Hans Petter Langtangen
Aslak Tveito
*Editors*

# Advanced Topics in Computational Partial Differential Equations

Numerical Methods
and Diffpack Programming

Springer

*Editors*

Hans Petter Langtangen
Aslak Tveito

Simula Research Laboratory
Martin Linges vei 17, Fornebu
P.O. Box
1325 Lysaker Norway
e-mail:
hpl@simula.no
aslak@simula.no
http://www.simula.no

Cataloging-in-Publication Data applied for

A catalog record for this book is available from the Library of Congress.

Bibliographic information published by Die Deutsche Bibliothek
Die Deutsche Bibliothek lists this publication in the Deutsche Nationalbibliografie;
detailed bibliographic data is available in the Internet at <http://dnb.ddb.de>.


Mathematics Subject Classification (2000): primary: 65M60, 65M06, 68U20, 60G60, 91B28, 60H15, 92C50; secondary: 65F10, 65M12, 65N06, 65N30, 65Y99, 73C02, 73E60, 76D05, 76M10, 76M20


ISSN 1439-7358
ISBN 3-540-01438-1 Springer-Verlag Berlin Heidelberg New York

Springer-Verlag Berlin Heidelberg New York
a part of Springer Science+Business Media

http://www.springer.de

© Springer-Verlag Berlin Heidelberg 2003
Printed in Germany

Cover Design: Friedhelm Steinen-Broo, Estudio Calamar, Spain
Cover production: *design & production*
Typeset by the authors using a Springer TeX macro package

Printed on acid-free paper     46/3111 – 5 4 3 2 1     SPIN 11324331

# Lecture Notes
# in Computational Science
# and Engineering

# 33

Springer
*Berlin*
*Heidelberg*
*New York*
*Hong Kong*
*London*
*Milan*
*Paris*
*Tokyo*

# Preface

This book is about solving partial differential equations (PDEs). Such equations are used to model a wide range of phenomena in virtually all fields of science and technology. In the last decade, the general availability of extremely powerful computers has shifted the focus in computational mathematics from simplified model problems to much more sophisticated models resembling intricate features of real life. This change challenges our knowledge in computer science and in numerical analysis.

The main objective of the present book is to teach modern, advanced techniques for numerical PDE solution. The book also introduces several models arising in fields like finance, medicine, material technology, and geology. In order to read this book, you must have a basic knowledge of partial differential equations and numerical methods for solving such equations. Furthermore, some background in finite element methods is required. You do not need to know Diffpack, although this programming environment is used in examples throughout the text. Basically, this book is about models, methods, and how to implement the methods. For the implementation part it is natural for us to use Diffpack as the programming environment, because making a PDE solver in Diffpack requires little amount of programming and because Diffpack has support for the advanced numerical methods treated in this book. Most chapters have a part on models and methods, and a part on implementation and Diffpack programming. The exposition is designed such that readers can focus only on the first part, if desired. The prerequisites for the present book (PDEs and numerics) as well as an introduction to Diffpack programming can be found in "Computational Partial Differential Equations – Numerical Methods and Diffpack Programming" by H. P. Langtangen.

The basic aim of each chapter is to introduce the reader to one advanced topic. The first none chapters address numerical methods and programming techniques, whereas the last five chapters are devoted to application of PDEs in diverse fields. Parts of this text have been used in a course at the University of Oslo over a period of four years. The idea is that such a course can now be adapted to the students' need by picking various chapters from this book. Most of the chapters in this text can be read independently, thus giving a lot of freedom in constructing new courses.

Parallel computing is introduced in the first chapter. No prior knowledge to this field is assumed, and we start by presenting some very basic examples. The concept of performance modeling is explained and basic steps towards programming parallel computers are introduced at an easy-to-read pace. Tools for parallel Diffpack computations are presented. The aim is to show that within this programming environment, parallelization of a code is easy, at least compared to traditional approaches like low-level MPI programming.

The second chapter is about overlapping domain decomposition methods. Such methods are extremely important in parallel computing and are a natural part of advanced courses in computational partial differential equations. Again, this chapter requires no prior knowledge of domain decomposition; the method is introduced through simple examples with accompanying software for numerical experiments. Thereafter, the general applicability of the domain decomposition strategy is emphasized, with a special focus on parallelization of PDE simulators. Diffpack has support for easy implementation of overlapping domain decomposition methods in a general PDE simulator. This feature provides an efficient way of adapting serial Diffpack simulators, developed without multi-processing computers in mind, to parallel computing environments. The chapter ends with detailed examples on how to parallelize stationary and time-dependent Diffpack simulators.

Chapter 3 contains instructions on how to equip a Diffpack simulator with multigrid methods. Thereafter the authors use the flexible multigrid toolbox in Diffpack to explore the method in different applications, including challenging problems with discontinuous coefficients, nonlinear problems, and systems of PDEs. In particular, it is easy to set up numerical experiments to find a favorable combinations of multigrid ingredients (smoothers, cycles, grid levels, etc.) in a particular application.

When solving systems of PDEs it is often an advantage to use different types of finite elements for different physical unknowns. Such techniques are usually referred to as mixed methods and play a fundamental role in the discretization of the Navier-Stokes equations and the system formulation of the Poisson equation. An easy-to-read introduction to mixed finite element discretization is provided in Chapter 4. There are two well-known difficulties with practical application of mixed finite element methods: (i) the implementation, especially on general 2D/3D unstructured grids, and (ii) efficient solution of the resulting indefinite linear systems. Diffpack has special support for treating both difficulties; programming with mixed finite elements is treated in Chapter 4, whereas efficient linear system solution is described in Chapter 5. The latter chapter addresses preconditioning of block matrices arising from systems of PDEs. One challenge in this field is to generalize well known optimality properties for discretized scalar partial differential equations. Another challenge is to develop efficient techniques for implementing optimal methods for block preconditioning.

Systems of PDEs are also the main topic of Chapter 6, where both fully implicit methods and various techniques for operator splitting are discussed. In particular, it is discussed how existing software for scalar equations can be combined to yield *fully implicit* solvers for a coupled system of equations. Examples from non-Newtonian fluid flow and oil reservoir simulations illustrate the implementation techniques.

Chapter 7 addresses stochastic PDEs. These are PDEs where some of the coefficients, boundary conditions, or initial conditions are uncertain and therefore modeled as stochastic quantities. The solutions of such equations then also become stochastic. The introductory literature to stochastic PDEs requires, to our knowledge, a firm background in mathematical analysis and integration theory and therefore seldom reaches the computational scientist or engineer. The authors of the chapters have therefore taken a less rigorous, more intuitive and computational approach to the subject. The purpose is to expose this important and consistent way of dealing with uncertainty to a larger audience. The chapter provides some simple examples of stochastic PDEs, where the solution can be achieved by analytical means. Thereafter, the authors address generic tools, mainly Monte Carlo simulation and perturbation methods, for numerical solution of stochastic PDEs in general. The Diffpack support for stochastic PDE computations is outlined and some simulators are explained in detail.

Chapter 8 describes how to operate Diffpack from Python scripts. The Python programming language has a simple and clear syntax, much like Matlab and Maple. Using Diffpack through Python is convenient and simplifies gluing Diffpack and, e.g., visualization packages. Calling complicated C++ codes like Diffpack from Python is in principle straightforward, but numerous technical details have to be resolved. In this chapter, we share the experiences in coupling Diffpack and Python, list recipes for carrying out the coupling, and describe how to take advantage of the Python interfaces.

Chapter 9 shows how to use numerical experiments to gain insight into the CPU efforts needed to solve PDE problems. The chapter contains parameterized results from huge collections of experiments, involving a range of models and methods: Poisson problems, fully nonlinear water waves, viscous fluid flow (the Navier-Stokes equations), two-phase porous media flow, finite element methods, finite difference methods, preconditioned Conjugate Gradient-like methods, multigrid methods, and so on.

Chapter 10 deals with a grand challenge problem in the computational sciences; computation of the electrical activity in the heart. Such electrical activity is the basis for Electrocardiographic (ECG) measurements, which provide a fundamental diagnostic tool of in medicine. The task of computing the effects of a single heart beat is tremendous and constitute an active research field internationally. This chapter introduces some basic principles and presents the most important models and numerical methods applied in this field.

Chapters 11 and 12 are about some basic problems in computational finance. It has been known for 30 years that partial differential equations can be used to model the fair price of financial derivatives such as stock options. In Chapter 11 we discuss several models in this field, and Chapter 12 focuses on solving the associated partial differential equations.

The finite element method is a fundamentally important tool when designing structures like ships, buildings, cars, and aircrafts. However, the formulation of the finite element method in structural analysis is seemingly quite different from the finite element method for solving partial differential equations in general (some refer to the former as the engineering finite element method and the latter as the mathematical finite element method). The purpose of Chapter 13 is show in detail that these two "schools" of finite element methods are mathematically equivalent. Popular structural elements, such as beams and plates, are described, and several examples on developing Diffpack simulators for elastic structures are presented.

Chapter 14 is about extrusion of aluminum. This is an important industrial forming process, typically for producing aluminum profiles for use in cars, for instance. The mathematical model used to model the extrusion process couples non-Newtonian flow, heat transfer, and elasticity. The authors describe the PDEs, numerical methods, the implementation in Diffpack, and some applications.

Establishing the evolution of sedimentary basins, which is a basic ingredient of oil exploration projects, is the topic of Chapter 15. Physically, this is a coupling of elastic deformations, porous media flow, and convective heat transfer in the Earth's crust over very long (geological) time scales. The authors present the geological, mathematical, and numerical models, along with an overview of the implementation. Extensive applications to sill intrusions in sedimentary basins are reported and discussed.

Many of the chapters make extensive use of software to illustrate mathematical, numerical, and implementational issues. The source code of this software is available for free from *www.diffpack.com/Book*.

*Acknowledgements.* This book has for a large part been prepared by members of the Scientific Computing Department at the Simula Research Laboratory; *www.simula.no/sc*. We want to express our gratitude to the members of the group for their enthusiasm with finishing this project. In particular, we would like to thank Dr. Xing Cai for his extraordinary technical contribution to the present volume and Dr. Are Magnus Bruaset for his very valuable feedback on many chapters. It is also a pleasure to thank our other friends and colleagues participating in various chapters of this book; both in writing and refereeing. It has been a pleasure to work together with you and we hope to be able to continue the collaboration. We would also like to thank Dr. Martin Peters for all encouragements, interesting discussions, and good advice in the process of completing this manuscript. Finally, we would like to thank the Research Council of Norway, the Department of Informatics and the Department of Mathematics at the University of Oslo, and the Simula Research Laboratory for providing substantial and long-term funding for our research. The editors also greatly acknowledge the financial support from the Norwegian Non-fiction Literature Fund.

*Fornebu, June 2003*

*Hans Petter Langtangen*                                          *Aslak Tveito*

# Table of Contents

**1 Parallel Computing** .................................... 1

    *X. Cai, E. Acklam, H. P. Langtangen, A. Tveito*

  1.1 Introduction to Parallel Computing ......................... 1

      1.1.1 Different Hardware Architectures ..................... 2

      1.1.2 The Message-Passing Programming Model ........... 2

      1.1.3 A Multicomputer Model ........................... 3

      1.1.4 Two Examples of Parallel Computing ............... 3

      1.1.5 Performance Modeling ............................. 5

      1.1.6 Performance Analysis of the Two Examples .......... 6

  1.2 A Different Performance Model ........................... 7

      1.2.1 The 2D Case ..................................... 8

      1.2.2 The 3D Case ..................................... 10

      1.2.3 A General Model ................................. 11

  1.3 The First MPI Encounter ............................... 11

      1.3.1 "Hello World" in Parallel ......................... 11

      1.3.2 Computing the Norm of a Vector ................... 12

  1.4 Basic Parallel Programming with Diffpack .................. 14

      1.4.1 Diffpack's Basic Interface to MPI ................... 14

      1.4.2 Example: The Norm of a Vector .................... 17

  1.5 Parallelizing Explicit FD Schemes ........................ 18

      1.5.1 Partitioning Lattice Grids ......................... 18

      1.5.2 Problem Formulation .............................. 20

      1.5.3 Hiding the Communication Cost .................... 22

      1.5.4 Example: The Wave Equation in 1D ............... 23

      1.5.5 How to Create Parallel FD Solvers in Diffpack ......... 26

      1.5.6 Tools for Parallel Finite Difference Methods ........... 27

      1.5.7 Example: Heat Conduction in 2D ................... 30

      1.5.8 The Sequential Heat Conduction Solver .............. 32

      1.5.9 The Parallel Heat Conduction Solver ................ 34

  1.6 Parallelizing FE Computations on Unstructured Grids ........ 37

      1.6.1 Parallel Linear Algebra Operations .................. 38

      1.6.2 Object-Oriented Design of a Parallel Toolbox .......... 42

      1.6.3 Major Member Functions of GridPartAdm ............ 46

      1.6.4 A Parallelization Example ......................... 49

      1.6.5 Storage of Computation Results and Visualization ..... 51

      1.6.6 Questions and Answers ............................ 52

  References ............................................... 55

## 2 Overlapping Domain Decomposition Methods ..... 57
*X. Cai*

2.1  Introduction ................................................. 57
2.2  The Mathematical Formulations ............................. 58
    2.2.1  The Classical Alternating Schwarz Method ............ 58
    2.2.2  The Multiplicative Schwarz Method ................... 59
    2.2.3  The Additive Schwarz Method ........................ 62
    2.2.4  Formulation in the Residual Form ................... 62
    2.2.5  Overlapping DD as Preconditioner ................... 63
2.3  A 1D Example ............................................... 64
2.4  Some Important Issues ...................................... 67
    2.4.1  Domain Partitioning ................................ 67
    2.4.2  Subdomain Discretizations in Matrix Form ........... 67
    2.4.3  Inexact Subdomain Solver ........................... 67
    2.4.4  Coarse Grid Corrections ............................ 68
    2.4.5  Linear System Solvers at Different Levels .......... 68
2.5  Components of Overlapping DD Methods ...................... 69
2.6  A Generic Implementation Framework ........................ 70
    2.6.1  The Simulator-Parallel Approach .................... 70
    2.6.2  Overall Design of the Implementation Framework ...... 71
    2.6.3  Subdomain Solvers .................................. 71
    2.6.4  The Communication Part ............................. 76
    2.6.5  The Global Administrator ........................... 76
    2.6.6  Summary of the Implementation Framework .......... 80
2.7  Parallel Overlapping DD Methods ........................... 80
2.8  Two Application Examples .................................. 82
    2.8.1  The Poisson1 Example ............................... 83
    2.8.2  The Heat1 Example .................................. 88
    2.8.3  Some Programming Rules ............................. 93
References ...................................................... 95

## 3 Software Tools for Multigrid Methods .............. 97
*K.-A. Mardal, G. W. Zumbusch, H. P. Langtangen*

3.1  Introduction ............................................... 97
3.2  Sketch of How Multilevel Methods are Implemented in Diffpack  99
3.3  Implementing Multigrid Methods ........................... 100
    3.3.1  Extending an Existing Application with Multigrid ..... 100
    3.3.2  Making Grid Hierarchies ............................ 104
    3.3.3  A Proper Test Example ............................. 105
3.4  Setting up an Input File .................................. 105
    3.4.1  Running a Plain Gaussian Elimination ............... 105
    3.4.2  Filling out MGtools Menu Items ..................... 106
3.5  Playing Around with Multigrid ............................ 107
    3.5.1  Number of Grids and Number of Iterations ........... 108
    3.5.2  Convergence Monitors .............................. 109

|       | 3.5.3 | Smoother | 110 |
|       | 3.5.4 | W Cycle and Nested Iteration | 113 |
|       | 3.5.5 | Coarse-Grid Solver | 116 |
|       | 3.5.6 | Multigrid as a Preconditioner | 117 |
|       | 3.5.7 | Additive Preconditioner | 119 |
| 3.6   |       | Equipping the Poisson2 Solver with Multigrid | 121 |
|       | 3.6.1 | Multigrid and Neumann or Robin Boundary Conditions | 122 |
|       | 3.6.2 | Debugging Multigrid | 124 |
|       | 3.6.3 | Domains With Geometric Singularities | 128 |
|       | 3.6.4 | Jumping Coefficients | 129 |
|       | 3.6.5 | Anisotropic Problems and (R)ILU smoothing | 133 |
| 3.7   |       | Systems of Equations, Linear Elasticity | 133 |
|       | 3.7.1 | Extending Elasticity1 with Multigrid | 134 |
|       | 3.7.2 | The $\lambda$-dependency | 136 |
| 3.8   |       | Nonlinear Problems | 136 |
|       | 3.8.1 | The Nonlinear Multigrid Method | 138 |
|       | 3.8.2 | The Full Approximation Scheme | 139 |
|       | 3.8.3 | Software Tools for Nonlinear Multigrid | 141 |
|       | 3.8.4 | Default Implementation of Some Components | 144 |
|       | 3.8.5 | Experiments with Nonlinear Multigrid | 145 |
|       | 3.8.6 | Nonlinear Multigrid for a Linear Problem | 147 |
|       | 3.8.7 | Nonlinear Multigrid Applied to a Non-Linear Problem | 148 |
| References | | | 150 |

## 4 Mixed Finite Elements ........................... 153
*K.-A. Mardal, H. P. Langtangen*

| 4.1 | | Introduction | 153 |
| 4.2 | | Model Problems | 154 |
|     | 4.2.1 | The Stokes Problem | 154 |
|     | 4.2.2 | The Mixed Poisson Problem | 154 |
| 4.3 | | Mixed Formulation | 155 |
|     | 4.3.1 | Weighted Residual Methods | 155 |
|     | 4.3.2 | Mixed Elements, Do We Really Need Them? | 158 |
|     | 4.3.3 | Function Space Formulation | 159 |
|     | 4.3.4 | The Babuska-Brezzi Conditions | 161 |
| 4.4 | | Some Basic Concepts of a Finite Element | 163 |
|     | 4.4.1 | A General Finite Element | 163 |
|     | 4.4.2 | Examples of Finite Element Spaces | 165 |
|     | 4.4.3 | Diffpack Implementation | 175 |
|     | 4.4.4 | Numbering Strategies | 177 |
| 4.5 | | Some Code Examples | 181 |
|     | 4.5.1 | A Demo Program for Special Versus General Numbering | 181 |
| 4.6 | | Programming with Mixed Finite Elements in a Simulator | 183 |
|     | 4.6.1 | A Standard Solver with Some New Utilities | 183 |
|     | 4.6.2 | A Simulator for the Stokes Problem | 186 |

4.6.3   Numerical Experiments for the Stokes Simulator . . . . . . . 191
4.6.4   A Simulator for the Mixed Poisson Problem . . . . . . . . . . 192
4.6.5   Numerical Experiments for the Mixed Poisson Simulator 195
References . . . . . . . . . . . . . . . . . . . . . . . . . . . . . . . . . . . . . . . . . . . 196

**5   Systems of PDEs and Block Preconditioning** . . . . . . 199
*K.-A. Mardal, J. Sundnes, H. P. Langtangen, A. Tveito*
5.1   Introduction . . . . . . . . . . . . . . . . . . . . . . . . . . . . . . . . . . . . . 199
5.2   Block Preconditioners in General . . . . . . . . . . . . . . . . . . . . . 200
    5.2.1   Operator Splitting Techniques . . . . . . . . . . . . . . . . . . . 201
    5.2.2   Conjugate Gradient-Like Methods . . . . . . . . . . . . . . . . 203
    5.2.3   Software Tools for Block Matrices
            and Numbering Strategies . . . . . . . . . . . . . . . . . . . . . . 205
    5.2.4   Software Tools for Block Structured Preconditioners. . . . 208
5.3   The Bidomain Equations . . . . . . . . . . . . . . . . . . . . . . . . . . . . 209
    5.3.1   The Mathematical Model . . . . . . . . . . . . . . . . . . . . . . . 209
    5.3.2   Numerical Method . . . . . . . . . . . . . . . . . . . . . . . . . . . . 211
    5.3.3   Solution of the Linear System . . . . . . . . . . . . . . . . . . . 212
    5.3.4   A Simulator for the Bidomain Model . . . . . . . . . . . . . . . 213
    5.3.5   A Simulator with Multigrid . . . . . . . . . . . . . . . . . . . . . 220
    5.3.6   A Solver Based on Operator Splitting . . . . . . . . . . . . . . 222
5.4   Two Saddle Point Problems . . . . . . . . . . . . . . . . . . . . . . . . . 226
    5.4.1   Stokes Problem . . . . . . . . . . . . . . . . . . . . . . . . . . . . . . 227
    5.4.2   Mixed Poisson Problem . . . . . . . . . . . . . . . . . . . . . . . . 231
References . . . . . . . . . . . . . . . . . . . . . . . . . . . . . . . . . . . . . . . . . . . 235

**6   Fully Implicit Methods for Systems of PDEs** . . . . . . 237
*Å. Ødegård, H. P. Langtangen, A. Tveito*
6.1   Introduction . . . . . . . . . . . . . . . . . . . . . . . . . . . . . . . . . . . . . 237
6.2   Implementation of Solvers for PDE Systems in Diffpack . . . . . . . 238
    6.2.1   Handling Systems of PDEs . . . . . . . . . . . . . . . . . . . . . . 238
    6.2.2   A Specific $2 \times 2$ System . . . . . . . . . . . . . . . . . . . . . . . 239
6.3   Problem with the Gauss–Seidel Method, by Example . . . . . . . . 240
    6.3.1   A System of Ordinary Differential Equations . . . . . . . . . 240
6.4   Fully Implicit Implementation . . . . . . . . . . . . . . . . . . . . . . . . 243
    6.4.1   A System of Linear Partial Differential Equations . . . . . . 244
    6.4.2   Implementation . . . . . . . . . . . . . . . . . . . . . . . . . . . . . . 244
    6.4.3   Systems of Non–Linear Partial Differential Equations . . . 247
6.5   Applications . . . . . . . . . . . . . . . . . . . . . . . . . . . . . . . . . . . . . 248
    6.5.1   A System of PDEs Modelling Pipeflow . . . . . . . . . . . . . . 248
    6.5.2   Comparison of the Gauss–Seidel
            and the Fully Implicit Method . . . . . . . . . . . . . . . . . . . 250
    6.5.3   Two–phase Porous Media Flow . . . . . . . . . . . . . . . . . . . 251
    6.5.4   Solver . . . . . . . . . . . . . . . . . . . . . . . . . . . . . . . . . . . . . 254

      6.5.5   Comparison of the Gauss–Seidel Method and the Fully
              Implicit Method .................................. 254
  6.6   Conclusion ............................................. 255
  References ................................................... 256

**7  Stochastic Partial Differential Equations** ........... 257
     *H. P. Langtangen, H. Osnes*
  7.1   Introduction ........................................... 257
  7.2   Some Simple Examples.................................. 259
      7.2.1   Bending of a Beam............................. 260
      7.2.2   Viscous Drag Forces........................... 264
      7.2.3   Heat Conduction in the Earth's Crust ............... 266
      7.2.4   Transport Phenomena ........................... 268
      7.2.5   Generic Problems .............................. 269
  7.3   Solution Methods ...................................... 271
      7.3.1   Monte Carlo Simulation ........................ 272
      7.3.2   Perturbation Methods .......................... 275
  7.4   Quick Overview of Diffpack Tools ........................ 277
  7.5   Tools for Random Variables ............................. 279
      7.5.1   Random Number Generation ...................... 279
      7.5.2   Description of Probability Distributions .............. 280
      7.5.3   Estimation of Statistics of Random Variables ......... 282
      7.5.4   Example: Simulation of a Stochastic Beam ............ 284
      7.5.5   Suggested Design of Stochastic PDE Simulators ....... 287
      7.5.6   Example: Stochastic Heat Conduction ............... 290
  7.6   Diffpack Tools for Random Fields ........................ 298
      7.6.1   Generation of Random Fields ...................... 298
      7.6.2   Statistical Properties ................................. 303
      7.6.3   Example: A Stochastic Poisson Equation.............. 304
  7.7   Summary.............................................. 307
  7.A   Transformation of Random Variables ...................... 308
      7.A.1   Transformation of a Single Random Variable .......... 308
      7.A.2   Transformation of Normal
              and Lognormal Random Variables .................. 311
      7.A.3   Other Variable Transformations ..................... 312
      7.A.4   Partial and Approximate Analysis
              of Random Transformations....................... 314
  7.B   Implementing a New Distribution ........................ 316
  References ................................................... 318

**8  Using Diffpack from Python Scripts** ............... 321
     *H. P. Langtangen, K.-A. Mardal*
  8.1   Introduction ........................................... 321
      8.1.1   The Advantages of High-Level Languages ............. 322
      8.1.2   Python Scripting and Diffpack...................... 323

|  | 8.1.3 | Example: Running a Diffpack Simulator from Python | 324 |
| 8.2 | | Developing Python Interfaces to C/C++ Functions | 325 |
|  | 8.2.1 | Wrapper Functions | 326 |
|  | 8.2.2 | Creating and Using a Module | 328 |
|  | 8.2.3 | How SWIG Simplifies Writing Wrapper Code | 328 |
|  | 8.2.4 | Writing Python Interfaces to Diffpack Relies on SWIG | 329 |
| 8.3 | | Compiling and Linking Wrapper Code with Diffpack | 329 |
|  | 8.3.1 | Makefiles | 330 |
|  | 8.3.2 | Summary of Creating the Interface | 332 |
|  | 8.3.3 | A Trivial Test Module | 333 |
|  | 8.3.4 | Common Errors in the Linking Process | 334 |
|  | 8.3.5 | Problems with Interfacing a Simulator Class | 335 |
|  | 8.3.6 | Python Interface, Version 1 | 335 |
|  | 8.3.7 | Version 2; Setting Input Parameters | 337 |
|  | 8.3.8 | Version 3; Using the C Preprocessor to Expand Macros | 340 |
|  | 8.3.9 | Version 4; Extending the Python Interface with Auxiliary Functions | 342 |
|  | 8.3.10 | SWIG Pointers | 343 |
| 8.4 | | Converting Data between Diffpack and Python | 346 |
|  | 8.4.1 | A Class for Data Conversion | 346 |
|  | 8.4.2 | Conversion between Vec and NumPy Arrays | 346 |
|  | 8.4.3 | Creating a Python Interface to the Conversion Class | 348 |
|  | 8.4.4 | Examples on Using the Conversion Class Interface | 349 |
|  | 8.4.5 | A String Typemap | 350 |
| 8.5 | | Building an Interface to a More Advanced Simulator | 351 |
|  | 8.5.1 | Computing Empirical Convergence Estimates | 351 |
|  | 8.5.2 | Visualization with Vtk | 355 |
| 8.6 | | Installing Python, SWIG etc. | 356 |
| 8.7 | | Concluding Remarks | 359 |
|  | | References | 359 |

# 9   Performance Modeling of PDE Solvers ............. 361

*X. Cai, A. M. Bruaset, H. P. Langtangen, G. T. Lines,*
*K. Samuelsson, W. Shen, A. Tveito, G. Zumbusch*

| 9.1 | | Introduction | 361 |
| 9.2 | | Model Problems | 363 |
|  | 9.2.1 | The Elliptic Boundary-Value Problem | 363 |
|  | 9.2.2 | The Linear Elasticity Problem | 365 |
|  | 9.2.3 | The Parabolic Problem | 366 |
|  | 9.2.4 | The Nonlinear Water Wave Problem | 366 |
|  | 9.2.5 | The Two-Phase Flow Problem in 2D | 368 |
|  | 9.2.6 | The Heart-Torso Coupled Simulations | 369 |
|  | 9.2.7 | The Species Transport Problem | 370 |
|  | 9.2.8 | Incompressible Navier-Stokes Equations | 371 |
| 9.3 | | Numerical Methods | 371 |

9.3.1   The Elliptic Problems ............................... 371
9.3.2   The HC2 Problem .................................. 372
9.3.3   The Nonlinear Water Wave Problem ................ 372
9.3.4   The TF2 Problem .................................. 372
9.3.5   The HT2 and HT3 Simulations ..................... 373
9.3.6   The AD3 Problem ................................. 373
9.3.7   The NS2 Problem.................................. 374
9.3.8   Solution Methods for Linear Systems ................ 374
9.3.9   Diffpack Implementation and Notation .............. 376
9.4   Total CPU Time Consumption .......................... 377
9.4.1   Establishing Performance Models ................... 378
9.4.2   The Software and Hardware Specification ............ 379
9.4.3   The Best Performance Model for Each Test Problem ... 380
9.4.4   The CPU Measurements and Performance Models ..... 380
9.4.5   Some Remarks About the Measurements............. 385
9.4.6   Measurements Obtained on Another Platform ........ 385
9.5   Solution of Linear Systems ............................. 386
9.5.1   Summary ........................................ 387
9.5.2   Measurements ................................... 387
9.5.3   Efficiency of the Linear Algebra Tools in Diffpack ...... 392
9.6   Construction of Linear Systems .......................... 392
9.6.1   The Process...................................... 392
9.6.2   Some Guidelines ................................. 394
9.6.3   The Mapped Laplace Equation in the WA3 Problem ... 395
9.6.4   Parabolic Problems ............................... 396
9.7   Concluding Remarks.................................... 397
References................................................... 398

**10   Electrical Activity in the Human Heart** ........... 401
    *J. Sundnes, G.T. Lines, P. Grøttum, A. Tveito*
10.1   The Basic Physiology ................................. 401
10.2   Outline of a Mathematical Model........................ 402
10.3   The Bidomain Model ................................. 404
10.3.1   A Continuous Model for the Heart Tissue ............. 404
10.3.2   Derivation of the Bidomain Equations ............... 405
10.4   A Complete Mathematical Model......................... 407
10.4.1   Boundary Conditions on the Heart-Torso Interface ..... 407
10.4.2   Boundary Conditions on the Surface of the Body. ...... 408
10.4.3   Summary of the Mathematical Problem .............. 408
10.5   Physiology of the Heart Muscle Tissue ................... 409
10.5.1   Physiology of the Cell Membrane ................... 409
10.5.2   The Nernst Potential .............................. 410
10.5.3   Models for the Ionic Current ....................... 412
10.5.4   Electric Circuit Model for the Membrane ............. 412
10.5.5   Channel Gating .................................. 413

10.5.6  The Hodgkin-Huxley Model ......................... 415
10.5.7  The Beeler-Reuter Model ........................... 415
10.5.8  The Luo-Rudy Model ............................... 416
10.5.9  A Different Model for Calcium Dynamics ............ 420
10.5.10 Structure of the Heart Tissue ...................... 425
10.6  The Numerical Method ................................... 427
10.6.1  Simplifications due to Boundary Conditions on $\partial H$ .... 427
10.6.2  Time Discretization ............................... 428
10.6.3  Discretization in Space ........................... 429
10.6.4  Calculation of the Conductivity Tensors ............ 431
10.6.5  Solution of the ODEs ............................. 431
10.7  Implementation ......................................... 433
10.7.1  Design ........................................... 433
10.7.2  The CommonRel Class ............................. 434
10.7.3  The Cells Class ................................... 436
10.7.4  The CellModel Class .............................. 436
10.7.5  The Parabolic Class ............................... 437
10.7.6  The Elliptic Class ................................. 437
10.8  Optimization of the Simulator .......................... 438
10.8.1  An Implicit ODE Solver .......................... 438
10.8.2  Local makeSystem Routines ....................... 440
10.8.3  Adaptivity ....................................... 440
10.9  Simulation Results ..................................... 442
10.9.1  Test Results on a Simple 3D Geometry ............. 443
10.9.2  Physiologically Correct Geometries ................. 445
10.10 Concluding Remarks ..................................... 447
References ...................................................... 448

# 11  Mathematical Models of Financial Derivatives ... 451

### O. Skavhaug, B. F. Nielsen, A. Tveito

11.1  Introduction ........................................... 451
11.2  Basic Assumptions ..................................... 453
11.2.1  Arbitrage–Free Markets ........................... 453
11.2.2  The Efficient Market Hypothesis ................... 454
11.2.3  Other Assumptions ............................... 454
11.3  Forwards and Futures ................................... 454
11.3.1  The Forward Contract ............................. 455
11.3.2  The Forward Price ................................ 455
11.3.3  The Future Contract .............................. 457
11.4  The Black-Scholes Analysis ............................. 457
11.4.1  Ito's Lemma ...................................... 459
11.4.2  Elimination of Randomness ........................ 460
11.4.3  The Black-Scholes Equation ....................... 460
11.5  European Call and Put Options ......................... 462
11.5.1  European Call Options ............................ 462

11.5.2 European Put Options .............................. 464
11.5.3 Put–Call Parity .................................. 465
11.5.4 Analytical Solutions ............................... 465
11.6 American Options ........................................ 466
11.6.1 American Put Options ............................ 467
11.7 Exotic Options .......................................... 468
11.7.1 Correlation–Dependent Options ..................... 468
11.7.2 Path–Dependent Options ........................... 471
11.8 Hedging ................................................ 474
11.8.1 The Delta Greek ................................. 474
11.8.2 The Gamma Greek ............................... 476
11.8.3 The Theta Greek ................................ 478
11.8.4 The Vega Greek ................................. 479
11.8.5 The Rho Greek .................................. 480
11.9 Remarks ............................................... 481
References ..................................................... 481

12   **Numerical Methods for Financial Derivatives** .... 483
     *O. Skavhaug, B. F. Nielsen, A. Tveito*
12.1 Introduction ............................................ 483
12.2 Model Summary ........................................ 484
12.3 Monte–Carlo Methods ................................... 488
12.4 Lattice Methods ......................................... 490
12.5 Finite Difference Methods ................................ 493
12.5.1 European Options ............................... 494
12.5.2 American Options ............................... 496
12.6 Finite Element Methods .................................. 497
12.6.1 Implementing a FEM Solver ...................... 498
12.6.2 Extensions ...................................... 504
References ..................................................... 505

13   **Finite Element Modeling of Elastic Structures** ... 507
     *T. Thorvaldsen, H. P. Langtangen, H. Osnes*
13.1 Introduction ............................................ 507
13.1.1 Two Versions of the Finite Element Method ........... 507
13.1.2 Two Element Concepts ............................ 509
13.2 An Introductory Example; Bar Elements .................... 510
13.2.1 Differential Equation Formulation ................... 511
13.2.2 Energy Formulation .............................. 516
13.3 Another Example; Beam Elements ......................... 518
13.3.1 Differential Equation Formulation ................... 519
13.3.2 Energy Formulation .............................. 521
13.4 General Three-Dimensional Elasticity ...................... 522
13.4.1 Differential Equation Formulation ................... 522
13.4.2 Energy Formulation .............................. 525

13.5   Degrees of Freedom and Basis Functions ..................... 527
    13.5.1  Bar Elements ....................................... 528
    13.5.2  Beam Elements...................................... 530
    13.5.3  Frame Elements .................................... 531
    13.5.4  DKT Plate Elements ................................ 532
13.6   Material Types and Elasticity Matrices...................... 534
    13.6.1  Linear Elastic, Isotropic Material ..................... 535
    13.6.2  Orthotropic Material; Fiber Composite ............... 535
13.7   Element Matrices in Local Coordinates...................... 537
    13.7.1  Elements with Linear, Isotropic Material Properties .... 537
    13.7.2  Elements with Orthotropic Material Properties ........ 539
    13.7.3  DKT Plate Elements ................................ 539
13.8   Element Load Vectors in Local Coordinates.................. 539
    13.8.1  Bar Elements ....................................... 540
    13.8.2  Beam Elements...................................... 542
    13.8.3  Frame Elements .................................... 544
    13.8.4  DKT Plate Elements ................................ 545
13.9   Element Matrices and Vectors in Global Coordinates.......... 546
    13.9.1  Bar, Beam, and Frame Elements..................... 546
    13.9.2  DKT Plate Elements ................................ 546
13.10  Element Forces, Stresses, and Strains ...................... 549
    13.10.1 Bar, Beam, and Frame Elements..................... 549
    13.10.2 DKT Plate Elements ................................ 550
13.11  Implementation of Structural Elements...................... 551
    13.11.1 Class StructElmDef ................................ 552
    13.11.2 Class BarElm ...................................... 553
    13.11.3 Class BeamElm .................................... 553
    13.11.4 Class IsoDKTElm .................................. 554
    13.11.5 Class OrthoDKTElm................................ 554
    13.11.6 Class StructElms .................................. 554
    13.11.7 How to Implement New Structural Elements .......... 555
13.12  Some Example Programs ................................... 556
    13.12.1 The Bar Element Simulator ......................... 557
    13.12.2 Indicators in the ANSYS Simulators ................. 560
13.13  Test Problems ........................................... 561
    13.13.1 Bar Elements in 1D ................................ 561
    13.13.2 Bar Elements in 2D ................................ 562
    13.13.3 Beam Elements in 2D; Three-Storey Framework ....... 567
    13.13.4 Bar and Frame Elements in 2D ...................... 569
    13.13.5 Twisting of a Square Plate; Isotropic Material ......... 570
    13.13.6 Simply Supported Fiber Composite Plate ............. 572
    13.13.7 Test Problems Using ANSYS Input Files ............. 573
    13.13.8 Test Problems Summary............................ 574
13.14  Summary................................................. 574
References ..................................................... 576

**14    Simulation of Aluminum Extrusion** .............. 577

*K. M. Okstad, T. Kvamsdal*

14.1   Introduction ............................................ 577
14.2   Mathematical Formulation ............................. 579
    14.2.1   Basic Definitions ..................................... 579
    14.2.2   Governing Equations ................................. 580
    14.2.3   Boundary Conditions ................................. 581
    14.2.4   Variational Formulation .............................. 583
14.3   Finite Element Implementation ........................ 585
    14.3.1   Time Discretization .................................. 585
    14.3.2   Spatial Discretization ............................... 586
    14.3.3   Global Solution Procedure and Mesh Movement ....... 586
14.4   Object-Oriented Implementation ....................... 587
    14.4.1   Introduction ........................................ 587
    14.4.2   Class Hierarchy for the Problem-Dependent Data ...... 588
    14.4.3   Class Hierarchy for the Numerical Solvers ............ 593
14.5   Numerical Experiments ................................ 600
    14.5.1   The Jeffery–Hamel Flow Problem .................... 600
    14.5.2   The Extrusion Problem .............................. 603
    14.5.3   Simulations of the Temperature ..................... 604
14.6   Concluding Remarks ................................... 604
References ................................................... 608

**15    Simulation of Sedimentary Basins** ............... 611

*A. Kjeldstad, H. P. Langtangen, J. Skogseid, K. Bjørlykke*

15.1   Introduction ............................................ 611
15.2   The Geomechanical and Mathematical Problem .......... 613
    15.2.1   Elastic Deformations ................................ 613
    15.2.2   Fluid Flow .......................................... 616
    15.2.3   Heat Transfer ....................................... 619
    15.2.4   Initial Conditions ................................... 620
    15.2.5   Boundary Conditions ................................. 622
15.3   Numerical Methods ..................................... 622
    15.3.1   Discretization Technique ............................. 623
    15.3.2   Nonlinear Solution Technique ........................ 623
    15.3.3   The Linear System ................................... 624
15.4   Implementing a Solver for a System of PDEs ........... 628
15.5   Verification ............................................ 630
    15.5.1   Cylinder with Concentric Circular Hole .............. 630
    15.5.2   One-Dimensional Consolidation ...................... 633
15.6   A Magmatic Sill Intrusion Case Study ................. 635
    15.6.1   Case Definition ..................................... 641
    15.6.2   Results and Discussion .............................. 643
15.7   Concluding Remarks ................................... 656
References ................................................... 657

# Chapter 1

# Parallel Computing

X. Cai[1,2], E. Acklam[3], H. P. Langtangen[1,2], and A. Tveito[1,2]

[1] Simula Research Laboratory
[2] Department of Informatics, University of Oslo
[3] Numerical Objects AS

**Abstract.** Large-scale parallel computations are more common than ever, due to the increasing availability of multi-processor systems. However, writing parallel software is often a complicated and error-prone task. To relieve Diffpack users of the tedious and low-level technical details of parallel programming, we have designed a set of new software modules, tools, and programming rules, which will be the topic of the present chapter.

## 1.1 Introduction to Parallel Computing

Parallel computing, also known as concurrent computing, refers to a group of independent processors working collaboratively to solve a large computational problem. This is motivated by the need to reduce the execution time and to utilize larger memory/storage resources. The essence of parallel computing is to partition and distribute the entire computational work among the involved processors. However, the hardware architecture of any multi-processor computer is quite different from that of a single-processor computer, thus requiring specially adapted parallel software. Although the message passing programming model, especially in terms of the MPI standard [3], promotes a standardized approach to writing parallel programs, it can still be a complicated and error-prone task. To reduce the difficulties, we apply object-oriented programming techniques in creating new software modules, tools, and programming rules, which may greatly decrease the user effort needed in developing parallel code. These issues will be discussed in this chapter within the framework of Diffpack.

The contents of this chapter are organized as follows. The present section contains a brief introduction to some important concepts related to parallel computing in general. Section 1.2 proposes a performance model that has advantages in analyzing and predicting CPU consumptions by complex parallel computations. Then, Section 1.3 is devoted to an introduction to the basic functionality of the MPI standard. Later on, Section 1.4 presents a high-level Diffpack interface to the MPI routines, whereas Section 1.5 concerns parallelizing Diffpack simulators that use explicit finite difference schemes. Finally, Section 1.6 covers the topic of parallelizing implicit finite element computations on unstructured grids.

### 1.1.1   Different Hardware Architectures

The hardware architectures of different multiple-processor systems span a wide spectrum. We may categorize multiple-processor systems from the angle of memory configuration. At one extreme, there are the so-called *shared memory* parallel computers. The most important feature of such parallel computers is that all the processors share a single global memory space, which is realized either at the hardware level or at the software level. At the other extreme of the hardware architecture spectrum, there are the so-called *distributed memory* parallel computers, where each processor has its own individual memory unit. All the processors are then connected through an interconnection network, by which communication between processors takes place. Using the network speed as a main criterion, we can further divide the distributed memory systems into sub-categories such as high-end massive parallel computers, low-cost PC clusters, networks of workstations, etc. Finally, in the middle of the hardware architecture spectrum, we find *distributed shared-memory* machines, which are in fact clusters of shared-memory parallel computers.

### 1.1.2   The Message-Passing Programming Model

The different types of hardware architecture have their impact on parallel software programming. Two major parallel programming models are thus so-called *shared memory* and *message passing* models, which arise from the shared memory and distributed memory hardware architectures, respectively. The shared memory programming model assumes that every process can directly access the global data structure of a parallel program. Each process is implicitly assigned to work on a portion of the global data. Moreover, work load partitioning and inter-process communication are hidden from the user. This programming model is well suited for static and regular data structures, leading normally to parallel execution at the loop level.

In the message passing model, however, it is necessary to partition a global problem into several smaller sub-problems and assign them to different processes. Each process has direct access only to its sub-problem data structure. In order for the processes to solve the global problem collaboratively, the user has to insert communication commands in the parallel program. Collaboration between the processes is achieved by *explicit message passing*, i.e., different processes communicate with each other in form of sending and receiving messages. We remark that message passing enables not only data transfer, but also synchronization, which refers to waiting for the completion of the involved processes, at certain pre-determined points during execution. As the message passing model is more flexible and can also be realized on any shared memory machine, we focus on this programming model. It has been applied in the development of the parallel Diffpack libraries, and will be assumed throughout the rest of this chapter.

*Remark.* Strictly speaking, the term *process* is different from the term *processor*. A process is an instance of a program executing autonomously, see e.g. [8, Ch. 2.2], whereas a processor is a hardware unit of a computer on which one or several processes can be executed. However, assuming the most common situation where only one process within a parallel computation is executed on one processor, we loosely use these two terms interchangeably in the rest of this chapter.

### 1.1.3    A Multicomputer Model

To simplify the description and analysis of the forthcoming parallel algorithms, we adopt the following model of a theoretical *multicomputer* (see e.g. [11, Ch. 1.2]), which is in fact an idealized model of distributed memory systems.

A multicomputer has $P$ identical processors, where every processor has its own memory. Each processor has control of its own local computation, which is carried out sequentially. The processors are interconnected through a communication network, so that any pair of two processors can exchange information in form of sending and receiving messages. Moreover, communication between any two processors is of equal speed, regardless of physical distance.

### 1.1.4    Two Examples of Parallel Computing

*Addition of Two Vectors.* Addition of two vectors is a frequently encountered operation in scientific computing. More precisely, we want to compute

$$\mathbf{w} = a\mathbf{x} + b\mathbf{y},$$

where $a$ and $b$ are two real scalars and the vectors $\mathbf{x}$, $\mathbf{y}$ and $\mathbf{w}$ are all of length $M$. In order to carry out the vector addition in parallel, it is necessary to first partition $\mathbf{x}$, $\mathbf{y}$ and $\mathbf{w}$ among the processors. That is, each processor creates its local storage to hold a portion of every global vector. Using an integer $p$, which has value between 0 and $P-1$, as the processor rank on a $P$-processor multicomputer, we will have for processor number $p$:

$$\mathbf{x}_p \in \mathbb{R}^{M_p}, \quad \mathbf{y}_p \in \mathbb{R}^{M_p}, \quad \mathbf{w}_p \in \mathbb{R}^{M_p}, \quad M = \sum_p M_p.$$

The above partitioning means that for every global index $i = 1, 2, \ldots, M$, there exists an index pair $(p, j)$, $0 \leq p < P$, $1 \leq j \leq M_p$, such that the correspondence between $i$ and $(p, j)$ is one-to-one. Note that if $M = PI$ for some integer $I$, a global vector can be divided into $P$ equally sized sub-vectors, each containing $I = M/P$ entries. (For the case where $M$ is not divisible by

$P$, we refer to Section 1.3.2.) A partitioning scheme that lets each sub-vector contain contiguous entries from the global vector can be as follows:

$$I = \frac{M}{P}, \quad p = \lfloor \frac{i}{I} \rfloor, \quad j = i - pI .$$

Note that we use integer division in the above formula to calculate the $p$ value.

Once the partitioning is done, addition of the two vectors can be carried out straightforwardly in parallel by

$$\mathbf{w}_p = a\mathbf{x}_p + b\mathbf{y}_p, \quad p = 0, 1, \ldots P - 1 .$$

*Remarks.* The partitioning introduced above is *non-overlapping* in that each entry of the global vectors $\mathbf{x}$, $\mathbf{y}$ and $\mathbf{w}$ is assigned to *only one* processor and $M = \sum_p M_p$. Furthermore, the local vector addition operation can be carried out by each processor, totally independent of other processors.

*Inner Product of Two Vectors.* Unlike the above vector addition operation, most parallel computations require inter-processor communication when being carried out on a multicomputer. An example is the parallel implementation of the inner-product between two vectors. More precisely, we have $\mathbf{x} \in \mathbb{R}^M$ and $\mathbf{y} \in \mathbb{R}^M$, and want to compute

$$c = \mathbf{x} \cdot \mathbf{y} = \sum_{i=1}^{M} x_i y_i .$$

Again, a partitioning of the global vectors $\mathbf{x}$ and $\mathbf{y}$ needs to precede the parallel inner-product operation. When each processor has the two sub-vectors ready, we continue with a local computation of the form:

$$c_p = \mathbf{x}_p \cdot \mathbf{y}_p = \sum_{j=1}^{M_p} x_{p,j} y_{p,j} .$$

Then, we can obtain the correct global result by

$$c = \sum_{p=0}^{P-1} c_p, \tag{1.1}$$

which can be realized by, e.g., asking every processor to send its local result $c_p$ to all the other processors. In this case, the parallel inner-product operation requires an all-to-all inter-processor communication, and we note that the correct global result $c$ will be available on every processor.

*Remark.* An alternative way of realizing (1.1) will be to designate a so-called *master* processor, which collects all the local results $c_p$ for computing $c$ and then broadcasts the correct global result to all the processors. Consequently, the above all-to-all communication is replaced by first an all-to-one communication and then a one-to-all communication.

### 1.1.5  Performance Modeling

We now study how the execution time of a parallel program depends on the number of processors $P$. Let us denote by

$$T = T(P)$$

the execution time for a parallel program when being run on $P$ processors. Performance analysis is to study the properties of $T(P)$, which mainly consists of two parts: arithmetic time and message exchange time. We remark that the above simple composition of $T(P)$ is a result of using the multicomputer model from Section 1.1.3. In the present section, we assume that the global problem size is fixed and thus only study how $T$ depends on $P$. Assuming also that computation and communication can not be performed simultaneously, we can come up with the following idealized model:

$$T(P) = \max_{0 \le p < P} I_p \tau_A + \max_{0 \le p < P} T_C^p, \tag{1.2}$$

where $I_p$ and $\tau_A$ are the number of arithmetic operations on processor number $p$ and the time needed by one single arithmetic operation, respectively. Moreover, $T_C^p$ is processor number $p$'s total communication time, which is spent on the involved message exchanges. More specifically, we have

$$T_C^p = \tau_C(L_1^p) + \tau_C(L_2^p) + \dots, \tag{1.3}$$

where $L_1^p$ is the length of message number 1, $L_2^p$ is the length of message number 2, and so on. Moreover, the terms $\tau_C(L)$ in (1.3) are used to model the exchange cost of a message of length $L$ between two processors:

$$\tau_C(L) = \tau_L + \xi L. \tag{1.4}$$

Here, $\tau_L$ is the so-called *latency*, which represents the startup time for communication, and $1/\xi$ is often referred to as the *bandwidth*, which indicates the rate at which messages can be exchanged between two processors.

Recall that one major motivation of parallel computing is to reduce the execution time. The quality of a parallel program can thus be indicated by two related quantities: *speed-up* $S(P)$ and *efficiency* $\eta(P)$. More precisely,

$$S(P) = \frac{T(1)}{T(P)}, \quad \eta(P) = \frac{S(P)}{P},$$

where $T(1) = \hat{I}\tau_A$ is the execution time of a purely sequential program. We note that $T(1)$ is free of communication cost and the following observation is valid:

$$\hat{I} \le \sum_{p=0}^{P-1} I_p \le P \left( \max_{0 \le p < P} I_p \right) \quad \Rightarrow \quad T(1) \le P T(P).$$

This observation also implies that

$$S(P) \leq P \tag{1.5}$$

and

$$\eta(P) \leq 1. \tag{1.6}$$

In practice, one normally has to be satisfied with $S(P) < P$ due to the communication cost. However, in certain special situations, *superlinear* speed-up, i.e., $S(P) > P$, can be obtained due to the cache effect. In other words, $\tau_A$ in this case effectively becomes smaller when $P$ is so large that local data fit into the processor cache, and, at the same time, the resulting efficiency gain exceeds the overhead communication cost.

### 1.1.6   Performance Analysis of the Two Examples

*Addition of Two Vectors.* For the example of adding two vectors in parallel, which was described in Section 1.1.4, we have

$$T(1) = 3M\tau_A \quad \text{and} \quad T(P) = \max_{0 \leq p < P} (3M_p\tau_A) = 3I\tau_A, \quad I = \max_p M_p.$$

Therefore

$$S(P) = \frac{T(1)}{T(P)} = \frac{3M\tau_A}{3I\tau_A} = \frac{M}{I}.$$

Consequently, when the computational load is perfectly balanced among the processors, i.e., $I = M/P$, we obtain perfect speed-up

$$S(P) = P$$

and efficiency $\eta(P) = 1$.

*Inner Product of Two Vectors.* For the example of computing the inner-product in parallel, the situation is different because inter-processor communication is involved. We see that

$$T(1) = (2M - 1)\tau_A$$

and

$$T(P) = \max_{0 \leq p < P} ((2M_p - 1)\tau_A + (P - 1)(\tau_C(1) + \tau_A)).$$

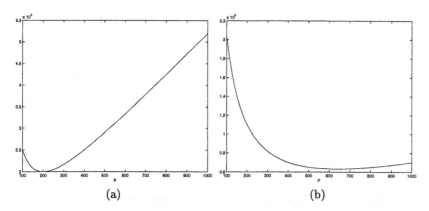

**Fig. 1.1.** Two plots of the function $T(P) = 2M/P + \gamma P$, where $\gamma = 50$. The value of $M$ is $10^6$ in (a) and $10^7$ in (b), respectively.

*Remark.* The part $(P - 1)(\tau_C(1) + \tau_A)$ is due to the communication and computation cost associated with $c = \sum_p c_p$. That is, every processor sends its local result $c_p$ to all the other processors. The global result $c$ will be calculated on every processor. However, we note that real applications will use so-called *collective communication* routines for computing $c = \sum_p c_p$, where the total cost is of order $\log P(\tau_C(1) + \tau_A)$; see [4, Ch. 8].

Let us introduce the factor

$$\gamma = \frac{\tau_C(1)}{\tau_A}.$$

Under the assumption that $P \ll M$ and the load balance is perfect, i.e., $\max_p M_p = M/P$, we arrive at

$$S(P) = \frac{T(1)}{T(P)} \approx \frac{2M\tau_A}{2\frac{M}{P}\tau_A + P\tau_C(1)} = \frac{2M\tau_A}{(2\frac{M}{P} + \gamma P)\tau_A} = \frac{P}{1 + \frac{\gamma P^2}{2M}}.$$

Consequently, to achieve perfect speed-up, i.e., $S(P) \approx P$, it is necessary to have

$$\frac{\gamma P^2}{2M} \ll 1,$$

which is either due to an extremely fast communication speed (very small $\gamma$), or due to very large long vectors (very large $M$), or both reasons. The effect of $M$ is illustrated by the two plots in Figure 1.1 It is also obvious that a small value of $\gamma$ is always favorable for achieving good speed-up.

## 1.2 A Different Performance Model

The performance model (1.2) given in the preceding section is of rather limited usage in practice, due to the following reasons:

1. It is extremely difficult to come up with an accurate count $I_p$ for all the involved arithmetic operations in complicated parallel computations.

2. It is not realistic to assume that all the different arithmetic operations take constant time $\tau_A$. For instance, one division of two floating point numbers may be 8 times as costly as one multiplication. Besides, for more complicated operations, such as the calculation of a sine function, it is difficult to determine the exact cost.

3. The cost model (1.4) for communication is only valid for one-to-one inter-processor communication and therefore not valid for the collective inter-processor communication that is frequently encountered.

In the following, we propose a different performance model for $T(M, P)$, where $M$ represents the global size of the computation. This performance model does not make explicit use of $I_p$, $\tau_A$, $\tau_L$ and $\xi$.

### 1.2.1    The 2D Case

Suppose we are interested in parallel computations where the amount of local computations per processor is linearly proportional to the number of subgrid points. In addition, the involved inter-processor communication can be of two types:

1. Exchanging long vectors that contain boundary data between immediate neighboring processors;

2. Every processor collectively sends and receives short messages from all the other processors, as needed during parallel calculation of inner-product.

Let us consider a 2D computational grid with $M$ grid points and a multi-computer with $P$ processors. For simplicity, we consider a 2D lattice having $\sqrt{M}$ points in each direction, i.e., the total number of grid points is

$$M = \sqrt{M} \times \sqrt{M}.$$

Let us also assume that we can organize the multicomputer in a two-dimensional array of

$$P = \sqrt{P} \times \sqrt{P}$$

processors.

On each processor, there are

$$\frac{\sqrt{M}}{\sqrt{P}} \times \frac{\sqrt{M}}{\sqrt{P}} = \frac{M}{P}$$

computational points. The boundary of each subdomain consequently consists of $\mathcal{O}(\sqrt{M}/\sqrt{P})$ points. Here, the symbol '$\mathcal{O}$' indicates that the number of points on the boundary is proportional to $\sqrt{M}/\sqrt{P}$. In other words, we do not distinguish between schemes that exchange values on one or several layers of boundary points.

For this 2D case, the overhead due to the first type of communication is of order $\mathcal{O}(\sqrt{M}/\sqrt{P})$. The overhead due to the second type of communication will be of order $\mathcal{O}(\log P)$, assuming use of an effective implementation of such collective communication (see [4, ch. 8]). Since the computation time is of order $\mathcal{O}(M/P)$, we argue that the overall execution time can be modeled by

$$T(M, P) = \alpha \frac{M}{P} + \beta \frac{\sqrt{M}}{\sqrt{P}} + \varrho \log P, \qquad (1.7)$$

where $\alpha$, $\beta$ and $\varrho$ are three constant parameters to be determined by applying e.g. the method of least squares (see [7, ch. 8]) to a series of measurements of $T$ for different values of $M$ and $P$.

*Remarks*

1. We have assumed that the cost of the first type of communication depends only on the amount of data to be exchanged between two processors. In other words, we have ignored latency in (1.7). Most likely, this is probably appropriate for properly scaled problems. However, if we keep $M$ fixed and increase $P$ to be close to $M$, the model will produce a poor approximation. In such a case the execution time $T$ will eventually be dominated by the overhead of latency.

2. Note also that model (1.7) does not consider the cost of waiting for a communication channel to be available. That is, we assume that any pair of processors can communicate with each other undisturbed. This assumption is valid for e.g. clusters of PC nodes that are connected through a high-speed switch.

3. We have also neglected the overhead of synchronization that will arise during the second type of communication, in case of poor computational load balance among the processors.

4. In practical computations, the convergence of iterative schemes may slow down as $M$ increases. We do not take this effect into account. A more realistic model can be formulated as:

$$T(M, P) = \alpha \frac{M^\delta}{P} + \beta \frac{\sqrt{M}}{\sqrt{P}} + \varrho \log P,$$

where $\delta \geq 1$.

*A Numerical Experiment.* We consider a 2D parallel computation that involves 400 Conjugate Gradient iterations. Based on a set of actual timing measurements, see Table 1.1, we have used the method of least squares to compute the parameters $\alpha$, $\beta$ and $\varrho$ to produce the following model:

$$T(M, P) = 1.086 \cdot 10^{-3} \frac{M}{P} + 6.788 \cdot 10^{-4} \frac{\sqrt{M}}{\sqrt{P}} + 2.523 \cdot 10^{-1} \log P. \quad (1.8)$$

**Table 1.1.** Actual timing measurements of a 2D parallel computation that involves 400 Conjugate Gradient iterations.

| $M$ | $P = 2$ | $P = 4$ | $P = 8$ | $P = 16$ |
|---|---|---|---|---|
| 40,401 | 21.61 | 10.81 | 5.96 | 3.57 |
| 90,601 | 49.43 | 24.85 | 12.73 | 7.14 |
| 160,801 | 87.67 | 45.07 | 23.06 | 12.21 |
| 251,001 | 136.36 | 69.32 | 35.67 | 18.10 |

Thereafter, we use an additional set of timing measurements of the *same* computation to check the accuracy of model (1.8). Note that this additional timing set was not used for estimating $\alpha$, $\beta$ and $\varrho$ in (1.8). In Table 1.2, the actual timing measurements are compared with the estimates, which are the italic-fonted numbers in parentheses. We observe that the estimates are fairly accurate.

**Table 1.2.** Comparison between the actual timing measurements and the estimates (in parentheses) produced by model (1.8).

| $M$ | $P = 2$ | $P = 4$ | $P = 8$ | $P = 16$ |
|---|---|---|---|---|
| 641,601 | 348.58 (*348.93*) | 175.55 (*174.92*) | 89.16 (*88.02*) | 45.39 (*44.68*) |

### 1.2.2   The 3D Case

We can argue similarly for the 3D case, as for the above 2D case. Suppose we have a computational lattice grid with $M = M^{1/3} \times M^{1/3} \times M^{1/3}$ points and the multicomputer has a three-dimensional array of $P = P^{1/3} \times P^{1/3} \times P^{1/3}$ processors.

On each processor, there are

$$\frac{M^{1/3}}{P^{1/3}} \times \frac{M^{1/3}}{P^{1/3}} \times \frac{M^{1/3}}{P^{1/3}} = \frac{M}{P}$$

computational points. The number of points on the boundary surface of one subdomain is of order

$$\frac{M^{1/3}}{P^{1/3}} \times \frac{M^{1/3}}{P^{1/3}} = \frac{M^{2/3}}{P^{2/3}}.$$

So the communication cost is of order $\mathcal{O}(\frac{M^{2/3}}{P^{2/3}})$, and the computation time is of order $\mathcal{O}(\frac{M}{P})$. Therefore, the overall execution time can be modeled by

$$T(M, P) = \alpha \frac{M}{P} + \beta \frac{M^{2/3}}{P^{2/3}} + \varrho \log P. \tag{1.9}$$

### 1.2.3   A General Model

In general, we can argue that $T(M, P)$ can be modeled by

$$T(M, P) = \alpha \frac{M}{P} + \beta \left(\frac{M}{P}\right)^{\frac{d-1}{d}} + \varrho \log P, \tag{1.10}$$

where $d$ is the number of spatial dimensions. The remarks given in Section 1.2.1 also apply to the models (1.9) and (1.10).

## 1.3   The First MPI Encounter

Recall that we have briefly explained the basic ideas of the message passing programming model in Section 1.1.2. The acronym MPI stands for *message passing interface* [3], which is a library specification of message passing routines. MPI has a rich collection of such routines and is the de-facto standard for writing portable message passing programs. This section aims to introduce the reader to basic MPI routines by showing a couple of simple MPI programs in the C language. A complete comprehension of the MPI routines is not strictly necessary, since those low-level routines will be replaced by more flexible high-level Diffpack counterparts in parallel Diffpack programs.

### 1.3.1   "Hello World" in Parallel

The most important thing to keep in mind when writing an MPI program is that the same program is to be executed, simultaneously, as multiple processes. These processes are organized, by default, in the context of a so-called MPI communicator with name MPI_COMM_WORLD. Inside MPI_COMM_WORLD, each process is identified with a unique rank between 0 and $P - 1$, where $P$ is the total number of processes. The processes can refer to each other by their ranks. The following short program demonstrates the MPI routines for determining the number of processes and the process rank, while also showing the most fundamental pair of MPI routines, namely MPI_Init and MPI_Finalize. Calls to these two routines must enclose any other other MPI routines to be used in an MPI program.

```
#include <stdio.h>
#include <mpi.h>

int main (int nargs, char** args)
{
  int size, my_rank;
  MPI_Init (&nargs, &args);
  MPI_Comm_size (MPI_COMM_WORLD, &size);
  MPI_Comm_rank (MPI_COMM_WORLD, &my_rank);
  printf("Hello world, I'm No.%d of %d procs.\n",my_rank,size);
  MPI_Finalize ();
  return 0;
}
```

The compilation of the above MPI program, assuming the file name `hello.c`, can be done by, e.g.,

```
mpicc hello.c
```

The resulting executable, say `a.out`, can be run by e.g. the following command:

```
mpirun -np 48 a.out
```

We note that the above `mpicc` and `mpirun` commands are standard in many MPI implementations. The execution result in each process printing out a message. More precisely, if the MPI program is executed as e.g. 48 processes, the process with rank 5 will print out the following message:

```
Hello world, I'm No.5 of 48 procs.
```

### 1.3.2   Computing the Norm of a Vector

Our next example concerns the computation of the Euclidean norm of a long vector $v = (v_1, \ldots, v_M)$. The vector is to be distributed among a number of processes, i.e., we cut up $v$ into $P$ pieces. Process number $p$ owns the local vector segment: $v_1^p, \ldots, v_{M_p}^p$, where $M_p$ is the number of vector components on process $p$. For instance, if vector $v$ is of length $4 \cdot 10^6$ and we have four processes available, we would typically set $M_p = 10^6$ and store $(v_1, \ldots, v_{10^6})$ in the process with rank 0, $(v_{10^6+1}, \ldots, v_{2 \cdot 10^6})$ in the process with rank 1, and so on.

In the general case, where we assume that $M$ may not be divisible by $P$, we can use a simple (but not optimal) workload partitioning strategy by assigning the remainder from $M/P$ to the last process. More precisely, we first set $r = (M \bmod P)$, and then calculate the length of the local subvectors by $M_p = (M - r)/P$ for $p < P - 1$, and $M_{P-1} = (M - r)/P + r$. Using integer division in C we can simply write

```
M_p = M/P;   r = M % P;
```

As for the implementation, we assume that the process with rank 0 is in charge of reading $M$ from some user input, e.g., the command line. Then the value $M$ is broadcast to all the other processes. The next task of the program is to create the local segment of the vector and fill the vector components with appropriate numbers. Assume for simplicity that the components of vector $v$ have e.g. values $v_i = 3i + 2.2, 1 \le i \le M$. Then, on process $p$, local component $v_j^p$ equals $3(j + pM_p) + 2.2, 1 \le j \le M_p$.

After having initialized the local vector segment, we compute the square of the norm of the current segment: $N_p = \sum_{j=1}^{M_p} (v_j^p)^2$. The next step is to add the $N_p$ values from all the processes and write out the global norm $\sqrt{\sum_p N_p}$

of the $v$ vector in one of the processes, e.g. process 0. We mention that a minor modification of the following implementation can be used to compute the inner-product between two vectors, described in Section 1.1.4.

```c
#include <stdio.h>
#include <mpi.h>
#include <malloc.h>
#include <math.h>

int main(int argc, char** argv)
{
  int P, my_rank, i, j;
  int M = 1000;  /* length of global vector, default 1000 */
  int M_p;     /* length of vector segment for this process */
  int r;       /* the remainder of the length (for last proc) */
  double local_norm2, global_norm2;
  double* vector;

  MPI_Init (&argc, &argv);
  MPI_Comm_size (MPI_COMM_WORLD, &P);
  MPI_Comm_rank (MPI_COMM_WORLD, &my_rank);

  /* only the master process should read input */
  if (my_rank==0 && argc>1)
    M = atoi(argv[1]);

  /* broadcast the length to the other processes */
  MPI_Bcast (&M, 1, MPI_INT, 0, MPI_COMM_WORLD);

  M_p = M/P;  r = M % P;

  /* the remaining components will be placed on the last process, so
     the remainder variable is set to 0 on all other processes. */
  if (my_rank < P-1)   r = 0;

  /* create the vector segment on this process */
  vector = (double*)malloc((M_p+r)*sizeof(double));

  /* initialize vector  (simple formula: vector(i) = 3*i + 2.2) */
  for (j = 0; j < M_p+r; j++) {
    i = j + 1 + my_rank * M_p; /* find global index */
    vector[j] = 3*i + 2.2;
  }

  /* Compute the _square_ of the norm of the local vector segment
     (cannot simply add local norms; must add square of norms...) */
  local_norm2 = 0.;
  for (j = 0; j < M_p+r; j++)
    local_norm2 += vector[j]*vector[j];

  /* let the master process sum up the local results */
  MPI_Reduce (&local_norm2,&global_norm2,1,MPI_DOUBLE,
              MPI_SUM,0,MPI_COMM_WORLD);

  /* let only the master process write the result to the screen */
  if (my_rank==0)
    printf("\nThe norm of v(i) = 3*i + 2.2, i = (1,...,%d) is %g\n",
```

```
          M, sqrt(global_norm2));

  free (vector);

  MPI_Finalize ();
  return 0;  /* successful execution */
}
```

In the above MPI program, MPI_Bcast is used to "broadcast" the value of M, which is an integer available on process 0, to all the other processes. Moreover, the MPI_Reduce command collects the local results local_norm2 from all the processes and calculates the global result global_norm2, which will be available on process 0. We remark that both the MPI commands need to be invoked on *all* the processes, even though the input or output value is only available on process 0.

## 1.4   Basic Parallel Programming with Diffpack

Developing a parallel program is always more complicated than developing a corresponding sequential program. Practice shows that transforming a parallel algorithm into running code is often a time-consuming task, mainly because the parallel programming tools are primitive. The standard message passing protocol in wide use today is MPI, but MPI programming tends to be notoriously error-prone due to the many low-level message passing details, which need to be taken care of by the programmer.

To increase the human efficiency in developing parallel computer codes, we should develop a software environment where the programmer can concentrate on the principal steps of parallel algorithms, rather than on MPI-specific details. A desired situation will be that a programmer can start with developing a sequential solver and then in just a few steps transform this solver to a parallel version. Realization of such a software environment is indeed possible and requires a layered design of software abstractions, where all explicit MPI calls are hidden in the most primitive layer, and where the interface to message passing tools is simple and adapted to the programming standard of sequential solvers. We have developed this type of software environment in Diffpack, and the usage of the environment will be explained in the present section.

### 1.4.1   Diffpack's Basic Interface to MPI

MPI routines normally have a long list of input/output arguments. The task of calling them can be considerably simplified by using the strengths of C++. That is, overloaded functions, classes, and dynamic binding can be used to build a generic and simple interface to a subset of MPI routines that is needed for solving partial differential equations in parallel. In Diffpack, this interface consists of a class hierarchy with base class `DistrProcManager` (distributed process manager), which gives a generic representation of all the commonly needed message passing calls in the Diffpack framework. Class `DistrProcManagerMPI` is one concrete subclass where all the virtual member functions of `DistrProcManager` are implemented using MPI calls. The reason for keeping such a class hierarchy is to make a switch of message passing protocols painless to the user. That is, we can e.g. make another subclass `DistrProcManagerPVM` in the parallel Diffpack library, so that the user can use the other message passing standard PVM [10] without having to change the source code of his parallel Diffpack application. Therefore the reader only needs to get familiar with the functionality of class `DistrProcManager`.

*The DistrProcManager Class.* A Diffpack programmer normally calls a member function belonging to the `DistrProcManager` class, instead of calling a primitive message passing routine directly. Diffpack has a global handle[1] variable with name `proc_manager`, which is available everywhere in a program. At the beginning of every parallel Diffpack simulation, the `proc_manager` variable is bound to an object of some subclass of `DistrProcManager` (see above). As usual in object-oriented programming, the programmer can always concentrate on the base class interface to a class hierarchy, so we will only discuss `DistrProcManager` in the following.

Overloading is an attractive feature of C++ that distinguishes between calls to functions having the same name but with different argument lists. This has been used in the `DistrProcManager` class to allow the user to call, for instance, the `broadcast` function with a variable to be broadcast as parameter. The C++ run-time system will automatically choose the appropriate `broadcast` function depending on the argument type. On the other hand, using MPI's `MPI_Bcast` function directly would have required the datatype, along with several other arguments, to be specified explicitly.

The ability to use default arguments in C++ is another feature that is used frequently in `DistrProcManager`. Most calls to MPI routines require a communicator object as argument. However, the average user is unlikely to use any other communicator than `MPI_COMM_WORLD`. By using this communicator as a default argument to all the functions in `DistrProcManager`, the users are allowed to omit this argument unless they wish to use a different communicator. By taking advantage of overloaded functions and arguments with default values, the interface to common MPI calls is significantly simplified by the `DistrProcManager` class. The main results are twofold: (i) novice programmers of parallel computers get started more easily than with pure MPI, and (ii) the resulting code is cleaner and easier to read.

In the same manner as overloaded versions of the `broadcast` function are built into `DistrProcManager`, other useful functions are also present in

---

[1] A Diffpack handle is a smart pointer with internal functionality for reference counting.

this class. These include, e.g., the functions for sending and receiving messages, both blocking (MPI_Send and MPI_Recv) and non-blocking (MPI_Isend and MPI_Irecv). Overloaded versions of the functions cover communication of real, integer, string, and Diffpack vector variables. For example, native Diffpack vectors like Vec(real) can be sent or received directly, without the need for explicitly extracting and communicating the underlying C-style data structures of the Diffpack vector classes.

*Some Important DistrProcManager Functions.* To give the reader an idea about programming with the DistrProcManager interface, we list some of its important functions (often in their simplest form):

```
int getNoProcs();        // return number of processes
int getMyId();           // return the ID (rank) of current process,
                         // numbered from 0 to getNoProcs()-1
bool master();           // true: this is the master process (ID=0)
bool slave();            // true: this is a slave process (ID>0)

void broadcast (real& r);   // send r to all the processes
void broadcast (int&  i);
void broadcast (String& s);
void broadcast (VecSimple(int)& vec);
void broadcast (VecSimple(real)& vec);

// blocking communications:
void send (real& r, int to = 0, int tag = 0);
void recv (real& r, int from,   int tag);
void send (int&  i, int to = 0, int tag = 0);
void recv (int&  i, int from,   int tag);
void send (VecSimple(real)& vec, int to = 0, int tag = 0);
void recv (VecSimple(real)& vec, int from,   int tag);

// non-blocking communications:
void startSend (real& r, int to = 0, int tag = 0);
void startRecv (real& r, int from,   int tag);
void startSend (int&  i, int to = 0, int tag = 0);
void startRecv (int&  i, int from,   int tag);
void startSend (VecSimple(real)& vec, int to = 0, int tag = 0);
void startRecv (VecSimple(real)& vec, int from,   int tag);
void finishSend ();
void finishRecv ();
```

We remark that using the so-called blocking versions of the send and receive functions imply that communication and computation can not take place simultaneously. In contrast, the non-blocking version allows overlap between communication and computation. It is achieved by first invoking a startSend routine, then issuing some computation routines, and finally calling finishSend.

## 1.4.2   Example: The Norm of a Vector

For the purpose of illustrating how one can use class `DistrProcManager` as a high-level MPI interface, we re-implement here our simple example from Section 1.3.2 concerning parallel computation of the norm of a vector.

```cpp
#include <MenuSystem.h>
#include <Vec_real.h>
#include <DistrProcManager.h>
#include <initDpParallelBase.h>

int main(int argc, const char** argv)
{
  initDpParallelBase (argc, argv);
  initDiffpack (argc, argv);

  int M;        // length of global vector
  int M_p;      // length of vector segment for this process
  int r;        // the remainder of the length (for last proc)

  // only the master process (with rank 0) should read input:
  if (proc_manager->master())
    initFromCommandLineArg ("-L", M, 1000);

  // broadcast the length to the other processes
  proc_manager->broadcast(M);

  const int P = proc_manager->getNoProcs();
  M_p = M/P;   r = M % P;

  // the remaining components will be placed on the last process, so
  // the remainder variable is set to 0 on all other processes.
  if (proc_manager->getMyId() < P-1)   r = 0;

  // create the vector segment on this process:
  Vec(real) vector(M_p + r);

  // initialize vector  (simple formula: vector(i) = 3*i + 2.2)
  int i,j;
  for (j = 1; j <= vector.size(); j++) {
    i = j + proc_manager->getMyId() * M_p; // find global index
    vector(j) = 3*i + 2.2;
  }

  // Compute the _square_ of the norm of the local vector segment
  // (cannot simply add local norms; must add square of norms...)
  real local_norm2 = sqr(vector.norm()); // or vector.inner(vector)
  real global_norm2;
  // let the master process (with rank 0) sum up the local results
  proc_manager->reduce(local_norm2, global_norm2, PDP_SUM);

  // let only the master process write the result to the screen:
  if (proc_manager->master()) {
    s_o << "\nThe norm of v(i) = 3*i + 2.2, i = (1,...,"
        << M << ")\n" << " is " << sqrt(global_norm2) << "\n\n";
  }
  closeDpParallelBase();
```

```
   return 0;  // successful execution
}
```

We note that there are no direct calls to MPI routines in the above parallel program. All the communications are invoked through the global Diffpack variable `proc_manager`. The first initialization statement of the program, which is `initDpParallelBase`, takes care of `MPI_Init`, `MPI_Comm_rank`, and `MPI_Comm_size`, after having internally instantiated the global variable `proc_manager`. In this example, `proc_manager` is pointing to an object of class `DistrProcManagerMPI`. At the end of the program, `MPI_Finalize` is invoked internally within the closing statement `closeDpParallelBase`. If more advanced functionality for parallelizing, e.g., implicit finite element codes (see Section 1.6) is desired, the `initDpParallelBase` and `closeDpParallelBase` pair should be replaced by `initDpParallelLA` and `closeDpParallelLA`. Note also that the call of `reduce` in the above program invokes a global collective reduction operation. A global summation is invoked in the current example, due to the input argument `PDP_SUM`, which is Diffpack's equivalent of MPI's `MPI_SUM`. The member function `reduce` must be called on every process, and the result is only available on the master process (with rank 0).

## 1.5    Parallelizing Explicit FD Schemes

The topic of this section is how to parallelize sequential Diffpack simulators that employ explicit finite difference schemes. We will present a set of programming rules and the associated classes specially designed for this type of parallelization.

### 1.5.1    Partitioning Lattice Grids

Assume that we want to solve a system of partial differential equations over some spatial domain $\Omega$, which is covered by a rectangular lattice grid. For such a case, the easiest way of partitioning the global computation is to decompose the global lattice grid with cutting lines or planes that are perpendicular to the coordinate axes. This type of domain partitioning ensures that the resulting subdomains $\Omega_s$, $s = 1, \ldots, D$, can also be covered by rectangular lattice subgrids. We note that this rectangular domain partitioning scheme can use fewer space dimensions than that of the global lattice grid. For example, we can partition a 2D global lattice grid with only cutting lines perpendicular to the $x$-axis. However, the resulting subdomain lattice grids will always have the original number of space dimensions.

It is obvious that the above decomposition of the global lattice grid gives rise to a partitioning of the global computation, so that one lattice subgrid is assigned to one process. An immediate observation is that the same explicit finite difference scheme can be applied to each lattice subgrid. However, the

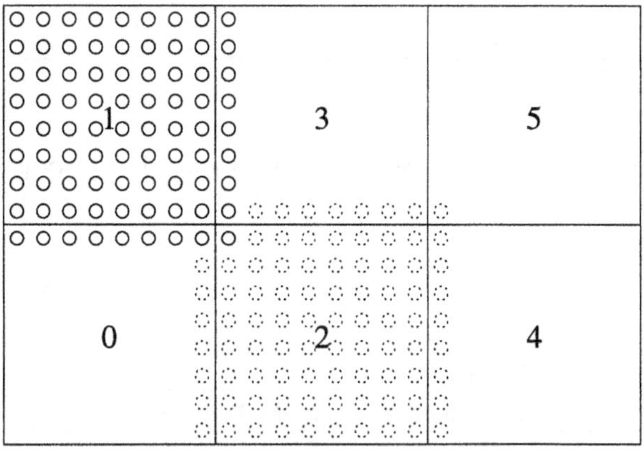

**Fig. 1.2.** An example of 2D overlapping subdomains.

grid point layers that lie beside the cutting lines or planes need special attention. This is because to update the point values in those layers, we need point values belonging to the neighboring subdomain(s). This can be done by exchanging arrays of point values between neighboring subdomains. A feasible data structure design is therefore to enlarge each lattice subgrid by one or several grid point layers beyond the cutting line or plane. For example, one grid point layer suffices for the standard second-order finite difference approximation $(u_{i-1} - 2u_i + u_{i+1})/\Delta x^2$ to second-order differentiation. The subdomains therefore become overlapping, see Figure 1.2 for an example of overlapping 2D subdomains.

After the enlargement of the data structure on each subdomain, we distinguish between two types of grid points, in addition to the genuine physical boundary points. These two types of grid points are: *computational points* whose values are updated by local computations, and *ghost boundary points* whose values come from the neighboring subdomains. Figure 1.3 shows an example of partitioning a global 1D lattice grid, where the vertical lines are the cutting lines. Each subdomain is enlarged with one ghost boundary point at each end. We have plotted grid points of the middle subdomain as circles above the axis, where empty circles represent the computational points and shaded circles represent the ghost boundary points. Note that the ghost boundary points of one subdomain are computational points of neighboring subdomains. Note also that the local computation will only traverse the computational points but make use of the values on the ghost boundary points. In effect, the data structure of each subdomain closely mimics that of the global domain, because we can view the ghost boundary points as having essential boundary conditions that are determined by the neighboring subdomains.

**Fig. 1.3.** An example of 1D lattice subgrids with ghost boundary points (shaded circles).

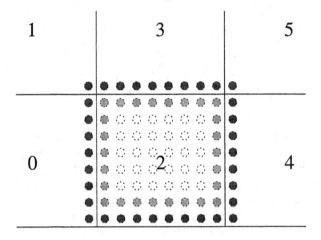

**Fig. 1.4.** The dark shaded grid points have offset +1, the light shaded points have offset 0, the clear grid points have offset -1 and less.

Another classification of the grid points can be convenient for dealing with communication and computations. Grid points at the boundary of a subdomain are said to have offset +1. The immediate neighboring points within the boundary are said to have offset 0, while the remaining interior points have offset -1 or less, as illustrated in Figure 1.4. The points with offset +1 are either part of the genuine physical boundary or they correspond to points with offset 0 on a neighboring subdomain. According to the above definitions of ghost boundary points and computational points, grid points with offset +1 are either ghost boundary points or physical boundary points, whereas grid points with zero or negative offset are computational points.

### 1.5.2    Problem Formulation

The most common application of explicit finite difference schemes is perhaps the solution of temporal evolution equations. This class of computation can be written as

$$u_i^+(x) = \mathcal{L}_i(u_1^-(x), \ldots, u_n^-(x)), \quad i = 1, \ldots, n, \qquad (1.11)$$

where $u_i(x)$ is a function of the spatial coordinates $x \in \Omega$, $\mathcal{L}_i$ is some spatial differential operator (possibly non-linear), and $n$ is the number of unknown functions in the underlying partial differential equation system. The superscript $+$ denotes an unknown function at the current time level, while the superscript $-$ refers to a function that is known numerically from previous time levels. Restriction of (1.11) to the subdomains is straightforward:

$$u_i^{s,+}(x) = \mathcal{L}_i(u_1^{s,-}(x), \ldots, u_{n_s}^{s,-}(x); g^-), \quad i = 1, \ldots, n_s, \ s = 1, \ldots, D.$$
$$(1.12)$$

The new solution over $\Omega_s$ is $u_i^{s,+}(x)$, while $u_i^{s,-}(x)$ are previously computed solutions over $\Omega_s$. Depending on the discretization of $\mathcal{L}_i$, previously computed functions $u_i^{q,-}(x)$ from *neighboring* subdomains $\Omega_q$, with $q$ in some suitable index set, are also needed when solving the subproblem on $\Omega_s$. This information is represented as the $g^-$ quantity in (1.12), which refers to the values on the ghost boundary points. In other words, the coupling between the subproblems is through previously computed functions, on the ghost boundary points, when we work with explicit temporal schemes. At a new time level, we can then solve the subproblems over $\Omega_s$, $s = 1, \ldots, D$ in parallel. Thereafter, we need to exchange the $g^-$ information between the subdomains before we proceed to the next time level. The mathematical and numerical details will become clearer when we study a specific initial-boundary value problem below.

One important observation now is that sequential software can be re-used for updating the values on the computational points on each subdomain. For exchanging data between neighboring subdomains, additional functionality needs to be developed. When using C++, Diffpack and object-oriented concepts, we can build the software for a parallel solver in a clean way, such that the debugging of e.g. communication on parallel computers is minimized. First, one develops a standard sequential solver class for the problem over $\Omega$ and tests this solver. Thereafter, one derives a subclass with additional general data structures for boundary information exchange. This subclass contains little code and relies on the numerics of the already well-tested base class for the global problem, as well as on a general high-level communication tool for exchange of boundary information.

The subclass solver will be used to solve problems on every subdomain $\Omega_s$, $s = 1, \ldots, D$. In addition, we need a manager class per subdomain that holds some information about the relations between the subdomains $\Omega_s$. The principle is that the managers have a global overview of the subproblems, while the subproblem solvers have no explicit information about other subproblem solvers. At any time, one can pull out the original sequential solver to ensure that its numerics work as intended. With this set-up, it is fairly simple to take an existing sequential solver, equip it with two extra small classes and turn it into a simulation code that can take advantage of parallel computers. This strategy may have a substantial impact on the development time of parallel simulation software.

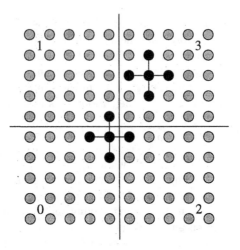

**Fig. 1.5.** Updating the value in subdomain 0 (points with offset 0) requires values from subdomains 1 and 2. The indicted value in subdomain 3 (point with offset -1) depends only on local information and can be updated without communicating any values from the neighbors.

### 1.5.3  Hiding the Communication Cost

When applying the finite difference scheme on a subdomain, we need boundary values from the neighboring subdomain at the points with offset +1. These values are used in the difference equations for the points with offset 0. See Figure 1.5 for the example of a five-point computational stencil. The main idea in the parallel algorithm is to communicate these boundary values while applying the finite difference scheme to all points with offset -1 or less. When the boundary values are received, one can apply the scheme to the points with offset 0 and the solution over a subdomain is complete.

On many parallel computers, computations and communication can be carried out simultaneously. In MPI, for instance, this is accomplished by non-blocking send and receive calls. We can therefore devise the following algorithm for a subdomain solver:

1. Send requested points to all neighboring subdomains (these points will be used as boundary conditions in the neighboring subdomain solvers).

2. Update the solution at all purely inner points (those with offset -1 and less).

3. Receive boundary values at points with offset +1 from all neighboring subdomains.

4. Update the solution at the points with offset 0.

For small grids we may have to wait for the messages to arrive, because the time spent on communication may exceed the time spent on computing inner points. The length of the message depends on the circumference of the grid, while the number of inner values to be computed depend on the area of the grid (cf. Section 1.2.1). Therefore, as the size of the grid grows, the time spent on computing will grow faster than the time spent on communication, and at some point the communication time will be insignificant. The speedup will then be (close to) optimal; having $P$ processors reduces the CPU time by approximately a factor of $P$.

### 1.5.4  Example: The Wave Equation in 1D

Consider the one-dimensional initial-boundary value problem for the wave equation:

$$\frac{\partial^2 u}{\partial t^2} = \gamma^2 \frac{\partial^2 u}{\partial x^2}, \quad x \in (0,1), \ t > 0, \tag{1.13}$$

$$u(0,t) = U_L, \tag{1.14}$$

$$u(1,t) = U_R, \tag{1.15}$$

$$u(x,0) = f(x), \tag{1.16}$$

$$\frac{\partial}{\partial t} u(x,0) = 0. \tag{1.17}$$

Let $u_i^k$ denote the numerical approximation to $u(x,t)$ at the spatial grid point $x_i$ and the temporal grid point $t_k$, where $i$ and $k$ are integers, $1 \le i \le n$ and $k \ge -1$ ($k = 0$ refers to the initial condition and $k = -1$ refers to an artificial quantity that simplifies the coding of the scheme). Assume here that $x_i = (i-1)\Delta x$ and $t_k = k\Delta t$. With the definition of the Courant number as $C = \gamma \Delta t / \Delta x$, a standard finite difference scheme for this equation reads

$$u_i^0 = f(x_i), \quad i = 1, \ldots, n,$$

$$u_i^{-1} = u_i^0 + \frac{1}{2} C^2 (u_{i+1}^0 - 2u_i^0 + u_{i-1}^0), \quad i = 2, \ldots, n-1,$$

$$u_i^{k+1} = 2u_i^k - u_i^{k-1} + C^2 (u_{i+1}^k - 2u_i^k + u_{i-1}^k), \quad i = 2, \ldots, n-1, \ k \ge 0,$$

$$u_1^{k+1} = U_L, \quad k \ge 0,$$

$$u_n^{k+1} = U_R, \quad k \ge 0.$$

A domain partitioning for the numerical scheme is easily formulated by dividing the domain $\Omega = [0,1]$ into overlapping subdomains $\Omega_s$, $s = 1, \ldots, D$, and applying the scheme to the interior points in each subdomain. More precisely, we define the subdomain grids as follows.

$$\Omega_s = [x_L^s, x_R^s], \quad x_i^s = x_L^s + (i-1)\Delta x, \ i = 1, \ldots, n_s.$$

24    X. Cai et al.

Here, $n_s$ is the number of grid points in $\Omega_s$. For the end points $x_L^s$ and $x_R^s$ of subdomain $\Omega_s$, we have

$$x_1^s = x_L^s, \quad x_{n_s}^s = x_R^s.$$

The physical boundaries are given by

$$x_L^1 = 0, \quad x_R^D = 1.$$

The scheme for $u_i^{k+1}$ involves a previously computed value on the grid point to the left $(u_{i-1}^k)$ and the grid point to the right $(u_{i+1}^k)$. We therefore need one grid cell overlap between the subdomains. This implies some relations between the start and end grid points of the subdomain, which are listed here (although they are not of any practical use in the implementation):

$$x_1^s = x_{n_{s-1}-1}^{s-1}, \ x_2^s = x_{n_{s-1}}^{s-1}, \quad x_{n_s-1}^s = x_1^{s+1}, \ x_{n_s}^s = x_2^{s+1},$$

for $s = 2, \ldots, D - 1$. The relations assume that the subdomains $\Omega_1, \Omega_2, \Omega_3$ and so on are numbered from left to right.

*Remark.* In multi-dimensional problems, a precise mathematical notation for the overlap is probably not feasible. Instead, we only know that there is some kind of overlap, and we need the principal functionality of evaluating $u$ in a neighboring subdomain at a specified point. Let $\mathcal{N}_s$ be an index set containing the neighboring subdomains of $\Omega_s$ (in 1D $\mathcal{N}_s = \{s-1, s+1\}$). In general, we need an interpolation operator $I(x, t; \mathcal{N}_s)$ that interpolates the solution $u$ at the point $(x, t)$, where $x$ is known to lie in one of the subdomains $\Omega_q, q \in \mathcal{N}_s$. We note that if the grid points in the subdomain grids coincide, $I(x, t; \mathcal{N}_s)$ just makes an appropriate copy.

Defining $u_i^{s,k}$ as the numerical solution over $\Omega_s$, we can write the numerical scheme for subdomain $\Omega_s$ as follows.

$$u_i^{s,0} = f(x_i), \quad i = 1, \ldots, n_s,$$

$$u_i^{s,-1} = u_i^{s,0} + \frac{1}{2}C^2(u_{i+1}^{s,0} - 2u_i^{s,0} + u_{i-1}^{s,0}), \quad i = 2, \ldots, n_s - 1,$$

$$u_i^{s,k+1} = 2u_i^{s,k} - u_i^{s,k-1} + C^2(u_{i+1}^{s,k} - 2u_i^{s,k} + u_{i-1}^{s,k}), \ i = 2, \ldots, n_s - 1, \ k \geq 0,$$

$$\text{if } x_1^s = 0, \ u_1^{s,k+1} = U_L, \quad \text{else} \quad u_1^{s,k+1} = I(x_1^s, t_{k+1}; \mathcal{N}_s), \quad k \geq 0,$$

$$\text{if } x_{n_s}^s = 1, \ u_{n_s}^{s,k+1} = U_R, \quad \text{else} \quad u_{n_s}^{s,k+1} = I(x_{n_s}^s, t_{k+1}; \mathcal{N}_s), \quad k \geq 0.$$

Apart from the $s$ superscript and the need to interpolate values between neighboring subdomains, this scheme is identical to the scheme for the global problem. In other words, we can re-use most of the sequential code for the original global problem, provided that this code can deal with an arbitrary subdomain and that we add the interpolation functionality. This latter feature involves communication between subdomains, implying communication between processors.

In the following, we present a plain C program for the one-dimensional wave equation problem and equip it with direct MPI calls for communicating boundary values. We remark that the lower bound of an array index in C starts at 0, instead of 1 in Diffpack. In the program, we also use an integer variable n_i, which has value equal to $n_s - 2$. Knowledge about the forthcoming details of the implementation is not required elsewhere in the chapter, so readers who want to concentrate on Diffpack's parallel programming interface can safely move on to the next section.

```c
#include <stdio.h>
#include <malloc.h>
#include <mpi.h>

int main (int nargs, char** args)
{
  int size, my_rank, i, n = 1001, n_i, lower_bound;
  double h, x, t, dt, gamma = 1.0, C = 0.9, tstop = 1.0;
  double *up, *u, *um, umax = 0.05, UL = 0., UR = 0.;
  MPI_Status status;

  MPI_Init (&nargs, &args);
  MPI_Comm_size (MPI_COMM_WORLD, &size);
  MPI_Comm_rank (MPI_COMM_WORLD, &my_rank);

  if (my_rank==0 && nargs>1)    /* total number of points in [0,1] */
    n = atoi(args[1]);

  if (my_rank==0 && nargs>2)    /* read in Courant number */
    C = atof(args[2]);

  if (my_rank==0 && nargs>3)    /* length of simulation */
    tstop = atof(args[3]);

  MPI_Bcast (&n, 1, MPI_INT, 0, MPI_COMM_WORLD);
  MPI_Bcast (&C, 1, MPI_DOUBLE, 0, MPI_COMM_WORLD);
  MPI_Bcast (&tstop, 1, MPI_DOUBLE, 0, MPI_COMM_WORLD);

  h = 1.0/(n-1);   /* distance between two points */

  /* find # of local computational points, assume mod(n-2,size)=0 */
  n_i = (n-2)/size;
  lower_bound = my_rank*n_i;

  up = (double*) malloc ((n_i+2)*sizeof(double));
  u  = (double*) malloc ((n_i+2)*sizeof(double));
  um = (double*) malloc ((n_i+2)*sizeof(double));

  /* set initial conditions */
  x = lower_bound*h;
  for (i=0; i<=n_i+1; i++) {
    u[i] = (x<0.7) ? umax/0.7*x : umax/0.3*(1.0-x);
    x += h;
  }
  if (my_rank==0) u[0] = UL;
  if (my_rank==size-1) u[n_i+1] = UR;
  for (i=1; i<=n_i; i++)
```

```
    um[i] = u[i]+0.5*C*C*(u[i+1]-2*u[i]+u[i-1]); /* artificial BC */
  dt = C/gamma*h;
  t = 0.;
  while (t < tstop) {      /* time stepping loop */
    t += dt;
    for (i=1; i<=n_i; i++)
      up[i] = 2*u[i]-um[i]+C*C*(u[i+1]-2*u[i]+u[i-1]);

    if (my_rank>0) {
      /* receive from left neighbor */
      MPI_Recv (&(up[0]),1,MPI_DOUBLE,my_rank-1,501,MPI_COMM_WORLD,
                &status);
      /* send left neighbor */
      MPI_Send (&(up[1]),1,MPI_DOUBLE,my_rank-1,502,MPI_COMM_WORLD);
    }
    else
      up[0] = UL;

    if (my_rank<size-1) {
      /* send to right neighbor */
      MPI_Send (&(up[n_i]),1,MPI_DOUBLE,my_rank+1,501,MPI_COMM_WORLD);
      /* receive from right neighbor */
      MPI_Recv (&(up[n_i+1]),1,MPI_DOUBLE,my_rank+1,502,
                MPI_COMM_WORLD,&status);
    }
    else
      up[n_i+1] = UR;

    /* prepare for next time step */
    um = u;  u = up;  up = um;
  }

  free (um); free (u); free (up);
  MPI_Finalize ();
  return 0;
}
```

### 1.5.5   How to Create Parallel FD Solvers in Diffpack

The process of building a parallel finite difference solver in Diffpack can be summarized as follows. First, one develops a standard sequential solver, using grid and field abstractions (GridLattice, FieldLattice etc, see [6, Ch. 1.6.5]). The "heart" of such a solver is the loops implementing the explicit finite difference scheme. These loops should be divided into two parts: loops over the inner points and loops over the boundary points. As we shall see later, this is essential for parallel programming. The solver must also be capable of dealing with an arbitrary rectangular grid (a trivial point when using Diffpack grids and fields).

We recall (cf. [6, Ch. 7.2]) that it is customary in Diffpack to assemble independent solvers for the different components in solving a system of partial differential equations. It is therefore useful to have a scan function in the solver that can either create the grid internally or use a ready-made external grid. We want the sequential solver to solve the partial differential equations on its own, hence it must create a grid. On the other hand, when it is a part of a parallel algorithm, the grid is usually made by some tool that has an overview of the global grid and the global solution process. This type of flexibility is also desirable for other data members of the solver (e.g. time parameters).

Having carefully tested a flexible sequential solver on the global domain, we proceed with the steps towards a parallel solver as follows:

1. Derive a subclass of the sequential solver that implements details specific to parallel computing. This subclass is supposed to be acting as subdomain solver. The code of the subclass is usually very short, since major parts of the sequential solver can be re-used also for parallel computing. Recall that this was our main goal of the mathematical formulation of the initial-boundary value problems over the subdomains. The subclass mainly deals with exchanging boundary values with neighboring subdomains.

2. Make a manager class for administering the subdomain solvers. The manager has the global view of the concurrent solution process. It divides the grid into subdomains, calls the initialization processes in the subdomain solver, and administers the time loop. The subdomain solver has a pointer to the manager class, which enables insight into neighboring subdomain solvers for communication etc. The code of the manager class is also short, mainly because it can be derived from a general toolbox for parallel programming of explicit finite difference schemes (see `ParallelFD` in Section 1.5.6).

### 1.5.6   Tools for Parallel Finite Difference Methods

*Flexible Assignment of Process Topology.* In many applications, the order in which the processes are arranged is not of vital importance. By default, a group of $P$ processes are given ranks from 0 to $P-1$. However, this linear ranking of the processes does not always reflect the geometry of the underlying numerical problem. If the numerical problem adopts a two- or three-dimensional grid, a corresponding two- or three-dimensional process grid would reflect the communication pattern better. Setting up the process topology can be complicated, and the required source code follows the same pattern in all the programs. In Diffpack, the topology functionality is offered by class `TopologyMPI`, which has a subclass `CartTopologyMPI`. By using `CartTopologyMPI` the user can organize the processes in a Cartesian structure with a row-major process numbering beginning at 0 as indicated in Figure 1.6. The `CartTopologyMPI` class also provides functions that return the rank of a process given its process coordinates and vice versa. We refer to the MPI

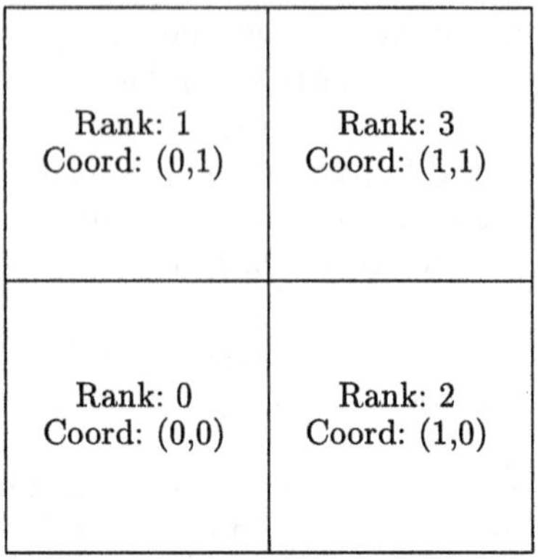

| | |
|---|---|
| Rank: 1<br>Coord: (0,1) | Rank: 3<br>Coord: (1,1) |
| Rank: 0<br>Coord: (0,0) | Rank: 2<br>Coord: (1,0) |

**Fig. 1.6.** Relation between coordinates and ranks for four processes in a ($2 \times 2$) process grid.

documentation (e.g. [3]) for more information on the topology concept and functionality.

For example, the user can use the following user-friendly input format

```
CART d=2 [2,4] [ false false ]
```

for setting up a two-dimensional Cartesian process grid with $2 \times 4$ processes, non-periodic in both directions. We remark that being non-periodic in one direction means that the "head" processes of that direction do not communicate with the "tail" processes. We refer to MPI's MPI_Cart_create, MPI_Cart_coords, and MPI_Cart_rank functions for more details.

We remark that some parallel computers do not completely follow the multiprocessor model from Section 1.1.3. The communication network may be asymmetric, in that a processor has large bandwidth connections to some processors but small bandwidth to others. It is thus important to notice the difference between the logical topology (also called virtual topology) of processes and the underlying physical layout of the processors and the network. A parallel computer may exploit the logical process topology when assigning the processes to the physical processors, if it helps to improve the communication performance. If the user does not specify any logical topology, the parallel computer will make a random mapping, which may lead to unnecessary contention in the communication network. Therefore, setting up the logical topology may give performance benefits as well as large benefits for program readability.

*The Grid and Process Handler: ParallelFD.* Class ParallelFD is the parallel toolbox for parallelizing Diffpack FD solvers. It contains functionality generally needed for solving problems using finite difference methods on parallel computers. The `ParallelFD` class uses the basic parallel class `DistrProcManager`, which supplies information such as rank of the process and total number of processes. It also holds the topology object from which we can obtain information related to the multidimensional topology.

One of the main tasks of the manager class is to divide the global grid into subdomains. This general problem is supported by class `ParallelFD`. The global grid is here divided in a manner such that all the local grids are approximately of the same size, ensuring that the load is evenly balanced between the processes. There are two versions of the function `initLocalGrid`, one for ordinary finite difference grids, and one for staggered grids. The latter version is more advanced and allows a different size of the overlap in the different space directions.

The `ParallelFD` class contains menu functionality for a `TopologyMPI` object and some very useful utility functions for computing the local grid for a processor

```
void initLocalGrid (GridLattice& local_grid,
                    const GridLattice& global_grid);
```

In addition, class `ParallelFD` has functionality for creating a global representation of the unknown field for visualization purposes (the global field can have coarser resolution than the local fields if there is not sufficient memory available). The function

```
void gatherFields(FieldLattice& local_res,
                  FieldLattice& global_res_u);
```

is then used to deliver a local part of a field to the global representation of the field.

*Communication Between Processes: The Points Classes.* A central issue in parallel algorithms for solving partial differential equations is to communicate values at the boundaries between subdomains. Some of the required functionality is common to a wide range of problems, and can hence be collected in a general class, here called `Points`, whereas other types of functionality depend on the particular solution method being used. The special functionality for explicit finite difference methods is collected in class `PointsFD`, which is a subclass of `Points`. The subclass `PointsFD` is mainly concerned with setting up the proper data structures for the information on boundary points.

The `Points` classes offer functionality for

– setting up data structures for the boundary points and values to be sent and received,

– sending boundary values to neighbors,

– receiving boundary values from neighbors.

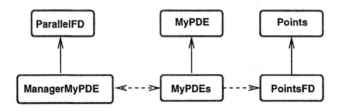

**Fig. 1.7.** Sketch of a sequential solver, its subclass, manager, and the toolboxes that these classes utilize. Solid arrow indicate class derivation ("is-a" relationship), whereas dashed arrows indicate pointers ("has-a" relationship).

Figure 1.7 shows a schematic design of a parallel Diffpack FD simulator. Assuming there exists an original sequential simulator MyPDE, the main work of parallelization consists in extending two subclasses from respectively MyPDE and ParallelFD. Subclass MyPDEs is to work as a subdomain solver and subclass ManagerMyPDE is a manager for communication related tasks. Note that the two subclasses are to work closely with PointsFD. We explain how these two subclasses can be written in the following text concerning the creation of a parallel heat conduction simulator.

### 1.5.7    Example: Heat Conduction in 2D

We consider a scaled heat conduction problem:

$$\frac{\partial u}{\partial t} = \frac{\partial^2 u}{\partial x^2} + \frac{\partial^2 u}{\partial y^2}, \quad (x,y) \in (0,1) \times (0,1), \ t > 0, \tag{1.18}$$

$$u(0,y,t) = 0, \quad t > 0, \tag{1.19}$$

$$u(1,y,t) = 0, \quad t > 0, \tag{1.20}$$

$$u(x,0,t) = 0, \quad t > 0, \tag{1.21}$$

$$u(y,1,t) = 0, \quad t > 0, \tag{1.22}$$

$$u(x,y,0) = f(x,y). \tag{1.23}$$

Following the notation used in the previous example, we define $u_{i,j}^k$ as the approximation to $u(x,y,t)$ on the spatial grid point $(x_i, y_j)$ at time step $t_k$, where the integers $i$, $j$ and $k$ are given by $i = 1, \ldots, n$, $j = 1, \ldots, m$ and $k \geq 0$, and we have $x_i = (i-1)\Delta x$, $y_j = (j-1)\Delta y$ and $t_k = k\Delta t$.

Discretization of the equations (1.18–1.23) using an explicit finite difference scheme gives

$$u_{i,j}^0 = f(x_i, y_j), \quad i = 1, \ldots, n, \; j = 1, \ldots, m,$$

$$u_{i,j}^{k+1} = u_{i,j}^k + \Delta t \left( [\delta_x \delta_x u]_{i,j}^k + [\delta_y \delta_y u]_{i,j}^k \right), \quad \begin{array}{l} i = 2, \ldots, n-1, \\ j = 2, \ldots, m-1, \end{array} \; k > 0,$$

$$u_{1,j}^{k+1} = u_{n,j}^{k+1} = 0, \quad j = 1, \ldots, m, \; k \geq 0,$$

$$u_{i,1}^{k+1} = u_{i,m}^{k+1} = 0, \quad i = 1, \ldots, n, \; k \geq 0,$$

where

$$[\delta_x \delta_x u]_{i,j}^k \equiv \frac{1}{\Delta x^2} \left( u_{i-1,j}^k - 2u_{i,j}^k + u_{i+1,j}^k \right),$$

$$[\delta_y \delta_y u]_{i,j}^k \equiv \frac{1}{\Delta y^2} \left( u_{i,j-1}^k - 2u_{i,j}^k + u_{i,j+1}^k \right).$$

We wish to split the domain $\Omega$ into $D$ subdomains $\Omega_s$, where $s = 1, \ldots, D$. However, since we now work with two-dimensional, rectangular, finite difference lattice grids, it is convenient to introduce a subdomain numbering using double indices $(p, q)$ in a "Cartesian" grid of $P \times Q$ subdomains. When we need to switch between a single index $s$ and the pair $(p, q)$, we assume that there is a mapping $\mu$ such that $s = \mu(p, q)$ and $(p, q) = \mu^{-1}(s)$. The subdomains can now be defined as follows.

$$\Omega_{p,q} = [x_L^p, x_R^p] \times [y_B^q, y_T^q]$$

$$x_i^p = x_L^p + (i - 1)\Delta x, \; i = 1, \ldots, n_p$$

$$y_j^q = y_B^q + (j - 1)\Delta y, \; j = 1, \ldots, m_q$$

$$x_1^p = x_L^p, \; x_{n_p}^p = x_R^p, \; y_1^q = y_B^q, \; y_{m_q}^q = y_T^q,$$

$$x_L^1 = 0, \; x_R^P = 1, \; y_B^1 = 0, \; y_T^Q = 1.$$

Here, we have $p = 1, \ldots, P$, $q = 1, \ldots, Q$ and $D = PQ$.

Our numerical scheme has a five-point computational stencil, i.e., for each point, we only need values on points that are one grid cell away. This implies that the optimal overlap between the subdomains is one grid cell. For $p = 2, \ldots, P - 1$ and $q = 2, \ldots, Q - 1$ we then have the relations

$$x_1^p = x_{n_{p-1}-1}^{p-1}, \quad x_2^p = x_{n_{p-1}}^{p-1}, \quad x_{n_p-1}^p = x_1^{p+1}, \quad x_{n_p}^p = x_2^{p+1},$$

$$y_1^q = y_{m_{q-1}-1}^{q-1}, \quad y_2^q = y_{m_{q-1}}^{q-1}, \quad y_{m_q-1}^q = y_1^{q+1}, \quad y_{m_q}^q = y_2^{q+1}.$$

If either $P = 1$ or $Q = 1$ we note that we have a one-dimensional partitioning of the domain.

The set of neighboring subdomains $\mathcal{N}_{p,q} = \mathcal{N}_{\mu^{-1}(s)}$ is given by

$$\mathcal{N}_{p,q} = \{(p + 1, q), (p - 1, q), (p, q + 1), (p, q - 1)\}.$$

In conformity with the one-dimensional case (Section 1.5.4), we denote the interpolation operator by $I(x, y, t; \mathcal{N}_{p,q})$ which interpolates the numerical solution at $(x, y, t)$, where $(x, y)$ is in one of the neighboring subdomains $\Omega_{k,l}$ of $\Omega_{p,q}$, i.e., $(k, l) \in \mathcal{N}_{p,q}$. The numerical equations for subdomain $\Omega_{p,q}$ then read

$$u_{i,j}^{p,q,k+1} = u_{i,j}^{p,q,k} + \Delta t([\delta_x \delta_x u^{p,q}]_{i,j}^k + [\delta_y \delta_y u^{p,q}]_{i,j}^k), \quad \begin{matrix} i = 2, \ldots, n_p - 1, \\ j = 2, \ldots, m_q - 1, \end{matrix} \quad k > 0,$$

$$u_{1,j}^{p,q,k+1} = I(x_1^p, y_j^q, t_{k+1}; \mathcal{N}_{p,q}), \quad j = 1, \ldots, m_q, \ k > 0,$$

$$u_{n_p,j}^{p,q,k+1} = I(x_{n_p}^p, y_j^q, t_{k+1}; \mathcal{N}_{p,q}), \quad j = 1, \ldots, m_q, \ k > 0,$$

$$u_{i,1}^{p,q,k+1} = I(x_i^p, y_1^q, t_{k+1}; \mathcal{N}_{p,q}), \quad i = 1, \ldots, n_p, \ k > 0,$$

$$u_{i,m_q}^{p,q,k+1} = I(x_i^p, y_{m_q}^q, t_{k+1}; \mathcal{N}_{p,q}), \quad i = 1, \ldots, n_p, \ k > 0.$$

For subdomains that border with the physical boundaries, the following updating schemes apply for the points that lie on the physical boundaries:

$$u_{1,j}^{1,q,k+1} = 0, \quad j = 1, \ldots, m_q, \ k \geq 0,$$

$$u_{n_P,j}^{P,q,k+1} = 0, \quad j = 1, \ldots, m_q, \ k \geq 0,$$

$$u_{i,1}^{p,1,k+1} = 0, \quad i = 1, \ldots, n_p, \ k \geq 0,$$

$$u_{i,m_Q}^{p,Q,k+1} = 0, \quad i = 1, \ldots, n_p, \ k \geq 0.$$

It is evident that a sequential program solving the global problem can be reused for solving the problem over each subdomain, provided that the domain can be an arbitrary rectangle and that we add functionality for communicating interpolated values between the subdomain solvers. If the subdomains have coinciding grid points, which is the usual case when working with basic finite difference methods, interpolation just means picking out the relevant grid point values from a neighboring subdomain.

## 1.5.8   The Sequential Heat Conduction Solver

Following standard Diffpack examples on creating finite difference solvers, see e.g. [6, Ch. 1.7], the heat equation simulator is typically realized as a C++ class with the following content.

```
class Heat2D
{
protected:
  Handle(GridLattice)  grid;      // uniform "finite difference" grid
  Handle(FieldLattice) u;         // the unknown field to be computed
  Handle(FieldLattice) u_prev;    // u at the previous time level
  Handle(TimePrm)      tip;       // time step etc.

  CPUclock clock;                 // for timings within the program
  real compute_time;
```

```
int i0, in, j0, jn;              // help variables for loops
real mu, nu;                     // help variables in the scheme

real initialField (real x, real y );    // initial condition func.

virtual void setIC ();
virtual void timeLoop ();
void computeInnerPoints ();
void computeBoundaryPoints ();
void solveAtThisTimeStep ();
void updateDataStructures () { *u_prev = *u; }
public:
Heat2D ();
virtual ~Heat2D () {}

void scan (GridLattice* grid_ = NULL, TimePrm* tip_ = NULL);
void solveProblem ();
virtual void resultReport ();

};
```

In the above class declaration, we have introduced two non-standard functions computeInnerPoints and computeBoundaryPoints, which separate the computations on inner and boundary points. This is an important split of the implementation of the scheme, because non-blocking communication calls can then be used to hide the cost of communication. Another non-standard feature is that scan can make use of an external grid and external time integration parameters.

The particular numerical problem solved by class Heat2D is

$$\frac{\partial u}{\partial t} = \frac{\partial^2 u}{\partial x^2} + \frac{\partial^2 u}{\partial y^2}$$

on the unit square $\Omega = [0, 1] \times [0, 1]$ with $u = 0$ on the boundary. The initial condition reads

$$u(x, y, 0) = \sin \pi x \sin \pi y \,.$$

The computeInnerPoints function typically takes the form

```
void Heat2D:: computeInnerPoints()
{
  int i, j;
  ArrayGenSel(real)& U  = u->values();
  ArrayGenSel(real)& Up = u_prev->values();
  for (j = j0+2; j <= jn-2; j++) {
    for (i = i0+2; i <= in-2; i++) {
      U(i,j) = Up(i,j) + mu*(Up(i-1,j) - 2*Up(i,j) + Up(i+1,j))
                       + nu*(Up(i,j-1) - 2*Up(i,j) + Up(i,j+1));
    }
  }
}
```

Here, i0, j0, in, and jn are precomputed start and stop indices for the loops over the grid points. We remark that the speed of the loop can be enhanced by working directly on the underlying C array as explained in [6, App. B.6.4]. As always, we emphasize that such optimizations should be performed after the code is thoroughly tested.

Notice that the i and j loops in computeInnerPoints touch neither the subdomain boundary points nor the ghost boundary points. Updating the solution at the subdomain boundaries takes place in computeBoundaryPoints:

```
void Heat2D:: computeBoundaryPoints()
{
  int i,j;
  ArrayGenSel(real)& U  = u->values();
  ArrayGenSel(real)& Up = u_prev->values();

  for (j = j0+1; j <= jn-1; j++) {
    U(i0+1,j) = Up(i0+1,j)
                + mu*(Up(i0,j) - 2*Up(i0+1,j) + Up(i0+2,j))
                + nu*(Up(i0+1,j-1) - 2*Up(i0+1,j) + Up(i0+1,j+1));
    U(in-1,j) = Up(in-1,j)
                + mu*(Up(in-2,j) - 2*Up(in-1,j) + Up(in,j))
                + nu*(Up(in-1,j-1) - 2*Up(in-1,j) + Up(in-1,j+1));
  }
  for (i = i0+1; i <= in-1; i++) {
    U(i,j0+1) = Up(i,j0+1)
                + mu*(Up(i-1,j0+1) - 2*Up(i,j0+1) + Up(i+1,j0+1))
                + nu*(Up(i,j0) - 2*Up(i,j0+1) + Up(i,j0+2));
    U(i,jn-1) = Up(i,jn-1)
                + mu*(Up(i-1,jn-1) - 2*Up(i,jn-1) + Up(i+1,jn-1))
                + nu*(Up(i,jn-2) - 2*Up(i,jn-1) + Up(i,jn));
  }

  // boundary values are never recalculated, hence the boundary
  // conditions will not be set at every time step
}
```

### 1.5.9   The Parallel Heat Conduction Solver

*The Subclass Solver.* The basic finite difference schemes for updating the solution at inner and boundary points are provided by the sequential solver, which is class Heat2D in our current example. Communication of boundary values between processors is the only additional functionality we need for turning the sequential solver into parallel code. We add the communication functionality in a subclass Heat2Ds. Its header is like this:

```
class Heat2Ds: public HandleId, public Heat2D
{
  Ptv(int) coords_of_this_process; // my processor coordinates
  Ptv(int) num_solvers;            // # of procs in each space dir
  int s;                           // my subdomain number is s
  VecSimple(int) neighbors;        // identify my neighbors
  ManagerHeat2D* boss;             // my manager (for global info)
```

```
  PointsFD bc_points;              // boundary communication points

  void initBoundaryPoints ();      // init the bc_points object
  void setBC1 ();                  // send request for boundary values
  void setBC2 ();                  // receive & insert boundary values
  void solveAtThisTimeStep();
public:
  Heat2Ds (ManagerHeat2D* boss_);
  ~Heat2Ds () {}

  void scan();
  virtual void resultReport();

  friend class ManagerHeat2D;

};
```

Class `Heat2Ds` inherits the functionality from class `Heat2D`, but extends it for parallel computing:

- We have some data members for representing the processor coordinates of the current process, identification of neighbors etc.
- A `PointsFD` structure holds the boundary points for communication (the points to send and receive and their values).
- `initBoundaryPoints` initializes the `PointsFD` structure.
- `setBC1` applies the `PointsFD` structure for sending the boundary values to the neighbors.
- `setBC2` applies the `PointsFD` structure for receiving the boundary values from the neighbors.

Looking at the following parts of the source code, we observe that there is a close relation between the program abstractions and the formulation of the parallel algorithm:

```
void Heat2Ds:: setBC1 ()
{
  bc_points.fillBoundaryPoints(u());
  bc_points.sendValuesToNeighbors(neighbors);
}

void Heat2Ds:: setBC2 ()
{
  bc_points.receiveValuesFromNeighbors(neighbors.size());
  bc_points.extractBoundaryPoints(u_prev());
}

void Heat2Ds:: solveAtThisTimeStep()
{
  setBC1();
  computeInnerPoints();
```

```
  setBC2();
  computeBoundaryPoints();
  updateDataStructures();
}
```

The initialization of the `PointsFD` structure goes as follows.

```
void Heat2Ds:: initBoundaryPoints ()
{
  bc_points.initBcPointList(coords_of_this_process, num_solvers, *u);
  bc_points.initBcComm(neighbors, *u, coords_of_this_process);
}
```

*The Manager Class.* The particular `ManagerHeat2D`, inheriting the common `ParallelFD` functionality and making use of the `Heat2Ds` class, will be quite simple and can look like this:

```
class ManagerHeat2D : public ParallelFD, public SimCase
{

  friend class Heat2Ds;

  Handle(Heat2Ds) solver;

  // hold global grid information:
  Handle(GridLattice) global_grid;

  // global variables for the coarse grid used to report the results
  Handle(GridLattice)  global_resgrid;
  Handle(FieldLattice) global_res_u;

  Handle(TimePrm) tip;
  Handle(SaveSimRes) database;

  VecSimple(int) neighbors;

  void scan (MenuSystem& menu);
  void define (MenuSystem& menu, int level = MAIN);
  void timeLoop ();

public:

  ManagerHeat2D ();
  virtual ~ManagerHeat2D () {}

  virtual void adm (MenuSystem& menu);
  void gatherData (FieldLattice& local_res);
  void solveProblem ();
  void resultReport ();

};
```

The computational core of the `ManagerHeat2D` class is simply carried out by invoking the `solveAtThisTimeStep` belonging to class `Heat2Ds`. More specifically, we have

```
void ManagerHeat2D:: timeLoop ()
{
  tip->initTimeLoop();
  while (!tip->finished()) {
    tip->increaseTime();
    solver->solveAtThisTimeStep();
  }
}
```

Finally, the task of collecting the subdomain solutions becomes also simple by calling the `ParallelFD::gatherFields` function. In other words, we have

```
void ManagerHeat2D :: gatherData (FieldLattice& local_res)
{
  ParallelFD::gatherFields(local_res, *global_res_u);

  if (proc_manager->master())
    database->dump (*global_res_u, tip.getPtr());
}
```

## 1.6   Parallelizing FE Computations on Unstructured Grids

Finite element methods are often associated with unstructured computational grids. Compared with parallelizing codes for explicit finite difference schemes, see Section 1.5, the task of parallelizing codes for computation on unstructured grids is more demanding. Many new issues arise, such as general grid partitionings, distribution of unstructured data storage, complex inter-processor communication etc. Parallelizing a sequential finite element simulator from scratch is therefore a complicated and error-prone process.

To provide Diffpack users with a straightforward approach to parallelizing their sequential finite element simulators, we have devised an add-on toolbox that hides the cumbersome parallelization specific codes. Instead, the toolbox offers high-level and user-friendly functionality, such that a Diffpack user needs only to insert a few lines of code into his original sequential simulator to transform it into a parallel one. The parallel toolbox ensures that the computation-intensive operations are run in parallel at the linear algebra level. The resulting parallel simulator will maintain the same portability and flexibility as its sequential counterpart.

This section gives an introduction to the add-on parallel toolbox. We start with a mathematical explanation of how different linear algebra operations can be parallelized in a divide-and-conquer style. Then, we continue with a brief description of how object-oriented programming enables a seamless coupling between the parallel toolbox and the huge library of existing sequential Diffpack codes. Thereafter, we focus on some of the major functions of the toolbox. Finally, the actual parallelization process is demonstrated by an example, before we give answers to some frequently asked questions about this parallelization approach. We mention that there exists another high-level parallelization approach, which incorporates the mathematical framework of domain decomposition methods. The discussion of this topic will be left to another chapter [1] in this book.

### 1.6.1   Parallel Linear Algebra Operations

Let us consider for instance an elliptic boundary value problem discretized on an unstructured finite element grid. Suppose the solution process involves an iterative linear system solver, then the most computation intensive operations viewed at the level of matrices and vectors are: (i) calculation of element matrices and vectors, and (ii) solution of the resulting linear system of equations. Roughly speaking, the parallelization starts with partitioning the global finite element grid into a set of smaller subgrids to be hosted by different processors of a parallel computer. It is important to note that global matrices and vectors *need not to be constructed physically*. It suffices to deal only with sub-matrices and sub-vectors that are associated with the subgrids. Neighboring processors need to exchange information associated with the nodes that are shared between them. However, the amount of information exchange is rather limited. It will be shown in the following that the linear algebra operations on the original global matrices and vectors can be achieved by operations on the sub-matrices and sub-vectors plus inter-processor communications. We also note that the following parallelization approach and its implementation are only valid for Diffpack finite element simulators using iterative linear system solvers, not direct solvers as e.g. Gaussian elimination.

*Partitioning Unstructured Finite Element Grids.* The essence of parallel computing is to distribute the work load among the processors of a parallel computer. For finite element computations, the work load distribution arises naturally when the entire solution domain is decomposed into a number of subdomains, as a processor only carries out computation restricted to its assigned subdomain. Partitioning a global finite element grid primarily concerns dividing all elements into subgroups. The elements of the global finite element grid $\mathcal{M}$ are partitioned to form a set of smaller subgrids $\mathcal{SM}_i$, $1 \leq i \leq P$, where $P$ is typically the number of available processors. By a *non-overlapping partitioning*, we mean that each element of $\mathcal{M}$ is to belong to only one of the subgrids. Element edges shared between neighboring subgrids constitute the so-called *internal boundaries*. The boundary of a subdomain is either entirely internal or partly internal and partly physical. We note that although every element of $\mathcal{M}$ belongs to a single subgrid after a non-overlapping partitioning, the nodes lying on the internal boundaries belong to multiple subgrids at the same time. In Figure 1.8, the nodes marked with a circle denote such nodes.

Ideally, the entire work load should be divided evenly among the available processors. Since the work load of a finite element computation is proportional to the number of elements, the subgrids should have approximately

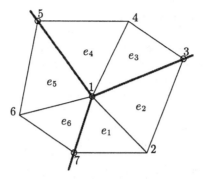

**Fig. 1.8.** An example of internal boundary nodes that are shared between neighboring subgrids. The figure shows a small region where three subgrids meet. Elements $e_1$ and $e_2$ belong to one subgrid, $e_3$ and $e_4$ belong to another subgrid, while $e_5$ and $e_6$ belong to a third subgrid. Nodes marked with a circle are internal boundary nodes.

the same number of elements. Due to the necessity of exchange of information associated with the internal boundary nodes, a perfect partitioning scheme should minimize both the number of neighbors and the number of internal boundary nodes for each subgrid. In this way, the communication overhead is minimized. General non-overlapping partitioning of unstructured finite element grids is a non-trivial problem. Here, we only mention that it is often transformed into a graph partitioning problem and solved accordingly, see e.g. [2,12].

*Distributed Matrices and Vectors.* When a non-overlapping partitioning of the unstructured global finite element grid is ready, one subgrid is to be held by one processor. The discretization of the target partial differential equation on one processor is done in the same way as in the sequential case, only that the discretization is restricted to the assigned subdomain. The assembly of the local stiffness matrix and right-hand side vector needs contribution *only* from all the elements belonging to the current subgrid. So far no inter-processor communication is necessary. Let us denote the number of nodes in subgrid $\mathcal{SM}_i$ by $N_i$. For any scalar partial differential equation to be discretized on subgrid $\mathcal{SM}_i$, we will come up with a linear system of equations

$$\mathbf{A}_i \mathbf{x}_i = \mathbf{b}_i,$$

where the local stiffness matrix $\mathbf{A}_i$ is of dimension $N_i \times N_i$, and the local right-hand side vector $\mathbf{b}_i$ and solution vector $\mathbf{x}_i$ have length $N_i$. The result of the parallel finite element solution process will be that $\mathbf{x}_i$ holds the correct solution for all the nodes belonging to $\mathcal{SM}_i$. The solution values for the internal boundary nodes are duplicated among neighboring subdomains. In comparison with $\mathbf{A}$ and $\mathbf{b}$, which would have arisen from a discretization done on the entire $\mathcal{M}$, entries of $\mathbf{A}_i$ and $\mathbf{b}_i$ that correspond to the *interior nodes* of $\mathcal{SM}_i$ are identical with those of $\mathbf{A}$ and $\mathbf{b}$. For entries of $\mathbf{A}_i$ and $\mathbf{b}_i$ that correspond to the internal boundary nodes, they are different from those of $\mathbf{A}$ and $\mathbf{b}$, because the contributions from the elements that lie in neighboring subgrids are missing. However, this does not prevent us from obtaining the correct result of different linear algebra operations that should have been carried out on the *virtual* global matrices and vectors. It is achieved by operating solely on local matrices and vectors plus using inter-processor communication, as the following text will show.

*Three Types of Parallel Linear Algebra Operations.* At the linear algebra level, the objective of the parallel solution process is to find the solution of the global system $\mathbf{Ax} = \mathbf{b}$ through linear algebra operations carried out only on the local systems $\mathbf{A}_i\mathbf{x}_i = \mathbf{b}_i$. We emphasize that global matrices and vectors have no physical storage but are represented virtually by local matrices and vectors that are held by the different processors. For iterative solution algorithms, it suffices to parallelize the following three types of global linear algebra operations:

1. Vector addition: $\mathbf{w} = \mathbf{u} + \alpha\mathbf{v}$, where $\alpha$ is a scalar constant;

2. Inner-product between two vectors: $c = \mathbf{u} \cdot \mathbf{v}$;

3. Matrix-vector product: $\mathbf{w} = \mathbf{Au}$.

*Parallel vector addition.* Assuming that values of $\mathbf{u}_i$ and $\mathbf{v}_i$ on the internal boundary nodes are correctly duplicated between neighboring subdomains, the parallel vector addition $\mathbf{w}_i = \mathbf{u}_i + \alpha\mathbf{v}_i$ is straightforward and needs no inter-processor communication (see Section 1.1.4).

*Parallel inner-product between two vectors.* The approach from Section 1.1.4 is to be used. However, the result of the local inner-product $c_i = \sum_j^{N_i} u_{i,j}v_{i,j}$ can not be added together directly to give the global result, due to the fact that internal boundary nodes are shared between multiple subgrids. Vector entries on those internal boundary nodes must be scaled accordingly. For this purpose, we denote by $\mathcal{O}_i$ the set of all the internal boundary nodes of $\mathcal{SM}_i$. In addition, each internal boundary node has an integer count $o_k$, $k \in \mathcal{O}_i$, which is the total number of subgrids it belongs to, including $\mathcal{SM}_i$ itself. Then we can get the adjusted local result by

$$\tilde{c}_i = c_i - \sum_{k \in \mathcal{O}_i} \frac{o_k - 1}{o_k} u_{i,k}v_{i,k} \, .$$

Thereafter, the correct result of $c = \mathbf{u} \cdot \mathbf{v}$ can be obtained by collecting all the adjusted local results in form of $c = \sum_i^p \tilde{c}_i$. This is done by inter-processor communication in form of an all-to-all broadcast operation.

*Parallel matrix-vector product.* We recall that a global matrix **A** does not exist physically, but is represented by a series of local matrices $\mathbf{A}_i$. These local matrices arise from a local assembly process that is restricted to each subdomain. The rows of $\mathbf{A}_i$ that correspond to the interior subgrid nodes are correct. A local matrix-vector product $\mathbf{w}_i = \mathbf{A}_i\mathbf{u}_i$ will thus give correct values of $\mathbf{w}_i$ for the interior subgrid nodes. However, the rows of $\mathbf{A}_i$ that correspond to the internal boundary nodes are only partially correct. For a specific internal boundary node, some of the elements that share this node belong to the neighboring subdomains. The contributions from the element matrices are thus distributed among the neighboring subdomains. Therefore, the correct value in $\mathbf{w}_i$ for an internal boundary node can only be obtained by adding up contributions from *all* the neighboring subdomains. For instance, in Figure 1.8, the nodes with global numbers 3,5,7 need contributions from two neighbors, while the node with global number 1 should add together contributions from all three neighbors. This requires inter-processor communication in form of two and two neighboring subdomains exchanging values on the relevant internal boundary nodes.

*Preconditioning and Overlapping Grid Partitioning.* Iterative solvers for linear systems often use preconditioners to speed up the convergence. Solving linear systems in parallel requires therefore parallel preconditioning operations. This is an important but tricky issue. We will leave the discussion of parallel multilevel preconditioners such as domain decomposition methods to a later chapter [1], and consider the other Diffpack preconditioners here. It should be pointed out immediately that many preconditioners such as SSOR relaxation and RILU are inherently sequential algorithms. A remedy is to just let each processor run the corresponding localized preconditioning operations, and then take some kind of average of the different values associated with each internal boundary node from neighboring subgrids, see e.g. [9]. However, a different and possibly weakened preconditioning effect may be expected.

Careful readers will notice, however, that even the intrinsically parallel Jacobi relaxation does not produce the correct result on a non-overlapping partitioning, because the internal boundary nodes will not be updated correctly. That is, the need for an *overlapping* partitioning of $\mathcal{M}$ arises. More specifically, the non-overlapping subgrids are extended slightly so that every node of $\mathcal{M}$, which does not lie on the physical boundary, is an interior node of at least one subgrid. Figure 1.9 shows an example of extending a non-overlapping partitioning and making it overlapping. A parallel Jacobi relaxation on an overlapping partitioning will be done as follows: each subdomain runs *independently* its local Jacobi relaxation, then local results on the internal boundary nodes are discarded and replaced by the correct results sent from the neighboring subdomains, which have those nodes as their interior grid nodes.

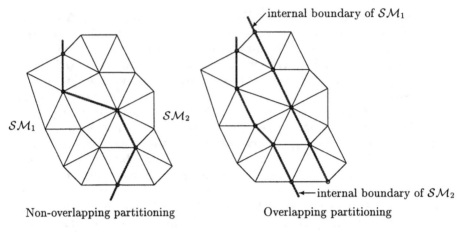

Fig. 1.9. Extension of a non-overlapping partitioning of a finite element grid into an overlapping partitioning.

### 1.6.2 Object-Oriented Design of a Parallel Toolbox

It is now clear that at the linear algebra level the global finite element computational operations can be replaced with localized operations plus necessary inter-processor communication. This is good news regarding the parallelization of Diffpack finite element computations, because we can directly utilize the existing sequential codes for carrying out all the local linear algebra operations. Among other things we are able to use the *same* Diffpack classes for storing the local matrices and vectors, and all the iterative methods for solving linear systems can be used in the same way as before. Of course, new codes need to be written to incorporate two major parallelization specific tasks: general grid partitioning and different forms of inter-processor communication. In this section, we present the design of a small add-on parallel toolbox that contains a number of new C++ classes capable of these two tasks. The add-on toolbox can be coupled seamlessly with the existing huge sequential Diffpack library, thereby allowing Diffpack users to parallelize their finite element computation in a straightforward way. In this connection, object-oriented programming techniques have proved to be successful.

*Coupling with the Sequential Diffpack Library.* One of our objectives is to maintain the comprehensive sequential library of Diffpack independently of the parallel toolbox. The only new class that we have introduced into the sequential Diffpack library is a pure interface class with name SubdCommAdm. The following simplified class definition gives an overview of its most important member functions:

```
class SubdCommAdm : public HandleId
{
```

```
protected:
  SubdCommAdm () {}
public:
  virtual ~SubdCommAdm () {}
  virtual void updateGlobalValues (LinEqVector& lvec);
  virtual void updateInteriorBoundaryNodes (LinEqVector& lvec);
  virtual void matvec (const LinEqMatrix& Amat,
                       const LinEqVector& c,
                             LinEqVector& d);
  virtual real innerProd (LinEqVector& x, LinEqVector& y);
  virtual real norm (Vec(real)& c_vec, Norm_type lp=l2);
};
```

We note that the constructor of class `SubdCommAdm` is a protected member function, meaning that no instance of `SubdCommAdm` can be created. In addition, all its member functions, which are virtual, have empty definitions. Therefore, it only serves as a pure interface class defining the name and syntax of different functions for inter-processor communication. Inside the parallel toolbox, we have developed a subclass of `SubdCommAdm` in order to implement such communication functions. Here, we mention that the member function `updateInteriorBoundaryNodes` will invoke an inter-processor communication, which ensures that all the internal boundary nodes to have correctly duplicated values among neighboring subgrids. In contrast, the `updateGlobalValues` function concerns overlapping partitionings, i.e., it is meant for all the overlapping nodes shared among neighboring subgrids. Moreover, the member functions `matvec`, `innerProd`, and `norm` are meant to carry out the parallel versions of the matrix-vector product, the inner product between two vectors, and the norm of a vector, respectively. These member functions work for both non-overlapping and overlapping partitionings.

It suffices for the rest of the sequential Diffpack library to only know how inter-processor communication can be invoked following the definition of class `SubdCommAdm`. A few of the linear algebra classes in the sequential Diffpack library were slightly extended to allow a connection with the add-on parallel toolbox. The rest of the sequential Diffpack library classes remain intact. For example, class `LinEqSystemPrec` is now equipped with the following two lines in its definition:

```
Handle(SubdCommAdm) comm_adm;
void attachCommAdm (const SubdCommAdm& adm_);
```

The member function `attachCommAdm` can be used to set the internal smart pointer `comm_adm` to a concrete object of type `SubdCommAdm`. Consequently, the member function `LinEqSystemPrec::matvec` is modified as follows:

```
void LinEqSystemPrec::matvec(const LinEqVector& c, LinEqVector& d)
{
  if (comm_adm.ok())
    comm_adm->matvec (*Amat, c, d);   // parallel computation
  else
    Amat->prod (c, d);                // sequential computation
```

```
    matvecCalls++;
  }
```

It can be observed that, during a sequential Diffpack simulation, the test

```
    if (comm_adm.ok())
```

should return a **false** value, so the parallel version of the matrix-vector product will not be invoked. The situation is the opposite during a parallel Diffpack simulation. It is also important to note that these if-tests are hidden from the user and have completely negligible overhead.

To relieve Diffpack users of the burden of inserting the **attachCommAdm** function in many places when parallelizing a sequential Diffpack simulator, the member function **LinEqAdm::attachCommAdm** makes sure that all the involved linear algebra classes call their respective **attachCommAdm** functions. So, for the application programmer, one explicit call of **LinEqAdm::attachCommAdm** is enough before the parallel solution process starts.

*The Parallel Toolbox.* The two main tasks of the toolbox are to provide Diffpack users with different grid partitioning methods and several high-level inter-processor communication functions. In addition, the toolbox also provides parallel versions of some of Diffpack's most frequently used linear algebra operations.

*A hierarchy of grid partitioning methods.* The purpose of grid partitioning is to provide subdomains with subgrids where local discretization can be carried out. Although the most common situation is to start with a global finite element grid and partition it into smaller subgrids, there can be cases where Diffpack users wish to let each subdomain create its local finite element grid directly either by reading an existing Diffpack **GridFE** file or meshing the subdomain based on sufficient geometry information. In order to incorporate different grid partitioning approaches, we have created a class hierarchy with a base class named **GridPart**, whose simplified definition is as follows.

```
    class GridPart
    {
    public:
      GridPart (const GridPart_prm& pm);
      virtual ~GridPart () {}
      virtual bool makeSubgrids ()=0;
      virtual bool makeSubgrids (const GridFE& global_grid)=0;
    };
```

We can see that **GridPart** is a so-called pure virtual class in C++, because both versions of the member function **makeSubgrids** are required to be implemented in a derived subclass. At run time, the real work of grid partitioning is done by one of the **makeSubgrids** functions in a specific subclass decided by the Diffpack user. The input argument to the constructor is an object of type GridPart_prm, which contains diverse grid partition-related parameters. The following subclasses of GridPart are already included in the hierarchy, which also allows a Diffpack user to extend it by developing new subclasses.

1. GridPartUnstruct. This class creates subgrids by partitioning an unstructured global finite element grid. The number of resulting subgrids can be arbitrary and is decided by the user at run time. Class GridPartUnstruct is in fact a pure virtual class itself. Different graph partitioning methods are implemented in different subclasses of GridPartUnstruct. One example is subclass GridPartMetis where the METIS algorithm [5] is implemented.

2. GridPartUniform. This class creates subgrids by a rectangular partitioning of a structured global grid. The subgrids will also be structured so that the user has more control over the partitioning. This simple and flexible partitioning method allows also a 2D or 3D structured global grid to be partitioned in fewer directions than its actual number of spatial dimensions.

3. GridPartFileSource. This class reads ready-made subgrids from Diffpack GridFE data files. All the grid data files should be of the format accepted by the standard Diffpack readOrMakeGrid function.

*The main control class.* We recall that SubdCommAdm is a pure interface class; all its inter-processor communication functions need to be implemented in a subclass before they can be used in parallel Diffpack finite element computations. For this purpose, we have created a subclass of SubdCommAdm in the parallel toolbox and named it GridPartAdm. Here, object-oriented programming sets SubdCommAdm as the formal connection between the existing sequential Diffpack library and the parallel toolbox, while GridPartAdm is the actual working unit. It should also be mentioned that class GridPartAdm not only implements the different inter-processor communication functions, but at the same time provides a unified access to the different methods offered by the GridPart hierarchy. In this way, it suffices for a user to only work with GridPartAdm during parallelization of a sequential Diffpack simulator, without making explicit usage of the details of GridPart and GridPart_prm etc. A simplified definition of class GridPartAdm is as follows.

```
class GridPartAdm : public SubdCommAdm
{
protected:
  Handle(GridPart_prm) param;
  Handle(GridPart) partitioner;
  bool overlapping_subgrids;

public:
  GridPartAdm ();
  virtual ~GridPartAdm ();
  void copy (const GridPartAdm& adm);
  static void defineStatic  (MenuSystem& menu, int level = MAIN);
  virtual void scan         (MenuSystem& menu);
```

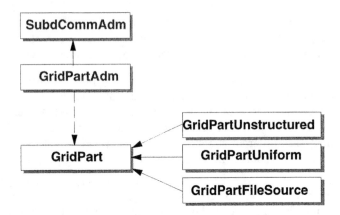

**Fig. 1.10.** The relationship between SubdCommAdm, GridPartAdm, and GridPart. Note that the solid arrows indicate the "is-a" relationship, while the dashed arrows indicates the "has-a" relationship.

```
    virtual int getNoGlobalSubds() {return param->num_global_subds;}
    virtual bool overlappingSubgrids(){return overlapping_subgrids;}
    virtual void prepareSubgrids ();
    virtual void prepareSubgrids (const GridFE& global_grid);
    virtual GridFE& getSubgrid ();
    virtual void prepareCommunication (const DegFreeFE& dof);
    virtual void updateGlobalValues (LinEqVector& lvec);
    virtual void updateInteriorBoundaryNodes (LinEqVector& lvec);
    virtual void matvec (const LinEqMatrix& Amat,
                         const LinEqVector& c,
                               LinEqVector& d);
    virtual real innerProd (LinEqVector& x, LinEqVector& y);
    virtual real norm (LinEqVector& lvec, Norm_type lp=l2);
  };
```

We can see from above that class GridPartAdm has smart pointers to objects of GridPart_prm and GridPart, through which it administers the process of grid partitioning. Inside the GridPartAdm::prepareSubgrids function, an object of GridPart is instantiated to carry out the grid partitioning. The actual type of the grid partitioner is determined *at run time* according to the user's input data. We refer to Figure 1.10 for a schematic diagram describing the relationship between SubdCommAdm, GridPartAdm, and GridPart.

### 1.6.3  Major Member Functions of GridPartAdm

During parallelization of most sequential Diffpack simulators, an object of GridPartAdm is all a user needs in order to access the desired functionality of the parallel toolbox. We will here give a detailed description of the major functions of GridPartAdm.

*Grid Partition Related Member Functions.* Since there are many parameters associated with the different grid partitioning approaches, the Diffpack MenuSystem class is heavily involved in this process. Thus, almost every parallel Diffpack simulator will need such a statement in its define function:

```
GridPartAdm:: defineStatic (menu, level+1);
```

Then, typically inside the corresponding scan function, the work of grid partitioning can be carried out as follows:

```
// read information from the input menu
gp_adm->scan (menu);
// assuming that grid is pointing to the global grid
gp_adm->prepareSubgrids (*grid);
// now let grid point to the subgrid
grid.rebind (gp_adm->getSubgrid());
```

Here, gp_adm is a Handle to a GridPartAdm object. The member function prepareSubgrids has the responsibility of carrying out the grid partitioning, after which the member function getSubgrid will return the reference of the resulting subgrid on every processor. The function prepareSubgrids makes use of different grid parameters that are supplied by the member function pair GridPartAdm::defineStatic and GridPartAdm::scan. In this way, the user can make a flexible choice among different grid partitioning methods at run time. The following are examples of parameter input files for three different ways of producing the subgrids:

1. General partitioning of an unstructured global finite element grid:

```
sub GridPart_prm
set grid source type = GlobalGrid
set partition-algorithm = METIS
set number overlaps = 1
ok
```

The input value GlobalGrid indicates that a global finite element grid is to be partitioned into an arbitrarily given number of subgrids. The number of subgrids is equal to the number of involved processors. For example a run-time command such as

```
mpirun -np 8 ./app
```

means that eight subgrids will be created, each residing on one processor. The above parameter input file also says that the method of the METIS package will be used to do the general grid partitioning and the resulting non-overlapping partitioning is to be extended to create an overlapping partitioning.

2. User-controlled rectangular partitioning of a uniform grid:

```
sub GridPart_prm
set grid source type = UniformPartition
set subdomain division = d=2 [0,1]x[0,1] [0:4]x[0:2]
set overlap = [1,1]
set use triangle elements = ON
ok
```

The above parameter input file implies that a global 2D uniform mesh, in form of `GridFE`, will be partitioned into $4 \times 2 = 8$ uniform local grids. We remark that the redundant geometry information (`[0,1]x[0,1]`) on the `set subdomain division` line is ignored by the partitioner. For this example, the parallel finite element application must be started on 8 processors, each taking care of one subdomain. The number of partitions is 4 in the $x$-direction and 2 in the $y$-direction. The resulting subgrids will all have 2D triangular elements, and one element overlap layer exists between the neighboring subgrids in both the $x$- and $y$-directions.

3. Creation of subgrids directly from grid data files:

```
sub GridPart_prm
set grid source type = FileSource
set subgrid root name = subd%02d.grid
ok
```

An assumption for using the above input file is that we have a series of grid data files of format `GridFE`, named `subd01.grid`, `subd02.grid`, and so on. Note also that for this grid partitioning approach, we have to use the version of the `GridPartAdm::prepareSubgrids` function that takes no `GridFE` input argument. This is because no global finite element grid exists physically. In other words, the work of grid partitioning is done by

```
gp_adm->scan (menu);
gp_adm->prepareSubgrids ();
grid.rebind (gp_adm->getSubgrid());
```

*Communication-Related Member Functions.* The communication pattern between neighboring subgrids that arise from a general grid partitioning is complex. However, for a fixed partitioning, whether overlapping or not, the communication pattern is fixed throughout the computation. The member function `prepareCommunication` thus has the responsibility of finding out the communication pattern *before* the parallel computation starts. One example of invoking the function is

```
gp_adm->prepareCommunication (*dof);
```

where dof is a `Handle` to a `DegFreeFE` object. That is, the function takes a Diffpack `DegFreeFE` object as the input argument. For each subdomain, this function finds out all its neighboring subdomains. Additionally, the function also finds out how to exchange information between each pair of neighboring subdomains. The storage allocation for the outgoing and incoming messages is also done in `prepareCommunication`. After that, the member function `attachCommAdm` belonging to class `LinEqAdm` should be called such as:

```
lineq->attachCommAdm (*gp_adm);
```

The design of the parallel toolbox has made the parallelization process extremely simple for the user. The above statement is sufficient for ensuring that necessary inter-processor communication will be carried out by the standard Diffpack libraries when solving the linear systems of equations. However, in certain cases a user needs to explicitly invoke one of the two following functions of `GridPartAdm` for inter-processor communication.

`updateGlobalValues` ensures that all the nodes that are shared between multiple subgrids have correctly duplicated values. For non-overlapping grid partitionings, this means that for a particular internal boundary node, the different values coming from the neighboring subgrids are collected and summed up. For overlapping grid partitionings, every internal boundary node throws away its current value and replaces it with a value provided by a neighboring subgrid, which has an interior node with the same coordinates. If an internal boundary node *lies in the interior* of more than one neighboring subgrid, an average of the values, which are provided by these neighboring subgrids, becomes the new value.

`updateInteriorBoundaryNodes` has the same effect for non-overlapping grid partitionings as `updateGlobalValues`. For overlapping grid partitionings, the function only applies to the internal boundary nodes, whereas values of the other nodes in the overlapping regions are unaffected.

The member functions `matvec`, `innerProd`, and `norm` of class `GridPartAdm` can be used to carry out the parallel version of respectively matrix-vector product, inner product between two vectors, and norm of a vector. Inside these functions, local linear algebra operations are carried out by invoking the corresponding sequential Diffpack functions before some inter-processor communication, see also Section 1.6.2.

### 1.6.4   A Parallelization Example

We consider a standard sequential Diffpack finite element simulator `Poisson1` from [6]. The parallelization is quite straightforward, so it is sufficient to insert parallelization specific codes directly into the original sequential simulator, enclosed in the preprocessor directive

```
#ifdef DP_PARALLEL_LA
...
#endif
```

In this way, we will be able to maintain a single code for both the sequential and parallel simulators.

The first thing to do is to insert the following line in the beginning of the header file:

```
#ifdef DP_PARALLEL_LA
#include <GridPartAdm.h>
#endif
```

The next thing to do is to include a handle of a `GridPartAdm` object in the definition of class `Poisson1`, i.e.,

```
#ifdef DP_PARALLEL_LA
   Handle(GridPartAdm) gp_adm;
#endif
```

Thereafter, three modifications remain to be done in the original sequential code. The first modification is to put inside the `define` function the following line:

```
#ifdef DP_PARALLEL_LA
   GridPartAdm::defineStatic ·(menu,level+1);
#endif
```

The second code modification is to let the `GridPartAdm` object produce the subgrid, i.e., inside the `scan` function:

```
#ifdef DP_PARALLEL_LA
   gp_adm.rebind (new GridPartAdm);
   gp_adm->scan (menu);
   gp_adm->prepareSubgrids (*grid);
   grid.rebind (gp_adm->getSubgrid());
#endif
```

The third required code modification is to add the following two lines of new code, after the `LinEqAdmFE` object is created and its `scan` is called:

```
#ifdef DP_PARALLEL_LA
   gp_adm->prepareCommunication (*dof);
   lineq->attachCommAdm (*gp_adm);
#endif
```

Here, the first of the above two new code lines invokes the member function `GridPartAdm::prepareCommunication`, which must precede all the subsequent inter-processor communications. The second line attaches the `GridPartAdm` object to `lineq`, which points to a Diffpack `LinEqAdmFE` object. As mentioned earlier in Section 1.6.2, the `LinEqAdmFE` object will then automatically invoke all the other needed `attachCommAdm` functions, including e.g. the one belonging to `LinEqSystemPrec`. In this way, necessary communication and synchronization will be enforced during the solution process later.

The final task is to insert

```
   initDpParallelLA (nargs, args);
```

as the first statement in the main program and insert

```
   closeDpParallelLA();
```

at the end.

### 1.6.5    Storage of Computation Results and Visualization

The storage of computation results during a parallel Diffpack simulation can be carried out exactly as before, e.g. through the usage of SaveSimRes. The only exceptions are the use of SaveSimRes::lineCurves and time series plot for a given spatial point. These should be avoided in a parallel Diffpack simulator. For each processor involved in a parallel Diffpack simulation, the global casename variable will be automatically suffixed with _p0000, _p0001, ..., for respectively processor number one, number two, and so on. That is, each processor has a unique casename variable and therefore operates with its own SimResFile. In multiple-loop computations, the processor rank is added in front of the multiple-loop identification number, like _p0000_m01, _p0001_m01.

*Four Useful Tools.* As explained above, each processor generates its own SimResFile during a parallel simulation. When the parallel simulation is done, the user will face a series of SimResFile databases, possibly after collecting them from different hard disks. To visualize such parallel simulation results, using e.g. Vtk, it is necessary to run the simres2vtk filter for each of the SimResFiles. In order to relieve the user's effort of having to filter *all* the SimResFiles one by one, we have devised a tool named distrsimres. This is a Perl script that automatically visits all the SimResFiles. An example of using distrsimres can e.g. be:

```
distrsimres simres2vtk -f SIMULATION -E -a
```

That is, the user just needs to prefix any simres2xxx command with the distrsimres script. Similarly, for a multiple-loop simulation, it is necessary to add the loop-number specification to the casename, like e.g.

```
distrsimres simres2vtk -f SIMULATION_m02 -E -a
```

The second tool is another Perl script with name distrsummary. It can be used to generate a single HTML summary file that contains links to all the casename_p00xx-summary.html files, which will be generated by a parallel simulation that uses Diffpack's automatic report generation functionality. The syntax is simply

```
distrsummary -f casename
```

The third tool is a compiled Diffpack application, named distr2glob, which can be used to "patch together" chosen field(s) that are stored distributedly in the different SimResFiles. To use distr2glob, the user specifies a global grid, typically a quite coarse grid of type GridLattice. Then distr2glob maps the distributed SimResFile data onto the specified global coarse grid. The result is a new SimResFile that contains the chosen field(s) on the global coarse grid. An example execution of distr2glob can be

```
distr2glob -gg coarse.grid -const 0 -f SIMULATION -r 1 -s -a
```

The effect is that the first scalar field (due to the -r option), which is stored distributedly in the different SimResFiles that share the root casename SIMULATION (due to the -f option), is mapped onto a global coarse grid given by the grid file coarse.grid (due to the -gg option) and stored eventually in a new SimResFile. In short, the user can use most of the options available for a standard simres2xxx filter, such as -r, -t, -A, and -E, to choose wanted subdomain field(s). In addition, the new option -gg takes a text string of the format acceptable by Diffpack's readOrMakeGrid command and the other new option -const reads a default value to be assigned to those grid points of the global coarse grid lying outside the original global solution domain.

The fourth tool is a new version of RmCase for parallel simulations. By specifying the value of casename such as SIMULATION or SIMULATION_m01, the user can remove all the files related to casename.

*A Simple GUI for Visualization.* We have mentioned that the distr2glob tool can allow the user to rapidly check the parallel data set against a global coarse grid. However, for more detailed visualization of the parallel computation results, we have created a new Perl script similar to vtkviz. The new Perl script is named distrvtkviz and its work is to run vtk with an input Vtk-Tcl script named ParaViewer.tcl. We recall that the work of vtkviz is to run vtk with Examiner.tcl as input. The new features of ParaViewer.tcl are that the user can freely add/remove sub-grid data set(s) and concentrate on a particular sub-grid, if desired. To use distrvtkviz, the only thing a user needs to do is to run the distrsimres simres2vtk command, before starting the GUI by issuing the command

```
distrvtkviz
```

### 1.6.6   Questions and Answers

1. *When should I use a non-overlapping partitioning of an unstructured finite element grid, when to use an overlapping partitioning?*
   In principle, using an overlapping partitioning is always a *safe* approach in that parallelization of certain operations can not achieve correct result on a non-overlapping partitioning. An example is parallel Jacobi relaxation. However, an overlapping partitioning results in more computation per processor. So it is advantageous to first check whether the two different partitions produce the same result on a coarse global finite element grid before running full-scaled simulations in parallel.

2. *How do I produce a 2D partitioning of a 3D uniform finite element grid?*
   Assume that a $100 \times 100 \times 100$ uniform mesh covering the unit cube is desired to be partitioned into 10 subgrids. The number of partitions in the $x-$, $y-$, and $z-$directions are 5, 2, 1, respectively. In addition, one layer of element overlap is desired between the neighboring subgrids. Then the following information can be used in the input parameter file:

```
sub GridPart_prm
set grid source type = UniformPartition
set subdomain division =d=3 [0,1]x[0,1]x[0,1]  [0:5]x[0:2]x[0:1]
set overlap = [1,1,0]
ok
```

3. *When should I use* `GridPartAdm::updateGlobalValues`?
   For non-overlapping partitionings, the two functions `updateGlobalValues` and `updateInteriorBoundaryNodes` of class `GridPartAdm` have exactly the same effect. For overlapping partitionings, some situations only require that internal boundary nodes of a subgrid should receive their correct values from neighboring subgrids, where those nodes are interior subgrid nodes. More precisely, the functionality of `updateInteriorBoundaryNodes` is a subset of that of `updateGlobalValues`. So `updateGlobalValues` is always safe to use but may introduce some unnecessary communication overhead.

4. *What do I do when I want to work with both scalar and vector fields in my Diffpack simulator?*
   Assume that all the scalar fields use one `DegFreeFE` object and all the vector fields use another `DegFreeFE` object, while the two `DegFreeFE` objects share the same `GridFE` object. In this situation, two objects of type `GridPartAdm` are needed, one for all the scalar fields, the other for all the vector fields. Since all the fields need to work on the same partitioning of the global finite element grid, only one `GridPartAdm` will have the responsibility of doing the grid partitioning, whereas the other `GridPartAdm` copies the partitioning.

```
Handle(GridPartAdm) adm1, adm2;
Handle(GridFE) grid;
Handle(DegFreeFE) dof1, dof2;
Handle(LinEqAdm) lineq1, lineq2;
// ....
adm1->scan (menu);
adm1->prepareSubgrids ();
grid.rebind (adm1->getSubgrid());
// ...
adm1->prepareCommunication (*dof1);
lineq1->attachCommdm (*adm1);
adm2->copy (*adm1);
adm2->prepareCommunication (*dof2);
lineq2->attachCommdm (*adm2);
```

5. *How do I parallelize a nonlinear PDE solver?*
   Carry out all the above steps, and invoke in addition:

```
nlsolver->attachCommAdm (*gp_adm);
```

6. *Is it possible to parallelize explicit FE schemes?*
   Yes. Let us look at `$NOR/doc/Book/src/fem/Wave1` for instance (see [6]). The main simulator class `Wave1` does not use `LinEqAdmFE`, but uses Gaussian elimination to solve the *diagonal* mass matrix M at each time level.

Because the entries of M associated with the internal boundary nodes are not correct, we need the following modification of M for each subgrid, once and for all:

```
for (i=1; i<=nno; i++) scratch(i) = M(i);
gp_adm->updateInteriorBoundaryNodes (scratch);
for (i=1; i<=nno; i++) M(i) = scratch(i);
```

Besides, an explicit call of GridPartAdm::updateInteriorBoundaryNodes is also necessary after each matrix-vector product involving the stiffness matrix K:

```
prod (scratch, K, u_prev->values());   // scratch = K*u^0
gp_adm->updateInteriorBoundaryNodes (scratch);
```

7. *Are there any functions of* SaveSimRes *that can not be used in parallel simulations?*
   The use of SaveSimRes::lineCurves and time series plot for a given spatial point should be avoided in parallel Diffpack simulations.

8. *How do I avoid unnecessary output from every processor?*
   Use the following if-test before e.g. an output statement to s_o:

```
if (proc_manager->master()) s_o << "t=" << tip->time();
```

   Recall that proc_manager is a global Diffpack handle variable of type DistrProcManager.

9. *Can I use the static functions of* ErrorNorms *as before?*
   Almost. Just replace all the static Lnorm functions belonging to ErrorNorms by the corresponding functions of a new class DistrErrorNorms, with a reference to a GridPartAdm object as an additional input argument. The only exception is the ErrorNorms::errorField, which should be used exactly as before. The following is an example

```
ErrorNorms::errorField (*uanal, *u, tip->time(), *error);
DistrErrorNorms::Lnorm  // fills the error norms
   (*gp_adm, *uanal, *u, tip->time(),
    L1_error, L2_error, Linf_error, error_itg_pt_tp);
```

10. *How do I find the global maximum value of a scalar field?*
    The basic idea is to first find the local maximum value and then use the functionality offered by DistrProcManager to find the global maximum value.

```
real global_max, local_max = sub_field.values().maxValue();
proc_manager->allReduce(local_max,global_max,PDP_MAX);
```

11. *Is it possible to solve a global linear system in parallel by Gaussian Elimination?*
    No, this is not possible for the current version of the parallel toolbox. Only parallel iterative solvers are supported. As the default answer to the basic method item in a LinEqSolver_prm sub-menu is GaussElim, the user has to specifically write, e.g.,

```
set basic method = ConjGrad
```

in an input file to the Diffpack MenuSystem.

# References

1. X. Cai. Overlapping domain decomposition methods. In H. P. Langtangen and A. Tveito, editors, *Advanced Topics in Computational Partial Differential Equations – Numerical Methods and Diffpack Programming.* Springer, 2003.
2. C. Farhat and M. Lesoinne. Automatic partitioning of unstructured meshes for the parallel solution of problems in computational mechanics. *Internat. J. Numer. Meth. Engrg.*, 36:745–764, 1993.
3. Message Passing Interface Forum. MPI: A message-passing interface standard. *Internat. J. Supercomputer Appl.*, 8:159–416, 1994.
4. I. Foster. *Designing and Building Parallel Programs.* Addison-Wesley, 1995.
5. G. Karypis and V. Kumar. Metis: Unstructured graph partitioning and sparse matrix ordering system. Technical report, Department of Computer Science, University of Minnesota, Minneapolis/St. Paul, MN, 1995.
6. H. P. Langtangen. *Computational Partial Differential Equations - Numerical Methods and Diffpack Programming.* Textbooks in Computational Science and Engineering. Springer, 2nd edition, 2003.
7. D.J. Lilja. *Measuring Computer Performance – A Pratitioner's Guide.* Cambridge University Press, 2000.
8. P.S. Pacheco. *Parallel Programming with MPI.* Morgan Kaufmann Publishers, 1997.
9. G. Radicati and Y. Robert. Parallel conjuget gradient-like algorithms for solving sparse nonsymmetric linear systems on a vector multiprocessor. *Parallel Computing*, 11:223–239, 1989.
10. V. Sunderam. PVM: A framework for parallel distributed computing. *Concurrency: Practice and Experience*, 2:315–339, 1990.
11. E.F. Van de Velde. *Concurrent Scientific Computing.* Springer-Verlag, 1994.
12. D. Vanderstraeten and R. Keunings. Optimized partitioning of unstructured finite element meshes. *Internat. J. Numer. Meth. Engrg.*, 38:433–450, 1995.

# Chapter 2

# Overlapping Domain Decomposition Methods

X. Cai[1,2]

[1] Simula Research Laboratory
[2] Department of Informatics, University of Oslo

**Abstract.** Overlapping domain decomposition methods are efficient and flexible. It is also important that such methods are inherently suitable for parallel computing. In this chapter, we will first explain the mathematical formulation and algorithmic composition of the overlapping domain decomposition methods. Afterwards, we will focus on a generic implementation framework and its applications within Diffpack.

## 2.1  Introduction

The present chapter concerns a special class of numerical methods for solving partial differential equations (PDEs), where the methods of concern are based on a physical decomposition of a global solution domain. The global solution to a PDE is then sought by solving the smaller subdomain problems collaboratively and "patching together" the subdomain solutions. These numerical methods are therefore termed as *domain decomposition* (DD) methods. The DD methods have established themselves as very efficient PDE solution methods, see e.g. [3,8]. Although sequential DD methods already have superior efficiency compared with many other numerical methods, their most distinguished advantage is the straightforward applicability for parallel computing. Other advantages include easy handling of global solution domains of irregular shape and the possibility of using different numerical techniques in different subdomains, e.g., special treatment of singularities.

In particular, we will concentrate on one special group of DD methods, namely iterative DD methods using *overlapping* subdomains. The overlapping DD methods have a simple algorithmic structure, because there is no need to solve special interface problems between neighboring subdomain. This feature differs overlapping DD methods from non-overlapping DD methods, see e.g. [3,8]. Roughly speaking, overlapping DD methods operate by an iterative procedure, where the PDE is repeatedly solved within every subdomain. For each subdomain, the artificial internal boundary condition (see Section 2.2) is provided by its neighboring subdomains. The convergence of the solution on these internal boundaries ensures the convergence of the solution in the entire solution domain.

The rest of the chapter is organized as follows. We start with a brief mathematical description of the overlapping DD methods. The mathematical description is followed by a simple coding example. Then, we continue

with some important issues and a discussion of the algorithmic structure of
the overlapping DD methods. Thereafter, we present a generic framework
where the components of the overlapping DD methods are implemented as
standardized and extensible C++ classes. We remark that Sections 2.2-2.6
do not specifically distinguish between sequential and parallel DD methods,
because most of the parallel DD methods have the same mathematical and
algorithmic structure as their sequential counterparts. We therefore postpone
until Section 2.7 the presentation of some particular issues that are only applicable to parallel DD methods. Finally we provide some guidance on how to
use the generic framework for easy and flexible implementation of overlapping
DD methods in Diffpack.

## 2.2   The Mathematical Formulations

In this section, we introduce the mathematical formulations of the overlapping DD methods. The mathematics will be presented at a very brief level,
so interested readers are referred to the existing literature, e.g. [3,8,10], for
more detailed theories.

### 2.2.1   The Classical Alternating Schwarz Method

The basic mathematical idea of overlapping DD methods can be demonstrated by the very first DD method: the classical *alternating Schwarz method*,
see [7]. This method was devised to solve a Poisson equation in a specially
shaped domain $\Omega = \Omega_1 \cup \Omega_2$, i.e., the union of a circle and a rectangle, as
depicted in Figure 2.1. More specifically, the boundary-value problem reads

$$-\nabla^2 u = f \quad \text{in } \Omega = \Omega_1 \cup \Omega_2,$$
$$u = g \quad \text{on } \partial\Omega.$$

The part of the subdomain boundary $\partial\Omega_i$, which is not part of the global
physical boundary $\partial\Omega$, is referred to as the artificial *internal boundary*. In
Figure 2.1, we see that $\Gamma_1$ is the artificial internal boundary of subdomain
$\Omega_1$, and $\Gamma_2$ is the artificial internal boundary of subdomain $\Omega_2$.

In order to utilize analytical solution methods for solving the Poisson
equation on a circle and a rectangle separately, Schwarz proposed the following *iterative* procedure for finding the approximate solution in the entire
domain $\Omega$. Let $u_i^n$ denote the approximate solution in subdomain $\Omega_i$, and $f_i$
denote the restriction of $f$ to $\Omega_i$. Starting with an initial guess $u^0$, we iterate
for $n = 1, 2, \ldots$ to find better and better approximate solutions $u^1$, $u^2$, and
so on. During each iteration, we first solve the Poisson equation restricted to
the circle $\Omega_1$, using the previous iteration's solution from $\Omega_2$ on the artificial

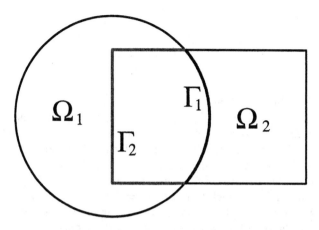

**Fig. 2.1.** Solution domain for the classical alternating Schwarz method.

internal boundary $\Gamma_1$:

$$-\nabla^2 u_1^n = f_1 \quad \text{in } \Omega_1,$$
$$u_1^n = g \quad \text{on } \partial\Omega_1\backslash\Gamma_1,$$
$$u_1^n = u_2^{n-1}|_{\Gamma_1} \quad \text{on } \Gamma_1.$$

Then, we solve the Poisson equation within the rectangle $\Omega_2$, using the latest solution $u_1^n$ on the artificial internal boundary $\Gamma_2$:

$$-\nabla^2 u_2^n = f_2 \quad \text{in } \Omega_2,$$
$$u_2^n = g \quad \text{on } \partial\Omega_2\backslash\Gamma_2,$$
$$u_2^n = u_1^n|_{\Gamma_2} \quad \text{on } \Gamma_2.$$

The two local Poisson equations in $\Omega_1$ and $\Omega_2$ are coupled together in the following way: the artificial Dirichlet condition on the internal boundary $\Gamma_1$ of subdomain $\Omega_1$ is provided by subdomain $\Omega_2$ in form of $u_2^{n-1}|_{\Gamma_1}$, and vice versa. It is clear that $u_2^{n-1}|_{\Gamma_1}$ and $u_1^n|_{\Gamma_2}$ may change from iteration to iteration, while converging towards the true solution. Therefore, in each Schwarz iteration, the two Poisson equations need to update the artificial Dirichlet conditions on $\Gamma_1$ and $\Gamma_2$ by exchanging some data. Note also that the classical alternating Schwarz method is *sequential* by nature, meaning that the two Poisson solves within each iteration must be carried out in a predetermined sequence, first in $\Omega_1$ then in $\Omega_2$. Of course, the above alternating Schwarz method can equally well choose the rectangle as $\Omega_1$ and the circle as $\Omega_2$, without any noticeable effects on the convergence.

### 2.2.2 The Multiplicative Schwarz Method

We now extend the classical alternating Schwarz method to more than two subdomains. To this end, assume that we want to solve a linear elliptic PDE of the form:

$$Lu = f \quad \text{in } \Omega, \tag{2.1}$$

$$u = g \quad \text{on } \partial\Omega, \tag{2.2}$$

where $L$ is some linear operator.

In order to use a "divide-and-conquer" strategy, we decompose the global solution domain $\Omega$ into a set of $P$ subdomains $\{\Omega_i\}_{i=1}^P$, such that $\Omega = \cup_{i=1}^P \Omega_i$. As before, we denote by $\Gamma_i$ the internal boundary of subdomain number $i$, i.e., the part of $\partial\Omega_i$ not belonging to the physical global boundary $\partial\Omega$. In addition, we denote by $\mathcal{N}_i$ the index set of neighboring subdomains for subdomain number $i$, such that $j \in \mathcal{N}_i \Rightarrow \Omega_i \cap \Omega_j \neq \emptyset$. We require that there is explicit overlap between each pair of neighboring subdomains. (The case of non-overlapping DD methods is beyond the scope of this chapter.) In other words, every point on $\Gamma_i$ must also lie in the *interior* of at least one neighboring subdomain $\Omega_j$, $j \in \mathcal{N}_i$.

When the set of overlapping subdomains is ready, we run an iterative solution procedure that starts with an initial guess $\{u_i^0\}_{i=1}^P$. The work of iteration number $n$ consists of $P$ sub-steps that must be carried out *in sequence* $i = 1, 2, \ldots, P$. For sub-step number $i$, the work consists in solving the PDE restricted to subdomain $\Omega_i$:

$$L_i u_i^n = f_i \quad \text{in } \Omega_i, \tag{2.3}$$

$$u_i^n = g \quad \text{on } \partial\Omega_i \backslash \Gamma_i, \tag{2.4}$$

$$u_i^n = \tilde{g}^* \quad \text{on } \Gamma_i. \tag{2.5}$$

Here, in a rigorous mathematical formulation, $L_i$ in (2.3) means the restriction of $L$ onto $\Omega_i$. However, for most cases the $L_i$ and $L$ operators have exactly the same form. The right-hand side term $f_i$ in (2.3) arises from restricting $f$ onto $\Omega_i$. The notation $\tilde{g}^*$ in (2.5) means an artificial Dirichlet condition on $\Gamma_i$. The artificial Dirichlet condition is updated by "receiving" the *latest* solution on $\Gamma_i$ from the neighboring subdomains. Different points on $\Gamma_i$ may use solution from different neighboring subdomains. More precisely, for every point $x$ that lies on $\Gamma_i$, we suppose it is also an interior point to the neighboring subdomains: $\Omega_{j_1}, \Omega_{j_2}, \ldots, \Omega_{j_m}$, where $j_1 < j_2 < \ldots < j_m$ with $j_k \in \mathcal{N}_i$ for $k = 1, \ldots, m$. Then, the latest approximate solution at $x$ from a particular subdomain, number $j_k$, will be used. That is, $u_{j_k}^n(x)$ is "received" as the artificial Dirichlet condition on $x$. Here, $j_k$ should be the maximal index among those indices satisfying $j_k < i$. In case of $i < j_1$, $u_{j_m}^{n-1}(x)$ will be used as the artificial Dirichlet condition on $x$. The derived method is the so-called *multiplicative Schwarz method*.

*Example: Multiplicative Schwarz on the Unit Square.* Suppose we want to partition $\Omega$, the unit square $(x, y) \in [0, 1] \times [0, 1]$, into four overlapping subdomains. This can be achieved by first dividing $\Omega$ into four non-overlapping

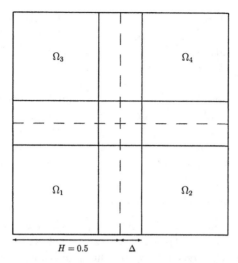

**Fig. 2.2.** An example of partitioning the unit square into four overlapping subdomains.

subdomains, each covering one quarter of the unit square. This is shown by the dashed lines in Figure 2.2. Then, each subdomain extends itself in both $x$- and $y$-direction by $\Delta$, resulting in four overlapping subdomains $\Omega_1$, $\Omega_2$, $\Omega_3$, and $\Omega_4$, bounded by the solid lines in Figure 2.2. The artificial internal boundary $\Gamma_i$ of each subdomain can be divided into three parts, two of them border with only one neighboring subdomain, while the third part borders with all the three neighboring subdomains. Let us take $\Omega_1$ for instance. The first part of $\Gamma_1$: $x = 0.5 + \Delta$, $0 \le y \le 0.5 - \Delta$ lies in the interior of $\Omega_2$, so the artificial Dirichlet condition on this part is updated by "receiving" the latest solution from $\Omega_2$. The second part of $\Gamma_1$: $0 \le x \le 0.5 - \Delta$, $y = 0.5 + \Delta$ lies in the interior of $\Omega_3$. The artificial Dirichlet condition on this part is thus updated by "receiving" the latest solution from $\Omega_3$. The remaining part of $\Gamma_1$, i.e., $0.5 - \Delta \le x \le 0.5 + \Delta$, $y = 0.5 + \Delta$, and $x = 0.5 + \Delta$, $0.5 - \Delta \le y \le 0.5 + \Delta$, lies in the interior of all the three neighbors. The artificial Dirichlet condition on this part, however, needs to be updated by "receiving" the latest solution from $\Omega_4$. This is because subdomain number 4 is the last to carry out its subdomain solve in a previous multiplicative Schwarz iteration.

Like the classical alternating Schwarz method, the multiplicative Schwarz method is also sequential by nature. This is because in each iteration the $P$ sub-steps for solving the PDE restricted onto $\Omega_i$, $i = 1, 2, \ldots, P$, must be carried out in a fixed order. More precisely, simultaneous solution of (2.3)-(2.5) on more than one subdomain is not allowed due to (2.5).

In order for the multiplicative Schwarz method to be able to run on a multi-processor platform, it is necessary to modify the method by using a multi-coloring scheme. More specifically, we use a small number of different colors and assign each subdomain with one color. The result of multi-coloring is that any pair of neighboring subdomains always have different colors. In a modified multiplicative Schwarz method, all the subdomains that have the same color can carry out the subdomain solves simultaneously. Therefore, multiple processors are allowed to work simultaneously in subdomains with the same color. It should be noted that the modified multiplicative Schwarz method is not so flexible as the additive Schwarz method (see below) when it comes to parallel implementation. The convergence speed of the modified multiplicative Schwarz method becomes slower compared with that of the standard multiplicative Schwarz method.

### 2.2.3    The Additive Schwarz Method

Another variant of overlapping DD methods, which inherently promotes parallel computing, is called *additive Schwarz method*. The difference between the multiplicative Schwarz method and the additive counterpart lies in the way how the artificial Dirichlet condition is updated on $\Gamma_i$. For the additive Schwarz method, we use

$$u_i^n = \tilde{g}^{n-1} \quad \text{on } \Gamma_i,$$

instead of (2.5). This means that the artificial Dirichlet condition on $\Gamma_i$ is updated in a Jacobi fashion, using solutions from all the relevant neighboring subdomains from iteration number $n-1$. Therefore, the subdomain solves in the additive Schwarz method can be carried out completely independently, thus making the method inherently parallel. In more precise terms, for every point $x$ that lies on $\Gamma_i$, we suppose it is also an interior point of the neighboring subdomains $\Omega_{j_1}, \Omega_{j_2}, \ldots, \Omega_{j_m}$, where $j_1 < j_2 < \ldots < j_m$ with $j_k \in \mathcal{N}_i$ for $k = 1, \ldots, m$. Then, the average value

$$\frac{1}{m} \sum_{k=1}^{m} u_{j_k}^{n-1}(x)$$

will be used as the artificial Dirichlet condition on $x \in \Gamma_i$. Although the additive Schwarz method suits well for parallel computing, it should be noted that its convergence property is inferior to that of the multiplicative Schwarz method. In case of convergence, the additive Schwarz method uses roughly twice as many iterations as that of the standard multiplicative Schwarz method. This is not surprising when the Schwarz methods are compared with their linear system solver analogues; multiplicative Schwarz is a block Gauss-Seidel approach, whereas additive Schwarz is a block Jacobi approach.

### 2.2.4    Formulation in the Residual Form

It is often more convenient to work with an equivalent formulation of the overlapping DD methods. More precisely, Equation (2.3) can be rewritten using a residual form:

$$u_i^n = u_i^{n-1} + L_i^{-1}(f_i - L_i u_i^{n-1}). \tag{2.6}$$

The above residual form is, e.g., essential for introducing the so-called coarse grid corrections (see Section 2.4.4). Moreover, by using a restriction operator $R_i$ from $\Omega$ to $\Omega_i$, whose matrix representation consists of only ones and zeros, we can rewrite, e.g., the additive Schwarz method more compactly as

$$u^n = u^{n-1} + \sum_{i=1}^{P} R_i^T L_i^{-1} R_i(f - Lu^{n-1}), \tag{2.7}$$

where $R_i^T$ is an interpolation operator associated with $R_i$. To understand the effect of $R_i$, suppose we want to solve a scalar PDE on a global grid with $M$ grid points. Let subgrid number $i$ have $M_i$ points, where point number $j$ has index $I_j^i$ in the global grid. Then, the corresponding matrix for $R_i$ is of dimension $M_i \times M$. On row number $j$, there is only one entry of "1" at column number $I_j^i$, the other entries are "0". The matrix form of $R_i^T$ is simply the transpose of that of $R_i$. However, in a practical implementation of an overlapping DD method, we can avoid physically constructing the global matrices and vectors. It is sufficient to work directly with the subdomain matrices/vectors, which together constitute their global counterparts. Thus, we do not need to know $I_j^i$ explicitly, only the correspondence between neighboring subdomains in terms of the shared grid points is necessary. In other words, the $R_i$ and $R_i^T$ operators are only necessary for the correctness of the mathematical formulation, they are not explicitly used in an implementation where all the global vectors are distributed as subdomain vectors.

Consequently, iteration number $n+1$ of the multiplicative Schwarz method can be expressed in the residual form, consisting of $P$ sequential sub-steps:

$$u^{n+\frac{1}{P}} = u^n + R_1^T L_1^{-1} R_1(f - Lu^n), \tag{2.8}$$

$$u^{n+\frac{2}{P}} = u^{n+\frac{1}{P}} + R_2^T L_2^{-1} R_2(f - Lu^{n+\frac{1}{P}}), \tag{2.9}$$

$$\dots$$

$$u^{n+1} = u^{n+\frac{P-1}{P}} + R_P^T L_P^{-1} R_P(f - Lu^{n+\frac{P-1}{P}}). \tag{2.10}$$

## 2.2.5    Overlapping DD as Preconditioner

So far, we have discussed the overlapping DD methods in the operator form using $L_i$, $R_i$ and $R_i^T$. When we discretize (2.1)-(2.2), by e.g. the finite element method (FEM), a global system of linear equations will arise in the following matrix-vector form:

$$\mathbf{Ax} = \mathbf{b}. \tag{2.11}$$

If the above global linear system is large, it is beneficial to apply an iterative method. Moreover, preconditioning is often necessary for achieving fast convergence. That is, we solve

$$\mathbf{BAx} = \mathbf{Bb}$$

instead of (2.11), where $\mathbf{B}$ is some preconditioner expressed in the matrix form. The preconditioner $\mathbf{B}$ should be close to $\mathbf{A}^{-1}$, so that the condition number of $\mathbf{BA}$ becomes much smaller than that of $\mathbf{A}$. An optimal situation will be that the condition number of $\mathbf{BA}$ is independent of $h$, the grid size. Then the number of iterations needed by a global Krylov subspace method for obtaining a desired convergence will remain a constant, independent of $h$. In fact, for certain PDEs, one Schwarz iteration with coarse grid correction (see Section 2.4.4) can work as such an optimal preconditioner. Using overlapping DD methods as preconditioners in such a way actually results in more robust convergence than using them as stand-alone solvers.

In the operator form, the result of applying preconditioner $B$ can be expressed by

$$v = Bw.$$

Typically, $w$ is a residual vector given by $w = b - Ax$ and $v$ is the resulting preconditioned correction vector.

We note that a Schwarz preconditioner, either additive or multiplicative, is no other than one Schwarz iteration with zero initial guess. More precisely, an additive Schwarz preconditioner produces $v$ by

$$v = \sum_{i=1}^{P} R_i^T L_i^{-1} R_i w. \tag{2.12}$$

For a multiplicative Schwarz preconditioner, it is difficult to express it in a single and compact formula as above. But its preconditioned vector $v$ can be obtained as $u^{n+1}$, if we set $f = w$ and $u^n = 0$ in (2.8)-(2.10). It is also worth noting that a multiplicative Schwarz preconditioner has better convergence property than its additive counterpart.

## 2.3    A 1D Example

In order to help the reader to associate the mathematical algorithm of the additive Schwarz method with concrete C++ programming, we present in the following a simple Diffpack program that applies the additive Schwarz method with two subdomains for solving a 1D Poisson equation: $-u''(x) = f$, where $f$ is some constant.

More specifically, we cover the 1D solution domain $\Omega = [0, 1]$ with a global uniform lattice grid having $n$ points, whose indices are from 1 to $n$. In addition, we introduce two indices $a$ and $b$, where $b < a$, such that points with indices $1, \ldots, a$ form the left subdomain and points with indices $b, \ldots, n$ form the right subdomain. Note the condition $b < a$ ensures that the two subdomains are overlapping.

In the following Diffpack program, where the simulator class has name DD4Poisson, the data structure for the subdomain solvers consists of

- two GridLattice objects: grid_1 and grid_2 for the left and right subdomain, respectively,

- two FieldLattice objects containing the subdomain solutions from the current Schwarz iteration: u_1 and u_2,

- two FieldLattice objects containing the subdomain solutions from the previous Schwarz iteration: u_1_prev and u_2_prev,

- two pairs of MatTri(real)/ArrayGen(real) objects: A_1 and b_1 for $\Omega_1$ and A_2 and b_2 for $\Omega_2$. They contain the subdomain linear systems.

The Diffpack program starts with preparing the subdomain grids and subdomain linear systems. Then, it creates the data structure for a global GridLattice object grid and a global FieldLattice object u. Thereafter, the program proceeds with a setUpMatAndVec function for filling up the values of the subdomain linear systems:

```
void DD4Poisson:: setUpMatAndVec()
{
  // set up matrices and vectors

  int i;
  const real h = grid->Delta(1);

  A_1.fill(0.0);
  A_2.fill(0.0);
  b_1.fill(0.0);
  b_2.fill(0.0);

  // left domain   (domain 1):
  A_1(1,0) = 1;
  b_1(1) = 0;
  for (i = 2; i < a; i++) {
    A_1(i,-1) = 1;   A_1(i,0) = -2;   A_1(i,1) = 1;
    b_1(i) =   -f*h*h;
  }
  A_1(a,0) = 1;
  // b_1(a) is updated for each iteration, in solveProblem()

  A_1.factLU();

  // right domain (domain 2):
  A_2(1,0) = 1;
  // b_2(1) is updated for each iteration, in solveProblem()
  for (i = 2; i < n-b+1; i++) {
    A_2(i,-1) = 1;   A_2(i,0) = -2;   A_2(i,1) = 1;
    b_2(i) = - f*h*h;
  }
```

```
A_2(n-b+1,0) = 1;
b_2(n-b+1) = 0;

A_2.factLU();
}
```

The main computational loop of the additive Schwarz method looks as follows:

```
for (int j = 2; j <= no_iter; j++) {
    // local solution in each subdomain, and update the global domain
    b_1(a) = u_2_prev.values()(a);
    // the input rhs vector is changed during the solution process
    // we send a copy of the b_1 vector to the function
    rhs_tmp = b_1;
    A_1.forwBackLU(rhs_tmp,u_1.values());   // Gaussian elimination
    b_2(1) = u_1_prev.values()(b);
    rhs_tmp = b_2;
    A_2.forwBackLU(rhs_tmp,u_2.values());   // Gaussian elimination

    updateValues();
}
```

In the above loop, the subdomain matrices A_1 and A_2 remain constant, so the LU factorization is carried out once and for all inside setUpMatAndVec, whereas only two calls of forwBackLU need to be invoked in each Schwarz iteration. Furthermore, the change of the right-hand side vectors b_1 and b_2 from iteration to iteration is small, i.e., only the last entry of b_1 and the first entry of b_2 need to be updated. The work of function updateValues is to store the current subdomain solutions for the next Schwarz iteration, while also updating the global solution field u. We mention that the points of the right subdomain grid, grid_2, have indices $b, \ldots, n$. This greatly simplifies the procedure of mapping the subdomain nodal values from u_2 to u, as is shown below.

```
void DD4Poisson:: updateValues()
{
    int i;

    // remember the subdomain solutions for the next iteration
    u_1_prev.values() = u_1.values();
    u_2_prev.values() = u_2.values();

    // update global domain values
    for (i = 1; i <= b; i++)
        { u.values()(i) = u_1.values()(i); }

    for (i = b+1; i < a; i++)
        { u.values()(i) = (u_1.values()(i) + u_2.values()(i))/2; }

    for (i = a; i <= n; i++)
        { u.values()(i) = u_2.values()(i); }
}
```

## 2.4    Some Important Issues

### 2.4.1    Domain Partitioning

The starting point of an overlapping DD method is a set of overlapping subdomains $\{\Omega_i\}_{i=1}^{P}$. There is, however, no standard way of generating $\{\Omega_i\}_{i=1}^{P}$ from a global domain $\Omega$. One commonly used approach is to start with a global fine grid $\mathcal{T}$ that covers $\Omega$ and consider the graph that corresponds to $\mathcal{T}$. By a graph we mean a collection of nodes, where some nodes are connected by edges. Note that nodes and edges in a graph are not the same as the grid points and element sides of $\mathcal{T}$. For the graph that arises from a finite element grid $\mathcal{T}$, each element is represented by a node, an edge exists between two nodes if the two corresponding elements are neighbors; see e.g. [4,9]. That is, the graph is a simplified topological representation of the grid, without considering the size and location of the elements. We can then use some graph partitioning algorithm to produce non-overlapping subdomains, such that every element belongs to a unique subdomain. Thereafter, a post-processing procedure is necessary to enlarge the subdomains and thereby ensure that every grid point on $\Gamma_i$ eventually also lies in the interior of at least one neighboring subdomain.

### 2.4.2    Subdomain Discretizations in Matrix Form

When the set of overlapping subdomains $\{\Omega_i\}_{i=1}^{P}$ and the associated subgrids $\mathcal{T}_i$ are ready, the next step for a DD method is to carry out discretization on each subgrid $\mathcal{T}_i$. An important observation about overlapping DD methods is that the subproblems arise from restricting the original PDE(s) onto the subdomains. Discretization on $\mathcal{T}_i$ can therefore be done straightforwardly, without having to first construct the global stiffness matrix $\mathbf{A}$ for $\Omega$ and then "cut out" the portion corresponding to $\mathbf{A}_i$.

Let us denote respectively by $\mathbf{A}_i^0$ and $\mathbf{f}_i^0$ the resulting local stiffness matrix and right-hand vector, without enforcing the artificial Dirichlet condition on $\Gamma_i$. Then, to incorporate the artificial Dirichlet condition, we need to remove all the non-zero entries on the rows of $\mathbf{A}_i^0$ that correspond to the grid points on $\Gamma_i$. In addition, the main diagonal entries for these rows are set to one. The modified subdomain matrix $\mathbf{A}_i$ remains unchanged during all the DD iterations. During DD iteration number $i$, the right-hand side vector $\mathbf{f}_i^n$ is obtained by modifying $\mathbf{f}_i^0$ on the entries that correspond to the grid points on $\Gamma_i$. Those entries will use values that are provided by the neighboring subdomains.

From now on we will use the matrix representations $\mathbf{A}_i$, $\mathbf{A}_i^{-1}$, $\mathbf{R}_i$, $\mathbf{R}_i^T$ for $L_i$, $L_i^{-1}$, $R_i$, $R_i^T$, respectively.

### 2.4.3    Inexact Subdomain Solver

From Section 2.2 we can see that the main computation of an overlapping DD method happens in the subdomain solves (2.3)-(2.5). Quite often, the size of the subgrids may still be large, therefore preventing an exact subdomain solution in form of $\mathbf{A}_i^{-1}$. This means that we have to resort to some iterative solution method also for the subdomain problems. That is, we use an inexact subdomain solver $\tilde{\mathbf{A}}_i^{-1}$, which approximates the inverse of $\mathbf{A}_i$ in some sense. However, the subdomain problems have to be solved quite accurately when overlapping DD methods are used as stand-alone iterative solvers. When an overlapping DD method is used as a preconditioner, it is sufficient to solve the subdomain problems only to a certain degree of accuracy. In the latter case, it is standard to use in the subdomain solves an iterative Krylov solver with a loose convergence criterion, or even one multigrid V-cycle [5]. This may greatly enhance the overall efficiency of the overlapping DD methods.

### 2.4.4   Coarse Grid Corrections

One weakness with the Schwarz methods from Section 2.2 is that convergence deteriorates as the number of subdomains increases, especially when an additive Schwarz method is used as a stand-alone iterative solver. This is due to the fact that each subdomain problem is restricted to $\Omega_i$. The global information propagation, in form of data exchange within the overlapping regions, happens only between immediate neighboring subdomains during one DD iteration. That is, the information of a change at one end of $\Omega$ needs several DD iterations to be propagated to the other end of $\Omega$.

To incorporate a more rapid global information propagation mechanism for faster convergence, we may use a coarse global grid $\mathcal{T}_H$ that covers the whole domain $\Omega$. Then an additional global coarse grid problem is solved in every DD iteration. Such overlapping DD methods with *coarse grid corrections* can be viewed as a variant of two-level multigrid methods, where the collaborative subdomain solves work as a special multigrid smoother, see e.g. [10]. For example, the additive Schwarz preconditioner equipped with a coarse grid solver can be expressed as:

$$\mathbf{v} = \sum_{i=0}^{P} \mathbf{R}_i^T \tilde{\mathbf{A}}_i^{-1} \mathbf{R}_i \mathbf{w}, \qquad (2.13)$$

where we denote by $\tilde{\mathbf{A}}_0^{-1}$ a solver for the global coarse grid problem, $\mathbf{R}_0$ and $\mathbf{R}_0^T$ represent the restriction and interpolation matrices describing the mapping between $\mathcal{T}$ and $\mathcal{T}_H$. In the later implementation, the matrices $\mathbf{R}_0$ $\mathbf{R}_0^T$ are not formed explicitly. Their operations are achieved equivalently by using a series of mapping matrices between $\mathcal{T}_i$ and $\mathcal{T}_H$. We also note that (2.13) allows inexact subdomain solvers.

### 2.4.5   Linear System Solvers at Different Levels

It is worth noticing that overlapping DD methods involve linear system administration on at least two levels (three if coarse grid corrections are used).

First of all, we need a *global* administration of the solution of the global linear system, e.g., calculating the global residual, updating the artificial Dirichlet conditions, and monitoring the convergence of the global solution. Secondly, we need a separate *local* administration for handling the local linear system restricted to each subdomain. In case an iterative subdomain linear system solver is chosen, it is normally sufficient to use a less strict subdomain stopping criterion than that used for monitoring the convergence of the global DD iterations. An important observation is that although linear solvers exist at both global and subdomain levels, there is no absolute need for constructing physically the global matrices and vectors. We can actually adopt a *distributed data storage* scheme, where we only store for each subdomain its subdomain matrix and vector. These subdomain matrices and vectors, which are simply parts of their global counterparts, can together give a *logical* representation of the global matrix and vector. More specifically, all the global linear algebra operations can be achieved by running local linear algebra operations on the subdomains, together with necessary data exchange between the subdomains. This allows us to avoid physical storage of the global matrices and vectors, thus making the transition from a sequential overlapping DD code to a parallel code quite straightforward.

## 2.5    Components of Overlapping DD Methods

Before moving on to the implementation of the overlapping DD methods, we find it important to view these methods at a higher abstraction level. This is for gaining insight into the different components that form the algorithmic structure.

We have pointed out in Section 2.4.1 that the starting point of an overlapping DD method is a set of overlapping subdomains. Based on this overlapping subdomain set, the basic building block of an overlapping DD method is its subdomain solvers. The subdomain solvers should be flexible enough to handle different shapes of subdomains and mixture of physical and artificial boundary conditions. They also need to be numerically efficient, because the overall efficiency of an overlapping DD method depends mainly on the efficiency of the subdomain solvers.

Due to the necessity of updating the artificial Dirichlet condition (2.5) during each DD iteration, data must be exchanged between neighboring subdomains. In order to relieve the subdomain solvers of the work of directly handling the data exchange, there is need for a so-called "data exchanger". This component has an overview of which grid points of $\mathcal{T}_i$ that lie on $\Gamma_i$, and knows which neighboring subdomain should provide which part of its solution along $\Gamma_i$. For every pair of neighboring subdomains, the task is to construct data exchange vectors by retrieving grid point values from corresponding locations of the local solution vector and then make the exchange.

In order to coordinate the subdomain solves and the necessary data exchanges, it is essential to have a global administration component. The tasks of this component include controlling the progress of DD iterations, monitoring the global convergence behavior, starting the subdomain solves, and invoking data exchange via the data exchange component.

As a summary, an overlapping DD method should have the following main components:

- Subgrid preparation component;
- Subdomain solver component;
- Data exchange component;
- Global administration component.

## 2.6     A Generic Implementation Framework

If there already exists a global solver for a PDE, then it should, in principle, also work as a subdomain solver in an overlapping DD method. This is due to the fact that the subdomain problems of an overlapping DD method are of the same type as the original global PDE problem. Of course, small modifications and/or extensions of the original global solver are necessary, such as for incorporating the artificial Dirichlet boundary condition (2.5).

Object-oriented programming is an ideal tool for code extensions. The objective of this section is thus to present a generic object-oriented implementation framework, where an existing Diffpack PDE solver can be easily extended for incorporating an overlapping DD method. The implementation framework is used to simplify the coding effort and promote a structured design.

### 2.6.1     The Simulator-Parallel Approach

The traditional way of programming overlapping DD methods is to implement from scratch all the involved algebra-level operations, like matrix-vector product, residual vector calculation etc. We use however a so-called *simulator-parallel* approach, which was first implemented in a generic fashion in [1] for implementing parallel DD solvers, but applies equally well to sequential DD solvers. This approach takes an existing PDE simulator as the starting point. The simulator is re-used as the subdomain solver component, after minor code modifications and/or extensions. Each subdomain is then assigned with such a subdomain solver. The computation coordination of the subdomain solvers is left to a global administrator, which is implemented at a high abstraction level close to the mathematical formulation of the overlapping DD methods.

The simulator-parallel approach can be used to program any overlapping DD code, which is achievable by the traditional linear-algebra level programming approach. This is because any object-oriented sequential simulator, which has data structure for local matrices/vectors, a numerical discretization scheme, and a linear algebra toolbox, is capable of carrying out all the operations needed in the subdomain solves. An advantage of the simulator-parallel approach is that each subdomain simulator "sees" the physical properties of the involved PDEs, which may be used to speed up the solution of the subproblems. Furthermore, a subdomain simulator may have independent choice of its own solution method, preconditioner, stopping criterion etc. Most importantly, developing DD solvers in this way strongly promotes code reuse, because most of the global administration and the related data exchange between subdomains can be extracted from specific applications, thereby forming a generic library.

### 2.6.2    Overall Design of the Implementation Framework

As we have pointed out in Section 2.5, the subdomain solvers are only one of the components that form the algorithmic structure of an overlapping DD method. The other components are normally independent of specific PDEs, thus enabling us to build a generic implementation framework where components are realized as standardized and yet extensible objects.

The implementation framework is made up of three parts. In addition to the subdomain solvers, the other two parts are a communication part and a global administrator. The communication part contains a subgrid preparation component and a data exchange component, as mentioned in Section 2.5. In this way, the data exchange component has direct access to the information about how the overlapping subgrids are formed. The information will thus be used to carry out data exchanges between neighboring subdomains.

When using the framework for implementing an overlapping DD method, the user primarily needs to carry out a coding task in form of deriving two small-sized C++ subclasses. The first C++ subclass is for modifying and extending an existing PDE simulator so that it can work as a subdomain solver. The second C++ subclass is for controlling the global calculation and coupling the subdomain solver with the other components of the implementation framework. The coding task is quite different from that of incorporating a multigrid solver into a Diffpack simulator, see [6], where a toolbox is directly inserted into an existing PDE simulator without having to derive new C++ subclasses. This seemingly cumbersome implementation process of an overlapping DD method is necessary due to the existence of linear system solvers at different levels, see Section 2.4.5. We therefore let the global administrator take care of the global linear system, while the subdomain solvers control the solution of the subdomain linear systems. Besides, we believe this coding rule promotes a structured implementation process with sufficient flexibility.

### 2.6.3    Subdomain Solvers

A typical Diffpack PDE simulator has functionality for doing discretization on a grid and solving the resulting linear system. In order for such an existing

Diffpack PDE simulator to be accepted by the implementation framework, it is important that the simulator is first "wrapped up" within a generic standard interface recognizable by the other generic components. We have therefore designed

```
class SubdomainFEMSolver : public virtual HandleId
```

that provides such a generic interface. The purpose of this class is to allow different components of the implementation framework to access the subdomain data structure and invoke the functions related to the subdomain solves. The main content of class SubdomainFEMSolver consists of a set of data object handles and virtual member functions with standardized names. It is also through these handles and functions that the generic implementation framework utilizes the data and functions belonging to an existing PDE simulator.

The most important internal variables of class SubdomainFEMSolver are the following data object handles:

```
Handle(GridFE)      subd_fem_grid;
Handle(DegFreeFE)   subd_dof;
Handle(LinEqAdm)    subd_lineq;
Handle(Matrix(real)) A_orig;
Handle(Matrix(real)) A_new;
Handle(Vec(real))   global_solution_vec;
Handle(Vec(real))   global_orig_rhs;
Handle(Vec(real))   global_residual_vec;
Handle(Vec(real))   solution_vec;
Handle(Vec(real))   orig_rhs;
Handle(Vec(real))   rhs_vec;
Handle(Vec(real))   residual_vec;
```

The above handles can be divided into three types. The first type of handles are to be bound to external data objects created by some other components of the generic implementation framework. The subd_fem_grid handle is such an example. More precisely, the subdomain solvers of the implementation framework are no longer responsible for constructing the subgrids, but receive them from the global administrator (see Section 2.6.5). It is important to note that the subgrid is the starting point for building up the data structure of a subdomain solver. The second type of handles are to be bound to external data objects living inside an existing PDE simulator, thus enabling the implementation framework to utilize those data objects. Handles subd_dof, subd_lineq, A_orig, and global_solution_vec belong to this second type. The rest of the above handles belong to the third type and are to be bound to some internal data objects, which are specially needed for carrying out a generic subdomain solution process.

When deriving a new subdomain solver, which is to be a subclass of both class SubdomainFEMSolver and an existing PDE simulator class, we normally only make direct use of the handles subd_fem_grid, subd_dof, subd_lineq, and global_solution_vec. The other handles of SubdomainFEMSolver are normally used within the implementation framework. For example, the matrix object that contains the local matrix $\mathbf{A}_i^0$ lies physically inside an existing PDE simulator. It is assumed to be physically allocated in the LinEqAdm object to which the subd_lineq handle is bound. We therefore do not need to directly access the A_orig handle, which is to be bound to the $\mathbf{A}_i^0$ matrix object inside a function named SubdomainFEMSolver::modifyAmatWithEssIBC. This function is normally invoked by the global administrator to produce the internal matrix object $\mathbf{A}_i$, which is pointed by the handles A_new.

We note that all the matrix and vector objects, which belong to one object of SubdomainFEMSolver, arise from the particular subgrid $\mathcal{T}_i$, not the global grid $\mathcal{T}$. That is, the prefix global rather reminds us of the fact that we need to operate with linear systems at two different levels. The first level concerns the logically existing global linear system. For example, global_residual_vec is used for monitoring the convergence of the global solution. The second level concerns the subdomain local linear system. For example, solution_vec represents $\mathbf{u}_i^n$ that is involved in every subdomain solve $\mathbf{A}_i \mathbf{u}_i^n = \mathbf{f}_i^n$. In other words, we avoid the physical storage of global matrices and vectors that are associated with the entire global grid $\mathcal{T}$. This is because any global linear algebra operation can be realized by local linear algebra operations plus necessary data exchange, as explained in Section 2.4.5.

The aforementioned function modifyAmatWithEssIBC is an important member function of class SubdomainFEMSolver. In addition, the other five important member functions of class SubdomainFEMSolver are:

```
virtual void initSolField (MenuSystem& menu) =0;
virtual void createLocalMatrix ();
virtual void initialize (MenuSystem& menu);
virtual void updateRHS ();
virtual void solveLocal ();
```

All these six member functions rarely need to be directly invoked by the user. Among them only initSolField needs to be *explicitly* re-implemented in a subclass. The task of this function is to make sure that the needed external data objects are constructed and the handles subd_dof, subd_lineq, and global_solution_vec are correctly bound to those external data objects. Besides, the member function createLocalMatrix, which is used for building $\mathbf{A}_i^0$ and $\mathbf{f}_i^0$, normally also requires re-implementation. For most cases, this can be done by simply using

```
FEM:: makeSystem (*subd_dof, *subd_lineq);
```

The other three member functions, like modifyAmatWithEssIBC, have a default implementation that seldom needs to be overridden in the subclass. The task of the member function initialize is to invoke the user re-implemented member function initSolField, in addition to preparing the internal data objects like residual_vec, solution_vec, rhs_vec, and orig_rhs. The member function updateRHS is normally used by the global administrator for updating the right-hand side vector $\mathbf{f}_i^n$, see Section 2.4.2, before solving the subdomain problem in iteration number $n$. Moreover, the member function `solveLocal` has the following default implementation:

```
void SubdomainFEMSolver:: solveLocal ()
{
  subd_lineq->solve (first_local_solve);
  first_local_solve = false;
}
```

where `first_local_solve` is an internal boolean variable belonging to class `SubdomainFEMSolver`. Since the subdomain matrix $\mathbf{A}_i$ does not change in the DD iterations, the `first_local_solve` variable thus gives this signal to the internal `LinEqAdm` object. We remark that `first_local_solve` is assigned with a `dpTrue` value inside every call of the `modifyAmatWithEssIBC` function, which follows every call of the `createLocalMatrix` function. We also remark that the user is responsible for deciding how the subdomain linear system (with an updated right-hand side $\mathbf{f}_i^n$) should be solved in each DD iteration. A typical choice can be a Krylov subspace solver using a moderate-sized tolerance in the stopping criterion.

Such a generic subdomain solver interface makes it easy to incorporate an existing PDE simulator into the generic implementation framework. Given an existing PDE simulator, say `MySolver`, we can construct a new subdomain solver that is recognizable by the generic implementation framework as follows:

```
class MySubSolver : public MySolver, public SubdomainFEMSolver
```

After re-implementing the virtual member functions such as `initSolField` and `createLocalMatrix`, we can readily "plug" the new subdomain solver `MySubSolver` into the generic implementation framework.

*Class CoarseGridSolver.* The data structure and functionality needed in coarse grid corrections are programmed as a small general class with name `CoarseGridSolver`. Inside class `SubdomainFEMSolver` there is a following handle

```
Handle(CoarseGridSolver) csolver;
```

which will be bound to a `CoarseGridSolver` object if the user wants to use coarse grid corrections. The other variables and member functions of class `SubdomainFEMSolver` that concern coarse grid corrections are:

```
Handle(DegFreeFE) global_coarse_dof;
virtual void buildGlobalCoarseDof ();
virtual void createCoarseGridMatrix ();
```

When coarse grid corrections are desired, the implementation framework equips *each* `SubdomainFEMSolver` object with an object of `CoarseGridSolver`. Of course, only one `CoarseGridSolver` object (per processor) needs to contain the data for e.g. $\mathbf{A}_0$. But every `CoarseGridSolver` object has the data structure for carrying out operations such as to map a subdomain vector associated with $\mathcal{T}_i$ to a global vector associated with $\mathcal{T}_H$, where the combined result from all the subdomains is a mapping between $\mathcal{T}$ and $\mathcal{T}_H$. The user only needs to re-implement the two following virtual member functions of SubdomainFEMSolver

```
virtual void buildGlobalCoarseDof ();
virtual void createCoarseGridMatrix ();
```

in a derived subclass. The purpose of the first function is to construct the global_coarse_dof object and fill it with values of essential boundary conditions, if applicable. The user should normally utilize the default implementation of SubdomainFEMSolver::buildGlobalCoarseDof, which constructs the global_coarse_dof object without considering the boundary conditions. The second function is meant for carrying out an assembly process for building the global coarse grid matrix $\mathbf{A}_0$. In the following, we list the default implementation in class SubdomainFEMSolver.

```
void SubdomainFEMSolver:: buildGlobalCoarseDof ()
{
    // menu_input is an internal MenuSystem pointer
    String cg_info = menu_input->get("global coarse grid");
    Handle(GridFE) coarse_grid = new GridFE;
    readOrMakeGrid (*coarse_grid, cg_info);

    const int cg_ndpn
        = menu_input->get("degrees of freedom per node").getInt();
    global_coarse_dof.rebind (new DegFreeFE(*coarse_grid, cg_ndpn));
}

void SubdomainFEMSolver:: createCoarseGridMatrix ()
{
    // user must redefine this function to include assembly
    if (!csolver->coarse_lineq->ok() ||
        !csolver->coarse_lineq->bl().ok()) {
      warningFP("SubdomainFEMSolver::createCoarseGridMatrix",
                "You should probably redefine this function!");
      return;
    }

    if (csolver->coarse_lineq.ok() &&
        csolver->coarse_lineq->getMatrixPrm().storage=="MatBand") {
      // carry out LU-factorization only once
      FactStrategy fstrat;
      LinEqSystemPrec& sys=csolver->coarse_lineq->getLinEqSystem();
      sys.allow_factorization = true;
      csolver->coarse_lineq->A().factorize (fstrat);
      sys.allow_factorization = false;
    }
}
```

Examples of how these two member functions can be re-implemented in a subclass of SubdomainFEMSolver can be found in Section 2.8.

### 2.6.4   The Communication Part

Data exchange between neighboring subdomains within the overlapping regions is handled by class CommunicatorFEMSP in the generic implementation framework. This class is derived as a subclass of GridPartAdm, see [2], so the functionalities for overlapping subgrid preparation and data exchange are automatically inherited.

We note that a sequential DD solver is simply a special case of a parallel DD solver. In a sequential DD solver all the subdomains reside on a single processor, and class CommunicatorFEMSP is capable of handling this special case. This part of the implementation framework seldom requires re-implementation and the user normally does not invoke it directly.

### 2.6.5   The Global Administrator

The global administrator of the implementation framework is constituted by several generic classes. The responsibility of the global administrator includes construction of the subdomain solvers and the communication part, choosing a particular DD method during a setting-up phase, and invoking all the necessary operations during each DD iteration.

*Class ParaPDESolver.* We recall that overlapping DD methods can work as both stand-alone iterative solution methods and preconditioners for Krylov subspace methods. By using the object-oriented programming techniques, we wish to inject this flexibility into the design of the global administrator. We have therefore designed class ParaPDESolver to represent a generic DD method.Two subclasses have also been derived from ParaPDESolver, where the first subclass BasicDDSolver implements an overlapping DD method to be used as a stand-alone iterative solver, and the second subclass KrylovDDSolver implements an overlapping DD preconditioner.

*Class ParaPDESolver_prm.* In order to allow the user to choose, *at run-time*, whether to use an overlapping DD method as a preconditioner or as a stand-alone iterative solver, we have followed the Diffpack standard and devised a so-called parameter class with name ParaPDESolver_prm. The most important member function of this simple parameter class is

```
virtual ParaPDESolver* create () const;
```

which creates a desired BasicDDSolver or KrylovDDSolver object at run-time. In addition to the type of the DD method, class ParaPDESolver_prm also contains other parameters whose values can be given through the use of MenuSystem. The most important parameters of class ParaPDESolver_prm are:

1. A flag indicating whether the overlapping DD method should be used as a preconditioner for a Krylov method, or as a stand-alone iterative solver.

2. A `LinEqSolver_prm` object for choosing a particular global Krylov method when DD is to work as a preconditioner, the maximum number of Krylov or DD iterations, and a prescribed accuracy.

3. A `ConvMonitorList_prm` object for choosing a global convergence monitor.

Below, we also list a simplified version of the `defineStatic` function that belongs to class `ParaPDESolver_prm`.

```
void ParaPDESolver_prm:: defineStatic(MenuSystem& menu, int level)
{
  String label = "DD solvers and associated parameters";
  String command = "ParaPDESolver_prm";
  MenuSystem::makeSubMenuHeader(menu,label,command,level,'p');
  menu.addItem (level,
                "domain decomposition scheme","dd-scheme",
                "class name in ParaPDESolver hierarchy",
                "KrylovDDSolver",
                "S/BasicDDSolver/KrylovDDSolver");
  LinEqSolver_prm::defineStatic (menu, level+1);
  ConvMonitorList_prm::defineStatic (menu, level+1);
}
```

*Class SPAdmUDC.* The main class of the global administrator is `SPAdmUDC`, which controls parameter input and the creation of a desired DD solver. In addition, class `SPAdmUDC` also controls the communication part and all the generic subdomain solvers. We remark that "UDC" stands for "user-defined-codes" and is used to indicate that the user has the possibility of re-implementing `SPAdmUDC`'s member functions and introducing new functions when he or she derives a new subclass. The simplified definition of class `SPAdmUDC` is as follows:

```
class SPAdmUDC : public PrecAction
{
  int num_local_subds;
  bool use_coarse_grid;
  bool additive_schwarz;
  Handle(MenuSystem) menu_input;

  VecSimplest(Handle(GridFE)) subd_fem_grids;
  VecSimplest(Handle(SubdomainFEMSolver)) subd_fem_solvers;
  Handle(ParaPDESolver_prm) psolver_prm;
  Handle(ParaPDESolver) psolver;
  Handle(CommunicatorFEMSP) communicator;

  virtual void setupSimulators ();
  virtual void setupCommunicator ();
  virtual void prepareGrids4Subdomains();
  virtual void prepareGrids4Subdomains(const GridFE& global_grid);
  virtual void createLocalSolver (int subd_id);
  virtual void init (GridFE* global_grid = NULL);
  // DD as stand-alone solver
  virtual void oneDDIteration (int iteration_counter);
  virtual void genAction    // DD used as preconditioner
    (LinEqVector&, LinEqVector&, TransposeMode);
```

```
    static  void defineStatic (MenuSystem& menu, int level = MAIN);
    virtual void scan          (MenuSystem& menu);
    virtual bool solve ();
};
```

Among the above member functions of SPAdmUDC, three of them are meant
to be invoked directly by the user:

1. The defineStatic function designs the layout of an input menu.

```
void SPAdmUDC:: defineStatic (MenuSystem& menu, int level)
{
    // choices such as whether DD is to be used as preconditioner
    ParaPDESolver_prm::defineStatic (menu, level);

    String label
      = "User-defined components of domain decomposition";
    String command = "SPAdmUDC";
    MenuSystem::makeSubMenuHeader(menu,label,command,level,'U');

    GridPartAdm::defineStatic (menu,level+1);  // subd grid creation

    menu.addItem (level,
                  "use additive Schwarz","additive-Schwarz",
                  "additive Schwarz","ON","S/ON/OFF");

    // concerning coarse-grid-correction
    menu.addItem (level,
                  "use coarse grid","use_coarse_grid",
                  "use coarse grid","OFF","S/ON/OFF");
    menu.addItem (level,
                  "global coarse grid","gcg",
                  "global coarse grid","global-coarse.grid","S");
    menu.addItem (level,
                  "degrees of freedom per node","ndpd",
                  "ndpn-global-coarse-grid","1","I1");
    menu.setCommandPrefix ("coarsest");
    LinEqAdm::defineStatic (menu,level+1); //for coarse grid correc.
    menu.unsetCommandPrefix ();
}
```

2. The scan function reads the input parameters and does the necessary
   initialization work.

```
void SPAdmUDC:: scan (MenuSystem& menu)
{
  menu_input.rebind (menu);

  psolver_prm.rebind (new ParaPDESolver_prm);
  psolver_prm->scan (menu);

  additive_schwarz = menu.get("use additive Schwarz").getBool();
  use_coarse_grid = text2bool (menu.get("use coarse grid"));
```

```
// other menu items are handled by 'SubdomainFEMSolver' later

init ();   // do all the initialization work
}
```

3. The `solve` function carries out the solution of a logically existing global linear system using a particular DD method.

```
bool SPAdmUDC:: solve ()
{
  return psolver->solve();
}
```

Inside the `scan` function, the initialization of all the components of a chosen DD method is done by the `init` function, which invokes in sequence: `prepareGrids4Subdomains`, `setupSimulators`, and `setupCommunicator`. The final work of the `init` function is to create the chosen DD method by

```
psolver.rebind(psolver_prm.create());
```

The solution of a virtual global linear system by the chosen DD method will be done by the virtual member function `ParaPDESolver::solve`. For class `BasicDDSolver` the `solve` function has the following algorithmic core:

```
bool BasicDDSolver:: solve ()
{
  // udc is of type SPAdmUDC*
  iteration_counter = 0;
  while (iteration_counter++<max_iterations && !satisfied())
    udc->oneDDIteration (iteration_counter);
  return convflag;
}
```

where the `oneDDIteration` function of class `SPAdmUDC` has the following simplified implementation:

```
void SPAdmUDC:: oneDDIteration (int /*iteration_counter*/)
{
  if (use_coarse_grid)
    coarseGridCorrection ();
  for (int i=1; i<=num_local_subdomains; i++) {
    local_fem_solvers(i)->updateRHS ();
    local_fem_solvers(i)->solveLocal ();
  }
  updateGlobalValues ();
}
```

We mention that it is possible for the user to re-implement the `oneDDIteration` function in a derived subclass to design a non-standard DD method as a stand-alone solver. We also mention that the `genAction` function does the work of a DD preconditioner, associated with class KrylovDDSolver. The function is invoked inside every iteration of a chosen global Krylov method. Re-implementation of genAction is also allowed in a derived subclass.

Unlike most member functions of class SPAdmUDC, the createLocalSolver function *must* be re-implemented in a derived subclass. The task of the function is to bind each entry of the subd_fem_solvers handle array to a newly defined subdomain solver, say MySubSolver. Therefore, the simplest re-implementation of createLocalSolver in a subclass of SPAdmUDC is

```
void MyUDC:: createLocalSolver (int local_id)
{
  subd_fem_solvers(local_id).rebind (new MySubSolver);
}
```

However, if the global administrator also wants to use some non-generic data objects or functions of the newly defined subdomain solver, it is necessary to introduce in the class definition of MyUDC an additional handle array

```
VecSimplest(Handle(MySubSolver)) subd_my_solvers;
```

Then a more advanced version of MyUDC::createLocalSolver can be as follows:

```
void MyUDC:: createLocalSolver (int local_id)
{
  MySubSolver* my_solver = new MySubSolver;
  subd_my_solvers.redim (num_local_subds);
  subd_my_solvers(local_id).rebind (*my_solver);
  subd_fem_solvers(local_id).rebind (*my_solver);
}
```

### 2.6.6   Summary of the Implementation Framework

To summarize, we present a schematic diagram of the generic implementation framework in Figure 2.3. In the diagram, the top two classes SPAdmUDC and ParaPDESolver constitute the global administrator. Class CommunicatorFEMSP constitutes the communication part, which has a direct connection to all the subdomain solvers. We note that we have omitted drawing the direct connection from SPAdmUDC to the subdomain solvers in the diagram.

## 2.7   Parallel Overlapping DD Methods

Although one major strength with the overlapping DD methods is their suitability for parallel computing, they can achieve superior numerical efficiency even on sequential computers. In particular, the content of Sections 2.2-2.6 applies for both a sequential and a parallel implementation. For sequential overlapping DD methods, all the subdomains reside on the same processor, and one object of class SubdomainFEMSolver is used for each subdomain. The

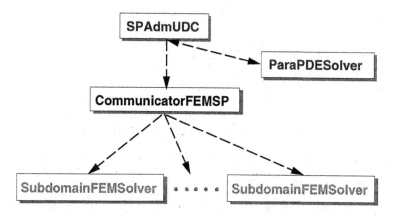

**Fig. 2.3.** A generic implementation framework for sequential overlapping DD methods. Note that the dashed arrows mean the "has a" relationship.

global iteration administration and data exchange are handled by a *single* set of class objects containing SPAdmUDC, ParaPDESolver_prm, ParaPDESolver, and CommunicatorFEMSP.

For running e.g. an additive Schwarz method on a multi-processor platform, the only difference is that every involved processor will use its own set of class objects for iteration administration and data exchange. Furthermore, one processor may have one or several subdomain solvers, depending on the relation between the number of subdomains and the number of processors. Since class CommunicatorFEMSP is derived from GridPartAdm, the situation where one processor is responsible for multiple subdomains can be handled by exactly the *same* Diffpack code as that for the situation of one subdomain per processor. By considering a sequential overlapping DD execution as a special case of a parallel overlapping DD execution, we can easily understand that the difference between the sequential code and the parallel code exists only in the parallel Diffpack library and is thus invisible to the user.

Following the implementation framework explained in the preceding section, every processor has a global coarse grid solver with complete data structure for $A_0$. Before each processor calculates the same coarse grid correction, *all* the subdomain solvers need to communicate with each other to form a global residual vector $w_0$ that is associated with the global coarse grid $\mathcal{T}_H$. Therefore, inter-processor communication after solving

$$A_0 u_0 = w_0 \tag{2.14}$$

is not necessary. However, the size of the coarse grid problem needs some attention. A finer $\mathcal{T}_H$ normally gives better overall convergence, but it may come at the cost of lower parallel efficiency, due to increased overhead associated with the coarse grid correction. So it is important to have a coarse grid problem that is considerably smaller than every subdomain problem. Under this assumption, it suffices to use a sequential solution procedure for (2.14).

In practice, it is unusual that a parallel DD solver demonstrates linear speed-up results, i.e., the execution time decreases linearly as a function of the number of processors. This is due to the overhead that consists of (a) computation not being parallelized, like the above mentioned way of treating the coarse grid correction, and (b) work that is only present in a parallel execution. The latter type of overhead again can be attributed to two sources:

1. overhead of communication, and

2. overhead of synchronization.

The communication overhead is determined by both the amount of information neighboring subdomains need to exchange and the number of neighbors each subdomain has. This is because the cost of sending/receiving a message between two processors roughly consists of two parts: a constant start-up cost and the message exchange cost that is proportional to the message size. So one of the requirements for the overlapping grid partitioning should be to keep the average number of neighbors small. Meanwhile, reducing the size of overlap between neighboring subdomains will also reduce the overhead, but may however affect the convergence speed. *Therefore, a good overlapping grid partitioning algorithm should provide the user with the freedom of adjusting the size of overlap to reach a compromise between fast convergence and low communication overhead.*

The synchronization overhead is primarily due to the difference in subgrid size. That is, some subdomains have more computation because they have more grid points. However, unbalanced local convergence may also destroy the work balance, even if the grid partitioning algorithm produces subgrids of an identical size (e.g. out from a structured global grid). For example, if a Krylov subspace method is used in the subdomain solves, different subdomains may then need different numbers of iterations to achieve the local convergence. This situation of unbalanced local convergence may change from one DD iteration to another. *Therefore, if applicable, a fixed number of multigrid V-cycles are optimal subdomain solvers with respect to both computational efficiency and synchronization overhead.*

## 2.8    Two Application Examples

In order to show the reader how an overlapping DD method can be incorporated into an existing PDE simulator, we present in this section two application examples. The first example is for solving a stationary PDE, whereas the second example is for solving a time-dependent PDE.

### 2.8.1   The Poisson1 Example

We use the standard Diffpack sequential Poisson equation solver Poisson1[1] as the starting point for this example. The coding effort needed for incorporating an overlapping DD method is in form of creating two new C++ classes: SubdomainPESolver as the new subdomain solver and Poisson1DD that controls the global computation.

*Definition of Class SubdomainPESolver.*

```
class SubdomainPESolver : public SubdomainFEMSolver, public Poisson1
{
protected:
  virtual void initSolField (MenuSystem& menu);
  virtual void createLocalMatrix ();
  virtual void buildGlobalCoarseDof ();
  virtual void createCoarseGridMatrix ();

public:
  SubdomainPESolver () {}
  virtual ~SubdomainPESolver () {}
  static  void defineStatic  (MenuSystem& menu, int level = MAIN);
};
```

*Implementation of Class SubdomainPESolver.*

```
void SubdomainPESolver:: initSolField (MenuSystem& menu)
{
  // subd_fem_grid should already be sent in by SPAdmUDC
  grid.rebind (SubdomainFEMSolver::subd_fem_grid.getRef());

  // read info (except for grid) from menu
  menu.setCommandPrefix ("subdomain");
  A.redim (grid->getNoSpaceDim());
  A.scan (menu.get ("A parameters"));
  // database is to be given by Poisson1DD
  FEM:: scan (menu);
  lineq.rebind (new LinEqAdmFE);
  lineq->scan (menu);
  menu.unsetCommandPrefix ();

  // allocate data structures in the class:
  u.rebind    (new FieldFE(*grid,"u"));     // allocate, field name "u"
  error.rebind(new FieldFE(*grid,"error"));// another scalar field
  flux.rebind (new FieldsFE(*grid,"flux"));// vector field

  dof.rebind    (new DegFreeFE(*grid, 1));
  linsol.redim (dof->getTotalNoDof());
  uanal.rebind (new Ell1AnalSol(this)); // functor for analy. solution

  // connection between SubdomainFEMSolver and SubdomainPESolver
  subd_lineq.rebind (*lineq);
```

---

[1] The Poisson1 solver can be found in the $NOR/doc/Book/src/fem/Poisson1/ directory of a standard Diffpack distribution.

```
  subd_dof.rebind (*dof);
  global_solution_vec.rebind (linsol);

  // fill the DegFreeFE object with ess. BC, must be done here
  Poisson1:: fillEssBC ();
}

void SubdomainPESolver:: createLocalMatrix ()
{
  Poisson1:: makeSystem (*dof, *lineq);
  linsol.fill (0.0);        // set all entries to 0 in start vector
  dof->insertEssBC (linsol);// insert boundary values in start vector
}

void SubdomainPESolver:: buildGlobalCoarseDof ()
{
  SubdomainFEMSolver:: buildGlobalCoarseDof ();
  GridFE& coarse_grid = global_coarse_dof->grid();
  global_coarse_dof->initEssBC ();
  const int c_nno = coarse_grid.getNoNodes();
  Ptv(real) pt;
  for (int i=1; i<=c_nno; i++)
    if (coarse_grid.BoNode(i)) {
      coarse_grid.getCoor (pt,i);
      global_coarse_dof->fillEssBC (i,g(pt));
    }
}

void SubdomainPESolver:: createCoarseGridMatrix ()
{
  FEM:: makeSystem (csolver->coarse_dof(), csolver->coarse_lineq());
  SubdomainFEMSolver:: createCoarseGridMatrix (); // default work
}

void SubdomainPESolver:: defineStatic (MenuSystem& menu, int level)
{
  menu.addItem (level,            // menu level: level+1 is a submenu
                "A parameters",   // menu command/name
                "A_1 A_2 ... in f expression", // help/description
                "3 2 1");         // default answer
  // submenus:
  LinEqAdmFE::defineStatic (menu, level+1);  // linear solver
  FEM::defineStatic (menu, level+1);         // num. integr. rule etc
}
```

*Some Explanations.* It is mandatory to redefine the initSolField function inside class SubdomainPESolver. First of all, it is needed for reading input from MenuSystem and constructing the subdomain data structure. Secondly, it is needed for coupling the external data objects with the generic data handles: subd_lineq, subd_dof, and global_solution_vec. The functions defineStatic and createLocalMatrix also need re-implementation in most applications. Thirdly, if coarse grid corrections are desired, the createCoarseGridMatrix function must be re-implemented. Part of this coding work can be done by re-using the default implementation from class SubdomainFEMSolver, see the above program. In addition, we also extend the default implementation of the `buildGlobalCoarseDof` function in this example, because we want to fill in the desired essential boundary conditions.

*Definition of Class Poisson1DD.*

```
class Poisson1DD: public SPAdmUDC
{
protected:
  Handle(SaveSimRes) database;   // used by all the local solvers
  VecSimplest(Handle(SubdomainPESolver)) subd_poisson_solvers;
  virtual void createLocalSolver (int local_id);
  virtual void setupSimulators ();

public:
  Poisson1DD () {}
  virtual ~Poisson1DD () {}

  virtual void adm    (MenuSystem& menu);
  virtual void define (MenuSystem& menu, int level = MAIN);
  virtual void solveProblem ();  // main driver routine
  virtual void resultReport ();  // write error norms to the screen
};
```

*Implementation of Class Poisson1DD.*

```
void Poisson1DD:: createLocalSolver (int local_id)
{
  SubdomainPESolver* loc_solver = new SubdomainPESolver;
  loc_solver->database.rebind (*database);

  subd_poisson_solvers.redim (num_local_subds);
  subd_poisson_solvers(local_id).rebind (*loc_solver);
  subd_fem_solvers(local_id).rebind (*loc_solver);
}

void Poisson1DD:: setupSimulators ()
{
  database.rebind (new SaveSimRes);
  database->scan (*menu_input, subd_fem_grids(1)->getNoSpaceDim());

  SPAdmUDC:: setupSimulators();  // do the default work
}

void Poisson1DD:: adm (MenuSystem& menu)
{
  define (menu);
  menu.prompt ();
  SPAdmUDC::scan (menu);
}

void Poisson1DD:: define (MenuSystem& menu, int level)
{
  SPAdmUDC:: defineStatic (menu,level+1);
  SaveSimRes:: defineStatic (menu, level+1);  // storage of fields
```

```
    menu.setCommandPrefix ("subdomain");
    SubdomainPESolver:: defineStatic (menu,level);   // for subd-solvers
    menu.unsetCommandPrefix ();
}

void Poisson1DD:: solveProblem ()
{
    // solution of the global linear system
    // DD either as stand-alone solver or precond
    SPAdmUDC:: solve ();

    // save results for later visualization
    SubdomainPESolver* loc_solver;
    for (int i=1; i<=num_local_subds; i++) {
      loc_solver = subd_poisson_solvers(i).getPtr();
      loc_solver->dof->vec2field (loc_solver->linsol, loc_solver->u());
      database->dump (loc_solver->u());
      loc_solver->makeFlux (loc_solver->flux(), loc_solver->u());
      database->dump (loc_solver->flux());
    }
}
```

*Some Explanations.* As class Poisson1DD now replaces class Poisson1 as the main control class, the administration of the SaveSimRes object is transferred from Poisson1 to Poisson1DD. Thus, a sub-menu for SaveSimRes is inserted into the define function of class Poisson1DD. The setupSimulators function is also modified to include the building-up of the SaveSimRes object. The re-implemented createLocalSolver function has also taken this to account. We note that class Poisson1DD has an additional array of subdomain simulator handles:

```
    VecSimplest(Handle(SubdomainPESolver)) subd_poisson_solvers;
```

This handle array allows direct access to the public member functions and data objects that belong to class Poisson1.

We can see that the solveProblem function for this particular example is very simple. The reason is that the scan function of class SPAdmUDC has taken care of all the initialization work, including building up the subdomain linear systems. Recall that the global linear system is virtually represented by all the subdomain linear systems, the overall solution process is simply invoked by

```
    SPAdmUDC:: solve ();
```

*The Main Program.* The following main program is primarily written for use on a parallel computer, this explains the appearance of the function pair initDpParallelDD and closeDpParallelDD, which belong to the parallel Diffpack library. However, as we have mentioned before, using the same code for a sequential DD solver can be achieved by simply running it on one processor.

```
#include <Poisson1DD.h>
#include <initDpParallelDD.h>

int main (int nargs, const char** args)
{
  initDpParallelDD (nargs, args);
  initDiffpack (nargs, args);
  global_menu.init ("Solving the Poisson equation","PDD approach");

  Poisson1DD udc;
  udc.adm (global_menu);
  udc.solveProblem ();
  udc.resultReport ();

  closeDpParallelDD();
  return 0;
}
```

*An Input File for MenuSystem.* An input file for the Diffpack MenuSystem may be composed as follows. Note that for this particular input file the overlapping DD method will be used as a preconditioner. To use it as a stand-alone iterative solver, it suffice to change the value of domain decomposition scheme to BasicDDSolver.

```
sub ParaPDESolver_prm
set domain decomposition scheme = KrylovDDSolver
sub LinEqSolver_prm
set basic method = ConjGrad
set max iterations = 200
ok
sub ConvMonitorList_prm
sub Define ConvMonitor #1
set #1: convergence monitor name = CMRelResidual
set #1: convergence tolerance = 1.0e-4
ok
ok
ok

sub SPAdmUDC
sub GridPart_prm
set grid source type = GlobalGrid
set global grid=P=PreproBox|d=2[0,1]x[0,1]|d=2e=ElmB4n2D[40,40][1,1]
set number overlaps = 1
ok
set use additive Schwarz = ON
set use coarse grid = ON
set global coarse grid=P=PreproBox|d=2[0,1]x[0,1]|d=2e=ElmB4n2D[4,4][1,1]
set degrees of freedom per node = 1
ok

set subdomain A parameters = 2 1
sub subdomain LinEqAdmFE
sub subdomain Matrix_prm
set subdomain matrix type = MatSparse
ok
```

```
sub subdomain LinEqSolver_prm
set subdomain basic method = ConjGrad
set subdomain max iterations = 100
ok
sub subdomain ConvMonitorList_prm
sub subdomain Define ConvMonitor #1
set subdomain #1: convergence monitor name = CMRelResidual
set subdomain #1: convergence tolerance = 1.0e-2
ok
ok
ok
ok
```

### 2.8.2   The Heat1 Example

We use the standard Diffpack sequential heat conduction simulator Heat1[2]
as the starting point. The purpose of this example is to show the additional
considerations that are required for incorporating an overlapping DD method
into a simulator that solves a time-dependent PDE. Similar to the previous
example, the coding effort is also in form of creating two new C++ classes:
SubdomainHeatSolver and Heat1DD.

*Definition of Class SubdomainHeatSolver.*

```
class SubdomainHeatSolver : public SubdomainFEMSolver, public Heat1
{
protected:
  virtual void initSolField (MenuSystem& menu);
  virtual void buildGlobalCoarseDof ();
  virtual void integrands (ElmMatVec& elmat, const FiniteElement& fe);

public:
  SubdomainHeatSolver () { make4coarsegrid = false; }
  virtual ~SubdomainHeatSolver () {}
  static  void defineStatic  (MenuSystem& menu, int level = MAIN);
  virtual void createLocalMatrix ();
  bool make4coarsegrid;
  virtual void createCoarseGridMatrix ();
};
```

*Implementation of Class SubdomainHeatSolver.*

```
void SubdomainHeatSolver:: initSolField (MenuSystem& menu)
{
  // subd_fem_grid should already be sent in by SPAdmUDC
  grid.rebind (SubdomainFEMSolver::subd_fem_grid.getRef());

  // read info (except for grid and tip) from menu
  menu.setCommandPrefix ("subdomain");
  assignEnum(error_itg_pt_tp,menu.get("error integration point type"));
```

---

[2] The Heat1 simulator can be found in the $NOR/doc/Book/src/fem/Heat1/ direc-
tory of a standard Diffpack distribution.

```
    diffusion_coeff = menu.get ("diffusion coefficient").getReal();
    FEM::scan (menu);
    uanal.rebind (new Heat1AnalSol (this));
    // 'database' and 'tip' are to be given by Heat1DD

    // fill handles with objects:
    u.      rebind (new FieldFE (*grid, "u"));
    u_prev.rebind (new FieldFE (*grid, "u_prev"));
    error. rebind (new FieldFE (*grid, "error"));
    flux.  rebind (new FieldsFE (*grid, "flux"));
    u_summary.attach (*u);  u_summary.init ();

    dof.rebind(new DegFreeFE(*grid,1));
    lineq.rebind(new LinEqAdmFE);
    lineq->scan(menu);
    linsol.redim(grid->getNoNodes());
    linsol.fill(0.0);
    menu.unsetCommandPrefix ();

    // connection between SubdomainFEMSolver and SubdomainHeatSolver
    subd_lineq.rebind (*lineq);
    subd_dof.rebind (*dof);
    global_solution_vec.rebind (linsol);

    // fill the DegFreeFE object with ess. BC, must be done here
    Heat1:: fillEssBC ();
}

void SubdomainHeatSolver:: buildGlobalCoarseDof ()
{
    SubdomainFEMSolver:: buildGlobalCoarseDof ();
    GridFE& coarse_grid = global_coarse_dof->grid();
    global_coarse_dof->initEssBC ();
    const int c_nno = coarse_grid.getNoNodes();
    Ptv(real) pt;
    for (int i=1; i<=c_nno; i++)
        if (coarse_grid.BoNode(i)) {
            coarse_grid.getCoor (pt,i);
            global_coarse_dof->fillEssBC (i,g(pt));
        }
}

void SubdomainHeatSolver:: integrands (ElmMatVec& elmat,
                                       const FiniteElement& fe)
{
    if (!make4coarsegrid)
        return Heat1::integrands (elmat, fe);

    const real detJxW = fe.detJxW();
    const int  nsd = fe.getNoSpaceDim();
    const int  nbf = fe.getNoBasisFunc();
    const real dt  = tip->Delta();
    const real t   = tip->time();
    real gradNi_gradNj;
    const real k_value = k (fe, t);
    int i,j,s;
    for(i = 1; i <= nbf; i++)
```

```
    for(j = 1; j <= nbf; j++) {
      gradNi_gradNj =0;
      for(s = 1; s <= nsd; s++)
        gradNi_gradNj += fe.dN(i,s)*fe.dN(j,s);

      elmat.A(i,j) +=(fe.N(i)*fe.N(j) +
                       dt*k_value*gradNi_gradNj)*detJxW;
    }
  // right-hand does not need to be filled here
}

void SubdomainHeatSolver:: createLocalMatrix ()
{
  Heat1:: makeSystem (*dof, *lineq);
}

void SubdomainHeatSolver:: createCoarseGridMatrix ()
{
  if (!make4coarsegrid) return;
  FEM::makeSystem (csolver->coarse_dof(), csolver->coarse_lineq());
  SubdomainFEMSolver::createCoarseGridMatrix (); // default work
}

void SubdomainHeatSolver:: defineStatic (MenuSystem& menu, int level)
{
  menu.addItem (level, "diffusion coefficient", "k", "1.0");
  menu.addItem (level, "error integration point type", "err_itg",
               "type of itg rule used in ErrorNorms::Lnorm",
               "GAUSS_POINTS", "S/GAUSS_POINTS/NODAL_POINTS");
  LinEqAdmFE::defineStatic(menu,level+1);
  FEM::defineStatic(menu,level+1);
}
```

*Definition of Class Heat1DD.*

```
class Heat1DD: public SPAdmUDC
{
protected:
  Handle(TimePrm)    tip;      // time discretization parameters
  Handle(SaveSimRes) database; // used by all the local subd-solvers

  VecSimplest(Handle(SubdomainHeatSolver)) subd_heat_solvers;
  virtual void createLocalSolver (int local_id);
  virtual void setupSimulators ();

  virtual void timeLoop ();
  virtual void solveAtThisTimeStep ();

public:
  Heat1DD () {}
  virtual ~Heat1DD () {}

  virtual void adm      (MenuSystem& menu);
  virtual void define (MenuSystem& menu, int level = MAIN);
  virtual void solveProblem ();  // main driver routine
  virtual void resultReport ();  // write error norms to the screen
};
```

*Implementation of Class Heat1DD.*

```
void Heat1DD:: createLocalSolver (int local_id)
{
  SubdomainHeatSolver* loc_solver = new SubdomainHeatSolver;
  loc_solver->tip.rebind (*tip);
  loc_solver->database.rebind (*database);

  subd_heat_solvers.redim (num_local_subds);
  subd_heat_solvers(local_id).rebind (loc_solver);
  subd_fem_solvers(local_id).rebind (loc_solver);
}

void Heat1DD:: setupSimulators ()
{
  // make 'tip' & 'database' ready for 'SPAdmUDC::setupSimulators'
  tip.rebind (new TimePrm);
  tip->scan(menu_input->get("time parameters"));
  database.rebind (new SaveSimRes);
  database->scan (*menu_input, subd_fem_grids(1)->getNoSpaceDim());

  SPAdmUDC:: setupSimulators();   // do the default work
}

void Heat1DD:: timeLoop()
{
  tip->initTimeLoop();

  // set the initial condition
  SubdomainHeatSolver* hsolver;
  int i;
  for (i=1; i<=num_local_subds; i++) {
    hsolver = subd_heat_solvers(i).getPtr();
    hsolver->setIC();
    database->dump (hsolver->u(), tip.getPtr(), "initial condition");
  }

  while(!tip->finished())
    {
      tip->increaseTime();
      solveAtThisTimeStep();
      for (i=1; i<=num_local_subds; i++) {
        hsolver = subd_heat_solvers(i).getPtr();
        hsolver->u_prev() = hsolver->u();
      }
    }
}

void Heat1DD:: solveAtThisTimeStep ()
{
  SubdomainHeatSolver* hsolver;
  int i;

  for (i=1; i<=num_local_subds; i++) {
    hsolver = subd_heat_solvers(i).getPtr();
    hsolver->fillEssBC ();
    hsolver->createLocalMatrix ();
    hsolver->modifyAmatWithEssIBC (); // must be called
```

```
  hsolver->dof->field2vec (hsolver->u(), hsolver->linsol);
  // only necessary to build the coarse grid matrix once
  if (use_coarse_grid && i==1) {
    hsolver->make4coarsegrid = true;
    hsolver->createCoarseGridMatrix ();
    hsolver->make4coarsegrid = false;
  }
}

SPAdmUDC:: solve ();    // solve the global linear system
if (proc_manager->master()) {
  s_o << "t=" << tip->time();
  s_o << oform("   solver%sconverged in %3d iterations\n",
               psolver->converged() ? " " : " not ",
               psolver->getItCount());
  s_o.flush();
}

for (i=1; i<=num_local_subds; i++) {
  hsolver = subd_heat_solvers(i).getPtr();
  hsolver->dof->vec2field (hsolver->linsol, hsolver->u());
  database->dump (hsolver->u(), tip.getPtr());
  hsolver->u_summary.update (tip->time());
  hsolver->makeFlux (hsolver->flux(), hsolver->u());
  database->dump (hsolver->flux(), tip.getPtr(),
                  "smooth flux -k*grad(u)");
}
}

void Heat1DD:: adm (MenuSystem& menu)
{
  define (menu);
  menu.prompt ();
  SPAdmUDC::scan (menu);
}

void Heat1DD:: define (MenuSystem& menu, int level)
{
  SPAdmUDC:: defineStatic (menu,level+1);
  menu.addItem (level, "time parameters", "step and domain for time",
                "dt =0.1 t in [0,1]");
  SaveSimRes::defineStatic (menu, level+1);

  menu.setCommandPrefix ("subdomain");
  SubdomainHeatSolver::defineStatic (menu,level); // for subd-solvers
  menu.unsetCommandPrefix ();
}

void Heat1DD:: solveProblem () { timeLoop(); }
```

*Some Explanations.* The main program and the MenuSystem input file are
similar to those from the previous example, so we do not list them here.
However, there are several new programming issues worth noticing. First,
the explicit call of

```
  u_prev->valueFEM(fe)
```

inside the original `Heat1::integrands` function makes it impossible to re-use the function in assembling the global coarse grid matrix. Therefore, we have to write a new version of the `integrands` function in class `SubdomainHeatSolver`, which makes use of a boolean variable `make4coarsegrid`, see the above code. Second, the `tip` handle is now controlled by class `Heat1DD`. This results in some extra code lines in the `createLocalSolver` and `setupSimulators` functions of class `Heat1DD`. Third, class `Heat1DD` also needs to re-implement the functions `timeLoop` and `solveAtThisTimeStep`. There is close resemblance between most parts of these two functions and their counterparts in class `Heat1`. Fourth, re-generation of the virtually existing global linear system at each time level needs some attention. It is achieved on each subdomain by first creating the subdomain matrix $\mathbf{A}_i^0$ and then modifying it into $\mathbf{A}_i$, see Section 2.4.2. More precisely, the following two code lines are used:

```
hsolver->createLocalMatrix ();
hsolver->modifyAmatWithEssIBC ();
```

Note that we have assumed the most general situation, so the $\mathbf{A}_i^0$ matrix is re-generated and modified into $\mathbf{A}_i$ at every time level. In case the existing simulator `Heat1` only needs to modify the right-hand side vector, e.g., due to time-independent coefficients, then the above call to the `modifyAmatWithEssIBC` function should be replaced with

```
if (tip->getTimeStepNo()==1)
  hsolver->modifyAmatWithEssIBC ();
else {
  hsolver->orig_rhs() = (Vec(real)&)hsolver->subd_lineq->b();
  hsolver->global_orig_rhs() = hsolver->orig_rhs();
}
```

### 2.8.3   Some Programming Rules

To help the reader to extend a standard Diffpack simulator with overlapping DD functionality, we have summarized the following programming rules:

1. Two new C++ classes need to be created. The first class, e.g. with name `NewSubdSolver`, is to work as the new subdomain solver and should therefore be a subclass of both `SubdomainFEMSolver` and an existing simulator class. The second class is to work as the new global administrator and should be a subclass of `SPAdmUDC`.

2. The new subdomain solver must re-implement the `initSolField` function, which replaces the `scan` function of the existing simulator class and binds the handles of `SubdomainFEMSolver` to the correct data objects. Inside the `initSolField` function, the `subd_fem_grid` handle can be assumed to have already been bound to a subgrid covering the subdomain. It is also important to initialize the `subd_dof` object with correct essential boundary conditions, inside the `initSolField` function if applicable.

3. The new subdomain solver normally needs to have a new `defineStatic` function, which designs the `MenuSystem` layout part that is related to the subdomain solver, by including the necessary menu items from the original `define/defineStatic` function.

4. Normally, the new subdomain solver also needs to re-implement the `createLocalMatrix` function by e.g. utilizing the original `integrands` function and the generic `FEM::makeSystem` function.

5. If coarse grid corrections are desired, the functions `buildGlobalCoarseDef` and `createCoarseGridMatrix` normally need to be re-implemented in the new subdomain solver. The re-implemented functions normally make use of their default implementation inside class `SubdomainFEMSolver`. The following are two examples:

```
void NewSubdSolver:: buildGlobalCoarseDof ()
{
    // construct the coarse grid and its Dof
    SubdomainFEMSolver:: buildGlobalCoarseDof ();
    // fill the Dof object with essential BC if any
}

void NewSubdSolver:: createCoarseGridMatrix ()
{
    FEM::makeSystem (csolver->coarse_dof(),csolver->coarse_lineq());
    SubdomainFEMSolver::createCoarseGridMatrix ();
}
```

6. The new global administrator class should normally introduce an additional array of handles:

```
VecSimplest(Handle(NewSubSolver)) subd_new_solvers;
```

This is for accessing the non-generic functions of `NewSubSolver`.

7. The `createLocalSolver` function should be re-implemented in the new global administrator class.

8. The new global administrator class should also take over the responsibility for, e.g.,

```
Handle(TimePrm) tip;
Handle(SaveSimRes) data;
```

Their corresponding menu items should be included in the `define` function of the new global administrator class. Normally, a slight modification of the `setupSimulators` function is also necessary. Re-implementing the default `SPAdmUDC::scan` function is, however, not recommended.

9. In case of solving a time-dependent PDE, the new global administrator class should also implement new functions such as `timeLoop` and `solveAtThisTimeStep`.

# References

1. A. M. Bruaset, X. Cai, H. P. Langtangen, and A. Tveito. Numerical solution of PDEs on parallel computers utilizing sequential simulators. In Y. Ishikawa et al., editor, *Scientific Computing in Object-Oriented Parallel Environment, Springer-Verlag Lecture Notes in Computer Science 1343*, pages 161–168. Springer-Verlag, 1997.

2. X. Cai, E. Acklam, H. P. Langtangen, and A. Tveito. Parallel computing in Diffpack. In H. P. Langtangen and A. Tveito, editors, *Advanced Topics in Computational Partial Differential Equations – Numerical Methods and Diffpack Programming*. Springer, 2003.

3. T. F. Chan and T. P. Mathew. Domain decomposition algorithms. In *Acta Numerica 1994*, pages 61–143. Cambridge University Press, 1994.

4. C. Farhat and M. Lesoinne. Automatic partitioning of unstructured meshes for the parallel solution of problems in computational mechanics. *Internat. J. Numer. Meth. Engrg.*, 36:745–764, 1993.

5. W. Hackbusch. *Multigrid Methods and Applications*. Springer, Berlin, 1985.

6. K.-A. Mardal, H. P. Langtangen, and G. Zumbusch. Multigrid methods. In H. P. Langtangen and A. Tveito, editors, *Advanced Topics in Computational Partial Differential Equations – Numerical Methods and Diffpack Programming*. Springer, 2003.

7. H. A. Schwarz. *Gesammelte Mathematische Abhandlungen*, volume 2, pages 133–143. Springer, Berlin, 1890. First published in Vierteljahrsschrift der Naturforschenden Gesellschaft in Zürich, volume 15, 1870, pp. 272–286.

8. B. F. Smith, P. E. Bjørstad, and W. Gropp. *Domain Decomposition: Parallel Multilevel Methods for Elliptic Partial Differential Equations*. Cambridge University Press, 1996.

9. D. Vanderstraeten and R. Keunings. Optimized partitioning of unstructured finite element meshes. *Internat. J. Numer. Meth. Engrg.*, 38:433–450, 1995.

10. J. Xu. Iterative methods by space decomposition and subspace correction. *SIAM Review*, 34(4):581–613, December 1992.

# Chapter 3

# Software Tools for Multigrid Methods

K.-A. Mardal[1,2], G. W. Zumbusch[3], and H. P. Langtangen[1,2]

[1] Simula Research Laboratory
[2] Dept. of Informatics, University of Oslo
[3] Institute for Applied Mathematics, University of Bonn

**Abstract.** This chapter provides a tutorial for the multigrid solver module in Diffpack. With the `MGtools` class or the `NonLinMGtools` class, a standard Diffpack finite element application code can be equipped with a flexible multigrid solver by adding about 10 lines of extra code. The applications covered here are the Poisson equation, more general elliptic problems with anisotropic or jumping coefficients and various boundary conditions, the equation of linear elasticity, and a nonlinear Poisson-like problem.

## 3.1 Introduction

When discretizing partial differential equations by the finite element, difference, or volume methods, one often ends up with a system of linear equations (hereafter called *linear system*). In time-dependent problems such linear systems must be solved at each time level. Quite frequently one experiences that the solution of the linear system is the bottleneck of the simulation, especially on large grids. The reason for this is easy to explain. Linear system solvers (often just called *linear solvers*) require work of the order of $n^\alpha$, where $n$ is the number of unknowns in the system and $\alpha \geq 1$. If $\alpha > 1$, which is the common case for Gaussian elimination, classical iterations, and Conjugate-Gradient-like methods [9], the solution of linear systems will dominate over the work required to set up the system (which is proportional to $n$). Table 3.1 shows the amount of work which is needed for some typical linear solvers applied to the Poisson equation in 2D (see [13]). The iterative methods need a stopping criterion, such that the overall accuracy is in the range of the discretization error, this is reflected by the $\log \epsilon$ term.

In the 70's and 80's, researchers constructed so-called *multi-level methods* having the nice property that $\alpha = 1$. These methods are characterized by employing a hierarchy of grids, or in other words, multi-level discretizations. Special cases of multi-level strategies are the multigrid and the domain decomposition methods. The current chapter is restricted to multigrid methods.

Looking at the whole family of linear solvers, one observes that direct methods, like Gaussian elimination and its variants, require no user-chosen parameters and the amount of work is known a priori, while iterative methods

**Table 3.1.** Complexity of different solvers for the 2D Poisson problem.

| Method | number of operations |
|---|---|
| banded Gaussian elimination | $O(n^2)$ |
| Jacobi iteration | $O(n^2 \log 1/\epsilon)$ |
| Conjugate Gradient method | $O(n^{3/2} \log 1/\epsilon)$ |
| Multigrid method | $O(n \log 1/\epsilon)$ |
| Full Multigrid | $O(n)$ |

usually involve user-chosen, problem-dependent parameters, and the amount of work is not known. It is also uncertain whether iterative methods will converge or not. The family of multigrid methods is the extreme part of this picture; it contains a wide range of user-chosen, problem-dependent parameters: the number and the type of grids, pre- and post-smoothing solvers, the number of pre- and post-smoothing iterations, the type of coarse-grid solver, the type of multigrid strategy (e.g. V-cycles versus W-cycles), and so on. A lot is known for elliptic problems, but the multilevel idea (and algorithm) is far more general. For most applications, appropriate values of these user-chosen parameters are not known, or at best, some rough guidelines from numerical analysis and previous experience exist. A few simple problems have been analyzed in detail, and the most effective multigrid strategies are then known.

To obtain an effective and working multigrid method the user must often carry out lots of experiments to tune the various parameters. This requires great flexibility in the implementation of the multigrid solver, which is exactly the advantage of the Diffpack multigrid toolbox. The purpose of this chapter is to show how easy it is to implement the (complicated) multigrid algorithm in an existing Diffpack finite element solver and to show how flexible this implementation is.

The required background for reading this chapter is knowledge of typical finite element solvers in Diffpack, like the `Poisson1`, `Poisson2`, and `Elasticity1` solvers from [9]. Furthermore, the reader must know the basics of multigrid algorithms, e.g., from [9, App. C.4.2]. In particular, we will use mathematical symbols from [9, App. C.4.2].

In any implementation of the multigrid method one needs the following operations and data structures:

- a hierarchy of grids that may be nested[1] or not,
- a hierarchy of linear systems,
- linear solvers for pre-smoothing,
- linear solvers for post-smoothing,

---

[1] Nested here means that each coarse grid element can be represented as a superposition of fine grid elements. For standard (isoparametric) elements this implies that the gridlines in the coarse grid are also in the fine grid.

- transfer operators (restriction and prolongation), used to transfer vectors from one grid to the next coarser or finer grid),
- a (sufficiently accurate) coarse grid solver.

Diffpack has basic building blocks for all these ingredients so implementing multigrid is, in principle, a matter of putting these building blocks together. The typical way of realizing the mentioned data structures and operations is to introduce a vector of grids, a vector of degrees of freedom handlers, DegFreeFE, a vector of coefficient matrices, a vector of solutions and right-hand sides, a vector of projections, a vector of smoothers (linear solvers), and so on. One example is VecSimplest(Handle(LinEqSolver)). As you will need the same data structures from problem to problem, it is an idea to collect them in a multigrid toolbox, which is called MGtools. The MGtools class also contains functions for initializing and operating on these data structures. As an application programmer, all you have to do is to include an MGtools object in your class and add about 10 lines with calls to MGtools functions! With the Diffpack menu system you have full flexibility in choosing the type of smoothers, the number of pre- and post-smoothing iterations, the type of multigrid algorithm, the type of cycling, the coarse-grid solver, and so on. Through examples and exercises in this chapter we shall demonstrate this great flexibility and how you can use it to learn about the behavior of multigrid in the application at hand.

There are many good books about multigrid methods. Books emphasizing practical issues are, e.g., [4,13], while, e.g., [2,7,8] cover the theory.

The software described in this chapter is mainly an interface and a re-design of the multilevel software tools described in [6,15].

## 3.2    Sketch of How Multilevel Methods are Implemented in Diffpack

Before we begin with the toolbox MGtools we give a brief description of what happens behind the curtain. It is safe to jump to the next section, if one only intends to employ MGtools or NonLinMGtools. Multigrid methods in Diffpack are implemented as subclasses of a multi-level base class MLSolver. The multigrid algorithms are then realized as different subclasses, Multigrid, AddMultigrid, NestedMultigrid, NonlinearMultigrid, and FASMultigrid. The different algorithms are described later.

Multilevel solvers can act as both basic iterative solvers and as preconditioners. This means that the MLSolver hierarchy should be accessible from Diffpack's linear solver hierarchy (LinEqSolver) and from the preconditioner hierarchy (Precond) as well. The glue between MLSolver and Diffpack's linear solvers and preconditioners is the classes MLIter and PrecML. MLIter is a subclass of BasicItSolver, the base class for all iterative solvers (and a subclass of LinEqSolver), and represents a general basic iterative solver based on multilevel techniques. PrecML is a subclass of Precond and represents a general preconditioner based on multilevel strategies. Both `MLIter` and `PrecML` have a `MLSolver` base-class pointer for accessing various run-time chosen multigrid methods in the `MLSolver` hierarchy. That is, the algorithms are implemented in subclasses of `MLSolver`, while `PrecML` and `MLIter` are small classes that use these algorithms.

Since multilevel solvers need access to problem-dependent information in the user's application code, the `MLSolver` class needs to access some virtual functions that the application code implements. The strategy for implementing such a feature is the standard one in Diffpack: the application code is derived from a base class with *user-dependent code*, here `MLSolverUDC`. In class `MLSolver` we then have a pointer to an `MLSolverUDC` object through which we can call appropriate information provided by the user. There are three central virtual functions in `MLSolverUDC` that the application code must implement:

- `solveSubSystem` for implementing a smoother (used for pre- and post-smoothing on an arbitrary grid level)

- `transfer` for implementing projections (i.e. restrictions and prolongations)

- `residual` for implementing the evaluation of the residual on an arbitrary grid level

Since these three functions are approximately the same for a wide range of applications, we have provided suitable default versions of the functions in class `MGtools`. This means that *an application code does not need to be derived from* `MLSolverUDC` or to provide `solveSubSystem`, `transfer`, and `residual` if one applies the simplified multigrid interface in `MGtools`. This is what we aim at in the present chapter.

## 3.3   Implementing Multigrid Methods

In this section we will extend the `Poisson1` problem to a proper test example for multigrid methods.

### 3.3.1   Extending an Existing Application with Multigrid

Let us choose one of the simplest finite element solvers from [9], class `Poisson1`, and equip it with a multigrid solver. The multigrid version of `Poisson1` is called `Poisson1MG`. The following steps are required.

1. Make a new directory `Poisson1MG` with this command:

   ```
   Mkdir -md Poisson1MG
   ```

2. Move to the `Poisson1MG` directory, take a copy of the `Poisson1` files and the `main.cpp` file and change the classname to `Poisson1MG`:

   ```
   Rename -texttoo Poisson1 Poisson1MG Poisson1.*
   ```

The -texttoo flag implies a substitution of Poisson1 by Poisson1MG in all files[2].

3. Include the MGtools.h file in the header file Poisson1MG.h:

```
#include <MGtools.h>
```

4. At the beginning of the declaration of the Poisson1MG class, the following (partly new) data structures must be included:

```
Vec(real)            linsol; // solution of linear system
int                  no_of_grids;
Handle(MGtools)      mgtools;
```

The linsol data structure was present in the original Poisson1 solver so it is a matter of moving the declaration to the top, to avoid problems when Poisson1MG cleans up all the data structures; linsol is attached to other structures and must exist until these are deleted from memory. C++ deletes class data members in the reverse order of their declaration.

5. The Poisson1MG.cpp file needs to include

```
#include <PrecML.h>
#include <MLIter.h>
```

6. In Poisson1MG::define you add the MGtools menu to the complete menu for your application, e.g., at the end of the function,

```
MGtools:: defineStatic (menu, level+1);
```

7. The MGtools utility has menu items for reading the grids (and specification of the hierarchy of grids) such that the gridfile item on the original Poisson1 menu can (and should) be removed. Recall to remove it in Poisson1MG::scan too, together with the associated readOrMakeGrid call. Instead, a tool GridCollector will compute all the grids we need, based on items in the input file. This is discussed in Section 3.3.2.

8. In Poisson1MG::scan, after having declared a reference menu to the MenuSystem object, you need to perform the following steps.

   (a) Start with binding the lineq handle to a new LinEqAdmFE object and then scan its menu items[3]:

   ```
   lineq->scan (menu);  // this call is also in Poisson1
   ```

   (b) Create the MGtools object:

   ```
   mgtools.rebind( new MGtools(*this));
   mgtools->attach(*lineq);
   mgtools->scan(menu);
   ```

---

[2] A similar substitution can be done with:
perl -pi.old   -e 's/Poisson1/Poisson1MG/g;' *.h *.cpp

[3] Make sure that you move the lineq statements to the top of scan; a frequent error is to get two lineq initializations in scan, where the latter one destroys the multigrid initializations.

(c) If you have chosen a basic solver (on the menu) to be of multilevel type, you need to insert the multilevel solver in `mgtools` as the solver in the `LinEqAdmFE` object:

```
if (lineq->getSolver().description().contains("multilevel"))
  {
    MLIter& ml = CAST_REF(lineq->getSolver(),MLIter);
    ml.attach(mgtools->getMLSolver());
  }
```

If the preconditioner is chosen to be of multilevel type, you can apply the same strategy and insert `mgtools`' multigrid solver as preconditioner in the `LinEqAdmFE` object:

```
if (lineq->getPrec().description().contains("multilevel"))
  {
    PrecML& ml = CAST_REF(lineq->getPrec(),PrecML);
    ml.attach(mgtools->getMLSolver());
  }
```

(d) Make the data structures (matrices, vectors, projections, smoothers etc.):

```
mgtools->initStructure();
```

The numbering of the matrices, vectors, and projections is determined by the `DegFreeFE` objects, which again are based on the grid hierarchy in `GridCollector`.

(e) Connect the `GridFE` and `DegFreeFE` handles to the corresponding data structures in `MGtools` at the finest grid level:

```
no_of_grids = mgtools->getNoOfSpaces();
grid.rebind (mgtools->getGrid (no_of_grids));
dof.rebind  (mgtools->getDof (no_of_grids));
```

(f) Perform grid manipulations, such as redefining boundary indicators, on *all* grids. Such manipulations are not necessary in the `Poisson1` code, but they appear in class `Poisson2`, the `Elasticity1` solver and other Diffpack simulators, see class `Poisson2MG::scan` for an example.

(g) The rest of `Poisson1MG::scan` proceeds as `Poisson1::scan`. If desired, one can print an overview of the generated grids, e.g., at the end of scan,

```
mgtools->print (s_o, 1 /* level 1 */);
```

(A higher value for the level also prints the linear systems etc. at each grid level.)

9. The `Poisson1MG::mgFillEssBC` uses a parameter reflecting the current grid level and the function is different from `Poisson1::fillEssBC`. The main difference is that we use the grid level parameter to extract the grid and the degree of freedom handler at the current grid level. Thereafter we proceed as in `Poisson1::fillEssBC` using these grid and degree of freedom handler objects. Here is the `mgFillEssBC` function:

```
void Poisson1MG:: mgFillEssBC (int grid_level)
{
   // extract references to the grid at the current level and to
   // the degree of freedom handler at the current level:
   GridFE&    current_grid = mgtools->getGrid(grid_level);
   DegFreeFE& current_dof  = mgtools->getDof(grid_level);

   // the rest of the functions is similar to
   //Poisson1::fillEssBC, except that we now
   // use current_grid and current_dof instead
   // of grid and dof

   current_dof.initEssBC ();
   const int nno = current_grid.getNoNodes();
   Ptv(real) x;
   for (int i = 1; i <= nno; i++) {
      // is node i subjected to any boundary indicator?
      if (current_grid.boNode (i)) {
         x = current_grid.getCoor (i);    // coord. of node i
         current_dof.fillEssBC (i, g(x)); // u=g at boundary
      }
   }
}
```

10. In `Poisson1MG::solveProblem`, we must replace the `makeSystem` call in `Poisson1::solveProblem` by a call to our `MGtools` object for making the systems at all grid levels:

    `mgtools->makeSystemMl();`

    The `mgtools->makeSystemMl()` function applies the `FEM*` pointer in class `MGtools` for calling up the functions `integrands` and `mgFillEssBC` (and `integrands4side` if it exists) in the application code when running the element-by-element assembly loop on all grid levels.

    The rest of the `solveProblem` is not affected by multigrid methods. In particular, the call `lineq->solve()` uses multigrid or any other Diffpack solver according to the user's choice on the menu!

We are now done with all the modifications of the original `Poisson1` code. You will see that very few additional complications occur when we move on to more advanced applications, so the amount of work involved in equipping any Diffpack simulator with multigrid is well described in the previously listed points.

One modification is of fundamental importance: Diffpack must know that the new `Poisson1MG` code applies an add-on module (the original `Poisson1` code applied core Diffpack libraries only). The safest way of ensuring correct linking of a Diffpack application that applies add-on modules like multilevel solvers or adaptivity is to create a new application directory by

`Mkdir -md Poisson1MG`

The -md option to Mkdir specifies compilation and linkage against add-on modules in the .cmake files. In addition initDpMultilevel (argc, argv) should be called after initDiffpack (argc, argv) in main.cpp.

The FAQ [9] (see index "add-on modules") explains how the Makefile detects that an application is to be linked to an add-on module.

Before we test the software and discuss how the multigrid methods can be used, we turn to how a grid hierarchy is constructed and the importance of a proper test example.

### 3.3.2   Making Grid Hierarchies

Normally, the grid hierarchy is computed by GridCollector, according to the information set in its menu interface. This is done by default in MGtools. The grid hierarchy will then be a nested refinement of the coarse grid, which is specified in the input file by the menu item gridfile. An example of a grid hierarchy specification is

```
sub GridCollector
  set refinement = [2,2]
  set no of grid levels = 4
  set gridfile = P=PreproStdGeom | DISK_WITH_HOLE a=1 b=3   \
  degrees=90   | d=2 e=ElmB4n2D [12,6] [1,1.3]
ok
```

In this example the coarse grid is a $12 \times 6$ partition of a disk with a hole. GridCollector will consist of 3 nested refinements of this coarse grid and the coarse grid itself (no of grid levels = 4). The grid on level $i$ is a $2 \times 2$ refinement of the grid on level $i - 1$ (refinement = [2,2]).

MGtools and GridCollector can also be made and filled with precomputed grid hierarchies, e.g,

```
gridcoll.rebind(new GridCollector());
gridcoll->setNoOfSpaces(no_of_spaces);
Handle(GridFE) grid_i;
for (int i=1; i<=no_of_spaces; i++) {
  grid_i.rebind(new GridFE());
  ...
  // make grid somehow
  ...
  gridcoll->attach(*grid_i,i);
}
mgtools->gridcoll.rebind(gridcoll);
...
mgtools->scan(menu);
mgtools->initStructure();
```

If this technique is used then GridCollector should be filled with grids and attached to MGtools before MGtools::scan is called. MGtools will then avoid the call to GridCollector::scan and the grids attached will be used as they are. In MGtools::initStructure the grids are used to initialize all the data structure (matrices, vectors, projections, etc. ), and the grids should not be modified later (without a new call to MGtools::initStructure).

### 3.3.3  A Proper Test Example

The Poisson1 simulator is a poor example for multilevel methods as it is. The problem is too "simple" to be a real stress test. The reason is that the solution of the problem is smooth and the start vector is zero, therefore the initial error is smooth. Hence, multigrid might work very well in this problem, even when the smoothers are not able to remove "all kinds of high frequency errors"[4]. To make sure that the multigrid algorithms work properly we should instead use a start vector that contains "all possible errors", with both low and high frequencies. This is usually done by starting with a random vector, which is done in Poisson1MG by,

```
linsol.randomize(-1, 1, 1200);
```

In fact, this test with a random start vector can be used even when the solution is zero (homogeneous boundary conditions and zero right-hand side). This latter test is often useful during debugging.

## 3.4  Setting up an Input File

In this section we shall learn how to steer the multigrid computation from the menu system.

### 3.4.1  Running a Plain Gaussian Elimination

The gridfile item on the original Poisson1 menu is now replaced by grid-handling items on the MGtools menu. Therefore, it may be instructive to first run a plain Gaussian elimination to see how the new multigrid framework affects the input file in a case where multigrid is *not* used as a solution method. An input file gausselim.i in Poisson1MG/Verify is meant to be a counterpart to Poisson1/Verify/test1.i and just generates 5 × 5 bilinear elements on the unit square:

```
set A parameters = 2 1

! default LinEqAdmFE is GaussElim on MatBand

sub MGtools
 sub GridCollector
  set gridfile = P=PreproBox | d=2 [0,1]x[0,1] |  \
                              d=2 e=ElmB4n2D  [5,5]  [1,1]
  ok
 ok
ok
```

The results of this execution should of course be the same as running Poisson1 with its Verify/test1.i input.

---

[4] The reader is encouraged to use zero as a start vector and check how various smoothers and relaxation parameters behave in multigrid. Almost any of the iterative methods implemented in Diffpack will work.

### 3.4.2    Filling out MGtools Menu Items

The next natural step is to create a simple input file for solving the test problem in Poisson1MG by multigrid as the linear solver.

Using multigrid as a basic solver is enabled by choosing basic method as MLIter on the LinEqSolver_prm menu and setting the preconditioner to PrecNone (no preconditioning):

```
sub LinEqAdmFE
 sub Matrix_prm
  set matrix type = MatSparse
 ok
 sub LinEqSolver_prm
  set basic method = MLIter
  set max iterations = 300
 ok
!plain multigrid, no preconditioning:
 sub Precond_prm
  set preconditioning type = PrecNone
 ok
ok
```

The setting of multigrid parameters is done on the MGtools menu, which has various submenus:

GridCollector for parameters related to the hierarchy of grids

MLSolver_prm for choosing the multigrid type (classname in Multigrid hierarchy and V vs. W cycle)

smoother LinEqSolver_prm for setting the linear solver that acts as pre- and post-smoother in the multigrid algorithm

coarse grid LinEqSolver_prm for choosing a coarse-grid solver

Proj_prm for choosing the type of projection (interpolation, restriction)

FieldCollector as a companion to GridCollector for mixed finite element methods

Let us now specify a simple multigrid algorithm:

- V-cycle with one SSOR iteration as smoother,
- SSOR as coarse-grid solver, and
- four grid levels with $2 \times 2$ bilinear elements as the coarse grid and a $2 \times 2$ subdivision of elements from one level to the next.

The corresponding MGtools menu might look like this:

```
sub MGtools
 sub GridCollector
  set no of grid levels = 4
  set refinement = [2,2]      ! subdivide each elm into 2x2
  set gridfile = P=PreproBox | d=2 [0,1]x[0,1] |
```

```
                          d=2 e=ElmB4n2D { [2,2] } [1,1]
ok
sub MLSolver_prm
 set multilevel method = Multigrid
 set cycle type gamma = 1    ! V cycle
ok
set sweeps = [1,1]  ! 1 pre- and 1 post-smoothing sweep (V1,1 cycle)
sub smoother LinEqSolver_prm
 set smoother basic method = SSOR
 ! no of iterations governed by the number of sweeps
ok
set coarse grid solver = false ! => SSOR on the coarse grid
ok
```

Referring to the notation in [9, App. C.4], the sweeps item equals $[\nu_q, \mu_q]$, where $q$ is the grid level, cycle type gamma is $\gamma_q$, and the no of grid levels is $K = 4$. You can find the complete input file for this example in Verify/test1.i.

If coarse grid solver is set to false, the smoother menu items is used to initialize the coarse grid solver. In this example it will therefore be SSOR. Switching to an alternative coarse-grid solver is easy, just set the coarse grid solver menu item on and fill out a submenu on the MGtools menu that we did not cover in the previous example:

```
set coarse grid solver = true  ! iterative solver
sub coarse grid LinEqSolver_prm
 set coarse grid basic method = SSOR
 set coarse grid max iterations = 20
ok
```

Notice that when coarse grid solver is true and nothing else is specified, the *default* Gaussian elimination solver is chosen. This means the factLU function in the chosen matrix format class. If Gaussian elimination with pivoting is required, one should set coarse grid solver to true and fill in the exact specification of the GaussElim procedure on the coarse grid LinEqSolver_prm menu.

## 3.5   Playing Around with Multigrid

Even with this simple Poisson1MG simulator we can do several interesting experiments with multigrid. To get some feeling for different components of the algorithm, we encourage you to do some tests on your own. Playing around with parameters will be useful, especially if you want to apply multigrid in more advanced problems.

The following sections will explore various aspects of multigrid by suggesting a number of exercises/examples that the reader can play around with. First we use standard multigrid as a basic iterative method. Then we investigate other variants of multigrid and use multigrid as a preconditioner for Conjugate Gradient-like (Krylov) methods.

### 3.5.1  Number of Grids and Number of Iterations

*The number of iterations.* Take a multigrid V-cycle with an exact coarse grid solver, one pre- and one post-smoothing step, use a relative residual termination criterion for some arbitrarily prescribed tolerance, and let the coarse grid consist of $2 \times 2$ bilinear elements on the unit square. That is, use the `Verify/test1.i` input file as starting point. A central question is how the number of iterations depends on the number of grids or the number of unknowns. The sensitivity to the number of grids can easily be investigated by changing the `no of grid levels` item to a multiple answer, either in the file (do it in a copy of `Verify/test1.i`!)

```
set no of grid levels = { 2 & 3 & 4 & 5 & 6 & 7 & 8 & 9}
```

or directly on the command line[5]

```
--no_of_grid_levels '{ 2 & 3 & 4 & 5 & 6 & 7 & 8 & 9 }'
```

The number of iterations in multigrid is written by the `Poisson1MG` code to the screen. Here is a sample from an experiment on a Unix machine:

```
unix> ./app --iscl --Default Verify/test1.i \
     --no_of_grid_levels '{ 2 & 3 & 4 & 5 & 6 & 7 & 8 & 9 }' > tmp.1
unix> grep 'solver converged' tmp.1
   solver converged in   5 iterations
   solver converged in   6 iterations
   solver converged in   6 iterations
   solver converged in   6 iterations
   solver converged in   6 iterations
   solver converged in   6 iterations
   solver converged in   6 iterations
   solver converged in   6 iterations
```

The total number of unknowns in these eight experiments are 25, 81, 289, 1098, 4225, 16641, 66049, and 263169 and it appears that the number of iterations is constant. The V-cycle has a convergence rate, $\gamma$, independent of $h$. Hence, the number of iterations, $k$, to achieve convergence can be determined from the relation, $\frac{\epsilon_k}{\epsilon_0} \le \gamma^k$, where $\epsilon_k = \|e_k\| = \|x - x_k\|$ and $\|\cdot\|$ is, e.g., the $l_2$-norm, $x$ is the exact numerical solution and $x_k$ is the approximated solution after $k$ iterations. Since, $\frac{\epsilon_k}{\epsilon_{k-1}}$ is fixed, the number of iterations $k = \frac{\log \frac{\epsilon_k}{\epsilon_0}}{\log \gamma}$ should be fixed. However, we do not measure the error in this experiment. The relative residual $\frac{\|r_k\|}{\|r_0\|}$ is used as a convergence criterion. Therefore we cannot conclude that multigrid has a convergence rate independent of $h$, but it is still a good indication. A more rigorous test is done in Section 3.6.2, where `MGtools` is extended with some debugging functionality. Nevertheless, we continue our discussion about the number or iterations.

A bounded number of iterations (for a fixed tolerance), independent of the number of unknowns $n$, means a bounded number of operations per unknown:

---

[5] A list of all command line options is generated by the command `DpMenu__HTML`

The operations per iteration sum up to some constant times the number of unknowns. This means that we are solving the equation system for $n$ unknowns in $\mathcal{O}(n)$ operations, which is (order) optimal.

This should not be confused with the optimality of the Full Multigrid method, considered in Section 3.5.4. Full Multigrid reaches the discretization error in $\mathcal{O}(n)$ iterations. The discretization error, $E_h$, for elliptic problems is on the form $E_h \leq Kh^2$, $h = \frac{1}{n^{1/d}}$, where $K$ is a constant independent of $h$, and $d$ is the number of space dimensions. Hence, if the numerical error should be comparable with the discretization error we can not use a fixed tolerance, but need to choose $\epsilon_k \leq K\frac{1}{n^{2/d}}$ and determine $k$. A straightforward calculation shows that,

$$k \geq \frac{\log\left(\frac{K}{\epsilon_0}\frac{1}{n^{2/d}}\right)}{\log\gamma} = \frac{\log n^{2/d}}{\log 1/\gamma} - \frac{\log\frac{\epsilon_0}{K}}{\log 1/\gamma}.$$

Therefore, with a stopping criterion suitable for practical purposes the multigrid method requires the work of $\mathcal{O}(n\log n)$ operations.

### 3.5.2    Convergence Monitors

In the previous experiment we saw that the number of iterations was constant even though we changed the number of unknowns (grid levels). The reason was that multigrid reduces the error by a constant rate, independent of the grid size. So how can we choose a convergence criterion that guarantees that the numerical solution is approximated with the same accuracy as the discretization error? Several criteria can be used. In the previous experiment we simply used the $\|r_k\|/\|r_0\|$. This is implemented as CMRelResidual. We can estimate the error in the energy norm by measuring the residual. The residual equation reads,

$$\mathbf{A}\mathbf{e}_k = \mathbf{r}_k.  \tag{3.1}$$

From this equation we derive,

$$(\mathbf{A}\mathbf{e}_k, \mathbf{e}_k) = (\mathbf{r}_k, \mathbf{e}_k) = (\mathbf{r}_k, \mathbf{A}^{-1}\mathbf{r}_k) \leq \|\mathbf{A}^{-1}\|\|\mathbf{r}_k\|^2.$$

The multigrid method is a fix point iteration, hence $\|\mathbf{x}_k - \mathbf{x}_{k-1}\|$ can be used. We have,

$$\|\mathbf{x}_k - \mathbf{x}_{k-1}\| \leq \|\mathbf{x}_k - \mathbf{x}\| + \|\mathbf{x} - \mathbf{x}_{k-1}\| \leq \gamma^{k-1}(\gamma+1)\|\mathbf{x} - \mathbf{x}_0\|.  \tag{3.2}$$

This criterion is implemented as CMAbsSeqSolution. One can also measure the convergence factor, $\rho$, in the energy norm,

$$(\mathbf{A}\mathbf{e}_k, \mathbf{e}_k) \leq \rho^k(\mathbf{A}\mathbf{e}_0, \mathbf{e}_0),  \tag{3.3}$$

where $\mathbf{A}$ is the matrix to be solved in Poisson1. $\mathbf{A}$ is a symmetric positive definite (SPD) matrix. However, the error $\mathbf{e} = \mathbf{x} - \mathbf{x}_k$ is obviously unavailable in general and therefore unusable as convergence criterion. Instead we can use the residual equation (3.1), to derive a suitable criterion for the residual. First of all we consider some general properties. Multigrid is a linear operator, meaning that it can be represented as a matrix. However, it is more efficiently implemented as an algorithm, which produces an output vector $\mathbf{v}_k$, given an input vector $\mathbf{r}_k$. We therefore introduce the algorithm as an operator $\mathbf{B}$, and $\mathbf{v}_k = \mathbf{Br}_k$ makes sense. It is known that $\mathbf{B}$ is spectrally equivalent with $\mathbf{A}^{-1}$, and spectral equivalence of $\mathbf{A}^{-1}$ and $\mathbf{B}$ is defined by,

$$c_0(\mathbf{A}^{-1}\mathbf{x}, \mathbf{x}) \leq (\mathbf{Bx}, \mathbf{x}) \leq c_1(\mathbf{A}^{-1}\mathbf{x}, \mathbf{x}), \quad \forall \mathbf{x}, \tag{3.4}$$

or

$$c_0(\mathbf{Ax}, \mathbf{x}) \leq (\mathbf{ABAx}, \mathbf{x}) \leq c_1(\mathbf{Ax}, \mathbf{x}), \quad \forall \mathbf{x}, \tag{3.5}$$

From (3.1), (3.3) and (3.5) we derive,

$$(\mathbf{Br}_k, \mathbf{r}_k) = (\mathbf{BAe}_k, \mathbf{Ae}_k) = (\mathbf{ABAe}_k, \mathbf{e}_k) \leq c_1(\mathbf{Ae}_k, \mathbf{e}_k). \tag{3.6}$$

The term $(\mathbf{Br}_k, \mathbf{r}_k)$ is already computed by the preconditioned Conjugate-Gradient method and is available at no cost. It is implemented in the convergence monitor `CMRelMixResidual`. The term `Rel` or `Abs` refers to the fact that the criterion is relative or absolute, respectively. These convergence monitors and several others are implemented in Diffpack, look up the man page (`dpman ConvMonitor`). We usually use a relative convergence criterion when we test the efficiency of multigrid, since the criterion will then be independent of the grid size.

*Experiments in 3D.* The code in the `Poisson1MG` works for 3D problems as well (cf. [9]). We can redo the previous experiments to see if the number of iterations $n$ is bounded, i.e., independent of the number of unknowns, also in 3D. Since $n$ grows faster (with respect to the number of levels) in 3D than in 2D, we only try 2, 3, 4, and 5 refinements on the unit cube. The relevant lines in `Verify/test1.i` that needs to be updated are three items on the `GridCollector` submenu:

```
set no of grid levels = { 2 & 3 & 4 & 5 }
set refinement = [2,2,2]    ! subdivide each elm into 2x2x2
set gridfile = P=PreproBox | d=3 [0,1]x[0,1]x[0,1] |
                             d=3 e=ElmB8n3D [2,2,2] [1,1,1]
```

These modifications are incorporated in the `Verify/test2.i` file. Running the `Poisson1MG` with the `test2.i` input shows that the number of iterations seems to be no worse than constant (but higher than in the 2D experiment). The multigrid method is in general an $\mathcal{O}(n \log n)$ operation algorithm in any number of space dimensions.

### 3.5.3  Smoother

*Effect of Different Smoothers.* In the previous examples (input files `test1.i` and `test2.i`) we used the SSOR method as smoother. What is the effect of other choices, like SOR and Jacobi iterations? This is investigated by editing `Verify/test1.i` a bit[6]

```
set no of grid levels = 6
set smoother basic method = { SSOR & SOR & Jacobi }
```

(The resulting file is `Verify/test3.i`.) The critical result parameters to be investigated are the number of iterations and the CPU time of the solver. Especially the latter is a relevant measure of the relative performance of the smoothers. The CPU time of the linear solver is written to the screen if you run the simulator with the command-line option `--verbose 1`:

```
unix> app --verbose 1 < Verify/test3.i > tmp.1
```

Again you will need to grep on the `tmp.1` file to extract the relevant information:

```
unix> egrep 'solver converged|solver_classname' tmp.1
```

The performance of Jacobi, Gauss-Seidel, SOR, and SSOR iterations deteriorates with increasing number of unknowns in the linear system when these methods are used as stand-alone solvers. In connection with multigrid, this is no longer true, but there are of course significant differences between the efficiency of various smoothing procedures in a multigrid context and in particular the relaxation parameter is very important. Choosing the "wrong" relaxation parameter may lead to poor performance, as we will see below.

*Influence of the Relaxation Parameter.* Our choice of SOR in the previous test actually means the Gauss-Seidel method, because the relaxation parameter in SOR is 1 by default. For the same reason we used SSOR with a unit relaxation parameter. One should notice that the optimal relaxation parameter for SOR and SSOR as smoothers differs from the optimal value when using SOR and SSOR as stand-alone iterative solvers. In this case under-relaxation rather than over-relaxation is appropriate. It is trivial to test this too:

```
set smoother basic method = { SSOR & SOR }
set smoother relaxation parameter =
                { 0.8 & 1.0 & 1.2 & 1.4 & 1.6 & 1.8 }
```

(`test4.i` contains these modifications.) Now two menu items are varied. To see the menu combination in run number 5, just look at `SIMULATION_m5.ml`. From the `*.ml` files we realize that the relaxation parameter is fixed while changing between SSOR and SOR (or in other words, the smoother has the fastest variation). A relaxation parameter around unity seems appropriate.

In the context of preconditioning we will see that symmetric smoothers can be necessary.

---

[6] It would be convenient to just take `test1.i` as input and give the smoother method on the command-line. However, the command-line option `--itscheme` is ambiguous. We are therefore forced to use file modifications.

*The Number of Smoothing Steps.* The number of smoothing steps is another interesting parameter to investigate:

```
set sweeps = { [1,1] & [2,2] & [3,3] & [4,4] }
```

How many smoothing steps are optimal? We can simply run the application with test1.i as input and use command-line arguments for the number of sweeps (and for increasing the CPU time by setting the number of grid levels to 7):

```
./app --iscl --Default Verify/test1.i --verbose 1 \
    --no_of_grid_levels 7 \
    --sweeps '{ [1,1] & [2,2] & [3,3] & [4,4] }' > tmp.1
```

The number of iterations decreases slightly with the number of sweeps, but recall that the work in each iteration increases with the number of sweeps. We have included the --verbose 1 option such that we can see the CPU time of the total multigrid solver. The CPU times point out that one sweep is optimal in this case.

Another open question is whether the number of pre- and post-smooth operations should be equal. Let us experiment with pre-smoothing only, post-smoothing only, and combined smoothing. We can either modify a copy of the test1.i

```
set sweeps = { [1,0] & [2,0] & [0,1] & [0,2] & [1,1] & [2,2] }
```

or use the command-line option:

```
--sweeps '{ [1,0] & [2,0] & [0,1] & [0,2] & [1,1] & [2,2] }'
```

With seven grid levels, two post-smoothings or one pre- and post-smoothing turned out to be the best choices. This is a little bit strange. The transfer operators (standard $L_2$ projection) are not perfect, high frequency errors are restricted to low frequency errors and may therefore pollute the coarse grid correction. It is therefore very important that the pre-smoother removes all high frequency error before transferring. These projection effects are usually called *aliasing*. A funny example of aliasing is apparent in old western movies, where the wheels seem to go backwards. This is simply a result of too coarse sampling rate of a high frequency phenomena. In general, because of aliasing, we need pre- smoothers (at least in theory). However, the above numerical experiment indicated that multigrid might very well work with only post-smoothings.

If you have a self-adjoint operator and want to construct a *symmetric* multigrid *preconditioner* (for a Conjugate-Gradient solver), you will have to use an equal number of pre- and post-smoothings and the pre-smoother and post-smoother should be adjoint, to obtain a symmetric preconditioner.

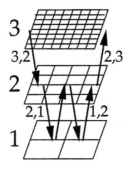

**Fig. 3.1.** Multigrid W-cycle

### 3.5.4   W Cycle and Nested Iteration

*Different Multigrid Cycles.* We specify the multigrid W-cycle and other cycles by introducing a *cycle* parameter, often denoted by $\gamma$ [9]. The value $\gamma = 1$ gives a V-cycle, whereas $\gamma = 2$ gives a W-cycle (see Figure 3.1 for an example on a W-cycle). The menu item cycle type gamma is used to set $\gamma$. Use test1.i as input and override $\gamma$ on the command line[7]:

```
unix> ./app --iscl --Default Verify/test1.i --verbose 1\
            --no_of_grid_levels 7 --gamma '{ 1 & 2 & 3 }' > tmp.1

unix> grep 'solver conv\|LinEqAdm' tmp.1
  LinEqAdm::solve: solve a linear system,  CPU time= 0.24
  solver converged in   6 iterations
  LinEqAdm::solve: solve a linear system,  CPU time= 0.31
  solver converged in   6 iterations
  LinEqAdm::solve: solve a linear system,  CPU time= 0.45
  solver converged in   6 iterations
```

The numerical experiments are done on an AMD Athlon 1800 MHz with 1 GB RAM.

The $\gamma$ parameter increases the complexity of the algorithm (the recursive calls of the multigrid routine). If you encounter convergence problems in an application, you can try a W-cycle multigrid or even $\gamma > 2$. Higher $\gamma$ values are usually used for more complicated grids or equations. For the current Poisson equation, a straightforward V-cycle is optimal.

*Nested Iteration.* Nested iteration, or full multigrid, or cascadic iteration[8] is based on the idea that a coarse grid solution may serve as a good start guess

---

[7] This is only possible as long as there are no --gamma command-line option from the simulator's own menu (or the Diffpack libraries for that sake). Adjusting the menu item in a file is always safe.

[8] This is a special case of the nested iteration, where no restriction is used and coarser grids are never revisited.

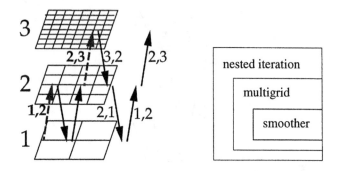

**Fig. 3.2.** Nested iteration, multigrid V-cycle

for a fine grid iteration. Hence, before the standard V- or W-cycle, a start vector is computed. This process starts by first computing an approximate solution on the coarsest grid. This solution is prolongated onto the next coarsest grid and a smoother is applied on this grid level. This process is repeated until the finest grid is reached, where the standard multigrid process starts. The right hand side on the coarsest grid is in Diffpack implemented as a restriction of the right hand side on the finest grid. It is cheaper to generate the right hand side on the coarsest grid, but this can not be done when the right hand side comes from a residual generated by Krylov solver. We have therefore chosen this implementation such that it is easy to use the nested iteration as a preconditioner for a Krylov solver.

We now want to run `NestedMultigrid`, which can be specified as the answer to the `multilevel method` menu item or the corresponding `--ml_method` command-line option. There is a parameter `nested cycles` (command-line option `--nestedcycles`) that controls the number of multigrid cycles before the solution is passed to the next finer grid as a start solution. We can include some values of this parameter:

```
./app --iscl --Default Verify/test1.i --verbose 1 \
     --nestedcycles '{ 1 & 2 & 3 }' --no_of_grid_levels 7 \
     --ml_method NestedMultigrid
```

A slight increase of the CPU time is seen as the `nested cycle` parameter increases.

It appears that the efficiency of nested multigrid is better than standard multigrid. In fact, nested multigrid is optimal, it reaches discretization error within $\mathcal{O}(n)$ iterations (see also page 109). Let us compare the two approaches directly and include a run-time plot of how the residual in the linear system decreases with the iteration number (the `run time plot` item on the `Define ConvMonitor #1` menu or the `--runtime_plot` command-line option):

```
./app --iscl --Default Verify/test1.i --verbose 1 \
     --no_of_grid_levels 7 --runtime_plot ON \
```

```
--ml_method '{ Multigrid & NestedMultigrid }'
```

The run-time plot of the residuals evolution during the multigrid algorithm is shown briefly on the screen, but the plots are also available in some standard Diffpack curveplot files[9]:

```
tmp.LinEqConvMonitorData.SIMULATION_m01.map
tmp.LinEqConvMonitorData.SIMULATION_m02.map
```

Each of the mapfiles contains only one curve, but we can plot both curves by, e.g.,

```
curveplot gnuplot \
    -f tmp.LinEqConvMonitorData.SIMULATION_m01.map \
    -f tmp.LinEqConvMonitorData.SIMULATION_m02.map \
    -r '.' '.' '.'  -ps r_nested_vs_std.ps
```

Figure 3.3 shows the resulting plot, where we clearly see that nested multigrid results in faster decrease in the absolute norm of the residual. (When preparing the plot in Figure 3.3, we edited the Gnuplot command file, which has the name .gnuplot.commands when produced by curveplot [9, App. B.5.1], to improve the labels, and then we loaded this file into Gnuplot to produce a new plot.)

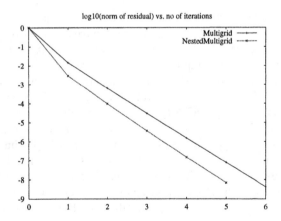

**Fig. 3.3.** Nested iteration vs. standard multigrid.

---

[9] Note that the filenames contain the string "tmp", which means that the files will be automatically removed by the Clean script [9].

### 3.5.5    Coarse-Grid Solver

*The Accuracy of the Coarse-Grid Solver.* On page 107 we outlined how easy it
is to switch from using Gaussian elimination as coarse-grid solver to using an
iterative method. Let us investigate SOR, SSOR, and Conjugate-Gradients
as coarse-grid solvers and how many iterations that are optimal (i.e., how
accurate the solution need to be on the coarse grid). Input file `test5.i` has
the relevant new menu items:

```
...
sub MGtools
  ...
  sub GridCollector
    ...
    set no of grid levels = 5
    set gridfile = P=PreproBox | d=2 [0,1]x[0,1]
                               | d=2 e=ElmB4n2D [8,8] [1,1]
    ...
  ok
  ...
  set coarse grid solver = true
  sub coarse grid LinEqSolver_prm
    set coarse grid basic method = { SOR & SSOR & ConjGrad }
    ! default relaxation parameter is 1.0
    set coarse grid max iterations = { 1 & 5 & 10 & 40 }
  ok
  ...
```

Since we are using an iterative solver, we need a coarse grid with some unknowns, at least more than 9. This is why we have specified $8 \times 8$ bilinear
elements for the coarsest grid. From the CPU-time values it appears that 10-
40 iterations have the best computational efficiency. That is, the coarse-grid
solver needs to be accurate, which also is a general result from the analysis
of multigrid methods. These experiments show that the Conjugate-Gradient
method is more efficient than SSOR, which is more efficient than SOR, when
then maximum number of coarse-grid iterations is small. The different solution methods result in approximately the same overall CPU time when the
optimal (a large) number of coarse-grid iterations is specified. It is important
to notice that the Conjugate-Gradient method is not a linear iteration and
does not fit into most theoretical analysis of multilevel methods. However,
the numerical experiments done here indicate that it may work anyway.

*The Optimal Coarse Grid and Number of Levels.* We now face the important
question of how to choose the coarse-grid partition and the number of grid
levels. Let us try the following combinations of grid levels and coarse-grid
partitions, designed such that the number of unknowns on the finest grid is
constant (16384):

```
levels    coarse-grid partition
  6              [4,4]
  5              [8,8]
```

| 4 | [16,16] |
| 3 | [32,32] |

These modifications have been incorporated in `test6a.i` to `test6d.i` (which are essentially minor edits of `test5.i` – a more efficient and convenient method of handling the input file is to embed one input file in a Perl script as explained in [9, Ch. 3.12.9]). As the coarse-grid solver we use SSOR with unit relaxation parameter, i.e., symmetric Gauss-Seidel iteration, and a maximum number of iterations of 40. The number of iterations is constant at 7, but the CPU-time increases dramatically as we go from a $16 \times 16$ to a $32 \times 32$ coarse grid. Thus, a very coarse coarse grid with many levels seems to be an optimal combination.

### 3.5.6    Multigrid as a Preconditioner

We now want to use multigrid as a preconditioner [9, App. C] instead of as a stand-alone iterative solver. We choose the Conjugate-Gradient algorithm to solve the Poisson equation. This algorithm requires a symmetric preconditioner and a symmetric matrix, which in a multigrid context means that the pre- and post-smoothing operators as well as the restriction and prolongation operators must be adjoint. Several methods are available:

- One way to satisfy the condition is to take a self-adjoint smoother like Jacobi iteration or symmetric Gauss-Seidel iteration (i.e. SSOR with unit relaxation parameter).

- Alternatively, use an non-symmetric smoother as a pre-smoother and its adjoint as a post-smoother. For example, take a Gauss-Seidel iteration (SOR) or a (R)ILU iteration with a node ordering $1, 2, \ldots, n$ as pre-smoother and the same method with a reversed node ordering $n, n - 1, \ldots, 1$ as post-smoother.

- Another alternative is to use an additive multigrid (see section 3.5.7) with a self-adjoint smoother like Jacobi iteration or symmetric Gauss-Seidel iteration.

A necessary requirement is that the number of pre- and post-smoothing steps must be equal.

Using multigrid as a preconditioner is pretty much the same as using it as a stand-alone solver, except that applying the preconditioner means a single multigrid iteration (e.g. one V-cycle). Technically in Diffpack, the preconditioner `PrecML` must have a `MLSolver`, or in the case of multigrid a `Multigrid`, attached. `PrecML` just passes the vectors `x` and `b` to `Multigrid` which does the computation. These steps were taken care of in the example of the simulator's `scan` function as we explained in Section 3.3.1.

*Specifying Multigrid as Preconditioner.* In the input file we basically change

```
sub LinEqAdmFE
  ...
  sub LinEqSolver_prm
    set basic method = ConjGrad
    set max iterations = 300
  ok
  sub Precond_prm
    set preconditioning type = PrecML
  ok
  ...
```

The input file `test7.i` runs a test with both multigrid and Conjugate-Gradients as basic solver and multigrid and the identity matrix as preconditioners. The parameters in the multigrid method (whether used as a basic iteration scheme or a preconditioner) are set on the `MGtools` menu.

**Table 3.2.** Comparison of Multigrid as a solver and as a preconditioner

| | MG | | CG/MG | |
|---|---|---|---|---|
| $h$ | #it. | CPU time | #it. | CPU time |
| $2^{-2}$ | 5 | 0.01 | 3 | 0.01 |
| $2^{-3}$ | 5 | 0.01 | 4 | 0.01 |
| $2^{-4}$ | 6 | 0.01 | 5 | 0.01 |
| $2^{-5}$ | 6 | 0.01 | 5 | 0.02 |
| $2^{-6}$ | 6 | 0.06 | 5 | 0.07 |
| $2^{-7}$ | 6 | 0.25 | 5 | 0.29 |
| $2^{-8}$ | 7 | 1.22 | 5 | 1.19 |
| $2^{-9}$ | 7 | 4.85 | 5 | 4.75 |
| $2^{-10}$ | 7 | 23.08 | 6 | 31.26 |

*Multigrid as a Preconditioner vs. a Basic Iteration Method.* In Table 3.2 we compare multigrid as a solver with multigrid as a preconditioner for the Conjugate Gradient method. The results are striking, there seems to be no point in using Krylov subspace acceleration for a highly efficient multigrid method as we have here (see also [13], Chapter 7.8). In general, multigrid preconditioners are more robust and in real-life applications it is often difficult to choose optimal multigrid parameters. While multigrid may reduce most parts of the error, other parts remain essentially unchanged. These are picked up by the Krylov method (see the Sections 3.6.4 and 3.7).

*Investigating the Effect of Different Smoothers.* Let us test the performance of different smoothers when multigrid is used as preconditioner. The `test8.i` file specifies Conjugate-Gradients as basic solver, multigrid V-cycle as the preconditioner, and the following parameters for the smoother and the number of smoothing sweeps:

```
set sweeps = { [1,1] & [0,1] }
sub smoother LinEqSolver_prm
  set smoother basic method = { Jacobi & SSOR & SOR }
```

The results are striking: using [0,1] as sweeps leads to divergence of the solver. This is in accordance with the requirement of an equal number of pre- and post-smoothing steps when multigrid is used as preconditioner, as we have already mentioned. The SOR smoother with one pre- and post-smoothing sweep converges well, despite being non-symmetric, but applies twice the CPU time and twice the number of Conjugate-Gradient iterations compared with the symmetric smoothers (Jacobi and SSOR). The typical behavior can be seen in Figure 3.4. Initially, we get nice convergence, but later we get stagnation or even divergence.

*Non symmetric Smoothers and Conjugate Gradient-like Algorithms.* If we apply a Conjugate Gradient-like method for non-symmetric linear systems, there is no requirement of a symmetric preconditioner, and we can play around with a wider range of smoothing strategies. The input file test9.i launches experiments with four basic solvers, BiCGStab, CGS, GMRES, and Conjugate-Gradients, combined with a highly non-symmetric smoother: two Gauss-Seidel sweeps as pre-smoothing and no post-smoothing.

```
sub LinEqSolver_prm
  set basic method = { BiCGStab & GMRES & CGS & ConjGrad }
...
sub ConvMonitorList_prm
  sub Define ConvMonitor #1
  ...
  set #1: run time plot = ON
...
set sweeps = [2,0]
sub smoother LinEqSolver_prm
  set smoother basic method = SOR
```

Both BiCGStab and CGS with two pre-smoothing Gauss-Seidel sweeps appear to be as efficient as Conjugate Gradients with a symmetric multigrid preconditioner (cf. the test8.i test). As expected, the symmetric Conjugate-Gradient solver stagnates in combination with the non-symmetric smoother, but the other algorithms behave well.

### 3.5.7  Additive Preconditioner

Additive multigrid refers to a strategy where the corrections on the different grid levels are run independently. Originally, this method was proposed in [14], and now it is often referred to as the BPX preconditioner (BPX) or "multilevel diagonal scaling" (MDS). The method played an important

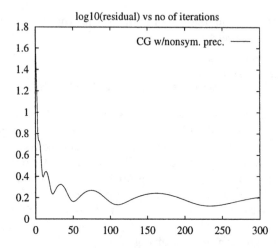

**Fig. 3.4.** The stagnation of a Conjugate-Gradient method with a multigrid V-cycle, combined with a non-symmetric smoother (two Gauss-Seidel pre-smoothing sweeps).

role for the proof of optimal complexity of multigrid, and the interpretation as additive multigrid was found later. Additive multigrid may diverge even for elliptic problems, but as a preconditioner, it is fairly efficient. The advantage of additive multigrid is that independent operations may serve as a source of parallelism, although the grid transfer operations, the restrictions and prolongations still are serial operations. These steps can be broken up into independent operations by splitting the computational domain into subdomains.

*Standard vs. Nested vs. Additive Multigrid.* We can start testing the additive multigrid method by running some experiments:

```
sub LinEqSolver_prm
  set basic method = ConjGrad
  ...
sub Precond_prm
  set preconditioning type = PrecML
  ...
sub MLSolver_prm
  set multilevel method = {AddMultigrid & Multigrid & NestedMultigrid}
  set cycle type gamma = 1  ! V cycle
  set nested cycles = 1
ok
set sweeps = [1,1]
sub smoother LinEqSolver_prm
  set smoother basic method = SSOR
```

Notice that additive multigrid is invoked by selecting the `AddMultigrid` multilevel method. Additive multigrid does not necessarily work as stand-alone solver.

## 3.6    Equipping the Poisson2 Solver with Multigrid

In this section we want to test the efficiency of multigrid on more general elliptic problems,

$$
\begin{aligned}
-\nabla \cdot [k\nabla u] &= f, \text{ in } \Omega, \\
u &= D_1, \text{ on } \partial\Omega_{E_1}, \\
u &= D_2, \text{ on } \partial\Omega_{E_2}, \\
u &= g, \text{ on } \partial\Omega_{E_3}, \\
-k\frac{\partial u}{\partial n} &= 0, \text{ on } \partial\Omega_N, \\
-k\frac{\partial u}{\partial n} &= \alpha u - U_0, \text{ on } \partial\Omega_R, \\
\partial\Omega &= \partial\Omega_{E_1} \cup \partial\Omega_{E_2} \cup \partial\Omega_{E_3} \cup \Omega_N \cup \partial\Omega_R.
\end{aligned}
\tag{3.7}
$$

We will investigate how multigrid behaves with respect to different boundary conditions, grids, and $k$-functions. This problem is implemented in `Poisson2` and described in [9, Ch. 3.5]. The code in the `Poisson2` solver is much larger and more complicated than the `Poisson1` simulator, but the modifications needed to incorporate multigrid are nearly the same. The only difference is that the `Poisson2` solver performs some grid manipulations (changing boundary indicators for instance), which now need to be implemented for each grid in the grid hierarchy.

We copy the `Poisson2` files and rename `Poisson2` to `Poisson2MG`. The various multigrid modifications explained for the `Poisson1` code are performed as explained previously. Additionally, the `Poisson2::scan` function performs some grid manipulations that must be repeated for each grid:

```
for (int i =1; i<= no_of_grids; i++)
  {
    grid.rebind(mgtools->getGrid(i));

    String redef   = menu.get ("redefine boundary indicators");
    if (!redef.contains("NONE"))  grid->redefineBoInds (redef);
    String addbn   = menu.get ("add boundary nodes");
    if (!addbn.contains("NONE"))  grid->addBoIndNodes (addbn, ON);
    addbn          = menu.get ("remove boundary nodes");
    if (!addbn.contains("NONE"))  grid->addBoIndNodes (addbn, OFF);
    String addmat = menu.get ("add material");
    if (!addmat.contains("NONE")) grid->addMaterial (addmat);
  }
```

Notice that `grid` is a handle to the finest grid, as it should be, at the end of this loop.

In addition, ensure that dof(no_of_grids) points to a DegFreeFE object from the MGtools and remove the statement

```
dof.detach().rebind (new DegFreeFE (*grid, 1)); \
// 1 for 1 unknown per node
```

If this statement is not removed then we will have two sets of DegFreeFEs on the finest grid, and the boundary conditions will probably end up in only one of them.

*Repeating the Poisson1MG Tests.* We can run the same test case as for the Poisson1MG code, i.e., we choose to use the Poi2sinesum subclass which implements the same boundary value problem, with analytical solution, as in class PoissonMG. However, simply running

```
./app --class Poi2sinesum --iscl --Default Verify/test1.i
```

with Poisson1MG's test1.i file gives results that differ from those generated by the Poisson1MG solver. The reason is due to the handling of the boundary conditions. In the Poisson2 solver, there are five different boundary indicators that influence the setting of boundary conditions, and we have simply applied the default set of four indicators, one for each side. This is incorrect, as all sides should be subject to indicator no. 3 (a function specifying Dirichlet conditions). We must therefore add the following instruction in the test1.i file:

```
set redefine boundary indicators = n=3
     names= dummy1 dummy2 boundary 1=() 2=() 3=(1 2 3 4)
```

Now Poisson2MG::scan gets the correct boundary indicator information and propagates it to all intermediate grids.

Looking at Poi2sinesum::scan we see that there is a comment block explaining how to carry out manipulations of the boundary indicators directly in the code. This manipulation adjusts the finest grid only, in a multigrid context. We must ensure that mapping of boundary indicators is applied to all grids, cf. Poisson2MG::scan. Some of the subclass solvers in Poisson2 and Poisson2MG manipulate boundary indicators by forcing special menu answers. This will in general result in correct extraction of grid information in Poisson2MG::scan and correct propagation to the hierarchy of grids.

The nice thing about Poisson2MG is the automatic generation of reports. We can tailor this report to empirical investigation of the multigrid method, but the implementation is left as an exercise.

### 3.6.1   Multigrid and Neumann or Robin Boundary Conditions

The Poisson2 simulator handles Dirichlet, Neumann, Robin boundary conditions, and a mixture of these. It is necessary to redefine the calcElmMatVec, integrands, and integrands4side functions derived from FEM. In multilevel methods we have a sequence of coefficient matrices and each should reflect the same boundary conditions. Thanks to the object-oriented design of Diffpack this happens automatically, since mgtools->makeSystemMl() is implemented roughly as,

```
for (int i=1 ; i< no_of_grids ; i++) {
    current_grid_level = i;
    fem->mgFillEssBC(i);
    fem->makeSystem(dof(i)(), mat(i)->mat(), b(i)->vec());
}
```

Notice that fem is a FEM pointer to the current simulator so fem->makeSystem will call the redefined functions in the simulator class, just as in a standard simulator.

Multigrid is robust and handles any boundary conditions. The following test confirms the efficiency with Robin conditions,

```
unix >app --class Poi2Robin --iscl --Default Verify/test1.i     \
--no_of_grid_levels '{2 & 3 & 4 & 5 & 6 & 7}'  | grep iterations
    solver converged in   6 iterations
    solver converged in   6 iterations
    solver converged in   6 iterations
    solver converged in   6 iterations
    solver converged in   6 iterations
    solver converged in   6 iterations
```

Notice that when we have Neumann conditions on the whole boundary, the solution is only determined up to a constant and the matrices are therefore singular. The Krylov solvers and the classical iteration methods like Jacobi and SSOR all handle this ambiguity. But the Gaussian elimination (LU-factorization) on the coarsest grid may be dangerous, depending on the right-hand side. One should therefore switch from Gaussian elimination to an iterative method or make the matrices non-singular. In order to make the solution unique one might impose an additional condition on $u$, e.g.

$$u_i = 0, \text{ for some nodal point } i,$$

or

$$\int_{\Omega_l} u = 0 .$$

The point-wise condition is however poorly approximated on the coarser grids and results in deterioration of the multigrid convergence. It is usually sufficient to switch to an iterative method on the coarsest grid. We can use, e.g., SSOR with enough iterations to get the desired accuracy. For a more general discussion on singular systems the reader is refereed to Chapter 12 in [7].

### 3.6.2   Debugging Multigrid

A common error when developing multigrid simulators is wrong boundary conditions on the coarser grids. All matrices should have the same boundary conditions. It is therefore convenient to make function to check that the coarser problems are properly set up. An easy way to do that is to calculate the solution on a coarser grid and plot it. Any error will then be quickly revealed. We make a function debugALevel(int grid_level), where we fetch the data structures from MGtools from the level grid_level, calculate a right hand side, calculate the solution and dump it to file. An appropriate function is

```
void Poisson1MGb:: debugALevel(int grid_level) {
  // fetch stuff:
  GridFE& grid_l = mgtools->getGrid(grid_level);
  grid_l.boundaryData().print(s_o);
  LinEqSystemStd& system_l = mgtools->getLinEqSystem(grid_level);
  DegFreeFE& dof_l = mgtools->getDof(grid_level);

  Handle(FieldFE) u_l; // the coarse grid solution
  u_l.rebind(new FieldFE(grid_l, oform("u_%d",grid_level)));

  // fetch the matrix and vectors:
  LinEqMatrix& A_l = system_l.A();
  LinEqVector& x_l= system_l.x();
  LinEqVector& b_l= system_l.b();

  // fill boundary conditions if
  // mgtools->makeSystemMl() is not called yet.
  mgFillEssBC(grid_level);

  // calculate the right hand side.
  makeSystem(dof_l,  b_l.getVec());

  // solve the system:
  FactStrategy fact; fact.fact_tp=LU_FACT;
  A_l.factorize(fact);
  A_l.forwBack(b_l, x_l);

  // dump to file.
  dof_l.vec2field(x_l.getVec(), *u_l);
  database->dump(*u_l);
}
```

In addition we may want to visualize the behavior of each multigrid iteration. We can do this by implementing a class MyMGtools derived from MGtools. The intention is to call up some functions that dump the solution. These functions, defined in Poisson2MG, should be called at strategic places in the multigrid algorithm. We are interested in dumping the solution before and after the smoothing on the finest grid. We therefore redefine solveSubSystem to call up some Poisson2MG functions before and after MGtools::solveSubSystem.

```
bool MyMGtools::solveSubSystem (LinEqVector& b, LinEqVector& x,
```

```
                                   int grid_level, StartVectorMode start,
                                   MLSolverMode mode)
{
  if (poisson_simulator_attached) {
    if ( grid_level == no_of_grids && turn_on ) {
      bool ret;
      poisson_simulator->beforeSolveSubSystem(mode);
      // The standard smoother
      ret = MGtools::solveSubSystem(b,x,grid_level, start, mode);
      poisson_simulator->afterSolveSubSystem(mode);
      return ret;
    }
    else {
      // The standard smoother
      return MGtools::solveSubSystem(b,x,grid_level, start, mode);
    }
  }
  else {
    errorFP("MyMGtools::solveSubSystem", "Poisson2MG not attached");
  }
  return false;
}
```

The poisson_simulator is here of type Poisson2MG, where we have defined the following functions,

```
void Poisson2MG:: beforeSolveSubSystem(MLSolverMode mode) {
  iterations++;
  u->setFieldname(oform("u_before_it%d", iterations));
  dof->vec2field(*linsol, *u);
  database->dump(*u);
}

void Poisson2MG:: afterSolveSubSystem(MLSolverMode mode) {
  u->setFieldname(oform("u_after_it%d", iterations));
  dof->vec2field(*linsol, *u);
  database->dump(*u);
}
```

Figure 3.5 displays one V-cycle, with the field dumped by the functions beforeSolveSubSystem and afterSolveSubSystem. Notice that if plain multigrid is used, then linsol is used within the algorithm on the finest grid. On the other hand if multigrid is used as a preconditioner, then multigrid computes the error correction and linsol can not be used to investigate multigrid so easily. If we are interested in vectors or matrices from coarser grids, then we can fetch them from MGtools as we did in debugALevel.

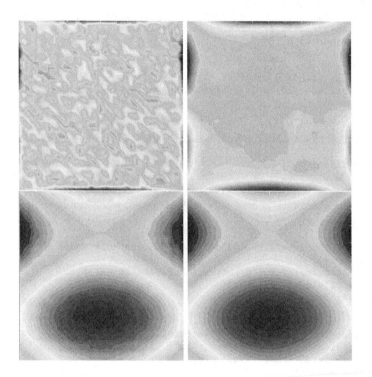

**Fig. 3.5.** The upper left picture shows the initial vector. It is a random vector that should contain "all possible" errors. In the upper right picture the solution after one symmetric Gauss-Seidel sweep is displayed. It is clear that the high frequency random behavior in the initial solution has been effectively removed. The picture down to the left shows the solution after the coarse grid correction. The smooth components of the solution have improved dramatically. In the last picture the solution after the post smoothing is displayed. The solution is now very close to the actual solution.

Another useful feature when debugging is to turn on --verbose 3 on the command line. One then obtains information about the residual on the coarser grid as well as several other things. A useful command may be of the form:

```
unix> app --verbose 3  --class Poi2Robin < Verify/test1.i  \
        | grep "residual in space 2"
The l2 norm of the residual in space 2 is 0.471954
The l2 norm of the residual in space 2 is 0.021194
The l2 norm of the residual in space 2 is 0.00118312
The l2 norm of the residual in space 2 is 7.24997e-05
The l2 norm of the residual in space 2 is 4.6181e-06
The l2 norm of the residual in space 2 is 3.00102e-07
```

In Poisson2MGanal the debugging functions are extended further. The final multigrid "stress-test" is to compute the error reduction after each cycle. We extend the afterSolveSubSystem function to do this,

```
void Poisson2MGanal:: afterSolveSubSystem(MLSolverMode mode) {
    Poisson2MG:: afterSolveSubSystem(mode);
    if (mode == SUBSPACE_BACK ) {
        ErrorNorms::Lnorm (*this,     // supplied function
                    *u,              // numerical solution
                    u->grid(),       // finest grid
                    false,           // same_grid=false
                    DUMMY,           // point of time
                    L1_error, L2_error, Linf_error, // error norms
                    error_itg_pt_tp, // point type for numerical integr.
                    0,               // relative order
                    this,            // FEM* solver
                    &energy_error);// compute error in energy norm

        H1_error = ErrorNorms::H1Norm
        (*this, // simulator is its own functor w/derivativePt
        *u, DUMMY, error_itg_pt_tp,
        0);     // relative integration rule
        s_o <<"L2 error at it. "<<iterations <<": "<<L2_error<<endl;
        s_o <<"H1 error at it. "<<iterations <<": "<<H1_error<<endl;
    }
}
```

This is mainly a cut and paste from Poisson2MGanal::resultReport. The error is measured after each cycle (mode == SUBSPACE_BACK). We can check the error rates by,

```
unix>app  --class Poi2sinesum --no_of_grid_levels 8  --iscl \
Verify/test1.i | grep L2
L2 error of u at iteration 1 is 0.110582
L2 error of u at iteration 2 is 0.00512824
L2 error of u at iteration 3 is 0.000312139
L2 error of u at iteration 4 is 8.51688e-05
L2 error of u at iteration 5 is 7.62436e-05
L2 error of u at iteration 6 is 7.58404e-05
```

We have a convergence rate at about 0.05 until we reach the discretization error. We obtain about the same rate if we use the $H^1$-norm.

### 3.6.3   Domains With Geometric Singularities

The multigrid theory in [2] applies to domains that are convex polygons. How does multigrid behave on a general domain? We test on a L-shaped domain, which is generated by the following grid generation command,

```
set gridfile = P=PreproStdGeom | BOX_WITH_BOX_HOLE \
d=2 [0,1]x[0,1] - [0,0.5]x[0,0.5] | d=2 e=ElmB4n2D [4,4] [1,1]
```

Simple mesh generation tools are described in [9, Ch. 3.5.2]. Note that the above `gridfile` command is used on the coarsest grid, the finer grids are uniform refinements of the coarsest. The resolution of the geometry is therefore limited by the coarsest grid. If the geometry of the problem is important and complex, one should instead make the grids manually and attach them to the `GridCollector` as described in Section 3.3.1. One might then end up with non-nested grids, but multigrid can be efficient and independent of $h$ if the grids have certain properties [2]. We investigate the behavior with respect to $h$,

```
unix>../app --class Poisson2MG --iscl   --Default  box1.i  \
--no_of_grid_levels
'{2 & 3 & 4 & 5 & 6}' | grep converged
solver converged in   6 iterations
solver converged in   6 iterations
solver converged in   6 iterations
solver converged in   6 iterations
solver converged in   6 iterations
```

Multigrid seems very robust, as it should in this case [3]. However, we should expect that it deteriorates for stronger geometrical singularities. In such problems the smoother will in general not be very efficient near the singularities. One cheap way to improve the convergence is to implement additional local smoothing in the neighborhood of the singularities. However, one can also often get $h$-independent behavior by switching to the W-cycle or Full Multigrid ([13, p. 174] or page 94 [8]).

To investigate convergence rates related to geometrical singularities, we make a geometry with a singularity dependent of the angle $D$ shown in Figure 3.6. Such grids can be made by Diffpacks interface to `Geompack`. An appropriate coarse grid generation command is

```
set gridfile = P=PreproStdGeom | DISK r=1 degree=355 | \
e=ElmT3n2D nel=26 resol=10
```

The angle $D$ (`degree`), is varied from $180^o - 359^o$, see Figure 3.6, but the convergence rate does not drop considerably. The number of iterations increases slightly from 8 to 11. There are four parameters r, `degree`, `nel`, and `resol` which should be related. The perl script `disk.pl` computes the number of elements according to the formula on page 306 in [9], and makes an inputfile `disk1.i` based on the test1 file `disk.test1`. Notice that in these experiments we have used nested grids and the geometrical resolution is therefore very coarse.

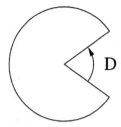

**Fig. 3.6.** Geometric singularity dependent on the angle D.

### 3.6.4   Jumping Coefficients

The class `Poi2flux` introduces a simulator with a possible discontinuous coefficient $k$ in the problem,

$$-\nabla \cdot (k\nabla u) = f, \tag{3.8}$$

where $k|_{\Omega_0} = k_0$, $k|_{\Omega_1} = k_1$, $\Omega = \Omega_0 \cup \Omega_1$. Such problems occur in groundwater flow and reservoir simulation, where $k$ is the permeability of the medium, consisting of rock and sand. For a strongly jumping coefficient the convergence will depend on the size of the jump and even divergence may occur. In such cases multigrid as a preconditioner is a more robust alternative. Prior to this simulation we have not edited the source code for `Poi2flux` at all. We decide to test `MLIter` vs. `ConjGrad/PrecML` with various $k_0$. The parameter $k_1$ is fixed to 1.

```
set k0 = {1 & 10 & 100 & 1000 }
```

We use the standard Unix-tool `tee` to redirect the output to both standard output and a file. By doing this we can grep on the essential information and save the output to a file at once.

```
unix> ./app --class Poi2flux < Verify/test4a.i | tee test4a.v \
 | grep converged
 solver converged in 145 iterations
 solver converged in 195 iterations
 solver converged in 201 iterations
 solver converged in 155 iterations
```

Multigrid as a solver seems to be a poor choice, but is rather independent of the $k_0$ value. In contrast the multigrid preconditioner seems pretty stable.

```
unix> ./app --class Poi2flux < Verify/test4b.i | tee test4b.v \
 | grep converged
 solver converged in  17 iterations
 solver converged in  17 iterations
 solver converged in  17 iterations
 solver converged in  17 iterations
```

The slow convergence when $k_0 = k_1 = 1$ is strange. However, thanks to
`debugALevel` the error is revealed when we look at the plots. We have implemented wrong boundary conditions on the coarser grids. In the source code
we see that the boundary indicators are redefined only on the finest grid.
We ensure that we redefine all the coarser grids by similar adjustment as in
`Poisson2MG`. We re-run the test and using `MLIter` we get,

```
unix> ./app --class Poi2flux < Verify/test4a.i | tee test4a.v \
 | grep converged
 solver converged in   6 iterations
 solver converged in  10 iterations
 solver converged in  12 iterations
 solver converged in  11 iterations
```

We see a slight dependency on the mesh size $h$, which is not apparent when
multigrid is used as a preconditioner,

```
unix> ./app --class Poi2flux  <Verify/test4b.i | tee test4b.v \
 | grep converged
 solver converged in   4 iterations
 solver converged in   6 iterations
 solver converged in   6 iterations
 solver converged in   6 iterations
```

The first thing that strikes us is that the multigrid preconditioner can be
rather effective with completely wrong boundary conditions on the coarser
grids. In general preconditioning is much more robust and therefore stand-
alone multigrid should always be used to validate the code. Any error or
none- optimal multigrid component will be much more apparent.

The results concerning the jumping coefficients have been much better
than expected. The convergence should in general depend on the jump and
on the geometry related to the jump. We check a somewhat harder and
slightly more interesting case by redefining the functions $k$ and $f$ according
to the Figure 3.7. This is implemented in the `Poi2flux2` class. In this example
multigrid even diverges, although the geometry and coefficients are rather
simple, we choose,

```
set k0 = {1 & 10 & 100 & 1000 }
```

```
unix> ./app  --class Poi2flux2  < Verify/test4a.i | tee test4a.v \
 | grep converged
 solver converged in  6 iterations
 solver converged in 10 iterations
 solver converged in 55 iterations
 solver not converged in 300 iterations
```

The problem here is that the smoother does not manage to reduce the error
sufficiently. Multigrid can still be used but requires more pre- and post sweeps
to reduce the error sufficiently. However, letting $k_0$ be 1000 and a V[10,10]-
cycle still leads to slow convergence (51 iterations) and it is therefore clear

that multigrid is unusable as a solution method, in general. However, it should be mentioned that multigrid still reduce some parts of the error. It is the error close to the jump that multigrid does not handle. We test multigrid as a preconditioner and obtain far better results,

```
unix> ./app  --class Poi2flux2  < Verify/test4b.i | tee test4b.v \
  | grep converged
  solver converged in   5 iterations
  solver converged in   8 iterations
  solver converged in  10 iterations
  solver converged in  13 iterations
```

It is also interesting to test how multigrid behaves in terms of $h$. In test4c.i we have set,

```
set k0 = {1000 }
...
set no of grid levels = {2 & 3 & 4 & 5 & 6 & 7   }
...
```

The output from the experiment is,

```
unix> ../app  --class Poi2flux2  < test4c.i | tee test4c.v \
  | grep converged
  solver converged in   5 iterations
  solver converged in  15 iterations
  solver converged in  13 iterations
  solver converged in  11 iterations
  solver converged in  13 iterations
  solver converged in  34 iterations
```

In the following, we will try to explain this odd behavior, but actually these experiments are beyond rigorous theoretical justification, at least to the author's knowledge. However, it still makes sense to do these numerical experiments, because it is a popular problem and the software allows it.

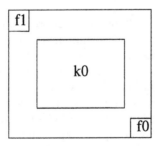

**Fig. 3.7.** This picture shows how the source $f$ and the permeability $k$ vary, $f$ is zero expect from the sink, f0, where $f = 1$, and the drain, f1, with $f = -1$. The value of $k$ in k0 vary, according to the input parameters, and $k = 1$ elsewhere.

*Multigrid as a Preconditioner vs. a Basic Iterative Method Revisited.* In the previous experiment stand-alone multigrid did not converge while the multigrid preconditioner was very efficient. Why did this happen? Is the preconditioner stable? Yes, with respect to $h$ it is, but it is dependent on $k$. To understand why the preconditioner is much more robust we use the terminology from Section 3.5.1. Multigrid is convergent if the reduction factor $\rho$ satisfy,

$$\rho = \|I - BA\| < 1. \tag{3.9}$$

On the other hand, the Conjugate gradient method is convergent as long as the preconditioned matrix $(BA)$ is SPD, in a suitable inner product[10]. However, the number of iterations $m$ will depend on the condition number of $BA$, $\kappa = \kappa(BA)$. In fact, the error bound,

$$\|e_m\|_A \leq \epsilon \|e_0\|_A,$$

is reached in at most $m$ iterations, where $m$ is

$$m = \frac{1}{2} \log \frac{2}{\epsilon} \sqrt{\kappa}.$$

This simple expression is derived in, e.g., [5] by assuming that $\epsilon \ll 1$. Hence, the number of iterations for the preconditioned Conjugate Gradient method to converge will be independent of $h$, if $\kappa$ is. This is more general than (3.9). In fact, we know that for (3.8) we have,

$$k_{min}(\nabla u, \nabla u) \leq (k\nabla u, \nabla u) \leq k_{max}(\nabla u, \nabla u), \tag{3.10}$$

where $k_{min} = \inf_{x \in \Omega} k(x)$ and $k_{max} = \sup_{x \in \Omega} k(x)$. This means that $B$ is spectrally equivalent to $A^{-1}$ even if $B$ is made as a preconditioner for the Poisson equation, $-\Delta u = f$, which we know how to solve efficiently. The condition number will then be proportional to $k_{max}/k_{min}$, but independent of $h$.

In the present case we did not use an approximation of $-\Delta^{-1}$ as preconditioner, but rather a multigrid iteration involving matrices with jumps. The number of iterations is also better than what is expected from the analysis above. This indicates that multigrid is efficient on most of the error, but some components of the error are not dealt with. These components are removed efficiently by the Conjugate-Gradient method. The above analysis, where the operator is compared with the standard Poisson equation is useful to determine the worst case. We can always use a preconditioner based on a multigrid sweep on the Poisson equation. However, it is usually more efficient to use a multigrid sweep on the operator at hand, even if multigrid as a stand-alone solver does not work at all.

---

[10] $B^{-1}$ in this case.

### 3.6.5    Anisotropic Problems and (R)ILU smoothing

Another interesting problem is the (axis parallel) anisotropic diffusion equation,

$$-k_0 u_{xx} - u_{yy} = f, \tag{3.11}$$

where, e.g., $k_0 \in [10^{-3}, 10^3]$. This problem is not implemented in the Poisson2 hierarchy, but we implement it in a class Poi2anis. It is well known that standard multigrid is not very efficient for such problems, but there are several ways to improve the convergence, such as semi coarsening and line smoothers (see [13]). We will consider the RILU algorithm implemented in Diffpack. RILU does not depend on a priori knowledge of the direction of the anisotropy.

We start with SSOR as the smoother and test different $k_0$, the input file is anis1.i, where we have

```
set k0 = {0.001 &  0.01 & 1 &  100 & 1000 }
```

```
unix>./app --class Poi2anis < Verify/anis1.i | grep conv
   solver converged in 285 iterations
   solver converged in  88 iterations
   solver converged in   6 iterations
   solver converged in  50 iterations
   solver converged in 125 iterations
```

We see that the performance deteriorates dramatically as we change the parameter $k_0$ away from unity. We redo the simulation using RILU as smoother.

```
unix>./app --class Poi2anis < Verify/anis2.i | grep conv
   solver converged in   4 iterations
   solver converged in   7 iterations
   solver converged in   5 iterations
   solver converged in   6 iterations
   solver converged in   3 iterations
```

The (R)ILU smoother seems to be stable with respect to the parameter $k_0$. This does not necessarily apply to anisotropy in 3D, but neither do line smoothers or semi-coarsening techniques.

## 3.7    Systems of Equations, Linear Elasticity

We want to extend the code to systems of equations. Take for example the equations of linear thermo-elasticity described in Chapter 5 in [9]:

$$\nabla(\lambda + \mu)\nabla \cdot \mathbf{u} + \nabla \cdot [\mu \nabla \mathbf{u}] = \nabla[\alpha(3\lambda + 2\mu)(T - T_0)] - \rho \mathbf{b}. \tag{3.12}$$

The unknown **u** is the displacement, and the other quantities are known. Instead of $\lambda$ and $\mu$ the Young's elasticity module E and the Poisson ration $\nu$ are often used. The relations between these parameters are given by

$$\lambda = \frac{E\nu}{(1+\nu)(1-2\nu)},$$

$$\mu = \frac{E}{2+2\nu}.$$

A good book about the theory of FEM, multigrid, and elasticity is [1].

### 3.7.1    Extending Elasticity1 with Multigrid

We now extend the `Elasticity1` simulator with multigrid. That is, we repeat the steps described in Section 3.3.1. In addition we also redefine boundary indicators and material types in the same way as the original code on all the grids.

```
int i;
for (i=1; i<= no_of_grids; i++) {
  grid.rebind(mgtools->getGrid(i))
  String redef  = menu.get ("redefine boundary indicators");
  if (!redef.contains("NONE"))  grid->redefineBoInds (redef);
  String addbn  = menu.get ("add boundary nodes");
  if (!addbn.contains("NONE"))  grid->addBoIndNodes (addbn, ON);
  addbn         = menu.get ("remove boundary nodes");
  if (!addbn.contains("NONE"))  grid->addBoIndNodes (addbn, OFF);
  String addmat = menu.get ("add material");
  if (!addmat.contains("NONE")) grid->addMaterial (addmat);
}
```

In `Elasticity1` the vector field of unknowns and the degrees of freedom are made by,

```
u.rebind (new FieldsFE (*grid, "u"));
dof.rebind (new DegFreeFE (*grid, nsd));
```

A vector field with `nsd` unknowns per node is made over the grid and the `DegFreeFE` is made accordingly. In `MGtools` the fields of unknowns are not strictly needed. If it is needed, e.g., when using non-isoparametric elements (see [12]), a `FieldCollector` object is used in the initialization of all the `DegFreeFE` objects. In this simulator we only use isoparametric elements and the `DegFreeFE` objects are constructed based on the grids on different levels and a parameter `nsd`, which is initialized with the following item in the input file.

```
set degrees of freedom per node = d
```

The value `d` should here be equal to the number of space dimensions for the `Elasticity1MG` simulator.

Our next task is to test the efficiency of multigrid on the linear elasticity problem.

*The Number of Iterations vs. Number of Unknowns.* We redo some of the test examples that follow the `Elasticity` simulator, but now with multigrid. We use the file `arch1.i` as the first test and investigate the behavior of multigrid compared with the RILU preconditioner. The `arch1.i` file is edited such that we have the following items,

```
set gridfile = P=PreproStdGeom | DISK_WITH_HOLE a=1 b=3 degrees=90 \
   | d=2 e=ElmB4n2D {[24,12] & [48,24] & [96,48] & [192,96]    \
   & [384,192]} [1,1.3]
set nu format = CONSTANT=0.25
...
set preconditioning type = PrecRILU
```

The corresponding multigrid input-file is `arch1mg.i`

```
set preconditioning type = PrecML
...
sub MGtools
...
set no of grid levels = {2 & 3 & 4 & 5 & 6 & 7}
set gridfile = P=PreproStdGeom | DISK_WITH_HOLE a=1 b=3 degrees=90 \
   | d=2 e=ElmB4n2D [12,6] [1,1.3]
```

In Table 3.3 the different methods are compared with respect to the CPU-time and the number of iterations, needed to satisfy a given convergence criterion. We see that the performance of CG/RILU deteriorates as the grid is refined, which is as we expected. In contrast both multigrid methods are pretty stable and converge in roughly the same number of iterations, independent of the number of unknowns. On the finest mesh, the fastest method, CG/MG is 15 times faster than the slowest, CG/RILU. We needed to use two pre- and post-smoothing sweeps to achieve convergence for the MG method, while CG/MG used only one pre- and post-smoothing sweeps. In this experiment the files `arch1mg.i` and `arch2mg.i` are used as input files for the multigrid simulator.

**Table 3.3.** Comparison of multigrid and RILU as preconditioners for CG, and plain multigrid. We used convergence criterion $\frac{\|r_k\|}{\|r_0\|} \leq 10^{-6}$, zero start vector, linear elements, $E = 100$, and $\nu = 0.25$.

|  | CG/RILU | | CG/MG | | MG | |
|---|---|---|---|---|---|---|
| unknowns | Time | # it | Time | # it | Time | # it |
| 650 | 0.01 | 38 | 0.06 | 24 | 0.06 | 19 |
| 2450 | 0.26 | 77 | 0.29 | 30 | 0.38 | 24 |
| 9506 | 2.34 | 160 | 1.33 | 33 | 2.11 | 26 |
| 37442 | 20.6 | 339 | 5.82 | 35 | 9.50 | 27 |
| 148610 | 173.2 | 731 | 24.84 | 37 | 39.1 | 27 |
| 592130 | 1516.4 | 1595 | 105.5 | 39 | 157.4 | 27 |

### 3.7.2    The λ-dependency

Some materials, like rubber, are nearly incompressible, it takes a lot of energy to make changes in the density. Such effects are characterized by large differences in the Lame constants,

$$\lambda \gg \mu.$$

When $\nu < 1/4$, then $\lambda < \mu$ and the elliptic nature of the equation dominates. The limit $\nu \to 1/2 \Rightarrow \lambda \to \infty$ means an incompressible material and (3.12) is no longer elliptic. Standard FEM is not appropriate in this case. Instead the mixed formulation described in [11] can be used, together with the efficient precondition techniques described in [12]. Other alternatives are reduced integration of the $\nabla(\nu + \lambda)\nabla \cdot \mathbf{u}$ term or nonstandard (nonconforming) FEM. The reduced integration technique is described in Chapter 6 in [9] and it is straightforward to extend the current simulator to include this.

We have tested the efficiency vs. $\lambda$ in Table 3.4. We experience a growth in the number of iterations when $\nu$ approaches $1/2$ for all methods. However, CG/MG seems remarkably stable compared with the other methods. Here we needed to use three pre- and post-smoothing sweeps for the plain multigrid method (MG), to get convergence for $\nu$ close to zero. On the other hand the CG/MG method converges fast with only one pre- and post-smoothing sweep. We used the `locking3.i` and `locking4.i` inputfiles in this experiment.

**Table 3.4.** Number of iterations vs. $\nu$.

| $\nu$ | # it,CG/RILU | # it, MG | # it, CG/MG |
|-------|--------------|----------|-------------|
| 0.0   | 315          | 20       | 35          |
| 0.1   | 323          | 21       | 34          |
| 0.25  | 339          | 27       | 35          |
| 0.3   | 355          | 30       | 37          |
| 0.4   | 1000+        | 51       | 55          |
| 0.45  | 1000+        | 1000+    | 97          |
| 0.49  | 1000+        | 1000+    | 439         |
| 0.499 | 1000+        | 1000+    | 1000+       |

## 3.8    Nonlinear Problems

In this section we consider multilevel methods for nonlinear problems. There are two different strategies to use. First one can use multigrid as a linear solver inside a nonlinear algorithm. This can be done by using the utilities developed in the previous sections. The second alternative is a generalization of the multilevel ideas into a nonlinear algorithm, which we will consider in this section. We introduce the model problem, a nonlinear elliptic equation,

$$\mathcal{L}(u) = \nabla \cdot (k(u)\nabla u) - f(u) = 0, \text{ in } \Omega,$$
$$u = g, \text{ on } \partial\Omega. \tag{3.13}$$

Notice that a linear problem with inhomogeneous right hand side is included in this definition. The problem is then $\mathcal{L}(u) = \nabla \cdot (k\nabla u) - f = 0$[11]. We assume that the equation (3.13) has at least one solution, so that the discretization makes sense. Hence, we assume that we have a family of well-posed nonlinear algebraic problems,

$$\mathcal{L}(\mathbf{u}_l) = 0, \text{ in } \Omega_l,$$
$$\mathbf{u}_l = \mathbf{g}_l, \text{ on } \partial\Omega_l. \tag{3.14}$$

at grid levels $l$ in $1, \ldots, n$.

Nonlinear multigrid employs nonlinear smoothers. The classical smoothers, Jacobi and Gauss-Seidel, can easily be extended to nonlinear solvers. The nonlinear Gauss-Seidel algorithm is described in Algorithm 3.81.

**Algorithm 3.81.**

---

*Nonlinear Gauss-Seidel*

for $i = 1, \ldots, n$:
    solve
        POINTWISE SMOOTHER
        $\mathcal{L}(u_1, \ldots, u_i', \ldots, u_n) = f_i$,
            with respect to $u_i'$,
    $u_i = u_i'$.

---

In some cases it is possible and efficient to implement the pointwise solver analytically, but in general this is difficult. Considering Algorithm 3.81 we see that the nonlinear Gauss-Seidel method is defined pointwise. For each pointwise calculation a nonlinear equation must be solved. This can be done by, e.g., the Newton-Raphson method. This pointwise linearization is usually used (in particular for finite difference (FDM) schemes), since it is cheaper in terms of memory. It is also probably more efficient that global linearization, since $\mathcal{L}$ is updated after each pointwise calculation. This is not yet implemented in Diffpack. In Diffpack we therefore linearize the global system by either Newton-Raphson or the Piccard iteration method and use standard linear smoothers. In other words, we use a global linearization technique. This technique can be seen as a mixture of the Gauss-Seidel and the Jacobi method, since the solution vector is updated during the smoothing while the

---

[11] We will also use a $f$ on the right-hand side. The reason will become clear in the next section.

linear system is not. The difference between this nonlinear smoother and a standard linear smoother is therefore the frequency of updating the linear system.

There are two fundamentally different versions of nonlinear multigrid algorithms, the nonlinear multigrid (NLMG) and the full approximation scheme (FAS) [7]. The main difference between these methods is that in NLMG the residual is transfered to the coarser grids, while in FAS the approximate solution and a right-hand side are transfered. To understand the difference we consider the nonlinear residual equation. Let $\mathbf{u}$ be the solution, $\mathbf{v}$ is an approximation, and $\mathbf{e} = \mathbf{u} - \mathbf{v}$ is the error. The nonlinear residual equation then reads

$$\mathcal{L}(\mathbf{u}) - \mathcal{L}(\mathbf{v}) = \mathbf{f} - \mathcal{L}(\mathbf{v}) = \mathbf{r}. \tag{3.15}$$

However, for nonlinear problems

$$\mathcal{L}(\mathbf{e}) \neq \mathcal{L}(\mathbf{u}) - \mathcal{L}(\mathbf{v}) = \mathbf{r}. \tag{3.16}$$

For linear problems we have two formulations of the residual equation,

$$\mathcal{L}\mathbf{e} = \mathbf{r}, \tag{3.17}$$

or

$$\mathcal{L}\mathbf{u} = \mathcal{L}\mathbf{v} + \mathbf{r}. \tag{3.18}$$

Both can be solved on the coarser grids. In (3.18) both the approximated solution $\mathbf{v}$ and the residual $\mathbf{r}$ must be transfered. On the other hand only the residual needs to be transfered in (3.17). This formulation is therefore cheaper. For nonlinear problems these two formulations turn into two different algorithms.

The two following sections describe these algorithms.

### 3.8.1    The Nonlinear Multigrid Method

On the coarser grids we need to solve an approximated residual equation:

$$\mathcal{L}(\mathbf{v}_l) = \mathbf{g}_l, \tag{3.19}$$

where $\mathbf{g}_l = \hat{\mathbf{f}}_l - s\,\mathbf{r}_l$, the residual $\mathbf{r}_l$ is defined in (3.15) and $\hat{\mathbf{f}}_l$ is obtained from the initially computed solution $\hat{\mathbf{u}}_l$, by $\hat{\mathbf{f}}_l = \mathcal{L}(\hat{\mathbf{u}}_l)$. Notice that a new parameter $s$ is introduced. The reason is that problem (3.19) may *not* be well posed, the solution only exists in a neighborhood of the (desired)[12] solution. It may therefore be necessary to introduce a damping parameter $s$ to ensure that $s\,\mathbf{r}_l$ is relatively close to zero. Since $\mathbf{r}_l$ is multiplied by $s$ after it is transfered

---

[12] Nonlinear problems may have more than one solution. To find the desired solution one has to start with values within the domain of attraction of the desired solution

back, the solution correction is divided by $s$ before it is transfered. A suitable and often used $s$ is

$$s_l = \sigma / \|\mathbf{r}_l\|, \tag{3.20}$$

where $\sigma$ is a real number to be chosen. For the convenience of the reader we list the algorithms.

### Algorithm 3.82.

---

*NLMG (The nonlinear multigrid method)*

l, grid level parameter
$\gamma$, cycle parameter
$\sigma$, damping parameter
k, nested cycles
given $\mathbf{u}_k^0$, compute $\mathbf{f}_k^0 = \mathcal{L}(\mathbf{u}_k^0)$, for $k = 0, .., l - 1$
$\mathbf{u}_l^3 = \Phi(0, \mathbf{f}_l, l)$
    APPLY PRE SMOOTHER
    $\mathbf{u}_l^1 = \mathcal{S}_1(\mathbf{u}_l, \mathbf{f}_l)$
    CALCULATE RESIDUAL
    $\mathbf{r}_l = \mathbf{f}_l - \mathcal{L}(\mathbf{u}_l)$
    RESTRICT THE RESIDUAL
    $\mathbf{r}_{l-1} = \mathbf{P}_{l,l-1}\mathbf{r}_l$
    CALCULATE DAMPING PARAMETER
    $s = \sigma(\mathbf{r}_l)$
    CALCULATE DEFECT
    $\mathbf{d}_{l-1} = \mathbf{f}_{l-1}^0 + s\,\mathbf{r}_{l-1}$
    INIT SOLUTION VECTOR
    $\mathbf{v}_{l-1} = \mathbf{u}_l^0$
    COARSE GRID SOLVER
    for $i = 1, \ldots, \gamma$ do
        $\Phi_{l-1}(\mathbf{v}_{l-1}, \mathbf{d}_{l-1}, l - 1)$
    PROLONGATE THE SOLUTION
    $\mathbf{u}_l^2 = \mathbf{u}_l + \mathbf{P}_{l-1,l}(\mathbf{v}_{l-1} - \mathbf{u}_l^0)/s$
    APPLY POST SMOOTHER
    $\mathbf{u}_l^3 = \mathcal{S}_2(\mathbf{u}_l, \mathbf{f}_l)$

---

### 3.8.2   The Full Approximation Scheme

In the FAS algorithm 3.83, the restriction operator works on the actual solution vector (left-hand side). It is not obvious that the same projection can be used. In fact, when we are using the finite element method they *cannot*. The projection operators developed for linear multigrid are the standard $L_2$ projection operators, and they enter the algorithms as follows:

$$\mathbf{P}_{l,l-1} : \mathbf{r}_l \to \mathbf{r}_{l-1}, \ (restriction),$$
$$\mathbf{P}_{l-1,l} : \mathbf{u}_{l-1} \to \mathbf{u}_l, \ (prolongation), \qquad (3.21)$$

where $\mathbf{u}_l$ is the solution vector and $\mathbf{r}_l$ is the residual or right-hand side. They *cannot* be used the other way around:

$$\mathbf{P}_{l,l-1} : \mathbf{u}_l \not\to \mathbf{u}_{l-1}. \qquad (3.22)$$

This is needed in the FAS method. The problem is related to the finite element method, in FDM the standard restriction and prolongation can be used. In FDM, $u_i = u(x_i)$ and $f_i = f(x_i)$, where $u_i$ and $f_i$ are the vectors in the linear system, whereas $u(\cdot)$ and $f(\cdot)$ are their representation (that can be evaluated pointwise). In the finite element method the right-hand side,

$$f_i^l = \int_\Omega f(x) N_i^l(x) d\Omega_l \ \approx \ f(x_i) V_{e(i,l)},$$

where $V_{e(i,l)}$ is the volume of element $e(i)$ on grid level $l$. Hence, $f_i$ is proportional to the element volume, which is dependent of $h$. On the other hand $u_i = u(x_i)$ and therefore independent of $h$. The restriction operator causes $b_{i,l-1} \sim b_{i,l} V_{e(i,l-1)}/V_{e(i,l)}$. Hence, this restriction can not be used, since $u_{i,l-1} \sim u_{i,l}$. In Diffpack appropriate operators for (3.22) are constructed simply by using the $L_2$ restriction operators and scaling them such that

$$u_{l-1}(x_i) = u_l(x_i), \ \text{if } u_l \text{ a constant function}.$$

If we choose $x_i$ as the nodal points in the coarse grid the scaling operator is uniquely determined.

**Algorithm 3.83.**

---

*FAS (The full approximation scheme)*

for $i = 1, \ldots, n$ do
   initialize $\mathbf{u}_l^0$ and $\mathbf{f}_l$
while not converged:
   $\mathbf{u}_l = \varPhi(\mathbf{u}_l, \mathbf{f}_l, l)$
where $\varPhi(\mathbf{u}_l, \mathbf{f}_l, l) = \mathbf{u}_l^3$
   APPLY PRE SMOOTHER
   $\mathbf{u}_l = \mathcal{S}_1(\mathbf{u}_l, \mathbf{f}_l)$
   RESTRICT SOLUTION
   $\mathbf{u}_{l-1} = \mathbf{P}_{l,l-1}\mathbf{u}_l$
   CALCULATE RESIDUAL
   $\mathbf{r}_l = \mathbf{f}_l - \mathcal{L}(\mathbf{u}_l)$
   RESTRICT THE RESIDUAL
   $\mathbf{r}_{l-1} = \mathbf{P}_{l,l-1}\mathbf{r}_l$
   CALCULATE DEFECT
   $\mathbf{d}_{l-1} = \mathcal{L}(\mathbf{u}_{l-1}) + \mathbf{r}_{l-1}$
   INIT COARSE START VECTOR
   $\mathbf{v}_{l-1} = \mathbf{u}_{l-1}$
   COARSE GRID SOLVER
   for $i = 1, \ldots, m$ do
      $\varPhi_{l-1}(\mathbf{u}_{l-1}, \mathbf{d}_{l-1}, l-1)$
   PROLONGATE THE SOLUTION
   $\mathbf{u}_l = \mathbf{u}_l + P_{l-1,l}(\mathbf{v}_{l-1} - \mathbf{u}_{l-1})$
   APPLY POST SMOOTHER
   $\mathbf{u}_l = \mathcal{S}_2(\mathbf{u}_l, \mathbf{f}_l)$

---

In Diffpack there are four different nonlinear multigrid solvers, based on the NLMG and FAS.

- NonlinearMultigrid, NLMG with $s = 1$
- NonlinearDampedMultigrid, NLMG
- NestedFASMultigrid, FAS
- FASMultigrid, FAS, $u_l^0 = 0$

### 3.8.3  Software Tools for Nonlinear Multigrid

Diffpack has a toolbox similar to `MGtools` for nonlinear multigrid methods. It is called `NonLinMGtools`. We will now describe in details how the simulator `NlElliptic` can be extended to the nonlinear multigrid solver `NlEllipticMG`.

1. Make a new directory `NlEllipticMG` with this command:

```
Mkdir -md NlEllipticMG
```

2. Move to the NlEllipticMG directory, take a copy of the NlElliptic files and change the filenames to NlEllipticMG:

```
Mvname NlElliptic NlEllipticMG NlElliptic.*
```

Also substitute NlElliptic by NlEllipticMG in all files:

```
perl -pi.old~~ -e 's/NlElliptic/NlEllipticMG/g;' *.h *.cpp
```

3. The following modifications is necessary in the header file.

   (a) Include the NonLinMGtools toolbox:

   ```
   #include <NonLinMGtools.h>
   ```

   (b) The simulator should be extended with the following data structures:

   ```
   Handle(NonLinMGtools) nlmgtools;
   int                   no_of_grids;
   ```

   (c) Additionally, some common data structures are used. These data structures are probably already in the header file (with different names).

   ```
   Handle(GridFE) grid;   // finite element grid
   Handle(DegFreeFE) dof; // mapping: nodal values <-> unknowns
   Handle(FieldFE) u;     // finite element field

   Handle(LinEqVector)     nonlin_solution;// solution
   Handle(LinEqVector)     linear_solution;// linear solution
   Handle(NonLinEqSolver) nlsolver;        // nonlinear solver
   Handle(NonLinEqSolver) current_nlsolver; // nonlinear solver
   ```

4. In NlEllipticMG::define the toolbox defines its own menu interface with a standard defineStatic function,

   ```
   NonLinMGtools:: defineStatic ( menu, level+1);
   ```

5. Create the NonLinMGtools object

   ```
   nlmgtools.rebind( new NonLinMGtools()) ;
   nlmgtools->attachFEM(*this);
   nlmgtools->attachNonLinEqSolverUDC(*this);
   nlmgtools->scan ( menu );
   nlmgtools->initStructure();
   nlmgtools->print(s_o);
   ```

NonLinMGtools needs access to the additional NonLinEqSolverUDC and FEM classes, from which NlEllipticMG is derived. Class NonLinEqSolverUDC is the standard interface for nonlinear problems and provides the function makeAndSolveLinearSystem, whereas FEM is attached because it provides makeSystem which is used to calculate the residual. The reader should consult the source code in Section 3.8.4 to make sure that these default implementations are appropriate.

6. We fetch and attach the data structures on the finest grid,

```
no_of_grids = nlmgtools->getNoOfSpaces();
nlsolver.rebind (nlmgtools->getNonLinMLSolver());
u.rebind( nlmgtools->getField(no_of_grids));
grid.rebind ( nlmgtools->getGrid ( no_of_grids ) ) ;
dof.rebind ( nlmgtools->getDof( no_of_grids ));
linear_solution.rebind (
    nlmgtools->getLinSmootherSolution(no_of_grids));
```

7. The nonlinear solution vector must be made and attached,

```
Handle(Vec(NUMT)) nl = new Vec(NUMT)(dof->getTotalNoDof());
nonlin_solution.rebind ( new LinEqVector(*nl));
nlmgtools->attachNonLinMLSolution(nonlin_solution());
```

8. The `fillEssBC` routine is replaced with a `mgFillEssBC` as described for linear multigrid.

9. The `integrands` function must fetch the field on the current grid level from `NonLinMGtools`,

```
int level = nlmgtools->getCurrentGridLevel();
u.rebind(nlmgtools->getField(level));
const real u_pt = u->valueFEM (fe);     // U (at present point)
```

10. In `solveProblem` we might want to insert a start vector since we need a start vector within the domain of attraction of the desired solution,

```
nlmgtools->initStartVector(a); //a is a NUMT
```

or

```
nlmgtools->initStartVector(field,level);
```

or a random start vector,

```
nlmgtools->initRandomStartVector();
```

where `field` is a `FieldFE` object and `level` is an integer indicating the grid level.

11. The `makeAndSolveLinearSystem` routine is implemented almost as usual. However, we need to fetch the data structures from `NonLinMGtools`, and additionally the right hand side is modified. An appropriate routine is as follows:

```
//fetch stuff
int level = nlmgtools->getCurrentGridLevel();
if (level == 0 ) level = no_of_grids;
DegFreeFE& dof = nlmgtools->getDof(level);
LinEqAdmFE& lineq = nlmgtools->getLinEqAdmFE(level);
FieldFE& field = nlmgtools->field(level)();
LinEqVector& ns = nlmgtools->getNonLinSmootherSolution(level);
LinEqVector& ls = nlmgtools->getLinSmootherSolution(level);
current_nlsolver.rebind ( nlmgtools->getNonLinEqSolver(level));
```

```
// copy most recent guess
dof.vec2field (ns.getVec(), field);

makeSystem (dof, lineq);
lineq.attach(ls);
//This is different from a standard makeAndSolverLinearSystem:
lineq.bl().add(lineq.bl(), nlmgtools->getLinRhs(level));

// init startvector (linear_solution) for iterative solver:
ls = ns;

lineq.solve();  // invoke a linear system solver
```

The complete source code is in

> src/multigrid/NlEllipticMG/NlEllipticMG.cpp

### 3.8.4    Default Implementation of Some Components

It is also useful to see the default implementation of some of the methods in NonLinMGtools, because it can be customized and optimized in a particular application. One example may be that there is no need to update the whole system, because parts of the matrix or the right-hand side are linear. These default functions may also be inappropriate for some applications where the standard makeSystem function is not used. One can then derive a subclass from NonLinMGtools and rewrite the functions, solveSubSystem, residual and matVec.

The smoother, $\mathbf{u}_l = \mathcal{S}_1(\mathbf{u}_l, \mathbf{f}_l)$, is implemented as

```
bool NonLinMGtools::solveSubSystem
  (LinEqVector& b,
   LinEqVector& x,
   int grid_level,
   StartVectorMode start,
   MLSolverMode)
{
  current_grid_level = grid_level;

  nonlin_solution(grid_level).rebind(x);
  nlsolver(grid_level)->attachNonLinSol(nonlin_solution(grid_level)());

  if (start==ZERO_START) nonlin_solution(grid_level)() = 0.0;
  dof(grid_level)->insertEssBC(nonlin_solution(grid_level)().getVec());

  linear_rhs(grid_level).rebind(b);

  nlsolver(grid_level)->solve();

  return true;
}
```

It is simply a matter of getting the right data structures and filling the boundary conditions before the call to the nonlinear smoother. The nonlinear smoother `nlsolver` will then call `makeAndSolverLinearSystem` in the simulator class inherited from `NonLinEqSolverUDC`.

Notice that when we write $\mathcal{L}(u)$ we include a possible inhomogeneous right-hand side, $\mathcal{L}(u) = \mathbf{A}(\mathbf{u})\mathbf{u} - \mathbf{b} = 0$. The residual of a linear system in Diffpack is defined as $\mathbf{r} = \mathbf{b} - \mathbf{A}(\mathbf{u})\mathbf{u}$, hence $-\mathcal{L}(u)$ is the residual. The function is implemented as `matVec`:

```
void NonLinMGtools::matVec
  (const LinEqVector& f, LinEqVector& x, int grid_level)
{
  current_grid_level = grid_level;

  dof(grid_level)->vec2field(x.getVec(), field(grid_level)());
  fem->makeSystem(dof(grid_level)(), lineq(grid_level)());

  lineq(grid_level)->getLinEqSystem().attach((LinEqVector&)x);
  lineq(grid_level)->getLinEqSystem().residual((LinEqVector&)f);
  Vec(NUMT)& ff = CAST_REF(f.vec(), Vec(NUMT) );
  ff.mult(-1.0);
}
```

The nonlinear multigrid algorithms also need to calculate the residual related to a particular right-hand side $\mathbf{f}_l$, on a given grid level, $\mathbf{r}_l = \mathbf{f}_l - \mathcal{L}(\mathbf{u}_l) = \mathbf{f}_l + \mathbf{b}_l - \mathbf{A}(\mathbf{u}_l)\mathbf{u}_l$.

```
void NonLinMGtools::residual
  (LinEqVector& b,
   LinEqVector& x,
   LinEqVector& r,
   int grid_level)
{
  current_grid_level = grid_level;
  dof(grid_level)->vec2field(x.getVec(), field(grid_level)());
  fem->makeSystem( dof(grid_level)(), lineq(grid_level)());
  lineq(grid_level)->getLinEqSystem().attach(x);
  lineq(grid_level)->getLinEqSystem().residual(r);
  r.add(r,b);
}
```

### 3.8.5    Experiments with Nonlinear Multigrid

*Model Problem 1.* We define two different test cases: (1) Standard Poisson equation and (2) The Bratu problem

$$
\begin{aligned}
1) \ -\Delta u &= 1 \quad \text{in } \Omega, \\
2) \ -\Delta u &= ce^u \text{ in } \Omega, \\
u &= 0 \quad \text{on } \partial\Omega, \text{ for all test cases}.
\end{aligned}
\tag{3.23}
$$

In the main program the user is able to choose between the three different problems, by the command line argument --case. Only the SuccessiveSubst linearization technique has been implemented. The linear system arising from NewtonRaphson linearization is more complicated and it is not clear that the smoothers will work.

*Setting up an Input File.* The nonlinear multilevel methods have a comprehensive set of parameters that can be varied and tuned through the menu systems interface.

The nonlinear multigrid method, which is named NonLinML and has a usual NonLinEqSolver_prm interface:

```
sub NonLinEqSolver_prm
set nonlinear iteration method = NonLinML
ok
```

The multilevel algorithm and its associated parameters are controlled with

```
sub MLSolver_prm
set multilevel method =    NonlinearDampedMultigrid
set sigma            =    1.0
ok
```

This interface is described earlier except for the sigma parameter, which is related to the damping parameter $s$ in Algorithm 3.82 by the relation,

$$s = \sigma/\|\mathbf{r}\| . \tag{3.24}$$

The smoother is also nonlinear and a linearization strategy and its associated parameters must be chosen.

```
sub smoother NonLinEqSolver_prm
set smoother nonlinear iteration method = NewtonRaphson
set smoother max nonlinear iterations =1
ok
```

The additional parameters of the linear solvers are controlled through a regular LinEqSolver_prm interface.

```
sub smoother LinEqSolver_prm
set smoother basic method = SSOR
set smoother max iterations = 1
ok
```

These are essentially all the parameters relevant for the linear multigrid. In nonlinear multigrid we have some additional features, which are included simply because the behavior of multigrid on nonlinear problems is much less understood. Also since the smoothers are already nonlinear, it seems tempting to include the Krylov family of nonlinear iterative methods with preconditioners. Hence, we have the following additional menus:

```
sub smoother Precond_prm
set smoother preconditioning type = PrecNone
```

One can, e.g, use the Conjugate Gradient method with a RILU preconditioner. A convergence criterion is chosen with:

```
sub smoother ConvMonitorList_prm
sub smoother Define ConvMonitor #1
set smoother #1: convergence monitor name = CMRelResidual
ok
ok
```

The same parameters are needed by the coarse grid solver.

```
sub coarse grid NonLinEqSolver_prm
set coarse grid nonlinear iteration method = SuccessiveSubst
ok
sub coarse grid LinEqAdmFE
sub coarse grid Matrix_prm
...
ok
sub coarse grid LinEqSolver_prm
set coarse grid basic method = GaussElim
ok
...
ok
```

### 3.8.6   Nonlinear Multigrid for a Linear Problem

The first test is to apply the nonlinear multigrid methods on the simplest linear problem, which we considered in Section 3.5. The NLMG method, described in Algorithm 3.82, is a generalization of the nested multigrid method, with a damping parameter $s$. The damping has no effect on the solution of the linear system, since we multiply the residual with $s$ and divide the solution correction by $s$.

We compare the efficiency of the different algorithms. In linear.i

```
unix>../app --case 1 < linear2.i | \
    grep "NonLinEq iter\|Method"
Method Multigrid
NonLinEq iter 5 (40)
Method NestedMultigrid
NonLinEq iter 5 (40)
Method NonlinearMultigrid
NonLinEq iter 4 (40)
Method NonlinearDampedMultigrid
NonLinEq iter 4 (40)
Method FASMultigrid
NonLinEq iter 5 (40)
```

The different algorithms have similar behavior for the linear problem.

### 3.8.7   Nonlinear Multigrid Applied to a Non-Linear Problem

The standard test example for nonlinear multigrid methods seems to be the Bratu problem (see [8,13]). This problem is known to have two solutions for $c < c^* \approx 6.8$, for $c = c^*$ there is a turning point with only one solution. If $c > c^*$ then there is no solution. The first solution with $c = 1$, has a maximum in $u(0.5, 0.5) \approx 0.078$, this solution is easily obtained. The other solution with a maximum $u(0.5, 0.5) \approx 7$ is hard to get, but can be obtained with a continuation method described in [13].

*Number of Iterations vs. Number of Nodes.* We now check the efficiency with respect to the number of nodes. We test `NonlinearDampedMultigrid` first.

```
unix>app --case 2 < Verify/noit.i | \
     grep 'Number of nodes\|NonLinEq iter'
Number of nodes 25
NonLinEq iter 5 (40)
Number of nodes 81
NonLinEq iter 5 (40)
Number of nodes 289
NonLinEq iter 6 (40)
Number of nodes 1089
NonLinEq iter 6 (40)
Number of nodes 4225
NonLinEq iter 5 (40)
Number of nodes 16641
NonLinEq iter 5 (40)
```

We run the same simulation with the FAS multigrid method,

```
unix>app --case 2 < Verify/noit2.i | grep 'Number of nodes\|NonLinEq iter
Number of nodes 25
NonLinEq iter 5 (40)
Number of nodes 81
NonLinEq iter 7 (40)
Number of nodes 289
NonLinEq iter 7 (40)
Number of nodes 1089
NonLinEq iter 7 (40)
Number of nodes 4225
NonLinEq iter 8 (40)
Number of nodes 16641
NonLinEq iter 7 (40)
```

Both methods converge in a number of iterations which is independent of the number of unknowns. The number of unknowns range from 25 to 16641 and the CPU time increase linearly (except from on the smallest grids).

*Stability Near the Turning Point.* Next we test the stability of the methods near the turning point $c = 6.8$. We use the `NonlinearDampedMultigrid` method. We set

```
set c = { 1 & 3 & 5 & 6 & 6.5 & 6.6 & 6.7 &  6.8 }
```

In the input file, `turning.i`.

```
unix> ./app --case 2 < Verify/turning | grep "NonLinEq iter"
NonLinEq iter 5 (80)
NonLinEq iter 6 (80)
NonLinEq iter 7 (80)
NonLinEq iter 10 (80)
NonLinEq iter 15 (80)
NonLinEq iter 19 (80)
NonLinEq iter 26 (80)
NonLinEq iter 81 (80)
```

The convergence gets slower as $c$ increases towards the turning point, and at $c = 6.8$ it does not converge in 80 iterations. The results with `FASMultigrid`, tested in `turning2.i`, are nearly the same.

*Accuracy of The Coarse Grid Solver.* For the linear multigrid we saw that the accuracy of the coarse grid solver was important. We therefore test how accurate the coarse grid solver has to be in the nonlinear case. This test is implemented in `coarse.i - coarse4.i`. We compare one nonlinear iteration, using one SSOR sweep on the linearized system, with an accurate coarse grid solver. We first test with $c = 1$,

```
unix> app --case 2 < Verify/coarse.i  | grep 'NonLinEq iter'
NonLinEq iter 5 (20)
NonLinEq iter 5 (20)
```

Both methods use the same number of iterations. The first result listed is with an accurate coarse grid solver. Also in the much harder case, with $c = 6.7$, the convergence is almost independent of the accuracy of the coarse grid solver,

```
unix> app --case 2 < Verify/coarse.i  | grep 'NonLinEq iter'
NonLinEq iter 25 (80)
NonLinEq iter 27 (80)
```

In both these cases we have used a very coarse grid, $2 \times 2$.

*SSOR vs. Conjugate Gradients.* Since we are using nonlinear iterative methods as smoothers, we might also want to try the nonlinear family of Krylov methods. We can for instance use a Conjugate Gradient method with SSOR or RILU as preconditioners. This is tested in the files `ssor.i`, `conjgrad.i`, and `conjgrad2.i`. All smoothers are comparable for $c = 1$, but when the nonlinear problem gets harder at $c = 6.7$, the preconditioned Conjugate Gradient method is better. The fastest method, Conjugate Gradient with RILU preconditioner, is in nearly 1.5 faster, in terms of CPU time, than plain SSOR smoothing.

*Continuation and FAS.* For many nonlinear problems it is necessary to use some kind of continuation to obtain convergence (see e.g. [9,13]). The FAS algorithms have an obvious advantage because the solution is available on each grid, and it is straightforward to use continuation within the multigrid algorithm. This means that most steps in the continuation method is done on the coarser grid. This can potentially speed up the continuation method significantly. For the NLMG algorithm the solution is only available on the fine grid. Only the equation for the error correction is known at coarser grids. It is therefore more difficult to design a continuation method that take advantage of intermediate steps on the coarser grids.

### Acknowledgements

The authors are grateful to Professor R. Winther for many useful discussions.

# References

1. D. Braess. *Finite elements; Theory, fast solvers, and applications in solid mechanics.* Cambridge University Press, 1997.
2. J. H. Bramble. *Multigrid Methods*, volume 294 of *Pitman Research Notes in Mathematical Sciences.* Longman Scientific & Technical, Essex, England, 1993.
3. J. H. Bramble, J. E. Pasciak, J. Wang, and J. Xu. Convergence estimates for multigrid algorithms without regularity assumptions. *Math. Comp.*, 57:23–45, 1991.
4. W. L. Briggs, V. E. Henson, and S. F. McCormick. *A Multigrid Tutorial.* SIAM Books, 2nd edition, 1996.
5. A. M. Bruaset. *A Survey of Preconditioned Iterative Methods.* Addison-Wesley Pitman, 1995.
6. A. M. Bruaset, H. P. Langtangen, and G. W. Zumbusch. Domain decomposition and multilevel methods in Diffpack. In P. Bjørstad, M. Espedal, and D. Keyes, editors, *Proceedings of the 9th Conference on Domain Decomposition.* Wiley, 1997.
7. W. Hackbusch. *Multi–Grid Methods and Applications.* Springer–Verlag, Berlin, 1985.
8. W. Hackbusch. *Iterative Solution of Large Sparse Systems of Equations.* Springer-Verlag, 1994.
9. H. P. Langtangen. Tips and frequently asked questions about Diffpack. Numerical Objects Report Series, Numerical Objects A.S., 1999. See URL *http://www.diffpack.com/products/faq/faq_main.html.*
10. H. P. Langtangen. *Computational Partial Differential Equations - Numerical Methods and Diffpack Programming.* Textbooks in Computational Science and Engineering. Springer, 2nd edition, 2003.
11. K.-A. Mardal and H. P. Langtangen. Mixed finite elements. In H. P. Langtangen and A. Tveito, editors, *Advanced Topics in Computational Partial Differential Equations – Numerical Methods and Diffpack Programming.* Springer, 2003.

12. K.-A. Mardal, J. Sundnes, H. P. Langtangen, and A. Tveito. Systems of PDEs and block preconditioning. In H. P. Langtangen and A. Tveito, editors, *Advanced Topics in Computational Partial Differential Equations – Numerical Methods and Diffpack Programming*. Springer, 2003.
13. U. Trottenberg, C. Oosterlee, and A. Schuller. *Multigrid*. Academic Press, 2001.
14. Xuejun Zhang. Multilevel Schwarz methods. *Numer. Math.*, 63(4):521–539, 1992.
15. G. W. Zumbush. Multigrid methods in Diffpack. *Technical report*, 1996.

# Chapter 4

# Mixed Finite Elements

K.-A. Mardal[1,2] and H. P. Langtangen[1,2]

[1] Simula Research Laboratory
[2] Department of Informatics, University of Oslo

**Abstract.** This chapter explains the need for mixed finite element methods and the algorithmic ingredients of this discretization approach. Various Diffpack tools for easy programming of mixed methods on unstructured grids in 2D and 3D are described. As model problems for exemplifying the formulation and implementation of mixed finite elements we address the Stokes problem for creeping viscous flow and the system formulation of the Poisson equation. Efficient solution of the linear systems arising from mixed finite elements is treated in the chapter on block preconditioning.

## 4.1 Introduction

In this chapter we will study two fundamental mathematical models in physics and engineering: the Stokes problem for slow (creeping) incompressible viscous fluid flow and the Poisson equation for, e.g., inviscid fluid flow, heat conduction, porous media flow, and electrostatics. The Stokes problem cannot be discretized by a straightforward Galerkin method as this method turns out to be unstable in the sense of giving non-physical oscillations in the pressure. Mixed finite element methods, however, result in a stable solution. In this chapter we use the term mixed finite elements, when the different physical unknowns utilize different finite element basis functions. In the Stokes problem, one example is quadratic elements for the velocity and linear elements for the pressure. The Poisson equation, on the other hand, does not need mixed finite element methods for a stable solution. However, if the main interest regards the gradient of the solution of the Poisson equation, the mixed formulation seems more natural because the gradient is one of the unknowns.

In the past, two particular difficulties have prevented widespread application of mixed finite element methods: (i) lack of convenient flexible implementations of the methods on general unstructured grids and (ii) the difficulty of constructing efficient solvers for the resulting linear systems. This chapter pays attention to the first issue, while a companion chapter [15] deals with the second. Actually, the methods and software from this chapter have applications far beyond the two model problems on which this exposition is focused. To the authors knowledge, Diffpack is at the time of writing the only software package that supports easy and flexible programming with mixed finite element methods on unstructured grids, coupled with state-of-the-art iterative schemes, like multigrid, for optimal solution of the linear systems [15].

## 4.2 Model Problems

### 4.2.1 The Stokes Problem

The Stokes problem can be formulated as follows:

$$-\mu \nabla^2 u + \nabla p = f \text{ in } \Omega, \quad \text{(equation of motion)}, \tag{4.1}$$

$$\nabla \cdot u = 0 \text{ in } \Omega, \quad \text{(mass conservation)}, \tag{4.2}$$

$$u = h \text{ on } \partial\Omega_E, \tag{4.3}$$

$$\frac{\partial u}{\partial n} + pn = 0 \text{ on } \partial\Omega_N. \tag{4.4}$$

Here, $u$ is the velocity of the fluid, $p$ is the pressure in the fluid, $f$ represents body forces and $n$ is the unit normal vector pointing out of $\Omega$. The boundary $\partial\Omega = \partial\Omega_E \cup \partial\Omega_N$, where $\partial\Omega_E$ and $\partial\Omega_N$ are the parts with essential and natural boundary conditions, respectively. The Stokes equation is the stationary linearized form of the Navier-Stokes equations and describes the creeping (low Reynolds number) flow of an incompressible Newtonian fluid.

### 4.2.2 The Mixed Poisson Problem

The Poisson equation

$$-\nabla \cdot (\lambda \nabla p) = g \text{ in } \Omega, \tag{4.5}$$

$$p = h \text{ on } \partial\Omega_E, \tag{4.6}$$

$$\frac{\partial p}{\partial n} = k \text{ on } \partial\Omega_N, \tag{4.7}$$

appears in many physical contexts. The boundary $\partial\Omega = \partial\Omega_E \cup \partial\Omega_N$, where $\partial\Omega_E$ and $\partial\Omega_N$ are the parts with essential and natural boundary conditions, respectively. The $\lambda$ is here assumed to be a scalar function such that $\lambda_{min} \leq \lambda(x) \leq \lambda_{max}$, where $\lambda_{min}$ and $\lambda_{max}$ are positive real numbers and $x \in \Omega$. One interpretation regards porous media flow, where (4.5) arises from a conservation equation combined with Newton's second law. The mass conservation equation reads $\nabla \cdot u = g$, where $u$ is some flux and $g$ denotes the injection or production through wells in groundwater flow or oil recovery. Newton's second law can be expressed by Darcy's law $u = -\lambda \nabla p$, where $\lambda$ denotes the mobility. This equation can be established as an average of the equation of motion in Stokes problem over a large number of pores. In porous media flow we are primarily interested in $u$, which is usually computed by solving the Poisson equation and computing $u = -\lambda \nabla p$ numerically. The numerical differentiation implies a loss of accuracy. The mixed formulation of the Poisson equation allows us to approximate the velocity $u$ as one of the unknowns. When applying mixed finite element methods the Poisson equation is reformulated as a system of partial differential equations (PDEs):

$$\frac{1}{\lambda}u + \nabla p = 0 \text{ in } \Omega \quad \text{(Darcy's law)}, \tag{4.8}$$

$$\nabla \cdot u = g \text{ in } \Omega \quad \text{(mass conservation)}, \tag{4.9}$$

$$p = h \text{ on } \partial\Omega_N, \tag{4.10}$$

$$\frac{\partial p}{\partial n} = u \cdot n = k \text{ on } \partial\Omega_E. \tag{4.11}$$

Notice that the essential boundary conditions in (4.6) appear as natural conditions in this formulation, while the natural conditions in (4.7) are essential.

Although the Stokes problem and the Poisson equation have seemingly similar mathematical structure, they require different types of mixed finite element methods. Knowing how to solve these two classes of problems should provide sufficient information to solve a wide range of systems of PDEs by mixed finite element methods.

The present chapter is organized as follows. In Section 4.3, we present the basic theory of mixed systems in an abstract setting. This abstract theory will introduce the Babuska-Brezzi conditions, i.e., the conditions that the mixed elements should meet. In Section 4.4, we present finite element spaces appropriate for solving our model problems and the corresponding software tools. We present the implementation of the simulators in Sections 4.5 and 4.6. Efficient iterative schemes, with multigrid, are described in [15]. The mixed finite element method is analyzed in [3,9].

## 4.3   Mixed Formulation

In this section we shall derive the finite element equations for our two model problems. First, we apply the weighted residual method [9] to the systems of PDEs and derive the resulting discrete equations. Thereafter, we present continuous mixed variational formulations, which can be discretized by introducing appropriate finite-dimensional function spaces [9]. The discretization of our problems needs to fulfill the discrete version of the Babuska-Brezzi conditions, which motivates the special finite elements presented in Section 4.4.

### 4.3.1   Weighted Residual Methods

*The Stokes Problem.* The starting point of a weighted residual formulation is the representation of the unknown scalar fields in terms of sums of finite element basis functions. In the Stokes problem we need to use different basis functions for the velocity components and the pressure. Hence, we may write

$$\boldsymbol{u} \approx \hat{\boldsymbol{v}} = \sum_{r=1}^{d} \sum_{j=1}^{n_v} v_j^r \boldsymbol{e}^r N_j, \qquad (4.12)$$

$$p \approx \hat{p} = \sum_{j=1}^{n_p} p_j L_j . \qquad (4.13)$$

We use $\boldsymbol{u}$ for the continuous vector field, while $\boldsymbol{v}$ is the vector consisting of the velocity values at the nodal points. The $N_j$ and $L_j$ denote the j-th basis functions for the velocity and the pressure, respectively. The $\{\boldsymbol{e}^r\}$ are the unit vectors in Cartesian coordinates and $d$ is the number of space dimensions. The number of nodes for the velocity and the pressure are $n_v$ and $n_p$, respectively. Notice that the nodal points of the velocity and pressure fields do not necessarily coincide. The unknowns $\{v_j^r\}$ and $\{p_j\}$ are represented as vectors,

$$\boldsymbol{v} = [v_1^1, v_2^1, \ldots, v_{n_v}^1, v_1^2, \ldots, v_{n_v}^2, \ldots, v_1^d, \ldots, v_{n_v}^d], \qquad (4.14)$$

$$\boldsymbol{p} = [p_1, p_2, \ldots, p_{n_p}] . \qquad (4.15)$$

The numbering used in (4.14) and (4.15) is only one of many possible numberings, which are described in Section 4.4.4.

Inserting the approximations $\hat{\boldsymbol{v}}$ and $\hat{p}$ in the equation of motion and the equation of continuity results in a residual since neither $\hat{\boldsymbol{v}}$ nor $\hat{p}$ are in general exact solutions of the PDE system. The idea of the weighted residual method is to force the residuals to be zero in a weighted mean. Galerkin's method is a version of the weighted residual method where the weighting functions are the same as the basis functions $N_i$ and $L_i$. Application of Galerkin's method in the present example consists of using $N_i$ as weighting function for the equation of motion and $L_i$ as weighting function for the equation of continuity. In this way we generate $dn_v + n_p$ equations for the $dn_v + n_p$ unknowns in $d$ space dimensions. The second-order derivative term in the equation of motion is integrated by parts. This is also done with the pressure gradient term $\nabla \hat{p}$. The resulting weighted residual or discrete weak form is as follows:

$$\int_\Omega \left( \mu \nabla \hat{v}^r \cdot \nabla N_i - \frac{\partial N_i}{\partial x^r} \hat{p} \right) d\Omega = \int_\Omega f^r N_i \, d\Omega, \quad r = 1, \ldots, d, \quad (4.16)$$

$$\int_\Omega L_i \nabla \cdot \hat{\boldsymbol{v}} \, d\Omega = 0 . \qquad (4.17)$$

Notice that we have assumed open boundary conditions, $\frac{\partial \boldsymbol{v}}{\partial n} + p\boldsymbol{n} = 0$. The notation $\hat{v}^r$ means the r-th component of $\hat{\boldsymbol{v}}$,

$$\hat{v}^r = \hat{\boldsymbol{v}} \cdot \boldsymbol{e}^r = \sum_i v_i^r N_i, \qquad (4.18)$$

with a similar interpretation for the other quantities with superscript $r$.

Inserting the finite element expressions for $\hat{v}$ and $\hat{p}$ in (4.16)–(4.17), gives the following linear system:

$$\sum_{j=1}^{n_v} A_{ij} v_j^r + \sum_{j=1}^{n_p} B_{ij}^r p_j = c_i^r, \quad i = 1, \ldots, n_v, \ r = 1, \ldots, d, \qquad (4.19)$$

$$\sum_{r=1}^{d} \sum_{j=1}^{n_v} B_{ji}^r v_j^r = 0 \quad i = 1, \ldots, n_p, \qquad (4.20)$$

where

$$\hat{v}^r = \sum_{j=1}^{n_v} v_j^r N_j, \qquad (4.21)$$

$$\hat{p} = \sum_{j=1}^{n_p} p_j L_j, \qquad (4.22)$$

$$A_{ij} = \int_\Omega \mu \left( \sum_{k=1}^{d} \frac{\partial N_i}{\partial x_k} \frac{\partial N_j}{\partial x_k} \right) d\Omega, \qquad (4.23)$$

$$B_{ij}^r = -\int_\Omega \frac{\partial N_i}{\partial x^r} L_j d\Omega, \qquad (4.24)$$

$$c_i^r = \int_\Omega f^r N_i d\Omega. \qquad (4.25)$$

*The Poisson Problem.* The expansions (4.12)–(4.13) are natural candidates when formulating a weighted residual method for the Poisson equation expressed as a system of PDEs (4.8)–(4.9). A Galerkin approach consists of using $N_i$ as weighting functions for (4.8) and $L_i$ as weighting functions for (4.9). The pressure gradient term in (4.8) is integrated by parts, resulting in the following system of discrete equations:

$$\sum_{j=1}^{n_v} A_{ij} v_j^r + \sum_{j=1}^{n_p} B_{ij}^r p_j = 0 \quad i = 1, \ldots, n_v, \ r = 1, \ldots, d, \qquad (4.26)$$

$$\sum_{r=1}^{d} \sum_{j=1}^{n_v} B_{ji}^r v_j^r = g_i \quad i = 1, \ldots, n_p, \qquad (4.27)$$

where

$$\hat{v}^r = \sum_{j=1}^{n_v} v_j^r N_j, \tag{4.28}$$

$$\hat{p} = \sum_{j=1}^{n_p} p_j L_j, \tag{4.29}$$

$$A_{ij} = \int_\Omega N_i N_j d\Omega, \tag{4.30}$$

$$B_{ij}^r = - \int_\Omega \frac{\partial N_i}{\partial x_r} L_j d\Omega, \tag{4.31}$$

$$g_i = \int_\Omega g L_i d\Omega. \tag{4.32}$$

### 4.3.2  Mixed Elements, Do We Really Need Them?

It is natural to ask whether mixed finite elements are actually needed. What will happen if one just discretizes the problem with a standard technique? One will then often experience non-physical pressure oscillations. The reason is that the pressure is not uniquely determined. This ambiguity comes in addition to the fact that the pressure is only determined up to a constant.

In this Section, the problem is presented from a purely algebraic point of view, to motivate that a naive discretization may lead to trouble (see also [19]). A typical discretization of the above described problems results in a matrix equation of the form:

$$\mathcal{A} \begin{bmatrix} v \\ p \end{bmatrix} = \begin{bmatrix} A & B^T \\ B & 0 \end{bmatrix} \begin{bmatrix} v \\ p \end{bmatrix} = \begin{bmatrix} f \\ g \end{bmatrix}. \tag{4.33}$$

When is $\mathcal{A}$ invertible? The matrix is of dimension $(n+m) \times (n+m)$, where $n$ is the length of the vector $v$ and $m$ is the length of $p$. $A$ is an $n \times n$ matrix, which is invertible for the problems considered here, and $B$ is $m \times n$. We may therefore multiply the first equation by $A^{-1}$ and obtain

$$v = A^{-1} f - A^{-1} B^T p. \tag{4.34}$$

We insert (4.34) in the second equation of (4.33) and get,

$$Bv = B(A^{-1} f - A^{-1} B^T p) = g,$$

or in other words, we must solve

$$-BA^{-1} B^T p = g - BA^{-1} f.$$

Hence, the system (4.33) is invertible if $BA^{-1}B^T$ is. The $BA^{-1}B^T$ is usually called the pressure Schur complement. Since $B$ is a $m \times n$ matrix it is not obvious that $BA^{-1}B^T$ is invertible. In fact, if the matrices come from a discretization where the velocity and the pressure are approximated with the same type of elements, then the pressure Schur complement is singular. This may result in pressure oscillations. The following algebraic analogy of the Babuska-Brezzi conditions ensure an invertible matrix $\mathcal{A}$,

$$v^T A v > 0, \quad \forall v \in Kernel(B), \quad (A \text{ is sufficiently invertible}), \qquad (4.35)$$

and

$$Kernel(B^T) = \{0\}. \quad (BA^{-1}B^T \text{ is invertible}). \qquad (4.36)$$

Although, it is not easy to deduce pressure oscillations from this algebraic discussion, it should be clear that a naive discretization does not necessarily lead to a well posed (at least non-singular) numerical problem.

We also supply the discussion with the following remark on standard linear elements approximation of fields with zero divergence. This is often refereed to as the locking phenomenon in elasticity theory (see, e.g., [2]).

*Remark: The Locking Phenomenon.* In the Stokes problem the approximate solution of the velocity should be *divergence free*, and it is not obvious that standard linear elements will give a good approximation under this constraint. In fact, an example of extremely poor approximation properties can be constructed with linear continuous elements and homogeneous Dirichlet boundary conditions. The only divergence free function satisfying these conditions is $\hat{v} = 0$, regardless of the mesh parameter $h$. (This result depends on the geometry of the triangulation, but it will often hold.) Hence, regardless of the equation we want to solve, the constraint, div $\hat{v} = 0$, enforces $\hat{v} = 0$.

These phenomena clearly shows that a careful discretization of our model problems is needed. We will now address the abstract properties that should be met by the mixed elements. The following conditions will ensure a well-posed discrete problem (i.e., there exist a unique solution depending continuously on the data).

### 4.3.3  Function Space Formulation

Both our problems can be viewed within the same abstract setting. Let $H$ and $P$ be suitable Hilbert spaces for $u$ and $p$, respectively. Then both these model problems can be formulated as follows:
Find $(u, p) \in H \times P$ such that

$$\begin{aligned} a(u, w) + b(p, w) &= (f, w), \quad \forall w \in H, \\ b(q, u) \phantom{+ b(p, w)} &= (g, q), \quad \forall q \in P. \end{aligned} \qquad (4.37)$$

The analysis of the mixed finite elements is related to this specific structure of the problem. We will need the following spaces:

- $L^2(\Omega) = \{\ w\ |\ \int_\Omega w^2 d\Omega < \infty\}$

- $H^k(\Omega) = \{\ w\ |\ \sum\limits_{|\alpha| \le k} \int_\Omega (D^\alpha w)^2 d\Omega < \infty\}$, where $k = |\alpha|$ and

$$D^\alpha w_i = \frac{\partial^{|\alpha|} w_i}{\partial x_1^{\alpha_1} \cdots \partial x_d^{\alpha_d}}$$

- $H(\mathrm{div}; \Omega) = \{\ w\ \in L^2(\Omega)\ |\ \nabla \cdot w \in L^2(\Omega)\}$

These spaces are Hilbert spaces of vector functions, the corresponding spaces for scalar functions, $L^2(\Omega)$ and $H^1(\Omega)$, are also used. Note that the derivatives are in distribution sense.

*The Stokes Problem.* The bilinear forms associated with the Stokes problem are:

$$a(u, w) = \int_\Omega \mu \nabla u \cdot \nabla w d\Omega, \tag{4.38}$$

$$b(p, w) = -\int_\Omega p \nabla \cdot w d\Omega. \tag{4.39}$$

$$\tag{4.40}$$

The product $\nabla u \cdot \nabla w$ is the "scalar tensor product"

$$\nabla u \cdot \nabla w = \sum_{i=1}^d \nabla u_i \cdot \nabla w_i = \sum_{i=1}^d \sum_{j=1}^d \frac{\partial u_i}{\partial x_j} \frac{\partial w_i}{\partial x_j}.$$

Suitable function spaces for this model problem are $H = H^1(\Omega)$ and $P = L^2(\Omega)$. The linear forms (the right-hand sides of (4.37)) are:

$$(f, w) = \int_\Omega f \cdot w\, d\Omega + \int_{\partial \Omega_N} \left(\mu w \cdot \frac{\partial u}{\partial n} - pn \cdot w\right) ds, \tag{4.41}$$

$$(g, q) = 0, \tag{4.42}$$

where $(\frac{\partial u}{\partial n} - pn)$ is known at the (parts of the) boundary with natural boundary condition.

*The Poisson Problem.* The bilinear forms in the mixed formulation of the Poisson equation read:

$$a(u, w) = \int_\Omega \lambda^{-1} u \cdot w\, d\Omega, \tag{4.43}$$

$$b(p, w) = \int_\Omega \nabla p \cdot w\, d\Omega = -\int_\Omega p \nabla \cdot w\, d\Omega + \int_{\partial \Omega_N} pw \cdot n. \tag{4.44}$$

The natural conditions here are the essential boundary conditions (4.6) As we see from (4.44), we have two possible choices for $b(\cdot, \cdot)$. The first alternative is $b(p, w) = (\nabla p, w)$. We must then require that $p \in H^1(\Omega)$ and $w \in L^2(\Omega)$. Another formulation is $b(p, w) = -(p, \nabla \cdot w) + \int_{\partial \Omega_N} pw \cdot n$, where we have used integration by parts. We must then require that $w \in H(\mathrm{div}; \Omega)$ and that $p \in L^2(\Omega)$. The difference between $H(\mathrm{div}; \Omega)$ and $H^1(\Omega)$ will be important in the construction of the finite elements and the preconditioner in [15]. The linear forms for the Poisson problem (the right-hand sides of (4.37)) are,

$$(f, w) = \int_{\partial \Omega_N} p n \cdot w ds, \tag{4.45}$$

$$(g, q) = \int_{\Omega} g \cdot q d\Omega. \tag{4.46}$$

### 4.3.4    The Babuska-Brezzi Conditions

In the previous section we saw that the fields $u$ and $p$ had different regularity requirements. The discrete analogy should reflect these differences and we therefore use elements designed for the problem at hand. Any combination of elements will *not* work. Let us assume that we have chosen two sets of basis functions $\{N_i\}$ and $\{L_i\}$ that span $H_h$ and $P_h$, respectively. We will in the following assume that $H_h \subset H$ and $P_h \subset P$. The norms in $H_h$ and $P_h$ can then be derived from $H$ and $P$, respectively. The Galerkin formulation of the problem can be written on the form:
Find $(\hat{v}, \hat{p}) \in H_h \times P_h$ such that

$$\begin{aligned} a(\hat{v}, m) + b(\hat{p}, m) &= (f, m), \quad \forall m \in H_h, \\ b(n, \hat{v}) &= (g, n), \quad \forall n \in P_h. \end{aligned} \tag{4.47}$$

This is just an abstract formulation of (4.19)-(4.20) and (4.26)-(4.27). The matrix equation of this problem is on the form (4.33). From the above definitions of the $H$ and $P$ we have that $a(\cdot, \cdot)$ and $b(\cdot, \cdot)$ are bounded for both model problems, i.e.,

$$a(u, v) \leq C\|u\|_H\|v\|_H, \quad \forall u, v \in H, \tag{4.48}$$

$$b(q, v) \leq C\|q\|_P\|v\|_H, \quad \forall v \in H, q \in P. \tag{4.49}$$

However, the two following conditions will in general not be fulfilled. The elements must be "designed" to meet them.

**Condition 1.** There exists a constant $\alpha$ (independent of h) such that $a(\cdot, \cdot)$ is $H$-elliptic on $H_h$:

$$a(\hat{w}, \hat{w}) \geq \alpha\|\hat{w}\|_H^2, \quad \forall \hat{w} \in H_h. \tag{4.50}$$

This property is trivially satisfied for the Stokes problem. Elements for the mixed Poisson problem are designed to meet this condition.

**Condition 2.** There exists a constant $\beta$ (independent of h) such that

$$0 < \beta := \inf_{\hat{q} \in P_h} \sup_{w_h \in H_h} \frac{b(\hat{w}, \hat{q})}{\|\hat{w}\|_H \|\hat{q}\|_P} . \tag{4.51}$$

The condition 2 is often called the Babuska-Brezzi condition or the inf-sup condition. These two conditions (in addition to (4.48) and (4.49)) are necessary for a well-posed discrete problem [3]. The algebraic analogies of the conditions 1 and 2 are (4.35) and (4.36), respectively. However, there is an important difference. The conditions are designed to ensure *optimal accuracy*. Hence, for $(\hat{v}, \hat{p})$ found by (4.47) we have,

$$\|u - \hat{v}\|_H + \|p - \hat{p}\|_P \le c(\inf_{v \in H_h} \|v - u\|_H + \inf_{q \in P_h} \|p - q\|_P) . \tag{4.52}$$

To get optimal accuracy, we must have $\beta > 0$ independently of $h$ (see, e.g., [3]). This is not needed to ensure that the algebraic system in (4.33) is invertible. The $\beta$ may then decrease towards zero as the grid is refined.

*Regularization of the Stokes Problem.* We can avoid the Babuska-Brezzi conditions by perturbing the problem. We restate the problem as: *find $(\hat{v}^\epsilon, \hat{p}^\epsilon)$ such that*

$$\begin{aligned}
a(\hat{v}^\epsilon, \hat{w}) + b(p_h^\epsilon, \hat{w}) &= (f, \hat{w}), \quad \forall \hat{w} \in H_h, \\
b(\hat{q}, \hat{v}^\epsilon) - \epsilon c(\hat{q}, \hat{p}^\epsilon) &= (g, \hat{q}), \quad \forall \hat{q} \in P_h,
\end{aligned} \tag{4.53}$$

where $\epsilon$ should be small. With this perturbation of the original problem, we get the non-singular matrix equation,

$$\mathcal{A} \begin{bmatrix} v^\epsilon \\ p^\epsilon \end{bmatrix} = \begin{bmatrix} A & B^T \\ B & -\epsilon M \end{bmatrix} \begin{bmatrix} v^\epsilon \\ p^\epsilon \end{bmatrix} = \begin{bmatrix} f \\ g + \epsilon d \end{bmatrix} . \tag{4.54}$$

There are mainly three methods used to construct $\epsilon M$, all based on perturbed versions of the equation of continuity,

$$\nabla \cdot v = \epsilon \nabla^2 p, \tag{4.55}$$

$$\nabla \cdot v = -\epsilon p, \tag{4.56}$$

$$\nabla \cdot v = -\epsilon \frac{\partial p}{\partial t} . \tag{4.57}$$

The approach (4.55) was derived with the purpose of stabilizing pressure oscillations and allowing standard grids and elements (see, e.g., [8]). The $\epsilon$ is usually $\alpha h^2$, where $\alpha$ can be tuned. This approach is usually preferred and does in fact satisfy a slightly more complicated version of the Babuska-Brezzi conditions, regardless of the choice of elements. The equations (4.56) and

(4.57) were not derived as stabilization methods, but were initiated from alternative physical and mathematical formulations of viscous incompressible flow. They are usually referred to as the penalty method and the artificial compressibility method, respectively [7]. The penalty method allows for elimination of the pressure and is usually used together with reduced integration. A Navier-Stokes solver using the penalty method is implemented in Diffpack and is described in [9]. The regularization techniques have relations to common operator splitting techniques [17].

## 4.4    Some Basic Concepts of a Finite Element

This Section presents the basics of a general finite element. We will focus on a definition that should be easy to use in a general implementation with mixed elements on a general unstructured mesh. This will be needed later when we implement the finite elements appropriate to our model problems. For further reading about the finite element method, see [2,3,10,9]. Detailed information on finite element programming in Diffpack can be found in [9,13].

### 4.4.1    A General Finite Element

Consider a spatially varying scalar field $f(x)$, where $x \in \Omega \subset \mathbb{R}^d$. The field $f(x)$ has the following approximate expansion $\hat{f}(x)$ in a finite element method:

$$f(x) \approx \hat{f}(x) = \sum_{j=1}^{n} \alpha_j N_j(x) \,. \tag{4.58}$$

If $f$ also depends on time, we write

$$f(x,t) \approx \hat{f}(x,t) = \sum_{j=1}^{n} \alpha_j(t) N_j(x) \,. \tag{4.59}$$

We refer to $N_j(x)$ as the basis functions in global coordinates and $\alpha_j$ are the *coefficients* in the expansion, or the global degrees of freedom of the discrete scalar field. The choice of basis functions $N_j(x)$ is problem dependent, and we "should" choose them based on the regularity requirements of the differential equation. The basis functions are piecewise polynomials and we can therefore deduce the following (see page 67 in [10]):

$$V_h \subset H^1(\Omega) \Leftrightarrow V_h \subset C^0(\Omega), \tag{4.60}$$
$$V_h \subset L^2(\Omega) \Leftrightarrow V_h \subset C^{-1}(\Omega), \tag{4.61}$$

where $C^0(\Omega)$ is the space of continuous functions and $C^{-1}(\Omega)$ is the space of discontinuous functions (with finite discontinuities). A vector element can often be made as a superposition of scalar elements.

Central to the finite element method is the partition of $\Omega$ into non-overlapping *subdomains* $\Omega_e$, $e = 1, \ldots, E$, $\Omega = \Omega_1 \cup \cdots \cup \Omega_E$. In each subdomain $\Omega_e$ we define a set of basis functions with support only in $\Omega_e$ and the neighboring subdomains. The basis functions are associated with a number of local degrees of freedom. A subdomain, along with its basis functions and degrees of freedom, defines a *finite element*, and the finite elements throughout the domain define a finite element space. The finite element space has a global number of degrees of freedom. These degrees of freedom can be the values of the unknown functions at a set of points (nodes). From an implementational point of view, a great advantage of the finite element method is that it can be evaluated locally and thereafter be assembled to a global linear system. We therefore focus on a local consideration of the finite element method. Similar definitions of finite elements and more information can be found in [2,10,9].

For each $\Omega_e$ there is a parametric mapping $M_e^{-1}$ from the physical subdomain to a reference subdomain:

$$\xi = M_e^{-1}(x), \quad x = M_e(\xi), \quad \xi \in \Omega_r \subset \mathbb{R}^d, \quad x \in \Omega_e \subset \mathbb{R}^d. \tag{4.62}$$

The $\Omega_r$ domain is often called the *reference domain*. The mapping of this reference domain to the physical coordinate system is defined by a set of geometry functions $G_i(\xi)$, i.e., $M_e$ is defined in terms of some $G_i$ functions. Normally, we associate the term *reference element* with the reference domain together with its degrees of freedom and basis functions.

**Definition 3.** An *isoparametric* element is a finite element for which the basis functions and the geometry functions coincide.

However, most mixed elements are not isoparametric. Diffpack therefore has a very general definition of a finite element. This definition is implemented in `ElmDef` which is the base class for all the elements.

**Definition 4.** A reference element is characterized by a reference subdomain $\Omega_r$ with a set of $n_g$ points $p_1, \ldots p_{n_g}$, referred to as geometry nodes, such that the mapping $M_e$ from the reference domain $\Omega_r$ onto the corresponding physical domain $\Omega_e$ has the form

$$x(\xi) = \sum_{k=1}^{n_g} G_k(\xi) x_k. \tag{4.63}$$

Here $x_k$ are the coordinates in physical space corresponding to the geometry node $p_j$. Moreover, the geometry function $G_j(\xi)$ has the property that $G_j(p_i) = \delta_{ij}$, where $\delta_{ij}$ is the Kronecker delta. The (transposed) Jacobi matrix element $J_{ij}$ of the mapping $M_e$ is then given by

$$J_{ij} = \sum_{k=1}^{n_g} \frac{\partial G_k(\xi)}{\partial \xi_i} x_k^j, \tag{4.64}$$

where $x_k^j$ is the $j$-th component of $x_k$.

If the element is non-isoparametric we must also specify the basis functions. We define a number $n_{bf}$ of basis functions $N_1^{ref}, \ldots N_{n_{bf}}^{ref}$ with $n_b$ associated nodes at points $q_1, \ldots q_{n_b}$ in the reference domain. These may in principle be chosen arbitrarily, designed for different purposes. A global basis function is then defined in terms of a mapping of the corresponding reference basis function. We can define a global element by using the geometry mapping $M_e$ in (4.62);

$$N_i^{glob}(x) = N_i^{ref}(M_e^{-1}(x)) = N_i^{ref}(\xi).$$

**Definition 5.** The restriction of a finite element scalar field, as in (4.58), to a finite element in the reference domain is given by

$$f(x)|_{\Omega_r} = \sum_{j=1}^{n_{bf}} \beta_j N_j(\xi), \quad \xi = (\xi_1, \ldots, \xi_d) \in \Omega_r.$$

Here $\beta_j$ are the local expansion coefficients, corresponding to the local degrees of freedom. The local basis functions are $N_j(\xi)$ and $n_{bf}$ is the number of local basis functions.

Finite element applications usually require the derivatives with respect to the physical (global) coordinates, $\partial N_i / \partial x_j$. The relation between derivatives with respect to local and global coordinates is given by

$$\frac{\partial N_i}{\partial \xi_j} = J_{ij} \frac{\partial N_i}{\partial x_j},$$

where $J_{ij}$ is given in (4.64).

The basis functions in the isoparametric elements are exactly the geometry functions, and the degrees of freedom are represented in the geometry nodes. This means that we get the physical elements (shape of the element and the expressions for the basis function in physical coordinates) by applying the parametric mapping $M_e(\xi)$ on the reference elements. Non-isoparametric elements might also be defined in terms of a mapped reference element. However, such elements might also have other non-equivalent definitions, as we will see for the elements used in $H(\mathrm{div}; \Omega)$.

### 4.4.2   Examples of Finite Element Spaces

This section describes some typical mixed elements and their implementation. We also list the error estimates for each element. These estimates are used later in the Sections 4.6.3 and 4.6.5 to verify the simulators. We assume that the triangulation is shape-regular and quasi-uniform. To clarify the notation introduced in the previous section we apply it to some well-known finite elements.

*The Linear Triangle.* The popular linear 2D triangular element is defined as follows. Let the reference domain be,

$$\Omega_r = \{(\xi_1, \xi_2) \mid 0 \le \xi_1 \le 1, \ \xi_2 \le \xi_1\}. \tag{4.65}$$

Moreover, $n_g = n_{bf} = n_b = 3$, $G_i = N_i$ and

$$N_1(\xi_1, \xi_2) = 1 - \xi_1 - \xi_2, \tag{4.66}$$
$$N_2(\xi_1, \xi_2) = \xi_1, \tag{4.67}$$
$$N_3(\xi_1, \xi_2) = \xi_2. \tag{4.68}$$

The geometry and basis function nodes are the corner points of the triangle; $p_1 = (0,0)$, $p_2 = (1,0)$, and $p_3 = (0,1)$. This element is named ElmT3n2D in Diffpack. The Figure 4.1 shows the element. The basis function nodes are indicated by the black circles. More details on the local side and node

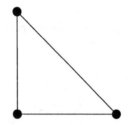

**Fig. 4.1.** Sketch of the 2D linear triangle element.

numbering can be found in [11].

*The 2D Piecewise Constant Element.* The 2D piecewise constant element over a triangle is defined by $\Omega_r$ in (4.65), $n_b = n_{bf} = 1$, $n_g = 3$, $N_1(\xi_1, \xi_2) = 1$ for all $\xi_1, \xi_2 \in \Omega_r$, while $G_i$ and $p_i$ are identical to the expressions for the standard linear triangle element, described above. The location of the basis function node is arbitrary, but an obvious choice is $q_1 = (1/3, 1/3)$. The name in Diffpack is ElmT3gn1bn2D, which refers to its basic properties. It is a triangle element, ElmT, with three geometry nodes, 3gn, and one basis function node, 1bn, in 2D, 2D. The Figure 4.2 shows the element.

As we stated in Section 4.3, our model problems require delicate combinations of finite element spaces, i.e, they must satisfy the Babuska-Brezzi conditions (2). Some appropriate mixed elements are presented here and we also discuss the implementation of these.

*Mixed Elements for the Stokes Problem.* We will now describe the mixed Stokes elements implemented in Diffpack. The Mini and Taylor-Hood elements satisfy the Babuska-Brezzi conditions (4.50)-(4.51) with $\boldsymbol{H} = \boldsymbol{H}^1(\Omega)$

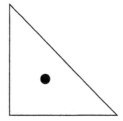

**Fig. 4.2.** Sketch of the 2D piecewise const element.

and $P = L^2(\Omega)$. The Crouzeix-Raviart element is non-conforming and the $\|\cdot\|_{H^1(\Omega)}$ norm has to be replaced with the element-wise norm $\|\cdot\|_{H^1_h(\Omega)}$ [5].

*The Taylor–Hood Element.* When solving Stokes problem or Navier-Stokes-type of equations for the pressure and velocity fields, mixed finite element methods are traditionally used. A possible choice is to let the geometry be described by six nodes in a triangle reference element. The velocity components may then use basis functions that coincide with the geometry functions $(n_g = n_b = n_{bf}, q_i = p_i,$ and $N_i = G_i)$. The pressure field, on the other hand, is based on an element where the geometry is the same as for the velocity field, but where the basis functions are linear. For such an element, we see that the basis function nodes are identical to a subset of the geometry function nodes, the corner nodes. The corresponding basis functions are the same as in the linear triangle element (4.66)-(4.68). Both the velocity and the pressure elements are continuous (more regularity than strictly needed for the pressure). For sufficiently regular $(u, p)$ we have the following approximation result for $k = 1, 2$ (e.g., [9]).

$$\|u - \hat{v}\|_{H^1(\Omega)} + \|p - \hat{p}\|_{L^2(\Omega)} \leq Ch^k(\|u\|_{H^{k+1}(\Omega)} + \|p\|_{H^k(\Omega)}), \quad (4.69)$$
$$\|p - \hat{p}\|_{H^1(\Omega)} \leq Ch^{k-1}(\|u\|_{H^{k+1}(\Omega)} + \|p\|_{H^k(\Omega)}) . (4.70)$$

If $\Omega$ is convex we have,

$$\|u - \hat{v}\|_{L^2(\Omega)} \leq Ch^{k+1}(\|u\|_{H^{k+1}(\Omega)} + \|p\|_{H^k(\Omega)}) . \quad (4.71)$$

The constant $C$ is here and in the following used as a generic constant independent of the mesh size $h$.

These elements are implemented in Diffpack as ElmT6n2D, these are continuous quadratic polynomials. The ElmT6gn3bn2D element are continuous linear polynomials. The same elements exist on quadrilaterals and is implemented as ElmB9n2D and ElmB9gn4bn2D. The Figure 4.3 shows the degrees of freedom in the element. In this and the following examples we have basis function nodes for both the pressure and the velocity. The black circles are velocity nodes, while the black squares denote nodes for both the pressure and the velocity.

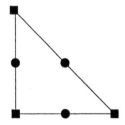

**Fig. 4.3.** Sketch of the 2D Taylor–Hood element; Quadratic velocity and linear pressure elements.

*The Mini Element.* Another approximation can be obtained by using the so-called Mini element [1]. The starting point is continuous linear elements for both the pressure and the velocity. This combination does not satisfy the Babuska-Brezzi conditions, but if we enrich the velocity space with the so-called "bubble" function, it does. Let the triangle be numbered such that $e_i$ are the edges at the opposite side of the vertex $v_i$. Let $\lambda_i$ be a linear function such that $\lambda_i(v_i) = 1$, while $\lambda_i(x_j) = 0 \; \forall x_j$ on $e_i$. The function $\lambda_1\lambda_2\lambda_3$ is the so called bubble function. The bubble function is zero on the edge of the element and has therefore support only in one element. The degrees of freedom are the nodes at the vertices and the center of the triangle. Hence, this element is simply a linear triangle element with one additional (bubble) basis function and the associated basis function node. Hence, $n_g = 3$, $n_{bf} = n_b = 4$. The basis functions in the reference element are

$$N_i(\xi_1, \xi_2) = G_i(\xi_1, \xi_2), \quad \text{for } i \leq 3, \tag{4.72}$$

$$N_4(\xi_1, \xi_2) = 27\,(1 - \xi_1 - \xi_2)\xi_1\xi_2, \tag{4.73}$$

where the geometry functions, $G_i$, are defined by the linear triangle element (page 166). The corresponding basis function nodes read,

$$q_i = p_i, \quad \text{for } i \leq 3, \tag{4.74}$$

$$q_4 = (1/3, 1/3)\,. \tag{4.75}$$

We have the following error estimate (e.g., [9]) for the Mini element:

$$\|\boldsymbol{u} - \hat{\boldsymbol{v}}\|_{\boldsymbol{H}^1(\Omega)} + \|p - \hat{p}\|_{L^2(\Omega)} \leq Ch(\|\boldsymbol{u}\|_{\boldsymbol{H}^2(\Omega)} + \|p\|_{H^1(\Omega)})\,. \tag{4.76}$$

If $\Omega$ is convex we have,

$$\|\boldsymbol{u} - \hat{\boldsymbol{v}}\|_{\boldsymbol{L}^2(\Omega)} \leq Ch^2(\|\boldsymbol{u}\|_{\boldsymbol{H}^2(\Omega)} + \|p\|_{H^1(\Omega)})\,. \tag{4.77}$$

We note that the approximation results rely on the linear polynomials, whereas the bubble functions give stability.

These elements are available in Diffpack as `ElmT3gn4bn2D` and should be used together with linear pressure elements, `ElmT3n2D`. The degrees of freedom is shown in Figure 4.4.

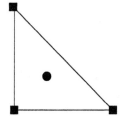

**Fig. 4.4.** Sketch of the 2D Mini velocity element and the linear pressure element.

**Definition 6.** A finite element space $V_h$ is a *conforming* approximation of $V$ if it is a subspace of $V$. A non-conforming finite element space approximation of $V$ is an outer approximation, that is, the finite element space is not a subspace of $V$.

*The Crouzeix-Raviart Elements.* The Crouzeix-Raviart element is linear and continuous at the midpoints of the triangle edges [5]. The midpoints are the only points where it is continuous. Hence, this element is not in $\boldsymbol{H}^1(\Omega)$. It is not conforming and $\boldsymbol{H}^1(\Omega)$ is therefore replaced with $\boldsymbol{H}_h^1(\Omega)$. The element has three basis functions and three associated basis function nodes, $n_g = n_{bf} = n_b = 3$. The basis functions in the reference element are,

$$N_1(\xi_1, \xi_2) = 1 - 2\xi_2,$$
$$N_2(\xi_1, \xi_2) = 2(\xi_1 + \xi_2) - 1,$$
$$N_3(\xi_1, \xi_2) = 1 - 2\xi_1,$$

and the corresponding basis function nodes are

$$q_1 = (1/2, 0),$$
$$q_2 = (1/2, 1/2),$$
$$q_3 = (0, 1/2).$$

The error estimates for this element are [5],

$$\|\boldsymbol{u} - \hat{\boldsymbol{v}}\|_{\boldsymbol{H}_h^1(\Omega)} \le Ch(\|\boldsymbol{u}\|_{\boldsymbol{H}^2(\Omega)} + \|p\|_{H^1(\Omega)}), \qquad (4.78)$$
$$\|p - \hat{p}\|_{L^2(\Omega)} \le Ch(\|\boldsymbol{u}\|_{\boldsymbol{H}^2(\Omega)} + \|p\|_{H^1(\Omega)}). \qquad (4.79)$$

If $\Omega$ is convex we have,

$$\|\boldsymbol{u} - \hat{\boldsymbol{v}}\|_{\boldsymbol{L}^2(\Omega)} \le Ch^2(\|\boldsymbol{u}\|_{\boldsymbol{H}^2(\Omega)} + \|p\|_{H^1(\Omega)}). \qquad (4.80)$$

It is implemented as `ElmT3gn3bn2D` and is used with the discontinuous constant pressure element `ElmT3gn1bn2D`. The degrees of freedom is shown in Figure 4.5. The white circle shows the pressure node. Note that this element only work with essential boundary conditions.

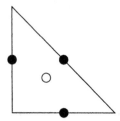

**Fig. 4.5.** Sketch of the 2D Crouzeix–Raviart element.

*Rannacher-Turek Elements.* The Rannacher-Turek elements [18,16] are the quadrilateral analogoue of the Crouzeix-Raviart elements. Several definitions are possible, but we have implemented it as mapped reference elements that are continuous in the midpoint on each side.

The error estimates are similar to the estimates for the Crouzeix-Raviart elements, since we have assumed a shape-regular and quasi-uniform triangulation. However, on a anisotropic mesh a global definition of the element is preferable and gives error estimates independent of the aspect ratio.

It is implemented as `ElmB4gn4bn2D` and should be used with discontinuous constant pressure elements `ElmB4gn1bn2D`.

*Mixed Elements for the Poisson Problem.* This section describes mixed elements suitable for the mixed Poisson problem. They satisfy the Babuska–Brezzi conditions (4.50)-(4.51) with $H = H(\text{div}; \Omega)$ and $P = L^2(\Omega)$. In $H(\text{div}; \Omega)$ continuity is only needed in the normal direction on the sides of the elements [3]. The following elements are designed to meet this requirement.

*The Raviart–Thomas Elements.* A popular choice of elements is the Raviart-Thomas element [20] of order 0 for the velocity approximation $\hat{v}$ and discontinuous piecewise constants for $\hat{p}$. The Raviart-Thomas elements are designed to approximate $H(\text{div}; \Omega)$. They are continuous only in the normal direction across the elements boundaries. The Raviart-Thomas elements of order 0 are on the form:

$$N_i = \begin{pmatrix} a + dx \\ b + dy \\ c + dz \end{pmatrix} \text{ in 3D and } N_i = \begin{pmatrix} a + dx \\ b + dy \end{pmatrix} \text{ in 2D}. \qquad (4.81)$$

The constants $a, b, c, d$ are determined such that

$$N_i \cdot n_j = \begin{cases} 1 & \text{if } i = j \\ 0 & \text{if } i \neq j \end{cases}. \qquad (4.82)$$

where $N_i$ are the basis functions and $n_j$ are the normal vectors associated with the side $i$ and $j$, respectively. This yields a global definition of the elements.

Using (4.81)-(4.82) to the reference element geometry we get the following basis functions in the reference element. The element has three basis functions and three associated basis function nodes, $n_g = n_{bf} = n_b = 3$. The $\xi_1$ components are,

$$N_1(\xi_1, \xi_2) = -\xi_1,$$
$$N_2(\xi_1, \xi_2) = -\sqrt{2}\,\xi_1,$$
$$N_3(\xi_1, \xi_2) = 1 - \xi_1,$$

and the $\xi_2$ components read,

$$N_1(\xi_1, \xi_2) = 1 - \xi_2,$$
$$N_2(\xi_1, \xi_2) = -\sqrt{2}\,\xi_2,$$
$$N_3(\xi_1, \xi_2) = -\xi_2 .$$

The basis function nodes are,

$$q_1 = (1/2, 0),$$
$$q_2 = (1/2, 1/2),$$
$$q_3 = (0, 1/2) .$$

While the previously mentioned elements is used such that the vector field is composed by scalar elements,

$$\hat{v} = \sum_i \mathbf{a}_i N_i,$$

the Raviart-Thomas element is a *vector* element in the sense that

$$\hat{v} = \sum_i a_i \mathbf{N}_i .$$

Moreover, these vector elements need to satisfy (4.82) globally, and this is not necessarily the case. We wish to define these elements in terms of a mapped reference element, because this is a very flexible strategy.

Following the basic concept of scalar elements we wish to use the definition for the reference element to map the reference element to the global element. However, the usual geometry mapping $M_e$ does *not* preserve the continuity in the normal direction and the mapped elements will therefore not be in $\mathbf{H}(\text{div}; \Omega)$. A slightly modified mapping does preserve the continuity in the normal direction, but unfortunately the mapping can not be defined componentwise by scalar elements. We remember the geometry mapping:

$$x = M_e(\xi).$$

In order to make elements that are continuous in the normal direction of the mapped reference element, we must assume that the geometry mapping is affine. Affine mappings can be expressed by

$$x = M_e(\xi) = x_e + B_e\xi.$$

for some matrix $B_e$ and a fixed point $x_e$ in physical space. The mapped reference element can then be defined as:

$$N(x) = \frac{1}{\det B_e} B_e \cdot N^{ref}(M_e^{-1}(x)). \tag{4.83}$$

This mapping is implemented in `MxMapping`. The element is a vector element and this is reflected in the code:

```
mfe.N(d,i);
mfe.dN(d,i,k);
```

Here, `MxFiniteElement` uses `MxMapping` to compute the basis function and its derivatives. Notice that this is in contrast to the standard use of vector elements composed by scalar elements,

```
mfe(d).N(i);
mfe(d).dN(i,k);
```

An example of use can be found in `PoissonMx.cpp`. To check whether the element requires the special mapping one can check the boolean variable `special_mapping` in `ElmDef` or `MxFiniteElement`. Vector fields based on these elements require a special initialization. This is described in Section 4.6.4

The 2D Raviart-Thomas reference element is shown in Figure 4.6 and we refer to [3,20] for more information. We have the following approximation properties [6]:

$$\|u - \hat{v}\|_{L^2(\Omega)} \le ch\|u\|_{H^1(\Omega))} \tag{4.84}$$

$$\|p - \hat{p}\|_{L^2(\Omega)} \le c(h\|p\|_{H^1(\Omega)} + h^2\|p\|_{H^2(\Omega)}). \tag{4.85}$$

Figure 4.6 shows the degrees of freedom in the element.

*A Mixed Poisson Element Which is Robust for the Stokes Problem.* All the previous elements discussed have been designed either for the Stokes problem or the mixed Poisson problem. We will now introduce a new element described in [16] which has good approximation properties in both cases. We will refer to this as the "robust" element, in lack of a better name. The polynomial space is defined by:

$$V(T) = \{v \in P_3^2 : \nabla \cdot v \in P_0, \quad (v \cdot n)|_e \in P_1, \quad \forall E(T)\},$$

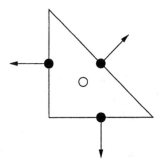

**Fig. 4.6.** Sketch of the 2D Raviart-Thomas element for the mixed formulation of the Poisson equation.

where $E(T)$ is the set of edges in triangulation. This element is designed such that the normal components on the faces are continuous, as is needed in $\boldsymbol{H}(\text{div}; \Omega)$. The tangential components are not continuous, but their mean value are. Hence, the element is conforming in $\boldsymbol{H}(\text{div}; \Omega)$, but non-conforming in $\boldsymbol{H}^1(\Omega)$. The element has nine basis functions and nine associated basis function nodes, $n_g = n_{bf} = n_b = 9$. The $\xi_1$ components of the basis functions in the reference element read,

$$N_1(\xi_1, \xi_2) = 2\xi_1 + 9\,\xi_1\xi_2 - 7.5\,\xi_1{}^2 + 6\,\xi_1{}^3 - 18\,\xi_2{}^2\xi_1,$$
$$N_2(\xi_1, \xi_2) = 6\,\xi_1 - 36\,\xi_1\xi_2 - 6\,\xi_1{}^2 + 36\,\xi_2{}^2\xi_1 + 24\,\xi_1{}^2\xi_2,$$
$$N_3(\xi_1, \xi_2) = 6\,\xi_1 - 51\,\xi_1\xi_2 - 13.5\,\xi_1{}^2 + 6\,\xi_1{}^3 + 54\,\xi_2{}^2\xi_1 + 48\,\xi_1{}^2\xi_2,$$
$$N_4(\xi_1, \xi_2) = \sqrt{2}\,(-7.5\,\xi_1 + 48\,\xi_1\xi_2 + 18\,\xi_1{}^2 - 9\,\xi_1{}^3 - 45\,\xi_2{}^2\xi_1 - 48\,\xi_1{}^2\xi_2),$$
$$N_5(\xi_1, \xi_2) = \sqrt{2}\,(-3\,\xi_1 + 18\,\xi_1\xi_2 + 9\,\xi_1{}^2 - 6\,\xi_1{}^3 - 18\,\xi_2{}^2\xi_1 - 24\,\xi_1{}^2\xi_2),$$
$$N_6(\xi_1, \xi_2) = \sqrt{2}\,(8.5\,\xi_1 - 36\,\xi_1\xi_2 - 24\,\xi_1{}^2 + 15\,\xi_1{}^3 + 27\,\xi_2{}^2\xi_1 + 48\,\xi_1{}^2\xi_2),$$
$$N_7(\xi_1, \xi_2) = -0.5 - 7\,\xi_1 + 2\,\xi_2 + 27\,\xi_1\xi_2 + 25.5\,\xi_1{}^2 - 18\,\xi_1{}^3 - 18\,\xi_2{}^2\xi_1 - 48\,\xi_1{}^2\xi_2,$$
$$N_8(\xi_1, \xi_2) = 6\,\xi_1 - 12\,\xi_1\xi_2 - 18\,\xi_1{}^2 + 12\,\xi_1{}^3 + 24\,\xi_1{}^2\xi_2,$$
$$N_9(\xi_1, \xi_2) = 1.5 - 3\,\xi_1 - 2\,\xi_2 + 15\,\xi_1\xi_2 - 4.5\,\xi_1{}^2 + 6\,\xi_1{}^3 - 18\,\xi_2{}^2\xi_1,$$

and the $\xi_2$ components are,

$N_1(\xi_1, \xi_2) = 1.5 - 2\,\xi_1 - 3\,\xi_2 + 15\,\xi_1\xi_2 - 4.5\,\xi_2{}^2 + 6\,\xi_2{}^3 - 18\,\xi_1{}^2\xi_2,$

$N_2(\xi_1, \xi_2) = -6\,\xi_2 + 12\,\xi_1\xi_2 + 18\,\xi_2{}^2 - 12\,\xi_2{}^3 - 24\,\xi_2{}^2\xi_1,$

$N_3(\xi_1, \xi_2) = 0.5 + 2\,\xi_1 - 7\,\xi_2 + 27.0\,\xi_1\xi_2 + 25.5\,\xi_2{}^2 - 18\,\xi_2{}^3 - 48\,\xi_2{}^2\xi_1 - 18\,\xi_1{}^2\xi_2,$

$N_4(\xi_1, \xi_2) = \sqrt{2}\,(8.5\,\xi_2 - 36\,\xi_1\xi_2 - 24\,\xi_2{}^2 + 15\,\xi_2{}^3 + 48\,\xi_2{}^2\xi_1 + 27\,\xi_1{}^2\xi_2),$

$N_5(\xi_1, \xi_2) = \sqrt{2}\,(-3\,\xi_2 + 18\,\xi_1\xi_2 + 9\,\xi_2{}^2 - 6\,\xi_2{}^3 - 24\,\xi_2{}^2\xi_1 - 18\,\xi_1{}^2\xi_2),$

$N_6(\xi_1, \xi_2) = \sqrt{2}\,(-7.5\,\xi_2 + 48\,\xi_1\xi_2 + 18\,\xi_2{}^2 - 9\,\xi_2{}^3 - 48\,\xi_2{}^2\xi_1 - 45\,\xi_1{}^2\xi_2),$

$N_7(\xi_1, \xi_2) = 6\,\xi_2 - 51\,\xi_1\xi_2 - 13.5\,\xi_2{}^2 + 6\,\xi_2{}^3 + 48\,\xi_2{}^2\xi_1 + 54\,\xi_1{}^2\xi_2,$

$N_8(\xi_1, \xi_2) = -6\,\xi_2 + 36\,\xi_1\xi_2 + 6\,\xi_2{}^2 - 24\,\xi_2{}^2\xi_1 - 36\,\xi_1{}^2\xi_2,$

$N_9(\xi_1, \xi_2) = 2\,\xi_2 + 9\,\xi_1\xi_2 - 7.5\,\xi_2{}^2 + 6\,\xi_2{}^3 - 18\,\xi_1{}^2\xi_2.$

The corresponding basis function nodes are,

$$q_1 = (1/4, 0),$$
$$q_2 = (2/4, 0),$$
$$q_3 = (3/4, 0),$$
$$q_4 = (3/4, 1/4),$$
$$q_5 = (2/4, 2/4),$$
$$q_6 = (1/4, 3/4),$$
$$q_7 = (0, 3/4),$$
$$q_8 = (0, 2/4),$$
$$q_9 = (0, 1/4).$$

We have the error estimates for the mixed Poisson equation,

$$\|\boldsymbol{u} - \hat{\boldsymbol{v}}\|_{L^2(\Omega)} \le Ch^2 \|\boldsymbol{u}\|_{H^2(\Omega)}, \tag{4.86}$$

$$\|\mathrm{div}\,(\boldsymbol{u} - \hat{\boldsymbol{v}})\|_{L^2(\Omega)} \le Ch \|\mathrm{div}\,\boldsymbol{u}\|_{H^1(\Omega)}, \tag{4.87}$$

$$\|p - \hat{p}\|_{L^2(\Omega)} \le Ch(\|p\|_{H^1(\Omega)} + h\|\boldsymbol{u}\|_{H^2(\Omega)}). \tag{4.88}$$

The corresponding estimates for Stokes problem reads,

$$\|\boldsymbol{u} - \hat{\boldsymbol{v}}\|_{H^1(\Omega)} \le Ch \|\boldsymbol{u}\|_{H^2(\Omega)},$$

$$\|p - \hat{p}\|_{L^2(\Omega)} \le Ch(\|p\|_{H^1(\Omega)} + \|\boldsymbol{u}\|_{H^2(\Omega)}).$$

The element has been implemented in Diffpack as `ElmT3gn9bn2Du` and `ElmT3gn9bn2Dv` and should be combined with the piecewise constant pressure elements, `ElmT3gn1bn2D`. Figure 4.7 shows the basis function nodes.

*Remark: Stokes Elements for the Poisson Problem.* The mixed elements for the Poisson problem considered above were such that $\boldsymbol{v} \in \boldsymbol{H}(\mathrm{div}; \Omega)$ and $p \in L^2(\Omega)$. One alternative approach is to seek an approximation where

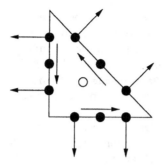

**Fig. 4.7.** The degrees of freedom of the robust element.

$v \in L^2(\Omega)$ and $p \in H^1(\Omega)$. Several of the above mentioned Stokes elements have continuous pressure elements and one might therefore suspect that these elements could be used. This is indeed true, at least for the Mini element and the Taylor–Hood element. Some numerical experiments in Section 4.6.5 show the behavior of these elements (c.f. also [16]). However, usually the highest level of accuracy is wanted in the velocity (or else one would not use the mixed formulation) and therefore this formulation is not often used.

### 4.4.3   Diffpack Implementation

The definition of the reference element is provided by a subclass of `ElmDef`. This definition is based on parameters (variables) in the base class `ElmDef` as well as virtual functions defined in `ElmDef`. For example, $n_b$, $n_{bf}$, and $n_g$ are integers in the class `ElmDef` having the names `nne_basis`, `nbf`, and `nne_geomt`, respectively. The virtual function `geomtFunc` defines the geometry functions $G_i(\xi)$, while the virtual function `basisFunc` defines the basis functions $N_i(\xi)$. The derivatives $\partial N_i / \partial \xi_j$ and $\partial G_i / \partial \xi_j$ are provided by the virtual functions `dLocBasisFunc` and `dLocGeomtFunc`, respectively. Notice that the derivatives refer to the reference coordinates $\xi$. It is the derivatives with respect to the physical coordinates that are of interest. Class `BasisFuncAtPt` is designed for evaluating and storing the geometry and basis functions, their global derivatives and the (transposed) Jacobi matrix determinant, at a particular $\xi$ point. This class relies on information from `ElmDef` and the global coordinates of the geometry nodes of the element.

Class `FiniteElement` is designed to offer the programmer easy access to the global element: the basis functions, their global derivatives, and the Jacobi determinant, evaluated at a point $\xi$ in the reference domain $\Omega_r$ (corresponding to a physical point $x$ in $\Omega_e$). Class `FiniteElement` is naturally based on a layered design where it gains its information from `BasisFuncGrid`, `ElmDef`, `BasisFuncAtPt`, and `GridFE` objects. In the case of mixed finite elements, one needs a `FiniteElement` object for each type of basis functions. This is provided by the class `MxFiniteElement`, which contains an array of pointers (handles) to `FiniteElement` objects and much of the same interface as class `FiniteElement`. Scalar finite element fields are represented by `FieldFE` and rely on information from `FiniteElement`, `BasisFuncGrid`, and `GridFE` objects. `FieldFE` objects are designed to represent the solution field and can be evaluated and differentiated at arbitrary points.

The initial version of Diffpack was written for isoparametric elements. Hence, most of the virtual functions in the `ElmDef` have a default implementation in class `ElmDef` for isoparametric elements (if the function will depend on the element shape, the default version assumes a box shape).

For instance, the Mini element is implemented in a subclasses `ElmT3gn4bn2D`, which is a subclass of `ElmT3n2D`, which is a subclass of the base class for all elements, `ElmDef`. In this case `ElmT3n2D` supply the information related to the geometry of the element. All that was needed to implement `ElmT3gn4bn2D` was the basis functions and the location of the basis function nodes.

The information on the elements, in particular the geometry nodes, their connectivity, and boundary indicators [9,4], is represented by class `GridFE` in Diffpack. Class `BasisFuncGrid` contains the basis function nodes, their connectivity, the boundary indicators at basis function nodes, etc. In addition, class `BasisFuncGrid` has a `GridFE` object (or rather a pointer or handle to such an object) such that a `BasisFuncGrid` object has complete information of all the geometry and basis function nodes and their relevant associated data.

When working with isoparametric elements, class `BasisFuncGrid` simply uses the `GridFE` object for looking up information on basis function nodes and degrees of freedom in scalar fields. For a user it is then only necessary to create a `GridFE` object. When other classes are initialized with a `GridFE` object, and they need information on the basis functions (class `FieldFE` is an example), it is assumed that the elements are isoparametric and a `BasisFuncGrid` object is easily constructed internally in these classes. None of the terms related to the distinction between basis and geometry functions need to be familiar to the user in the isoparametric case. In particular, class `BasisFuncGrid` is not apparent at all, see for example [9].

If non-isoparametric elements are applied, the user must allocate and initialize a `BasisFuncGrid` for each scalar field. In this way, the extra complexity associated with the details of non-isoparametric elements is only visible when it is really needed. One should notice that this design goal is readily achieved due to our usage of abstract data types and object-oriented programming.

The `GridFE` and `BasisFuncGrid` classes organize the *global* topology of the geometry and basis function information. The (local) definition of the basis functions are provided by an `ElmDef` subclass.

### 4.4.4  Numbering Strategies

Having performed the mixed finite element discretization, we end up, as usual, with a system of linear algebraic equations. The book-keeping of element degrees of freedom and linear system degrees of freedom is non-trivial in mixed methods and is described in the following.

*The Special Numbering.* It is assumed that isoparametric elements are used and that there are $m$ unknowns per node and a total of $n$ nodes. In other words, one needs to use the same element type for all the unknown fields that enter a system of partial differential equations. Problems for which this is a suitable approach, involve the Navier equations of elasticity, the incompressible Navier-Stokes equations treated by a penalty function approach, and simultaneous (implicit) solution of pressures and concentrations (saturations) in multi-phase porous media flow. The geometry and basis function nodes coincide, and it is natural to number the degrees of freedom of the linear system in this sequence:

$$u_1^1, u_1^2, \ldots, u_1^m, u_2^1, \ldots, u_2^m, \ldots, u_n^1, \ldots, u_n^m, \tag{4.89}$$

where $u_i^j$ is the degree of freedom of scalar field no. $j$ at node $i$. Given a global node $i$ and a scalar field number $j$, the global degree of freedom number is $m(i - 1) + j$. The DegFreeFE class in Diffpack takes care of computing the global degree of freedom number according to this formula.

At the element level, the structure of the special numbering is the same as at the global level. That is, if $i$ is a local node and $j$ is the field number, the associated *local* degree of freedom of the merged fields is $m(i - 1) + j$. This information is very important since the local numbering is fundamental when setting up the elemental matrix and vector contributions in the integrands functions in Diffpack. The book [9] contains some relevant examples. As an example on the special numbering, consider the $2 \times 2$ grid of bilinear elements

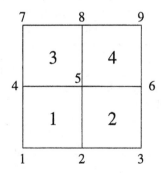

**Fig. 4.8.** Sketch of a $2 \times 2$ grid, with bilinear elements, and the corresponding numbering of elements and nodes.

in Figure 4.8. If there is only one scalar field, the degrees of freedom numbering in the linear system naturally follows the nodal numbering. A measure of
the associated matrix bandwidth could be the largest difference between two
degree freedom numbers in the same element. Here this is 4 (e.g. $5 - 1 = 4$
in the first element). With two unknown scalar fields, $u^{(1)}$ and $u^{(2)}$, there
are two ways of structuring the degrees of freedom in the linear system. The
suggestion above results in

$$u_1^1, u_1^2, u_2^1, u_2^2, \ldots, u_n^1, u_n^2 . \tag{4.90}$$

and depicted in Figure 4.9. Instead of numbering the local degrees of freedom

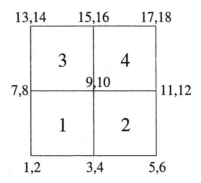

**Fig. 4.9.** The special numbering of degrees of freedom in the linear system arising
from two unknown scalar fields over the grid in Figure 4.8.

at each node consecutively, we could first number all the degrees of freedom
of scalar field one and then number the degrees of freedom of scalar field two:

$$u_1^1, u_2^1, \ldots, u_n^1, u_1^2, \ldots, u_n^2 . \tag{4.91}$$

The numbering on a $2 \times 2$ grid is displayed in Figure 4.10. The impact of the
two numberings on the bandwidth of the coefficient matrix should be clear:
9 in the first case and 13 in the second case. On a grid with $q \times q$ elements
the corresponding figures read $2q + 5$ and $q^2 + 3q + 3$!

*The General Numbering.* The general numbering is based on the following
strategy. At the element level, a field-by-field numbering like (4.91) is used
for the element degrees of freedom,

$$u_1^1, u_2^1, \ldots, u_n^1, u_1^2, \ldots, u_n^2, \ldots, u_1^m, \ldots, u_n^m .$$

where now $n$ is the number of nodes in the element and $m$ is the number of
unknown scalar fields. More generally, if we have $n_j$ degrees of freedom for
field no. $j$, we order the unknowns like this:

$$u_1^1, u_2^1, \ldots, u_{n_1}^1, u_1^2, \ldots, u_{n_2}^2, \ldots, u_1^m, \ldots, u_{n_m}^m .$$

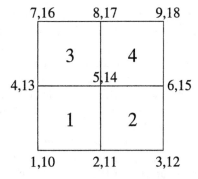

**Fig. 4.10.** A possible numbering of degrees of freedom in the linear system arising from two unknown scalar fields over the grid in Figure 4.8.

The corresponding *global* numbering is constructed from a simple (and general) algorithm that yields a reasonable small bandwidth: For each element we run through each local degree of freedom and increase the corresponding global number by one, provided the local degree of freedom has not been given a global number in a previously visited element. Such numbering of the degrees of freedom applied to a single scalar field is exemplified in Figure 4.11. In element 1 we go through the local nodes 1-4 and assign corresponding

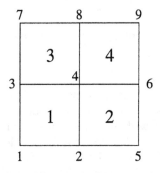

**Fig. 4.11.** The general numbering applied to a single scalar field over the grid in Figure 4.8.

global degree of freedom numbers 1-4. In element 2, the first local degree of freedom was already treated in element 1. The next local degree of freedom (local node no. 2) has not been treated before and can therefore be given the global degree of freedom number 5. The reader is encouraged to continue the algorithm and understand how the rest of the degree of freedom numbers arise.

Let us extend the general number example on a 2 × 2 grid to the case where we have two scalar fields as unknowns. In the first element we then run through the local degrees of freedom 1-4 of the first scalar field and generate corresponding global numbers 1-4. Then we run through the four degrees of freedom of the second scalar field and assign them to the global numbers 5-8. Proceeding with element two, the two nodes on the left have already been treated in element 2 so the second degree of freedom of scalar field no. 1 is assigned the global number 9, while the fourth degree of freedom of scalar field no. 1 corresponds to the global number 10. The second scalar field contributes with two new degrees of freedom, 11 and 12. Also in this case the reader should understand the rest of the global degree of freedom numbers. Figure 4.12 shows the numbering.

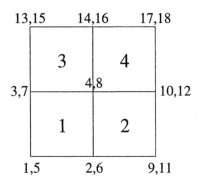

**Fig. 4.12.** The general numbering applied to two scalar fields over the grid in Figure 4.8.

Finally, we consider an example involving different number of degrees of freedom in different fields. Again we focus on the grid in Figure 4.8, but now we have two scalar fields with bilinear elements and one scalar field with piecewise constant elements. The location of the degree of freedom for the constant value in an element could be the centroid. (This example could correspond to bilinear elements for a vector field and piecewise constant elements for a scalar field.) Figure 4.13 shows the complete numbering.

As one can see, the general numbering requires quite some book-keeping. This is performed by the `DegFreeFE` object. However, the programmer must explicitly deal with the *local* degrees of freedom numbering when setting up the element matrix and vector, but this is quite simple, as one applies either the special numbering or the field-by-field numbering on the element level. The complicated details arise when going from the element to the global degrees of freedom numbering, but class `DegFreeFE` hides the book-keeping from the programmer. Some program examples appear later and illustrates the usage of various Diffpack tools.

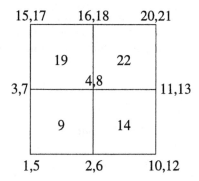

**Fig. 4.13.** The general numbering applied to two scalar fields with 9 nodes and one scalar field with 4 nodes over the grid in Figure 4.8.

## 4.5 Some Code Examples

In this Section we will build simulators appropriate for mixed systems, step by step. We begin with a simple demonstration program for the numbering schemes in Section 4.4.4 for scalar and vector fields. Then the usage of `BasisFuncGrid` and special vs. general numbering in a `Poisson1`-like (cf. [9]) solver is shown. The next step concerns a simple version of a Stokes-problem solver before we end up with a sophisticated block/multigrid solver for the Stokes problem.

### 4.5.1 A Demo Program for Special Versus General Numbering

Section 4.4.4 presents a series of examples, Figures 4.8–4.13, on various degree of freedom numbering strategies. Here we shall make a simple demo program that produces these numberings. The program is found in

```
src/mixed/numbering/numbering.cpp
```

Assume we have a finite element grid stored in a `GridFE` object `grid`. Making a `DegFreeFE` object directly from the grid, with one unknown per node,

```
DegFreeFE dof (grid, 1);
```

results in a standard numbering of the unknowns, which coincides with the nodal numbering (cf. Figure 4.8). If we have two fields over the grid, i.e., two unknowns per node, we just alter the second parameter to the `DegFreeFE` constructor:

```
DegFreeFE dof (grid, 2);
```

The resulting degree of freedom numbering corresponds to the special numbering (cf. Figure 4.9).

The general numbering requires a slightly different initialization procedure of the `DegFreeFE` object. Now we must explicitly define a `BasisFuncGrid` for the basis functions over the grid and tie a finite element field to this `BasisFuncGrid` object:

```
BasisFuncGrid f_grid (grid);
FieldFE f (f_grid, "f");
```

Notice that the field will here be isoparametric. Having the field `f` we can initialize the `DegFreeFE` object:

```
DegFreeFE dof (f);
```

The corresponding numbering of the degrees of freedom is shown in Figure 4.11. Notice that sending a field, instead of a grid, as argument to a `DegFreeFE` constructor implies the general numbering of the degrees of freedom.

With two scalar fields, having the same number of degrees of freedom, the fields must be attached to a `FieldsFE` collector prior to initializing the degree of freedom handler:

```
BasisFuncGrid u_grid (grid);
FieldFE u (u_grid, "u");
FieldFE v (u_grid, "v");
FieldsFE collection (2, "coll");
collection.attach (u, 1);
collection.attach (v, 2);

DegFreeFE dof (collection);
```

The degree of freedom numbering is depicted in Figure 4.12.

Our final example concerns two fields with isoparametric elements, like `u` and `v` above, plus one field having (possibly) non-isoparametric elements. In other words, we address the general case with several fields and different elements in the different fields. The set-up is the same:

```
BasisFuncGrid u_grid (grid);
FieldFE u (u_grid, "u");
FieldFE v (u_grid, "v");

BasisFuncGrid p_grid (grid);
p_grid.setElmType (p_elm);    // can change to non-isop. elm
FieldFE p (p_grid, "p");

FieldsFE collection (3, "coll");
collection.attach (u, 1);
collection.attach (v, 2);
collection.attach (p, 3);

DegFreeFE dof (collection);
```

Figure 4.13 displays the associated degree of freedom numbering.

## 4.6    Programming with Mixed Finite Elements in a Simulator

Programming with mixed finite elements is closely related to programming with isoparametric elements. The structure of the class that represents the simulator is the same in the two cases. The basic differences are

- Class FEM is replaced by its subclass MxFEM.
- Class FiniteElement is replaced by MxFiniteElement, which contains a set of FiniteElement objects, each corresponding to the element type of a field involved in the partial differential equations being solved.
- integrands is replaced by

  ```
  void integrandsMx (ElmMatVec& elmat, MxFiniteElement& mfe)
  ```

- Similarly, integrands4side is replaced by integrands4sideMx, makeSystem is replaced by makeSystemMx (with a slightly different argument list).
- The programmer must explicitly declare a BasisFuncGrid or

  ```
  VecSimplest(BasisFuncGrid)
  ```

  for each field type that corresponds to a primary unknown.
- All the BasisFuncGrid objects must be explicitly created and initialized by the programmer (but the initialization is trivial, just call a function with the element type for each field). The creation of DegFreeFE, FieldFE, and FieldsFE objects must be based on a BasisFuncGrid object as input rather than on a straight GridFE object.

### 4.6.1    A Standard Solver with Some New Utilities

Consider the Poisson1 class from [9, ch. 3.1]. Now we change the test problem a bit such that we have a numerical solution that coincides with the exact solution regardless of the number of elements. The test problem for this purpose reads $\nabla^2 u = 0$ in $[0, 1]^d$, with $\partial u/\partial n = 0$ on the boundary, except at $x_1 = 0$ where $u = 0$ and at $x_1 = 1$ where $u = 1$. The exact solution is then given as $u(x_1, \ldots, x_d) = x_1$. This should be reproduced by any grid.

The test problem is implemented in class Laplace1 whose directory is found in src/mixed/Laplace1. This class is a slight edit of Poisson1 where we basically have thrown away the flux computations, fixed the domain, and changed the analytical solution and the fillEssBC function. Of course, the define and scan functions are significantly changed.

Our first task is to extend class Laplace1 to allow for a general node numbering, just for the purpose of explaining how one introduces BasisFuncGrid objects in a solver and how the DegFreeFE object must be initialized. The following modifications to class Laplace1 must be performed:

- inclusion of a `BasisFuncGrid` object in the class:

      Handle(BasisFuncGrid) bgrid;

- definition of a menu item for the type of numbering
- creation of the unknown field from the `BasisFuncGrid` object[1]:

      bgrid.rebind (new BasisFuncGrid (*grid));
      u.rebind (new FieldFE (*bgrid, "u"));

- creation of the `DegFreeFE` object in two ways, depending on whether the user has chosen the special or general numbering:

      if (numbering == "special")
        dof.rebind (new DegFreeFE (*grid, 1));
      else
        dof.rebind (new DegFreeFE (*u));

- of reasons to be explained later, the `fillEssBC` function needs to be rewritten

These extra statements are conveniently placed in a subclass of `Laplace1`:

    class Laplace1GN : public Laplace1
    {
    public:
      Handle(BasisFuncGrid) bgrid;
      Laplace1GN () : Laplace1() {}
      ~Laplace1GN () {}
      virtual void define (MenuSystem& menu, int level = MAIN);
      virtual void scan ();
      virtual void fillEssBC ();
    };

The source code is placed in a subdirectory `Laplace1GN` of `Laplace1`[2].

The redefinition of `fillEssBC` is perhaps not obvious. If we use the familiar constructions for setting boundary conditions, e.g.,

    if (grid->boNode (i, 1))      // i=geometry node, bo.ind.=1
      dof->fillEssBC (i, 1.0);

a wrong solution is computed in the case of a general numbering. The reason is that using the node counter i in `dof->fillEssBC(i,1.0)` *implies that the numbering of the unknowns in the linear system coincides with the numbering of the geometry nodes.* Instead of the node number i we should use the *global degree of freedom number* corresponding to this node[3]. This number is in general computed by

---

[1] The field has a `BasisFuncGrid` which contains a `GridFE` object, making the access to all grid information complete from a field object.

[2] We have run `AddMakeSrc ..` to tell the makefile for `Laplace1GN` that the source depends on files in the parent directory.

[3] The numbering of the geometry nodes and the linear system degrees of freedom with a general numbering differ even when there is only one unknown scalar field.

```
idof = dof->fields2dof (i, 1);
```

Moreover, we should for the case of generality in the non-isoparametric case ask `BasisFuncGrid` if a basis function node is subject to the boundary condition:

```
if (bgrid->essBoNode (i, 1))
   dof->fillEssBC (idof, 1.0);
```

With isoparametric elements this can be simplified to

```
if (grid->boNode (i, 1))
   dof->fillEssBC (idof, 1.0);
```

i.e., asking the geometry nodes for essential boundary conditions. (If there is only one unknown per node, we have `idof=i`, of course.)

The next natural extension of class `Laplace1GN` is `Laplace1Mx` that allow for the mixed versions of `FEM` and `FiniteElement`. This class is essentially a merge of classes `Laplace1` and `Laplace1GN`, where `FEM` and `FiniteElement` are replaced by `MxFEM` and `MxFiniteElement`. In addition, we need a `FieldsFE` collection of all fields that enter as unknowns in the linear system. Notice that changing the elements even in the standard formulation of the Poisson problem has an effect. The Crouzeix-Raviart element is often used in to discretize the pressure in porous media flow, because of better results than standard elements.

The new parts of the code contain the construction of a `Handle(FieldsFE)` collection object, which is used in many of the mixed finite element utilities:

```
collection.rebind (new FieldsFE (1/*no of fields*/, "collection"));
collection->attach (*u, 1);
dof.rebind (new DegFreeFE (*collection));
```

The `collection` object is needed when making the linear system:

```
makeSystemMx (*dof, *lineq, *collection, 1);
```

The final argument 1 represents the field number in `collection` whose grid should be used for the numerical integration. (Usually, this should be the field that has the `BasisFuncGrid` with the highest order elements.)

The `integrands` routine is replaced by `integrandsMx` and works with an array of `FiniteElement` objects, one entry for each unknown field in the `collector` collection of fields. Here things are as simple as possible, i.e., only one unknown field.

```
void Laplace1Mx:: integrandsMx (ElmMatVec& elmat,
    const MxFiniteElement& mfe)
{
  int i,j,q;
  if (mfe.size() != 1)
    errorFP("Laplace1Mx:: integrandsMx",
            "Wrong size of MxFiniteElement");
```

```
    const FiniteElement& fe = mfe(1);
    const real detJxW = mfe.detJxW();
    ...
}
```

The rest of the routine is identical to Laplace1::integrands.

### 4.6.2   A Simulator for the Stokes Problem

The linear system arising from the Stokes problem can be written on the form (4.19)–(4.20). In our first implementation of the Stokes problem we shall assemble a single merged linear system, suitable for direct methods. The linear system is then not of the form (4.19)–(4.20), because the degrees of freedom and equations are merged according to the general numbering. However, at the element level, the discrete equations are on the form (4.19)–(4.20) with $n_v$ being the number of velocity nodes and $n_p$ the number of pressure nodes in an element.

The simulator is realized as a class Stokes. Its basic contents are, as usual, handles to a GridFE, LinEqAdmFE, DegFreeFE, FieldFE, and SaveSimRes objects. In addition we need some new data structures for mixed methods: BasicFuncGrid objects for specifying basis function nodes in mixed elements and FieldsFE objects for collecting unknown fields for book-keeping of the global set of degrees of freedom in the DegFreeFE object.

An outline of the class Stokes is listed next.

```
class Stokes : public MxFEM
{
protected:
  Handle(GridFE)         grid;     // underlying finite element grid
  Handle(BasisFuncGrid)  v_grid, p_grid;
  Handle(FieldFE)        p;          // Pressure
  Handle(FieldsFE)       v;          // Velocity
  Handle(FieldsFE)       coll;       // (v_1, ..., v_d, p)
  Handle(DegFreeFE)      dof;
  Vec(real)              linsol;
  Handle(LinEqAdmFE)     lineq;
  Handle(SaveSimRes)     database;

  Ptv(real) v_inlet;             // velocity at inlet
  Test_case test_case;

  Handle(FieldFE)        error;

  // used to partition scan into manageable parts:
  void scanGrid();
  virtual void scanData();
  virtual void initFieldsEtc();
  virtual void numbering();

  // standard functions:
  virtual void fillEssBC ();
  virtual void fillEssBC4V (DegFreeFE& dof);
```

```
virtual void fillEssBC4P (DegFreeFE& dof, int P);

virtual void calcElmMatVecMx
   (int e, ElmMatVec& elmat, MxFiniteElement& mfe);

virtual void integrandsMx
   (ElmMatVec& elmat, const MxFiniteElement& mfe);
void makeSystemMx
   (
   DegFreeFE&  dof,
   LinEqAdmFE& lineq,
   FieldsFE&   all_unknowns,
   int         itgrule_manager,    // one of the all_unknowns comp.
   bool        compute_A = true,
   bool        compute_RHS = true
   );


public:

   Handle(FieldsFunc)      usource;
   Stokes ();
  ~Stokes ();

   virtual void adm (MenuSystem& menu);
   virtual void define (MenuSystem& menu, int level = MAIN);
   virtual void defineData (MenuSystem& menu, int level = MAIN);
   virtual void scan ();

   virtual void solveProblem ();
   virtual void saveResults ();
   virtual void resultReport ();
};
```

The coll field contains all the unknown scalar fields and is required in some mixed finite element routines. The v field is simply the velocity field as a *vector field* suitable for dumping to a simres database and used later for visualization purposes.

The first non-trivial part of the code in class Stokes concerns initialization of the data structures. First we compute the grid, then the basis function grids, followed by the fields, then the degree of freedom handler, and finally the linear system. The element used in each basis function grid can be set on the menu. A typical initialization of a basis function grid object is then like

```
v_grid.rebind (new BasisFuncGrid (*grid));
String v_elm_tp = menu.get ("v element");
v_grid->setElmType (v_grid->getElmType(), v_elm_tp);
```

with similar code for p_grid. The string v element is read from the input file, where the Crouzeix-Raviart velocity elements and the constant pressure element is specified as,

```
set u element = ElmT3gn3bn2D
set p element = ElmT3gn1bn2D
```

The field objects for the velocity and the pressure are created from the basis function grids:

```
v.rebind (new FieldsFE (*v_grid, nsd, "v"));
p.rebind (new FieldFE (*p_grid, "p"));
```

The next step is to create a collector of all unknown vector and scalar fields.

```
coll.rebind (new FieldsFE (nsd+1,"collector"));
for (int k=1; k<=nsd; k++)
  coll->attach (v()(k), k);
coll->attach (*p, nsd+1);
dof.rebind (new DegFreeFE (*coll));
```

In the last line the `dof` is made from the `FieldsFE` object `coll`. This causes the general numbering to be used.

A central function is `fillEssBC`. The following code segment assigns the essential boundary conditions (by calling up functors for the exact $v$ at the boundary):

```
const int v_nno = v_grid->getNoNodes();
int    idof, i;
Ptv(real) x (nsd), values(nsd);
dof.initEssBC ();

// boundary indicators: Assume the default ones from PreproBox

// we assume that the v_x and v_y basis func nodes coincides such
// that we can use the same loop for all velocity components

for ( i=1; i<= v_nno; i++) {
  if ( v_grid->essBoNode(i)) {
    v_grid->getCoor(x,i);
    v_anal->valuePt(values,x);
    for (int d=1; d<= nsd; d++) {
      idof = dof.fields2dof(i,d);
      dof.fillEssBC(idof, values(d));
    }
  }
}
```

In this example we have split the `fillEssBC` function in two functions, called `fillEssBC4V` and `fillEssBC4P`. The reason is that we want to reuse these functions in the block solvers [15] that are derived from this class. The statements should be obvious from the previous material on the `Laplace1GN` and `Laplace1Mx` solvers.

The heart of any Diffpack finite element simulator is the `integrands` function. In the present case this routine is more complicated than in scalar PDE problems. The discrete equations (4.19)–(4.20) interpreted at the element level have the suitable form for direct implementation in an `integrands` routine, as we have already stated in Section 4.4.4 that the numbering of equations and degrees of freedom at the element level is always of the field-by-field form. We exemplify this in 2D, while the following code also works in 3D. The element equations consist of $n_v$ equations from the $x$ components of the equation of motion, then $n_v$ equations from the $y$ component (these two correspond to $r = 1$ and $r = 2$ in (4.19)). The last $n_p$ equations stem from the continuity equation (4.20). The unknowns are the $n_v$ nodal values of $\hat{v}^1$, the $n_v$ nodal values of $\hat{v}^2$, and the $n_p$ nodal values of $\hat{p}$. The element equations can be partitioned like this in 2D:

$$
\begin{pmatrix} A & 0 & B^1 \\ 0 & A & B^2 \\ (B^1)^T & (B^2)^T & 0 \end{pmatrix}
\begin{pmatrix} v_x \\ v_y \\ p \end{pmatrix}
=
\begin{pmatrix} c^{(1)} \\ c^{(2)} \\ 0 \end{pmatrix} ,
$$

with $A = \{A_{ij}\}$, $B^r = \{B^r_{ij}\}$, $(B^r)^T = \{B^r_{ji}\}$ and $c^i = \{c^i_i\}$. In `integrands` it is convenient to generate four block matrices: the upper left matrix, corresponding to the Laplace term $\nabla^2 v$,

$$
\begin{pmatrix} A & 0 \\ 0 & A \end{pmatrix} .
$$

The upper right matrix, corresponding to the pressure term $-\nabla p$,

$$
\begin{pmatrix} B^1 \\ B^2 \end{pmatrix} .
$$

The lower left matrix, corresponding to the divergence term $\nabla \cdot v$,

$$
\left( (B^1)^T \; (B^2)^T \right) .
$$

Finally, the lower right matrix, which is $\mathbf{0}$. With the above formulas we are ready to present the gory details of the `integrands` function:

```
void Stokes:: integrandsMx (ElmMatVec& elmat,
                            const MxFiniteElement& mfe)
{
  const int nsd   = mfe.getNoSpaceDim();
  const int nvxbf = mfe(1).getNoBasisFunc();
  const int nvbf  = nvxbf*nsd;
  const int npbf  = mfe(nsd+1).getNoBasisFunc();
  const real detJxW = mfe.detJxW();
  real   nabla;
  Ptv(NUMT) values(nsd);
  Ptv(NUMT) x(nsd);
  mfe.getGlobalEvalPt(x);
  if (usource.ok()) usource->valuePt(values,x,DUMMY);
  else values.fill(0.0);
  int    i,j,k,d,s;
  int ig,jg;

  // upper left block matrix, the term -Laplace(u)*mfe(U).N(i)
```

```
for (d = 1; d <= nsd; d++){
  for (i = 1; i <= nvxbf; i++) {
    ig = (d-1)*nvxbf+i;
    for (j = 1; j <= nvxbf; j++) {
      jg = (d-1)*nvxbf+j;
      nabla = 0;
      for (k = 1; k <= nsd; k++)
        nabla += mfe(d).dN(i,k)*mfe(d).dN(j,k);
      elmat.A(ig,jg) += nabla*detJxW;
    }
    elmat.b(ig) += values(d)*mfe(d).N(i)*detJxW;
  }
}

// upper right block matrix, the term (dp/dx)*mfe(U).N(i)
for (s = 1; s <= nsd; s++)
  for (i = 1; i <= nvxbf; i++) {
    ig = (s-1)*nvxbf+i;
    for (j = 1; j <= npbf; j++) {
      jg = nvbf+j;
      elmat.A(ig,jg) -= mfe(s).dN(i,s)*mfe(nsd+1).N(j)*detJxW;
    }
  }

// lower left block matrix, the term du/dx*mfe(P).N(i),
// eq.no. nubf+i
for ( d=1 ; d<= nsd; d++)
  for (i = 1; i <= npbf; i++) {
    ig = nvbf+i;
    for (j = 1; j <= nvxbf; j++) {
      jg = (d-1)*nvxbf+j;
      elmat.A(ig,jg) -= mfe(nsd+1).N(i)*mfe(d).dN(j,d)*detJxW;
    }
  }
}
```

Notice that both mfe(d) hold the $N_i$ functions and their derivatives (because
d <= nsd), whereas mfe(nsd+1) holds the $L_i$ functions. The program administration takes place in solveProblem:

```
void Stokes:: solveProblem ()
{
  fillEssBC ();
  makeSystemMx (*dof, *lineq, *coll, 1);
  linsol.fill(0.0);
  lineq->solve();
  dof->vec2field (linsol, *coll);

  database->dump(*v);
  database->dump(*p);
}
```

The contents of solveProblem follows the set-up from the examples in [9],
except that we call makeSystemMx and use the coll collection of fields, containing both v and p, both in makeSystemMx and when storing the solution of
the linear system (linsol) back in the fields (dof->vec2field).

At last, we compute the $L_2$, $L_1$ and $L_\infty$ norms of the error in both velocity and pressure. This is accomplished with some functions defined in `ErrorNorms`,

```
ErrorNorms::Lnorm (*v_anal,       // supplied function (see above)
                   *v,            // numerical solution
                   DUMMY,         // point of time
                   L1_error, L2_error, Linf_error, // error norms
                   GAUSS_POINTS); // point type for numerical integr.

s_o <<oform("L2 error of v=%12.5e \n", L2_error);
ErrorNorms::Lnorm (*p_anal,       // supplied function (see above)
                   *p,            //  numerical solution
                   DUMMY,         // point of time
                   L1_error, L2_error, Linf_error, // error norms
                   GAUSS_POINTS); // point type for numerical integr.

s_o <<oform("L2 error of p=%12.5e \n", L2_error );
```

The complete source code is found in

```
src/mixed/Stokes
```

Notice that efficient preconditioners are described in the companion chapter [15] and the source code is located in the subdirectory `block`. The multigrid preconditioners require the multigrid module [14].

### 4.6.3  Numerical Experiments for the Stokes Simulator

In this section we investigate the behavior of the error in terms of the mesh size $h$. We manufacture a non-polynomial smooth solution, simply by choosing some solutions $(u, p)$ and then compute $f$. Low-order polynomial solutions are avoided because these often lead to super convergence. Notice that $u$ must be divergence free and that $\int_\Omega p\, d\Omega = 0$. The manufactured solutions and the corresponding data read,

$$u = \begin{pmatrix} -\sin(xy)x \\ \sin(xy)y \end{pmatrix},$$

$$p = \cos(xy) - a,$$

$$a = \int_\Omega \cos(xy)\, d\Omega,$$

$$\mu = 1,$$

$$f = -\Delta u + \nabla p.$$

We use this solution instead of a physical relevant problem, to verify the implementation. It is recommended to always do such study of error behavior to eliminate bugs before advancing to more challenging physical problems. We measure the errors,

$$\varepsilon^v_{L^2(\Omega)} = \|u - \hat{v}\|_{L^2(\Omega)}, \tag{4.92}$$

$$\varepsilon^p_{L^2(\Omega)} = \|p - \hat{p}\|_{L^2(\Omega)}. \tag{4.93}$$

$$\tag{4.94}$$

The estimated errors are listed in Table 4.1. The Crouzeix-Raviart element and the Mini element show second order accuracy for the velocity and first order accuracy for the pressure, as expected from (4.76)-(4.77) and (4.79)-(4.80). The Taylor-Hood elements are better for smooth solutions, we get third order for the velocity and second order for the pressure, again in complete agreement with (4.69)-(4.71). The results listed below are made with

| $\hat{v}$-element | error\\$h$ | $2^{-2}$ | $2^{-3}$ | $2^{-4}$ | $2^{-5}$ |
|---|---|---|---|---|---|
| Crouzeix-Raviart | $\varepsilon^v_{L^2(\Omega)}$ | 9.00e-3 | 2.54e-4 | 6.66e-4 | 1.70e-4 |
| Crouzeix-Raviart | $\varepsilon^p_{L^2(\Omega)}$ | 7.72e-2 | 3.03e-2 | 1.25e-2 | 5.65-3 |
| Taylor-Hood | $\varepsilon^v_{L^2(\Omega)}$ | 2.99e-3 | 3.73e-4 | 4.65e-5 | 5.80e-6 |
| Taylor-Hood | $\varepsilon^p_{L^2(\Omega)}$ | 1.13e-2 | 1.71e-3 | 3.54e-4 | 8.60e-5 |
| Mini | $\varepsilon^v_{L^2(\Omega)}$ | 6.76e-3 | 1.64e-3 | 3.99e-4 | 9.86e-5 |
| Mini | $\varepsilon^p_{L^2(\Omega)}$ | 3.82e-1 | 8.66e-2 | 1.69e-2 | 5.53e-3 |

**Table 4.1.** Error obtained with different elements.

the solver described in [15]. That is, we have used the `SymMinRes` solver with a multigrid block preconditioner. It is worth noting that using a Krylov method, like `SymMinRes`, without a preconditioner leads to *very* slow convergence. Moreover, a standard preconditioner, e.g., RILU, actually often results in breakdown, because the system is indefinite. Instead block preconditioners, together with multigrid techniques from [14], should be used. In fact, optimal preconditioners are constructed in [15].

### 4.6.4  A Simulator for the Mixed Poisson Problem

We will now describe a mixed Poisson simulator, which is very similar to the Stokes simulator, but where we need to use some other elements. We have mentioned earlier that the Raviart-Thomas element and the robust element are implemented in Diffpack. Typical Stokes elements can also be used, but then lower accuracy of the velocity is obtained. The $H(\text{div}; \Omega)$ elements are *vector elements*, in the sense that the degrees of freedom correspond to vectors. These elements require special initialization. Therefore, we use a boolean variable `special_mapping` to distinguish vector elements from the other elements. The initialization of the velocity fields reads,

```
if (special_mapping) {
  u_x.rebind (new FieldFE (*u_bx, "u_x"));
  u_y.rebind (new FieldFE (*u_by, u_x->values (), "u_y"));
} else {
  u_x.rebind (new FieldFE (*u_bx, "u_x"));
  u_y.rebind (new FieldFE (*u_by, "u_y"));
}
```

In the case of vector elements, the same vector of unknowns, `u_x->values()`, is used both in the `u_x` field and the `u_y` field. However, these fields can be used as usual, e.g., a collection of all the fields is made as:

```
fields_all.rebind (new FieldsFE (3, "u-collector"));
fields_all->attach (*u_x, 1);
fields_all->attach (*u_y, 2);
fields_all->attach (*p, 3);
```

The `fields_all` object is suitable for post-processing. In the simulator for the Stokes problem we based the `DegFreeFE` on `coll` which is similar to the `fields_all` object made here. However, because `u_x` and `u_y` share the same unknowns we would then end up with too many degrees of freedom. Instead only one of the velocity fields is used to initialize the `DegFreeFE`.

```
fields_numbering.rebind (new FieldsFE (2, "u-collector"));
fields_numbering->attach (*u_x, 1);
fields_numbering->attach (*p, 2);
dof_numbering.rebind (new DegFreeFE (*fields_numbering));
```

The degrees of freedom in the Raviart-Thomas and the robust element are vectors (normal and tangential) rather than scalar values, and therefore the `fillEssBC` routine is slightly different. We need to know the normal direction on a side, as well as the value in a given point. To achieve this we make a loop that goes through all local nodes in all elements,

```
void PoissonMx:: fillEssBC4V (DegFreeFE& dof) {
... // initialization
  if (special_mapping) {

    for (e=1; e<= no_elms; e++) {
      nl = v_grid.getNoNodesInElm(e);
      for (l=1; l<= nl; l++) {
        idof = dof.loc2glob(e,l);
        dof.dof2fields(idof, i, f );
        if (v_grid.essBoNode (i)) {
          v_grid.getCoor (x, i);
          dof.fillEssBC (idof, uanalcomp->valuePt (x,e,l));
        }
      }
    }
  }
```

The element number and the local node can be used to compute the normal vector by `MxMapping`. This is done in the `uAnalVecComp` functor holding a pointer to the analytical solution `uAnal` functor `anal_funcs`. The `uAnal` contains the analytical solution in vector form, while the `uAnalVecComp` computes the value in either the normal or tangential direction. The code reads,

```
real uAnalVecComp:: valuePt (const Ptv (real) & p,
                int elm_no, int loc_node, real t ) {
  Ptv (real) values;
  anal_funcs->valuePt (values, p);
  MxFiniteElement mfe (data->fields_all ());
```

```
mfe.setItgRuleManager (1);
mfe.refill (elm_no);
MxMapping & mxmapping = mfe.map ();
Ptv(real) dummy(2); dummy = 0;
Ptv(real)& vector = dummy;

if (data->special_mapping) {
  if (data->u_bx->getElmType () == "ElmT3gn3bn2Du") {
    int side = loc_node;
    vector = mxmapping.getNormalVector (side);
    return values.inner(vector);
```

The `integrandsMx` function also needs to be modified slightly. When using vector elements, the basis functions `mfe.N(1,i)` and `mfe.N(2,i)` correspond to the same entry in the element matrix. We manage this by using an additional integer D, which is set to one in case of vector elements, while it is set to d (as usual) for standard elements. The $(\text{div }\hat{v}, L_i)$ term then reads,

```
// lower left block matrix, the term du/dx*mfe(P).N(i), eqn. nubf+i
for ( d=1 ; d<= nsd; d++) {
  if ( special_mapping ) D = 1;
  else D = d;
  for (i = 1; i <= npbf; i++)
    for (j = 1; j <= nuxbf; j++)
      elmat.A(nUbf+i,(D-1)*nuxbf+j) +=
            mfe.N(3,i)*mfe.dN(d,j,d)*detJxW;
}
```

A suitable `makeSystemMx` is,

```
makeSystemMx (*dof_numbering, *lineq, *fields_all, 1);
```

This variant of `makeSystemMx` makes a matrix based on the numbering of `dof_numbering`, but base the mixed elements on `fields_all`.

We can estimate the error with a varity the functions in `ErrorNorms`. Functions suitable for our purpose are,

```
ErrorNorms::HdivNorm(*uanal, *fields_u, Hdiv_error_norm,
            L2_part_error, div_part_error, inf_error, DUMMY, false);
ErrorNorms::Lnorm(*panal, *p, DUMMY, L1_error_of_p,
            L2_error_of_p, Linf_error_of_p);
```

The complete source code is found in

```
src/mixed/PoissonMx
```

Notice that efficient preconditioners are described in the companion chapter [15] and the source code is located in the subdirectory `block`. The multigrid preconditioners require the multigrid module [14].

### 4.6.5 Numerical Experiments for the Mixed Poisson Simulator

To verify the program we construct a simple problem with a known analytical solution,

$$\lambda = 1, \tag{4.95}$$

$$p = -\cos(\pi x)\cos(\pi y), \tag{4.96}$$

$$v = \nabla p, \tag{4.97}$$

$$f = \nabla \cdot v. \tag{4.98}$$

We have changed the sign of $p$ to obtain symmetry. Krylov solvers that utilize the symmetry, like SymMinRes and Symmlq, are much more efficient than the methods that do not, e.g, BiCGStab and Orthomin. In addition to the norms of the errors in (4.92)-(4.93), we also measure the $H(\mathrm{div};\Omega)$ norm,

$$\varepsilon^v_{H(\mathrm{div};\Omega)} = \|u - \hat{v}\|_{H(\mathrm{div};\Omega)}, \tag{4.99}$$

The Table 4.2 shows the errors associated with the Raviart-Thomas element, the robust element, and the Mini element. Both the Raviart-Thomas element and the robust element are designed for this problem and we get linear convergence of the velocity error in $H(\mathrm{div};\Omega)$ and $L^2(\Omega)$, whereas the pressure converges linearly in $L^2(\Omega)$, as expected from (4.84)-(4.85) and (4.86)-(4.88). On the other hand, the Mini element approximation does not converge in $H(\mathrm{div};\Omega)$. However, the velocity shows linear convergence in $L^2(\Omega)$ and we have quadratic convergence of the pressure in $L^2(\Omega)$ (c.f [16]).

| $\hat{v}$-element | error$\backslash h$ | $2^{-2}$ | $2^{-3}$ | $2^{-4}$ | $2^{-5}$ |
|---|---|---|---|---|---|
| Raviart-Thomas | $\varepsilon^v_{H(\mathrm{div};\Omega)}$ | 1.00e-1 | 5.05e-2 | 2.53e-2 | 1.26e-2 |
| Raviart-Thomas | $\varepsilon^v_{L^2(\Omega)}$ | 6.72e-2 | 3.36e-2 | 1.68e-2 | 8.39e-3 |
| Raviart-Thomas | $\varepsilon^p_{L^2(\Omega)}$ | 1.60e-2 | 7.69e-3 | 3.74e-3 | 1.85e-3 |
| Robust | $\varepsilon^v_{H(\mathrm{div};\Omega)}$ | 6.41e-2 | 3.22e-2 | 1.62e-2 | 8.08e-3 |
| Robust | $\varepsilon^v_{L^2(\Omega)}$ | 5.81e-3 | 1.49e-3 | 3.79e-4 | 9.65e-5 |
| Robust | $\varepsilon^p_{L^2(\Omega)}$ | 1.48e-2 | 7.34e-3 | 3.66e-3 | 1.83e-3 |
| Mini | $\varepsilon^v_{H(\mathrm{div};\Omega)}$ | 1.26e+0 | 1.32e+0 | 1.35e+0 | 1.36e+0 |
| Mini | $\varepsilon^v_{L^2(\Omega)}$ | 4.82e-2 | 2.42e-2 | 1.22e-2 | 6.09e-3 |
| Mini | $\varepsilon^p_{L^2(\Omega)}$ | 2.89e-2 | 8.75e-3 | 2.58e-3 | 7.42e-4 |

Table 4.2. Error obtained with different elements.

### Acknowledgements

The authors are grateful to Professor R. Winther for many useful discussions.

# References

1. D.N. Arnold, F. Brezzi, and M. Fortin. A stable finite element method for the stokes equations. *Calcolo*, 21:337–344, 1984.
2. S. C. Brenner and L. R. Scott. *The Mathematical Theory of Finite Element Methods*. Springer-Verlag, 1994.
3. F. Brezzi and M. Fortin. *Mixed and Hybrid Finite Element Methods*. Springer-Verlag, 1991.
4. A. M. Bruaset and H. P. Langtangen. A comprehensive set of tools for solving partial differential equations; Diffpack. In M. Dæhlen and A. Tveito, editors, *Mathematical Models and Software Tools in Industrial Mathematics*, pages 61–90. Birkhäuser, 1997.
5. M. Crouzeix and P.A. Raviart. Conforming and non–conforming finite element methods for solving the stationary stokes equations. *RAIRO Anal. Numér*, 7:33–76, 1973.
6. R. S. Falk and J. E. Osborn. Error estimates for mixed methods. *R.A.I.R.O. Numerical Analysis*, 14:249–277, 1980.
7. C. A. J. Fletcher. *Computational Techniques for Fluid Dynamics, Vol I and II*. Springer Series in Computational Physics. Springer-Verlag, 1988.
8. L. P. Franca, T. J. R. Hughes, and R. Stenberg. Stabilized finite element methods. In M. D. Gunzburger and R. A. Nicolaides, editors, *Incompressible Computational Fluid Dynamics; Trends and Advances*. Cambridge University Press, 1993.
9. V. Girault and P. A. Raviart. *Finite Element Methods for Navier-Stokes Equations*. Springer-Verlag, 1986.
10. C. Johnson. *Numerical solution of partial differential equations by the finite element method*. Studentlitteratur, 1987.
11. H. P. Langtangen. Tips and frequently asked questions about Diffpack. World Wide Web document: Diffpack v1.4 Report Series, SINTEF & University of Oslo, 1996. *http://www.nobjects.com/diffpack/reports*.
12. H. P. Langtangen. *Computational Partial Differential Equations - Numerical Methods and Diffpack Programming*. Textbooks in Computational Science and Engineering. Springer, 2nd edition, 2003.
13. H. P. Langtangen. Details of finite element programming in Diffpack. The Numerical Objects Report Series #1997:9, Numerical Objects AS, Oslo, Norway, October 6, 1997. See *ftp://ftp.nobjects.com/pub/doc/NO97-09.ps.gz*.
14. K.-A. Mardal, H. P. Langtangen, and G.W. Zumbusch. Software tools for multigrid methods. In H. P. Langtangen and A. Tveito, editors, *Advanced Topics in Computational Partial Differential Equations – Numerical Methods and Diffpack Programming*. Springer, 2003.
15. K.-A. Mardal, J. Sundnes, H.P Langtangen, and A. Tveito. Systems of PDEs and block preconditionering. In H. P. Langtangen and A. Tveito, editors, *Advanced Topics in Computational Partial Differential Equations – Numerical Methods and Diffpack Programming*. Springer, 2003.
16. K.-A. Mardal, X.-C. Tai, and R. Winther. Robust finite elements for Darcy-Stokes flow. *SIAM Journal on Numerical Analysis*, 40:1605–1631, 2002.
17. R. Rannacher. Finite element methods for the incompressible Navier-Stokes equation. 1999. *http://www.iwr.uni-heidelberg.de/sfb359/Preprints1999.html*.
18. R. Rannacher and S. Turek. A simple nonconforming quadrilateral stokes element. *Numer. Meth. Part. Diff. Equ*, 8:97–111, 1992.

19. P. A. Raviart. Mixed finite element methods. In D. F. Griffiths, editor, *The Mathematical Basis of Finite Element Methods*. Clarendon Press, Oxford, 1984.

20. P. A. Raviart and J. M. Thomas. A mixed finite element method for 2-order elliptic problems. *Matematical Aspects of Finite Element Methods*, 1977.

21. S. Turek. *Efficient Solvers for Incompressible Flow Problems*. Springer, 1999.

# Chapter 5

# Systems of PDEs and Block Preconditioning

K.-A. Mardal[1,2], J. Sundnes[1,2], H. P. Langtangen[1,2], and A. Tveito[1,2]

[1] Simula Research Laboratory
[2] Department of Informatics, University of Oslo

**Abstract.** We consider several systems of PDEs and how block preconditioners can be formulated. Implementation of block preconditioners in Diffpack is in particular explained. We emphasize object-oriented design, where a standard simulator is developed and debugged before being extended with efficient block preconditioners in a derived class. Optimal preconditioners are applied to the Stokes problem, the mixed formulation of the Poisson equation, and the Bidomain model for the electrical activity in the heart.

## 5.1 Introduction

In this chapter we will see how block preconditioners can be utilized in Diffpack for three different model problems. The chosen model problems are fundamental problems in engineering, medicine, and science and need to be solved in real-life situations, implying very large scale simulations. In such situations the algorithm for solving the matrix equation is of vital importance, and this motivates the use of highly efficient block preconditioners.

A model for the electrical activity in the heart is described in [14]. We will consider a simplified form of the equations, given by

$$\frac{\partial v}{\partial t} = \nabla \cdot (\sigma_i \nabla v) + \nabla \cdot (\sigma_i \nabla u_e), \ x \in \Omega, \ t > 0,$$
$$\nabla \cdot (\sigma_i \nabla v) + \nabla \cdot ((\sigma_i + \sigma_e) \nabla u_e) = 0, \ x \in \Omega, \ t > 0,$$

where $v$ is the transmembrane potential, $u_e$ is the extracellular potential, and $\sigma_i$ and $\sigma_e$ are the intra- and extracellular conductivities, respectively. The two other problems, described in [10], are the Stokes problem and the mixed Poisson problem. It is well known how efficient preconditioners may be constructed for these problems (see e.g. [13,1]). The Stokes problem, modeling creeping Newtonian flow, reads

$$-\mu \Delta v + \nabla p = f \text{ in } \Omega \text{ (equation of motion)},$$
$$\nabla \cdot v = 0 \text{ in } \Omega \text{ (mass conservation)},$$

where $v$ is the velocity, $p$ is the pressure, $\mu$ is the viscosity, and $f$ denotes the body forces. The last example is the mixed Poisson problem,

$$v + \lambda \nabla p = 0 \text{ in } \Omega \quad \text{(Darcy's law)},$$
$$\nabla \cdot v = g \text{ in } \Omega \quad \text{(mass conservation)},$$

where $v$ is the velocity, $p$ is the pressure, $\lambda$ is a mobility parameter, and $g$ is a source term. These equations are discretized by the finite element method. In the Stokes problem and the mixed Poisson problem we use mixed elements.

With a suitable numbering of the unknowns the discretization of all the equations given above may be written as

$$\begin{bmatrix} A & B \\ C & D \end{bmatrix} \begin{bmatrix} x \\ y \end{bmatrix} = \begin{bmatrix} b \\ c \end{bmatrix}. \tag{5.1}$$

It is this particular block structure that motivates the preconditioners under considerations. We want to utilize the structure of the problems to construct efficient preconditioners. In fact, for all our model problems it is possible to obtain optimal preconditioners by applying simple algebraic operator splitting techniques to (5.1). An optimal preconditioner leads to a number of iterations that is bounded independent of the number of unknowns. This implies that the amount of computational work is proportional to the number of unknowns. This is (order) optimal.

The remainder of this paper is organized as follows. Section 1 introduces block preconditioners from an algebraic point of view and describes some software tools in Diffpack. In Section 2 the Bidomain equations are considered, and an optimal preconditioner is developed together with appropriate software tools. In Section 3 we extend the simulators presented in [10] to utilize block solvers.

## 5.2    Block Preconditioners in General

In this section we will try to motivate block preconditioners from an algebraic point of view. We first consider the Jacobi method on block form, which is often used to solve coupled systems of PDEs. We then discuss some basic properties of the Jacobi method and motivate why and how it can be used as a preconditioner. We discuss the Jacobi method instead of the slightly faster Gauss-Seidel, because the Jacobi iteration is symmetric and can therefore be used as a preconditioner for, e.g., the Conjugate-Gradient method. Then we briefly consider Conjugate-Gradient-like methods and see how they can be combined with these block preconditioners. The two last sections deal with software tools to utilize these preconditioners in existing simulators. Readers interested in the mathematical details of iterative methods, e.g., the Conjugate-Gradient method, multigrid methods, and classical iterations, are refereed to, e.g., Hackbusch [8]. Krylov solvers in general, for non-symmetric and indefinite matrices, are dealt with in, e.g., Bruaset [3].

## 5.2.1    Operator Splitting Techniques

Operator slitting techniques are widely used for coupled systems of PDEs, and there are many different techniques that are used for different PDE problems. Often the operator splitting is used in time, resulting in a splitting on the PDE level. One of the main advantages with these techniques is that a complicated system of coupled PDEs is reduced to simpler equations like the Poisson equation, convection-diffusion equations, etc. For such equations a fast solver like multigrid can be employed. Additionally well tested software for such problems can be reused.

However, here we will consider techniques on the linear algebra level, but we should bear in mind that the fast solvers should and will also be used in this approach. The reader interested in the mathematical details is refereed to [8]. Given an algebraic system on the form,

$$\mathcal{A} \begin{bmatrix} x \\ y \end{bmatrix} = \begin{bmatrix} A & B \\ C & D \end{bmatrix} \begin{bmatrix} x \\ y \end{bmatrix} = \begin{bmatrix} b \\ c \end{bmatrix}, \tag{5.2}$$

the simplest operator splitting technique is the Jacobi method,

$$x^{k+1} = A^{-1}(b - By^k), \tag{5.3}$$
$$y^{k+1} = D^{-1}(c - Cx^k). \tag{5.4}$$

Let $\mathcal{D}$ be the block-wise diagonal of $\mathcal{A}$,

$$\mathcal{D} = \begin{bmatrix} A & 0 \\ 0 & D \end{bmatrix}. \tag{5.5}$$

The convergence of the Jacobi iteration is determined by the spectral radius of $\mathcal{I} - \mathcal{D}^{-1}\mathcal{A}$, $\rho(\mathcal{I} - \mathcal{D}^{-1}\mathcal{A})$, where $\mathcal{I} = diag(I, I)$. A necessary condition for convergence is that

$$\rho(\mathcal{I} - \mathcal{D}^{-1}\mathcal{A}) \le \delta < 1. \tag{5.6}$$

Hence, the standard results of point-wise Jacobi apply also to the block version of Jacobi's method. For coupled systems of PDEs it is often not easy to verify (5.6) a priori. The convergence/divergence determined by $\delta$, is usually found experimentally, and quite often the Jacobi method diverges. The Jacobi is also known to be a rather slow iterative method, hence it might be necessary to iterate quite a number of times.

A natural question to ask is how accurate $A^{-1}$ and $D^{-1}$ need to be. Can we use cheap approximations of these matrices, $\tilde{A}^{-1}$ and $\tilde{D}^{-1}$, and speed up the simulation time? A naive inexact version of block Jacobi (5.3)-(5.4) reads,

$$x^{k+1} = \tilde{A}^{-1}(b - By^k), \tag{5.7}$$
$$y^{k+1} = \tilde{D}^{-1}(c - Cx^k). \tag{5.8}$$

However, we see that this approach is dangerous, since we then solve a perturbed system

$$\tilde{A}\begin{bmatrix} \tilde{x} \\ \tilde{y} \end{bmatrix} = \begin{bmatrix} \tilde{A} & B \\ C & \tilde{D} \end{bmatrix}\begin{bmatrix} \tilde{x} \\ \tilde{y} \end{bmatrix} = \begin{bmatrix} b \\ c \end{bmatrix}. \tag{5.9}$$

Instead one should use the consistent version of the inexact Jacobi method[1]

$$x^{k+1} = x^k + \tilde{A}^{-1}(b - Ax^k - By^k), \tag{5.10}$$
$$y^{k+1} = y^k + \tilde{D}^{-1}(c - Cx^k - Dy^k). \tag{5.11}$$

The convergence will then be determined by

$$\rho(\mathcal{I} - \tilde{D}^{-1}\mathcal{A}). \tag{5.12}$$

Note that this quantity is even harder to estimate than $\delta$. The method (5.10)-(5.11) can also be seen as a Richardson method preconditioned with $\tilde{D}^{-1}$. Introducing a damping parameter $\tau$,

$$x^{k+1} = x^k + \tau\tilde{A}^{-1}r_x^k, \tag{5.13}$$
$$y^{k+1} = y^k + \tau\tilde{D}^{-1}r_y^k, \tag{5.14}$$

where the residual $[r_x^{k+1}, r_y^{k+1}]$ is defined by

$$r_x^k = b - Ax^k - By^k, \tag{5.15}$$
$$r_y^k = c - Cx^k - Dy^k. \tag{5.16}$$

We can always make the iteration convergent by adjusting[2] $\tau$ such that

$$\rho(\mathcal{I} - \tau\tilde{D}^{-1}\mathcal{A}) < 1. $$

However, $\tau$ must then usually be found and tuned experimentally for optimal performance. A too small $\tau$ will reduce the efficiency, while a too large $\tau$ will lead to divergence.

Although we have here focused on the block Jacobi method, a large number of other operator splitting methods exist. Variants that are similar to block Jacobi are the block versions of Gauss-Seidel, SOR, and SSOR, which may all be obtained by modifying the preconditioner. On this $2 \times 2$ block system we replace $\mathcal{D}^{-1}$ with the following $\mathcal{B}_{GS}^{-1}$ to obtain the block Gauss-Seidel method,

$$\mathcal{B}_{GS}^{-1} = \begin{bmatrix} A^{-1} & 0 \\ -D^{-1}CA^{-1} & D^{-1} \end{bmatrix}. \tag{5.17}$$

The following discussion will not only concern with the block-diagonal preconditioner leading to the block Jacobi method, but with preconditioners in

---

[1] This is what Hackbusch [8] calls the second normal form of a consistent linear iteration. Inexact Jacobi is used to generate $\tilde{D}^{-1}$.

[2] It can be done by setting $\tau = \frac{1}{\|\tilde{D}^{-1}\mathcal{A}\|}$.

general. Hence, the notation $\tilde{\mathcal{D}}^{-1}$ is replaced with $\mathcal{B}$. For symmetric problems the condition number $\kappa(\mathcal{B}\mathcal{A})$ determines the efficiency of the algorithm and whether it converges or not. In practice the estimates of these parameters are almost always computed. For the Stokes problem the Uzawa algorithm or the more general Pressure Schur Complement framework (see [16]) is often used. We will see that similar reasoning can be used here. In the next section we will consider more robust methods, that in principle always are convergent. Still, the convergence rate depends depends strongly on $\kappa(\mathcal{B}\mathcal{A})$.

## 5.2.2   Conjugate Gradient-Like Methods

Although there exists a large variety of iterative solvers that may utilize preconditioners based on operator splitting, we will focus on a family of such methods that is particularly interesting. This is the group of Conjugate Gradient-like methods or Krylov Subspace methods. There has been developed a rich set of Krylov solvers for different purposes, many of which are implemented in Diffpack. Some of the most commonly used are ConjGrad for symmetric and positive definite (SPD) systems, Symmlq and SymMinRes for symmetric and indefinite problems, and general purpose methods like Orthomin, GMRES, BiCGStab. The reader is referred to [3] for a general discussion of Krylov solvers and iterative solvers in general, and [4,9] for a presentation of the Krylov solvers implemented in Diffpack.

The Krylov solvers do not need a $\tau$ to be chosen to be convergent. However, the efficiency of an appropriate Krylov solver is to a large extent dependent on the condition number of the coefficient matrix $\mathcal{A}$, $\kappa(\mathcal{A})$. We have that the number of iterations is proportional to $\sqrt{\kappa(\mathcal{A})}$ for the Conjugate Gradient method[3]. For standard second order elliptic problems in 2D the condition number is typically $\mathcal{O}(h^{-2})$, where $h$ is the grid size parameter, and the performance of the Krylov solver will therefore deteriorate as the grid is refined. For sparse matrices arising from the discretization of PDEs, the memory required to store the matrix and vectors is $\mathcal{O}(n)$, where $n$ is the number of unknowns in the linear system. The basic operations in Krylov solvers are matrix-vector products, vector additions, and inner products, which are all $\mathcal{O}(n)$ operations. The overall performance is therefore $\mathcal{O}(n^{3/2})$, since the number of iterations is $\sqrt{h^{-2}} \sim n^{1/2}$.

In matrices arising from discretizations of PDEs the condition number increases as the resolution of the grid increases. We are not satisfied with being able to solve the problem on one particular grid. In four years when we have a four times faster computer with four times as much memory[4], we

---

[3] some assumptions must be made, see Section 5.3.3

[4] A popular version of Moore's law predicts that the processor speed double at least every other year. However, it seems to be slowing down, a doubling is more likely to happen every two year. It is also becoming apparent that the speed of the CPU is not the only factor that determine the total CPU time. The latency

want a four times as accurate solution in the same amount of time. Hence, if we want efficient solution methods we need to consider a family of problems,

$$\mathcal{A}_h \begin{bmatrix} x_h \\ y_h \end{bmatrix} = \begin{bmatrix} A_h & B_h \\ C_h & D_h \end{bmatrix} \begin{bmatrix} x_h \\ y_h \end{bmatrix} = \begin{bmatrix} b_h \\ c_h \end{bmatrix}, \tag{5.18}$$

where $h$ is the most important parameter for the condition number. Our ultimate goal is to construct methods that will solve the problem in $\mathcal{O}(n)$ operations, independently of the grid size $h$. As mentioned above, this is commonly called an *optimal method*. We are considering iterative methods, hence, the demands on a optimal solver can be summarized as follows[5]

A. The number of operations in each iteration should be $\mathcal{O}(n)$ .

B. The memory requirement should be $\mathcal{O}(n)$.

C. The number of iterations needed to reach convergence should be bounded independently of the mesh.

As mentioned earlier, the convergence of a Krylov solver for symmetric systems can be estimated in terms of the condition number of $\mathcal{A}_h$ and C is therefore not satisfied. A common way to speed up the Krylov solver is therefore to use a preconditioner, essentially meaning that we multiply the system by a matrix $\mathcal{B}_h$ to obtain another system with the same solution,

$$\mathcal{B}_h \mathcal{A}_h \begin{bmatrix} x \\ y \end{bmatrix} = \mathcal{B}_h \begin{bmatrix} b \\ c \end{bmatrix}. \tag{5.19}$$

The preconditioner $\mathcal{B}_h$ should be a well designed operator, because the efficiency of the Krylov solver then depends on $\kappa(\mathcal{B}_h \mathcal{A}_h)$. The preconditioner $\mathcal{B}_h$ does not need to be formed as a matrix explicitly, it can also be an algorithm like, e.g., multigrid. The only action that is needed is its evaluation on a vector, $v_h = \mathcal{B}_h u_h$. The demands given above can then be stated as properties of the preconditioner $\mathcal{B}_h$:

D. The application of $\mathcal{B}_h$ to a vector should require $\mathcal{O}(n)$ operations.

E. The memory required to store $\mathcal{B}_h$ should be $\mathcal{O}(n)$.

F. $\kappa(\mathcal{B}_h \mathcal{A})$ should be "small" and bounded independently of $h$.

Item F, in the case of symmetric matrices, is often stated as

$$\kappa(\mathcal{B}_h \mathcal{A}_h) = \frac{\max_i |\lambda_i|}{\min_i |\lambda_i|} \le C, \tag{5.20}$$

---

of the RAM is also very important. In fact, for the problems we consider here, which are very memory intensive, the CPU usually spend 90% of the time waiting for lookups in RAM, see [5].

[5] For a time-dependent problem, the number of operations should be proportional to the number of unknowns at each time-step multiplied by the number of time-steps.

where $\lambda_i$ are the eigenvalues and $C$ is a constant that is independently of the discretization parameter $h$. We will then say that $\mathcal{B}_h \mathcal{A}_h \sim I$. Notice that the above condition (5.20) differ slightly from the spectral equivalence that is usually wanted for elliptic operators. It is needed for the indefinite systems in Section 5.4.

The "operator splitting" technique in (5.10)-(5.11) can be reused directly. Using the matrix representation in (5.5) with inexact subsolvers, the preconditioner in (5.19) reads

$$\mathcal{B}_h = \begin{bmatrix} \tilde{A}^{-1} & 0 \\ 0 & \tilde{D}^{-1} \end{bmatrix}. \tag{5.21}$$

This is the simplest form of the block preconditioners. Nevertheless it will be the focus of this chapter.

### 5.2.3  Software Tools for Block Matrices and Numbering Strategies

Equation (5.1) is just an ordinary matrix equation,

$$\mathcal{A}u = b,$$

with a particular numbering. That is, we have two unknowns $x$ and $y$ numbered block wise giving the solution vector,

$$u = [x_1, x_2, \ldots, x_n, y_1, y_2, \ldots, y_m].$$

In general $x$ and $y$ define different fields which may be defined on different grids. One example is the Stokes problem which requires mixed elements, e.g., bilinear elements for $y$ and biquadratic for $x$. Diffpack has a general numbering strategy capable of numbering several different fields simultaneously. However, a straightforward use of this numbering (as was explained in [10]) results in a matrix where the block structure is lost. We want to keep the system on block form, implying that the matrix $B$ in (5.1) needs a numbering consistent with $x$ column-wise, while the row numbering must be the same as in $y$. Hence, $B$ is a $n \times m$ matrix and $n$ may be unequal to $m$. This is the case for the Stokes and the mixed Poisson problem. Diffpack has several tools for assembling and solving linear system on block-structured form. The coupling of unknown fields (FieldFE) and their degree of freedom in the linear system (LinEqSystem) is handled by a DegFreeFE object, and in case of block-structured systems, we use two DegFreeFE objects for each block matrix. DegFreeFE has two numbering strategies, the special and the general, we need both. In the first simulators for the Bidomain model it is sufficient to use isoparametric elements and the special numbering can be used. In Section 5.4 we need to use mixed elements, and we must then use the general numbering.

Here is a list of the tools used to facilitate the handling of block-structured systems.

- Each matrix in $\mathcal{A}$ needs two DegFreeFE objects to handle the numbering. One is used for the column-wise numbering and one for the row-wise.
- Class ElmMatVecs holds a matrix of ElmMatVec objects, where each object of type ElmMatVec contains a block matrix and a block vector at the element level.
- Class SystemCollector administers the computation of the element block matrices and vectors, as well as the modification of the matrices and vectors due to essential (Dirichlet) boundary conditions. The class contains handles to an ElmMatVecs object, all the DegFreeFE objects, and functors [9] to hold the integrands for each block matrix and vector. The assembly of element contributions in LinEqAdmFE::assemble relies on a SystemCollector object.

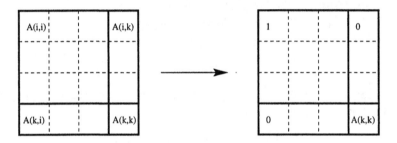

**Fig. 5.1.** Enforcing boundary condition subject to node i.

Linear systems in Diffpack support block structures. That is, the classes LinEqAdm, LinEqSolver, and LinEqSystem and their subclasses always work on LinEqMatrix and LinEqVector objects. LinEqMatrix is a MatSimplest (usually $1 \times 1$) of pointers (handles) to Matrix objects. The LinEqVector has a similar structure. This means that the Diffpack Krylov solvers can be used on the mixed system without modifications.

LinEqSystemPrec is a subclass of LinEqSystem designed for preconditioned linear systems. To handle block preconditioners, this class is extended by the subclass LinEqSystemBlockPrec. It has the same interface as LinEqSystemPrec plus some more functions associated with the block preconditioner. We refer to the manual page for the full interface. The use of LinEqSystemBlockPrec is exemplified in the source code in Chapter 5.3.4.

```
//Make a 2x2 matrix of integrand-functors
integrands.redim(2,2);
integrands(1,1).rebind( new Integrand11(this));
integrands(2,1).rebind( new Integrand21(this));
integrands(1,2).rebind( new Integrand12(this));
integrands(2,2).rebind( new Integrand22(this));
```

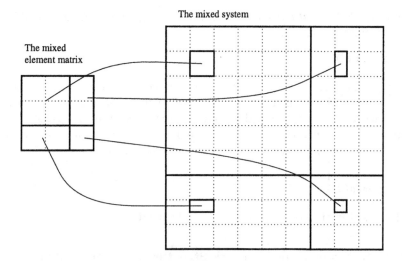

**Fig. 5.2.** The assembly process of the mixed system.

```
//Make ElmMatVecCalcStd functor to be used from makeSystem
calcelm.rebind (new ElmMatVecCalculator());


//Block matrix of ElmMatVec's
elmats.rebind(new ElmMatVecs());
//SystemCollectorFEM handles the assembly
//the block element matrices into the
//block matrix
elmsys.rebind(new SystemCollectorFEM());

//redim and attach datastructure:
elmsys->redim(2,2);
for (int i=1; i<= 2; i++)
  for (int j=1; j<= 2; j++)
    if (integrands(i,j).ok())
      elmsys->attach(integrands(i,j)(),i,j);

elmsys->attach(*calcelm);

elmsys->attach(*u_dof,1,true);
elmsys->attach(*u_dof,1,false);

elmsys->attach(*v_dof,2,true);
elmsys->attach(*v_dof,2,false);

//lineq is initialized from the DegFreeFE's attached
//to elmsys
lineq->initAssemble(*elmsys);

//attach the ElmMatVec's
elmats->attach(*elmsys);
elmsys->attach(*elmats);
```

*Assembly of the right hand side* Each of the `integrandMx` functors could have done calculations like

```
elmat.b(i) += ...
```

on the right hand side, as is common in a Diffpack `integrands` function. This would lead to a distributed calculation of the right hand side in several `IntegrandCalc` objects. Instead we have chosen to use the `IntegrandCalc` objects associated with the diagonal to handle the computation of the element vector. This gives cleaner and more efficient code. Hence, only

```
integrands(1,1)
integrands(2,2)
```

will be used in the computation of the right hand side.

### 5.2.4   Software Tools for Block Structured Preconditioners

As indicated above we will in this paper focus on block diagonal preconditioners, which may be written as

$$
\mathcal{B} = \begin{bmatrix} M & 0 \\ 0 & N \end{bmatrix}, \tag{5.22}
$$

where $M$ and $N$ are operators that are constructed in some way reflecting the structure of the linear system matrix $\mathcal{A}$. The software is not restricted to $2 \times 2$ block systems, any block system can be used. In this chapter we will however stick to $2 \times 2$ systems, for simplicity. In Diffpack block preconditioners can be made by defining the menu interface `MenuSystem`,

```
menu.setCommandPrefix("M prec");
Precond_prm ::defineStatic(menu, level+1);
menu.unsetCommandPrefix();
menu.setCommandPrefix("N prec");
Precond_prm::defineStatic(menu, level+1);
menu.unsetCommandPrefix()
```

scan the parameters,

```
menu.setCommandPrefix("M prec");
M_prec_prm.rebind( Precond_prm::construct());
M_prec_prm->scan(menu);
menu.unsetCommandPrefix ();
menu.setCommandPrefix("N prec");
N_prec_prm.rebind( Precond_prm::construct());
N_prec_prm->scan(menu);
menu.unsetCommandPrefix ();
```

and attach the preconditioners to the linear system, represented by the class `LinEqSystemBlockPrec`:

```
LinEqSystemBlockPrec& lins = CAST_REF(lineq->getLinEqSystem(),
      LinEqSystemBlockPrec);
lins.attach(*M_prec_prm,1);
lins.attach(*N_prec_prm,2);
```

`LinEqSystemBlockPrec` supply the routines

```
virtual void applyLeftBlockDiagPrec
(
  const LinEqVector& c,
        LinEqVector& d,
        TransposeMode tpmode = NOT_TRANSPOSED
);

// apply right block preconditioner Cright, d=Cright*c
virtual void applyRightBlockDiagPrec
(
  const LinEqVector& c,
        LinEqVector& d,
        TransposeMode tpmode = NOT_TRANSPOSED
);
```

which are used by the Krylov solvers. The user should make sure that the matrix, the vector and the corresponding preconditioner are all constructed using the same `DegFreeFE`-object. This will ensure that the numbering of all components are consistent.

This completes our general discussion of block preconditioners and the corresponding software tools. In the remainder of the paper we will demonstrate how the proposed solution method may be applied to specific problems.

## 5.3    The Bidomain Equations

As a first introduction to problems that lead to block structured linear systems, we consider a set of partial differential equations modeling the electrical activity in the heart. Solving these equations enables us to simulate the propagation of the electrical field that causes the cardiac cells to contract and thus initiates each heartbeat. Measurements of this electrical field on the surface of the body are known as electrocardiograms or ECG measurements, and are an important tool for identifying various pathological conditions. Performing realistic computer simulations may increase our understanding of this process, and may thus be a valuable tool for research and educational purposes.

### 5.3.1    The Mathematical Model

To obtain a set of partial differential equations describing the propagation of the electrical field, we introduce the quantities $u_i$ and $u_e$, denoting the intracellular and extracellular potential, respectively. To avoid treating the complex cellular structure of the tissue explicitly, we consider the heart muscle to be a continuum consisting of both intracellular and extracellular space. The potentials $u_i$ and $u_e$ are then volume averaged quantities defined at every point in the domain. To easily relate the electrical potentials to ionic currents crossing the cell membranes, which are extremely important for the propagation of the potentials, we introduce the transmembrane potential $v$, defined as $v = u_i - u_e$. If we further denote the intra- and extracellular conductivities by $\sigma_i$ and $\sigma_e$, respectively, the dynamics of the electrical potentials are governed by the following equations,

$$\chi C_m \frac{\partial v}{\partial t} + \chi I_{\text{ion}}(v, s) = \nabla \cdot (\sigma_i \nabla v) + \nabla \cdot (\sigma_i \nabla u_e), \quad x \in \text{Heart}(\Omega),$$
$$\nabla \cdot ((\sigma_i + \sigma_e)\nabla u_e) = -\nabla \cdot (M_i \nabla v), \qquad\qquad x \in \text{Heart}(\Omega).$$

This is the Bidomain model, introduced by Geselowitz [6] in the late 70s. The term $I_{\text{ion}}$ is the transmembrane ionic current, $C_m$ is the capacitance of the cell membrane, and $\chi$ is a parameter relating the area of the cell membrane to the tissue volume. A detailed derivation of these equations is given in [14]. The vector $s$ in the ionic current term $I_{\text{ion}}$ is a state vector characterizing the conductivity properties of the cell membrane. Realistic models of cardiac cells require this state vector to be determined from a complex set of ordinary differential equations. The solution of these ODEs contributes significantly to the complexity of solving the Bidomain equations numerically. Simulations based on the complete equations are presented in [14]. In the present section we want the Bidomain problem to serve as a simple model problem to introduce the concepts of block systems and block preconditioning. To obtain a problem that is more suitable for this purpose we disregard the ionic current term $I_{\text{ion}}$ and assume that $\chi C_m = 1$. We then get the simplified problem

$$\frac{\partial v}{\partial t} = \nabla \cdot (\sigma_i \nabla v) + \nabla \cdot (\sigma_i \nabla u_e), \quad x \in \Omega, \tag{5.23}$$

$$\nabla \cdot (\sigma_i \nabla v) + \nabla \cdot ((\sigma_i + \sigma_e)\nabla u_e) = 0, \quad x \in \Omega. \tag{5.24}$$

For the present study, we assume that the heart muscle is immersed in a non-conductive media, so that no current leaves the heart tissue. The current is assumed to be given by Ohms law, so this gives

$$n \cdot (\sigma_i \nabla u_i) = 0,$$

and

$$n \cdot (\sigma_e \nabla u_e) = 0,$$

on the surface of the heart, which defines the boundary conditions for (5.23)-(5.24). By utilizing that $u_i = v + u_e$, we may express these conditions in terms of our primary variables $v$ and $u_e$. Combining this with equations (5.23) and

(5.24) gives the following initial-boundary value problem

$$\frac{\partial v}{\partial t} = \nabla \cdot (\sigma_i \nabla v) + \nabla \cdot (\sigma_i \nabla u_e), \quad x \in \Omega, \ t > 0 \tag{5.25}$$

$$\nabla \cdot (\sigma_i \nabla v) + \nabla \cdot ((\sigma_i + \sigma_e)\nabla u_e) = 0, \quad x \in \Omega, \ t > 0 \tag{5.26}$$

$$n \cdot (\sigma_i \nabla v + \sigma_i \nabla u_e) = 0, \quad x \in \partial\Omega, \ t > 0 \tag{5.27}$$

$$n \cdot (\sigma_e \nabla u_e) = 0, \quad x \in \partial\Omega, \ t > 0 \tag{5.28}$$

$$v(x,0) = v_0(x), \quad x \in \Omega, \ t = 0, \tag{5.29}$$

where $v_0$ is a given initial distribution for the transmembrane potential.

## 5.3.2   Numerical Method

*Time Discretization* We let $\Delta t$ be the (constant) time step and denote by $v^l$ and $u^l$ approximations to $v$ an $u_e$, respectively, at time step $t_l$. More precisely we have

$$v^l(x) \approx v(x, l\Delta t),$$
$$u^l(x) \approx u_e(x, l\Delta t).$$

A time discretization of equation (5.25) may be obtained from a simple implicit Euler scheme, leading to the following equations to be solved at each time step,

$$\frac{v^l - v^{l-1}}{\Delta t} = \nabla \cdot (\sigma_i \nabla v^l) + \nabla \cdot (\sigma_i \nabla u^l), \quad x \in \Omega, \tag{5.30}$$

$$\nabla \cdot (\sigma_i \nabla v^l)\nabla \cdot ((\sigma_i + \sigma_e)\nabla u^l) = 0, \quad x \in \Omega. \tag{5.31}$$

To simplify the notation we introduce the inner products

$$a_i(v, \phi) = \Delta t \int_\Omega \sigma_i \nabla v \cdot \nabla \phi dx, \tag{5.32}$$

$$a_{i+e}(v, \phi) = \Delta t \int_\Omega (\sigma_i + \sigma_e)\nabla v \cdot \nabla \phi dx. \tag{5.33}$$

If we now define a suitable function space $V(\Omega)$, a weak formulation of equations (5.30)-(5.31) is to find $v^l, u^l \in V(\Omega)$ which satisfies

$$(v^l, \phi) + a_i(v^l, \phi) + a_i(u^l, \phi) = (v^{l-1}, \phi), \quad \forall \phi \in V(\Omega), \tag{5.34}$$

$$a_i(v^l, \phi) + a_{i+e}(u^l, \phi) = 0, \quad \forall \phi \in V(\Omega), \tag{5.35}$$

where $(\cdot, \cdot)$ denotes the $L^2$ inner product. All boundary integral terms in the weak equations vanish because of the homogeneous boundary conditions (5.27)-(5.28).

We now introduce a discrete subspace $V_h(\Omega)$ of $V(\Omega)$, spanned by basis functions $\phi_i(x), i = 1, \ldots, n$. We approximate $v^l$ and $u^l$ by

$$v^l(x) \approx \sum_{j=1}^{n} \phi_j(x)v_j, \qquad (5.36)$$

$$u^l(x) \approx \sum_{j=1}^{n} \phi_j(x)u_j, \qquad (5.37)$$

where $\phi_i(x), i = 1, \ldots, n$ is a suitable set of basis functions defined in $\Omega$. For the application of the finite element method, the basis functions are usually piecewise polynomials defined over a grid $\Omega_h$. The refinement of the grid is characterized by the discretization parameter $h$. We have $n \sim h^{-d}$, where $d$ is the number of physical space dimensions. Applying a standard Galerkin finite element discretization to the weak formulation (5.34)-(5.35), we obtain a system of linear equations given by

$$\sum_{j=1}^{l} (\phi_j, \phi_i)v_j + \sum_{j=1}^{l} a_i(\phi_j, \phi_i)v_j \qquad (5.38)$$

$$+ \sum_{j=1}^{l} a_i(\phi_j, \phi_i)u_j = (v^{l-1}, \phi_i), \quad \text{for } i = 1, \ldots, n, \qquad (5.39)$$

$$\sum_{j=1}^{l} a_i(\phi_j, \phi_i)v_j + \sum_{j=1}^{l} a_{i+e}(\phi_j, \phi_i)u_j = 0, \quad \text{for } i = 1, \ldots, n. \qquad (5.40)$$

In matrix form, this system may be written as

$$A \begin{bmatrix} \mathbf{v} \\ \mathbf{u} \end{bmatrix} = \begin{bmatrix} A & B \\ B & C \end{bmatrix} \begin{bmatrix} v \\ u \end{bmatrix} = \begin{bmatrix} \alpha \\ 0 \end{bmatrix}, \qquad (5.41)$$

where the matrix blocks are defined by

$$A_{ij} = (\phi_i, \phi_j) + a_i(\phi_i, \phi_j), \qquad (5.42)$$
$$B_{ij} = a_i(\phi_i, \phi_j), \qquad (5.43)$$
$$C_{ij} = a_{i+e}(\phi_i, \phi_j), \qquad (5.44)$$
$$\alpha_i = (v^l, \phi). \qquad (5.45)$$

### 5.3.3   Solution of the Linear System

It is of course possible to solve the linear system (5.41) without considering the block structure of the problem. Direct and iterative methods exist that will handle this problem fairly well. We may, however, improve the efficiency significantly by utilizing our knowledge of the system structure. The focus of this section is the construction of a highly efficient block preconditioner for this system.

*A Preconditioner* In this section we will consider the linear system to be solved at each time step as a stationary problem. We do not make any assumptions on the right hand side, but consider (5.41) with an arbitrary right hand side. The problem is given by

$$\mathcal{A} \begin{bmatrix} v \\ u \end{bmatrix} = \begin{bmatrix} A & B \\ B^T & C \end{bmatrix} \begin{bmatrix} v \\ u \end{bmatrix} = \begin{bmatrix} a \\ b \end{bmatrix}, \tag{5.46}$$

where $a$ and $b$ are arbitrary vectors. The reason for this is simply that the preconditioner needs to be able to work on any right hand side, since we do not have any a priori information about the behavior of the residual. Our suggested preconditioner for this linear system is

$$\mathcal{B}_h = \begin{bmatrix} \widehat{(A^{-1})} & 0 \\ 0 & \widehat{(C^{-1})} \end{bmatrix},$$

where $\widehat{(\cdot)}$ denotes multigrid approximations to the respective inverse. Since the blocks $A$ and $C$ are similar to the discrete Laplace operator, it is well known from multigrid theory that such approximations can be made that are spectrally equivalent with the true inverse, see e.g. [17]. Hence, the preconditioner is spectrally equivalent with the exact Jacobi preconditioner.

*The Solver* Since our linear system is symmetric and positive definite, we use the conjugate gradient method to solve the preconditioned system. If we denote the start vector by $x^0$ and the exact solution $x$, the initial error is given by $e^0 = x - x^0$. It can be shown that a reduction of the initial error with a factor $\epsilon \ll 1$ is achieved after at most

$$\frac{1}{2} \ln \frac{2}{\epsilon} \sqrt{\kappa} \tag{5.47}$$

iterations, see e.g. [3], where $\kappa$ is the condition number of $\mathcal{BA}$. The condition number of $\mathcal{A}$ without preconditioning is $\kappa \sim h^{-2}$, and the number of iterations is thus proportional to $h^{-1}$. Preliminary analysis indicates that the preconditioner $\mathcal{B}$ introduced in the previous section is spectrally equivalent to the inverse of $\mathcal{A}$, and the condition number of the preconditioned system is hence bounded independently of $h$. According to (5.47), the number of iterations is thus independent of $h$, i.e., independent of the number of unknowns $n$.

### 5.3.4    A Simulator for the Bidomain Model

We first make a simulator for the Bidomain model without considering the block structure, primarily for debugging purposes. The complete source code for this simulator is located in `src/doc/mixed/Heart`.

```
class Heart : public FEM
{
public:
  Handle(GridFE)         grid;    // underlying finite element grid
  Handle(FieldFE)        u;
  Handle(FieldFE)        u_prev;
  Handle(FieldFE)        v;
  Handle(DegFreeFE)      dof;
  Handle(TimePrm)        tip;
  Vec(real)              linsol;
  Handle(LinEqAdmFE)     lineq;
  Handle(SaveSimRes)     database;
  MatSimple(real)  sigma_i;
  MatSimple(real)  sigma_e;
  bool dump;


  Handle(FieldFE)          error;

  // used to partition scan into manageable parts:
  virtual void sigma_scan();
  void scanGrid();
  virtual void initFieldsEtc();

  virtual void timeLoop();
  virtual void solveAtThisTimeStep();
  virtual void setIC();

  // standard functions:
  virtual void fillEssBC ();
  virtual void fillEssBC4U (DegFreeFE& dof, int U);
  virtual void fillEssBC4V (DegFreeFE& dof, int V);

  virtual void calcElmMatVec
     (int e, ElmMatVec& elmat, FiniteElement& fe);

  virtual void integrands
     (ElmMatVec& elmat, const FiniteElement& fe);


public:

  Handle(FieldFunc)    usource;
  Handle(FieldFunc)    uic;
  Heart ();
  ~Heart ();

  virtual void adm (MenuSystem& menu);
  virtual void define (MenuSystem& menu, int level = MAIN);
  virtual void scan ();

  virtual void solveProblem ();
  virtual void saveResults ();
  virtual void resultReport ();
};
```

This simulator should be thought of as the main simulator. It is in charge of initialization of all the datastructure related to the physical properties of the model problem. One should also use this simulator to validate the model and verify new simulators that utilize more efficient solution techniques, like the block and multigrid solvers that will be presented below. To ease the process of debugging the simulator, we also introduce a sub-class named `HeartAnalytic` which is intended purely for debugging purposes. The class solves the problem with very simple boundary- and initial conditions, and compares the numerical solution with analytical results. This simulator is also useful to find appropriate convergence criteria.

It is evident from the header file that the structure of this simulator class is not very different from the simplest Diffpack simulators introduced in [9]. The major difference is that we have two unknown fields, and one `DegFreeFE` object, based on the grid and the number of unknowns per node,

```
dof.rebind (new DegFreeFE (*grid, 2));
```

The fields of unknowns are made accordingly,

```
coll.rebind(new FieldsFE(*grid, "coll"));
u.rebind (coll()(1));
v.rebind (coll()(2));
```

Hence, the fields are made as vector fields, similar to the way they are made in the `Elasticity1` simulator described in [9]. When we use the special (or the general) numbering technique the block structure of the matrices will be lost. Usually one wants to avoid this block structure because it results in a very large bandwidth of the matrices and for factorization methods small bandwidths are desirable. We employ these numbering strategies within each block, such that each block has a very small bandwidth.

To be able to reuse the `fillEssBC` routine in the block simulators we split it into two functions:

```
fillEssBC4U(DegFreeFE& dof, int U)
fillEssBC4V(DegFreeFE& dof, int V)
```

The main idea is that a `DegFreeFE& dof` and a field number U is sufficient to fill the boundary conditions with `dof.fields2dof(· , U)`. This can be done in both the block simulators, multigrid, and this simulator. Consult `src/mixed/Heart/HeartAnalytic.cpp` for the complete code.

The `HeartAnalytic` class is used to verify that the simulator produces correct results. A case with very simple geometry and initial conditions is considered, and the numerical results are compared to the analytical solution. The next step is to develop a simulator that utilizes the block structure in the solution algorithm. The purpose of this simulator is of course to produce the exact same results as the general `Heart` class, but to speed up the solution process by applying specialized block preconditioners. We first make a simple block structured simulator `HeartBlocks`, without any multigrid functionality. The process of debugging the simulators is substantially simplified by this incremental development, as we limit the amount of new functionality introduced in each step.

```
class HeartBlocks : public HeartAnalytic
{
public:
  Handle(Precond_prm)         prec_prm1;
  Handle(Precond_prm)         prec_prm2;
  Handle(FieldsFE)            u_coll;
  Handle(FieldsFE)            v_coll;
  Handle(DegFreeFE)           u_dof;
  Handle(DegFreeFE)           v_dof;
  Handle(LinEqVector)         linsol_block;
  Integrands_type             integrand_type;
  int nsd;

public:
  HeartBlocks();
 ~HeartBlocks();

  MatSimplest(Handle(IntegrandCalc)) integrands;
  Handle(ElmMatVecCalc)    calcelm;
  Handle(ElmMatVecs)       elmat;    // element level blocks
  Handle(SystemCollector) elmsys;  // main utility for block adm.

  virtual void adm (MenuSystem& menu);
  virtual void timeLoop();
  virtual void define (MenuSystem& menu, int level = MAIN);
  virtual void scan ();
  void initBlocks();
  void initBlockDofs();
  virtual void solveAtThisTimeStep();
  virtual void solveProblem();
  virtual void resultReport ();
  virtual void fillEssBC();
  virtual void makePreconditioner();
  virtual void makePreconditioner4U();
  virtual void makePreconditioner4V();

  void makeBlockSystem
  (
   LinEqAdmFE&        lineq,
   SystemCollector&   elmsys,
   FieldsFE&          coll,
   ElmMatVecs&        elmat,
   int                itg_man
  );

  void makeSystem
  (
    DegFreeFE&         dof,
    FieldsFE&          coll,
    IntegrandCalc&     integrand,
    ElmMatVecCalc&     emv,
    Matrix(NUMT)&      A,
```

```
    int              itgrule_man
);

};
```

The `LinEqAdmFE` handle `lineq` introduced in the main simulator `Heart` is a general structure for linear systems, which is capable of handling block structured systems. The default version has only one block, but in this case we want to construct a $2 \times 2$ block system corresponding to the two primary unknowns. The call

```
lineq.rebind (new LinEqAdmFE(2,2));
```

in the `scan` function handles this basic initialization. This call constructs a `LinEqSystemBlockPrec` with a $2 \times 2$ block structure of the matrices, preconditioners and vectors. The preconditioners are attached as described in Section 5.2.3.

The main simulator class `Heart` was derived from the base class `FEM` which offers a virtual function `integrands`, which is used to make the linear system in `makeSystem`. This is the common way to construct the linear system in a Diffpack simulator. However, Diffpack also offers other possibilities, which are more suitable for handling block structured systems. Instead of redefining the provided `integrands` function, we may introduce the integrands as derived subclasses of `IntegrandCalc`, which offers the integrands (`ElmMatVec& elmat, const FiniteElement& fe`), similar to `FEM`. The base class `FEM` then offers general `makeSystem` routines like,

```
virtual void makeSystem
    (DegFreeFE& dof, FieldsFE& coll, IntegrandCalc* integrand,
    ElmMatVecCalc& emv, Matrix(NUMT)& A),
```

which only uses the datastructures submitted in the function call. This is in fact the function we will use to make the algebraic preconditioners, as we will see later.

For a block system we need several integrand classes, all sub-classes of `IntegrandCalc`, which are typically organized in a 2-dimensional array.

```
MatSimplest(Handle(IntegrandCalc)) integrands;
```

The matrix is initialized according to the structure of the system, which in our case is a $2 \times 2$ system:

```
integrands.redim(2,2);
integrands(1,1).rebind( new Integrand11(this));
integrands(2,1).rebind( new Integrand21(this));
integrands(1,2).rebind( new Integrand12(this));
integrands(2,2).rebind( new Integrand22(this));
```

Each element in `integrands` is derived of `IntegrandCalc` and needs to define an `integrands` function. In addition they usually contain a pointer to a `Heart` object, submitted as this, to be able to fetch various data from the main simulator.

The four integrands functions will correspond to the four distinct parts of the integrands function implemented in the Heart class. The following one, corresponding to the block $C$ in (5.41), is a typical function for Poisson type equations (compare with the Poisson1 example ).

```
void Integrand22::integrands
  (ElmMatVec& elmat, const FiniteElement& fe)
{
  real nabla;
  real detJxW = fe.detJxW();
  int nbf = fe.getNoBasisFunc();
  int nsd = fe.getNoSpaceDim();
  int i,j,k;
  real dt = data->tip->Delta(); //data is a Heart pointer
  for (i = 1; i <= nbf; i++) {
    for (j = 1; j <= nbf; j++) {
      nabla = 0;
      for (k = 1; k <= nsd; k++)
          nabla += (data->sigma_e(k,k)+data->sigma_i(k,k))
                    *fe.dN(i,k)*fe.dN(j,k);
      elmat.A(i,j) += dt*nabla*detJxW;
    }
    elmat.b(i) += 0.0;
  }
}
```

Since all the calcElmMatVec functions are identical in this application they are all represented as one ElmMatVecCalculator object. Class ElmMatVecCalculator is derived from ElmMatVecCalc, which supply functions for calculating the element matrices. The following is a typical declaration of such a function, further details are found in the manual pages for the FEM class [9].

```
virtual void calcElmMatVec
  (
  int             elm_no,
  ElmMatVec&      elmat,
  FiniteElement&  fe,
  IntegrandCalc&  integrand,
  FEM*            fem
  );
```

We also need some temporary data structures to store the element matrices and administer the computation:

```
elmsys.rebind(new SystemCollector());
elmat.rebind(new ElmMatVecs());
```

ElmMatVecs is simply a matrix of ElmMatVec objects, in the present case a $2 \times 2$ system of element matrices. This class also enforces the essential boundary conditions in the different element matrices simultaneously.

The `SystemCollector` object `elmsys` is the manager that connects the block structured `LinEqMatrix` to the `DegFreeFE` objects and the classes containing the integrands for the system. It should be fed with a consistent set of `DegFreeFEs` and `IntegrandCalc` objects.

```
elmsys->redim(2,2);
for (int i=1; i<= 2; i++)
  for (int j=1; j<= 2; j++)
    if (integrands(i,j).ok())
      elmsys->attach(integrands(i,j)(),i,j);

elmsys->attach(*calcelm);

elmsys->attach(*u_dof,1,true);
elmsys->attach(*u_dof,1,false);

elmsys->attach(*v_dof,2,true);
elmsys->attach(*v_dof,2,false);

lineq->initAssemble(*elmsys);
elmat->attach(*elmsys);
elmsys->attach(*elmat);
elmsys->attach(this);
```

The assembly of the linear system is performed by a function corresponding to the `makeSystem` call in common Diffpack simulators,

```
makeBlockSystem(*lineq,*elmsys,*coll,*elmat,1);
```

A suitable definition of this function can be found in `src/mixed/Heart/block/HeartBlocks.cpp`.

Having now constructed the block linear system, we need to build the preconditioners. We will use the block Jacobi preconditioner introduced in sections 5.2.1 and 5.2.2 , but as stated above we will not consider multigrid preconditioners in this version of the simulator. We will instead use simple algebraic preconditioners to approximate the block inverses. The matrices needed by the preconditioner are assembled using the `IntegrandCalc` subclasses defined earlier. The following code segment handles the assembly of the preconditioner corresponding to block $A$ in (5.41).

```
Handle(IntegrandCalc) integrand;
LinEqSystemBlockPrec& lins =
    CAST_REF(lineq->getLinEqSystem(), LinEqSystemBlockPrec);
MenuSystem& menu = SimCase::getMenuSystem();
integrand = new Integrand11(this);

Handle(Matrix_prm(NUMT)) mat_prec_prm;
...
//initialize mat_prec_prm somehow
...
Handle(Matrix(NUMT)) mat_prec;
mat_prec= mat_prec_prm().create();

makeSystem(u_dof(), *coll, *integrand, *calcelm, *mat_prec, 1);
```

The assembled matrix is attached to the preconditioner and then to the preconditioned linear system represented by the `LinEqSystemBlockPrec lins`.

```
Handle(PrecBasis) prec_basis = new PrecBasis();
prec_basis->attach(*mat_prec);
prec_prm1->print(s_o);
lins.attach(*prec_prm1, *prec_basis,1);
```

Numerical results for this simulator are shown in Table 5.1. We have used the Conjugate Gradient method with RILU block preconditioners. We see that the number of iterations increases when the problem size increases. Because the cost of one iteration is proportional to the number of unknowns, the work of solving the problem is $\mathcal{O}(n^p)$, with $p > 1$.

**Table 5.1.** Numerical results for the Bidomain solver with RILU block preconditioning.

| Unknowns | Iterations | p |
|----------|------------|-----|
| 882 | 34 | - |
| 3362 | 39 | 1.1 |
| 13122 | 56 | 1.3 |
| 51842 | 86 | 1.3 |
| 206082 | 150 | 1.3 |

### 5.3.5  A Simulator with Multigrid

The final step in the process of speeding up the simulator is to replace the simple algebraic preconditioners with efficient multigrid algorithms. The Multigrid tools in Diffpack are documented in [11] and the reader is referred to this chapter for more details concerning the `MGtools` class. In the present section we will see how the toolbox can be utilized to easily employ multigrid for block systems. The `HeartBlocksMG` simulator is a derived sub-class of the `HeartBlocks` class. In our problem we need two `MGtools` objects, one for each block on the diagonal, and we thus introduce the vector `VecSimplest(Handle(MGtools)) mg`. We also need to rewrite some of the functions in the `HeartBlocks`. The complete class declaration is as follows.

```
class HeartBlocksMG : public HeartBlocks
{
 public:
  int no_of_grids;
  VecSimplest(Handle(MGtools)) mg;
  Handle(FieldCollector) fieldcoll2;

  HeartBlocksMG();
```

```
~HeartBlocksMG();

    virtual void define (MenuSystem& menu, int level = MAIN);
    virtual void initFieldsEtc();
    virtual void scan ();
    virtual void solveAtThisTimeStep();
    virtual void fillEssBC(SpaceId space);
    virtual void mgFillEssBC(SpaceId space);
    virtual void getDataStructureOnGrid(SpaceId space);
    virtual void resultReport();
    virtual void makePreconditioner();
    virtual void saveResults();
    virtual void setIC();
};
```

As usual in Diffpack we have an extensive menu-interface, which makes it easy to test various parameters. We use the command prefixes block1 and block2 to allow separate menu input for the two blocks.

```
menu.setCommandPrefix("block1");
Precond_prm ::defineStatic(menu, level+1);
menu.unsetCommandPrefix();
menu.setCommandPrefix("block2");
Precond_prm::defineStatic(menu, level+1);
menu.unsetCommandPrefix();
```

The MGtools menu is also initialized according to the Diffpack standard presented in [9],

```
MGtools::defineStatic(menu,level+1);
```

In the scan function the preconditioners are initialized and attached to the linear system.

```
menu.setCommandPrefix("block1");
prec_prm1.rebind( Precond_prm::construct());
prec_prm1->scan(menu);
menu.unsetCommandPrefix ();
menu.setCommandPrefix("block2");
prec_prm2.rebind( Precond_prm::construct());
prec_prm2->scan(menu);
menu.unsetCommandPrefix ();


LinEqSystemBlockPrec& lins =
    CAST_REF(lineq->getLinEqSystem(), LinEqSystemBlockPrec);
lins.attach(*prec_prm1,1);
lins.attach(*prec_prm2,2);
```

We also need to initialize the VecSimplest holding the MGtools. The length of the vector is set and each MGtools object is initialized according to the procedure described in [11].

```
mg.redim(2);
for (int i=1; i<= 2; i++){
```

```
if ( i== 1){
  mg(i).rebind(new MGtools(*this, oform("mg toolbox",i)));
}
else{
  //use the same grid hierarchy as mg(1) in mg(2):
  mg(i).rebind(new MGtools(*this,mg(1)->gridcoll(),
    oform("mg toolbox",i)));
}
mg(i)->attach(lineq());
mg(i)->scan(menu); //make grid etc.
mg(i)->initStructure();
if ( lins.getPrec(i).description().contains("multilevel")){
  PrecML& p = CAST_REF(lins.getPrec(i), PrecML);
  p.attach(mg(i)->getMLSolver());
}
```

Both MGtools objects are now initialized properly and attached to the linear system lineq. The only remaining task in the construction of the pre-conditioners is to make the linear systems for the multigrid algorithms.

```
void HeartBlocksMG:: makePreconditioner() {
    int i;
    integrand_type = PRECONDITIONER;
    for (i=1; i<= 2; i++) {
      mg(i)->makeSystemMl(*integrands(i,i), *calcelm, true);
    }
}
```

The code segments presented here include all the major steps in the process of including the block multigrid preconditioner in the simulator. Because the simulator was already prepared for block linear systems and block pre-conditioning, the addition of the multigrid preconditioner was fairly straightforward. The source code for the simulator is found in src/mixed/Heart/block/ in the files HeartBlocksMG.h and HeartBlocksMG.cpp.

Numerical results for the simulator with multigrid are shown in Table 5.2. We see that the number of iterations does not increase when the problem size increases. Because the number of iterations is constant and the cost of one iteration is proportional to the number of unknowns, the total work for one simulation is $\mathcal{O}(n)$.

### 5.3.6    A Solver Based on Operator Splitting

The block structured preconditioners presented in this paper are all based on operator splitting techniques. These techniques may also be used as solvers for the linear system. The simplest operator splitting algorithm is the block Jacobi method given by (5.3)-(5.4). A very similar algorithm is the block

**Table 5.2.** Numerical results for the Bidomain solver with multigrid block preconditioning.

| Unknowns | Iterations |
|----------|------------|
| 882      | 17         |
| 3362     | 17         |
| 13122    | 16         |
| 51842    | 15         |
| 206082   | 15         |

Gauss-Seidel iteration, defined by

$$x^{k+1} = x^k + A^{-1}(b - Ax^k - By^k) \tag{5.48}$$
$$= A^{-1}(b - By^k), \tag{5.49}$$
$$y^{k+1} = y^k + D^{-1}(c - Cx^{k+1} - Dy^k) \tag{5.50}$$
$$= D^{-1}(c - Cx^{k+1}). \tag{5.51}$$

Here we have inverted $A$ and $D$ exactly. This leads to the simplified expression.

The major work of one iteration is the solution of the linear sub-systems (5.49) and (5.51) and the efficiency of the algorithm depends on our ability to solve these linear systems efficiently. For these systems it is possible to use multigrid directly as a solver, by simply repeating the multigrid sweeps until convergence is reached. A slightly different approach is to use multigrid as preconditioner for the conjugate gradient method. We want to implement the new simulator as a sub-class of the HeartBlocksMG class described in the previous section, with as much reuse as possible, and thus the second approach seems like a good choice. To maximize the possibility for reuse, we employ the Gauss-Seidel method on linear algebra level. This means that we assemble the global linear system first, before it is decomposed into the sub-systems in (5.49)-(5.51).

A different, and perhaps more common, approach would be to split the equations on the PDE level and assemble the two linear systems separately. The main advantage of our approach is that we are able to reuse the code from the HeartBlocksMG simulator. Other advantages include a simple and efficient update of the right hand sides for each iteration, and that the residual for the complete linear system is easily available, to be used as a convergence test for the Gauss-Seidel iterations. The main disadvantage of the chosen approach is that the assembly of the complete linear system is more costly than assembling the two smaller sub-systems, and because of the need to store all the matrix blocks, the memory requirement is also increased.

Most of the datastructures and functions of the HeartBlocksMG class are reused unaltered in the new HeartBlocksGS simulator. The main addition to the class is that we need two new LinEqAdmFE objects to handle the linear systems (5.49)-(5.51). These linear systems are initialized and preconditioners attached by the following code segment in the scan function.

```
lineq1.rebind (new LinEqAdmFE(EXTERNAL_STORAGE));
lineq2.rebind (new LinEqAdmFE(EXTERNAL_STORAGE));

//use the same menu input for both lineqs
lineq1->scan(menu);
lineq2->scan(menu);

LinEqSystemPrec& lins1 =
    CAST_REF(lineq1->getLinEqSystem(), LinEqSystemPrec);
lins1.attach(*prec_prm1);

LinEqSystemPrec& lins2 =
    CAST_REF(lineq2->getLinEqSystem(), LinEqSystemPrec);
lins2.attach(*prec_prm1);

if ( lins1.getPrec().description().contains("multilevel")){
  PrecML& p = CAST_REF(lins1.getPrec(), PrecML);
  p.attach(mg(1)->getMLSolver());
}

if ( lins2.getPrec().description().contains("multilevel")){
  PrecML& p = CAST_REF(lins2.getPrec(), PrecML);
  p.attach(mg(2)->getMLSolver());
}
```

In addition to the additions made to the scan function, we need to change the function solveAtThisTimeStep to incorporate the new solution procedure. After assembling the global system, the following do-loop handles the Gauss-Seidel iterations.

```
do {
  //extract the u blocks as a separate system:
  Matrix(real)& block12 = lineq->Al().mat(1,2);

  block12.prod(linsol_block->getVec(2), rhs1);

  rhs1.add(lineq->bl().getVec(1),'-',rhs1);

  lineq1->attach(lineq->Al().mat(1,1),
      linsol_block->getVec(1), rhs1);

  //attach preconditioner for u:
  mg(1)->attach(lineq1->Al().mat(), no_of_grids);
  mg(1)->attachSol(lineq1->xl().vec(), no_of_grids);
  mg(1)->attachRhs(lineq1->bl().vec(), no_of_grids, false);
  //solve the linear system for u:
  lineq1->solve();

  //extract the v blocks as a separate system
  Matrix(real)& block21 = lineq->Al().mat(2,1);

   block21.prod(linsol_block->getVec(1), rhs2);
   rhs2.add(lineq->bl().getVec(2),'-',rhs2);
```

```
lineq2->attach(lineq->Al().mat(2,2),
    linsol_block->getVec(2), rhs2);

//attach preconditioner for v:
mg(2)->attach(lineq2->Al().mat(), no_of_grids);
mg(2)->attachSol(lineq2->xl().vec(), no_of_grids);
mg(2)->attachRhs(lineq2->bl().vec(), no_of_grids, false);
//solve the linear system for v
lineq2->solve();                     // solve linear system

lineq->getLinEqSystem().residual(res);
r = res.norm();
its ++;
} while (r/r0 > tol);
```

We use the relative residual r/r0 of the complete system to monitor
the convergence. The same residual was used for the block preconditioned
solvers described previously, and the performance of the simulators should
thus be directly comparable. The source code for this simulator is found in
src/mixed/Heart/block in the files HeartBlocksGS.h and HeartBlocksGS.cpp.
Results from numerical experiments with the Gauss-Seidel based simulator
are presented in Table 5.3. With the given structure of the two sub-problems
and the results achieved in the previous section, we expect multigrid to be an
optimal preconditioner for solving the two sub-problems with the Conjugate-
Gradient method. This is confirmed by the non-increasing iteration count for
these sub-problems when the problem size increases. It is not quite as obvious
that the number of Gauss-Seidel iterations is also independent of the problem size, but this is confirmed by the presented results. The present solution
method is thus $\mathcal{O}(n)$, exactly as we had for the block-preconditioned solver
with Conjugate Gradient iterations. Still, experiments indicate that the block
preconditioned solver is significantly faster, with CPU times approximately
25% of the Gauss-Seidel based solver.

**Table 5.3.** Numerical results for the Bidomain solver with multigrid-based Gauss-
Seidel iterations.

| Unknowns | CG its u system | CG its v system | GS iterations |
|---|---|---|---|
| 882 | 6 | 6 | 19 |
| 3362 | 6 | 5 | 16 |
| 13122 | 7 | 5 | 14 |
| 51842 | 7 | 6 | 12 |
| 206082 | 7 | 6 | 10 |

## 5.4   Two Saddle Point Problems

In the next sections we will consider two well known saddle point problems; the Stokes problem and the mixed formulation of the Poisson equation. These problems may both be written on the form

$$\mathcal{A}_h \begin{bmatrix} v \\ p \end{bmatrix} = \begin{bmatrix} A & B^T \\ B & 0 \end{bmatrix} \begin{bmatrix} v \\ p \end{bmatrix} = \begin{bmatrix} f \\ g \end{bmatrix}. \tag{5.52}$$

This is a special case of (5.2) where $C$ is $0$ and $C^{-1}$ is undefined. Therefore we cannot use the Jacobi method as preconditioner. But as we will see, the preconditioner will be block diagonal and is constructed similarly. These systems are indefinite and the discretization of these model problems require mixed elements as discussed in [10].

How should this preconditioner be constructed? Such systems have been studied in [13] and we follow their conclusion. The best preconditioner is of course $\mathcal{A}^{-1}$, which is obviously not an option. Our mixed system is indefinite, but we see that if we use the best preconditioner possible, $\mathcal{A}^{-1}$, we get a positive definite system. Hence, a natural question to ask, is whether the preconditioned system should be indefinite or definite. We seek an "approximation of the inverse" and an approximate transformation from an indefinite to a positive system may be dangerous. We may end up with eigenvalues close to zero, causing a breakdown of the basic iterative method. The cure is to make a preconditioner such that the preconditioned system is also an indefinite problem. Guided by [13] we consider block preconditioners on the form $\mathcal{B} = diag(M, N)$ such that (5.19) has the following form

$$\mathcal{B}_h \mathcal{A}_h = \begin{bmatrix} MA & MB^T \\ NB & 0 \end{bmatrix}. \tag{5.53}$$

The preconditioners $M$ and $N$ are chosen positive definite and therefore the preconditioned system will still be indefinite. The best possible such saddle point problem is

$$\mathcal{B}\mathcal{A} = \begin{bmatrix} I & Q^T \\ Q & 0 \end{bmatrix}, \text{ where } QQ^T = I. \tag{5.54}$$

This system has eigenvalues, $1, 1/2 \pm 1/2\sqrt{5}$ (both positive and negative). Hence, the preconditioner is not similar to the inverse of the coefficient matrix. To construct a system similar to (5.54), we should have block operators such that,

$$MA \sim I, \tag{5.55}$$

$$NBMB^T \sim I. \tag{5.56}$$

If we are able to construct blocks with these properties, we will bound the eigenvalues of $BA$ above and away from zero, independent of the mesh size $h$.

*Pressure Schur Complement methods.* In [16] they have formulated a general framework, a basic iteration on the Pressure Schur Complement, where the Projection method, Pressure correction methods, Uzawa methods, and the Vanka smoother are special cases. Such a framework can be developed by using the block utilities in Diffpack in a similar way that is done in Section 5.3.6.

### 5.4.1   Stokes Problem

*Mathematical Model* The Stokes problem fits elegantly in the setup described in the previous section. We briefly recapture the equations from [10],

$$-\mu\Delta v + \nabla p = f \text{ in } \Omega, \quad \text{(equation of motion)}, \tag{5.57}$$

$$\nabla \cdot v = 0 \text{ in } \Omega, \quad \text{(mass conservation)}, \tag{5.58}$$

$$v = h \text{ on } \partial\Omega_E, \tag{5.59}$$

$$\frac{\partial v}{\partial n} + pn = 0 \text{ on } \partial\Omega_N. \tag{5.60}$$

*Numerical Method* We remember the discrete coefficient operator for the Stokes problem,

$$\begin{bmatrix} A & B^T \\ B & 0 \end{bmatrix} = \begin{bmatrix} -\mu\Delta_h & \nabla_h \\ \nabla_h^T & 0 \end{bmatrix}. \tag{5.61}$$

To obtain a well posed discrete and continuous problem the Babuska-Brezzi condition have to be fulfilled [7], this means that we need to use the mixed elements described in [10]. Examples of elements are the Taylor-Hood, Mini and Crouzeix-Raviart.

*Solution of the Linear System* We can choose $\mathbf{M}$ as an efficient preconditioner for $A$ since $\mathbf{M}A \sim I$. Such preconditioner can be made by multigrid because $A$ is an elliptic operator. The Schur complement $\mathbf{BMB}^T$ is already well conditioned, thus $N$ can be chosen as $I$ (the mass matrix). We present some numerical experiments of this preconditioner at the end of this section, similar experiments are shown in [13].

*A Simulator for Stokes problem* In [10] a Stokes simulator, using mixed finite elements, was developed. We will now extend this simulator to utilize efficient block preconditioners. Most of the steps to do this are the same as for the Bidomain problem presented in the previous section, but we then only used isoparametric elements. For the Stokes problem and the mixed Poisson problem discussed later, we need to use mixed elements or else (5.52) will be singular[6] [2]. The transition from isoparametric elements to mixed is first of

---

[6] Notice that the pressure is only determined up to a constant and the system is therefore singular with a one dimensional kernel. However, when not using certain mixed elements, the kernel will be larger.

all a query-replace,

FEM, FiniteElement, integrand, makeSystem, IntegrandCalc

are replaced with

MxFEM, MxFiniteElement, integrandMx, makeSystemMx, IntegrandCalcMx.

In addition, it is important to know that the *general numbering* strategy of DegFreeFE is used instead of the *special numbering* which is used for isoparametric elements. The way DegFreeFE is constructed will determine which strategy it will use. A DegFreeFE made like,

```
GridFE grid;
DegFreeFE dof;
...
//make grid somehow
...
dof.rebind(new DegFreeFE(grid, 2)); //dof is made
```

results in the special numbering, whereas, the general numbering is used if DegFreeFE is made like,

```
FieldsFE fields;
DegFreeFE dof;
fields_grid.rebind (new BasisFuncGrid (*grid));
String fields_elm_tp = menu.get ("fields element");
fields_grid->setElmType (fields_grid->getElmType(), fields_elm_tp);
fields.rebind (new FieldsFE (*fields_grid, nsd, "fields"));

dof.rebind(new DegFreeFE(fields)); //dof is made
```

The general numbering will be used even though isoparametric elements are used in FieldsFE. We turn on the mixed element utilities in MGtools by

```
MGtools::useFieldCollector(true,no_fields);
```

where no_fields is an integer. The fields of unknowns will then be collected in a FieldCollector object managed by the MGtools. When needed, the fields should be fetched from there,

```
Handle(FieldsFE) v_coll;
v_coll.rebind( mg->fieldcoll->getFields(no_of_grids));
```

In contrast to HeartBlocksMG we made the fields ourselves by

```
coll.rebind(new FieldsFE(*grid, "coll"));
u.rebind (coll()(1));
v.rebind (coll()(2));
u_prev.rebind (new FieldFE (*grid, "u_prev"));
```

The implementation of the multigrid tools for the velocity preconditioner is more or less the same as for the Bidomain problem and we therefore skip the details. The source code can be found in `src/mixed/Stokes/block`.

However, now we also must make a preconditioner similar to $I$, the mass matrix. It can be misleading to use the term identity when using finite element methods. However, it is the natural counterpart to the identity in finite difference methods. We therefore describe the term in greater detail. Assume we have a function $f$ and we want a finite element representation, $\sum_j u_j N_j$, of this function, then $u_j$ is determined by

$$(\sum_j u_j N_j, N_i) = f_i = (f, N_i). \qquad (5.62)$$

Hence, we see that $u_i$ and $f_i$ are scaled differently. Inside the integration the coefficient $u_j$ is multiplied roughly with $N_j^2$, whereas $f_i$ is $f$ multiplied by $N_i$. We make a preconditioner to account for this different scaling, namely the inverse of a lumped mass matrix. To do this we derive a preconditioner from FEM and `PrecUserDefProc`,

```
class PressurePrec : public PrecUserDefProc, public MxFEM {
  public:
  StokesBlocks* solver;
  Handle(Precond)     mass_prec;
  Handle(Precond_prm) mass_prm;
  Handle(Matrix_prm(NUMT)) mat_prm;
  Handle(DegFreeFE)   p_dof;
  Handle(Matrix(NUMT))            mass;
  bool                inited;

  PressurePrec(Precond_prm& prm):PrecUserDefProc(prm) {
     inited= false;
  }
  ~PressurePrec() {}

  virtual void scan(MenuSystem& menu);
  static void defineStatic(MenuSystem& menu, int level);
  virtual void init();
  virtual void attachPDof(DegFreeFE& dof) { p_dof.rebind(dof); }
  virtual void makeMassPrec();
  virtual void apply(
      const LinEqVector& c,
      LinEqVector& d,
      TransposeMode tpmode = NOT_TRANSPOSED
  );

  bool ok() { return inited; }

};
```

Any derived class of `PrecUserDefProc` must supply an `apply` function, which in our case simply is,

```
    void PressurePrec:: apply ( const LinEqVector& c_, LinEqVector& d_,
```

```
                              TransposeMode tpmode )
    {
      if (!inited) init();
      mass->forwBack (( Vector(NUMT)&)c_.vec(), d_.vec());
    }
```

The `init` function should make the mass matrix and factorize it.

```
    void PressurePrec:: init(){
      makeMassPrec();
      inited = true;
    }
```

The mass matrix is made from the finite element fields related to the pressure, a `DegFreeFE` object p_dof containing the sufficient information and is fetched from the `Stokes` class. The mass matrix is made as follows,

```
    void PressurePrec:: makeMassPrec() {
      if (p_dof->noEssBC()) {
        warningFP("PressurePrec:: makeMassPrec",
                  "p_dof has no essential bc.");
      }
      mass.rebind(new MatDiag(NUMT)(p_dof->getTotalNoDof()));
      mass()=0.0;
      Handle(IntegrandMassP) integrand = new IntegrandMassP();
      integrand->P = 1;
      FEM::makeSystemMx(*p_dof, integrand.getPtr(), *mass);
      FactStrategy fact; fact.fact_tp=LU_FACT;
      mass->factorize(fact);
    }
```

We make a diagonal mass matrix for fast inversion. It is sufficient to use the diagonal version since this matrix only should deal with the scaling in (5.62). The following `integrands` function makes the mass matrix.

```
    void IntegrandMassP::integrandsMx
      (ElmMatVec& elmat, const MxFiniteElement& mfe)
    {
      const int nsd = mfe.getNoSpaceDim();
      const int npbf = mfe(P).getNoBasisFunc();

      const real detJxW = mfe.detJxW();
      int    i,k,j;
      for (i = 1; i <= npbf; i++){
        for (j = 1; j <= npbf; j++){
          elmat.A(i,i) += mfe(P).N(i)*mfe(P).N(j)*detJxW;
        }
      }
    }
```

*Numerical Experiments* In this section we test the efficiency of different block preconditioners. We fix $\mathbf{N}$ as the inverse of a lumped mass matrix. The comparison of $M$ made by multigrid and RILU is shown in Table 5.4. We have used the Krylov method SymMinRes (Minimum residual method for symmetric and indefinite problems), with a random start vector. The convergence criterion is set as,

$$\frac{(\mathcal{B}r_k, r_k)}{(\mathcal{B}r_0, r_0)} \le 10^{-4},$$

where $r_k$ is the residual at iteration $k$ and $\mathbf{B}$ is the preconditioner. This convergence monitor is implemented in Diffpack as CMRelMixResidual. The multigrid preconditioner is constructed as a V-cycle with SSOR with relaxation parameter 1.0 as smoother. On the coarsest grid we use Gauss elimination. As expected, the multigrid preconditioner ensures convergence in (roughly) a fixed number of iterations, while the efficiency of the RILU preconditioner deteriorates as the number of unknowns increases. Similar results are obtained with the Crouzeix-Raviart and the Mini element (see the examples that follow the source code.).

**Table 5.4.** Numerical results for the Stokes problem with Taylor-Hood elements.

| Unknowns | Iterations with RILU | Iterations with MG |
|---|---|---|
| 187 | 29 | 21 |
| 659 | 47 | 23 |
| 2467 | 59 | 22 |
| 9539 | 66 | 20 |
| 37507 | 116 | 20 |
| 148739 | 136 | 20 |

### 5.4.2  Mixed Poisson Problem

*Mathematical Model* The mixed Poisson problem is

$$\mathbf{v} - \lambda \nabla p = 0 \text{ in } \Omega \quad \text{(Darcy's law)}, \tag{5.63}$$

$$\nabla \cdot \mathbf{v} = g \text{ in } \Omega \quad \text{(mass conservation)}. \tag{5.64}$$

The sign of $p$ is changed to obtain a symmetric system with more efficient solution algorithms. More details on this problem and its implementation in Diffpack can be found in [10].

*Numerical Method* The coefficient operator is,

$$\mathcal{A}_h = \begin{bmatrix} A & B^T \\ B & 0 \end{bmatrix} = \begin{bmatrix} I_h & -\nabla_h \\ \nabla_h^T & 0 \end{bmatrix}. \tag{5.65}$$

Examples of elements that can be used are the Raviart-Thomas (see, e.g., [2,10]) and a new element [10,12], which we referred to as the robust element, in lack of a better name. Some usual Stokes elements, with continuous pressure elements, can also be used at the expense of loss of accuracy in the velocity field.

*Solution of the Linear System* Analogous with the preconditioner made in Section 5.4.1 we can choose $M$ as the inverse of a lumped mass matrix, since $A = I$. The preconditioner $N$ should then be made such that $N B M B^T \sim I$ and this means that $N \sim \Delta^{-1}$. This preconditioner can be used in connection with the Mini and Taylor-Hood, because the pressure elements are continuous. We have implemented this and numerical experiments are shown in the end of this section.

However, we are primarily interested in using discontinuous pressure elements, since the elements of particular interest are the Raviart-Thomas and the robust element. These elements are formulated as subspaces of $H(div; \Omega) \times L^2(\Omega)$. The point is that the gradient on the pressure is transfered to a divergence on the velocity. Applying the $\Delta$-operator on these pressure elements is not straightforward.

One alternative is to use the auxiliary space technique [18]. This technique allows us to make a multigrid preconditioner based on standard continuous linear elements and employ it on the piecewise constant elements. We refer to [18] for the mathematical description and the requirements on the operators involved. Notice also that the numerical experiments done in this section have not been verified theoretically.

Let $C$ and $L$ be the spaces of piecewise constant elements and continuous linear elements, respectively. Furthermore, let $P : L \to C$ be an interpolation or projection operator, and $P^*$ is the adjoint operator with respect to the $L_2$ inner product. The preconditioner is then,

$$L_C = S_C + P L_L P^*. \tag{5.66}$$

The basic ingredients are

- A multigrid preconditioner for continuous linear elements, $L_L$.
- Transfer operators mapping between the two different element spaces, $P$ and $P^*$.
- A smoother for the piecewise constants, $S_C$.

The multigrid preconditioner made for the the Mini and the Taylor–Hood elements can be reused. The smoother is implemented as the inverse of a mass matrix multiplied with $\tau h^2$. This is similar to the $N$-preconditioner we made for the Stokes problem (except that we multiply with $\tau h^2$). The transfer operators are made in a class `Interpolator`, which implements a general $L_2$ projection. The $L_2$ projection involves the making and inversion of mass matrices. This might seem like overkill, but the mass matrix has a small condition number and can be inverted sufficiently accurate quite efficiently. A few Conjugate Gradient iterations, with SSOR as the preconditioner, is usually sufficient. In our case the mass matrix is diagonal, because of the piecewise constants. Therefore, we can simply use one SSOR iteration to invert the mass matrix exactly.

The NCPressurePrecMG implements the above described preconditioner. We will comment the key points. The NCPressurePrecMG is a subclass of PressurePrecMG, which is a multigrid preconditioner similar to the one we implemented for the Stokes problem.

The new software in this preconditioner is the $L_2$ projection, Interpolator, which we will describe below. The Interpolator has a standard menu interface and can be made as follows,

```
interpolator.rebind(new Interpolator());
interpolator->scan(menu);
```

Additionally, it has to be initialized with the fields that we will transfer between.

```
interpolator->init(
   solver->p_dof().fields()(1),
   mgtools_laplace->getDof(no_of_grids).fields()(1));
```

These two steps are all that must be done to initialize the $L_2$ projection. We now apply this projection. The mapping from the piecewise constant elements onto the continuous linear element is then,

```
void NCPressurePrecMG::cons2linear(LinEqVector& from,
  LinEqVector& to)
{
   interpolator->interpolate(
     from.getVec(), DUAL, FIRST,
     to.getVec(), DUAL, SECOND);
}
```

The enum variable DUAL means that a right-hand side is transfered. A left-hand side transfer operation use the NODAL flag. The FIRST and SECOND flags mean that the vector from and to corresponds to the first and second field used in the initialization, Interpolator::init. The mapping from linear to constant elements is similar,

```
void NCPressurePrecMG::linear2cons(
  LinEqVector& from, LinEqVector& to)
{
   interpolator->interpolate(
     from.getVec(), NODAL, SECOND,
     to.getVec(), NODAL, FIRST);
}
```

The complete preconditioner in (5.66) is implemented as

```
void NCPressurePrecMG:: apply (
   const LinEqVector& c_,
   LinEqVector& d_,
   TransposeMode tpmode )
{
  if  (!inited) init();

  // init vectors
  LinEqVector& c_not_const = (LinEqVector&)c_;
  c->attach(c_not_const.getVec(1));
  d_ = 0.0;
  d->attach(d_.vec(1));
  tmp_vec->vec(1).fill(d->getVec(1));

  //lapace prec on linear elements
  cons2linear(*c, *lrhs);
  laplace_prec->apply(*lrhs, *llhs);
  linear2cons(*llhs, *tmp_vec);

  //smoother
  ssor_vec.fill( d->getVec());
  mass->SSOR1it(d->vec(), ssor_vec, c->vec(), 1.0);
  d->mult(tau*h_char*h_char);

  //add together
  d->add(*tmp_vec, *d);
}
```

There are several other alternatives to construct optimal preconditioners for
the mixed Poisson problem (see, e.g., [15,1]).

*Numerical Experiments* We now show the results of some numerical experiments that document the efficiency of the preconditioner. We have tested the
Taylor-Hood, Mini, Raviart-Thomas, and the robust elements. Notice that
the results with the Raviart-Thomas and the robust elements have not been
verified theoretically.

**Table 5.5.** Numerical results for the mixed Poisson problem with Taylor-Hood
elements.

| Unknowns | Iterations with RILU | Iterations with MG |
|----------|----------------------|---------------------|
| 187      | 77                   | 51                  |
| 659      | 139                  | 54                  |
| 2467     | 235                  | 54                  |
| 9539     | 305                  | 57                  |
| 37507    | 455                  | 50                  |
| 148739   | 619                  | 53                  |

**Table 5.6.** Numerical results for the mixed Poisson problem with Raviart-Thomas elements.

| Unknowns | Iterations with RILU | Iterations with MG |
|---|---|---|
| 88 | 63 | 41 |
| 336 | 132 | 49 |
| 1312 | 221 | 51 |
| 5184 | 374 | 56 |
| 20608 | 487 | 55 |
| 82176 | 645 | 56 |

The number of iterations to achieve the convergence criterion

$$\|\mathcal{B}r_k\|/\|\mathcal{B}r_0\| \leq 10^{-6}$$

with the Taylor-Hood element is shown in Table 5.5. We obtain similar behavior with the Mini element, where we use the same preconditioner. The number of iterations needed with the RILU preconditioner increases as the number of unknowns increases. The multigrid preconditioner ensures a fixed convergence rate independent of the number of unknowns. The input files for the Taylor-Hood experiment is b_th.i and mg_th.i, while the Mini element is tested with the b_mini.i and the mg_mini.i input files.

The preconditioner for the Raviart-Thomas and the robust elements is on the form (5.66). Table 5.6 shows the results with the Raviart-Thomas element. It seems that the preconditioner is about as efficient as in the case of the Taylor-Hood element. The robust element behaves the same way. The Raviart-Thomas element is tested in b_rt.i and mg_rt.i, while the robust element is tested in b_new.i and mg_new.i. The complete source code for the block preconditioned mixed Poisson problem is in src/mixed/PoissonMx/block.

# References

1. D. N. Arnold, R. S. Falk, and R. Winther. Preconditioning in H(div) and applications. *Math. Comp. 66*, 1997.
2. F. Brezzi and M. Fortin. *Mixed and Hybrid Finite Element Methods*. Springer-Verlag, 1991.
3. A. M. Bruaset. *A Survey of Preconditioned Iterative Methods*. Addison-Wesley Pitman, 1995.
4. A. M. Bruaset and H. P. Langtangen. Object-oriented design of preconditioned iterative methods in Diffpack. *Transactions on Mathematical Software*, 23:50–80, 1997.
5. Craig C. Douglas, Jonathan Hu, Markus Kowarschik, and Ulrich Rude. Cache based algorithms. In *MGNet Virtual Proceedings 2001*, http://www.mgnet.org/mgnet-cm2001.html, 2001.
6. D. B. Geselowitz and W. T. Miller. A bidomain model for anisotropic cardiac muscle. *Annals of Biomedical Engineering*, 11:191–206, 1983.

7. V. Girault and P. A. Raviart. *Finite Element Methods for Navier-Stokes Equations*. Springer-Verlag, 1986.
8. W. Hackbusch. *Iterative Solution of Large Sparse Systems of Equations*. Springer-Verlag, 1994.
9. H. P. Langtangen. *Computational Partial Differential Equations - Numerical Methods and Diffpack Programming*. Textbooks in Computational Science and Engineering. Springer, 2nd edition, 2003.
10. K.-A. Mardal and H. P. Langtangen. Mixed finite elements. In H. P. Langtangen and A. Tveito, editors, *Advanced Topics in Computational Partial Differential Equations - Numerical Methods and Diffpack Programming*. Springer, 2003.
11. K.-A. Mardal, H. P. Langtangen, and G.W. Zumbusch. Software tools for multigrid methods. In H. P. Langtangen and A. Tveito, editors, *Advanced Topics in Computational Partial Differential Equations - Numerical Methods and Diffpack Programming*. Springer, 2003.
12. K.-A. Mardal, X.-C. Tai, and R. Winther. Robust finite elements for Darcy-Stokes flow. *SIAM Journal on Numerical Analysis*, 40:1605–1631, 2002.
13. T. Rusten and R. Winther. A preconditioned iterative method for saddlepoint problems. *SIAM J. Matrix Anal.*, 1992.
14. J. Sundnes, G. T. Lines, P. Grøttum, and A. Tveito. Electrical activity in the human heart. In H. P. Langtangen and A. Tveito, editors, *Advanced Topics in Computational Partial Differential Equations - Numerical Methods and Diffpack Programming*. Springer, 2003.
15. P. S. Vassilevski T. Rusten and R. Winther. Interior penalty preconditioners for mixed finite element approximations of elliptic problems. *Mathematics of Computation*, 1996.
16. S. Turek. *Efficient Solvers for Incompressible Flow Problems*. Springer, 1999.
17. C. Oosterlee U. Trottenberg and A. Schuller. *Multigrid*. Academic Press, 2001.
18. J. Xu. The auxiliary space method and optimal multigrid preconditioning techniques forunstructured grids. *Computing*, 56:215–235, 1996.

# Chapter 6

# Fully Implicit Methods for Systems of PDEs

Å. Ødegård[1], H. P. Langtangen[1,2], and A. Tveito[1,2]

[1] Simula Research Laboratory
[2] Department of Informatics, University of Oslo

**Abstract.** Operator splitting is a common method for solving systems of partial differential equations. This is particularly the case if the system is composed of separate equations for which suitable software already exists. In such cases, operator splitting combined with explicit time–stepping is a straight forward approach. However, for some systems, implicit time–stepping is to be preferred because of better stability properties. The problem of combining existing codes in a fully *implicit* manner is much harder than the explicit case. The purpose of the present chapter is to discuss some examples illustrating the possibilities of combining existing codes in an implicit manner.

## 6.1 Introduction

We focus on a general implementation technique where a solver for a PDE is an object that can be combined with other solver objects in order to easily construct software for solving systems of PDEs. Using this technique, we make use of existing individual PDE solvers to simplify the discretization of a system of PDEs. Descriptions of boundary and initial conditions in the existing solvers can also be reused. The first step in this direction was taken in [1] and developed further by Langtangen and Munthe in [3, Ch. 7.2]. One major advantage of this method is its flexibility. It is particularly easy to assemble a solver for a system of PDEs based on independent solvers for each equation in the system. A potential problem of the methods presented in these works is that operator splitting algorithms are used to solve systems of PDEs. This means that the PDEs are solved in sequence, one PDE at a time, wrapped with some kind of iteration. A problem with this approach is that it is not unconditionally stable. Also, the numerical solution will in general depend on the order in which the equations are solved. For simple problems, a stability condition may be derived, but this is not the case for complicated, non–linear problems. We are hence urged to seek other methods lacking these unfortunate properties but still inviting an object-oriented approach. Fully implicit methods are known to be more reliable than operator splitting methods, and the main goal of this chapter is therefore to establish a framework for implementation of fully implicit solvers for systems of PDEs. This framework retains the flexibility and extensibility of implementating the operator splitting approach.

All the examples in this paper are implemented in Diffpack [2,3]. Below, we give a brief review of how solvers for PDEs can be combined, to form a solver for a system of equations, using the operator splitting approach. Next we recall some basic properties of this method, showing why the method may fail. An example is included where Gauss–Seidel iteration used in an operator splitting context fails. Finally we describe the new method, which implements flexible and extensible fully implicit solvers, based on independent solvers for each PDE in the system. Some applications where the new method is implemented are described.

## 6.2    Implementation of Solvers for PDE Systems in Diffpack

### 6.2.1    Handling Systems of PDEs

We consider systems of differential equations on the form

$$L_1(u, v) = f, \tag{6.1}$$
$$L_2(u, v) = g, \tag{6.2}$$

on some domain $\Omega$, and with some boundary condition specified on $\partial\Omega$. We assume properties on the operators $L_1$ and $L_2$ such that a finite element method can be applied in the discretization. Examples of such systems are given below. In Diffpack we can implement flexible solvers for such problems, using operator splitting algorithms. We can formulate this algorithm as an iteration on the equation level as follows:

1. Set $u$ and $v$ to proper initial functions.
2. Solve (6.1) with respect to $u$, regarding $v$ as a given function.
3. Solve (6.2) with respect to $v$, regarding $u$ as a given function.
4. Repeat steps 2 and 3 until some convergence criterion is fulfilled.

The method using both $u$ and $v$ in steps 2 and 3 from the previous iteration is referred to as the Jacobi algorithm. If the most recently available values of $u$ and $v$ are used, we call it Gauss–Seidel algorithm. See [3, Ch. 7.2, App. C].

Given $u^n$, $v^n$, we can formulate one iteration in the Jacobi algorithm:

1. Compute $u^{n+1}$ such that $L_1(u^{n+1}, v^n) = f$.
2. Then compute $v^{n+1}$ such that $L_2(u^n, v^{n+1}) = g$.

Similarly, given $u^n$, $v^n$, we can formulate one iteration in the Gauss–Seidel method:

1. Compute $u^{n+1}$ such that $L_1(u^{n+1}, v^n) = f$.
2. Compute $v^{n+1}$ such that $L_2(u^{n+1}, v^{n+1}) = g$.

We observe that both methods reduce the problem of solving one system of equations to solving two scalar equations. Solvers for such scalar equations are easily implemented in Diffpack. Assume that two solver classes are implemented for the two scalar equations, as described in [3, Ch. 3.1, 7.2]. Most functions in the basic solvers are defined as virtual functions, such that we can derive a subclass for each solver where only the coupling between the equations are implemented. Usually, this requires at least the integrands function to be re–implemented. Then, we can introduce a control unit which holds one object of each solver class, and solves the system through either the Jacobi or the Gauss–Seidel iterations. The control unit is implemented as a new class. The objects to each solver class in the control unit should be of the derived types, not the original solver classes.

## 6.2.2    A Specific 2 × 2 System

As a specific case of (6.1)-(6.2), we consider[1], $u = u(\mathbf{x}), v = v(\mathbf{x})$,

$$
\begin{aligned}
-\nabla^2 u + u + v &= f, \quad \mathbf{x} \in \Omega, \\
-\nabla^2 v + u + v &= g, \quad \mathbf{x} \in \Omega, \\
\frac{\partial u}{\partial n} &= 0, \quad \mathbf{x} \in \partial\Omega, \\
\frac{\partial v}{\partial n} &= 0, \quad \mathbf{x} \in \partial\Omega.
\end{aligned}
\tag{6.3}
$$

Let $(p, q)$ denote $\int_\Omega pq \, d\Omega$. A standard Galerkin finite element method [3, Ch. 2], gives

$$
\begin{aligned}
\sum_{j=1}^{n}(\nabla N_j, \nabla N_i)u_j + \sum_{j=1}^{n}(N_j, N_i)u_j + \sum_{j=1}^{n}(N_j, N_i)v_j = (f, N_i), \quad i = 1, \ldots, n, \\
\sum_{j=1}^{n}(\nabla N_j, \nabla N_i)v_j + \sum_{j=1}^{n}(N_j, N_i)u_j + \sum_{j=1}^{n}(N_j, N_i)v_j = (g, N_i), \quad i = 1, \ldots, n,
\end{aligned}
\tag{6.4}
$$

where

$$
\hat{u} = \sum_{j=1}^{n} u_j N_j \quad \text{and} \quad \hat{v} = \sum_{j=1}^{n} v_j N_j
$$

are finite element approximations of $u$ and $v$. Further, $N_j$ are the basis functions in the finite element discretization and $n$ is the number of basis functions.

---

[1] Note that this system can be decoupled into two completely independent scalar problems by solving for $e = u - v$, hence it is used here for the purpose of illustration only.

Given the initial guess on the form $(u^{(0)}, v^{(0)})$, the Gauss–Seidel iteration for this system can the be formulated as

$$\sum_{j=1}^{n}(\nabla N_j, \nabla N_i)u_j^{(k+1)} + \sum_{j=1}^{n}(N_j, N_i)u_j^{(k+1)} = -\sum_{j=1}^{n}(N_j, N_i)v_j^{(k)} + (f, N_i),$$

$$\sum_{j=1}^{n}(\nabla N_j, \nabla N_i)v_j^{(k+1)} + \sum_{j=1}^{n}(N_j, N_i)v_j^{(k+1)} = -\sum_{j=1}^{n}(N_j, N_i)u_j^{(k+1)} + (g, N_i).$$

$$(6.5)$$

The above two linear systems can be solved in sequence with a control class and two solver objects as briefly outlined above.

## 6.3   Problem with the Gauss–Seidel Method, by Example

For simple systems of linear PDEs as the example in Section 6.2.2, convergence of the Jacobi and Gauss–Seidel iteration algorithms to the correct solution can be proven by well known theories from linear algebra [3, Ch. C.1]. However, real–life applications of PDEs tend to be extremely complicated and consequently simple conditions guaranteeing convergence are not available. Therefore, a realistic scenario is that we do not know a priori whether or not a simple splitting scheme like the Jacobi or Gauss–Seidel algorithm will work. We present an example illustrating this in the next section.

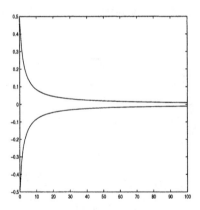

**Fig. 6.1.** The exact solution of (6.6).

### 6.3.1   A System of Ordinary Differential Equations

We will now apply the methods described above for a non–linear system of ordinary differential equations. The purpose of this example is to show that the simple splitting schemes may fail. Consider the system

$$y' = z^2 \qquad y(0) = -\varepsilon,$$
$$z' = -y^2 \qquad z(0) = \varepsilon. \tag{6.6}$$

If we add the two equations together, we get

$$(y+z)' = z^2 - y^2 = (y+z)(z-y). \tag{6.7}$$

Let $u = y + z$ and $v = z - y$. It then follows from (6.7) that

$$u' = uv.$$

Since $u(0) = -\epsilon + \epsilon = 0$, we have $u' = uv = 0$ at $t = 0$. Hence $u$ remains zero for all $t > 0$ and thus $y = -z$. We can then reformulate (6.6) as

$$y' = y^2, \qquad y(0) = -\varepsilon,$$
$$z' = -z^2, \qquad z(0) = \varepsilon. \tag{6.8}$$

Both equations can be solved analytically, and the solutions are

$$y(t) = \frac{-\epsilon}{1 + t\epsilon}, \tag{6.9}$$
$$z(t) = \frac{\epsilon}{1 + t\epsilon}. \tag{6.10}$$

A plot of the analytical solutions for $t$ between 0 and 100 is given in Figure 6.1. Note that both $y$ and $z$ are monotone functions.

The Gauss–Seidel method can be formulated in two ways for the *original* system (6.6), depending on the order in which we solve the two equations; either

$$y_{n+1} = y_n + \Delta t z_n^2 \qquad y_0 = -\varepsilon,$$
$$z_{n+1} = z_n - \Delta t y_{n+1}^2 \qquad z_0 = \varepsilon, \tag{6.11}$$

or

$$z_{n+1} = z_n - \Delta t y_n^2 \qquad z_0 = \varepsilon,$$
$$y_{n+1} = y_n + \Delta t z_{n+1}^2 \qquad y_0 = -\varepsilon. \tag{6.12}$$

The two versions of Gauss–Seidel give the solution curves plotted in Figure 6.2a and 6.2b, respectively, with $\Delta t = 0.0014$. We observe that none of them gives a correct solution, and also that they produce slightly different results. In Figure 6.2c, we have used a smaller value of $\Delta t$. This shows that we can improve the approximation, but still both $y(t)$ and $z(t)$ blow up when $t$ tends to infinity.

We now consider the Jacobi method. For system (6.6) the method is given by

$$y_{n+1} = y_n + \Delta t z_n^2 \qquad y_0 = -\varepsilon,$$
$$z_{n+1} = z_n - \Delta t y_n^2 \qquad z_0 = \varepsilon. \tag{6.13}$$

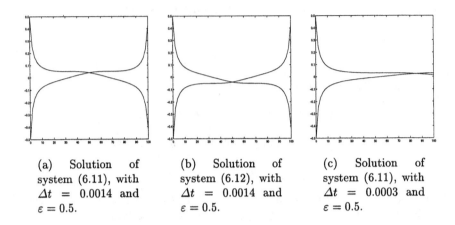

(a) Solution of system (6.11), with $\Delta t = 0.0014$ and $\varepsilon = 0.5$.

(b) Solution of system (6.12), with $\Delta t = 0.0014$ and $\varepsilon = 0.5$.

(c) Solution of system (6.11), with $\Delta t = 0.0003$ and $\varepsilon = 0.5$.

**Fig. 6.2.** The solution of (6.6) approximated with Gauss–Seidel's method.

Due to the initial data given in (6.6), a required stability condition for the Jacobi method is $\Delta t \varepsilon - 1 < 0$. If this is fulfilled, the iteration gives the solution in Figure 6.3a. If the stability condition is broken, we get something totally different, as we can see in Figure 6.3b, so the stability condition is crucial. Given that the stability condition is fulfilled, the Jacobi algorithm seems to be a better method than the Gauss–Seidel method for this particular example, since we get an approximation close to the exact solution. It is generally believed that the Gauss–Seidel method yields more accurate solutions than the Jacobi method, but as seen in this example, this is not always the case. Since the Gauss–Seidel algorithm generates solutions that lack the basic characteristics of the analytical solutions, it is important to use this type of splitting with care.

The third method we consider is the fully implicit iteration where both equations are solved simultaneously, i.e.,

$$y_{n+1} = y_n + \Delta t z_{n+1}^2 \qquad y_0 = -\varepsilon,$$
$$z_{n+1} = z_n - \Delta t y_{n+1}^2 \qquad z_0 = \varepsilon. \tag{6.14}$$

This method produces a good approximation to the exact solution, as we can see in Figure 6.4.

This example illustrates that simple splitting–techniques may lead to wrong solutions or slow convergence as well as strict stability conditions. However, a fully implicit approach may produce an adequately accurate solution.

## 6.4   Fully Implicit Implementation

Both the Jacobi and the Gauss–Seidel approach on the equation level are especially flexible and easy to implement in Diffpack, because we can benefit from object–oriented techniques. We have also seen above that we need to consider fully implicit methods for systems, since these generally are more stable. In order to implement fully implicit methods in the same flexible way as Gauss–Seidel or Jacobi methods, we first need to examine the differences between these two classes of methods.

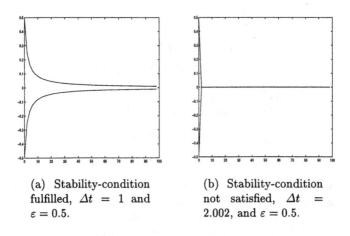

(a) Stability-condition fulfilled, $\Delta t = 1$ and $\varepsilon = 0.5$.

(b) Stability-condition not satisfied, $\Delta t = 2.002$, and $\varepsilon = 0.5$.

**Fig. 6.3.** The solution of (6.6) approximated with Jacobi's method.

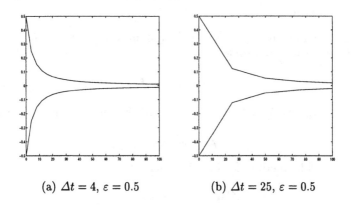

(a) $\Delta t = 4$, $\varepsilon = 0.5$

(b) $\Delta t = 25$, $\varepsilon = 0.5$

**Fig. 6.4.** The solution of (6.6) approximated with an implicit method.

### 6.4.1  A System of Linear Partial Differential Equations

We want to implement a fully implicit solver for a system of linear partial differential equations and compare this with the Gauss–Seidel method presented above in Section 6.2.1. We show that the implicit method can be implemented in the same flexible manner as the Gauss–Seidel method. Consider the system

$$-\nabla^2 u + u + v = f,$$
$$-\nabla^2 v + u + v = g. \tag{6.15}$$

Let matrix $\mathbf{A}$ have components $A_{i,j} = (\nabla N_i, \nabla N_j)$ and $\mathbf{M}$ have components $M_{i,j} = (N_i, N_j)$. The finite element form of system (6.15) can then be written with block matrices as

$$\begin{bmatrix} \mathbf{A}+\mathbf{M} & \mathbf{M} \\ \mathbf{M} & \mathbf{A}+\mathbf{M} \end{bmatrix} \begin{bmatrix} \hat{\mathbf{u}} \\ \hat{\mathbf{v}} \end{bmatrix} = \begin{bmatrix} \mathbf{F} \\ \mathbf{G} \end{bmatrix}, \tag{6.16}$$

where $F_i = (f, N_i)$ and $G_i = (g, N_i)$.

Remark that the Gauss–Seidel iteration (6.5) may be written with block matrices as

$$[\mathbf{A} + \mathbf{M}]\,\hat{\mathbf{u}}^{(k+1)} = -\mathbf{M}\hat{\mathbf{v}}^{(k)} + \mathbf{F},$$
$$[\mathbf{A} + \mathbf{M}]\,\hat{\mathbf{v}}^{(k+1)} = -\mathbf{M}\hat{\mathbf{u}}^{(k+1)} + \mathbf{G}. \tag{6.17}$$

We observe that the Gauss–Seidel approach and the implicit approach on the block matrix form only differ in how the terms of the linear system are arranged. We also note that the implicit method is a one–step procedure, while the Gauss–Seidel iteration is a two–step procedure.

### 6.4.2  Implementation

Recall that we implemented a solver for system (6.15) in Section 6.2.2, based on a sub–solver for the problem

$$-\nabla^2 u + u = f. \tag{6.18}$$

Using the matrices $\mathbf{A}$ and $\mathbf{M}$, we can write the finite element form of this problem as

$$[\mathbf{A} + \mathbf{M}]\hat{\mathbf{u}} = \mathbf{F}. \tag{6.19}$$

The Gauss–Seidel method for (6.15) is therefore

$$[\mathbf{A} + \mathbf{M}]\,\hat{\mathbf{u}}^{(k+1)} = -\mathbf{M}\hat{\mathbf{v}}^{(k)} + \mathbf{F}, \tag{6.20}$$
$$[\mathbf{A} + \mathbf{M}]\,\hat{\mathbf{v}}^{(k+1)} = -\mathbf{M}\hat{\mathbf{u}}^{(k+1)} + \mathbf{G}. \tag{6.21}$$

Now, we can clearly see that each equation in this system is similar to (6.19), with one term added at the right side in each equation.

The implicit approach applied to system (6.15) leads to the linear system (6.16). If we compare (6.16) with (6.20)–(6.21), we observe that the methods again only differ in how the terms are arranged.

As soon as the system (6.16) is assembled into an object of the class LinEqAdm, the system can be solved by the utilities for linear systems in Diffpack. Hence, the only problem left for the fully implicit method is how to assemble the system into this object, in such a way that the flexibility of the approach from the Gauss–Seidel method is retained. That is, we still want to use one sub–solver for each equation and a control unit. The most obvious difference from the Gauss–Seidel method, is that we now need to assemble a linear system in the control unit. We also need to modify the integrands function in the solver classes, because each solver now have to contribute two blocks into the coefficient matrix in (6.16). Below we present the integrands function in the solver classes derived for the fully implicit solver. Note that the name is changed to integrandsMx. This reflects the fact that we assemble a system for two or more unknown fields in one operation, and therefore are using the tools for Mixed Finite Element methods in Diffpack:

```
Simulator::integrandsMx (
   ElmMatVec& elmat
   const MxFiniteElement& mfe)
{
   const int nubf = mfe(1).getNoBasisFunc();
   const int nvbf = mfe(2).getNoBasisFunc();
   const int nsd = mfe.getNoSpaceDim();
   Ptv(real) x (nsd)
   fe.getGlobalEvalPt(x);
   real f_value = f(x);
   real nabla;
   for( int i=1; i < nubf; i++){
      //Left Block, -\nabla^2u+u
      for( int j=1; j < nubf; j++){
         nabla = 0;
         for( int k=1;k < nsd; k++)
            nabla += mfe(1).dN(i,k)*mfe(1).dN(j,k);
         elmat.A(i,j) += (nabla +
            mfe(1).N(i)*mfe(1).N(j))*mfe.detJxW();
      } // Right block, v
      for( int j=1; j < nvbf; j++)
         elmat.A(i,nubf+j) += mfe(1).N(i)*mfe(2).N(j)*
            mfe.detJxW();
      // Right hand side
      elmat.b(i) += mfe.N(i)*f_value*mfe.detJxW();
   }
}
```

The difference between system (6.16) and equations (6.20)–(6.21) is that the blocks off the main diagonal in the coefficient matrix in (6.16) has moved into the right hand side of equations (6.20)–(6.21). This is also the only real difference between the integrandsMx function above and the integrands functions used in a Gauss–Seidel method.

In order to complete this description, we must describe the control unit for the fully implicit approach. The difference from the control unit described for the Gauss–Seidel method in [3, Ch. 7.2], is that we now want to assemble and solve a linear system inside it. Hence, we need a object for administering the linear system and a vector for that of storing the solution. Otherwise, the content of the control unit is similar to the control unit described before.

The assembly process of the linear system is usually taken care of by functions from the Diffpack library, e.g., the function MakeSystem in the class FEM. But the existing functions can only assemble the linear system from one solver, while we need to assemble the linear system from an arbitrary number of attached sub–solvers. This is taken care of in a library class, SystemFEM. This library class has a VecSimplest (SimulatorSystem) array named Solvers where pointers to each solver class in the system will be attached, and a function makeSystemMx which assembles a linear system from the attached solver classes. The SimulatorSystem is a library class which ensures a common interface for the solver classes. Hence, solver classes must be derived from this class.

The SystemFEM class is derived from another class in the Diffpack library, called MxFEM, which handles mixed finite element methods. Consequently, the function which assembles the linear system is calles makeSystemMx. For the same reason you will see other methods with the MX suffix in this framework. Below we sketch a control class for a fully implicit solver:

```
class FImanager : public SystemFEM
{
// Variables:
    SimulatorSystem Sim_u; // A subsolver
    SimulatorSystem Sim_v; // Another subsolver
    GridFE* grid;
    FieldsFE* coll_of_fields; // Basically a vector of Fieldpointers
    LinEqAdm* lineq;
    Vec linsol;

// Methods:
    virtual void define (MenuSystem &menu, int level = MAIN);
    virtual void scan (MenuSystem &menu);
    virtual void fillEssBC();
    solveProblem;
}
```

The solveProblem function in the above class has access to two solver classes, and relies on the tailored makeSystemMx to assemble the linear system. Basically, this function initializes the Solvers array of solvers in the SystemFEM base class, calls the function makeSystemMx, and then solves the linear system that was assembled inside the LinEqAdmFE object:

```
void Solver::solveProblem(){
    Solvers.redim(2);            // Set number of solverclasses
    Solvers(1).rebind(Sim_u());// Attach solver 1
    Solvers(2).rebind(Sim_v());// Attach solver 2
```

```
fillEssBC();
makeSystemMx(lineq, coll); // Assemble the linear system into
                           // the LineEqAdm object lineq
lineq->solve();
dof->vec2field (linsol, coll); // Map the linear solution onto the
collection of Fields
}
```

We have seen that the Gauss–Seidel solver and the fully implicit solver require the same information. We have also seen that the fully implicit method can be implemented in a similar flexible manner as the Gauss–Seidel method.

### 6.4.3   Systems of Non–Linear Partial Differential Equations

The Gauss–Seidel algorithm for solving systems of PDEs can be applied in the non–linear case, hence we want to investigate whether the fully implicit method can be implemented in a similar flexible manner as the Gauss–Seidel method also in the non–linear case. We solve all non–linear problems with the Newton–Raphson method: Given a non–linear system $F(u) = 0$, the Newton iteration is given by

$$\frac{\partial F}{\partial u}(u^{(i-1)})r_u^{(i)} = -F(u^{(i-1)}), \tag{6.22}$$

where $\partial F/\partial u$ is the Jacobi matrix and $r_u^{(i)}$ is the correction of $u$, that is $r_u^{(i)} = u^{(i)} - u^{(i-1)}$. Further, $u^{(i)}$ is a series of solutions generated by the Newton–Raphson iteration.

First, we describe why the Gauss–Seidel approach is easy to implement. Let

$$F_1(u, v) = 0 \quad \text{and} \quad F_2(u, v) = 0 \tag{6.23}$$

be two non–linear differential equations, where the dependency is removed by setting $v$ to some known function in the first equation and $u$ to some know function in the second. Assume that a Diffpack solver for each problem exists. The implementation of a nonlinear solver in Diffpack is similar to that of linear solvers. The implementation of nonlinear solvers is discussed in [3, Ch. 4.2].

The coupled system is given by

$$\begin{aligned} F_1(u, v) &= 0, \\ F_2(u, v) &= 0. \end{aligned} \tag{6.24}$$

The Gauss–Seidel method for this problem is given by

$$\begin{aligned} F_1(u_k, v_{k-1}) &= 0, \\ F_2(u_k, v_k) &= 0. \end{aligned} \tag{6.25}$$

The corresponding Newton iterations on the discrete problem is given by

$$
\begin{aligned}
\frac{\partial F_1}{\partial u}(u_k^{(i-1)}, v_{k-1}) r_u^{(i)} &= -F_1(u_k^{(i-1)}, v_{k-1}), \\
\frac{\partial F_2}{\partial v}(u_k, v_k^{(i-1)}) r_v^{(i)} &= -F_2(u_k, v_k^{(i-1)}).
\end{aligned}
\tag{6.26}
$$

These are the same Jacobi–matrices that are already present in the sub-solvers we depend on. Hence, it is simple to implement a Gauss–Seidel solver for the system, in the same manner as before. We just need to reimplement the method named `integrands`, introducing the new coupling between the two equations.

A fully implicit solver applies Newton–Raphsons method to (6.24) directly. The Newton iteration on the corresponding discrete system is given by

$$
\begin{bmatrix} \frac{\partial F_1^{(i-1)}}{\partial u} & \frac{\partial F_1^{(i-1)}}{\partial v} \\ \frac{\partial F_2^{(i-1)}}{\partial u} & \frac{\partial F_2^{(i-1)}}{\partial v} \end{bmatrix} \begin{bmatrix} r_u^{(i)} \\ r_v^{(i)} \end{bmatrix} = \begin{bmatrix} -F_1^{(i-1)} \\ -F_2^{(i-1)} \end{bmatrix},
\tag{6.27}
$$

where $F_n^{(i-1)} = F_n(u_k^{i-1}, v_k^{i-1})$ and $r_x^{(i)} = x^{(i)} - x^{(i-1)}$ is the residual of $x$ where $x$ can be $u$ and $v$. We observe that the blocks on the main diagonal of the coefficient matrix are the Jacobi matrices involved in the Gauss–Seidel solver, but the two off–diagonal blocks are new. This shows a difference between the linear and the non–linear case. In the linear case when we go from the Gauss–Seidel method to the fully implicit method, we just need to move some terms from one side in the equation to the other. For the non–linear problem we need to compute a full Jacobi matrix for the coupled system, which includes blocks that are neither present in the Gauss–Seidel solver nor in the stand-alone sub–solvers. We also observe, however, that the linear system we need to solve in each Newton iteration, is very similar to (6.16). Hence, we can implement a fully implicit solver in the non–linear case, similarly as we have done in the linear case. The only exception is that the control–unit now must be a non–linear solver.

## 6.5   Applications

The purpose of this section is to demonstrate how the fully implicit, yet object–oriented, implementation sketched above can be applied to two test cases. For both cases, sub–solvers are already implemented in Diffpack and we want to reuse these codes as building blocks in a fully implicit solver for the system of PDEs.

### 6.5.1   A System of PDEs Modelling Pipeflow

Consider a model for Newtonian pipeflow. The model is studied in Chapter 7.2 in [3], where a Gauss–Seidel algorithm is implemented. We describe how we can extend this into a fully implicit solver.

The model is given by

$$\nabla \cdot [\mu \nabla w] = -\beta,$$
$$\nabla^2 T = -\kappa^{-1} \mu \dot{\gamma}^2,$$
$$w = 0 \text{ on } \partial\Omega,$$
$$T = T_0 \text{ on } \partial\Omega_1, \ T = T_1 \text{ on } \partial\Omega_2. \tag{6.28}$$

Here,

$$\mu = \mu_T(T) \mu_w(\dot{\gamma}),$$

and

$$\dot{\gamma} = \sqrt{(w_{,x})^2 + (w_{,y})^2},$$

$\mu_T$ and $\mu_w$ are given functions, describing the physical relations, and $\partial\Omega_1 \cup \partial\Omega_2 = \partial\Omega$. In our test–case, we solve (6.28) on the unit square. The part $\partial\Omega_2$ of the boundary is taken as the side $y = 1.0$ in the square, and we choose $T_0 = 0.0$ and $T_1 = 1.0$.

We solve (6.28) by a standard Galerkin finite element method. Suppose that $N_i(x)$, $i = 1, \ldots n$ is a set of finite element basis functions. The discrete equations then take the form

$$F_{w;i} = \int_\Omega (\mu \nabla w \cdot \nabla N_i - \beta N_i) d\Omega \qquad = 0,$$
$$F_{T;i} = \int_\Omega (\nabla T \cdot \nabla N_i - \kappa^{-1} \mu \dot{\gamma} N_i) d\Omega \quad = 0. \tag{6.29}$$

Let $(w_k, T_k)$ be the latest approximation. An iteration in the Gauss–Seidel method is then given as

1. Compute $w_{k+1}$ such that $F_w(w_{k+1}, T_k) = 0$.
2. Compute $T_{k+1}$ such that $F_T(w_{k+1}, T_{k+1}) = 0$.

This is the outer iteration in the solution algorithm. Since both sub–problems are nonlinear, we apply the Newton iteration to each problem as an inner iteration. Given the approximation $(w_{k+1}^{(i)}, T_k)$, we have to solve the following system as a step in the inner iteration for step 1 above:

$$\partial F_w(w_{k+1}^{(i)}, T_k) r_w^{(i)} = -F_w(w_{k+1}^{(i)}, T_k), \tag{6.30}$$

where $r_w^{(i)} = w^{(i+1)} - w^{(i)}$, and $\partial F_w$ is the Jacobi matrix. The inner iteration of step 2 above has a similar expression. Note that solvers for step 1 and step 2 in the Gauss–Seidel loop already exist.

Let us turn to the fully implicit case. The linear system of the $i$'th step in the Newton iteration, which we need to assemble, is given by

$$\begin{bmatrix} \frac{\partial F_w^{(i-1)}}{\partial w} & \frac{\partial F_w^{(i-1)}}{\partial T} \\ \frac{\partial F_T^{(i-1)}}{\partial w} & \frac{\partial F_T^{(i-1)}}{\partial T} \end{bmatrix} \begin{bmatrix} r_w^{(i)} \\ r_T^{(i)} \end{bmatrix} = \begin{bmatrix} -F_w^{(i-1)} \\ -F_T^{(i-1)} \end{bmatrix}. \tag{6.31}$$

where $r_x^{(i)} = x^{(i)} - x^{(i-1)}$, $x = w, T$. The expressions for the derivatives depend on the choice of $\mu_w$ and $\mu_T$. For a particular choice of these functions, see [3, Ch. 7.2].

The fully implicit solver for this problem is built on top of the stand-alone solvers for the two equations, see [3, Ch. 7.2]. We have to derive a new class as a solver class for each equation. These new solver classes must also be derived from `SimulatorSystem`, as described above, with a function called `integrandsMx`, where the element contributions to the system (6.31) are implemented. In [3, Ch. 7.2], a new class for common relations is implemented for the Gauss–Seidel method. This class holds functions and variables which are common for both solvers in the problem. We can reuse this class in the fully implicit solver. At last we need to implement the control unit, as a combination of the fully implicit solver outlined in Section 6.4.1 and the non–linear solver discussed in [3, Ch. 4.2].

### 6.5.2   Comparison of the Gauss–Seidel and the Fully Implicit Method

We have done some simple experiments to compare the performance of the two methods presented. These experiments show that the fully implicit method performs slightly better for this problem.

The results of the experiments are summarized in Table 6.1. In these computations,

$$\mu_T(T) = \mu_0 e^{-\alpha(T-T_0)}, \qquad \mu_w(\dot{\gamma}) = \dot{\gamma}^{n-1}. \tag{6.32}$$

The table gives the number of iterations for both the Gauss–Seidel method and the fully implicit method. These iteration counts are associated with two values of $n$: $n = 0.8$, $n = 1.0$, with $\alpha = 1.0$ and $\mu_0 = 0.5$. The problem is solved on the unit square. The discretization uses a grid of $800 \times 800$ elements. The stopping criterion used for the Gauss–Seidel iteration is

$$\frac{|r_w|_2 + |r_T|_2}{|w|_2 + |T|_2} < \epsilon, \tag{6.33}$$

where $r_w$ and $r_T$ are the corrections of the momentum and energy fields, respectively, and $\epsilon$ is a number used as tolerance. The norm used is the discrete $L^2$ norm: Given a vector with $n$ components, $\mathbf{u} = (u_1, \ldots, u_n)$,

$$|\mathbf{u}|_2 = \left[ \frac{1}{n} \sum_{i=1}^{n} u_i^2 \right]^{1/2}.$$

The same criterion is also used in the fully implicit method. For both methods, we have set the stopping criterion to $\epsilon = 10^{-2}$ for the outer iterations. In the fully implicit method, conjugate gradient method with RILU preconditioner is used to solve the linear system in each Newton iteration. The

**Table 6.1.** Performance comparison. For the fully implicit case, *iter* is the number of Newton iterations. For Gauss–Seidel, *iter* is outer iterations.

| pow.law exponent: | $n = 0.8$ | | $n = 1.0$ | |
|---|---|---|---|---|
| | iter. | time | iter. | time |
| Fully Implicit | 4 | 4342.7s | 3 | 3487.6s |
| Gauss–Seidel | 3 | 7747.8s | 3 | 5995.8s |

relaxation parameter is set to 0.7 and the tolerance for the conjugate gradient iteration is $\epsilon = 10^{-3}$. In the Gauss–Seidel method, the tolerance for the Newton iterations in the sub–problems is also $\epsilon = 10^{-3}$, while $\epsilon = 10^{-4}$ is used for the tolerance in the conjugate gradient iteration used to solve the linear systems in each Newton iteration. The same RILU preconditioner is used in this case. Table 6.1 also gives the CPU time used for the computation in each case [2]. As these results suggest, the fully implicit method is preferable at least for grid sizes and parameters comparable to what we have used. On the other hand, the Gauss–Seidel method requires less memory because we do not need to store matrices and vectors for the full coupled system, so this method can be used to solve larger systems than the fully implicit method.

### 6.5.3   Two–phase Porous Media Flow

We consider a simplified model of two–phase flow from oil reservoir simulation,

$$s_t + \mathbf{v} \cdot \nabla(f(s)) = \nabla \cdot (D\nabla s), \tag{6.34}$$
$$-\nabla \cdot (\lambda(s)\nabla p) = q, \tag{6.35}$$
$$\mathbf{v} = -\lambda(s)\nabla p. \tag{6.36}$$

Here, $s$ is the saturation of water in the reservoir, $p$ is the pressure, and $\mathbf{v}$ is the velocity vector field. The source term $q$ represents the wells in the reservoir and $D$ is a diagonal matrix with small positive entries. The parameters are defined as:

$$\lambda_w = K\frac{k_w(s)}{\mu_w},$$
$$\lambda_o = K\frac{k_o(s)}{\mu_o}, \tag{6.37}$$
$$\lambda = \lambda_w + \lambda_o,$$
$$f(s) = \frac{\lambda_w}{\lambda}.$$

---

[2] All computations are performed on a dual AMD Athlon MP1800 workstation supplied with 1GB memory, running Diffpack under the Linux operating system.

A derivation of this model can be found in, e.g, Peaceman[4], where the basic properties of the model are discussed.

We set the absolute permeability $K = 1$ and the viscosity of water and oil $\mu_w = 1$ and $\mu_o = 4$, respectively. These parameters are not necessarily realistic, they are used only as an example. The relative permeabilities are $k_w(s) = s^2$ for water and $k_o(s) = (1-s)^2$ for oil. The boundary condition for the pressure equation (6.35) is no flow on the entire boundary, $\mathbf{v} \cdot \mathbf{n} = 0$. For the saturation field the initial condition is given as

$$s_0(x) = 1, \quad x \in \Omega_I, \qquad s_0(x) = 0 \text{ elsewhere}. \tag{6.38}$$

Here, $\Omega_I$ is the domain of the injection wells. Similarly, the domain of the production wells is $\Omega_P$. The boundary condition for the saturation equation (6.34) is

$$\frac{\partial s(x)}{\partial n} = 0, \ x \in \partial\Omega. \tag{6.39}$$

We assume that there exists a Gauss–Seidel type solver for the problem (6.34)-(6.36). The solver for the saturation equation (6.34) is implicit in time. Since this equation is non–linear, we use the Newton–Raphson method to solve the discrete problem obtained with the Galerkin finite element method.

The weak form of (6.34) is

$$(s_t, \varphi) + (\mathbf{v} \cdot \nabla(f(s)), \varphi) + (D\nabla s, \nabla\varphi) = 0 \tag{6.40}$$

where $\varphi$ is a trial function. We approximate the time derivative with an implicit difference method, hence we can give a semi–discrete form of (6.40):

$$(s^{n+1}, \varphi) - (s^n, \varphi) + \Delta t[(\mathbf{v} \cdot \nabla(f(s^{n+1})), \varphi) + (D\nabla s^{n+1}, \nabla\varphi)] = 0. \tag{6.41}$$

Assume that $\{N_i\}_{i=1}^n$ is a set of finite element basis functions, and define discrete approximation of pressure and saturation as

$$\mathbf{s} = \sum_j s_j N_j, \quad \mathbf{p} = \sum_j p_j N_j \tag{6.42}$$

respectively. In order to achieve good accuracy and stability of the Galerkin finite element method for this problem, we use the shock–capturing streamline diffusion method, that is, we choose

$$W_i = N_i + \delta(\mathbf{v} \cdot \nabla N_i f'(s))$$

as the weighting functions. We also add an residual–dependent artificial diffusion term $\hat{\varepsilon}$, which gives the following finite element method: Find $s_h \in V_h$ such that

$$(s_h^{n+1}, W_i) - (s_h^n, W_i)$$
$$+ \Delta t[(\mathbf{v} \cdot \nabla(f(s_h^{n+1})), W_i) + (D\nabla s_h^{n+1}, \nabla W_i) + (\varepsilon \nabla s_h^{n+1}, \nabla N_i)] = 0 \tag{6.43}$$

for all $N_i, W_i, i = 1, \ldots, n$. For a description of this method, see Samuelsson [5].

The parameters $\delta$ and $\hat{\varepsilon}$ are element wise constant functions which depend on the residual of the solution $s_h$. They are chosen as follows:

$$\hat{\varepsilon} = \sqrt{D^2 + \tilde{\varepsilon}} - D, \qquad \delta = \max\left(\frac{h}{2|f'(s_h)\mathbf{v}|} - \bar{\varepsilon}, 0\right),$$

where

$$\tilde{\varepsilon} = 1.5h^2|\mathbf{v} \cdot \nabla(f(s_h)) - \nabla \cdot (D\nabla s_h)|, \qquad \bar{\varepsilon} = \frac{\sum_{i=1}^{2} d_i(v_i f'(s_h))^2}{|\mathbf{v} f'(s_h)|^4}.$$

Here, $d_i$ are the diagonal entries of $D$ and $h$ is the smallest length scale of the elements.

If we omit the non–linearities in $W_i$, the Jacobian matrix $J$ for (6.43) becomes

$$\begin{aligned}
J_{i,j} = (N_j, W_i) &+ \Delta t[(\mathbf{v} \cdot \nabla s_h^{n+1} f''(s_h^{n+1})N_j \\
&+ \mathbf{v} \cdot \nabla N_j f'(s_h^{n+1}) - \lambda'(s_h^{n+1})N_j \nabla p \cdot \nabla(f(s_h^{n+1})), W_i) \\
&+ (D\nabla N_j, \nabla W_i) + (\hat{\varepsilon}\nabla N_j, \nabla N_i)].
\end{aligned} \quad (6.44)$$

In the fully implicit solver, we want to solve for both $s$ and $p$ simultaneously. This problem is again non–linear, hence we use the Newton–Raphson method to solve it. In addition to the derivatives of the saturation equation with respect to $s_j$, we also need to compute the derivatives of the pressure equation and the derivatives of the saturation equation with respect to $p_j$.

The complete system we want to solve is

$$F = \binom{F^s}{F^p} = 0,$$

where $F^s$ is given in (6.43). The form of $F^p$ is simply

$$F^p = (\lambda(s)\nabla p, \nabla N_i) - (q, N_i).$$

A step in the Newton–Raphson method for this problem can be written as

$$\begin{bmatrix} \frac{\partial F^s}{\partial s_h} & \frac{\partial F^s}{\partial p_h} \\ \frac{\partial F^p}{\partial s_h} & \frac{\partial F^p}{\partial p_h} \end{bmatrix} \begin{bmatrix} r^s \\ r^p \end{bmatrix} = -\begin{bmatrix} F^s \\ F^p \end{bmatrix}. \quad (6.45)$$

The entries of $\partial F^s/\partial s_h$ in the Jacobian equal those given for the single saturation equation (6.44). The remaining blocks of the Jacobian have entries

$$\frac{\partial F_i^s}{\partial p_j} = -(\Delta t \lambda(s_h^{n+1})\nabla N_j \cdot \nabla(f(s_h^{n+1})), W_i), \quad (6.46)$$

$$\frac{\partial F_i^p}{\partial s_j} = (\lambda'(s_h^{n+1})N_j \nabla p_h, \nabla N_i)i, \quad (6.47)$$

$$\frac{\partial F_i^p}{\partial p_j} = (\lambda(s_h^{n+1})\nabla N_j, \nabla N_i), \quad (6.48)$$

**Table 6.2.** Performance comparison for twophase–flow with the source term $q = 1.0$ and various gridpartitions.

| Grid | $200 \times 200$ | $400 \times 400$ | $800 \times 800$ |
|---|---|---|---|
| Fully implicit | 141.63s | 1073.2s | 5885.9s |
| Gauss–Seidel | 538.53s | 1225.8s | 3816.7s |

if we still omit non–linearities in $W_i$.

### 6.5.4   Solver

The solver for the fully implicit method is built on top of the Gauss–Seidel solver, exactly as we did in the pipeflow example in Section 6.5.1. Again, we derive solver classes for the the sub–solvers, where some functions are reimplemented.

The coupled system (6.34)-(6.36) is both time–dependent and non–linear. The fully implicit solver therefore needs initial values for each field. For the saturation field initial values are available, since the saturation equation is time–dependent, while the situation is slightly complicated for the pressure, where no initial values are given in the sub–solver. We can now benefit from the object–oriented approach, since the sub–solver for the pressure equation in the fully implicit solver is derived from a stand–alone sub–solver. Hence, there is a function inside this class, which solves the pressure equation. The natural choice for an initial condition, is then to compute the pressure, based on the initial state for the saturation. This is implemented in a function called setIC in the new pressure sub–solver. Since the function $\lambda(s)$ in the pressure equation is connected with the saturation field in the Gauss–Seidel solver, we just have to ensure that the saturation field is initialized with the initial state (6.38) before this function is called from the control unit.

### 6.5.5   Comparison of the Gauss–Seidel Method and the Fully Implicit Method

To compare the performance of the two solution methods, we have run some simple experiments. For these experiments, the solution domain is the unit square in two dimensions. One injection well and one production well are placed in opposite corners. Both wells are square shaped with area $0.05^2$. In Table 6.2, we compare the CPU time of the methods for different grid resolutions and in Table 6.3 the rates of injection and production in the wells are varied on fixed grid size. The flow rate $q$ given in the table is the integral of the flow function over the area of the well.

**Table 6.3.** Performance comparison for twophase–flow with grid $400 \times 400$ and various settings for the source term $q$.

| $q$ | 1.0 | 2.0 |
|---|---|---|
| Fully implicit | 1073.2s | 1144.5s |
| Gauss–Seidel | 1225.8s | 2343.6s |

The convergence criterion used, for both the fully implicit method and the outer iteration in the Gauss–Seidel method, is:

$$\frac{|\mathbf{r}_{s,p}|_2}{|\mathbf{x}_{s,p}|_2} < \epsilon$$

where

$$\mathbf{x}_{s,p} = \begin{pmatrix} \mathbf{s} \\ \mathbf{p} \end{pmatrix},$$

and $\mathbf{r}_{s,p}$ is the correction vector between the two subsequent solutions $\mathbf{x}_{s,p}$. In our tests, we have set the tolerance in the outer iteration to $\epsilon = 10^{-3}$. The time–step used in these tests is $10^{-5}$, and we compute the solution for $t \in [0, 5.0 \cdot 10^{-5}]$. The linear system in each Newton iteration is solved with the bi–conjugate gradient method with a RILU preconditioner, to a precision of $\epsilon = 10^{-4}$, in both the fully implicit method and the saturation solver in the Gauss–Seidel method. The pressure equation in the Gauss–Seidel method is solved with conjugate gradients. We use the same tolerance $\epsilon = 10^{-4}$ for this solver.

The results presented in these tables suggest that the fully implicit method is preferable when the rate of injection and production in the wells increase. We also observe a problem with the fully implicit method when the grid size grows. For large systems, the Gauss–Seidel method is preferable. The main reason is that the non–linear system we need to solve in the fully implicit solver approximately grows by a factor of four compared with the non–linear system in the Gauss–Seidel solver, where only the saturation equation is non–linear.

## 6.6   Conclusion

In this chapter, we have presented a framework for implementation of solvers for systems of PDEs, based on fully implicit methods. The reason for doing this is that a simple Gauss–Seidel splitting algorithm for some problems may produce a wrong solution, as we have seen in an example. We have explained how such an implementation can be done in a similar flexible manner as the Gauss–Seidel methods. We have also seen in a few examples that such an implementation can perform as good as a Gauss–Seidel method, but this will be problem dependent.

Based on solvers for each sub–problem in the system, the user will implement a fully implicit solver by deriving a solver class from the original solver and the `SimulatorSystem` library class for each original solver. In this new solver class, the coupling between the equations will be introduced. Usually, the `integrandsMx` function is the most important function to reimplement, while functions for initial and boundary conditions may be reused. Further, a control unit must be implemented as described. While this is still some functionality to implement, code from the original solvers can be reused. Since the solvers for each equation have been through a debugging phase, this will increase the reliability of the new solver. The presented framework easily extends to systems with more than two equations.

The framework presented may in the future be developed further into a system for automatic generation of solvers for systems of PDEs. The coupling between the equations should be defined in some high level language, to ease the development of system solvers further. Combined with other frameworks and functionality available in the Diffpack environment, this may further be developed into a Problem Solving Environment (PSE) for the solution of partial differential equations.

# References

1. E. J. Holm, A. M. Bruaset, and H. P. Langtangen. Increasing the reliability and efficiency of numerical software development. In A. M. Bruaset, E. Arge, and H. P. Langtangen, editors, *Modern Software Tools for Scientific Computing*. Birkhuser, 1997.
2. Diffpack website. http://www.diffpack.com.
3. H. P. Langtangen. *Computational Partial Differential Equations - Numerical Methods and Diffpack Programming*. Textbooks in Computational Science and Engineering. Springer, 2nd edition, 2003.
4. Donald W. Peaceman. *Fundamentals of numerical reservoir simulations*. Elsevier Scientific Publishing company, 1977.
5. Klas Samuelsson. *Adaptive Algorithm for Finite Element Methods approximating Flow Problems*. PhD thesis, Royal Institute of Technology, Department of Numerical Analysis and Computing Science, Stockholm, Sweden, 1996.

# Chapter 7

# Stochastic Partial Differential Equations

H. P. Langtangen[1,2] and H. Osnes[3,4]

[1] Simula Research Laboratory
[2] Dept. of Informatics, University of Oslo
[3] Dept. of Mathematics, University of Oslo
[4] Det norske Veritas

**Abstract.** The purpose of this chapter is to give an introduction to stochastic partial differential equations from a computational point of view. The presented tools provide a consistent quantitative way of relating uncertainty in input to uncertainty in output for PDE-based models. We first give an analytical treatment of some stochastic differential equation model problems. These problems concerns deflection of a beam with random loading and material strength, heat conduction in random media, and pollution transport with random advection. Later, we develop Diffpack simulators for solving the model problems numerically. Two numerical solution methods are addressed: Monte Carlo simulations and perturbation methods. The main tools for generating and estimating stochastic variables and fields are outlined, and we show a suggested design of stochastic PDE simulators, which makes it easy to equip a standard sequential Diffpack simulator with stochastic treatment of uncertain input data.

## 7.1   Introduction

A problem of fundamental importance when using mathematical models for scientific or engineering investigations, is to determine how the uncertainty in the input data affects the uncertainty in the output data. A possible way of attacking this problem is to impose a perturbation of a subset of the input data and then compute the resulting perturbation in significant output parameters from the model. This type of analysis, often called deterministic sensitivity analysis, indicates a level of accuracy of the model. Such information is useful for decision makers.

A deficiency of deterministic sensitivity analysis is that, in reality, large perturbations of the input data are less likely to occur compared with small perturbations. Thus, we now face a problem to determine an appropriate size of the perturbations. The most rational way to treat this problem is to establish a statistical distribution of the input data, deduced from, e.g., physical experiments or Bayesian-like subjective reasoning, and then compute the corresponding statistics of the output data. Such an approach is able to give accurate and rational measures of the uncertainty in the results produced by the mathematical model.

Roughly speaking, we can say that deterministic sensitivity analysis is an efficient approach to obtain qualitative information on the propagation of uncertainty in a mathematical model, while the statistical approach is a more computationally expensive analysis method aimed at giving accurate numerical estimates of the uncertainty of the results in a particular problem. Many physical or financial problems cannot be described realistically without modeling some input data statistically. Uncertainty in input is then quantified, and it is natural to see how this uncertainty propagates through the model.

Partial differential equations (PDEs) where some or all of the input data are treated as stochastic quantities are referred to as stochastic partial differential equations (SPDEs). The solutions of such equations will also be stochastic quantities. Recall that input data to PDEs consist of the domain, the coefficients in the equations, as well as initial and boundary conditions.

In a few engineering branches, use of models based on partial stochastic differential equations is well established as part of the engineering routine. This is particularly the case when building structures and pricing financial derivatives. Pricing of financial products related to stock markets is to a large degree based on models involving stochastic ordinary differential equations. In the field of structural analysis, environmental loads from wind, waves, and currents are often modeled as stochastic variables, processes, or fields. Material properties represent another source of uncertainty that is often represented as stochastic variables or fields. Many design formulas in daily use by engineers are in fact derived from stochastic models, although the pure application of such formulas seldom require any probabilistic thinking. One aim of establishing efficient computational methods for stochastic differential equations is to be able to improve simple design formulas.

Many of the analysis methods for stochastic ordinary and partial differential equations developed in the field of structural engineering are applicable to other scientific branches, either directly or in a modified form. Such transfer of knowledge should indeed occur. In our opinion, the quality of decisions based on numerical simulations can be significantly improved by introducing stochastic elements in the equations. Important application areas include, e.g., heat conduction and fluid flow in geological (porous) media, shallow water waves with uncertain bottom topography, pollution transport under stochastic atmospheric conditions, and viscous flow in uncertain geometries.

Numerical solution of stochastic differential equations is orders of magnitude more time consuming than the solution of the corresponding deterministic differential equations. The information from a stochastic analysis is also much more comprehensive than its deterministic counterpart. The amount of computer time needed depends on the desired amount of statistical information about the output data. For example, determination of the mean and standard deviation of the solution requires considerably less CPU time than the determination of the probability distribution.

A rough rational estimate of the uncertainty in the mean value is provided by the standard deviation. For more accurate uncertainty estimates involving probabilities, for example for decision reliability calculations, the complete distribution must be computed. There are several methods available for calculating the mean, the standard deviation, and the distribution of output quantities in SPDEs. The present chapter deals with both analytical and numerical techniques.

The field of SPDEs is comprehensive. The engineering branch of the field has been mostly interested in methods for practical computations. Up to now, Monte Carlo techniques have usually been considered too costly and the vast research has concentrated on perturbation methods of some kind [7–9,16]. For reliability computations involving very small probabilities, variable transformations based on efficient numerical integration (actually carried out by the FORM/SORM optimization algorithm [17]) have been a popular approach. Considerable effort in the stochastic engineering community has also been devoted to the study of random fields [13,25] and feasible generation of (Gaussian) random fields with a prescribed correlation structure. An important physical application of SPDEs is flow and transport in geological structures, e.g., oil recovery and groundwater contamination [1,8,5,6]. The properties of geological structures show a random nature, and random fields constitute an important tool for description and analysis. Since random fields can be measured at wells, it is common to enforce the field to have fixed values at some spatial points. This is obtained by a method called kriging, or equivalently, conditional Gaussian random fields [4,13].

In structural analysis and material science, uncertain knowledge about the material or the loading environment calls for stochastic models (implying SPDEs). These application areas have made major contributions to solution techniques for SPDEs. Environmental loads on vibrating, linear structures can be described in terms of a randomly excited system of *ordinary* differential equations. Effective methods of analysis have been developed for this type of applications [18]. Stochastic finite element methods addressing SPDEs were first formulated for structural analysis, i.e., the equations of elasticity. This class of methods is based on perturbation methods, typically with the variance as the small perturbation parameter. The stochastic finite element approach of this kind can be adapted to many other areas [20].

SPDEs have received quite some attention in mathematics, which might give an impression that dealing with SPDEs requires a comprehensive mathematical background. The situation is, as in the field of deterministic PDEs, that much insight can be gained by a computational approach. This is what we aim at in the present chapter.

## 7.2   Some Simple Examples

When encountering a new topic, such as stochastic partial differential equations, it is advantageous to play around with some simple examples that can be analyzed in detail. Here we shall look at differential equations whose analytical solution is easy to obtain, i.e., we have symbolic expressions relating input data to output data. Using results from transformations of random variables, as listed in Appendix 7.A, we can establish stochastic properties of the output data given stochastic properties of the input data. Hopefully, this provides a feeling for what a stochastic differential equation is about and how it differs from a deterministic differential equation. In general, it is not possible to work out the stochastic properties of the output data analytically, as we do in the present section. Therefore, we focus on more generally applicable numerical methods in the rest of the chapter.

Readers who do not have a solid background in basic probability theory are recommended to read through Appendix 7.A before studying the forthcoming examples.

### 7.2.1   Bending of a Beam

Beams are important components in buildings, bridges, and ships, thus making the analysis of stress and deflection of beams central in civil engineering. Because the loads on the beams and the properties of the materials in the beams are often uncertain quantities, stochastic analysis of beam deformations becomes highly relevant.

We shall here look at a simple cantilever beam subject to an end load as depicted in Figure 7.1. Assuming that the standard beam theory [2,11]

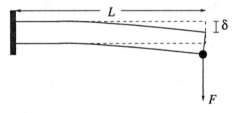

**Fig. 7.1.** Deflection of a cantilever beam of length $L$ subject to an end force $F$.

applies to our beam problem, the governing differential equation reads

$$\frac{d^2}{dx^2}\left(EI(x)\frac{d^2u}{dx^2}\right) = q(x)\,. \tag{7.1}$$

In this equation, $u(x)$ is the deflection of the beam, $E$ is Young's modulus, reflecting the elastic material properties of the beam, $I(x)$ is a geometric property of the cross section of the beam, and $q(x)$ is a distributed load along the beam in the direction of the deflection. In our example of a cantilever beam, $q = 0$ and the discrete end load is modeled through the boundary conditions. At the left end, $x = 0$, the deflection is zero. Moreover, the beam is clamped, which implies that $u'(0) = 0$. At the right end, $x = L$, there is no moment, $EIu''(L) = 0$, while the shear force $-EIu'''(L)$ equals the applied load $F$.

Equation (7.1) and its associated boundary conditions constitute a stochastic boundary value problem if any of the input data $(E, I, L, F)$ are uncertain and modeled as stochastic quantities. Consequently, the solution $u(x)$ will then also be a stochastic quantity. That is, for a given $x$, $u(x)$ is a random variable. Since we expect the function $u(x)$ to be smooth, it is clear that two random variables $u(x_1)$ and $u(x_2)$ are dependent, and we expect any covariance of $u(x_1)$ and $u(x_2)$ to depend on the distance $|x_1 - x_2|$.

In the present example we can find a closed-form solution $u(x)$ of the boundary value problem, as the differential equation can easily be integrated four times in the case $I$ is a constant. The integration constants are thereafter determined from the boundary conditions, yielding the solution

$$u(x; E, I, F, L) = \frac{Fx^2}{6EI}(3L - x). \tag{7.2}$$

Of particular interest, in an engineering context, is the moment $M$ in a cross section of the beam:

$$M(x; E, I, F, L) = -EI\frac{d^2u}{dx^2} = F(x - L). \tag{7.3}$$

The existence of such closed-form analytical solutions relating the input data directly to the output data is useful when analyzing stochastic differential equation problems. The solution might be simple enough so that one can apply random variable transformations and derive analytical expressions for the statistics of the output data. We shall demonstrate such procedures below. When an analytical random variable transformation becomes too complicated, a closed-form solution of the differential equation is still advantageous, because we can simply apply Monte Carlo simulations to obtain approximate statistics of the output data. In the case the differential equation must be solved numerically, calling up a solver for each Monte Carlo sample makes the entire statistical analysis much more expensive.

*Stochastic Load.* Suppose the load $F$ is modeled as a stochastic variable. We assume that $F$ is normally distributed with mean $\mu_F$ and standard deviation $\sigma_F$. This can be compactly expressed by the notation $F \sim N(\mu_F, \sigma_F)$. The normal distribution is convenient from a hand-calculation point of view, but it can often lead to unphysical values. This might be the case here; $F$ is assumed to act downwards, but the probability of a load acting upwards is nonzero as long as $F$ is normally distributed. For practical purposes, however, a normal distribution is an appropriate approximation if the probability of $F < 0$ is sufficiently small.

The statistics of $u$ is found by resolving the following question: Given the distribution of $F$, what is the distribution of $u(x) = \alpha(x)F$, where $\alpha(x) =$

$x^2(3L - x)/(6EI)$ is a known number? From basic probability theory [3,22] we know that if $F \sim N(\mu_F, \sigma_F)$, $u$ is also normally distributed, because a linear transformation preserves normality. Hence, from page 309 we get that

$$u = \alpha(x)F \sim N(\alpha(x)\mu_F, \alpha(x)\sigma_F).$$

In particular, the end deflection $\delta$ (occurring for $x = L$) becomes

$$\delta \sim N(\alpha(L)\mu_F, \alpha(L)\sigma_F), \quad \alpha(L) = \frac{L^3}{3EI}.$$

The standard deviation $\alpha(x)\sigma_F$ is an immediate measure of uncertainty. We can also use the knowledge about the distribution of $\delta$ to find, for instance, the probability $P$ that $\delta$ is inside some safe interval $[a, b]$ is

$$P = \int_a^b f_\delta(\tau)d\tau = \Phi(b) - \Phi(a),$$

where $\Phi(\delta)$ is the cumulative normal distribution function and $f_\delta$ is the associated probability density. Since $u(x) = \alpha(x)F$, we realize that $u(x) \sim N(\alpha(x)\mu_F, \alpha(x)\sigma_F)$ for any $x \in [0, L]$.

Another central quantity is the moment. The maximum value $m$ of the moment, occurring at the end ($x = L$) of the beam equals $FL$. Consequently, $m \sim N(L\mu_F, L\sigma_F)$ since $F \sim N(\mu_F, \sigma_F)$.

*Stochastic Material.* The material property $E$ must often be modeled as an uncertain parameter. Assigning, for example, a normal distribution with mean $\mu_E$ and standard deviation $\sigma_E$ to $E$, the distribution of the maximum deflection $\delta = \beta/E$, where $\beta = FL^3/(3I)$, is less trivial to find by analytical means. However, the statistics of $\beta/E$ can easily and effectively be found by Monte Carlo simulation. This will be demonstrated in Section 7.5.4.

In the case $E$ is a lognormal random variable with parameters $\lambda_E$ and $\zeta_E$, the general product transformation (7.57) in Section 7.A.2 can be applied to find the expectation and standard deviation of $\delta$. The expression $\delta = \beta/E$ can be written on the form (7.57) with

$$b = F\frac{L^3}{3I}, \quad a_1 = -1,$$

implying that $\delta$ is lognormally distributed with parameters

$$\lambda_\delta = \ln\left[\frac{FL^3}{3I}\right] - \lambda_E, \quad \zeta_\delta = \zeta_E.$$

The product transformation $\delta = \beta/E$ also allows us to use results for Beta, F, and Gamma distributed variables from Appendix 7.A.3. Provided $E$ follows any of these three distributions, $\delta$ achieves the same distribution, but with different parameters. An understanding of how to construct the statistics of $\delta$, when $E$ is a lognormal variable, is sufficient to readily apply the formulas in Appendix 7.A.3 to derive the distribution of $\delta$ when $E$ is a Beta, F, or Gamma distributed variable.

*Stochastic Load and Material.* Treating both the load $F$ and the material property $E$ as stochastic variables yields another example of relevance. The maximum deflection $\delta$ can then be written as $\gamma F/E$, where $\gamma = L^3/(3I)$. Again, Monte Carlo simulation represents a simple way of determining the distribution of $\delta$, once the distributions of $F$ and $E$ are given.

The product transformation of lognormally distributed variables can also be used to compute the statistics of $\delta$ when both $E$ and $F$ are independent, lognormal variables. In the transformation formula (7.57) in Section 7.A.2 we can then put $X_1 = E$, $X_2 = F$, $\lambda_1 = \lambda_E$, $\lambda_2 = \lambda_F$, $\zeta_1 = \zeta_E$, $\zeta_2 = \zeta_F$, $a_1 = -1$, $a_2 = 1$, and $b = L^3/(3I)$. The result is that $\delta$ is a lognormal random variable with parameters

$$\lambda_\delta = \ln\left[\frac{L^3}{3I}\right] - \lambda_E + \lambda_F, \quad \zeta_\delta^2 = \zeta_E^2 + \zeta_F^2.$$

*Approximate Expectation and Variance.* In the present model example, we work with a problem where we know the exact relation between the input and output statistics. Of this reason, we can investigate the accuracy of various *approximate* approaches for deriving the expectation $E[\delta]$ and variance $\text{Var}[\delta]$. These approaches are relevant when the distribution of the input parameters are unknown and/or when fast computational procedures for the expectation and variance are desired.

Let us write $\delta = g(E, F) = FL^3/(3EI)$. We assume that $E$ and $F$ are stochastic quantities with mean values $\mu_E$ and $\mu_F$. Their standard deviations are $\sigma_E$ and $\sigma_F$. We could also prescribe a covariance of $E$ and $F$, but in the present physical application this covariance is naturally zero; there is no reason to think that the load on a (linearly elastic) beam depends on the material properties of the beam.

Section 7.A.4 presents expressions for the first-order approximations of the expectation and variance of $\delta$ as well as the second-order approximation of the expectation. In these expressions, we need the derivatives of $g$ with respect to the variables $E$ and $F$:

$$\frac{\partial g}{\partial E} = -\frac{FL^3}{3E^2 I}, \quad \frac{\partial g}{\partial F} = \frac{L^3}{3EI},$$

$$\frac{\partial^2 g}{\partial E^2} = \frac{2FL^3}{3E^3 I}, \quad \frac{\partial^2 g}{\partial F^2} = 0, \quad \frac{\partial^2 g}{\partial E \partial F} = \frac{-L^3}{3E^2 I}.$$

The first-order approximation of the expectation of $\delta$ is simply

$$E[\delta]' = \frac{\mu_F L^3}{3I\mu_E},$$

i.e., the analytical formula for $\delta$ with just the mean values inserted for the stochastic input data. The second-order approximation reads

$$E[\delta]'' = \frac{\mu_F L^3}{3I\mu_E} + \frac{\mu_F L^3}{3\mu_E^3 I}\sigma_E^2.$$

The first-order approximation of $\mathrm{Var}[\delta]$ reads

$$\mathrm{Var}[\delta]' = \left(\frac{L^3}{3\mu_E I}\right)^2 \left(\sigma_F^2 + \frac{\mu_F^2}{\mu_E^2}\sigma_E^2\right).$$

We have in the latter two formulas used that $E$ and $F$ are uncorrelated. Note that in the case $E$ and $F$ are lognormally distributed, we could derive the exact probability distribution of $\delta$.

The accuracy of the first-order perturbation method for the beam is illustrated in Figure 7.2, where the approximations for the expectation and standard deviation of $\delta$ are compared with the exact quantities. In both cases it is presumed that $F$ and $E$ are lognormally distributed with the same expectation and standard deviation. It is seen that the perturbation method offers reasonable accuracy for $\sigma_F = \sigma_E < 0.2$. Figure 7.5 on page 287 shows how the probability density function of $\delta$ appears when $E$ is lognormal and $F$ is normal.

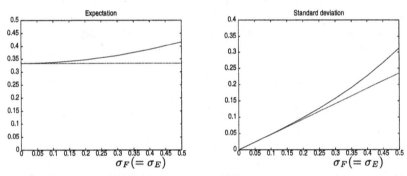

**Fig. 7.2.** Comparison of first-order perturbation approximations (dashed line) and exact values (solid line) of the expectation and standard deviation of $\delta$ as functions of $\sigma_F$ ($= \sigma_E$). The random variables $F$ and $E$ are both lognormally distributed with expectations equal to unity. Left: Expectation. Right: Standard deviation.

### 7.2.2    Viscous Drag Forces

The drag force $D$ on a body moving through or standing in a fluid with velocity $U$ can be modeled as $D = \alpha U^2$. How will uncertainties in $U$ propagate through this model? In the case $U$ can be viewed as a lognormal, Beta, F, or Gamma distributed variable, we can derive the probability distribution of $D$ using formulas from Appendices 7.A.2 and 7.A.3.

1. Lognormal distributed variables: If $U$ is lognormal with parameters $\lambda_U$ and $\zeta_U$, $D$ is also lognormal with parameters $\lambda_D = \ln\alpha + 2\lambda_U$ and $\zeta_D = 2\zeta_U$.

2. Using the Beta distribution:

$$U \sim b(u; a, b, h, C) \quad \Rightarrow \quad D = \alpha U^2 \sim b(d; a, b, h/2, \alpha C^2).$$

3. Using the F distribution:

$$U \sim f(u; a, b, k, \Delta) \quad \Rightarrow \quad D = \alpha U^2 \sim f(d; a, b, k/2, \alpha \Delta^2).$$

4. Using the Gamma distribution:

$$U \sim g(u; a, h, A) \quad \Rightarrow \quad D = \alpha U^2 \sim g(d; a, h/2, \alpha A^2).$$

The directory src/stoch/drag contains an application dragpdfs, which offers a graphical user interface for adjusting the mean and standard deviation of $U$, in the case $U$ is lognormally distributed, and viewing the probability density functions of $U$ and $D$. This gives a visual impression of how the uncertainties propagate through the formula $D = \alpha U^2$. Figure 7.3 displays two different examples.

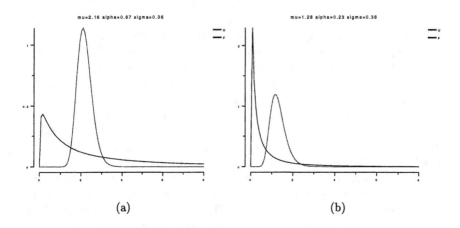

(a)                              (b)

**Fig. 7.3.** Graphs of the probability density of a lognormal variable $U$ and a lognormal variable $D = \alpha U^2$ in two different parametric cases. The "mu", "sigma", and "alpha" quantities in the plot title corresponds to E[$U$], StDev[$U$], and $\alpha$.

### 7.2.3   Heat Conduction in the Earth's Crust

Stationary heat conduction through a geological medium like the earth's crust can be modeled by the two-point boundary value problem [15, ch. 2.1]

$$\frac{d}{dx}\left(k(x)\frac{du}{dx}\right) = s(x), \quad 0 < x < L, \tag{7.4}$$

$$u(0) = U_0, \tag{7.5}$$

$$-k(L)u'(L) = -Q. \tag{7.6}$$

Here, $x$ is a coordinate, which is zero at the earth's surface and which is increasing with the crust depth. The primary unknown is the temperature $u(x)$. The thickness of the crust is set to $L$. The function $k(x)$ is the heat conduction coefficient, whereas the function $s(x)$ denotes heat generation by radioactivity. The temperature is known at the earth's surface and represented by $U_0$. Under the crust we assume that we know the net inflow of heat, here set equal to $Q$. Geological media are often changing in a random fashion so modeling the heat conduction properties $k(x)$ as a stochastic quantity (stochastic field) is often required.

At this point it is appropriate to distinguish between random variables, random processes, and random fields. Given a real number $x$, $k(x)$ is a random variable, just as the outcome when flipping a coin. However, if we look at a different $x$, say $x + h$, $k(x + h)$ is a random variable, but it might be highly dependent on $k(x)$ if $x$ and $x + h$ are close and $k(x)$ is a smooth function. For example, if we know $k(x)$, we can approximately predict $k(x+h)$ by $k(x) + k'(x)(x + h)$ (the first two terms in a Taylor series) when $h$ is small. Knowledge of $k(x)$ hence implies knowledge of $k(x+h)$. Of course, the dependence of $k(x + h)$ on $k(x)$ generally decreases as $h$ grows.

Since functions $k(x)$ give us an infinite number of possibly dependent stochastic variables, for each $x$, a special mathematical structure, called *random fields*, has been formulated. Roughly speaking, a random field is a spatial function $k(x)$ with a probability distribution for the value of $k$, given $x$, and a correlation structure between $k(x_1)$ and $k(x_2)$ for two arbitrary points $x_1$ and $x_2$. Here, $x$ can be a point in 1D, 2D, 3D, or higher dimensions if appropriate. Functions of time, say $f(t)$, are also random fields according to this brief explanation, but the uni-dimensional direction of time has made it natural to develop specialized mathematics. Such time-dependent stochastic $f(t)$ functions are then called *random processes*. A difference between random fields and processes occurs when conditioning a function value on a neighboring value; in a random process only neighboring values from the past are relevant, whereas in a random field neighbors in all spatial directions are relevant. This difference makes random fields more complicated than random processes. Stochastic ordinary differential equations have solutions depending on time, and the solutions are therefore random processes. Stochastic partial differential equations, which are addressed in this chapter, have solutions depending on space (and possibly time), and random fields becomes an important topic.

Let us return to the heat conduction example, where we now assume that $k(x)$ is a random field. Also in this example we can find a closed-form solution $u(x)$, a fact that simplifies the treatment of the stochastic problem. Assuming for simplicity that $s = 0$, we can integrate (7.6) and use the boundary conditions to arrive at

$$u(x) = Q \int_0^x [k(\tau)]^{-1} d\tau + U_0 \,. \tag{7.7}$$

Suppose the geological medium consists of $n$ layers, each with a constant heat conductivity $k_i$, $i = 1, \ldots, n$. Layer number $i$ occupies the region $[x_i, x_{i+1}]$. In this case our $k(x)$ is represented by a one-dimensional *discrete* random field, in the meaning of a finite number of possibly dependent random variables $k_i$, and we can easily carry out the integration of $1/k(x)$. The result becomes

$$u(x; k_1, \ldots, k_n) = U_0 + Q \left( (x - x_i) k_i^{-1} + \sum_{j=1}^{i-1} (x_{j+1} - x_j) k_j^{-1} \right), \tag{7.8}$$

provided that $x$ is in layer number $i$. The parameters $k_1, \ldots, k_n$ are naturally taken as either correlated or uncorrelated random variables. The heat conductivity $k(x)$ is hence a piecewise constant stochastic field, and the temperature $u(x; k_1, \ldots, k_n)$ also becomes a stochastic field.

The mean and variance of $u$ are easily found from (7.8) using the basic rules for expectation and variance of a sum of stochastic variables:

$$E[u] = U_0 + Q \left( (x - x_i) E[k_i^{-1}] + \sum_{j=1}^{i-1} (x_{j+1} - x_j) E[k_j^{-1}] \right), \tag{7.9}$$

$$Var[u] = Q^2 (x - x_i)^2 Var[k_i^{-1}] +$$
$$2Q^2 \sum_{j=1}^{n-1} (x - x_i)(x - x_j) Cov[k_i^{-1}, k_j^{-1}] +$$
$$Q^2 \sum_{\ell=1}^{i-1} \sum_{j=1}^{i-1} (x_{j+1} - x_j)(x_{\ell+1} - x_\ell) Cov[k_j^{-1}, k_\ell^{-1}]. \tag{7.10}$$

In case the $k_1, \ldots, k_n$ are independent stochastic variables, $k_1^{-1}, \ldots, k_n^{-1}$ are also independent, and the second part of the expression for $Var[u]$ vanishes because

$$Cov[k_i^{-1}, k_j^{-1}] = 0, \quad i \neq j \,.$$

The third part simplifies since $\text{Cov}[k_j^{-1}, k_\ell^{-1}] = 0$ for $j \neq \ell$. We then get

$$\text{Var}[u] = Q^2 \left( (x - x_i)^2 \text{Var}[k_i^{-1}] + \sum_{j=1}^{i-1} (x_{j+1} - x_j)^2 \text{Var}[k_j^{-1}] \right). \quad (7.11)$$

In the present case, each layer is considered homogeneous with no internal random variation. Because two neighboring geological layers can have quite different properties, the assumption that the $k_i$ stochastic variables are independent is relevant.

There is no simple way of prescribing a distribution of $k_i$ and then deriving the distribution of $u \sim \sum_j c_j k_j^{-1}$ analytically (the $c_j$ coefficients follow from (7.8)). Figures 7.6–7.11 on pages 295–297 in Section 7.5.6 show some statistics of $k$ and $u$ based on Monte Carlo simulation.

### 7.2.4   Transport Phenomena

Various (one-dimensional) transport phenomena can be modeled by the following linear advection equation

$$\frac{\partial c}{\partial t} + \frac{\partial}{\partial x}(vc) = 0, \quad 0 < x < L, \ t > 0. \quad (7.12)$$

Here, $v$ is the transport velocity, $x$ is the coordinate along the medium through which the transport phenomenon occurs, and $t$ is time. The function $c(x, t)$ is the primary unknown which can represent, e.g., a concentration in pollution transport or temperature in heat transfer. The solution of (7.12) with the initial condition

$$c(x, 0) = g(x), \quad 0 < x < L \quad (7.13)$$

can be expressed as

$$c(x, t) = g(x - vt), \quad 0 < x < L, \ t > 0. \quad (7.14)$$

Assume that the advective velocity $v$ is modeled as a random variable with the probability density function $f_V(v)$. We now write the solution as $c(x, t; v)$ to explicitly show that $c$ depends on the random variable $v$. With the initial condition from (7.13), it is possible to express the statistical moment of order $m$ about zero as

$$\text{E}[c^m] = \int [c(x, t; v)]^m f_V(v) dv = \int [g(x - vt)]^m f_V(v) dv. \quad (7.15)$$

The integral limits in (7.15) are determined by the probability density function, i.e., the domain in which $v$ is defined. The corresponding moments about the mean $c_\mu = \text{E}[c]$ are defined as

$$M_m \equiv \text{E}[(c - c_\mu)^m].$$

Notice that $E[c^m]$ and $M_m$ are functions of $x$ and $t$. It should be noted that the moments are smooth functions even if the corresponding deterministic solutions for a fixed $v$ contain discontinuities [12,21].

To gain some insight into the nature of the moments, consider the frequently arising case with discontinuous initial data: $g(x) = 1 - H(x)$, $H(x)$ being the Heaviside function[1]. Then

$$E[c(x,t;v)] = 1 - F_V(x/t) \tag{7.16}$$

and

$$\mathrm{Var}[c(x,t;v)] \equiv M_2 = (1 - F_V(x/t))\, F_V(x/t), \tag{7.17}$$

while the third and fourth central moments are

$$M_3 = c_\mu - 3c_\mu^2 + 2c_\mu^3, \quad M_4 = c_\mu - 4c_\mu^2 + 6c_\mu^3 - 3c_\mu^4. \tag{7.18}$$

In (7.16) and (7.17), $F_V(v)$ is the cumulative distribution corresponding to $f_V(v)$. In other words, the expected front is smeared in the same way as the cumulative distribution is, and the variance is peak formed. Furthermore, the level set

$$\{x(t) \,|\, E[c(x,t)] = \hat{c}\}$$

moves with the velocity $F_V^{-1}(1 - \hat{c})$, which for the "middle point" $\hat{c} = 1/2$ (in the discontinuity) equals the median of $v$. This is also the velocity of the peak of $\mathrm{Var}[c(x,t)]$. The third moment has two extrema that move with velocities $F_V^{-1}(0.5 \pm 1/\sqrt{12})$, whereas the fourth moment has two peaks moving with velocities that depend even more on the tails of $f_V(v)$. In Figure 7.4 the first four moments (as well as the corresponding deterministic solution) of $c$ are shown at time $t = 0.5$ for $v \sim N(1, 0.15)$.

That the solution, here moments, of a stochastic partial differential equation is smoother than the solution of the deterministic counterpart is clearly demonstrated in this example. The velocity is uncertain, implying that different realizations will have different front locations at a fixed point in time. These different front locations give a smooth average, and in a similar way, smooth higher-order moments. Of course, one could, in principle, take advantage of this smoothness in numerical methods for SPDEs. This issue is addressed in [21].

## 7.2.5   Generic Problems

For software development purposes, it is convenient to formulate a generic form of the various problems that arise from stochastic differential equations. Let the function $g$ denote the mapping from stochastic input data to stochastic output data in the problem. Although we have worked with analytical expressions for $g$ in the previous section, just to illustrate the effect of

---

[1] $H(x) = 1$ for $x > 0$, $H(0) = 1/2$, and $H(x) = 0$ for $x < 0$.

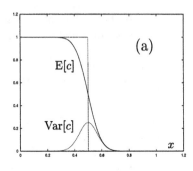 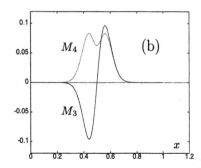

**Fig. 7.4.** The first four moments, along with the corresponding deterministic solution, of $c$ at time $t = 0.5$ for a normally distributed velocity; $v \sim N(1, 0.15)$. Figure (a): Expectation (solid line), variance (dotted line) and deterministic solution (dashed line) of $c$. Figure (b): Third (solid line) and fourth moment (dashed line) of $c$.

stochasticity in PDEs, we assume from now on that $g$ must be evaluated by solving some PDE, or system of PDEs, numerically.

The simplest case of SPDEs is where we have some stochastic variables collected in a vector $r$ as input data and we compute one scalar $T$ as a key output result in the problem. Of course, $T$ is derived from the solution of the differential equation and is a stochastic variable. We write

$$T = g(r). \tag{7.19}$$

One example appears in Section 7.2.1, where $r = (F, E)$ consists of the load and material properties and $T$ is the deflection $\delta$ of the beam, $\delta = \max_{0 \le x \le L} |u(x)|$. We also mention that instead of a scalar stochastic variable $T$ one can let $T$ be a vector of stochastic variables, i.e., one derives more than one scalar measure of the solution field. In the heat conduction example, $r = (k_1, \ldots, k_n)$ and $T$ is the continuous $u(x)$ field. In practice we compute $u$ at discrete points, giving a vector $T$ of stochastic variables.

In the most general case, the stochastic input data are stochastic space-time fields, and the solution also becomes a stochastic space-time field. We can write this relation in the compact form

$$T(x, t) = g(r(x, t)).$$

Both $r$ and $T$ can be stochastic vector fields, i.e., collections of stochastic scalar fields. One example is stochastic elasticity, where $T$ is the displacement vector field, or the stress tensor, and $r$ is material property fields (elastic parameters, thermal expansion coefficient) or a force or temperature loading.

The statistics $T$ can be of various types. In the case of a scalar random variable $T$ we are interested in the mean, the standard deviation, perhaps third and fourth order moments (especially the quantities skewness and kurtosis, which can be derived from the third and fourth order moments). Moreover, the complete distribution of $T$ is of interest. Random fields or processes $T$ are more difficult to estimate. The mean is straightforward to estimate, for example, as the average of each sample field. The variance and higher-order moments can be estimated by summing up independent field realizations. The spatial or temporal correlation between points in a field reflects how smooth or how "randomly varying" the function is. This can be estimated by summing up products of $T$ evaluated at two different spatial or temporal points. The amount of data generated is huge so one often restricts the attention to correlations along a line in space, for instance.

## 7.3   Solution Methods

There are three major categories of solution methods for SPDE problems on the form $T = g(r)$:

1. *Variable transformations.* This is an analytical tool that is useful for some SPDEs where the associated deterministic PDE has a simple closed-form solution. The method is exemplified in Section 7.2, and the transformation tools we use are listed in Appendix 7.A.

2. *Monte Carlo simulation (MCS).* This is a simple and widely applicable, but also computationally intensive, method to be described in Section 7.3.1.

3. *Perturbation methods.* Problems with small variances in the input data can be analyzed with perturbation methods. The SPDE is transformed to a hierarchy of recursive, deterministic PDEs describing the statistics of the solution. More information is provided in Section 7.3.2.

Perturbation based methods [7–9,16] are typically used to predict approximations to the expectation and standard deviation (and possibly covariances) of the response. Higher-order moments as well as the entire probability distributions must normally be obtained by MCS. Perturbation methods, which are applicable in the context of both analytical and numerical solution strategies, are usually very efficient for small and intermediate problems. However, for problems that involve a large number of random parameters perturbation methods require a large number of differential equations to be solved. In such cases MCS based methods are often preferred. MCS can also be required when the complete statistical distribution is of interest and the behavior of the system is far from normal (Gaussian). MCS methods are usually applied in a numerical context.

### 7.3.1   Monte Carlo Simulation

Let us consider an SPDE problem on the form $T = g(r)$, where $r$ is a collection of random input quantities and $T$ is a collection of random output quantities. The principles of Monte Carlo simulation are simple:

1. Generate $n$ random samples of $r$, $r = r_1, \ldots, r_n$.

2. Compute $n$ corresponding random samples of $T$, $T = T_1, \ldots, T_n$, where $T_i = g(r_i)$ for $i = 1, \ldots, n$.

3. Use statistical estimation techniques to derive statistics for $T$, based on the samples $T_1, \ldots, T_n$.

As a simple example on the relation $T = g(r)$, one can think of $\delta = \gamma F / E$, where $r = (F, E)$ and $T = \delta$. We can then use MCS to find, e.g., the expectation of $\delta$:

1. Draw $n$ random numbers $E_1, \ldots, E_n$ from the probability distribution of $E$. Draw $n$ numbers $F_1, \ldots, F_n$ from the probability distribution of $F$.

2. Compute $\delta_i = \gamma F_i / E_i$ for $i = 1, \ldots, n$.

3. Use the well-know estimator

$$\bar{\mu}_\delta = \frac{1}{n} \sum_{i=1}^{n} \delta_i$$

for computing an approximation $\bar{\mu}_\delta$ to $\mu_\delta = \mathrm{E}[\delta]$ from $n$ random samples $\delta_1, \ldots, \delta_n$.

Observe that once we have the samples $\delta_1, \ldots, \delta_n$, it is straightforward to compute other types of statistics of $\delta$. The time-consuming part of the method is, in an SPDE context, to evaluate the $g$ function since this normally means solving a PDE numerically.

A central issue when running MCS is the number of random samples we need, i.e., the size of $n$. The uncertainty of the quantity we want to estimate governs the choice of $n$. In case we want to estimate $\mathrm{E}[\delta]$, we know that $\bar{\mu}_\delta$ is a sum of random variables, which will, according to the central limit theorem, tend towards a normal distribution with expectation $\mathrm{E}[\delta]$ and standard deviation $\mathrm{StDev}[\delta]/\sqrt{n}$ as $n$ grows. Since $\mathrm{StDev}[\delta]$ is unknown, we can use the estimated standard deviation of the samples instead,

$$\bar{\sigma}_\delta = \sqrt{\frac{1}{n-1} \sum_{i=1}^{n} (\delta_i - \bar{\mu}_\delta)^2} \, .$$

The statistic

$$\bar{\mu} = \frac{\bar{\mu}_\delta - \mathrm{E}[\delta]}{\bar{\sigma}_\delta / \sqrt{n}} ,$$

is known to follow a Student's $t$ distribution with $n - 1$ degrees of freedom. A confidence interval for $\bar{\mu}$ can easily be formulated, but for MCS applications, $n$ is usually larger than 100, and Student's $t$ distribution can then be approximated by a normal distribution. The confidence interval can thus be expressed as

$$[\bar{\mu}_\delta - \Delta\bar{\mu}_\delta, \bar{\mu}_\delta + \Delta\bar{\mu}_\delta], \quad \Delta\bar{\mu}_\delta = \Phi^{-1}(\alpha/2)\frac{\bar{\sigma}_\delta}{\sqrt{n}},$$

where $1 - \alpha$ is the probability that $\bar{\mu}$ is within the interval. The uncertainty in $\bar{\mu}_\delta$ is hence $\Delta\bar{\mu}_\delta$. The convergence rate of MCS is easily seen to be behave like $n^{-1/2}$. This is slower than we are used to in numerical methods for deterministic problems[2]. In the numerical illustration in Section 7.5.4, we have $\Delta\bar{\mu}_\delta \sim 6 \cdot 10^{-3}$ after 1000 simulations and $\Delta\bar{\mu}_\delta \sim 6 \cdot 10^{-4}$ after 100000 simulations.

When using the MCS method, probability distributions of the stochastic input parameters must be prescribed. In case of dependent input quantities, the joint probability distribution must be known. This is often difficult. However, for jointly normally distributed variables it is sufficient to prescribe the means and a covariance matrix. Instead of the covariance matrix one often works with the normalized correlation matrix and a vector of standard deviations.

The generation of realizations of a single stochastic variable is in general easy. There exist numbers of random number generators. Most of these generate realizations of uniformly distributed variables $u_i$, which are easily transformed to any distribution $F$ by $F^{-1}(u_i)$, where $F$ is the desired cumulative distribution function. On the other hand, generation of vectors of random variables or random fields is more complicated.

Let $v$ be a vector of dependent, normally distributed, random variables. The associated correlation matrix is denoted by $R$. That is,

$$R_{ij} = \frac{\text{Cov}[v_i, v_j]}{\text{StDev}[v_i]\text{StDev}[v_j]}.$$

To generate a sample of the $v$ vector one first generates a sample of normalized, independent Gaussian variables $z$. The correct correlation among these variables is obtained by the transformation

$$y = L_0 z,$$

where $L_0$ is the lower triangular Cholesky decomposition of $R$:

$$R = L_0 L_0^T. \tag{7.20}$$

---

[2] Think of solving a 1D PDE like $u''(x) = f$ by a finite difference method of order $h^2$, $h$ being the cell size. Doubling the work, which implies halving $h$, reduces the error by a factor of four. To reduce the error in the Monte Carlo method by a factor of four, we need to increase the amount of work by a factor of 16.

Scaling $y$ by the standard deviations and adding the mean of $v$ gives a sample of $v$. That is, we compute $Sy + c$, where $S$ is diagonal with the standard deviations on the diagonal, $S_{ii} = \text{StDev}[v_i]$, and $c$ is the expectation of $v$. Notice that when $n$ is large, computing the Cholesky factorization of $R$ is expensive; the work is of order $n^3$.

The same procedure can be used when $v$ is a vector of discrete random variables in a random field [23]. In this case, the size of $v$ can be very large. Although the Cholesky factorization only has to be performed once, a practical limit on the number of representative field values emerges because of the time-consuming Cholesky decomposition and the large storage requirements of the matrices $R$ and $L_0$.

Lots of other generation methods for random fields exist, for example several sequential algorithms [19], where the structure of correlations are simplified considerably, and turning band methods [24], which require some kind of isotropic conditions in the random field.

The second step in the MCS method, the solution of the mathematical model for a realization of the stochastic input parameters, is not a subject of the present chapter (because it is the topic of the rest of this book and the one by Langtangen [15]). We therefore pay our attention to the next step, the procedure of estimating statistical properties of the response. For simplicity, we assume that the stochastic problem to be solved can be expressed in the form (7.19), and that $n$ realizations $r_i$, $i = 1, \ldots, n$, of the stochastic input variable $r$ are generated. For each realization of $r$ the function $g$ is evaluated, and the result yields a realization of the response $T$, that is $T_i = g(r_i)$. The expectation and standard deviation of $T$ are approximated by

$$\bar{\mu}_T = \frac{1}{n}\sum_{i=1}^{n} T_i = \frac{1}{n}\sum_{i=1}^{n} g(r_i)$$

$$\bar{\sigma}_T = \left\{ \frac{1}{n-1}\sum_{i=1}^{n}(T_i - \mu_T)^2 \right\}^{\frac{1}{2}},$$

respectively, while estimates of the third and fourth order moments about the mean, $\bar{M}_T^3$ and $\bar{M}_T^4$, are given by

$$\bar{M}_T^m = \frac{1}{N-m+1}\sum_{i=1}^{n}(T_i - \mu_T)^m, \quad m = 3, 4. \tag{7.21}$$

Also the probability distribution of $T$ can be estimated, at present by a plain histogram approach. The interval of possible values for the response is divided into a suitable number of subintervals (bins) say 20-100, depending on the size of the interval and the number of realizations to be generated. The number of realizations belonging to each cell is recorded, and the probability density is approximated by the relative frequency in each subinterval (bin).

In the context of a stochastic output field $u(x)$ the statistical moments and the probability distributions for the representative (nodal or element) field values can be obtained in the same manner as for the stochastic variable $T$. However, for random fields also the covariances might be of interest. When using the MCS method covariances can be estimated by

$$\text{Cov}[u(\boldsymbol{x}_i), u(\boldsymbol{x}_j)] \approx \frac{1}{n-1} \sum_{k=1}^{n} (u_k(\boldsymbol{x}_i) - \text{E}[u(\boldsymbol{x}_i)])\,(u_k(\boldsymbol{x}_j) - \text{E}[u(\boldsymbol{x}_j)])\,,$$

(7.22)

for $i, j = 1, \ldots, e$, where $e$ is the number of field values. The entries

$$\text{Cov}[u(\boldsymbol{x}_i), u(\boldsymbol{x}_j)]$$

in (7.22) are in fact elements in an $e \times e$ covariance matrix, and for large fields the calculation of all these quantities can be very time consuming and storage demanding. To reduce the consumption of computational resources, it is usual to restrict the calculation of covariances to certain lines or points in the field.

### 7.3.2    Perturbation Methods

To illustrate the perturbation method, we start by considering the simple beam example from section 7.2.1 and analyze it analytically. Let us investigate the maximum deflection $\delta$ of the beam due to a stochastic end load $F$. From (7.2) it is seen that $\delta$ is given by

$$\delta = \frac{FL^3}{3EI}\,.$$

(7.23)

In the perturbation method, the stochastic input parameters are divided into their mean values and stochastic deviations from the mean, i.e., we write

$$F = \mu_F + \hat{F}, \quad \text{E}[\hat{F}] = 0\,.$$

We can then view $\delta$ as a function of the stochastic variable $\hat{F}$:

$$\delta = \delta(\hat{F}) = \frac{(\mu_F + \hat{F})L^3}{3EI}\,.$$

The basic strategy in perturbation methods is to express the response as a Taylor series around the mean $\text{E}[F]$ with respect to the deviation $\hat{F}$ from the mean. This means that the method is suited for problems where the standard deviations of the input variables are small. A first-order Taylor series in the present example yields (a subscript 0 denotes evaluation at $\hat{F} = 0$)

$$\delta = \delta|_0 + \left.\frac{d\delta}{d\hat{F}}\right|_0 \hat{F}$$

(7.24)

$$= \frac{\mu_F L^3}{3EI} + \frac{L^3}{3EI}\hat{F}\,.$$

Based on this series we obtain the following approximations for the expectation and the standard deviation of the deflection:

$$\mu_\delta = \frac{\mu_F L^3}{3EI}, \tag{7.25}$$

$$\sigma_\delta = \frac{L^3}{3EI}\sigma_{\hat{F}},$$

where $\sigma_{\hat{F}} = \text{StDev}[\hat{F}] = \text{StDev}[F] = \sigma_F$. In the present simple example with a single stochastic input parameter, linearly related to the response, it is seen that the results obtained from the first-order perturbation method are exact. (Recall that $E[\delta]$ and $\text{Var}[\delta]$ can be calculated from (7.23).)

Let us now extend the example by considering both $F$ and $E$ as stochastic variables, which are divided into mean values ($\mu_F$ and $\mu_E$) and stochastic deviations from the mean ($\hat{F}$ and $\hat{E}$):

$$\delta(\hat{F}, \hat{E}) = \frac{(\mu_F + \hat{F})L^3}{3I(\mu_E + \hat{E})}.$$

Expanding $\delta$ into a first-order Taylor series reads (subscript 0 means that the quantity is to be evaluated for $\hat{E} = \hat{F} = 0$):

$$\delta = \delta|_0 + \left.\frac{\partial \delta}{\partial \hat{F}}\right|_0 \hat{F} + \left.\frac{\partial \delta}{\partial \hat{E}}\right|_0 \hat{E}, \tag{7.26}$$

$$= \frac{\mu_F L^3}{3\mu_E I} + \frac{L^3}{3\mu_E I}\hat{F} - \frac{\mu_F L^3}{3\mu_E^2 I}\hat{E},$$

leading to the following approximations for the expectation and the standard deviation of $\delta$:

$$\mu_\delta = \frac{\mu_F L^3}{3\mu_E I}, \tag{7.27}$$

$$\sigma_\delta = \frac{L^3}{3\mu_E I}\left\{\sigma_{\hat{F}}^2 + \left(\frac{\mu_F}{\mu_E}\right)^2 \sigma_{\hat{E}}^2 - 2\frac{\mu_F}{\mu_E}\text{Cov}[\hat{F}, \hat{E}]\right\}^{\frac{1}{2}}$$

$$= \frac{L^3}{3\mu_E I}\left\{\sigma_{\hat{F}}^2 + \left(\frac{\mu_F}{\mu_E}\right)^2 \sigma_{\hat{E}}^2\right\}^{\frac{1}{2}}.$$

Recall that $\text{Cov}[\hat{F}, \hat{E}] = 0$ since it is unlikely that uncertainties in material properties are related to uncertainties in environmental loadings. In the expansion (7.26) several higher order terms in $\hat{E}$ and coupling terms in $\hat{F}\hat{E}$ are neglected. Therefore, the first-order approximations (7.27) for the statistical behavior of $\delta$ is not exact, in contrast to the previous example with $F$ as a single stochastic input parameter.

Let us now consider the more general problem (7.19). The stochastic input field/vector $r$ still contains $e$ representative values. In the context of the perturbation method all $r_i$, for $i = 1, \ldots, e$, are divided into mean values $(\mu_{r_i})$ and stochastic deviations from the mean $(\hat{r}_i)$. We start by approximating $T$ as a first-order Taylor series in the $\hat{r}_i$ variables

$$T = g|_0 + \sum_{i=1}^{e} \left.\frac{\partial g}{\partial \hat{r}_i}\right|_0 \hat{r}_i .$$

The subscript 0 implies evaluation at the point $\hat{r}_1 = \hat{r}_2 = \ldots = \hat{r}_e = 0$. Terms of equal order in the stochastic input parameters, $\hat{r}_i$, are grouped and solved or evaluated separately. This leads to a single zero-order expression

$$g|_0 = 0, \tag{7.28}$$

and $e$ first-order problems

$$\left.\frac{\partial g}{\partial \hat{r}_i}\right|_0 = 0, \quad i = 1, \ldots, e. \tag{7.29}$$

Note that the evaluation of $g$ and its derivatives in (7.28) and (7.29) generally requires the solution of $e+1$ differential equations. From these evaluations and the series expansion (7.28) it now follows that the first-order approximations for the expectation and standard deviation of $T$ are given by

$$\mu_T = g|_0 \tag{7.30}$$

and

$$\sigma_T = \left\{ \sum_{i=1}^{e} \sum_{j=1}^{e} \left( \left.\frac{\partial g}{\partial \hat{r}_i}\right|_0 \left.\frac{\partial g}{\partial \hat{r}_j}\right|_0 \mathrm{Cov}[\hat{r}_i, \hat{r}_j] \right) \right\}^{\frac{1}{2}}, \tag{7.31}$$

respectively.

## 7.4    Quick Overview of Diffpack Tools

After having discussed several aspects and solution strategies for stochastic partial differential equations, it is time to give an overview of the Diffpack functionality applicable to stochastic problems, which is the topic of the rest of this chapter. The Diffpack tools for generating and analyzing stochastic quantities are tailored to the needs when solving SPDEs. That is, the tools are focused on generating and estimating scalar stochastic variables, vectors of stochastic variables, and large-scale stochastic fields. For more general statistical applications, involving small to moderate data sets, we recommend to use problem solving environments like S-Plus or SAS. You can easily dump Diffpack data in tabular form and import them in S-Plus for further analysis. However, the large data sets often generated when solving SPDEs make it advantageous to update estimators and then throw the samples away during each realization of input data and the response. This is a different analysis strategy than normally used in statistics, but highly useful and often necessary when dealing with the large amount of data produced in SPDE problems.

Diffpack has several class hierarchies and stand-alone classes for generating and estimating stochastic quantities. We provide a quick overview of the most important classes, before explaining them in more detail on the following pages.

- RandGen: generation of random variables.

- ProbDistr_prm: parameters for initializing objects in the ProbDistr hierarchy.

- ProbDistr: representation of probability distributions. Subclasses:

    1. UniformDistr: uniform distribution.

    2. NormalDistr: normal distribution.

    3. LognormalDistr: lognormal distribution.

    4. DetermDistr: deterministic numbers.

- RandFieldDescr: description of Gaussian random fields.

- CorrelModel: base class for description of correlation models.

- RandFieldGen: base class for random field generation. Generation classes:

    1. RandFieldGenGaussCond: generation of conditional Gaussian fields.

    2. Draw2D, Draw2D_s1, Draw_s2, Draw2D_matrix: generation of Gaussian random fields on lattice grids.

    3. RandFieldGenPiWisConst: generation of piecewise constant fields.

- EmpiricalStatistics: base class for estimation. Subclasses:

    1. ExpecVar: mean and variance.

    2. Moments: mean, variance, skewness, and kurtosis (first four moments).

    3. Extremes: minimum and maximum values in a sample.

    4. Histogram: histogram and probability density approximations.

    5. ExpecVarFieldFE, ExpecVarFieldLattice: mean and variance of random fields.

    6. MomentsFieldFE: mean, variance, skewness, and kurtosis of finite element random fields.

    7. CorrAlongLine: correlation of a random field along a line through the grid.

    8. TestOfNormality: tests if a random vector is normally distributed.

In addition to describing these tools we also present some stochastic Diffpack applications:

- Stochastic maximum beam deflection problem, introduced in Section 7.2.1.
- Stochastic heat conduction problem in 1D, introduced in Section 7.2.3.
- Stochastic groundwater flow mode, covered in Section 7.6.3.

## 7.5   Tools for Random Variables

Let us start the description of Diffpack tools for SPDEs by presenting the basic functionality for random variables and probability distributions. Our aim is to explain how to choose different probability distributions for the stochastic quantities, how to generate realizations of random variables and how to estimate statistical properties.

### 7.5.1   Random Number Generation

The fundamental class for generating realizations of random variables is the RandGen class. It contains a handle[3] to an object of class UnifRandGen, which is an abstract base class for generating uniformly distributed variables on $(0, 1)$. At the time of this writing, there are three such generation methods, implemented as subclasses of UnifRandGen.

1. Class UnifRandGenSimple offers a "quick and dirty" random generator, where the seed $s$ is updated according to

$$s = (s \cdot 8121 + 28411) \bmod 134456,$$

and the corresponding random number is taken as

$$r = \text{double}(s)/134456.$$

2. Class UnifRandGenOpSystem calls the standard random number generator offered by the operating system, i.e., the C functions rand and srand.

3. Class UnifRandGenKnuth is a high-quality random number generator suggested by Donald Knuth (a subtractive method).

Class RandGen is the main Diffpack tool for random number generation. From an application point of view, you need to know that the class has a uniform random number generator, a constructor for initialization, and functions for returning random variables according to various distributions:

```
class RandGen
{
    Handle(UnifRandGen) urg; // uniform random number generator
    public:
    RandGen
```

---
[3] Handles in Diffpack are smart pointers, see [15].

```
    (const String& rg,      // type of random generator (classname)
     int          seed);    // seed to random generator

  real gaussian (ProbabilityTrans_type trans_tp,
                 real mean, real stdev);
  real exponential (real lambda);
  real gamma (int n, real lambda);
  real Weibull (real lambda, real beta);
  real uniform  ()  { return urg->draw(); }
  real generate (ProbDistr& distribution);
};
```

The constructor needs a specification of the random number generator to be used, and a seed. The random number generator is specified by one of the previously listed names of subclasses in the `UnifRandGen` hierarchy.

The `gaussian` function draws random numbers from the normal distribution. The parameter `trans_tp` denotes the type of transformation from uniformly distributed variables that is to be used to generate Gaussian variates. An appropriate value is BOX_MULLER, reflecting the Box-Müller method,

$$g_1 = -2\ln u_1 \cos 2\pi u_2, \quad g_2 = -2\ln u_1 \sin 2\pi u_2,$$

where $u_1$ and $u_2$ are two independent uniformly distributed random variables on $(0, 1)$, and $g_1$ and $g_2$ are the corresponding Gaussian variates.

The `exponential`, `gamma`, and `Weibull` functions draw numbers from the exponential, gamma, and Weibull distributions, with densities

$$f_X(x) = \lambda \exp(-\lambda x) \quad \text{(exponential)},$$

$$f_X(x) = \lambda^n x^{n-1} \exp(-\lambda x)/\Gamma(n) \quad \text{(gamma)},$$

$$f_X(x) = \lambda^\beta \beta x^{\beta-1} \exp(-(\lambda x)^\beta) \quad \text{(Weibull)}.$$

The `uniform` function returns uniformly distributed random numbers between 0 and 1. Finally, we have the general `generate` function, which takes a specification of a probability distribution (explained in Section 7.5.2) and draws a number according to this distribution, i.e., it returns $F^{-1}(u)$, where $u$ is uniformly distributed on $(0, 1)$ and $F^{-1}$, the inverse of the cumulative distribution function.

### 7.5.2 Description of Probability Distributions

Some of the most commonly used probability distributions are implemented in an hierarchy with class `ProbDistr` as (abstract) base class. There are subclasses for

- the uniform distribution (`UniformDistr`),

- the normal (Gaussian) distribution (`NormalDistr`),

- the lognormal distribution (`LogNormalDistr`), and

– the delta-function distribution (DetermDistr), implying that deterministic numbers can be treated as a special case of stochastic variables[4].

More subclasses can be added using the set-up in Appendix 7.B.

Each subclass offers methods for evaluation of the probability density function, the cumulative distribution function, and the inverse of the cumulative distribution. A typical class in this hierarchy looks like this:

```
class UniformDistr : public ProbDistr
{
public:
  UniformDistr (const ProbDistr_prm& pm_) : ProbDistr(pm_) {}

  static real density (real x, real a=0, real b=1);
  static real cum     (real x, real a=0, real b=1);
  static real cumInv  (real p, real a=0, real b=1);

  virtual real f    (real x);
  virtual real F    (real x);
  virtual real Finv (real p);

  virtual real expec ();  // expectation
  virtual real stdev ();  // standard deviation
};

// minimum outline of the base class:
class ProbDistr
{
public:
  Handle(ProbDistr_prm) pm;
  ProbDistr (const ProbDistr_prm& prm) { pm.rebind(prm); }
  ...
};
```

Looking at a typical subclass in the ProbDistr hierarchy, we see that there are three static functions density, cum, and cumInv for computing the probability density, the cumulative distribution, and the inverse of the latter. In addition, there are three seemingly similar *virtual* functions f, F, and Finv, computing the same quantities as density, cum, and cumInv, respectively. The static versions can be called without an object. Therefore, they often contain the parameters of the distributions as arguments. The parameters are left out in the virtual versions of the functions, because these functions are called through an object, and the object inherits the parameters via a ProbDistr_prm object pm. Very often, the virtual functions are implemented just as plain calls to the corresponding static functions, with parameters from pm inserted as arguments.

The ProbDistr base class holds the parameters of the distribution in a ProbDistr_prm object:

---

[4] This allows you to turn stochasticity on and off in a simulator; just declare all possible variables as ProbDistr handles and choose, at run time, the ones to be stochastic and deterministic.

```
class ProbDistr_prm : public HandleId
{
public:
  // used instead of a constructor (Diffpack standard):
  static ProbDistr_prm* construct ();
  virtual ProbDistr* create () const;

  String         distribution;  // name of distribution
  VecSimple(real) parameters;    // parameters in a distribution

  real    operator () (int i) const;  // return parameter no i
  void    scan (Is is);
  ...
};
```

To create a ProbDistr_prm object, one never applies the new operator, but calls the construct function [14]:

```
Handle(ProbDistr_prm) p_prm = ProbDistr_prm::construct();
```

This is standard in Diffpack and essential if you want to extend parameter class hierarchies on your own.

In the case of a normal distribution, i.e. class NormalDistr, there are two parameters, and the length of the class member parameters array is 2. The first parameter is the mean value, whereas the second one is the standard deviation. Each class in the ProbDistr hierarchy has its own convention for numbering the parameters in the distribution, and the parameters are always stored in the parameters array. The parameter class offers a nice functionality for specifying a distribution and its parameters; the scan function reads initialization strings on the form X(p1,p2,...), where X is a name of a distribution (that is, a name in the ProbDistr hierarchy), and p1, p2, and so on are the associated parameters. For example, NormalDistr(2.1,0.4) means the normal distribution with mean 2.1 and standard deviation 0.4. There are short forms for the class names too: Normal for NormalDistr, Lognormal for LognormalDistr, and so on. That is, Normal(2.1,0.4) is the same as NormalDistr(2.1,0.4).

Having filled a ProbDistr_prm object with the proper data, the corresponding ProbDistr object is created by calling ProbDistr_prm's create function:

```
Handle(ProbDistr) p = p_prm->create();
```

We provide a complete example illustrating the use of the ProbDistr_prm and ProbDistr classes several places later in this chapter.

### 7.5.3   Estimation of Statistics of Random Variables

There are several classes available for estimating various statistical properties of a random variable. All these methods work on a set of realizations of the variable and are implemented as subclasses in the EmpiricalStatistics hierarchy. The base class keeps track of the number of samples and a logical name of the random variable to be estimated. Before an estimation process can start, the programmer must initialize the estimation object by calling the init function. For each realization of the variable we call the update function with the realization of the stochastic quantity as argument. Intermediate or final estimates are available from special functions in the subclasses.

The subclass ExpecVar computes the expectation and the standard deviation (or variance) of a random variable. Besides init and update functions, the class contains the functions getExpec for computing the expectation and getStDev for computing the standard deviation.

Class Moments is a subclass extension of ExpecVar, where also the skewness and kurtosis (or third- and fourth-order moments) can be estimated. Another class derived from ExpecVar, Extremes, computes the extremes (maximum and minimum values) from the set of realizations.

The class Histogram can be used for estimating the probability density (we remark that more sophisticated density estimation methods are often superior to the plain histogram approach in class Histogram). To initialize a Histogram object, the user must specify the number of uniform cells (bins) and provide an estimate of the minimum and maximum value of the realizations. The realizations are taken into account by calling the update function with the realization value as the argument. Finally, several plots can be made (bar plot, density plot, cumulative function plot) using the standard CurvPlotFile functionality in Diffpack [15]. An estimate of the cumulative probability distribution is available from the real-valued function cumulDistr.

It is also possible to test whether a set of realizations may stem from a normally distributed random variable or not. In the TestOfNormality class we have implemented the *Pearson's test for goodness of fit*, see a textbook on basic statistics ([3], for instance), for a vector of stochastic variables. This functionality can, of course, be used for a single random variable by regarding it as a vector of unit length.

```
class TestOfNormality : public EmpiricalStatistics
{
 public:
   TestOfNormality (int nvariables);
   ~TestOfNormality () {}

   void redim (int nvariables);
   void init ();  // if no sensible expec/stdev is available
   void init (const Vec(real)& expec, const Vec(real)& stdev);
   void update (const Vec(real)& realizations);
   void calcTestStatistic (Vec(real)& chi_sqr_test_statistic,
             const Vec(real)& expec, const Vec(real)& stdev);
};
```

An object of this class should be initialized by calling the init function with two arguments of type Vec(real) defining preliminary estimates of expectations and standard deviations of the random variables. These estimates are crucial in the construction of cells in which the different realizations of the variables are placed. For each realization we have to call the update function with the realization vector as the argument. The results of the test are obtained from the calcTestStatistic function containing three arguments of type Vec(real). The last two arguments are the (finally) estimated expectations and standard deviations, while the values of the Chi-square variables, which determine if the hypothesis of normally distributed variables can be rejected, are provided as the first argument.

The estimator classes have various functions for returning estimated quantities. These functions differ between the classes, and we refer to the man pages for exact information about what is estimated and how the relevant functions are called. The example in the next subsection shows how to call functions for estimating the mean, standard deviation, skewness, kurtosis, minimum, maximum, as well as histogram plots.

### 7.5.4 Example: Simulation of a Stochastic Beam

In the present section we describe a program for simulating the statistics of the beam model explained in Section 7.2.1. This program demonstrates the basic usage of the random number generator, the probability distribution classes (ProbDistr and ProbDistr_prm), and estimator classes (Moments, ExpecVar, Extremes, and Histogram). We assume that the load $F$ and the material property $E$ are mutually independent stochastic variables. The aim is to compute estimates of the statistics of the maximum deflection $\delta = \text{const} \cdot F/E$, using Monte Carlo simulation.

The algorithm is simple and easy to implement. Inside a loop from one to the given number of realizations we draw two random numbers, one for $E$ and one for $F$, compute $\delta$, and add the $\delta$ value to estimator objects for $\delta$.

We recommend to implement such stochastic simulators as a class. The basic variables in the present class consists of

- a random number generator (RandGen object),
- ProbDistr_prm and ProbDistr objects for specifying and holding distributions of $E$ and $F$,
- the number of Monte Carlo simulations,
- estimator objects: mean, standard deviation, skewness, and kurtosis (all provided by class Moments), extreme values (Extremes), and histogram (Histogram),
- a curve plot handler (CurvePlotFile), needed for plotting the histogram.

The functions include scan for reading input and initializing the class members and loopMonteCarlo for running the Monte Carlo simulation. The class can then be defined as

```
class StochasticBeam
{
```

```
    RandGen                 generator;
    Handle(ProbDistr_prm)   E_prm, F_prm;
    Handle(ProbDistr)       E_d, F_d;
    real                    L, I;              // beam parameters
    int                     nsimulations;

    Moments                 delta_moments;
    Histogram               delta_pdf;
    Extremes                delta_maxmin;
    CurvePlotFile           curveplot;
public:
    StochasticBeam () {}
    ~StochasticBeam () {}

    void scan ();
    void loopMonteCarlo ();
};
```

The scan function reads ProbDistr_prm initialization strings from the command line and initializes parameter objects (E_prm and F_prm) for the $E$ and $F$ quantities:

```
String E_d_str, F_d_str;
initFromCommandLineArg ("-E", E_d_str, "LogNormalDistr(1,0.2)");
initFromCommandLineArg ("-F", F_d_str, "LogNormalDistr(1,0.2)");

E_prm.rebind (ProbDistr_prm::construct());
F_prm.rebind (ProbDistr_prm::construct());

E_prm->scan (E_d_str);
F_prm->scan (F_d_str);
```

The next step in scan is to initialize the probability distribution objects E_d and F_d from the parameter objects:

```
E_d.rebind (E_prm->create());
F_d.rebind (F_prm->create());
```

The constructions so far follow the usual Diffpack conventions for programming with parameter objects and associated class hierarchies.

The estimator objects are initialized by setting a name of the quantity to be estimated and thereafter calling the init function:

```
delta_moments.setResponseName("delta");
delta_moments.init();
delta_maxmin.setResponseName("delta");
delta_maxmin.init();
```

Histogram objects need a slightly more comprehensive initialization, as we must provide a good initial guess of the minimum and maximum values of the realizations[5]:

---

[5] One can run a few simulations (say 100) and use the extreme of the response, combined with a suitable "safety factor", to estimate delta_min and delta_max.

```
real delta_min = 0, delta_max = 2; // guess
int nintervals=100;
delta_pdf.redim (delta_min, delta_max, nintervals);
delta_pdf.setResponseName ("delta");
delta_pdf.setXaxisName ("max deflection");
delta_pdf.init ();
```

The Monte Carlo loop is now straightforwardly implemented. We make a loop over the number of realizations. Inside the loop, we use `generator` to draw a random number for $E$ and for $F$. Thereafter we compute $\delta$, and then we add $\delta$ to the statistical estimator objects. The code can take this form:

```
void StochasticBeam:: loopMonteCarlo ()
{
  real E, F, delta;   // realizations of E, F, and delta
  delta_moments.init();
  delta_maxmin.init();
  delta_pdf.init();

  for (int i=1; i<=nsimulations; i++) {
    E = generator.generate (*E_d);  // draw input data
    F = generator.generate (*F_d);
    delta = F*L*L*L/(3*E*I);        // calculate response

    // update estimators:
    delta_moments.update (delta);
    delta_maxmin.update (delta);
    delta_pdf.update (delta);

    // write out 20 intermediate results,
    // plus results after the last simulation
    if (i % int(nsimulations/20) == 0 || i == nsimulations) {
      /* write delta_moments.getExpec() etc. */
      ...
      delta_pdf.makeBarPlot (curveplot);
      delta_pdf.makeDensityPlot (curveplot);
    }
  }
}
```

The relevant files associated with this program are found in `src/stoch/sbeam`. Having compiled the application, one can run (say) 10000 realizations, with $E$ and $F$ as normal and lognormal variables:

```
./app -n 10000 -E 'Normal(1.0,0.2)' -F 'LogNormal(1.0,0.2)'
```

Inside the Monte Carlo loop we create 20 intermediate plots of the histogram. Using Diffpack's `curveplot` facility, it is easy to generate an animation of how the histogram evolves throughout the simulation. The relevant command, using Gnuplot for generating the plot, is [15]:

```
curveplot gnuplot -f tmp_delta.map -r 'density' '.' '.' \
          -animate -o 'set yrange [0:7];' -fps 1
```

You can replace **gnuplot** by, e.g., **matlab** (specific Matlab commands for fixing the $y$ axis must then be inserted in the -o option).

Figure 7.5 shows the density, based on the histogram, after 1000 and 10000 realizations. This plot was obtained by

```
curveplot gnuplot -f tmp_delta.map \
           -r 'density' 'delta' '1000' -ps sbeam.ps
```

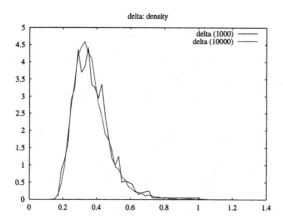

**Fig. 7.5.** Density plots produced by class **Histogram** for $\delta = \text{const} \cdot F/E$. The two curves correspond to 1000 and 10000 realizations, respectively. $E$ is lognormally distributed and $F$ is Gaussian.

### 7.5.5   Suggested Design of Stochastic PDE Simulators

The simple **StochasticBeam** class from the previous subsection employs a solution method and set-up that can be applied to stochastic partial differential equations in general. Instead of computing a formula for $\delta$, one could think of solving the differential equation for the deflection of a beam, and then compute $\delta$ is the maximum deflection. In such cases it is advantageous to separate the deterministic problem from the stochastic one. One can implement a standard deterministic simulator class called **MySimulator**, for instance:

```
class MySimulator : public SimCase
{
protected:
  // data
  ...
public:
  virtual void scan();
  virtual void solveProblem();
```

```
    virtual void resultReport();
    ...
};
```

Class `MySimulator` can be any type of simulator; the PDEs and the discretization methods are irrelevant when discussing Monte Carlo simulation. A trivial
example is a class solving the equations for the deflection of a beam and computing its maximum deflection:

```
class Beam
{
protected:
  real E, I, F, L;  // beam parameters
public:
  virtual void scan();  // read E, I etc. from somewhere
  virtual void solveProblem();
  virtual void resultReport();
  real delta ();          // return maximum deflection
};
```

A stochastic version of the simulator can be implemented in a subclass,
say `MySimulatorStoch`. This class contains random number generation and
estimator tools as well as a `scan` function, a function for Monte Carlo simulation, and also an extended version of `resultReport`, if desired.

```
class MySimulatorStoch : public MySimulator
{
  // random generator
  // probability distributions for various variables
  // estimators
  int nsimulations; // no of Monte Carlo simulations
public:
  virtual void scan();  // read input and init data
  virtual void loopMonteCarlo();
  virtual void resultReport();
};
```

The `loopMonteCarlo` function generates realizations of the stochastic input
data, calls the appropriate functionality in the base class for computing the
desired output data, and updates estimators for the output data. As an example of a `MySimulatorStoch`-type of class, we can create

```
class BeamStoch : public Beam
{
  RandGen                 generator;
  Handle(ProbDistr_prm)   E_prm, F_prm;
  Handle(ProbDistr)       E_d, F_d;
  int                     nsimulations;

  Moments                 delta_moments;
  Histogram               delta_pdf;
  Extremes                delta_maxmin;
  CurvePlotFile           curveplot;
```

```
public:
  BeamStoch () {}
  ~BeamStoch () {}

  void scan ();
  void loopMonteCarlo ();
};
```

The only difference from the previously presented class StochasticBeam is that we now call delta in class Beam for computing the maximum deflection and that we rely on class Beam for setting the beam parameters $L$ and $I$. The scan function hence looks like this:

```
void BeamStoch:: scan ()
{
  Beam::scan();
  // statements as in StochasticBeam::scan(), except L=1 and I=1
};
```

The physical parameters E and F in the base class are set in loopMonteCarlo function:

```
void BeamStoch:: loopMonteCarlo ()
{
  real d;  // realizations of delta
  delta_moments.init();
  delta_maxmin.init();
  delta_pdf.init();

  for (int i=1; i<=nsimulations; i++) {

    // generate input data:
    Beam::E = generator.generate (*E_d);
    Beam::F = generator.generate (*F_d);

    // compute response
    d = Beam::delta();

    // update estimators:
    delta_moments.update (d);
    delta_maxmin.update (d);
    delta_pdf.update (d);

    // write out intermediate results
    ...
  }
}
```

In more complicated problems, where the base class simulator and the stochastic version both can be quite large, the separation of the deterministic and the stochastic problems helps to make the software easier to maintain and extend.

### 7.5.6   Example: Stochastic Heat Conduction

We shall now develop a simulator for the physical problem outlined in Section 7.2.3. Following the set-up from the previous section, we first make a class for solving the deterministic problem:

```
class Heat
{
protected:
  real                 Q, U0;  // parameters in the PDE problem
  Handle(GridFE)       grid;
  Handle(FieldFE)      u;
  Handle(FieldPiWisConst) k;
public:
  Heat () {}
  virtual ~Heat () {}
  virtual void scan ();
  virtual void solveProblem ();
  virtual void resultReport () {}
};
```

Since we can solve the problem analytically, we just evaluate the closed-form solution (7.7) in the solveProblem function instead of solving the differential equation numerically. The integration is performed element by element to evaluate $u(x)$ at the nodal points:

```
void Heat:: solveProblem ()
{
  // integrate to obtain nodal values:
  const int nel = grid->getNoElms();
  u->values()(1) = U0;
  int j; real h;
  for (j = 1; j <= nel; j++) {
    h = grid->getCoor(j+1,1) - grid->getCoor(j,1); // element size
    u->values()(j+1) = u->values()(j) + Q*h/k->valueElm(j);
  }
}
```

The input to this deterministic simulator is a lattice grid, which is then transformed to a finite element grid, and a set of intervals to describe the geological layers. These layers are represented as *materials* in the finite element grid. The material concept is well supported in Diffpack, and a piecewise constant $k$ field over the different materials can be realized by a FieldPiWisConst object k. Let us introduce the following command-line arguments for specifying the grid and layers in our deterministic simulator:

```
-grid "d=1 [0,1] [1:51]" -layers "[0,0.1] [0.1,0.8] [0.8,1]"
```

This defines a 1D grid on $[0, 1]$ with 51 nodes and three materials: material 1 in $[0, 0.1]$, material 2 in $(0.1, 0.8]$, and material 3 in $(0.8, 1]$.

Based on these ideas, the scan function for reading input data and initializing the data structures can take the following form:

```
void Heat:: scan ()
{
    String lattice_grid;   // grid for u(x) evaluations:
    initFromCommandLineArg ("-grid", lattice_grid, "d=1 [0,1] [1:11]")
    grid.rebind (new GridFE());
    grid->scanLattice (lattice_grid);

    // -layers "[0,0.2] [0.2, 0.5] [0.5, 1];"
    // i.e. x_1=0, x_2=0.2, x_3=0.50, x_4=1.0 (Diffpack set syntax)
    String layers_str;
    initFromCommandLineArg ("-layers", layers_str, "[0,1];");
    SetSimplest(String) layers_set;
    layers_set.scan (layers_str);
    int i;
    for (i = 1; i <= layers_set.getNoMembers(); i++) {
        s_o << aform("material %d is %s\n", i, layers_set(i).c_str());
        grid->addMaterial (aform("no=%d %s", i, layers_set(i).c_str()));
    }

    u.rebind (new FieldFE (*grid, "u"));
    k.rebind (new FieldPiWisConst (*grid, false, "k"));
    k->fill(1.0);   // just for testing

    initFromCommandLineArg ("-U0", U0, 0.0);
    // if k is about 1, let u be about x, which gives Q=1:
    initFromCommandLineArg ("-Q", Q, 1.0);
}
```

One can test the Heat class separately from the stochastic simulator, which is a great advantage, especially in more complicated problems, where the deterministic simulator may involve intricate physics and numerics.

The stochastic version of our heat conduction simulator is realized as a subclass HeatStoch. In this subclass, we introduce the following items:

— a random field generator for piecewise constant fields, where the constants are independent random variables: RandFieldGenPiWisConst,

— an estimator object (MomentsFieldFE) for the expectation, standard deviation, skewness, and kurtosis of the numerical temperature field,

— an estimator object for the correlation of the temperature field along a line in the grid (CorrAlongLine),

— an estimator object for the probability density of the temperature field at a fixed node in the grid (Histogram),

— an object for finding the extreme value of the temperature (Extremes).

In C++, the class is declared as follows.

```
class HeatStoch : public Heat
{
protected:
    RandFieldGenPiWisConst          k_stoch;
    int                             nsimulations;
```

```
// hold mean, stdev, and moments of u(x):
MomentsFieldFE                         u_moments;
CorrAlongLine                          u_corr;

// histogram of u at one point u_pdf_pt (u_pdf_node):
Histogram                              u_pdf;
real                                   u_pdf_pt;
int                                    u_pdf_node;
Extremes                               u_max;

CurvePlotFile                          curveplot;

public:
  HeatStoch () {}
  ~HeatStoch () {}

  void scan ();
  void loopMonteCarlo ();
};
```

The `RandFieldGenPiWisConst` class has a `scan` function, which reads a set of of probability distribution descriptions, one description for each material in the grid. Each description follows the `ProbDistr_prm::scan` syntax, and the collection of descriptions follows the Diffpack set syntax (elements separated by blanks and a semicolon at the end of the set). An initialization string for probability distributions for $k_1$, $k_2$, and $k_3$ in a grid with three materials could read

```
"LogNormal(1.0,0.2) Normal(1.5,0.9) Normal(9,2) ;"
```

The `HeatStoch::scan` function can now be written like this:

```
void HeatStoch:: scan ()
{
  Heat::scan ();

  String k_str; // pass on to RandFieldGenPiWisConst::scan
  initFromCommandLineArg ("-k", k_str, "LogNormal(1.0,0.2);");
  k_stoch.scan (k_str, *grid, "k");
  k_stoch.setGenerator ("UnifRandGenKnuth", 1.24 /*seed*/);

  u_moments.redim (*u);

  // read the point where a histogram of u is desired:
  real distance; bool exact;
  initFromCommandLineArg ("-u_pdf_pt", u_pdf_pt,
     grid->getCoor(int(grid->getNoNodes()/2),1));   // midnode pt
  Ptv(real) x(1); x(1) = u_pdf_pt;
  u_pdf_node = grid->nearestPoint (x, distance, exact);

  initFromCommandLineArg ("-n", nsimulations, 40);

  u_moments.setResponseName ("u");
  u_moments.init ();
  u_max.setResponseName ("u");
```

```
      u_max.init ();

      // initialize the CorrAlongLine object u_corr:
      Ptv(real) corr_pt(1), start(1), stop(1);
      corr_pt(1) = grid->getCoor(u_pdf,node,1);
      start(1) = grid->getCoor(1,1);
      stop(1) = grid->getCoor(grid->getNoNodes(),1);
      u_corr.scan (corr_pt, start, stop, grid->getNoNodes(), "u");
      u_corr.init ();

      // u is approx U0 + x*Q/k, assume k and x are around 1, and
      // introduce a safety factor of about 50%:
      k_stoch.generate();
      real kmin, kmax;
      k_stoch.getField().minmax (kmin, kmax);
      real u_min=U0*0.5, u_max=Q/kmin*1.5; int nintervals=100;
      u_pdf.redim (u_min, u_max, nintervals);
      u_pdf.setResponseName("u");
      u_pdf.setXaxisName ("max deflection");
      u_pdf.init ();
      curveplot.open ("tmp_u");
    }
```

The heart of this MCS solver is the `loopMonteCarlo` function, which is very similar to the `loopMonteCarlo` function in (the mathematically much simpler) class `StochasticBeam`:

```
    void HeatStoch:: loopMonteCarlo ()
    {
      u_moments.init();
      u_max.init();
      u_pdf.init();
      u_corr.init();

      int i;  real umin, umax;
      for (i=1; i<=nsimulations; i++) {
        k_stoch.generate();              // generate k sample
        k.rebind (k_stoch.getField());   // insert k sample
        Heat::solveProblem ();           // compute u

        // update estimators:
        u_moments.update (*u);
        u->minmax (umin, umax);
        u_max.update (umax);
        u_pdf.update (u->values()(u_pdf_node));
        u_corr.update (*u, *u);

        // write out 20 intermediate results,
        // plus results after the last simulation
        if (i % int(nsimulations/20) == 0 || i == nsimulations) {
          String label = aform("u%04d",i); // annotate no of samples
          SimRes2gnuplot::makeCurvePlot (*u, curveplot, "u samples",
                                          label, " ");
          label = aform("k%04d",i);
          SimRes2gnuplot::makeCurvePlot (*k, curveplot, "k samples",
                                          label, " ");
          u_pdf.makeBarPlot (curveplot);
```

```
        u_pdf.makeDensityPlot (curveplot);
      }
    }
    String s = aform("%d simulations", nsimulations);
    String Es = "E[u]   " + s;  String Ss = "StDev[u]   " + s;
    SimRes2gnuplot::makeCurvePlot (u_moments.getExpec(),
                            curveplot, Es, "E[u]", s);
    SimRes2gnuplot::makeCurvePlot (u_moments.getStDev(),
                            curveplot, Ss, "StDev[u]", s);
    u_corr.dump (curveplot,
              u_moments.getExpec(), u_moments.getExpec());
  }
```

You can find the above code in the directory src/stoch/heat.

Let us experiment with a case with three materials, all having lognormally distributed heat conduction properties. An appropriate command for running the simulator is then

```
./app -n 10000 \
-k "LogNormal(1.0,0.8) LogNormal(8.0,1.0) LogNormal(0.8,0.1)" \
-grid "d=1 [0,1] [1:51]" \
-layers "[0,0.5] [0.5,0.7] [0.7,1]" \
-u_pdf_pt 0.8 -U0 0.0 -Q 1.0
```

This command specifies 10000 simulations. The simulator dumps 20 samples of $u$ and $k$. We can plot the samples writing[6]

```
curveplot gnuplot -f tmp_u.map -r 'u samples' '.' '.' \
          -o 'set nokey;' -ps heat_u_samples.ps
curveplot gnuplot -f tmp_u.map -r 'k samples' '.' '.' \
          -o 'set nokey;' -ps heat_u_samples.ps
```

The results appear in Figures 7.6 and 7.7.

We can visualize some key statistical quantities, e.g., the probability density of $u$ at $x = 0.8$, E[u], StDev[u], and Cov[u(0.8), u(x)]:

```
curveplot gnuplot -f tmp_u.map -r 'density' 'u' '10000'
curveplot gnuplot -f tmp_u.map -r '.' 'E\[u\]' '.'
curveplot gnuplot -f tmp_u.map -r '.' 'StDev\[u\]' '.'
curveplot gnuplot -f tmp_u.map -r 'covar' '.' '.'
```

The quantities are displayed in Figures 7.8–7.11.

---

[6] The set nokey command turns off curve legends. This is useful here as there are 20 curves.

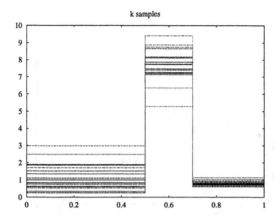

**Fig. 7.6.** 20 samples of the piecewise constant random field $k$. The three constants are lognormally distributed with means 1, 8, and 0.8, and standard deviations 0.8, 1, and 0.1, from left to right.

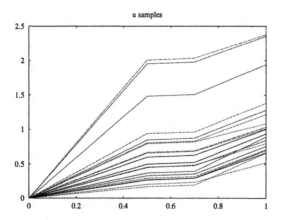

**Fig. 7.7.** Samples of the temperature field $u$, which are solutions of (7.4)–(7.6). The 20 samples correspond to the 20 samples of $k$ shown in Figure 7.6.

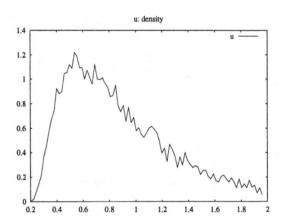

**Fig. 7.8.** The estimated probability density of the temperature $u$ at the node nearest the point $x = 0.8$.

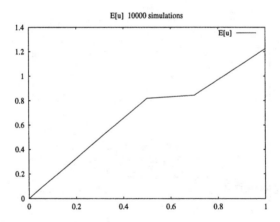

**Fig. 7.9.** The estimated expectation of the random temperature field $u$.

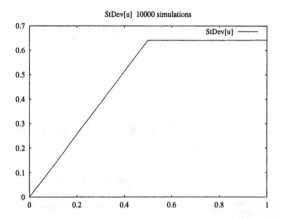

**Fig. 7.10.** The estimated standard deviation of the random temperature field $u$.

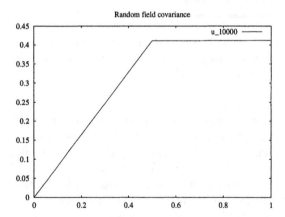

**Fig. 7.11.** The estimated correlation of $u(0.8)$ and $u(x)$ for $0 \le x \le 1$.

## 7.6    Diffpack Tools for Random Fields

### 7.6.1    Generation of Random Fields

The methods in Diffpack for dealing with random fields include

1. generation of realizations of random fields in various types of grid,

2. kriging,

3. estimation of statistical properties, and

4. test of normality of random fields.

The implemented functionality reflects the tools that are required by numerical solution methods for SPDEs.

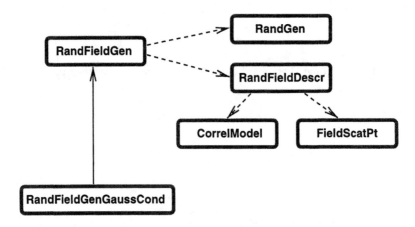

**Fig. 7.12.** A sketch of class RandFieldGen, its subclass and its data members. Solid lines indicate class derivation ("is-a" relationship), whereas dashed lines indicate data members ("has-a" relationship).

Figure 7.12 provides an overview of the central classes that come into play when generating random fields. A random field generator (RandFieldGen) has a random number generator (RandGen) and a random field description (RandFieldDescr). The latter object holds a correlation model (CorrelModel) and optionally a set of scattered points (FieldScatPt) for fixing the field values (kriging or conditional Gaussian fields). As an example of specific random field generators, we show the subclass RandFieldGenGaussCond for computing Gaussian random fields conditioned on a set of points with fixed (measured) values.

The class RandFieldGen acts as base class for a hierarchy of methods for generating discrete, spatial, random fields. The core of the class can briefly be sketched as follows:

```
class RandFieldGen
{
protected:
  Handle(RandGen) realiz;   // random number generator
  Handle(RandFieldDescr) description;
public:
    virtual void generate (FieldWithPtValues& field) =0;
    void transform2lognormal (FieldWithPtValues& field);
    virtual void scan (MenuSystem& menu);
    virtual void defineStatic (MenuSystem& menu, int level = MAIN);
};
```

An important data member is the RandFieldDescr object containing the description of the random field to be generated. This class will be explained later. The central function in class RandFieldGen is generate, which fills a scalar field with generated field values at each point. Since the argument FieldWithPtValues is a base class for fields over lattice grids (FieldLattice) and finite element fields (FieldFE), the generate function can be applied in both finite difference and finite element solvers. The transform2lognormal function transforms a Gaussian field to its lognormal counterpart, simply by applying the exponential function to each field value (cf. footnote on page 309). The define and scan functions make it possible to use the menu system for setting the input data necessary for initializing a random field generator.

The RandFieldDescr class contains the important input data for random field generators, mainly the description of the field: the mean and standard deviation values and the correlation structure. The class looks like this:

```
class RandFieldDescr
{
public:
  int nsd;                    // no of space dim.

  // Gaussian fields:
  real mean, stdev;           // unconditional mean and std.dev.
  bool lognormal_field;       // lognormal vs. normal field

  Handle(CorrelModel) corr_model;
  // storage of prescribed (deterministic) points from input:
  Handle(FieldScatPt) constraints;

  real spatialCorrelation // correl. between points x1 and x2
        (const Ptv(real)& x1, const Ptv(real)& x2);

  void scan (MenuSystem& menu);
  void defineStatic (MenuSystem& menu, int level = MAIN);
};
```

The distribution of the random field values is either Gaussian, or lognormal (which is just a point-by-point transformation of a Gaussian field). The user must provide a correlation model, represented here as a class in the CorrelModel hierarchy.

As an option, one can specify a set of points and corresponding deterministic (known) values of the field. This information is used for methods based on kriging, where random fields are fixed at measurement values at certain points. Such fields are also known as *conditional Gaussian fields*.

The CorrelModel hierarchy specifies models for the spatial correlation $C(r)$ in a random field, $r$ being the distance between two points. Expressed by covariances of the field values $f(x_1)$ and $f(x_2)$ at two points $x_1$ and $x_2$, $C(r)$ reads

$$C(r) = \frac{\text{Cov}[f(x_1), f(x_2)]}{\text{StDev}[f(x_1)]\text{StDev}[f(x_2)]}, \quad r = \|x_1 - x_2\|.$$

Note that $|C(r)| \leq 1$. Because only the distance between two points are taken into account, $C(r)$ can only be used to prescribe correlations in isotropic random fields.

At the time of this writing, four correlation models have been implemented as subclasses of CorrelModel. Class NoCorrelModel can be used to specify no correlation, $C(r) = 0$. Class ExponentialCorrelModel implements the model

$$C(r) = \exp\left(-\sum_{j=1}^{d}\left(\frac{x_j}{I_j}\right)^2\right),$$

whereas class SepExponentialCorrelModel offers

$$C(r) = \prod_{j=1}^{d}\exp\left(-\left(\frac{|x_j|}{I_j}\right)\right).$$

A generalization of this correlation function is implemented in a class with name GenExponentialCorrelModel:

$$C(r) = \prod_{j=1}^{d}\exp\left(-\left(\frac{|x_j|}{I_j}\right)^{a_j}\right).$$

The parameter $I_j$ is referred to as the integral scale in the random fields literature, and $d$ denotes the number of space dimensions. If we call the spatialCorrelation function in class RandFieldDescr through one of its subclasses, for computing the correlation between two points, this particular function in the CorrelModel hierarchy is called with its argument $r$ equal to the distance between the two points.

As we have pointed out, methods for generating (realizations of) random fields are implemented as subclasses of RandFieldGen. The first method to be discussed is provided by the RandFieldGenGaussCond class. The class generates (log-) normally distributed fields which may be conditioned on measurements from several locations (kriging). The generation method in this class, provided by the generate function, includes the following steps:

1. computation of conditional properties,

2. Cholesky-factorization of the covariance matrix,

3. generation of independent normally distributed variables for each point in the field and transformation of the variables (by using the Cholesky-factorized covariance matrix).

Step 1 and 2 are performed initially.

Figure 7.13 shows realizations of a 2D lognormally distributed random field, $T$, with parameters $\lambda_T = 1$ and $\zeta_T = 0.8$. The underlying Gaussian field has an exponential correlation function, and the field values are plotted along a prescribed line. In figures (a) and (c) high-frequency oscillations are observed. This is due to the fact that the correlation scale is small ($I = 0.1$) in these cases. On the other hand, in figures (b) and (d) a much larger integration scale is applied ($I = 2.5$). This means that the area in which the field values are correlated to each other is much larger than in the former case. Thus, the field is now relatively smoothly varying. Finally, while the realizations in (a) and (b) are unconditioned, the realizations in (b) and (d) are conditioned on the values 0.82 and 9.03, located at 2 and 8, respectively.

Another class of methods for generating random fields is provided by subclasses of the Draw2D class, where Draw2D is derived from RandFieldGen. The data structures are motivated by efficient Markov methods[7] [19]. These methods require uniform meshes with square cells. Therefore, Draw2D works with objects of type GridLattice and FieldLattice. These objects are created and initialized by the Draw2D::scan function. In addition, a new pure virtual generate function without arguments is introduced. Its implementation in the subclasses is supposed to fill the FieldLattice object with values. However, the general interface to an object of the class FieldWithPointValues is still provided through a call to the generate function with an argument of type FieldWithPointValues. The latter function now calls the new generate function, before the values from the FieldLattice object are interpolated into the desired FieldWithPointValues format. This means that one can generate a realization of a random field on a fine GridLattice grid and then use a coarser grid for SPDE calculations, provided one has a satisfactory method for upscaling the fine-grid field (this is a difficult problem, which has been subject to intensive research over the last decade).

The methods are based on the Markov property, which means that the statistical properties at a certain location is defined by the values of some neighboring points. This assumption is exactly fulfilled in several one-dimensional problems, but in higher dimensions it is only an approximation. Two 2D versions of the methods are implemented in the classes Draw2D_s1 and Draw2D_s2. In the former class, the generated values are conditioned on the corner points

---

[7] The C++ implementation in Diffpack is based on Fortran software developed at the Norwegian Computing Center (NR). NR has some rights in the C++ code, see the header of the Draw2D*.h files.

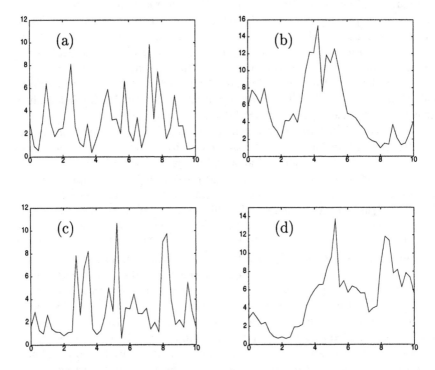

**Fig. 7.13.** Realization of a lognormally distributed 2D random field, $T$, with parameters $\lambda_T = 1$ and $\zeta_T = 0.8$, and an exponential correlation model for the underlying Gaussian field. The field is plotted along a prescribed line. (a): Correlation scale $I = 0.1$; (b): correlation scale $I = 2.5$; (c): field conditioned on the values $(2, 0.82)$ and $(8, 9.03)$, with correlation scale $I = 0.1$; (d): field conditioned on the values $(2, 0.82)$ and $(8, 9.03)$, with correlation scale $I = 2.5$.

in different squares of the domain, while the structure of conditioning is more complicated in the latter class. For reasons of efficiency, both methods require $2^n + 1$ (for integer $n$) nodal points in each spatial direction. To simplify the validation of these approximate methods, the accurate, but inefficient, matrix method described earlier is implemented for the present, more restrictive, two-dimensional problems in the class Draw2D_matrix.

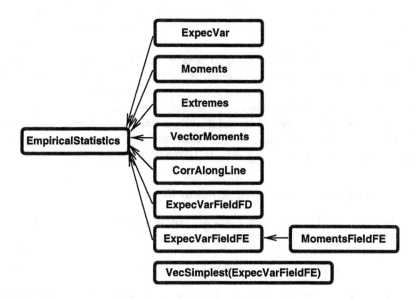

**Fig. 7.14.** A sketch of classes for empirical estimation of statistical properties. Solid arrows indicate class derivation ("is-a" relationship). Two other subclasses of EmpiricalStatistics, Histogram and TestOfNormality, have been omitted from the figure.

### 7.6.2  Statistical Properties

Diffpack has various functionality available for estimating statistical properties of random fields. The method for estimating the expectation and standard deviation of a discrete random field is simple; it is just a matter of estimating the expectation and standard deviation of each point value. Higher-order moments are handled by the same approach.

Following the class structure in the the classes ExpecVar and Moments for random variables we have developed the class ExpecVarFieldFE and its subclass MomentsFieldFE for estimating statistical moments of random fields of type FieldFE. The functionality of these classes is similar to the structure in the corresponding classes for a single random variable. However, there are of course some differences. Instead of representing the statistical properties as real numbers, they are now handled as objects of type `FieldFE`. Naturally, the argument to the `update` function is an object of the class `FieldFE`. In order to deal with stochastic fields represented as `FieldLattice` objects, a class `ExpecVarFieldLattice` is implemented similarly. In time-dependent problems one needs to estimate the statistics of a discrete random field at several time points. For this purpose, one can use a `VecSimplest` version of the field estimation classes mentioned, for example, `VecSimplest(ExpecVarFieldFE)`.

The covariance of a random field (or two different fields) for separation distances between a main point and points along a line can be estimated by the `CorrAlongLine` class (which is also a subclass of `EmpiricalStatistics`). This `CorrAlongLine` class has `init` and `update` functions as well as functions for plotting the covariance along lines. Note that the complete covariance information of a random field is very expensive to compute and store; it is much more efficient to estimate the covariance along lines.

The functionality provided by the `Histogram` and `TestOfNormality` classes can be easily applied in conjunction with random fields, when we look at point values. A sketch of estimation classes in Diffpack is given in Figure 7.14.

### 7.6.3   Example: A Stochastic Poisson Equation

The Diffpack tools for stochastic PDEs are now demonstrated in the following problem, modeling groundwater flow. The PDE reads

$$\nabla \cdot (T\nabla H) = 0. \tag{7.32}$$

Here, $T$ is a random conductivity field (which equals the permeability divided by the viscosity coefficient for the fluid) and $H$ denotes hydraulic head, which is obtained by subtracting the hydrostatic contribution from the actual pressure in the fluid. Equation (7.32) may be derived by combining Darcy's law and the mass balance equation.

Since $T$ is stochastic, $H$ also becomes stochastic. A general simulator, where the SPDE is solved by Monte Carlo simulation, has been developed. There is a class `Head` for solving equation (7.32) in the deterministic case ($T$ given as a realization) by the finite element method. The deterministic simulator has the subclass `HeadStoch` that implements the basic steps in the Monte Carlo method, e.g.,a `loopMonteCarlo` procedure and a modified `scan` routine, as well as some additional data required when dealing with stochastic problems, see the `HeatStoch` class introduced above. The authors believe that such a modular design is crucial for reducing the programming efforts and increasing the reliability when considering more complicated problems defined by systems of SPDEs. It also makes it quite simple to introduce stochastic effects in an existing deterministic simulator and solve the resulting SPDEs by Monte Carlo methods.

The SPDE (7.32) has also been solved numerically using a first-order perturbation approximation. As stated above, the conductivity $T$ is regarded as a random field. Without serious restrictions, we assume that it may be written in the form $T = \exp(Y)$, where $Y = \mu_Y + Y'$. Here, $\mu_Y$ is the expectation of $Y$ and, $Y'$ is the stochastic deviation from from the mean. In the following we will focus on the expectation and the covariance structure of $H$.

The perturbation method to be presented herein relies on a standard first-order Taylor series expansion of the SPDE (7.32). We start by expanding the transmissivity in a series with powers of the perturbation $Y'$ around $\mu_Y$. Only the first two terms are retained. The procedure results in the following equation:

$$\nabla \cdot \{\exp(\mu_Y)\,(1+Y')\,\nabla H\} = 0. \tag{7.33}$$

The solution of the SPDE (7.33) is accomplished by a finite element discretization and a first-order perturbation expansion of $H$, resulting in approximations to the expectation and the covariance structure of the stochastic head field. Letting $N_i(x)$ denote piecewise linear finite element basis functions, the random $H(x)$ field is approximated by

$$H \approx \sum_{i=1}^{N} h_i N_i(x), \tag{7.34}$$

where $h_i$ is the approximate head value at node $i$, and $N$ denotes the total number of nodes. Employing a Galerkin method to (7.33) (with corresponding boundary conditions) we arrive at the system

$$Kh = p, \tag{7.35}$$

where $K$ is the coefficient (stiffness) matrix, $h = \{h_1, h_2, \ldots, h_N\}^T$ is the vector of approximate nodal values, and the right hand side vector $p$ contains contributions due to the boundary conditions.

Provided that the expectation $\mu_Y = E[Y]$ is constant on each element, the contribution to $K$ from element $e$, denoted $K^{(e)}$, can be expressed as follows:

$$
\begin{aligned}
K^{(e)} &= \int_{D^{(e)}} B^{(e)^T} \exp(\mu_Y^{(e)})\,(1+Y')\,B^{(e)}\,dD \\
&= \exp(\mu_Y^{(e)}) B^{(e)^T} B^{(e)} \int_{D^{(e)}} (1+Y')\,dD \\
&= \exp(\mu_Y^{(e)}) B^{(e)^T} B^{(e)} \left( \mathrm{area}\left(D^{(e)}\right) + Y^{(e)} \right) \\
&= K_0^{(e)} + Y^{(e)} \Delta K_0^{(e)}.
\end{aligned}
\tag{7.36}
$$

Here, $B^{(e)}$ is a matrix containing first derivatives of the test functions belonging to element $e$, $\mathrm{area}\left(D^{(e)}\right) = \int_{D^{(e)}} dD$, and $Y^{(e)} = \int_{D^{(e)}} Y'\,dD$ is a stochastic variable. Because the basis functions are piecewise linear, $B^{(e)}$ is constant. The global coefficient matrix $K$ may now be written as

$$K = \sum_{e=1}^{E} \left( K_0^{(e)} + Y^{(e)} \Delta K_0^{(e)} \right)$$

$$= K_0 + \sum_{e=1}^{E} Y^{(e)} \Delta K_0^{(e)}, \tag{7.37}$$

where $E$ is the total number of elements. One should here notice that $K_0$ is the standard finite element discretization of the (possibly variable coefficient) Laplace operator when the multiplicative stochastic part of the input field $T$ is neglected. We also observe that $K$ is a function of the random variables $Y^{(1)}, Y^{(2)}, \ldots, Y^{(E)}$. Thus, from (7.35) it follows that also the head vector depends on these random variables. We expand $h$ in a first-order series around $E[Y^{(e)}] = 0$,

$$h \approx \hat{h} = h_0 + \sum_{e=1}^{E} \frac{\partial h}{\partial Y^{(e)}} Y^{(e)}, \tag{7.38}$$

where $\hat{h}$ is the first-order approximation of the stochastic vector $h$. The expressions (7.37) and (7.38) are inserted into (7.35), and terms of equal order in $Y^{(e)}$ are grouped and solved separately. This leads to the zero-order problem

$$K_0 h_0 = p. \tag{7.39}$$

From (7.38) and (7.39) it is seen that the approximate expectation of $H$ is obtained by solving the flow equation with $\exp(\mu_Y)$ as a conductivity field. Furthermore, there are $E$ first-order problems

$$K_0 \frac{\partial h}{\partial Y^{(e)}} = -\Delta K_0^{(e)} h_0, \quad e = 1, \ldots, E. \tag{7.40}$$

All deterministic boundary conditions are supposed to be fulfilled by the zero-order solution. Thus, the right hand side of (7.40) is set to zero for all nodes with prescribed head values. It is recognized that the $E + 1$ equation systems (7.39) and (7.40) contain the same coefficient matrix $K_0$. Moreover, the matrix is efficiently computed by integrating the element contributions $K_0^{(e)}$ analytically.

From the solutions of the linear systems (7.39) and (7.40) we can easily construct the first two moments of the approximate stochastic head field. To first order, the expectation is given by

$$E[\hat{h}] = h_0, \tag{7.41}$$

whereas the covariance becomes

$$\text{Cov}[\hat{h}, \hat{h}] = \text{E}[(\hat{h} - h_0)(\hat{h} - h_0)^T]$$

$$= \text{E}[(\sum_{e_1=1}^{E} \frac{\partial h}{\partial Y^{(e_1)}} Y^{(e_1)})(\sum_{e_2=1}^{E} \frac{\partial h}{\partial Y^{(e_2)}} Y^{(e_2)})^T]$$

$$= \sum_{e_1=1}^{E} \sum_{e_2=1}^{E} \frac{\partial h}{\partial Y^{(e_1)}} \left(\frac{\partial h}{\partial Y^{(e_2)}}\right)^T \text{E}[Y^{(e_1)} Y^{(e_2)}]. \qquad (7.42)$$

Using the definition of $Y^{(e)}$ the last expectation above may be written as

$$\text{E}[Y^{(e_1)} Y^{(e_2)}] = \int_{D^{(e_1)}} \int_{D^{(e_2)}} \text{E}[Y'(\boldsymbol{x}_{e_1}) Y'(\boldsymbol{x}_{e_2})] dD \ dD$$

$$= \int_{D^{(e_1)}} \int_{D^{(e_2)}} \text{Cov}[Y'(\boldsymbol{x}_{e_1}), Y'(\boldsymbol{x}_{e_2})] dD \ dD. \qquad (7.43)$$

The integration of the log conductivity covariance must normally be carried out by numerical techniques.

To summarize, the present technique yields first-order approximations to the head expectation and covariance without the common discretization of the random input field $Y$ in terms of finite elements. The method relies solely on specification of the mean and covariance structure of $Y$. Normally, the finite element mesh used for $Y$ has to be much finer than the one needed for the response. With the present approach one can use a fairly coarse mesh designed to resolve the moments of $H$. However, the integration of the covariance of $Y'$ will then need higher-order quadrature rules.

## 7.7  Summary

We have in this chapter explained how to deal with PDE models where some of the input data are given statistical descriptions. The solution will then be a stochastic field in space (and possibly time). Our simplest examples concerned models with known analytical solutions in symbolic form. If some of the input parameters are described as stochastic variables, the solution is a stochastic field, whose expression is obtained by the standard rules of stochastic variable transformations (Appendix 7.A). When numerical solution of the PDE(s) are required, we have presented two solution methods for handling stochastic input and output data: (i) Monte Carlo simulation, which is a general but computationally intensive approach, and (ii) perturbation methods, which are computationally fast but limited to small standard deviations. Both methods involve repeated solution of the underlying deterministic PDE(s). Hence, a deterministic solver can in principle be reused in the stochastic case. Diffpack has extensive support for such reuse, and the relevant tools for solving stochastic PDEs with Diffpack have been presented. We have also presented complete simulators for stochastic problems.

*Acknowledgement.* The authors thank Dr. Lars Holden at the Norwegian Computing Center (NR) for giving us NR's Fortran software for random field generation and for allowing us to distribute a C++ version of this software along with Diffpack (restricted to non-commercial use). Dr. Wen Chen, Ola Skavhaug, and Sylfest Glimsdal have contributed with corrections and constructive criticism of the present manuscript.

# 7.A   Transformation of Random Variables

A mathematical model with uncertainty can be viewed as a transformation of stochastic input data to stochastic output data. From probability theory we have some results regarding transformation of random variables. These are particularly useful in the analysis of simple stochastic differential equations, where the solution of the deterministic version of the differential equations is a closed-form mathematical expression. The purpose of the present section is to review the basic results regarding transformation of random variables.

## 7.A.1   Transformation of a Single Random Variable

Suppose we work with a mathematical model transforming a stochastic input parameter $X$ to a stochastic output parameter $Y$ through a function $g(X)$, whose mathematical expression is known,

$$Y = g(X) . \tag{7.44}$$

If $g$ is a monotonically increasing function of its argument, the probabilities expressed in either variables must be the same, that is,

$$P[Y \leq y] = P[X \leq x] \tag{7.45}$$

for $y = g(x)$. In terms of cumulative distribution functions $F_X(x)$ and $F_Y(y)$ of the two variables, (7.45) can be written

$$F_Y(y) = F_X(x) = F_X[g^{-1}(y)] .$$

Differentiating both sides of this expression with respect to $y$ results in a relation between the probability densities:

$$f_Y(y) = f_X[g^{-1}(y)] \frac{dg^{-1}}{dy} ,$$

where $f_Y$ and $f_X$ are the probability density functions corresponding to $F_X$ and $F_Y$ ($f_X = dF_X/dx$, $f_Y = dF_Y/dy$). Since such density functions cannot be negative, we take the absolute value of $dg^{-1}/dy$:

$$f_Y(y) = f_X[g^{-1}(y)] \left| \frac{dg^{-1}}{dy} \right| . \tag{7.46}$$

The inverse function $g^{-1}(y)$ can be multiple valued. That is, inverting $y = g(x)$ results in $x_i = g_i^{-1}(y)$ for $i = 1, \ldots, q$. As an example, think of $y = x^2$, resulting in $x_1 = g_1(y) = \sqrt{y}$ and $x_2 = g_2(y) = -\sqrt{y}$. In the case of a multiple-valued $g$, (7.46) is generalized to

$$f_Y(y) = \sum_{j \in J} f_X[g_j^{-1}(y)] \left| \frac{dg_j^{-1}}{dy} \right|. \tag{7.47}$$

The index set $J$ contains the indices $i = 1, \ldots, q$ for which $f_X(x_i)$ is defined and is a positive number. Equation (7.47) provides us with the *complete* information about the statistics of $Y$ given the statistics of $X$.

*Example: Linear Transformation.* When $Y = g(X)$ is a linear transformation, $Y = aX + b$, the distribution of $Y$ is the same as the distribution of $X$, just with other parameters:

$$f_Y(y) = \frac{1}{a} f_X[(y - b)/a].$$

For example, if $X$ is normally distributed with parameters $\mu_X$ and $\sigma_X$, $Y$ is normally distributed with parameters $a\mu_X + b$ and $a\sigma_X$.

*Example: Quadratic Transformation.* With a quadratic relationship, $Y = cX^2$, the inverse transformation $x = g^{-1}(y)$ has two roots:

$$g_1^{-1}(y) = \sqrt{y/c}, \quad g_2^{-1}(y) = -\sqrt{y/c}.$$

The derivative of $g^{-1}$ becomes

$$\frac{dg_j^{-1}}{dy} = (-1)^{j+1} \frac{1}{2\sqrt{cy}}, \quad j = 1, 2,$$

and then we get from (7.47),

$$f_Y(y) = \left( f_X(\sqrt{y/c}) + f_X(-\sqrt{y/c}) \right) \frac{1}{2\sqrt{cy}}, \tag{7.48}$$

provided $f_X$ is defined and nonzero at the two roots. Notice that we have taken the absolute value of the negative derivative of $g_2^{-1}$ ($y = cx^2$ is always positive).

Suppose $X$ is lognormally distributed[8] with parameters $\lambda_X$ and $\zeta_X$,

$$f_X(x) = \frac{1}{\zeta_X \sqrt{2\pi}} \frac{1}{x} \exp\left[ -\frac{1}{2} \left( \frac{\ln x - \lambda_X}{\zeta_X} \right)^2 \right], \quad x \geq 0. \tag{7.49}$$

---

[8] If $\ln X$ is normally distributed, then $X$ is lognormally distributed.

The parameters $\lambda_X$ and $\mu_X$ are related to the expected value $\mu_X$ and standard deviation $\sigma_X$ through

$$\lambda_X = \mathrm{E}[\ln X] = \ln \mu_X - \frac{1}{2}\zeta_X^2 \tag{7.50}$$

and

$$\zeta_X^2 = \mathrm{Var}[\ln X] = \ln\left[1 + \left(\frac{\sigma_X}{\mu_X}\right)^2\right]. \tag{7.51}$$

The inverse of (7.50)–(7.51) is useful when the parameters $\lambda_X$ and $\zeta_X$ are given and one wants to know the expectation and standard deviation of $X$, so we provide the formulas here for reference:

$$\mathrm{E}[X] = \exp\left(\lambda_X + \zeta_X^2/2\right), \tag{7.52}$$
$$\mathrm{Var}[X] = \mathrm{E}[X]^2\left(\exp\left(\zeta_X^2\right) - 1\right). \tag{7.53}$$

The probability density $f_X$ is not defined at the negative root $-\sqrt{y/c}$ so the density of $Y$ becomes

$$f_Y(y) = \frac{1}{\zeta_X\sqrt{2\pi}} \frac{1}{\sqrt{y/c}} \exp\left[-\frac{1}{2}\left(\frac{\ln\sqrt{y/c} - \lambda_X}{\zeta_X}\right)^2\right] \frac{1}{2\sqrt{cy}},$$

$$= \frac{1}{2\zeta_X\sqrt{2\pi}\,y} \exp\left[-\frac{1}{2}\left(\frac{\ln y - (\ln c + 2\lambda_X)}{2\zeta_X}\right)^2\right], \quad y \geq 0.$$

We see that $Y = cX^2$ is also lognormally distributed, but with parameters $\ln c + \lambda_X$ and $2\zeta_X$.

In the case $X$ is normally distributed with mean $\mu_X$ and standard deviation $\sigma_X$, $f_X$ can be evaluated at both roots and the probability density of $Y$ reads

$$f_Y(y) = \frac{1}{\sigma_X\sqrt{2\pi}} \frac{1}{2\sqrt{cy}} \left\{\exp\left[-\frac{1}{2}\left(\frac{+\sqrt{y/c} - \mu_X}{\sigma_X}\right)^2\right] + \right.$$

$$\left. \exp\left[-\frac{1}{2}\left(\frac{-\sqrt{y/c} - \mu_X}{\sigma_X}\right)^2\right]\right\},$$

for $-\infty < y < \infty$.

*Remark: Multi-Variable Random Transformations.* Transforming a vector $\boldsymbol{X} = (X_1, \ldots, X_n)$ of random variables to $\boldsymbol{Y} = (Y_1, \ldots, Y_n)$ through a vector-valued function $g$: $\boldsymbol{Y} = g(\boldsymbol{X})$, results in the following relation between the probability density $f_{\boldsymbol{Y}}$ of $\boldsymbol{Y}$ and the density $f_{\boldsymbol{X}}$ of $\boldsymbol{X}$:

$$f_{\boldsymbol{Y}}(y_1, \ldots, y_n) = f_{\boldsymbol{X}}(x_1, \ldots, x_n)|\det J|,$$

where $\det J$ is the determinant of the Jacobian of the transformation $g$. The usefulness of this relation is limited since the functional transformation $g$ is often difficult to invert (a system of nonlinear algebraic equations must be solved analytically).

### 7.A.2    Transformation of Normal and Lognormal Random Variables

*Sum of Independent Normally Distributed Variables.* Let $X_1, \ldots, X_n$ be independent normally distributed random variables with expectation values $\mu_1, \ldots, \mu_n$ and standard deviations $\sigma_1, \ldots, \sigma_n$. A general linear combination of these variables,

$$Y = b + \sum_{i=1}^{n} a_i X_i, \tag{7.54}$$

can be shown to be normally distributed with parameters

$$\mu_Y = b + \sum_{i=1}^{n} \mu_i, \tag{7.55}$$

$$\sigma_Y^2 = \sum_{i=1}^{n} a_i^2 \sigma_i^2. \tag{7.56}$$

*Product of Independent Lognormally Distributed Variables.* Consider a product $Y$ of independent lognormally distributed variables $X_1, \ldots, X_n$,

$$Y = b \prod_{i=1}^{n} X_i^{a_i}, \tag{7.57}$$

where $a_i$, $i = 1, \ldots, n$, are prescribed exponents. We denote the associated lognormal parameters by $\lambda_1, \ldots, \lambda_n$ and $\zeta_1, \ldots, \zeta_n$. Taking the logarithm of (7.57) yields

$$\ln Y = \ln b + \sum_{i=1}^{n} a_i \ln X_i.$$

We can now apply the result regarding transformation of a sum of independent normally distributed variables, because $\ln X_i$ is normal with mean $\lambda_i$ and standard deviation $\zeta_i$, when $X_i$ is lognormal with parameters $\lambda_i$ and $\zeta_i$. Hence, $Y$ in (7.57) is a lognormal variable with parameters

$$\lambda_Y = \ln b + \sum_{i=1}^{n} a_i \lambda_i, \tag{7.58}$$

$$\zeta_Y^2 = \sum_{i=1}^{n} a_i^2 \zeta_i^2. \tag{7.59}$$

Many formulas in engineering involve products of parameters so the transformation (7.57) arises frequently.

*The Central Limit Theorem.* When $n$ in (7.54) approaches infinity, one can prove that $Y$ approaches a normal distribution regardless of the distribution of the $X_i$ variables. The only requirement is that $X_1, \ldots, X_n$ are independent. This important result is known as the Central Limit Theorem. The concept can be extended to products of independent random variables as in (7.57), when $n \to \infty$. The distribution of the response $Y$ then approaches a lognormal distribution.

## 7.A.3   Other Variable Transformations

*Beta and F Distributed Variables.* There exist special results, which can be handy for stochastic analysis of engineering formulas, based on transformations of generalized Beta and F distributed variables. We briefly review some basic formulas and refer to Gran [10] for a complete list of distribution functions, moments, parameter estimation, and engineering applications.

The Beta distribution reads

$$b(x; a, b, h, C) = \frac{|h|}{B(a,b)C} \left(\frac{x}{C}\right)^{ah-1} \left[1 - \left(\frac{x}{C}\right)^{b-1}\right], \quad 0 \leq x \leq C, \quad (7.60)$$

where $B(a, b)$ is the complete Beta function, which may be expressed by gamma functions $\Gamma(x)$ through

$$B(a, b) = \frac{\Gamma(a)\Gamma(b)}{\Gamma(a + b)}. \quad (7.61)$$

The statistical moment $M_n = \mathrm{E}[X^n]$ is

$$M_n = C^n \frac{\Gamma(a + b)\Gamma(a + n/h)}{\Gamma(a + b + n/h)\Gamma(a)}. \quad (7.62)$$

The probability of a generalized F distributed variable can be written

$$f(x; a, b, k, \Delta) = \frac{|k|}{B(a,b)\Delta} \frac{\left(\frac{x}{\Delta}\right)^{ak-1}}{\left[1 + \left(\frac{x}{\Delta}\right)^k\right]^{a+b}}, \quad x \geq 0. \quad (7.63)$$

The corresponding moments read

$$M_n = \mathrm{E}[X^n] = \Delta^n \frac{\Gamma(a + n/k)\Gamma(b - n/k)}{\Gamma(a)\Gamma(b)}. \quad (7.64)$$

For the special case $a = b = 1$, a simpler formula can be derived:

$$M_n = \Delta^n \frac{\frac{n\pi}{k}}{\sin \frac{n\pi}{k}}. \quad (7.65)$$

The generalized F distribution reduces to well-known distributions for special choices of the parameters $a$, $b$, $k$, and $\Delta$: $f(x; n_1/2, n_2/2, 1, n_1/n_2)$ is a standard F distribution with degrees of freedom $n_1$ and $n_2$, $f(x; 1/2, 1/2, 2, \Delta)$ is

Cauchy's distribution, $f(x; 1/2, n/2, 2, \sqrt{n})$ is Student's t distribution with $n$ degrees of freedom.

We have three basic results regarding transformation of Beta and F distributed variables.

1. Let $X$ be Beta distributed with parameters $a$, $b$, $h$, and $C$:

$$X \sim b(x; a, b, h, C).$$

A power transformation preserves the distribution, i.e.,

$$Y = rX^s, \quad Y \sim b(y; a, b, h/s, rC^s). \tag{7.66}$$

2. Let $U$ be distributed according to the generalized F distribution:

$$U \sim f(u; a, b, k, \Delta).$$

A new variable $V = pU^q$ is also F distributed,

$$V = pU^q, \quad V \sim f(v; a, b, k/q, p\Delta^q). \tag{7.67}$$

3. Given

$$X \sim b(x; a, b, h, C), \quad U \sim f(u; a, b, k, \Delta),$$

then we have

$$\left(\frac{X}{C}\right)^h = \frac{\left(\frac{U}{\Delta}\right)^k}{1 + \left(\frac{U}{\Delta}\right)^k} \tag{7.68}$$

and

$$\left(\frac{U}{\Delta}\right)^k = \frac{\left(\frac{X}{C}\right)^h}{1 - \left(\frac{X}{C}\right)^h}. \tag{7.69}$$

These formulas are useful for stochastic analysis of, e.g., differential equations involving solutions on the (scaled) form $(x^p/(1+x^p))^q$ and $(x^p/(1-x^p))^q$.

*Gamma Distributed Variables.* The Gamma distribution has three parameters and the probability density function can be written

$$g(x; a, h, A) = \frac{|h|}{\Gamma(a)A} \left(\frac{x}{A}\right)^{ah-1} \exp\left(-\left(\frac{x}{A}\right)^h\right), \quad x \geq 0. \tag{7.70}$$

The associated moments are

$$M_n = E[X^n] = A^n \frac{\Gamma(a + n/h)}{\Gamma(a)}, \quad a + n/h \neq 0, -1, -2, \ldots.$$

The Gamma distribution contains many well-known distributions as special cases: the Rayleigh distribution ($a = 1$, $h = 2$, $A = \sqrt{2}\sigma$), the Maxwell distribution ($a = 3/2$, $h = 2$, $A = \sqrt{2}\sigma$), the exponential distribution ($a = 1$, $h = 1$), the two-parameter Weibull distribution ($a = 1$, $h > 0$), and the lognormal distribution ($a \to \infty$, $h^{-1} = \sqrt{a}\sigma$, $A = a^{-1/h}\exp(\bar{x})$, with mean $\bar{x}$ and standard deviation $\sigma$. In the case $a = 1$, $h = \bar{x}/\sigma \gg 1$, and $A = \bar{x}$, the Gamma distribution is an approximation of the one-sided normal distribution.

Given a Gamma distributed variable, $X \sim g(x; a, h, A)$, the related variable $Y = rX^s$ is also Gamma distributed:

$$Y = rX^s, \quad Y \sim g(x; a, h/s, rA^s). \tag{7.71}$$

Some of the transformation results can be extended to power products of two independent variables. We refer to Gran [10] for details.

### 7.A.4  Partial and Approximate Analysis of Random Transformations

In this subsection we shall mention some methods for computing the expectation and standard deviation of a random variable $Y$ related to random variables $X_1, \ldots, X_n$ whose mean values and standard deviations are known, and perhaps their covariances as well. That is, we do not find the complete probability density of $Y$; we only manage to compute some basic statistics. This is often a highly relevant setting as the complete (joint) distribution of stochastic input variables in a mathematical model is seldom known. Means and standard deviations are much easier to estimate and hence more frequently available.

*Linear Combination of Random Variables.* Suppose $Y$ is linearly related to $X_1, \ldots, X_n$,

$$Y = b + \sum_{i=1}^{n} a_i X_i.$$

The distributions of $X_1, \ldots, X_n$ are not known, but we have values for the mean values $\mu_i$, standard deviations $\sigma_i$, and their covariances $\mathrm{Cov}[X_i, X_j]$. Applying the expectation operator to the sum of the random variables, we obtain

$$\mathrm{E}[Y] = b + \sum_{i=1}^{n} a_i \mu_i. \tag{7.72}$$

Computing the variance implies more algebra, but the result becomes

$$\mathrm{Var}[Y] = \sum_{i=1}^{n} \sum_{j=1}^{n} a_i a_j \mathrm{Cov}[X_i, X_j]$$

$$= \sum_{i=1}^{n} a_i^2 \sigma_i^2 + \sum_{\substack{i=1 \\ i \neq j}}^{n} \sum_{j=1}^{n} a_i a_j \mathrm{Cov}[X_i, X_j]. \tag{7.73}$$

Notice that the last double sum in (7.73) vanishes if $X_1, \ldots, X_n$ are independent random variables.

*Approximate Analysis of Nonlinear Transformations.* Let us now work with a general nonlinear relationship between $X_1, \ldots, X_n$ and $Y$:

$$Y = g(X_1, \ldots, X_n).$$

By performing a Taylor expansion of $g$ around the mean values, we can apply the analysis from the previous paragraph. A second-order Taylor series expansion reads

$$Y \approx g(\mu_1, \ldots, \mu_n) + \sum_{i=1}^n (X_i - \mu_i)\frac{\partial g}{\partial X_i} + \frac{1}{2}\sum_{i=1}^n \sum_{j=1}^n (X_i - \mu_i)(X_j - \mu_j)\frac{\partial^2 g}{\partial X_i \partial X_j}.$$

$$(7.74)$$

The derivatives of $g$ are to be evaluated at the mean values $\mu_1, \ldots, \mu_n$.

Using the linear terms in (7.74) only, we can easily compute the first-order approximation of the expectation of $Y$:

$$\mathrm{E}[Y]' = g(\mu_1, \ldots, \mu_n). \tag{7.75}$$

The prime symbol signifies a first-order approximation of $\mathrm{E}[Y]$. The calculated expression for $\mathrm{E}[Y]'$ means that if we have a relation $Y = f(\boldsymbol{X})$ in a physical application, and regard $\boldsymbol{X}$ as stochastic, the first-order approximation of the mean of $Y$ is obtained by evaluating $f$ at the mean of $\boldsymbol{X}$.

A second-order approximation of the expectation of $Y$ is easily derived from (7.74), but it involves the covariances of the input variables $X_1, \ldots, X_n$:

$$\mathrm{E}[Y]'' = g(\mu_1, \ldots, \mu_n) + \frac{1}{2}\sum_{i=1}^n \sum_{j=1}^n \frac{\partial^2 g}{\partial X_i \partial X_j}\mathrm{Cov}[X_i, X_j]. \tag{7.76}$$

With uncorrelated input variables, the second-order approximation to $\mathrm{E}[Y]$ simplifies to

$$\mathrm{E}[Y]'' = g(\mu_1, \ldots, \mu_n) + \frac{1}{2}\sum_{i=1}^n \sum_{j=1}^n \frac{\partial^2 g}{\partial X_i^2}\sigma_i^2, \tag{7.77}$$

with $\sigma_i$ being the standard deviation of $X_i$, $i = 1, \ldots, n$.

An approximation of the variance of $Y$ can also be found from (7.74). We have to use only the linear part of (7.74) to obtain an expression for $\mathrm{Var}[Y]$, which requires information about the covariances of $X_1, \ldots, X_n$ only:

$$\mathrm{Var}[Y]' = \sum_{i=1}^n \sum_{j=1}^n \frac{\partial g}{\partial X_i}\frac{\partial g}{\partial X_j}\mathrm{Cov}[X_i, X_j]. \tag{7.78}$$

In the case the $X_i$ variables are uncorrelated, the result simplifies to

$$\mathrm{Var}[Y]' = \sum_{i=1}^n \left(\frac{\partial g}{\partial X_i}\right)^2 \mathrm{Var}[X_i]. \tag{7.79}$$

*Implicitly Defined Transformations.* The results for $E[Y]'$, $E[Y]''$, and $\text{Var}[Y]'$ enable computations of basic statistics of response variables in mathematical models involving differential equations when the function $g$ is known as an explicit formula such that the derivatives of $g$ can be computed. This is normally the case if we solve the differential equations analytically. Unfortunately, numerical solution methods only evaluate $g$ at discrete points $x_1, \ldots, x_n$. For such implicitly defined transformations, available through some "black-box" computer code, we can estimate derivatives of $g$ by finite differences. For example, a centered difference around the mean reads

$$\left. \frac{\partial g}{\partial X_i} \right|_{\mu_1, \ldots, \mu_n} \approx \frac{1}{2h\sigma_i} \left( g(x_1, \ldots, x_i + h\sigma_i, \ldots, x_n) - g(x_1, \ldots, x_i - h\sigma_i, \ldots, x_n) \right).$$

Another approach is to differentiate the differential equation to obtain an equation for $\partial g / \partial X_i$, $i = 1, \ldots, n$. This is often not possible if $Y$ is a quantity derived from the solution of the differential equation, for example, the maximum value of the solution as in Section 7.2.1.

## 7.B    Implementing a New Distribution

It is possible to implement a new class in the `ProbDistr` hierarchy without touching the existing source code files or recompiling them. The new distribution class is placed in a separate file in the programmer's library or application directory. In addition, one needs to extend the `ProbDistr_prm` class such that the Diffpack libraries become aware of the new subclass of `ProbDistr`. The recipe is described in general terms in the Diffpack FAQ, question "How can I extend a class hierarchy in Diffpack?".

Suppose you want to add the Weibull distribution to the `ProbDistr` hierarchy. The relevant formulas for the density, the cumulative distribution, and its inverse are

$$f(x) = \lambda^\beta \beta x^{\beta-1} \exp\left(-(\lambda x)^\beta\right), \quad x \geq 0 \tag{7.80}$$

$$F(x) = 1 - \exp\left(-(\lambda x)^\beta\right), \tag{7.81}$$

$$F^{-1}(p) = \frac{1}{\lambda} \left(-\ln(1-p)\right)^{1/\beta}. \tag{7.82}$$

We implement these functions as `density`, `cum`, and `cumInv`, respectively, in a new class `WeibullDistr`:

```
class WeibullDistr : public ProbDistr
{
public:
  WeibullDistr (const ProbDistr_prm& pm_) : ProbDistr(pm_) {}

  static real density (real x, real lambda, real beta);
  static real cum     (real x, real lambda, real beta);
  static real cumInv  (real p, real lambda, real beta);

  virtual real f     (real x);
  virtual real F     (real x);
```

```
   virtual real Finv (real p);

   virtual real expec    ();
   virtual real stdev    ();
};
```

The virtual functions f, F, and Finv simply call the static versions density, cum, and cumInv with pm()(1) as the first parameter, lambda ($\lambda$), and pm()(2) for the second parameter, beta ($\beta$). That $\lambda$ is the first parameter in pm is a convention introduced by the designer of the class. The expectation and the standard deviation of the Weibull distribution are implemented in expec and stdev, respectively. The code related to class WeibullDistr is implemented in the files WeibullDistr.h and WeibullDistr.cpp, found in the src/stoch/Weibull directory.

We also need to define a new parameter class (say) ProbDistr2_prm, which is a subclass extension of class ProbDistr_prm. Class ProbDistr2_prm will provide all the features of class ProbDistr_prm, in addition to being aware of the WeibullDistr class. According to the Diffpack FAQ, the new parameter class takes the form

```
class ProbDistr2_prm : public ProbDistr_prm
{
protected:
   virtual ProbDistr_prm* clone () { return new ProbDistr2_prm(); }
   friend void initDiffpack (int, const char**, bool);

   ProbDistr2_prm ();
public:
   virtual ProbDistr* create () const;
};
```

The corresponding implementation in ProbDistr2_prm.cpp reads

```
ProbDistr* ProbDistr2_prm::create () const
{
   ProbDistr* ptr = NULL;
   // test first for classes managed by ProbDistr2_prm:
   if (distribution == "WeibullDistr")
     ptr = new WeibullDistr (*this);
   else
     // rely on create() in the base class:
     ptr = ProbDistr_prm::create();

   return ptr;
}

ProbDistr2_prm:: ProbDistr2_prm ()
  : ProbDistr_prm()
{ subclasses.append("WeibullDistr"); }
```

To register the new class in the compiled Diffpack libraries, you need to copy an initDiffpack.h file, for instance, from

```
$NOR/dp/src/libs/dpVtk/s2vtk/initDiffpack.cpp
```

to the application directory. In the `initDiffpack` function you need to insert the following registration statements:

```
// register ProbDistr2_prm as official ProbDistr_prm:
ProbDistr_prm::registerPrmSubclass (*new ProbDistr2_prm());
```

These statements can be placed at the end of the `initDiffpack` function. You need, of course, to insert an include,

```
#include <ProbDistr2_prm.h>
```

in this file.

We are now in the position of creating a simple program for drawing random numbers from the Weibull distribution. This is accomplished by calling `RandGen`'s `generate` function, which requires a `ProbDistr` object as argument. This reference points to a `WeibullDistr` subclass object in the `ProbDistr` hierarchy, and the creation of the subclass object makes use of a `ProbDistr_prm` object.

```
#include <ProbDistr2_prm.h>
#include <RandGen.h>
#include <empirstat.h>
#include <MenuSystem.h>

int main (int argc, const char* argv[])
{
  initDiffpack (argc, argv);
  RandGen                 generator("UnifRandGenKnuth",0.12345);
  Handle(ProbDistr_prm) R_prm;
  Handle(ProbDistr)     R_d;

  R_prm.rebind(ProbDistr2_prm::construct());
  R_prm->scan ("Weibull(0.2,1.0)");
  R_d.rebind (R_prm->create());

  int nsimulations;
  initFromCommandLineArg ("-n", nsimulations, 40);
  real R;

  for (int i=1; i<=nsimulations; i++) {
    R = generator.generate(*R_d);
    s_o << aform("%5d R=%g\n",i,R);  // write realization
  }
}
```

All the relevant files for this demo are found in `src/stoch/Weibull`.

# References

1. A. Bellin, S. Salandin, and A. Rinaldo. Simulation of dispersion in heterogeneous porous formations: Statistics, first-order theories, convergence of computations. *Water Resour. Res.*, 28:2211–2227, 1992.

2. P. P. Benham, R. J. Crawford, and C. G. Armstrong. *Mechanics of Engineering Materials*. Longman, 2nd edition, 1996.
3. G. K. Bhattacharyya and R. A. Johnson. *Statistical Concepts and Methods*. Wiley, 1977.
4. G. Dagan. Stochastic modeling of groundwater flow by unconditional and conditional probabilities. i: Conditional simulation and the direct problem. *Water Resour. Res.*, 18:813–833, 1982.
5. G. Dagan. Solute transport in heterogeneous porous formations. *J. Fluid Mech.*, 145:151–177, 1984.
6. G. Dagan. Theory of solute transport by groundwater. *Ann. Rev. Fluid Mech.*, 19:183–215, 1987.
7. G. Deodatis and M. Shinozuka. The weighted integral method. ii: Response variability and reliability. *ASCE J. Engrg. Mech.*, 117:1865–1877, 1991.
8. M. D. Dettinger and J. L. Wilson. First order analysis of uncertainty in numerical models of groundwater flow, Part 1. Mathematical development. *Water Resour. Res.*, 17:149–161, 1981.
9. R. G. Ghanem and P. D. Spanos. *Finite Elements: A Spectral Approach*. Springer-Verlag, 1991.
10. S. Gran. *A Course in Ocean Engineering*. Elsevier, 1992.
11. R. C. Hibbeler. *Mechanics of Materials*. Prentice Hall, 4th edition, 1999.
12. H. Holden and N. H. Risebro. Stochastic properties of the scalar buckley-leverett equation. *SIAM J. Appl. Math.*, 51:1472–1488, 1991.
13. E. H. Isaaks and R. M. Srivastava. *An Introduction to Applied Geostatistics*. Oxford University Press, 1989.
14. H. P. Langtangen. Tips and frequently asked questions about Diffpack. Numerical Objects Report Series, Numerical Objects A.S., 2002. *http://www.diffpack.com/products/faq/faq_main.html*.
15. H. P. Langtangen. *Computational Partial Differential Equations – Numerical Methods and Diffpack Programming*. Texts in Computational Science and Engineering. Springer, 2nd edition, 2003.
16. W. L. Liu, A. Mani, and T. Belytschko. Finite element methods in probabilistic mechanics. *Probab. Engrg. Mech.*, 2:201–213, 1987.
17. H. O. Madsen, S. Krenk, and N. C. Lind. *Methods of Structural Safety*. Prentice-Hall, 1986.
18. D. E. Newland. *An Introduction to Random Vibrations and Spectral Analysis*. Longman, 1984.
19. H. Omre, K. Sølna, and H. Tjelmeland. Simulation of random functions on large lattices. In *Proceedings Geostatistics Tróia '92*, pages 179–199, Tróia, Italy, 1992.
20. H. Osnes and H. P. Langtangen. An improved probabilistic finite element method for stochastic groundwater flow. *Advances in Water Resources*, 22:185–195, 1998.
21. H. Osnes and H. P. Langtangen. A study of some finite difference schemes for a unidirectional stochastic transport equation. *SIAM J. Sci. Comput.*, 19(3):799–812, 1998.
22. S. M. Ross. *Introduction to Probability Models*. Academic Press, 6th edition, 1997.
23. L. Smith and R. A. Freeze. Stochastic analysis of steady state groundwater flow in a bounded domain. 2. Two-dimensional simulations. *Water Resour. Res.*, 15(6):1543–1559, 1979.

24. A. F. B. Tompson, R. Ababou, and L. W. Gelhar. Implementation of the three-dimensional turning bands random field generator. *Water Resour. Res.*, 25:2227–2243, 1989.

25. E. Vanmarcke. *Random Fields*. MIT Press, 1983.

# Chapter 8

# Using Diffpack from Python Scripts

H. P. Langtangen[1,2] and K.-A. Mardal[1,2]

[1] Simula Research Laboratory
[2] Department of Informatics, University of Oslo

**Abstract.** Diffpack is a comprehensive software library for solving partial differential equations. For the experienced user and C++ programmer, Diffpack offers lots of functionality, which simplify the development of new simulators. However, the nature of Diffpack/C++ programming is more detailed and cumbersome than programming in environments like Matlab and Maple. Coupling of Diffpack with other packages at the C++ level is possible, but requires quite some work and knowledge. Operating Diffpack through high-level Python scripts may meet these shortcomings. The Python language is very powerful, yet easy and very convenient to use, and provides a syntax and working style close to that of Matlab and Maple.

Coupling Diffpack/C++ and Python is a non-trivial task. However, there is a tool, SWIG, which provides the possibility to automate the coupling of C/C++ and Python such that Python scripts can call C/C++ functions and operate directly on the C/C++ data structures. Applying SWIG to a large unit of software such as Diffpack faces many technical challenges. The present chapter introduces new tools in Diffpack that enable application of SWIG to generate Python interfaces to Diffpack simulators in an almost automatic way. We present some step-by-step examples on equipping Diffpack simulators with Python interfaces. We also show more complete applications where a controlling Python script runs a Diffpack simulator interactively, modifies its data, and visualizes the data with the aid of the Vtk package. This example demonstrates how Python scripts may act as a glue between Diffpack and other software packages, allowing data to be efficiently sent back and forth as pointers. With the information in this chapter, the reader should be able to steer any Diffpack simulator from a Python script.

## 8.1 Introduction

Development of Diffpack simulators implies programming in C++. With all the high-level, ready-made classes in the Diffpack libraries and object-oriented programming techniques, the code in a Diffpack simulator is shorter, less detailed, and appear at a higher abstraction level than traditional numerical simulators written in C or Fortran. C++ programming and the Diffpack libraries constitute powerful tools for numerical software development. However, any programmable environment requires a user to *program* to accomplish a task, and programming in C++ is time consuming and error-prone, although the authors claim it to be more effective than traditional Fortran and C coding. This chapter outlines techniques for lifting Diffpack programming up to a higher and more convenient level.

### 8.1.1  The Advantages of High-Level Languages

Flexible *high-level*[1] computer languages and environments have gained significant popularity in recent years. Perl, Python, R/S-Plus, Matlab, Maple, Mathematica, and Gauss are some examples. Most of these tools are interactive, where the user issues simple, high-level commands to accomplish a task. The commands may have lots of parameters, but most of the parameters have sensible default values. Hence, beginners are not exposed to advanced concepts. A task may typically be accomplished in 10+ commands, which may be placed in a file, often referred to as a *script*, and equipped with loops, variables, etc., if desired.

The success of the mentioned languages and environments is very much due to the fact that the work appears to be at a higher level than traditional programming in compiled languages (i.e., Fortran, C, and C++). Variables are not declared with a type, hence an argument to a function can be used for many different types. Moreover, nested heterogeneous data structures are constructed on the fly, without the comprehensive object-oriented set-up in Java and C++, or explicit pointer management as in C.

All details about memory management are hidden in high-level languages. Data structures are created and extended when needed, and they can be passed efficiently in and out of functions by references. The run-time system will automatically deallocate structures that are no longer in use.

Functions may have keyword arguments for easy use and flexible setting of default values. As an example, a function `plot` can be called with three keyword arguments:

```
plot(func=g, xaxis=[0,10], xlabel='time')
```

Each argument has a name, or keyword, and a value (e.g., the first argument has the name `func` and the value `g`). The declaration of this `plot` function may have 30 keyword arguments, each with a default value. In the exemplified `plot` call we just set three of the arguments, relying on default values for the rest. This feature is heavily used in environments like Matlab and Maple and contributes to user-friendliness; novice users set a minimum of arguments, whereas experienced users can provide lots of arguments to achieve a higher degree of control.

High-level languages do not require time-consuming compilation and linking of programs, thus making the correct-and-test cycle during debugging very effective. Extensive safety checks are performed at run time, and in the case of Python, very informative error messages are issued. We also mention

---

[1] These languages are also referred to as *very high-level* languages; different authors use different scales.

that a high-level language like Python can be run interactively, command by command, a feature that has made Matlab popular. In a nutshell, scientists and engineers become more productive with these types of languages and environments.

Traditional numerical software, written in Fortran, C, or C++, can be equipped with scripting interfaces in high-level languages such that the software can be operated from the high-level language, with a syntax and working habit in the style of, e.g., Matlab. We remark that scripting means different things to different people. Here we use the term for a programming style, where the tasks to be done are expressed in a few very powerful statements in a high-level language.

### 8.1.2    Python Scripting and Diffpack

The purpose of the present chapter is to give some ideas for how you can equip a Diffpack simulator with a high-level scripting interface. We have chosen to use the Python language for scripting, because Python has a very clean syntax and because Python has a lot of convenient tools for numerical computing. However, there is no difficulty in using Perl, Tcl, Ocaml, Ruby, Java, Mzscheme, or Guile instead of Python; the tool we use to generate the scripting interfaces supports all these languages.

The simplest way of creating a scripting interface to Diffpack is to run the simulator as a stand-alone program and feed data into Diffpack through files or command-line arguments. A script is then used as a layer between Diffpack and the user. This script can offer a simpler-to-use command-line interface, e.g., with more high-level and tailored commands than what the Diffpack simulator offers through its menu system. The script can also provide a tailored interface through a graphical user interface (GUI). Maximum flexibility is gained by editing the script code on the fly, or issuing the statements in the script interactively. Numerous such scripts were developed as a part of the jump-start applications for Diffpack. Many of them are described in [6]. One example on a scripting interface to Diffpack can be found in src/app/SqueezeFilm/coupling/gui.py (see Chapter 7.1 in [6]). This is a GUI tailored for studying a fluid-structure interaction problem. The GUI allows convenient setting of physical parameters in the problem and offers tailored visualization of key results.

Creating tailored, scripting interfaces to stand-alone Diffpack simulators and visualization programs is easy if the basics of a language like Python is known. The book [4] explains the tools and programming techniques. However, more flexible interfaces can be developed if we are not limited by running the Diffpack simulator as a stand-alone program. That is, we would like to operate the data structures and functions in the simulator directly from Python. This is also possible, but technically more complicated and constitute the topic of the present chapter.

Debugging is one example where direct calls to Diffpack and operations on Diffpack data structures are beneficial. For example, in the search for errors we may investigate the application while it runs instead of post-processing data written to file. A Python script may act as a (tailored) debugger, allowing us to dump quantities when the need arises, call functions, examine new data, etc. Modifying the script is done on the fly; there is no time-consuming compilation and linking process.

Python has various modules like Numeric [1,9], Scientific [12], SciPy [13], Vtk [16], MayaVi [8], PyMat [7,10], and Gnuplot [11], which may make our interface to Diffpack simulators powerful:

- Numeric and other modules from the Numerical Python package (often called NumPy) extend Python with array data structures and efficient array and linear algebra operations.

- Scientific and SciPy offer various numerical functionality, such as non-linear least squares, statistics, elementary functions, etc., in Python programs.

- Vtk is a visualization system for 2D and 3D scalar and vector field, and MayaVi is a Python layer on top of Vtk with high-level scripting and graphical interfaces.

- With PyMat, Python arrays can be sent back and forth between the script and Matlab, enabling use of Matlab for computations or graphics.

- Gnuplot is a plotting package, mainly for curves and 2D data points.

All these software packages must, of course, be installed on the computer system. Section 8.6 gives some information on how to do this. The present chapter covers the technology of coupling the mentioned packages to Diffpack such that data can be transferred as desired between software components, via Python. Safe transfer normally means copying data, at least for the novice, but if you know what you are doing, very efficient large-scale data transfer, via pointers only, can be performed. This opens up the possibility to create sophisticated, easy-to-use, and efficient problem solving environments.

### 8.1.3  Example: Running a Diffpack Simulator from Python

Let us show a simple example on a useful Python script for studying convergence of a heat equation solver (the standard Heat1 solver [6]). The scripts performs a multiple loop[2] *outside* Diffpack; we want to reduce the grid spacing *and* the time step simultaneously in a series of experiments, and this is not possible with native Diffpack multiple loops (all combinations of grid and time spacings would then be run). The script may look like this:

---

[2] A multiple loop in Diffpack runs all combinations of multiple values of some input parameters.

```
from Heat1 import *    # load scripting interface to Heat1
import math
menu = MenuSystem()
menu.init("Python Interface", "cool huh")
heat = Heat1()
heat.define(menu)

L2_errors = []
dt = 0.1
n  = 8
for i in range(5):    # 5 refinements in space and time
  n = n*2
  grid_str = "P=PreproBox | d=2 [0,1]x[0,1] \
    | d=2 e=ElmB4n2D div=[%d,%d] grading=[1,1]" % (n,n)
  menu.set("gridfile", grid_str)
  dt = dt/4.0;   dx = 1.0/n
  tip_str = "dt=%g [0,3]" % dt
  menu.set("time parameters", tip_str)
  heat.scan()
  heat.solveProblem()
  heat.resultReport()
  L2_errors.append((dx, dt), heat.L2_error)
```

With a couple of extra statements we can grab the solution field from Diff-pack and transfer it to Vtk or Matlab for visualization, or plot the list of discretization errors as a curve in Gnuplot. As will be shown in Section 8.5, we can also easily introduce a least-squares fit of an error estimate in the Python script.

As we have outlined, a typical Python script can be used for preparing Diffpack input data, performing several simulations, visualizing data, and computing convergence rates. The high-level statements in such scripts actually offer users a new interface to Diffpack simulation, which tends to be simpler to work with than writing the corresponding administering programs directly in C++. In a way, the scripting interface gives a high-level access to Diffpack with many of Matlab's attractive features, such as a clean syntax, powerful commands, interactive computing, and a gentle learning curve.

We have written this chapter for a reader who is familiar with Python and Diffpack programming. One possible source for learning Python, especially if you are a computational scientist, is [4], from which Chapters 2 and 3 is a prerequisite for the material to follow. It will be an advantage to have some basic knowledge of Python-C/C++ integration, e.g. from [4, Ch. 4].

The tools described in this chapter come with the Diffpack 4.1 distribution in the $NOR/dp/etc/swig directory.

## 8.2   Developing Python Interfaces to C/C++ Functions

Python has a well-defined C API [15] for gluing Python and code written in C, C++, or Fortran[3]. However, one has to write or generate gluing code.

---

[3] In the following we speak about C code as a short term for C, C++, and Fortran code.

In fact, for each C function to be called from Python, one has to write a function that converts Python objects into the arguments needed by the C function. These conversion functions are usually referred to as *wrapper functions*. SWIG (Simplified Wrapper and Interface Generator) [14] is a tool for generating these wrapper functions automatically. We will in this section briefly describe why we need wrapper functions, what they look like, and what SWIG can do.

### 8.2.1   Wrapper Functions

In Python everything is an object, and the object may contain any type of data. Hence, the statement `d=3` makes a Python object `d` containing a C integer with the value 3. However, at any time we may change `d`, e.g., `d='some text'`. Now `d` contains a string instead. When sending Python objects to a C function we therefore need to check the data type in the object before we can make use of it in the C function. If the argument is different from what the C function expects, an error message is printed out[4] and the C function is never called (since this would probably result in a segmentation fault). These actions are accomplished by a *wrapper* function. A simple example will explain what wrapper functions are about. This information is rather technical and not required for further reading of the chapter. Hence, the impatient reader can therefore safely jump to Section 8.3 after having gotten the short story: Wrapper functions are tedious and boring to write, but SWIG can generate them automatically, more or less directly from the C code we want to interface.

Our first example on wrapper functions is pretty simple. We want to call a C function for drawing normally distributed random numbers,

```
extern double gauss (double mean, double stdev);
```

The call is to be performed from Python, i.e., we want to write something like

```
#!/usr/bin/env python
from draw import gauss
m = 1.0; s = 0.2
for i in range(1000):
    r = gauss(m, s)
    print "r=", r
```

To this end, the Python code must call a wrapper function, written in C, where the contents of the `m` and `s` objects are analyzed and the double precision floating-point numbers are extracted and stored in straight C `double` variables. Then, the wrapper function can call our `gauss` function. Since the

---

[4] Technically, an exception (of type `TypeError` in this example) is raised. Unless the exception is handled in the calling code, an error message is written to standard output and Python aborts the execution.

gauss function returns a `double`, we need to convert this `double` to a Python object that can be returned to the calling Python code and stored in the object `r`. A suitable wrapper function may in this case look like this:

```
static PyObject *_wrap_gauss(PyObject *self, PyObject *args) {
    PyObject *resultobj;
    /* temporary C variables needed by gauss  */
    double arg1;  double arg2;  double result;

    /*  Convert the args to suitable C data: */
    if(!PyArg_ParseTuple(args, (char *)"dd:gauss", &arg1, &arg2))
      return NULL;

    /* Call the C function gauss:  */
    result = (double) gauss(arg1, arg2);

    /* make a Python object from the double variable result:  */
    resultobj = PyFloat_FromDouble(result);
    return resultobj;
}
```

Every wrapper function has a similar look. In a nutshell, there is a lot of code around the principal statement where we call our C function (here `gauss`). Let us now go through this wrapper function line by line.

- `static PyObject *_wrap_gauss(PyObject *self, PyObject *args)`
  A wrapper function takes two or three input arguments of type `PyObject*`. The `args` object contains the function arguments, whereas `self` contains the object instance (`obj.somefunc(args)` may be a typical general call). Additionally, a third object with a dictionary of keyword arguments is optional. All objects in Python are "derived" from `PyObject`.

- `double arg1; double arg2; double result;`
  These are temporal C data needed in the C function call.

- `if(!PyArg_ParseTuple(args,(char *)"dd:gauss",&arg1,&arg2))`
  The Python objects are converted to C data by `PyArg_ParseTuple` or a similar function. The `PyArg_ParseTuple` function takes in one Python object supposed to be a tuple holding all arguments (`args`) transferred to the function, a format string, and a list of C variables. If the tuple contains the data indicated by the format string, the data will be stored in the C variables, and `PyArg_ParseTuple` returns a true value. If not, it will return false and set an appropriate exception. The convention is then that the wrapper function returns a `NULL` pointer.

- `result = (double) gauss(arg1, arg2);`
  The C function is called.

- `resultobj = PyFloat_FromDouble(result);`
  The output `result` is converted to a Python object.

- `return resultobj;`
  The Python object is returned to the calling script.

### 8.2.2    Creating and Using a Module

The wrapper function must be compiled, here with a C compiler. We must also compile the file with the `gauss` function in C. The object code of the `gauss` function (and all functions called from `gauss`) must then be linked with the wrapper code to form a *shared library* file (recognized by the `.so` filename extension on Unix machines). The shared library file forms the core of a Python module, which one can import to call up the C code from Python.

The following compilation commands illustrate how we can compile our wrapper code for the `draw` module and link it to form a shared library file `draw.so` on a Unix system:

```
gcc -c draw.c -o draw.o
gcc -c -fpic  draw_wrap.c  -I. -I.. \
   -DHAVE_CONFIG_H -I$SOFTWARE/include/python2.2  \
   -I$SOFTWARE/lib/python2.2/config

gcc -shared  draw_wrap.o  draw.o -o draw.so
```

The environment variable SOFTWARE appears in the compilation command, because we have installed all the software (binaries, headerfiles, and libraries) relative to the directory whose name is reflected by this variable. More about this can be found in Section 8.6 where the installation is described. The shared library module `draw.so` can be loaded into Python using the `import` statement. From Python, it is impossible to distinguish between a pure Python module or a shared library module consisting of C code. If the name of the shared library module is `draw`, you write

```
from draw import gauss
```

in the Python code and call `gauss` as if it were written in Python:

```
r = gauss(1.0, 0.1)
```

The code in this example is in the `src/pydp/draw` directory.

### 8.2.3    How SWIG Simplifies Writing Wrapper Code

As we have tried to demonstrate, even in a very simple example, the writing of wrapper functions requires knowledge of details of the C interface to Python. In other words, we need to know how Python objects are manipulated in C code. This is well explained in Beazley's book [2], in the official Python electronic documentation, and in Lutz [6]. Unfortunately, writing wrapper code is a big job; each C function you want to call from a Python script must have an associated wrapper function. Such manual work is boring and error-prone. Luckily, tools have been developed to automate most of this manual work. We shall here, contrary to what is normally done in textbooks covering Python-C interfacing, focus on using SWIG for automatic generation of wrapper code.

Basically, SWIG needs a list of the C or C++ functions we want to call from Python. This list appears in an *interface file*. In the current example, such an interface file could look as follows:

```
/* file: draw.i */
%module draw
%{
#include "draw.h"
%}
/* list the function we want to call */
double gauss(double mean, double stdev);
```

The interface file consists of C syntax and SWIG-specific syntax (keywords preceded by the percentage sign). Here we have defined a module `draw`, the interface needs access to the C header file `draw.h` to compile, and the final line lists the function we want to call from Python. Running SWIG with this interface file as a command-line argument automatically generates a wrapper function, similar to the previously shown wrapper function, `_wrap_gauss`. The remaining tasks are then to compile the wrapper code, the C code containing `gauss` and its associated functions, and link the pieces together in a shared library module.

### 8.2.4   Writing Python Interfaces to Diffpack Relies on SWIG

Calling your Diffpack simulator from Python means writing a large number of quite complicated wrapper functions. Fortunately, SWIG is a tool that automates the generation of wrapper functions, also in the context of a comprehensive library such as Diffpack. However, the application of SWIG to Diffpack code requires some rules and tools to obtain a semi-automatic procedure. With the information in the present chapter, it should be possible, with minor extra effort and knowledge, to generate Python interfaces to all of your Diffpack simulators.

Although we explain the most basic issues of SWIG and its usage, the reader should have quick access to the well-written SWIG manual [3]. As soon as one starts developing serious Python interfaces to Diffpack, we highly recommend reading the SWIG manual.

We should mention that there are other alternatives to SWIG for generating wrapper code. The SWIG homepage [14] contains links to alternative projects. To a limited extent we have tested other tools. Our experience so far is that SWIG seems to be the most mature and well-documented tool for generating wrapper code for C++ packages.

## 8.3   Compiling and Linking Wrapper Code with Diffpack

The purpose of our first coupling of Diffpack and Python is to make a module out of a Diffpack simulator. However, the module does not contain any data or functions; the goal is just to successfully compile and link the module. Simple C libraries are easily handled by SWIG [14] , whereas complicated C++ codes, such as Diffpack, usually require some extra "hacks" to define the interface and successfully compile and link a Python-loadable module. That is why we use a section in the this chapter on the technicalities of compiling and linking wrapper code with Diffpack.

Our example in the present section concerns creating an interface to the Heat1D finite difference heat equation solver. This simulator comes with the Diffpack distribution ($NOR/doc/Book/src/fdm/Heat1D). The example utilizing SWIG is located in src/pydp/Heat1D in the directory tree associated with the present book.

### 8.3.1   Makefiles

Compilation and linking of Diffpack rely on makefiles. SWIG also offers make-files for compiling and linking wrapper code and the C/C++ library we want to interface. Hence, we need to merge the Diffpack and SWIG makefiles.

We keep the Diffpack application and the SWIG interface in separate di-rectories, thereby ensuring that the Diffpack simulator can run without the Python interface. It is natural to let the SWIG makefile handle the compi-lation of the interface code. However, the SWIG makefile needs information about where to find Diffpack header files and libraries. This information is contained in the Diffpack makefile.

We have developed a script called MkDpSWIGMakefile, which creates a Makefile suitable for compiling the SWIG-generated wrapper code and link-ing this with the Diffpack application to form a shared library module. You can create a Makefile for the SWIG interface by running, e.g.,

```
MkDpSWIGMakefile  MODULE=Heat1D DPDIR=.. MODE=opt
```

The first argument is the name of the module, the second is the directory where the Diffpack application to link with is located, and the third is the usual optimization Diffpack flag. We remark that the MkDpSWIGMakefile script comes with Diffpack.

The MkDpSWIGMakefile script first moves to the Diffpack application direc-tory and runs

```
Make MODE=opt DPDIR=.. link2dp > DpMakefile.defs
```

The link2dp target in Diffpack's makefiles writes out the most important def-initions needed for compiling and linking Diffpack. This output from running Make is redirected to a file DpMakefile.defs in the SWIG interface directory. Here is a sketch of a typical DpMakefile.defs file:

```
# DP_DIR          : where the Diffpack application is located
DP_DIR            := ..
```

```
# DP_INCLUDEDIRS   : where the Diffpack include files are
DP_INCLUDEDIRS     := -I. -I/work/NO/bt/include ...

# DP_OBJS          : where the Diffpack object files to be linked
#                    with this application are
DP_OBJS            := ../.Heat1D.o ../.main.o

# DP_CXXFLAGS       : C++ compiler flags for Diffpack compilation
DP_CXXFLAGS        := -Dgpp_Cplusplus -O ...

# DP_LIBS           : Diffpack libraries to link with
DP_LIBS            := -lbt2_gui -larr1 ...

# DP_SYSLIBS        : system libraries needed by Diffpack
DP_SYSLIBS         := -lf2c -lm -lXt -lX11

# DP_LDFLAGS        : flags for linking, including -L paths
DP_LDFLAGS         := -L. -L/usr/X11/lib -L/work/NO/bt/lib/Linux/opt

# DP_CXX            : C++ compiler for Diffpack
DP_CXX             := g++

# DP_LD             : program for linking Diffpack applications
DP_LD              := g++
```

SWIG, Python, and Diffpack must be compiled with the same compiler, which means that DP_CXX and DP_LDD in Diffpack need to be the same as CXX and CXXSHARED in SWIG.

The MkDpSWIGMakefile script generates a file Makefile for compiling the wrapper code and linking it with Python and the Diffpack application. The file DpMakefile.defs is included in this Makefile. Here is an example on how the SWIG interface Makefile looks like:

```
# include definitions of Diffpack Makefile variables:
include DpMakefile.defs

# SWIG-specific make variables:
LIBS =
LDPATH =


SWIG       = swig
SWIGOPT    = -c -shadow -I. -I.. \
             -I$(NOR)/dp/etc/swig $(DP_INCLUDEDIRS)
INTERFACE  = Heat1D.i
TARGET     = Heat1D

# use the template makefile that was generated by SWIG (v1.3):
SWIGMAKEFILE = $(SWIGSRC)/Examples/Makefile

# Build a Python dynamic module
# (note that CXXUF=-D... can be used to turn on macros)
python::
        checkDpMode.py MODE=$(MODE) DPDIR=$(DP_DIR)
        $(MAKE) -f '$(SWIGMAKEFILE)' INTERFACE='$(INTERFACE)' \
        SWIG='$(SWIG)' SWIGOPT='$(SWIGOPT) $(CXXUF)' \
```

```
        INCLUDES='$(INCLUDES) -I.. $(DP_INCLUDEDIRS)' \
        LIBS='$(DP_OBJS) $(DP_LIBS) $(DP_SYSLIBS) $(LIBS) \
            $(DP_LDFLAGS) -lswigpy -L$(SOFTWARE)/lib ' \
        CFLAGS='$(DP_CXXFLAGS) $(CXXUF)' TARGET='$(TARGET)' \
        python_cpp

clean::
        rm -f *_wrap* *.so *.o $(OBJ_FILES)  *~

# dump the preprocessor command to be used with Diffpack files:
dpprepro::
        @echo $(DP_CXX) $(DP_CXXFLAGS) $(CXXUF) -I.. -E -DSWIG
```

Notice that an environment variable SWIGSRC is used to locate the root directory of the SWIG source code; you must set this variable (typically in a start-up file like .bashrc on Unix).

The Makefile can work with Diffpack applications that are distributed in many directories. If you just specify the "main" Diffpack directory, i.e., where the main.cpp file for the Diffpack application is located, running Diffpack's link2dp makefile target will detect all files belonging to the application and generate the correct DpMakefile.defs file.

*Remark.* The MkDpSWIGMakefile cannot make portable makefiles. That is, DpMakefile.defs and Makefile must be regenerated when you move your SWIG interface to a new platform or file system. Therefore, MkDpSWIGMakefile writes a one-line script configure.sh containing the command you used to run MkDpSWIGMakefile. On a new platform, all you have to do is execute

    ./configure.sh

and MkDpSWIGMakefile is rerun with the correct parameters, i.e., with the correct name of the module and the correct location of the Diffpack code.

### 8.3.2  Summary of Creating the Interface

Making a Python interface to Heat1D consists in

1. creating a subdirectory swig of Heat1D,

2. running MkDpSWIGMakefile in the swig subdirectory,

3. defining the Python interface to Heat1D in a SWIG interface file Heat1D.i,

4. running SWIG on the interface file to generate wrapper code, followed by compiling and linking a shared library module with the name _Heat1D.so.

The last three tasks are usually done by the make program. The name of the module file is in this case _Heat1D.so. SWIG generates a Python script Heat1D.py[5], which imports _Heat1D.so. The file Heat1D.py is to be imported in a Python script calling up Diffpack. That is, we load the Diffpack simulator into Python by writing

---

[5] That is, SWIG with the -shadow option makes Heat1D.py.

```
import Heat1D
```

This statement causes Python to search for the module file `Heat1D.py` in the directories contained in the `sys.path` list and load the file if it is found. You can add directories to this list by specifying them in the `PYTHONPATH` environment variable. By default, the current directory is a part of `sys.path` so Python will find the `Heat1D` module, available as the file `Heat1D.py`. If you want a Python script located in an arbitrary directory to load this module, you need to move the `_Heat1D.so` and `Heat1D.py` files to one of the directories in the `sys.path` list or, alternatively, add the directory where the files reside to the `PYTHONPATH` environment variable. In case you get an `ImportError` exception when importing `Heat1D`, make sure that one of the directories in `sys.path` really contains `_Heat1D.so` and `Heat1D.py`. Printing `sys.path` is easily done with

```
python -c 'import sys; print sys.path'
```

### 8.3.3   A Trivial Test Module

Our first module actually contains neither data nor functions. The purpose is just to test the makefile set up, compilation and linking, and loading of the module. We create a subdirectory `swig-empty` and build the test module there. The definition of an empty module can be accomplished by this SWIG interface file:

```
%module Heat1D
%{
#include <Heat1D.h>
%}
```

The only thing going on in this file is the definition of the module name (`Heat1D`) and inclusion of the simulator's header file. This header file will be compiled by SWIG, hence we can check compilation of some Diffpack code in a SWIG context, but no wrapper code is generated. This is because we do not define any functions or data in the interface.

The next step is to compile the Diffpack application. This is done in the `Heat1D` directory, e.g.,

```
cd ..
Make MODE=opt
```

Thereafter, we run SWIG in the `swig-empty` directory, compile the wrapper code, and link it with `Heat1D.o` and the rest of the Diffpack libraries. All these tasks are performed in the `Makefile` generated by the `MkDpSWIGMakefile` script. We simply write

```
cd swig-empty
./configure.sh    # or run the MkDpSWIGMakefile command
make
```

A successful compilation and linking results in a file _Heat1D.so. We can check what the module contain as follows:

```
python -c 'import Heat1D; print dir(Heat1D)'
```

The output becomes

```
['_Heat1D', '__builtins__', '__doc__', '__file__', '__name__',
'_newclass', '_object', '_swig_getattr', '_swig_setattr', 'types']
```

That is, there are no functions or data from Diffpack, only some standard SWIG/Python variables and functions.

### 8.3.4    Common Errors in the Linking Process

A very common mistake when building shared library modules is to forget to link some pieces, which results in undefined symbols. For example, if we leave out the Diffpack libraries in the definition of DP_LIBS (see the file DpMakefile.defs on page 331), the compilation and linking appears successful, but the import statement raises an ImportError exception:

```
ImportError: ./_Heat1D.so: undefined symbol: global_menu
```

To fix such errors, you need to locate the library where the code related to global_menu is defined, and add it to the LIBS= specification in the Makefile. Utilities like nm, ldd, and readelf can be used to locate undefined symbols. For example,

```
find $NOR -name '*.a' -exec nm -o {} \; | grep global_menu
```

reveals that the symbol is defined in libbt2.a (provided you know how to interpret the output of nm).

It may be hard to track down the definition of missing symbols. For instance, some symbols may only be defined when certain compilation flags are defined. During debugging it is usual to use Make MODE=nopt to make Diffpack carry out internal consistency and safety checks. If the Diffpack application is compiled with MODE=opt while the interface is compiled with MODE=nopt we do not experience any problems during the compilation and linking. However, importing the modul in Python we get

```
>>> import Heat1D
>>> Traceback (most recent call last):
>>>    File "<stdin>", line 1, in ?
>>>      File "Heat1D.py", line 2, in ?
>>>         import Heat1D
>>>         ImportError: ./_Heat1D.so:
>>>            undefined symbol: hasClassType__C8HandleId
```

In this case we have used all the libraries we should, the error is that we have used incompatible compilation commands and therefore linked with the wrong version of the libraries. Only the `nopt` version of the library `libarr1.a` defines the `hasClassType__C8HandleId` symbol. (The name of this symbol reveals how member functions in C++ classes get mangled when viewed from C; in this example the symbol represents the `HandleId::hasClassType` function, which is defined only in the `nopt` compilation mode in Diffpack.)

### 8.3.5  Problems with Interfacing a Simulator Class

In the previous section we tested that Python, SWIG, and Diffpack worked properly together. That is, the `Makefile` is constructed such that all required include files and libraries are found and used. We always recommend you to perform such a step before trying to create a useful Python interface to your Diffpack simulator.

The next step is to try to build a useful Python interface to the `Heat1D` simulator. We will

1. make the functions defined by `Heat1D` accessible from Python,

2. make all variables `public` and thereby accessible from Python,

3. comment away or modify some difficult stuff,

4. expand macros such that SWIG is able to wrap them,

5. extend the simulator with some functionality useful in Python.

### 8.3.6  Python Interface, Version 1

We start by creating a new directory, `swig-v0`, which is a subdirectory of `Heat1D`, where the Diffpack application is located. This first version is not intended to work, but to illustrate the typical way of working with SWIG. We describe solutions to the problems we face as we go along; these fixes are implemented in `swig-v1`. The starting point is simply to run SWIG on the header file to see what SWIG do not like. This is done by making the following SWIG interface file `Heat1D.i`:

```
%module Heat1D
%{
/* necessary header files to compile the wrapper code: */
#include <Heat1D.h>
%}

/* make a Python interface to everything in ../Heat1D.h: */
%include Heat1D.h
```

We run `MkDpSWIGMakefile` to build a `Makefile`. Trying to compile the code by `make` results in numerous error messages. SWIG does not eat the Diffpack code right away. The problems are related to the Diffpack macros,

```
Handle(GridLattice) grid;   // 1D grid
Handle(TimePrm)      tip;   // time step etc.
MatTri(real)         A;     // the coefficient matrix
ArrayGen(real)       b;     // the right-hand side
```

SWIG works perfectly when these macro constructions in `Heat1D.h` are removed out. There are different ways of doing this, which we explain in the forthcoming paragraphs.

*Using Preprocessor Directives to Leave out Code.* The simplest solution is to enclose the problematic statements inside `#ifndef SWIG` directives in the `Heat1D.h` file:

```
class Heat1D
{
  ...
#ifndef SWIG
  Handle(GridLattice)  grid;
  ...
  ArrayGen(real)       b;
#endif
  ...
};
```

When SWIG reads the file, `SWIG` is defined, and everything inside the `#ifndef SWIG` directive is left out. During compilation of the simulatior, `SWIG` is not defined, and the code is included such that the simulator object files become complete. We will come back to this approach.

*Manual Declaration of the Class Interface.* As an alternative to `#ifndef SWIG` we can omit

```
%include Heat1D.h
```

in the SWIG interface file `Heat1D.i` and instead write up (i.e., copy and paste) the desired class definition. In this definition we only need to include the data members and the functions we want to use in the interface. Problematic lines can simply be left out. In the present case we could include a class definition like the following one:

```
class Heat1D
{
public:
  real               theta;    // time discretization parameter
  int                n,q;      // grid has n=q*q points
  real               h;        // cell size: h=1/(q-1)^2
  int                bc_tp;    // boundary condition at x=1
  int                ic_tp;    // initial condition
  int                problem_tp; // 1: test with analyt. sol.

public:
  Heat1D () {}
```

```
virtual void scan    ();
virtual void solveProblem ();
virtual void resultReport ();
};
```

This approach is taken in `Heat1D/swig-v1`.

*Problems with Initialization of Diffpack Simulators.* Given that we have made the interface and the `Makefile` as described above we should be able to build a Python interface. The contents of the resulting module can be checked as follows:

```
>>> import Heat1D
>>> dir(Heat1D)
['Heat1D', 'Heat1DPtr', 'Heat1Dc', '__builtins__', '__doc__',
'__file__', '__name__']
>>> dir(Heat1D.Heat1D)   # contents of the simulator class
['__del__', '__doc__', '__getattr__', '__getmethods__',
'__init__', '__module__', '__repr__', '__setattr__',
'__setmethods__', 'resultReport', 'scan', 'solveProblem']
```

Unfortunately, we cannot yet run the simulator successfully:

```
>>> h = Heat1D.Heat1D()
>>> h.scan()

>>>>> Handling a fatal exception:  getCommandLineOption reports:
You must call initDiffpack at the beginning of main!
```

The problem is that all Diffpack programs must start with an `initDiffpack` call. Initialization statements of this type can be placed in the SWIG interface file as an `%init` block:

```
%init%{
  const char* p[] = { "Calling Diffpack from Python" };
  initDiffpack(1, p);
%}
```

Note that the `initDiffpack(1,p)` call is a quick hack as we do not transfer any command-line arguments to `initDiffpack`.

### 8.3.7   Version 2; Setting Input Parameters

The `simulator.scan()` functions examine the command-line options, but in our case, the SWIG interface file generates an `initDiffpack` call where any user-provided command-line arguments are ignored. Hence, we cannot use the command line as Diffpack users are used to in our Python interface. Instead, it is tempting to set, e.g., the grid resolution directly in the interface:

```
>>> from Heat1D import *
>>> simulator = Heat1D()
>>> simulator.scan()
>>> simulator.n = 120
>>> simulator.solveProblem()
>>> Segmentation fault
```

To track down this segmentation fault, we can run the Python interpreter under a standard debugger. This is a useful debugging technique for Python-Diffpack interface problems:

```
unix> gdb python
(gdb) run
>>>
>>> simulator = Heat1D.Heat1D()
>>> simulator.scan()
Running problem 1 with b.c. 1, 11 grid points and theta=0.5
>>> simulator.n = 120
>>> simulator.solveProblem()
Program received signal SIGSEGV, Segmentation fault.
[Switching to Thread 1024 (LWP 23011)]
0x405c6d51 in GridLattice::getPt () from ./_Heat1D.so
(gdb) where
#0  0x405c6d51 in GridLattice::getPt () from ./_Heat1D.so
#1  0x40596e86 in Heat1D::solveAtThisTimeStep () from ./_Heat1D.so
```

The cause of the error is easily explained. We allocate data structures in scan, based on command-line input or default values. In the present case, default values are used. Then we change n; if it is changed to a larger value than the default value, loops over grid points may typically be too long and cause indices out of bounds. In other words, the n value of 120 is no longer compatible with the allocated data structures.

The remedy to the problem described in the previous paragraph is to split the scan function into two parts: one for loading command-line arguments into class data and one for allocating data structures. The simplest version of the split may be performed right before the assignment to h in Heat1D::scan. The first part of the original scan function can be put in a function scan, whereas the rest can form a function allocate.

We mention here that the splitting of the scan function into reading input and allocating data structures is only necessary for Diffpack simulators that do not utilize the menu system. As we will show later, the menu system can be operated from the Python script, and no rewrite of the simulator is required to form a Python interface where all input variables can be set as desired.

In the file src/pydp/Heat1D/Heat1D.cpp we have introduced a preprocessor variable (macro) PYINTERFACE to indicate whether the original Heat1D::scan is to be used or if the split version with scan and allocate should be brought into play. (The PYINTERFACE variable must be used in Heat1D.h and main.cpp as well.) To enable all options to be set from Python, we introduce a char* string tip_str in the class to hold the TimePrm initialization string. We also shuffle all initFromCommandLine calls to scan.

To compile Heat1D for a Python interface, we should run

```
Make MODE=opt CXXUF=-DPYINTERFACE
```

The wrapper code and shared library module must thereafter be generated by

```
make CXXUF=-DPYINTERFACE
```

Now we can change n, boundary conditions, initial condition, etc., and run
several problems consecutively. Here is a sample session:

```
from Heat1D import *
h = Heat1D()
h.scan()
h.n = 4                     # 4 grid points
h.problem_tp = 2            # no source term in the PDE
h.bc_tp = 2                 # u_x(1,t)=0
h.allocate()
h.solveProblem()
h.n = 64
h.allocate()
h.solveProblem()
h.tip_str = "dt=0.1 [0,1]"
print h.tip_str
h.allocate()
h.solveProblem()
```

*Using Typedefs to Handle Macros.* If we really want to work with the grid,
field, matrix, and vector objects in the Python interface, we can define some
typedefs for the macros. This can be done in the interface file:

```
%module Heat1D
%{
#include <Heat1D.h>
typedef Handle(GridLattice)   Handle_GridLattice;
typedef Handle(FieldLattice)  Handle_FieldLattice;
typedef Handle(TimePrm)       Handle_TimePrm;
typedef MatTri(real)          MatTri_real;
typedef ArrayGen(real)        ArrayGen_real;
%}
```

The new names, like ArrayGen_real, are then used in the class declaration in
the Heat1D.h header file or a copy of the class declaration in the Heat1D.i
interface file. The latter strategy is followed in the file[6]

```
src/pydp/Heat1D/swig-v2/Heat1D.i
```

That is, we have made an additional directory swig-v2 for this version of the
interface file. Next we shall present a more general solution to the problems
with the macros: We simply let the C preprocessor expand the macros before
SWIG processes the source files.

---

[6] Also this file contains a preprocessor directive with a macro PYINTERFACE.

### 8.3.8    Version 3; Using the C Preprocessor to Expand Macros

We are now close to a successful Python interface to Diffpack simulators, but the building of the interface file demands quite some manual work, which we should try to automate through scripts. The most challenging part of the automation is perhaps to replace macros and other C preprocessor constructs by suitable **typedefs**. Instead of approaching this challenge, we apply a simpler and more general strategy: We run the C preprocessor on the header files and grab the class declaration from the macro-expanded output. This output can be placed together with proper Diffpack-specific initialization constructs in an automatically generated SWIG interface file. A script MkDpSWIGInterface (included in Diffpack) takes the module name and a list of classes as command-line arguments and builds an appropriate interface file. In the present case we can create a new SWIG directory swig-v3, move to this directory, and run

```
MkDpSWIGInterface MODULE=Heat1D CLASSES=Heat1D
```

The arguments specify that we want to make a Python module Heat1D consisting of the C++ class Heat1D. This command generates a Heat1D.i file, whose contents is listed in the following:

```
%module Heat1D
%{
/* necessary header files to compile the wrapper code: */
#include <Heat1D.h>
%}

%init%{
  const char* p[] =  { "Calling Diffpack from Python" };
  initDiffpack(1, p);
%}

/*
class Heat1D code with macros expanded and C comments
removed by running the C preprocessor on the header file:
*/

class Heat1D
{
public:
  Handle_GridLattice grid;
  Handle_FieldLattice u;
  Handle_FieldLattice u_prev;
  Handle_TimePrm tip;
  double theta;
  double diff;

  MatTri_double A;
  ArrayGen_double b;
  ...
public:
```

```
Heat1D () {}
virtual ~Heat1D () {}
virtual void scan ();
...
};
```

Such an interface file is found in the directory src/pydp/Heat1D/swig-v3.

We can now write make and the wrapper code is compiled and linked with the Diffpack simulator to form a Python-loadable module Heat1D. Its usage goes as follows:

```
>>> from Heat1D import *
>>> simulator = Heat1D()
>>> simulator.scan()
>>> simulator.allocate()
>>> simulator.solveProblem()
```

In between these commands, the Diffpack application writes some output to the screen. To summarize, with three commands: make, MkDpSWIGMakefile and MkDpSWIGInterface, we have built a Python interface to a Diffpack simulator. The interface file also contains some definitions of the handles like, e.g., Handle_GridLattice, which will be useful later.

*Remark.* Running the C preprocessor on the C++ header files is in principle easy, as one can just take the standard compilation command and add (for most compilers) the -E flag and redirect the output to a file. In practice, we generate a special C++ source code file for every header file. The source code file contains an inclusion of the header file only. We let MkDpSWIGMakefile generate an extra target dpprepro, which prints an appropriate Diffpack compilation command for these special source code files to the screen. To the string generated by make dpprepro we add the -E flag and redirect the output to file. Regular expressions are then used to grab the text between class someclassname and the closing }; in the class declaration. At present, it is important that the closing }; is written exactly as shown here and that it appears at the beginning of a line.

The MkDpSWIGInterface script can, to a large extent generate ready-made interface files, but for more comprehensive Diffpack simulators you need to adjust the automatically generated interface file. Such operations require an understanding of both SWIG and the problems with the Diffpack simulator interfaces as outlined above. Additionally, you have to make the simulator such that it handles to be modified interactively.

*Issues Regarding Copying of Data Members.* In class Heat1D we have the matrix and vector in the linear system as class members:

```
MatTri(real)    A;     // the coefficient matrix
ArrayGen(real)  b;     // the right-hand side
```

Suppose we have a SWIG-generated interface to class ArrayGen(real). If we in Python issue the commands

```
p = Heat1D()
a = p.b
```

the resulting object a will be a complete copy of b and not a pointer to
b. Diffpack arrays may be large, and we usually want to avoid a costly copy
operation. We leave this issue now, but the desired pointer copying can easily
be obtained by using Handle's instead of objects in the simulator:

```
Handle_MatTri_real      A;    // the coefficient matrix
Handle_ArrayGen_real    b;    // the right-hand side
```

This requires, unfortunately, quite some simulator code to be altered. (For
example, A(i,0) must be replaced by A()(i,0) or a MatTri(real)& reference
can be introduced to reference *A right before the subscripting of the matrix
starts.) The bottom line is that we should use declare Handles, not objects,
in simulator classes if we want to enable an efficient data transfer between
Python scripts and Diffpack simulators.

*Potential Problems with Missing Copy Constructors.* If we have an instance
of, and not only a pointer to, a Diffpack or user-defined object of type X, the
wrapper code will perform something like

```
X *arg2;
...
arg1->x = *arg2;
```

where x is the name of the X instance in the simulator class. This assignment
requires an operator= function in class X. In case class X does not contain
an operator= function, C++ will automatically generate it. The generated
function performs a simple memberwise copy. This may cause many problems,
from compilation errors to memory management errors. Running the Python
interpreter in a debugger can reveal problems with such assignments, and the
remedy is to implement a proper operator= function.

### 8.3.9   Version 4; Extending the Python Interface
       with Auxiliary Functions

An important point is that SWIG is able to make Python objects of all C
data types. As previously demonstrated, we can change, e.g., the integer n.
However, if we want to modify or read the contents of Diffpack classes, we
need to make special access functions. Such functions can be added in the
interface file or in a separate file included in the interface file.

SWIG has an %extend directive, which can be used to extend the Python
interface with auxillary functions. Suppose we are interested in setting and
getting the values of the primary unknown field u in the Heat1D simulator.
Additionally, we need to know the size of the underlying vector of grid point
values in the field. The relevant functions can be included as member functions of class Heat1D, as seen from Python, by the statements

```
%extend Heat1D {
    int u_size() { return self->u->getNoValues(); }

    double u_val_get(int i) { return self->u->valueIndex(i); }
    void u_val_set(int i, double a) { self->u->valueIndex(i) = a; }
}
```

in the SWIG interface file. Here, `self` is the simulator (a variable of type `Heat1D*`). From this pointer we can access its functions and variables in a standard C++ fashion. The `%extend` section above has been added to the interface file generated by `MkDpSWIGInterface`. This new interface file is located in the directory `src/pydp/Heat1D/swig-v4`.

After having compiled and linked the module, we can use the three new functions from Python:

```
n = h.u_size()

for i in range(1,n+1):
    print "u(%4d)=%g" % (i, h.u_val_get(i))

for i in range(1,n+1):
    h.u_val_set(i, 4.0)
```

Following this recipe, it is easy to make an interface to everything that is accessible from `Heat1D`. However, it is needless to say that the code segment above is not very efficient; one should avoid explicit loops over long arrays in Python.

## 8.3.10   SWIG Pointers

SWIG gives us access to all primitive C data types. User-defined classes as well as Diffpack classes can be accessed by special functions, as we have explained. But what about accessing Diffpack objects directly from Python? We may try `Heat1D::u`, which is of type `Handle(FieldLattice)`:

```
>>> from Heat1D import *
>>> h = Heat1D()
>>> h.scan(); h.allocate()
>>> print h.u
_38841608_p_Handle_FieldLattice
>>> print type(h.u)
<type 'str'>
```

Here, `u` is a string, which is not what we would like to have. From the Python point of view, `h.u` acts a pointer to a Diffpack object, and SWIG represents this pointer as a string containing information about the type of object it points to. The string with pointer information is very useful, though. Here is an example where we add some new methods taking the string with pointer information as argument in functions and performing some desired operations:

```
%extend Heat1D  {
  void fill (Handle_FieldLattice u, double a) {
     u->values() = a;
  }
  void Print (Handle_FieldLattice u) {
     u->print(s_o);
  }
}
```

Notice that the print functionality is given the name `Print` with capital P;
using `print` would give a name clash with Python's built-in `print` command.
The `fill` and `Print` functions can be used as expected in Python,

```
# fill u_prev
h.fill(h.u_prev, 13.0)
# u_prev = 13 is now used in the next time-step
h.solveAtThisTimeStep()
h.Print(h.u)
```

The functionality is defined in `src/pydp/Heat1D/swig-v4/Heat1D.i`.

`MkDpSWIGInterface` extends the `Heat1D` interface with some functionality
for the handles:

```
class Handle_FieldLattice

{
protected:
  FieldLattice* classptr;
  bool checkPtr () const;

public:
  Handle_FieldLattice ();
  Handle_FieldLattice (const FieldLattice& p);
  Handle_FieldLattice (FieldLattice* p);
  Handle_FieldLattice (const Handle_FieldLattice& p);
 ~Handle_FieldLattice ();

  void rebind (const FieldLattice* pc);
  void rebind (const FieldLattice& p);
  Handle_FieldLattice& detach ();

  void operator = (const Handle_FieldLattice& h);
  void operator = (const FieldLattice& p);
  void operator = (const FieldLattice* p);

  bool ok () const;
  bool ok (const char* message) const;

  const FieldLattice* operator -> () const;
  const FieldLattice& operator () () const;
  const FieldLattice& operator * () const;
  const FieldLattice* getPtr () const;
  const FieldLattice& getRef () const;

  FieldLattice* operator -> ();
  FieldLattice& operator () ();
```

```
    FieldLattice& operator * ();
    FieldLattice* getPtr ();
    FieldLattice& getRef ();

    FieldLattice** getPtrAdr ();

    bool operator == (const Handle_FieldLattice& h) const;
    bool operator != (const Handle_FieldLattice& h) const;
    bool operator < (const Handle_FieldLattice& h) const;
    bool operator > (const Handle_FieldLattice& h) const;
};
```

This interface not very usefull in itself. We can fetch, e.g., the FieldLattice object, but since SWIG does not know anything about FieldLattice we can not access its functionallity. Looking at the contents of this object, using dir, shows that it is basically a Python string (it does not contain any FieldLattice functions and data):

```
>>> h = Heat1D()
>>> ... #initialize and solve
>>> u1 = h.u()
>>> dir(u1)
['__add__', '__class__', '__contains__', '__delattr__', '__eq__',
'__ge__', '__getattribute__', '__getitem__', '__getslice__', '
__gt__', '__hash__', '__init__', '__le__', '__len__', '__lt__',
'__mul__', '__ne__', '__new__', '__reduce__', '__repr__', '__rmul__',
'__setattr__', '__str__', 'capitalize', 'center', 'count', 'decode',
'encode', 'endswith', 'expandtabs', 'find', 'index', 'isalnum', '
isalpha', 'isdigit', 'islower', 'isspace', 'istitle', 'isupper',
'join', 'ljust', 'lower', 'lstrip', 'replace', 'rfind', 'rindex',
'rjust', 'rstrip', 'split', 'splitlines', 'startswith', 'strip',
'swapcase', 'title', 'translate', 'upper']
```

We have developed a module DP, which contains the FieldLattice functionallity. Therefore, by importing this module, we have access to the FieldLattice functionallity also in our simulator:

```
>>> import DP
>>> u2 = h.u()
>>> dir(u2)
['__class__', '__del__', '__delattr__', '__dict__', '__getattr__',
'__getattribute__', '__hash__', '__init__', '__module__', '__new__'
, '__reduce__', '__repr__', '__setattr__', '__str__',
'__swig_getmethods__' , '__swig_setmethods__', '__weakref__', 'add',
'apply', 'attach', 'cast2FieldLattice', 'copyWithoutGhostBoundary',
'derivativeFEM', 'derivativePt', 'empty', 'exchangeValues',
'fictBoundary', 'fill', 'getGridBase', 'getGridWithPts', 'getInfo',
'getNoPoints', 'getNoValues', 'getPt', 'ghostBoundary', 'grid',
'loadData', 'minmax', 'mult', 'ok', 'pack', 'redim', 'scale', 'scan',
'this', 'thisown', 'type_info', 'unloadData', 'unpack', 'unscale',
'value', 'valueFEM', 'valueIndex', 'valuePoint', 'valuePt', 'values',
'values0', 'valuesVec', <class 'DP.DP.FieldLatticePtr'>]
```

At present, DP does not contain all the functionallity in Diffpack, only some of the basic classes that are useful in the simulators described in this chaphter.

## 8.4   Converting Data between Diffpack and Python

We have shown how to perform operations on data in C++. However, it would also be convenient to fetch data from Diffpack computations in Python. For computational efficiency and for easy data transfer to other numerical Python modules, the data should be stored in NumPy arrays. This implies that we need to grab Diffpack array data and wrap them in NumPy arrays.

A NumPy array is implemented as a C struct, and for details we refer to the NumPy C API described in [1]. For now it suffices to know that the NumPy C struct holds the dimensions of the array as well as a pointer to a contiguous memory segment with the array entries. Our implementation of the mapping between Diffpack arrays and NumPy arrays is carried out in a tailored C++ class, here called a *filter*. For safety, we copy data between Python and Diffpack. It would be easy to let the NumPy array and Diffpack array point to the same memory segment, but strange errors may occur if both Python and Diffpack think they own the same data; this may lead to deleting memory segments twice (and the well-known trouble with tracking down the associated strange behavior of the program).

### 8.4.1   A Class for Data Conversion

We have created a class `dp2pyfilters` containing several functions for converting Diffpack arrays to NumPy arrays and back again. The class also contains functions for converting Diffpack fields to Vtk structured and unstructured grids.

The header file `dp2pyfilters.h` of the filter reads

```
class dp2pyfilters
{
public:
  dp2pyfilters();
  ~dp2pyfilters() {}

  void py2dp(PyObject* array, Vec(real)& vec);
  PyObject* dp2py(const Vec(real)& vec);
  PyObject* dp2vtk(const FieldFE& field);
  PyObject* dp2vtk(const FieldsFE& field);
  PyObject* dp2vtk(const FieldLattice& field);
};
```

### 8.4.2   Conversion between Vec and NumPy Arrays

We see that all NumPy arrays are represented as `PyObject*` pointers. Hence, when performing operations in C++ with such objects we should check that the `PyObject*` pointer actually points to a NumPy array[7]. As an example,

---

[7] We could have used `PyArrayObject*` pointers instead, but it is even safer to include explicit `PyObject*` type checking in the C++ code.

consider the py2dp function for converting a NumPy array to a Diffpack Vec vector:

```
void dp2pyfilters:: py2dp(PyObject* py_obj, Vec(real)& vec)
{
  if (!PyArray_Check(py_obj)) {
    PyErr_SetString(PyExc_TypeError, "Not a NumPy array");
    return;
  }
  else {
    PyArrayObject* array = (PyArrayObject*)
      PyArray_ContiguousFromObject(py_obj, PyArray_DOUBLE, 1,1);
    if (array == NULL) {
      PyErr_SetString(PyExc_TypeError,
                      "The NumPy array is not one-dim.");
      return;
    }
    int size = array->dimensions[0];
    vec.redim(size);
    for (int i=1; i<= size; i++) {
      vec(i) = *(double*)(array->data + array->strides[0]*(i-1));
    }
  }
}
```

We first check that the Python object is a NumPy array (PyArrayObject) with PyArray_Check(py_obj). If not, an exeption is raised:

```
PyErr_SetString(PyExc_TypeError, "Not a NumPy array");
```

The PyArray_ContiguousFromObject function checks that the data are contiguous, one-dimensional, and of type double, otherwise it creates such an array from the data if possible. If all is well, we can redim the Diffpack vector and copy the data.

The inverse filter from Diffpack to Python (dp2py) converts a Diffpack Vec object to a NumPy array:

```
PyObject*
dp2pyfilters:: dp2py(const Vec(real)& vec)
{
  int dim = vec.size();
  PyArrayObject* array = (PyArrayObject*) \
          PyArray_FromDims(1, &dim, PyArray_DOUBLE);
  for (int i=1; i<=vec.size(); i++) {
    *(double*)(array->data + array->strides[0]*(i-1)) = vec(i);
  }
  return PyArray_Return(array);
}
```

We create a NumPy array of proper size with PyArray_FromDims. If we want to use the same underlaying data for both the Diffpack vector and the Python array, we can use PyArray_FromDimsAndDataAndDescr.

When working with NumPy arrays from C and C++, we need an explicit initialization statement import_array() before performing any computations. This statement is naturally placed in the constructor,

```
dp2pyfilters:: dp2pyfilters () { import_array(); }
```

### 8.4.3   Creating a Python Interface to the Conversion Class

To create a Python Interface to class dp2pyfilters, we must write an interface
file dp2pyfilters.i:

```
%module dp2pyfilters
%{
#include <dp2py.h>
#include <Python.h>
#include <Numeric/arrayobject.h>
%}

%init%{
  const char* p[] = { "Calling Diffpack from Python" };
  initDiffpack(1, p);
%}

class dp2pyfilters {
  public:
    dp2pyfilters();
    ~dp2pyfilters();
    void py2dp(PyObject* array, Vec_double& vec);
    PyObject* dp2py (const Vec_double& vec);
    PyObject* dp2vtk(const FieldFE& field);
    PyObject* dp2vtk(const FieldsFE& field);
    PyObject* dp2vtk(const FieldLattice& field);
};
```

We may mention that we have, in the dp2py directory, also copied the header
files defining the classes FieldFE and Vec in Diffpack, run MkDpSWIGInterface
on the files, and SWIG'ed also these classes.

To employ the dp2py method to an array of nodal field values, we need to
access the underlying vector in Diffpack field objects. This is easily accomplished by making a function, here u_vec, in the SWIG interface file:

```
%extend Heat1D  {
  Vec_double& getVec (Handle_FieldLattice f)
    { return self->f->values(); }
}
```

In a Python script we can then run

```
from Heat1D import *
h = Heat1D()
...
filter = dp2pyfilters.dp2pyfilters()
u = filter.dp2py(h.getVec(h.u))
```

The u variable is now a standard NumPy array.

The complete set of details on how to make a Python interface to the
dp2pyfilters code is found in the directory src/pydp/dp2py.

### 8.4.4    Examples on Using the Conversion Class Interface

The dp2pyfilters class enables us to grab data from Diffpack and pass them to a plotting program like Gnuplot. There is a Python module Gnuplot that plots NumPy arrays with the aid of the Gnuplot plotting program. Here is the simple code at the scripting level:

```
from Heat1D import *
h = Heat1D()
h.scan()
h.solveProblem()
h.resultReport()

import Gnuplot, Numeric
import dp2pyfilters
filter = dp2pyfilters.dp2pyfilters()
u = filter.dp2py(h.u_vec())

g = Gnuplot.Gnuplot(debug=1, persist=1)
# let x be just the indices of u...
x = Numeric.arange(1,len(u)+1,1, typecode=Numeric.Float)
d = Gnuplot.Data(x, u, title="Solution obtained from Diffpack",
                 with='lines')
g.plot(d)
```

Similarly, we can use Matlab from Python with NumPy arrays through the pymat module. The Matlab interface in Python is very elegant. We can put Python arrays into the Matlab workspace with the put function. Similarly, we can get Matlab arrays with get. Moreover, we can make any type of string and evaluate it in Matlab as a Matlab command with eval. Hence, with these three functions we have complete access to all the Matlab functionality! Here is a simple example:

```
from Heat1D import *
h = Heat1D()
h.scan()
h.solveProblem()
h.resultReport()

import dp2pyfilters
filter = dp2pyfilters.dp2pyfilters()
u = filter.dp2py(h.u_vec())

import pymat  # import Matlab
eng = pymat.open()
pymat.put(eng, "u" , u)
pymat.eval(eng, "plot(u)")

import time
time.sleep(4); pymat.close(eng)  # wait 4s before we kill Matlab
```

### 8.4.5    A String Typemap

Diffpack applies its own type `String` for strings. SWIG sees `String` objects as any other objects and do not know of the implicit conversion that Diffpack has between C strings `char*` and `String` objects (the conversion requires knowledge of the constructors of class `String`). When we wanted our `Heat1D` simulator to store the input string for initializing the `TimePrm` object from the command line, we had to include a `char*` pointer in the `Heat1D` class and perform low-level C string actions in order to make it possible to adjust this string from Python. We would like to use a `String` object instead in class `Heat1D`. This requires a mapping between Diffpack `String` and Python strings (via `char*` C strings).

The mapping can be accomplished in two ways: (i) we can SWIG class `String`, or parts of its interface, or (ii) we can make a *typemap*. Suppose we SWIG parts of class `String`. We can then work with the `Heat1D::tip_str` string as follows in Python:

```
h.tip_str = String('dt=0.01 [0,10]')
print tip_str.c_str()
```

This works well, but typemaps make a more seamless integration of Diffpack and Python strings.

Typemaps are SWIG's powerful way to support any kind of mapping from a C/C++ object to a Python object and back again. For instance, the filters made in the previous section could be made as typemaps. However, we chose to implement them as a filter in C++ code instead, to emphasize that a copy takes place. Additionally, a `Vec` object in Diffpack has different functionality than a NumPy array. However, when it comes to strings we have chosen to make a typemap, because

- strings are small (in common Diffpack use); the copy procedure is cheap,

- strings are often used; the filter invocation is boring to use repeatedly,

- we do not need Diffpack's `String` facilities in addition to Python's (much richer) string functionality.

Python offers a function `PyString_AsString`, which takes in a `PyObject*` and returns a `char*`. Diffpack strings have a constructor for taking `const char*` as input. The typemap from Diffpack to Python strings can be implemented as

```
%typemap(in) const String& {
        String* s = new String(PyString_AsString($input));
        $1 = s;
}
```

The `$input` variable is a Python string, which is converted to the output variable `$1` of specified type `const String&`. We can then write

```
h.tip_str = "dt=0.01 [0,10]"
```

in the Python interface to Diffpack; this is the most attractive syntax.

The inverse typemap employs `PyString_FromString`, which returns a Python object (`$result`) from `char* Strings::c_str()`:

```
%typemap(out) String {
        $result = PyString_FromString($input.c_str());
}
```

We refer to the SWIG manual for a thorough explanation of typemap construction. Our purpose here is just to outline that short typemap codes can automate the conversion of the different string representations in Python and Diffpack. Return values of type `String` from Diffpack functions can now be directly stored in Python strings.

## 8.5   Building an Interface to a More Advanced Simulator

With the techniques and tools explained in the previous section, we are now able to build a Python interface to a standard Diffpack finite element simulator in a straightforward way. The Diffpack simulator `Heat1` is sufficiently general, from a software point of view, to be a proof of the concept. The `Heat1` simulator solves an $n$-dimensional heat equation,

$$\frac{\partial u}{\partial t} = \nabla \cdot (k\nabla u) + f, \tag{8.1}$$

in an arbitrary domain (see [6] for details). In `Heat1` the computed solution is compared with a known solution and the error is computed. To partially verify this simulator we check that the numerical error rates are in agreement with the theoretical. This is done by running the experiments in a Python script.

The least squares method, which is implemented in Hinsen's `Scientific` Python package [12], is used to computed the error rates. Like most numerical modules in Python, `Scientific` assumes that data are stored in NumPy arrays.

Additionally, any verification and validation process needs appropriate visualization. For this purpose we could employ Vtk since it has a Python interface. However, MayaVi [8] is an easy-to-use data visualizer, written in pure Python and built upon Vtk, which is more convenient to call up in our script.

### 8.5.1   Computing Empirical Convergence Estimates

The `Heat1` simulator solves the heat equation with an implicit backward Euler scheme in time and the finite element method in space. We therefore assume that the error $e$ behaves like

$$e \equiv \|u - \hat{u}\| \leq ah^b + c(\Delta t)^d, \tag{8.2}$$

where $u$ is the known solution, $\hat{u}$ is the numerical solution, $h$ is a characteristic element length, $\Delta t$ is the time step, and the constants $a$, $b$, $c$, and $d$ are unknown. These constants are independent on the discretization parameters $h$ and $\Delta t$. A goal is to have a balance of different sources of error, which here means that we want to find $\Delta t$ and $h$ such that $ch^b$ and $c(\Delta t)^d$ are of the same size.

**Table 8.1.** $L_2$ norm of the error in terms of $h$ and $\Delta t$, when solving a 2D heat equation (**Heat1**) by a backward Euler scheme in time and bilinear finite elements in space.

| $\Delta t \backslash h$ | $2^{-2}$ | $2^{-3}$ | $2^{-4}$ | $2^{-5}$ | $2^{-6}$ |
|---|---|---|---|---|---|
| $2^{-6}$ | 2.70e-2 | 4.04e-3 | 1.21e-2 | 1.42e-2 | 1.47e-2 |
| $2^{-7}$ | 3.36e-2 | 2.81e-3 | 5.29e-3 | 7.33e-3 | 7.85e-3 |
| $2^{-8}$ | 3.73e-2 | 6.60e-3 | 1.46e-3 | 3.50e-3 | 4.01e-3 |
| $2^{-9}$ | 3.93e-2 | 8.63e-3 | 5.85e-4 | 1.45e-3 | 1.97e-3 |
| $2^{-10}$ | 4.03e-2 | 9.68e-3 | 1.64e-3 | 3.99e-4 | 9.09e-4 |
| $2^{-11}$ | 4.08e-2 | 1.02e-2 | 2.17e-3 | 1.38e-4 | 3.73e-4 |
| $2^{-12}$ | 4.11e-2 | 1.05e-2 | 2.44e-3 | 4.08e-4 | 1.02e-4 |
| $2^{-13}$ | 4.12e-2 | 1.06e-2 | 2.58e-3 | 5.44e-4 | 3.40e-4 |

We write a simple Python function that performs a single simulation based on the input parameters $\Delta t$, $T$, and $h$. This function administers the simulator through the **MenuSystem**.

```
def run(dt, T, dx):
    """run Heat1 simulator and return L2 error"
    xnodes = int(1/dx)
    grid_str = "P=PreproBox | d=2 [0,1]x[0,1] | \
    d=2 e=ElmB4n2D div=[%d,%d] grading=[1,1]" % (xnodes, xnodes)
    menu.set("gridfile", grid_str)
    time_str = "dt =%e t in [0,%e]" % (dt, T)
    menu.set("time parameters", time_str)
    heat.scan()
    heat.solveProblem()
    heat.resultReport()
    return heat.L2_error # L2_error is public variable in Heat1
```

The basic input parameters we vary is the time step and the space resolution. These are **double** variables, which are converted to strings by Python and thereafter passed to Diffpack's menu system. The strings **grid_str** and **time_str** are then passed to the **MenuSystem** by the **menu.set** calls. The menu system can load data from many sources, including the **get** function, which is very convenient in a script. This means that we can "program" the menu system. The menu variable is an instance of type MenuSystem and must be created in the script, e.g. by

```
menu = MenuSystem()
```

The DP module wraps MenuSystem, FieldFE, GridFE, and other useful Diffpack classes such that these are available in Python.

We are now in the position to make a simple multiple loop:

```
dts = [pow(2,-x) for x in range(6,12)]    # time steps
dxs = [pow(2,-x) for x in range(2,7)]     # nodes in x dir
T = 2.0

for dt in dts:
  for dx in dxs:
    L2_error = run(dt, T, dx)
    data.append(((dx, dt), L2_error))
```

The results of such a series of simulations is presented in Table 8.1. We then did a least square approximation of the data to determine the constants in the error estimate (8.2). It turned out that $a$ was negative! There are no bugs here, so let us briefly explain the unexpected result. A closer look at the Table 8.1 reveals that the error actually increases as $\Delta t$ is reduced if the grid is coarse. The same effect happens when refining the grid and $\Delta t$ is "too big". In a nonlinear least-squares fit, this "non-asymptotic" behavior of the error gives irrelevant values in the asymptotic error formula. We should therefore discard values associated with large $h$ and/or $\Delta t$. One solution is to perform the fit for the last column, last row, and the diagonal of Table 8.1. A better multiple loop therefore collects the three mentioned types of data:

```
for dt in dts:
  for dx in dxs:
    if dx == dxs[-1] or dt == dts[-1] or dt == dx*dx:
      L2_error = run(dt, T, dx)
      if dt == dts[-1]:
        data1.append(((dx), L2_error))
      if dx == dxs[-1]:
        data2.append(((dt), L2_error))
      if dt*dt == dx ):
        data3.append(((dx, dt), L2_error))
```

We use the least square method implemented in the Scientific package to compute the $a, b, c$, and $d$ in (8.2).

```
parameters, error = leastSquaresFit(model, guess, data)
```

The first parameter, model, is the function to be fitted. It will be called with two parameters. The first containing the fit parameters and the second contains the data (see, e.g., the error function below). The nonlinear least square method needs an initial guess, which is provided by the second parameter guess. The last parameter data is the actual gridsize-error data from numerical experiments. The data are built up as a list of tuples. Each tuple contains two elements. The first element is a tuple of the independent variables, the second contain the number that the model should fit. This is exactly how we stored the data (data1, data2, and data3) in the previous multiple loop. The return value of the function is two tuples. The first containing the estimated parameters, the second containing the quality of the fit.

The next step is to do define functions for our expected error behavior. We define three functions. First we define the total error function as assumed in (8.2):

```
def error(parameters, data):
    a, b, c, d = parameters
    h, dt = data
    a1 = a*Numeric.power(h, b)
    a2 = c*Numeric.power(dt,d)
    return Numeric.add(a1, a2)
```

Additionally, we make two auxiliary functions corresponding to the cases where either the $h$ or the $\Delta t$ part totally dominates the error during refinement. These functions read,

```
def h_error(parameters, data):
    a, b = parameters
    h = data
    a1 = a*Numeric.power(h, b)
    return a1

def dt_error(parameters, data):
    c, d = parameters
    dt = data
    a1 = c*Numeric.power(dt, d)
    return a1
```

Finally, we perform the least-squares approximation. The code reads

```
from Scientific.Functions.LeastSquares import leastSquaresFit

# fit a and b:
initial_guess = (1.0, 2.0)
fit = leastSquaresFit(h_error, initial_guess , data1)
print "Fitted parameters:", fit[0],  " Fit error:", fit[1]

# fit c and d:
initial_guess = ( 1.0, 1.0)
fit = leastSquaresFit(dt_error, initial_guess , data2)
print "Fitted parameters:", fit[0], " Fit error:", fit[1]

# fit a, b, c, and d:
initial_guess = (1.0, 2.0, 1.0, 1.0)
fit = leastSquaresFit(error, initial_guess , data3)
print "Fitted parameters:", fit[0], " Fit error:", fit[1]
```

The complete Python script is found in `src/pydp/Heat1/swig/estimate.py`.

### 8.5.2   Visualization with Vtk

Vtk comes with a Python interface. Unfortunately, to the authors knowledge, there is no documentation about how this is made, but it is not made with SWIG. Hence, we have to deal with the source code. The general interface is defined in `vtkPythonUtils.h`, and the reader is encouraged to check this out. In our case we want to make a `vtkUnstructuredGrid` object in Python[8]. In `vktPythonUtils.h` there is a function that converts any `vtkObject*` pointer to `PyObject*`:

```
PyObject *vtkPythonGetObjectFromPointer(vtkObject *ptr)
```

This function is applicable for turning all Vtk objects (subclasses of `vtkObject`) into Python objects[9]! Diffpack comes with a filter from `FieldFE` objects to the corresponding unstructured field objects of type `vtkUnstructuredGrid` in Vtk. This filter is accessed as `SimRes2vtk::field2vtk` (a static Diffpack function). Making a conversion from a `FieldFE` object to the Python version of Vtk's unstructured field object is therefore easy:

```
PyObject*
dp2pyfilters:: dp2vtk(const FieldFE& field) {
  vtkUnstructuredGrid* vtk_grid = NULL;
  SimRes2vtk::field2vtk(field, vtk_grid);
  vtk_grid = SimRes2vtk::u_grid;
  PyObject* py_obj=vtkPythonGetObjectFromPointer((vtkObject*)vtk_grid);
  return py_obj;
}
```

In the `Heat1` simulator we have a handle to the `FieldFE` objects: `Handle_FieldFE`. We therefore have to extend our Python interface with functionality to access the `FieldFE&` reference of the solution field directly:

```
%extend Heat1 {
  FieldFE& u()
    { return self->u.getRef(); }
}
```

Alternatively, we can make something more general, enabling us to fetch whatever field we like:

```
%extend Heat1 {
  FieldFE& getFieldFE (Handle_FieldFE f)
    { return f->u.getRef(); }
}
```

---

[8] The file `vtkUnstructuredGridPython.cxx` is also worth checking out.

[9] We also mention that SWIG uses the same strategy. It has a function `SWIG_NewPointerObj` that converts a C pointer to a Python object. The inverse mapping is supplied by `SWIG_ConvertPtr` that makes a C pointer from a Python object. These functions are frequently used in the wrapping code.

We will now show an example of usage. We could call Vtk directly through its Python interface. However, there is a higher-level interface written for Vtk, called MayaVi [8], which offers simplified commands for visualization and also a GUI. We assume that a simulation is done such that we have access to the computed field via the simulator function u.

```
# convert FieldFE to a Vtk field
u = heat.u()
import dp2pyfilters, vtk, mayavi
filter = dp2pyfilters.dp2pyfilters()
vtk_field = filter.dp2vtk(u)

# open mayavi and load the vtk field
v= mayavi.mayavi()
v.open_vtk_data(vtk_field)

# apply various filters to vtk_field
m = v.load_module('SurfaceMap', 0)
a = v.load_module('Axes', 0)
a.axes.SetCornerOffset(0.0) # configure the axes module
o = v.load_module('Outline', 0)
f = v.load_filter('WarpScalar', config=0)
config_file = open("test2.config")
f.load_config(config_file)

v.Render()  # visualize!
```

The complete script is found in the file src/pydp/Heat1/swig/visualize.py.

## 8.6   Installing Python, SWIG etc.

The major difficulty with launching environments of the type described in this chapter is that a range of different software packages are needed. You do not only need a working SWIG, Diffpack, Python, Vtk, MayaVi, Tcl/Tk, NumPy, Scientific, Gnuplot, Matlab, PyMat, etc., but the versions of these packages must be compatible. To ease the installation of the required components described in the present chapter, we have made a collection of scripts for downloading the source codes, packing them out, compiling and linking libraries, and installing the libraries locally on a part of the disk where the user has the necessary permissions. Normally, a system administrator is supposed to do this job, but system administrators are extremely busy people, so impatient scientists simply have to take over the software installation process.

The idea is to put all external software in a local directory, say the location $HOME/software. You must then define an environment variable $SOFTWARE, which points to this directory. Standard environment variables, such as PATH, and LD_LIBRARY_PATH, must be extended such that Unix searches correctly for the software installed locally on your account. Here is a suitable initialization segment for your .bashrc file:

```
export SOFTWARE=$HOME/software
```

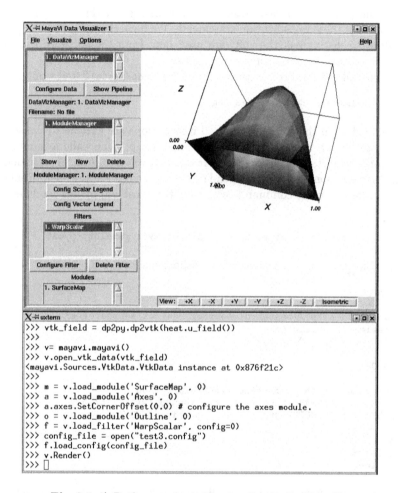

**Fig. 8.1.** A Python session with visualization in MayaVi.

```
export LD_LIBRARY_PATH=$SOFTWARE/lib:$LD_LIBRARY_PATH
export PATH=$SOFTWARE/bin:$PATH
export MANPATH=$SOFTWARE/man:$MANPATH
export PYTHONPATH=$SOFTWARE/src/Vtk/Wrapping/Python:\
$SOFTWARE/lib/vtk:$PYTHONPATH
export SWIGSRC=$SOFTWARE/src/SWIG-1.3.17
```

If you follow the set-up for installing software as suggested in [4], you already have most of these initialization statements. We have chosen to use the prefix SOFTWARE to distinguish the software used here from the software from [4], mainly because there are different versions.

To install the required software, go to src/pydp/install and run

```
./all-install 2>& 1 |  tee log
```

The output from the compilations and installations are written to the screen as well as to the file log[10]. The all-install script then calls a number of installation scripts, such as python_install, swig_install, etc., in a particular order, because some packages depend on others. Check the log file if problems occur. The following command tests that the most important modules are available:

```
python -c "import Tkinter, Scientific, mayavi, gnuplot, pymat"
```

Finally, we show an example on an installation script, swig-install:

```
#!/bin/sh -x

lwp-download http://belnet.dl.sourceforge.net/sourceforge/\
swig/swig-1.3.19.tar.gz
tar xzf swig-1.3.19.tar.gz

cd SWIG-1.3.19

./configure --with-python=$SOFTWARE/bin/python \
--with-tcl=$SOFTWARE/bin --with-tcllib=$SOFTWARE/lib/ \
--with-tclincl=$SOFTWARE/include --prefix=$SOFTWARE

make
make install
make clean
```

These installation scripts are not bullet-proof, but at least they may serve as a todo list. They also include the URLs to the source code.

---

[10] all-install 2>&1 means temporarily connecting the stderr of the command to the same "resource" as the shell's stdout. The stdout is redirected both to screen and the file log by tee.

## 8.7   Concluding Remarks

In this chapter we have shown how to build Python interfaces to Diffpack simulators. Several examples have hopefully demonstrated the potential power of such interfaces, especially when it comes to integrating Diffpack with other software packages such as Gnuplot, Vtk/MayaVi, Matlab, and numerical Python tools. The basic recipe for building Python interfaces to C++ codes via SWIG is simple, but there are numerous technical details that must be resolved when interfacing a huge piece of software such as Diffpack. We have tried to explain these technical issues in the present chapter, but complete information on how we have built the interfaces is contained in the software associated with this chapter.

This is our first step towards building Python interfaces to Diffpack. We will in the future most likely come up with more fine-tuned tools for simplifying, or even automating, the creation of such interfaces. We will also to a much larger extent explore the benefit of scripting interfaces to Diffpack. However, we still think that making Python interfaces to concrete applications is feasible for Diffpack users and that doing this will increase the reliability and efficiency simulation work.

### Acknowledgements

The authors are grateful to Magne Westlie and Ola Skavhaug for many improvements of the manuscript and code.

## References

1. D. Ascher, P. F. Dubois, K. Hinsen, J. Hugunin, and T. Oliphant. Numerical Python. Technical report.
2. D. Beazley. *Python Essential Reference.* New Riders Publishing, 2000.
3. D. Beazley et. al. Swig 1.3 Development Documentation. Technical report.
4. H. P. Langtangen. *Python Scripting for Scientific Computing.* Springer. Book in preparation.
5. H. P. Langtangen. *Computational Partial Differential Equations – Numerical Methods and Diffpack Programming.* Textbook in Computational Science and Engineering. Springer, 2nd edition, 2003.
6. M. Lutz. *Programming Python.* O'Reilly, second edition, 2001.
7. Matlab software package. *http://www.mathworks.com.*
8. MayaVi software package. *http://mayavi.sourceforge.net.*
9. Numerical python software package. *http://sourceforge.net/projects/numpy.*
10. Pymat Python-Matlab interface. *http://claymore.engineer.gvsu.edu/ steriana/Python.*
11. Python-gnuplot interface. *http://gnuplot-py.sourceforge.net.*
12. Scientificpython software package. *http://starship.python.net/crew/hinsen.*
13. SciPy software package. *http://www.scipy.org.*
14. Swig software package. *http://www.swig.org.*

15. G. van Rossum. Extending and embedding. Part of the electronic Python documentation at *http://www.python.org/doc.*

16. Vtk software software package. *http://www.kitware.com.*

# Chapter 9

# Performance Modeling of PDE Solvers

X. Cai[1,2], A. M. Bruaset[1,3], H. P. Langtangen[1,2], G. T. Lines[1,2],
K. Samuelsson[4], W. Shen[5], A. Tveito[1,2], and G. Zumbusch[6]

[1] Simula Research Laboratory
[2] Department of Informatics, University of Oslo
[3] Numerical Objects AS
[4] Chalmers University of Technology, Sweden
[5] SISSA, Italy
[6] Institute for Applied Mathematics, University of Bonn, Germany

**Abstract.** In this chapter, we collect actual CPU time measurements of a number of prototypical PDE simulators for solving the Poisson equation, the linear elasticity equation, the heat conduction equation, the equations of nonlinear water waves, the incompressible Navier-Stokes equations, and many more. We show how these measurements can be used to establish performance models of the form: $t = C n^{\alpha}$, which describes the CPU consumption as a function of the size of a discretized problem. The models can, in particular, quantify the efficiency of some standard numerical methods for solving the linear system in a particular problem. Different numerical methods can thus be compared, and we also hope that the performance models may form a basis for roughly estimating the consumption of CPU time by more complicated PDE simulators.

## 9.1 Introduction

The quality of any numerical software for solving partial differential equations (PDEs) is determined by the interplay between accuracy and efficiency. The ultimate question is: *How can a PDE simulator achieve a prescribed level of accuracy in the shortest amount of time?* Of course, we do not intend to give a complete answer to the question above. Rather, the main content of this chapter consists in a collection of statistical information on the CPU time consumed by some prototypical PDE simulators. The primary objective is to establish, through analyzing the CPU measurements, a set of performance models for some standard numerical methods when being applied in these prototypical PDE simulators. The measurements are obtained by running a series of simulations for each test problem, while we vary two factors: the size of the discretized problem and the solution method for the involved system of linear equations. Basically, we assume that for each solution method, the CPU consumption model is of the form:

$$t = C n^{\alpha},$$

where $n$ indicates the size of a discretized problem. The constants $C$ and $\alpha$ can be computed by fitting the model for a series of measurements, using the method of least squares, see Section 9.4.1. All the PDE simulators involved have been developed by using Diffpack [8,16].

The present chapter also aims at providing readers with a detailed study of different methods for solving linear systems. For example, one can easily compare the efficiency of multigrid methods with that of conjugate gradient methods, as functions of $n$. We also discuss some implementation issues that are relevant for achieving good efficiency in PDE software in general.

The contents of this chapter are organized as follows. Section 9.2 gives the mathematical formulations of a collection of model problems involving PDEs. Then, we briefly describe in Section 9.3 the numerical methods to be used for each of the model problems. Thereafter, in Section 9.4, we list the total CPU time consumption for every simulation. We also establish for each test problem performance models that depend on the problem size and the chosen method for solving the linear systems. In Sections 9.5 and 9.6, we analyze the CPU measurements in detail by examining the two major parts of each simulation: The CPU time for constructing the linear systems and the CPU time for solving them. In addition, we discuss some implementation issues that are of importance for improving the efficiency of PDE simulators.

Before we present the details of the test problems and numerical methods in Sections 9.2 and 9.3, we first list in Tables 9.1 and 9.2 the identifiers that will be used. These two tables give a quick overview of the test problems and methods we deal with in this chapter.

**Table 9.1.** Identifiers for the test problems.

| Identifier | Problem description |
|:---:|:---|
| PE2 | 2D Poisson equation |
| PE3 | 3D Poisson equation |
| EB2 | 2D elliptic boundary-value problem with variable coefficients |
| EB3 | 3D elliptic boundary-value problem with variable coefficients |
| DEB2 | 2D elliptic boundary-value problem with discontinuous coefficients |
| EL2 | 2D linear elasticity equation |
| HC2 | 2D heat conduction equation |
| WA2 | 2D nonlinear water wave equation |
| WA3 | 3D nonlinear water wave equation |
| TF2 | 2D two-phase flow problem |
| HT2 | 2D heart-torso coupled simulation |
| HT3 | 3D heart-torso coupled simulation |
| AD3 | 3D linear advection-diffusion equation |
| NS2 | 2D incompressible Navier-Stokes equations |

**Table 9.2.** Identifiers for the solution methods for linear systems.

| Identifier | Solution method | Reference(s) |
|:---:|:---|:---:|
| *BG* | Gauss elim. on banded matrix | [11] |
| *J* | Jacobi iterations | [23], [25] |
| *GS* | Gauss-Seidel iterations | [23], [25] |
| *CG* | Conjugate gradient method | [9], [15] |
| *IPCG* | CG + ILU prec. | [3], [12] |
| *MPCG* | CG + MILU prec. | [3], [12] |
| *RPCG* | CG + RILU prec. ($\omega = 0.9$) | [3], [12] |
| *FPCG* | CG + fast Fourier trans. prec. | [22] |
| *NMG* | Nested multigrid cycles | [13] |

## 9.2   Model Problems

In this section, we present the mathematical formulations of a collection of model problems. For each of them, one or several test problems are specified.

### 9.2.1   The Elliptic Boundary-Value Problem

We consider a second-order boundary-value problem of the form:

$$
\begin{aligned}
-\nabla \cdot (K\nabla u) &= f &&\text{in } \Omega, \\
u &= u_D &&\text{on } \Gamma_1, \\
\frac{\partial u}{\partial n} &= u_N &&\text{on } \Gamma_2,
\end{aligned}
\tag{9.1}
$$

where $\Omega \subset \mathbb{R}^d$, $d = 2, 3$. The boundary $\partial\Omega$ consists of two disjointed parts, such that $\Gamma_1 \cup \Gamma_2 = \partial\Omega$ and $\Gamma_1 \cap \Gamma_2 = \emptyset$. In (9.1), we have $K > 0$ as a scalar function defined in $\Omega$.

The simplest case of model problem (9.1) is the Poisson equation with homogeneous Dirichlet boundary conditions:

$$
\begin{aligned}
-\nabla^2 u &= f &&\text{in } \Omega, \\
u &= 0 &&\text{on } \partial\Omega.
\end{aligned}
$$

*The 2D Poisson equation* (PE2). For this 2D test problem, the solution domain $\Omega$ is the unit square, i.e., $\Omega = [0, 1]^2$. The right-hand side has the form:

$$
f(x, y) = e^{\sin(2\pi xy)}.
$$

*The 3D Poisson equation* (PE3). The solution domain $\Omega$ for this 3D test problem is the unit cube, i.e., $\Omega = [0, 1]^3$. The right-hand side has the form

$$
f(x, y, z) = e^{\sin(2\pi xyz)}.
$$

*The 2D elliptic boundary-value problem with variable coefficients* (EB2). The solution domain $\Omega$ and the right-hand side $f(x,y)$ are the same as in the PE2 problem. We also apply the same homogeneous Dirichlet boundary conditions $u = 0$ on the entire boundary. The scalar function $K(x,y)$ is given by

$$K(x,y) = 1 + xy + (xy)^2 .$$

*The 3D elliptic boundary-value problem with variable coefficients* (EB3). The boundary conditions, the solution domain $\Omega$, and the right-hand side $f(x,y,z)$ are the same as in the PE3 problem. The scalar function $K(x,y,z)$ has the following expression:

$$K(x,y,z) = 1 + xyz + (xyz)^2 .$$

*The 2D elliptic boundary-value problem with discontinuous coefficient* (DEB2). The original 2D solution domain $\Omega$, depicted in Figure 9.1, has four curved boundaries. In addition, the coefficient in (9.1) contains discontinuities. More precisely, we have

$$K(x,y) = \begin{cases} \delta_i & \text{for } (x,y) \in \Omega_i, \; i = 1,2,3, \\ 1.0 & \text{for } (x,y) \in \Omega \backslash \Omega_1 \backslash \Omega_2 \backslash \Omega_3, \end{cases}$$

where $\delta_i = 10^{-i}$, $i = 1,2,3$. The boundary condition valid on the entire boundary is homogeneous Neumann, i.e., we have $\partial u / \partial n = 0$ on $\partial \Omega$. The right-hand side $f$ has the form:

$$f(x,y) = \begin{cases} 1 \text{ for } (x,y) \in [0.1625, 0.2] \times [0.2375, 0.275], \\ -1 \text{ for } (x,y) \in [0.8, 0.8375]^2, \\ 0 \quad \text{elsewhere,} \end{cases}$$

which satisfies the condition

$$\int_\Omega f dx = 0.$$

Note that the special forms of $K$ and $f$ can be used to model the pressure equation associated with a simplified oil reservoir. In such a case, three different geological regions are represented by the discontinuities in $K$, and the right-hand side $f$ models one injection well and one production well.

By a domain imbedding technique, see [2,18], we introduce a regularization parameter $\epsilon$, such that we solve

$$-\nabla \cdot (K_\epsilon \nabla u_\epsilon) = f$$

in a larger and *rectangular* domain $\Omega_\epsilon = [-0.1, 1.1]^2$. The coefficient $K_\epsilon$ in the enlarged domain $\Omega_\epsilon$ is

$$K_\epsilon(x,y) = \begin{cases} K(x,y) & \text{for } (x,y) \in \Omega, \\ \epsilon & \text{for } (x,y) \in \Omega_\epsilon \backslash \Omega . \end{cases}$$

For this test problem, we have chosen $\epsilon = 10^{-9}$. To ensure a unique solution, we may introduce an additional requirement $\int_{\Omega_\epsilon} u_\epsilon = 0$.

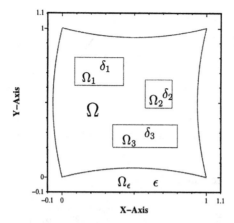

**Fig. 9.1.** A non-rectangular solution domain and its enlarged rectangular domain after domain imbedding.

### 9.2.2   The Linear Elasticity Problem

The displacement field $\mathbf{U} = (u_i)_{i=1}^{d}$ of a homogeneous isotropic elastic body can be modeled by the following linear vector PDE:

$$-\mu\Delta\mathbf{U} - (\mu+\lambda)\nabla(\nabla\cdot\mathbf{U}) = \mathbf{f}, \tag{9.2}$$

where $\mathbf{f}$ is a given vector function representing external load per volume. In (9.2) $\mu$ and $\lambda$ are elasticity constants. The boundary conditions are normally of two types: either the stress vector is prescribed, or the displacement is known, see e.g. [16, Ch. 5.1].

*The 2D linear elasticity problem* (EL2). Model problem (9.2) is to be solved in a 2D domain, which is a quarter of a hollow disk, see Figure 9.2, where the inner and outer radii are 1 and 2, respectively. The stress vector is prescribed on the entire boundary, except on $\Gamma_1$ and $\Gamma_2$, see Figure 9.2. On $\Gamma_1$, we have $u_1 = 0$ and the second component of the stress vector is prescribed. On $\Gamma_2$, we have $u_2 = 0$ and the first component of the stress vector is prescribed. More specifically, the stress vector on the boundary is given in form of $\boldsymbol{\sigma}\cdot\mathbf{n}$, where

$$\boldsymbol{\sigma} = \begin{pmatrix} \lambda(3x^2y + 4y^4) + 6\mu x^2 y & \mu x^3 \\ \mu x^3 & \lambda(3x^2y + 4y^4) + 8\mu y^3 \end{pmatrix}$$

is the stress tensor and $\mathbf{n}$ is the unit outward normal vector. The expressions are constructed to allow a simple analytical solution for the problem.

The elasticity constants $\mu$ and $\lambda$ have values as:

$$\mu = \frac{E}{2(1+\nu)}, \quad \lambda = \frac{\nu E}{(1+\nu)(1-2\nu)}, \quad \text{where } E = 10^5, \ \nu = 0.25\,.$$

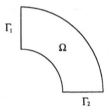

**Fig. 9.2.** The solution domain for the linear elasticity problem.

The external load per volume is

$$\mathbf{f} = \begin{pmatrix} -6\mu xy - 6(\mu + \lambda)xy \\ -12\mu y^2 - (\mu + \lambda)(3x^2 + 12y^2) \end{pmatrix} .$$

### 9.2.3    The Parabolic Problem

We consider the following parabolic initial-boundary value problem:

$$\begin{aligned}
\frac{\partial u}{\partial t} &= \nabla \cdot (K\nabla u) + f & &\text{in } \Omega \times (0, T], \\
u &= u_D & &\text{on } \Gamma \times (0, T], \\
\frac{\partial u}{\partial n} &= u_N & &\text{on } \partial\Omega \backslash \Gamma \times (0, T], \\
u(\mathbf{x}, 0) &= u^0(\mathbf{x}) & &\text{in } \Omega .
\end{aligned}$$

*The 2D heat conduction equation* (HC2). Model problem (9.3) is to be solved in the spatial domain $\Omega = [0, 1]^2$ and in the time interval between 0 and $T = 1$. With $K = 1$, the governing equation takes a simpler form:

$$\frac{\partial u}{\partial t} = \nabla^2 u + f .$$

We have chosen

$$f(x, y, t) = e^t(\tilde{x}\tilde{y} + 2\tilde{x} + 2\tilde{y}), \quad \tilde{x} = x(1 - x), \quad \tilde{y} = y(1 - y) .$$

The initial condition is set as

$$u(x, y, 0) = xy(1 - x)(1 - y) = \tilde{x}\tilde{y},$$

and we assume that homogeneous Dirichlet boundary conditions apply on the entire boundary.

### 9.2.4    The Nonlinear Water Wave Problem

Fully nonlinear free surface waves can be modeled by standard potential theory. Due to a coordinate transformation that will be used in the following solution approach, we denote by $\bar{x}, \bar{y}, \bar{z}$ the physical coordinates. The velocity potential $\varphi(\bar{x}, \bar{y}, \bar{z}, t)$ and the free surface elevation $\eta(\bar{x}, \bar{y}, t)$ are the primary unknowns. The mathematical model consists of a coupled system of PDEs. Under the standard assumption of incompressible flow without vorticity, the system of PDEs takes the following form:

$$\nabla^2 \varphi = 0 \quad \text{in the water volume,} \tag{9.3}$$

$$\eta_t + \varphi_{\bar{x}} \eta_{\bar{x}} + \varphi_{\bar{y}} \eta_{\bar{y}} - \varphi_{\bar{z}} = 0 \quad \text{on the free surface,} \tag{9.4}$$

$$\varphi_t + \frac{1}{2}(\varphi_{\bar{x}}^2 + \varphi_{\bar{y}}^2 + \varphi_{\bar{z}}^2) + g\eta = 0 \quad \text{on the free surface.} \tag{9.5}$$

Here, (9.3) is the Laplace equation, and equations (9.4) and (9.5) are referred to as the kinematic and dynamic boundary conditions, respectively. We refer to [24] for a detailed mathematical description.

We confine ourselves to a water volume of the form:

$$\overline{\Omega}(t) = \{(\bar{x}, \bar{y}, \bar{z}) \,|\, (\bar{x}, \bar{y}) \in \Omega_{\bar{x}\bar{y}}, -H \le \bar{z} \le \eta(\bar{x}, \bar{y}, t)\}, \tag{9.6}$$

where $\partial \varphi / \partial n = 0$ on the whole boundary except on the free surface. The time dependence of $\overline{\Omega}$ means that the Laplace equation (9.3) has to be solved in a dynamic physical domain. However, by introducing a time-dependent transformation:

$$x = \bar{x}, \quad y = \bar{y}, \quad z = \left(\frac{\bar{z} + H}{\eta + H} - 1\right) H, \tag{9.7}$$

we can instead solve an elliptic boundary-value problem of the form (9.1) in a *fixed* computational domain at each time step. The time-dependent coefficient matrix $K$ of the elliptic problem reads

$$K(x, y, z, t) = \frac{1}{H} \begin{bmatrix} \eta + H & 0 & -(z+H)\eta_x \\ 0 & \eta + H & -(z+H)\eta_y \\ -(z+H)\eta_x & -(z+H)\eta_y & \dfrac{H^2 + (z+H)^2(\eta_x^2 + \eta_y^2)}{\eta + H} \end{bmatrix}. \tag{9.8}$$

We note that $K$ is symmetric and positive definite, and $K$ is well-defined provided that the condition $|\eta| < H$ is satisfied. For more information on the transformation and $K$, we refer to [6]. Because the $\bar{x}$- and $\bar{y}$-coordinates are the same as the $x$- and $y$-coordinates after the transformation, we will drop notation $\bar{x}, \bar{y}$ in the rest of the report and simply use $x, y$ instead. Thus, the new system takes the following form:

$$\nabla \cdot (K \nabla \varphi) = 0 \quad \text{in } \Omega = \Omega_{xy} \times [-H, 0],$$

$$\eta_t + \varphi_x \eta_x + \varphi_y \eta_y - \varphi_z \frac{H}{\eta + H} = 0 \quad \text{on the free surface,}$$

$$\varphi_t + \frac{1}{2}\left(\varphi_x^2 + \varphi_y^2 + \left(\frac{\varphi_z H}{\eta + H}\right)^2\right) + g\eta = 0 \quad \text{on the free surface.}$$

*The 2D nonlinear water wave equation* (WA2). In this 2D test problem, the wave motion is restricted to the $(x, z)$ spatial coordinates. We solve the 2D wave system in the time interval $0 < t \leq T = 8$, where the spatial domain is

$$(x, z) \in [0, L] \times [-H, 0], \quad L = 160, \ H = 70.$$

The initial conditions on the free surface are

$$\eta(x, 0) = \eta^0(x),$$
$$\varphi(x, \eta^0(x), 0) = \varphi_z(x, \eta^0(x), 0) = 0,$$

where the initial form of the free surface is given by

$$\eta^0(x) = \frac{729}{16} \left[ \left(\frac{x}{L}\right)^2 \left(\frac{x - L}{L}\right)^4 - \frac{1}{105} \right].$$

*The 3D nonlinear water wave equation* (WA3). For this 3D test problem, the time interval of interest is chosen to be $0 \leq t \leq T = 4$. The spatial domain is

$$(x, y, z) \in [0, L_1] \times [0, L_2] \times [-H, 0], \quad L_1 = L_2 = 80, \ H = 50,$$

which is bounded by the free water surface on the top and solid walls $(\partial \varphi / \partial n = 0)$ on the rest part of the boundary. The initial conditions on the free surface are

$$\eta(x, y, 0) = \eta^0(x, y),$$
$$\varphi(x, y, \eta^0(x, y), 0) = \varphi_z(x, y, \eta^0(x, y), 0) = 0,$$

where the initial form of the free surface is expressed by

$$\eta^0(x, y) = \left(-0.9 \cos\left(\frac{\pi x}{L_1}\right) + \cos\left(\frac{2\pi x}{L_1}\right)\right) \times \left(1 - 0.9 \cos\left(\frac{\pi y}{L_2}\right) + \cos\left(\frac{2\pi y}{L_2}\right)\right).$$

## 9.2.5    The Two-Phase Flow Problem in 2D

We consider a simple model of two-phase (oil and water) flow in oil reservoir simulation:

$$s_t + \mathbf{v} \cdot \nabla(f(s)) = 0 \quad \text{in } \Omega \times (0, T], \tag{9.9}$$
$$-\nabla \cdot (\lambda(s)\nabla p) = q \quad \text{in } \Omega \times (0, T], \tag{9.10}$$
$$\mathbf{v} = -\lambda(s)\nabla p.$$

In the above system of PDEs, $s$ and $p$ are the primary unknowns, which represent the saturation of water and the pressure distribution, respectively. Equation (9.9) is referred to as the saturation equation, which is a hyperbolic

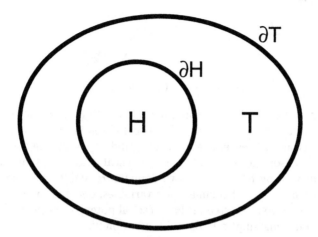

**Fig. 9.3.** A simplified 2D solution domain of the coupled torso and myocardium simulations.

conservation law, and (9.10) is referred to as the pressure equation, which is a second-order elliptic boundary-value problem. Here, $\mathbf{v}$ is the total velocity field, $\lambda$ is the mobility, and $f$ is the flux function. The source term $q$ in (9.10) represents injection and production wells. The boundary condition is $\mathbf{v} \cdot \mathbf{n} = 0$, which ensures no flow through the boundary.

As a 2D test problem, denoted by TF2, we consider the model with

$$f(s) = \frac{s^2}{s^2 + a(1-s)^2}, \qquad \lambda(s) = s^2 + a(1-s)^2,$$

where $a = 0.5$. The domain $\Omega$ is chosen to be the unit square and one injection well is located at $(0,0)$, while one production well is located at $(1,1)$. The contributions of the wells to the source term are treated as Dirac delta functions with proper strength, i.e., a positive value for the injection well and a negative value for the production well. The test problem is to be simulated in the time interval $(0,T]$ with $T = 0.1$. The initial condition for $s$ is $s^0$, where $s^0 = 1$ at the injection well, and $s^0 = 0$ elsewhere.

### 9.2.6   The Heart-Torso Coupled Simulations

We use the Bi-Domain model (see e.g. [10,14]) for modeling the electrical potential in the human body $\Omega = H \cup T$, where $H$ denotes the heart and $T$ denotes the torso exterior to the heart, see Figure 9.3. The mathematical model inside $H$ consists of two PDEs and one ordinary differential equation

(ODE) system as follows:

$$C\chi\frac{\partial v}{\partial t} + \chi I_{\text{ion}}(v, s) - \nabla \cdot (M_i \nabla v) = \nabla \cdot (M_i \nabla u_e), \quad (9.11)$$

$$\nabla \cdot ((M_i + M_e)\nabla u_e) = -\nabla \cdot (M_i \nabla v), \quad (9.12)$$

$$\frac{\partial s}{\partial t}(x) = F(t, s(t, x), v(t, x); x). \quad (9.13)$$

In the above system, $v$ is the membrane potential given by the relation $v = u_i - u_e$, where $u_i$ and $u_e$ denote the electrical potential inside and outside of the heart muscle cells respectively, with $M_i$ and $M_e$ being the corresponding conductivity tensors. Moreover, $C$ and $\chi$ are scalar constants, and $I_{\text{ion}}$ is the ionic current passing between the two domains $H$ and $T$. The ionic current depends on $v$ and a set of cellular state variables, denoted by $s$.

Since we are also interested in the electrical potential on the body surface, we use the following elliptic PDE valid in region $T$:

$$\nabla \cdot (M_T \nabla u_T) = 0. \quad (9.14)$$

In the following text, we denote by HT2 and HT3 heart-torso coupled simulations in 2D and 3D, respectively.

### 9.2.7   The Species Transport Problem

We consider the following advection-diffusion-reaction equation with initial and boundary conditions:

$$\frac{\partial C}{\partial t} = \nabla \cdot (\mathbf{D}\nabla C) - \mathbf{v} \cdot \nabla C - kC^2 \quad \text{in } \Omega \times (0, T], \quad (9.15)$$

$$C = C_0 \quad \text{on } \Gamma_1 \times (0, T], \quad (9.16)$$

$$\frac{\partial C}{\partial n} = 0 \quad \text{on } \Gamma_2 \times (0, T], \quad (9.17)$$

$$C(\mathbf{x}, 0) = 0 \quad \text{in } \Omega. \quad (9.18)$$

Here, $C(\mathbf{x}, t)$ is a solute concentration in a flow field with velocity $\mathbf{v}$, $\mathbf{D}$ is the hydrodynamic dispersion tensor, $\Omega$ is the domain of interest, and $\Gamma_1$ and $\Gamma_2$ denote the partition of the boundary: $\partial\Omega = \Gamma_1 \cup \Gamma_2$, $\Gamma_1 \cap \Gamma_2 = \emptyset$.

*The 3D linear advection-diffusion test problem* (AD3). With $k = 0$ equation (9.15) transforms into a linear equation. In this test problem, we simply prescribe a constant velocity $\mathbf{v} = (1, 0, 0)^T$. The anisotropic $\mathbf{D}$ tensor is taken to be constant, more precisely,

$$\mathbf{D} = \begin{bmatrix} 0.01 & 0 & 0 \\ 0 & 0.001 & 0 \\ 0 & 0 & 0.001 \end{bmatrix}.$$

Moreover, we set $T = 1$, $\Omega$ is a hypercube: $(x, y, z) \in [0, 1] \times [0, 0.2] \times [0, 0.1]$, and $\Gamma_1$ is the plane $x = 0$ where we have $C_0 = 100$.

### 9.2.8    Incompressible Navier-Stokes Equations

We consider the following incompressible Navier-Stokes equations:

$$\frac{\partial \mathbf{u}}{\partial t} + \mathbf{u} \cdot \nabla \mathbf{u} = -\frac{1}{\varrho} \nabla p + \nu \nabla^2 \mathbf{u} + \mathbf{b}, \tag{9.19}$$

$$\nabla \cdot \mathbf{u} = 0. \tag{9.20}$$

Here, $\mathbf{u}$ and $p$ are the primary unknowns representing the fluid velocity and pressure distribution. In addition, $\varrho$, $\nu$ and $\mathbf{b}$ denote the density, kinematic viscosity, and external body forces, respectively. The system (9.19)-(9.20) is to be supplemented with suitable boundary and initial conditions; see e.g. [17].

*The 2D Navier-Stokes test problem* (Ns2). The solution domain is the unit square, and we have

$$\varrho = 100, \quad \nu = 10^{-6}, \quad \mathbf{b} = 0.$$

The simulation is to be carried out within the time interval $0 < t \le 0.5$ with initial conditions as $p = 0$ and $\mathbf{u} = 0$.

## 9.3    Numerical Methods

In this section, we first describe briefly how to discretize the model problems presented in the preceding section. Thereafter, we concentrate on the solution of the linear systems involved in the simulations. Finally, we discuss some implementation issues in Diffpack.

### 9.3.1    The Elliptic Problems

We use finite element discretization (see e.g. [16, Ch. 2]) in the numerical solution of all the test problems derived from the second-order elliptic model problems (9.1) and (9.2). This involves primarily numerical calculation of the integrals of the weak formulation at the element level. The contributions are then added in an assembly process to construct a global system of linear equations. Due to the simple shapes of the solution domains, uniform grids have been applied. (We note that test problem DEB2 is solved in the imbedded domain of a rectangular shape.) Linear triangular elements are used in the 2D test problems, while trilinear elements are used in the 3D test problems.

For test problems PE2, PE3, EB2, EB3, and EL2, the finite element discretization results in a system of linear equations:

$$\mathbf{A}\mathbf{x} = \mathbf{b}, \tag{9.21}$$

where $\mathbf{A}$ is a sparse, symmetric, and positive definite matrix. The vector $\mathbf{x}$ contains the unknown values at the grid points. It is well known that different methods for solving the linear system result in different computational performance. This issue will be discussed later in Section 9.3.8. For test problem DEB2 in particular, $\mathbf{A}$ is only defined up to a constant due to the boundary condition. Although the singularity of $\mathbf{A}$ makes the Gaussian elimination an inappropriate solution method, the performance of the iterative methods does not seem to be affected in any way.

### 9.3.2   The HC2 Problem

For the HC2 problem, we apply the finite element method for the spatial discretization and the fully implicit Euler scheme in the temporal discretization. At each time level, the main computational task reduces to the solution of the following system of linear equations:

$$\mathbf{A}\mathbf{u}^l = \mathbf{b}(\mathbf{u}^{l-1}). \tag{9.22}$$

Here, $\mathbf{u}$ is the vector of nodal values of the approximate solution $u$, and the superscript $l$ denotes the time level. For the HC2 problem, $\mathbf{A}$ is independent of $\mathbf{u}$ and $t$, so it needs only to be computed once. We refer to Section 9.6 for the details of constructing $\mathbf{A}$. Linear 2D triangular elements of uniform size are used in the finite element discretization.

### 9.3.3   The Nonlinear Water Wave Problem

The fixed computational domain for the mapped Laplace equation is a rectangle for WA2 and a hypercube for WA3, respectively. This enables a uniform meshing, such that every finite element is of the same size. To be more specific, we use linear elements in 2D and trilinear elements in 3D. The free surface boundary conditions are discretized by a standard centered finite difference scheme. The values of $\eta$ and $\varphi|_{z=0}$, which will be used in the solution of the mapped Laplace equation, are updated at each time level in a leapfrog manner. Most of the computational effort is spent on the solution of the mapped Laplace equation, which is discretized by the finite element method. In other words, a system of linear equations in form of (9.21) needs to be solved at each time level. We mention that in this case a direct Poisson solver based on the fast Fourier transform (FFT) becomes an optimal preconditioner for the conjugate gradient method.

### 9.3.4   The TF2 Problem

The TF2 problem is solved by the Implicit-Pressure-Explicit-Saturation (IM-PES) algorithm. This is a sequential algorithm where the saturation and pressure equations are solved separately on discrete time levels. The algorithm reads:

At time level $t^l$, given $s^l$, do the following iteration:

I: Solve the pressure equation to get $p^l$, using $s^l$ in the mobility;

II: Compute the velocity $\mathbf{v}^l$ based on $p^l$;

III: Solve the saturation equation explicitly to get $s^{l+1}$ on the next time level, using $\mathbf{v}^l$.

In step I, the pressure equation is solved by the standard finite element method with linear elements, as mentioned above in Section 9.3.1. The saturation equation in step III is solved explicitly by an un-split second-order Godunov scheme (cf. [4,7,19]). The time stepping is adaptive where the CFL condition for the Godunov scheme is always satisfied. This results in a non-uniform temporal discretization.

### 9.3.5   The HT2 and HT3 Simulations

For the heart-torso coupled simulations, we use a series of uniformly distanced time levels $t_l = l\Delta t$, and an implicit scheme for discretizing the $\frac{\partial v}{\partial t}$ term in (9.11). The spatial discretization is done using the finite element method with linear elements. The numerical algorithm consists of three sub-steps at every time level, assuming that $u^l, v^l$, and $s^l$ are known at time $t_l$.

1. Compute $s^{l+1}$ by solving the ODE system:

$$\frac{\partial s}{\partial t} = F(t, s, v; x), \quad s(t_l, x) = s^l \text{ and } v(t_l, x) = v^l, \quad \text{for } t \in (t_l, t_{l+1}].$$

2. Compute $v^{l+1}$ in $H$ by finite element solution of

$$C\chi \frac{v^{l+1} - v^l}{\Delta t} + \chi I_{\text{ion}}(v^l, s^{l+1}) = \nabla \cdot (M_i \nabla v^{l+1}) + \nabla \cdot (M_i \nabla u_e^l).$$

3. Compute $u^{l+1}$ in $\Omega = H \cup T$, i.e. $u_e^{l+1}$ in $H$ and $u_T^{l+1}$ in $T$, by finite element solution of an elliptic PDE composed by (9.12) and (9.14).

The discretized PDEs are solved by the multigrid method, while the ODE system is solved by a Runge-Kutta method with adaptive time stepping. We refer to [21] for more details. The simulations are carried out on unstructured finite element grids with triangular elements in 2D and tetrahedral elements in 3D.

### 9.3.6   The AD3 Problem

For the AD3 problem, we use the finite element method with trilinear elements of uniform size. The discretization in time is based on the fully implicit Euler scheme. The resulting discrete equations take the form:

$$\mathbf{A}\,\mathbf{c}^l = \mathbf{b}(\mathbf{c}^{l-1})$$

at each time level. Here, **c** is the vector of nodal values of $C$, and the superscript $l$ denotes the time level. The coefficient matrix **A** is independent of **c** and $t$, whereas the right-hand side vector **b** can be updated by a simple matrix-vector product (involving $\mathbf{c}^{l-1}$). Therefore, the linear system can be generated very efficiently at each time level, without any need for a new finite element assembly. Thus, the CPU time will mostly be spent on the solution of the linear system. We apply the BiCGStab method with Jacobi (diagonal) preconditioning. The result from the previous time level is used as the start vector. The iteration is stopped when the discrete $L^2$-norm (see Section 9.3.8 for definition) of the preconditioned residual, divided by the discrete $L^2$-norm of **b**, is less than a prescribed tolerance of $10^{-10}$.

### 9.3.7    The Ns2 Problem

We consider a fast finite element Navier-Stokes solver, see [16, Ch. 6.5], which uses the technique of operator splitting and consists of the following sub-steps at each time level:

$$\mathbf{k}^{(1)} = -\Delta t(\mathbf{u}^l \cdot \nabla \mathbf{u}^l - \nu \nabla^2 \mathbf{u}^l), \tag{9.23}$$

$$\hat{\mathbf{u}}^l = \mathbf{u}^l + \mathbf{k}^{(1)}, \tag{9.24}$$

$$\mathbf{k}^{(2)} = -\Delta t(\hat{\mathbf{u}}^l \cdot \nabla \hat{\mathbf{u}}^l - \nu \nabla^2 \hat{\mathbf{u}}^l), \tag{9.25}$$

$$\mathbf{u}^* = \mathbf{u}^l + \frac{1}{2}(\mathbf{k}^{(1)} + \mathbf{k}^{(2)}), \tag{9.26}$$

$$\nabla^2 p^{l+1} = \frac{\varrho}{\Delta t}\nabla \cdot \mathbf{u}^*, \tag{9.27}$$

$$\mathbf{u}^{l+1} = \mathbf{u}^* - \frac{\Delta t}{\varrho}(\nabla p^{l+1} - \varrho \mathbf{b}). \tag{9.28}$$

We note that the work of (9.23)-(9.25) and (9.28) is simply explicit updating, provided that we use a lumped diagonal form to represent the mass matrix in the finite element discretization. For the pressure equation (9.27), we use the conjugate gradient method for solving the linear system at each time level. The stopping criterion is that the $L^2$-norm of the residual is reduced by a factor of $10^5$.

### 9.3.8    Solution Methods for Linear Systems

Systems of linear equations arise in the solution process of all the test problems. Different solution methods result in different computational efficiency. Table 9.3 contains a heuristic comparison of several methods with respect to computational cost, when being applied to the elliptic model problem (9.1). We remark that computing **A** and **b** for setting up a linear system requires work of order $(N)$, where $N$ denotes the total number of unknowns.

    Generally, the methods for solving linear systems can be divided into two categories: direct methods and iterative methods. However, direct methods, with the banded Gaussian ($BG$) elimination as a classical example, are

**Table 9.3.** Comparison of different solution methods in respect of computational cost; $N$ denotes the total number of unknowns. These bounds apply to linear systems arising from discretization on regular grids.

| Method | 2D | 3D |
|---|---|---|
| Banded Gauss Elim. | $\mathcal{O}(N^2)$ | $\mathcal{O}(N^{7/3})$ |
| Nested Dissection | $\mathcal{O}(N^{3/2})$ | $\mathcal{O}(N^2)$ |
| Jacobi | $\mathcal{O}(N^2)$ | $\mathcal{O}(N^2)$ |
| Gauss-Seidel | $\mathcal{O}(N^2)$ | $\mathcal{O}(N^2)$ |
| Conjugate Gradient | $\mathcal{O}(N^{3/2})$ | $\mathcal{O}(N^{4/3})$ |
| CG + MILU | $\mathcal{O}(N^{5/4})$ | $\mathcal{O}(N^{7/6})$ |
| CG + FFT | $\mathcal{O}(N \log N)$ | $\mathcal{O}(N \log N)$ |
| Multigrid | $\mathcal{O}(N)$ | $\mathcal{O}(N)$ |

rarely suitable for large-scale simulations. Iterative methods are asymptotically more favorable with respect to both CPU time and storage requirements. For a survey of these iterative methods, we refer to [5]. In these methods, we start with an initial guess $\mathbf{x}^0$ and generate a sequence of approximations $\{\mathbf{x}^k\}$, which hopefully converges toward the true solution $\mathbf{x}$. Traditional iterative methods based on splittings of the coefficient matrix $\mathbf{A}$ can be represented by the Jacobi ($J$) and Gauss-Seidel ($GS$) iterations. For a symmetric and positive definite $\mathbf{A}$, the conjugate gradient ($CG$) method, from the family of Krylov subspace methods, is the best choice. Krylov subspace methods can be preconditioned to speed up the convergence. One particularly simple preconditioning scheme is known as "incomplete LU-factorization" (ILU) [3,12]. On hypercube geometries, direct Poisson solvers based on FFT are very often used as the preconditioners. Recently, new iterative methods such as domain decomposition [20] and multigrid [13] have emerged as more robust preconditioners or efficient stand-alone iterative methods. Table 9.2 has summarized all the solution methods to be used in the simulations.

The stopping criterion is another important component of the iterative methods. That is, we need to continue the iterations until a prescribed tolerance of e.g. the residual is reached. There are several possible choices of stopping criteria:

- $Sc0(\epsilon)$: The relative stopping criterion monitoring the deviation from the reference solution $\dfrac{\|\mathbf{x} - \mathbf{x}_k\|}{\|\mathbf{x}\|} < \epsilon$. We remark that this criterion is rarely useful in practice.

- $Sc1(\epsilon)$: The relative stopping criterion monitoring the relation between the latest residual and the initial residual, in e.g. the discrete $l^2$-norm (see below for definition):

$$\frac{\|\mathbf{b} - \mathbf{A}\mathbf{x}_k\|_{l^2}}{\|\mathbf{b} - \mathbf{A}\mathbf{x}_0\|_{l^2}} < \epsilon.$$

– $Sc2(\epsilon)$: The absolute stopping criterion monitoring the discrete $L^2$-norm of the residual:

$$\|\mathbf{b} - \mathbf{A}\mathbf{x}_k\|_{L^2} < \epsilon, \quad \text{where } \|\mathbf{g}\|_{L^2} \equiv \sqrt{\frac{1}{N}\sum_{i=1}^{N} g_i^2}.$$

– $Sc3(\epsilon)$: The absolute stopping criterion monitoring the discrete $l^2$-norm of the residual:

$$\|\mathbf{b} - \mathbf{A}\mathbf{x}_k\|_{l^2} < \epsilon, \quad \text{where } \|\mathbf{g}\|_{l^2} \equiv \sqrt{\sum_{i=1}^{N} g_i^2}.$$

In general, the chosen stopping criterion and tolerance level should be related to the size of the discretization error. This is because we want the error, which is caused by an incomplete solution of the system of linear equations, to be negligible in comparison with the discretization error. The stopping criterion should therefore be grid dependent, typically of the form $C\,h^\alpha$, where $\alpha$ is the spatial order of the method and the constant $C$ is tuned such that $C$ is significantly less than the corresponding constants in the spatial (and temporal) error estimate(s). However for simplicity, we have used a fixed tolerance level well below the discretization error in the simulations. This is satisfactory if the monitored quantity, e.g., the residual, decreases at an approximately constant rate.

For most of our test problems, we have observed that the above stopping criteria behave similarly. Table 9.4 demonstrates the results concerning the number of iterations of the preconditioned CG method, when applied in test problem EB2. It should be noted here that $Sc0(\epsilon)$ is of little practical interest, because it requires the solution we are seeking. The problem with $Sc3(\epsilon)$ is that it does not mimic a norm for continuous functions. Thus, we regard $Sc1(\epsilon)$ and $Sc2(\epsilon)$ as the most appropriate alternatives. More specifically, we apply $Sc1(\epsilon)$ for stationary problems: PE2, PE3, EB2, EB3, DEB2; and $Sc2(\epsilon)$ for dynamic problems: HC2, WA2, WA3 and TF2. The reason for not applying $Sc1(\epsilon)$ for dynamic problems is that it may become unnecessarily strict as $\Delta t$ decreases, because a smaller $\Delta t$ results in a better start vector for every time level, and thereby a smaller value of the discrete $L^2$-norm of the initial residual.

### 9.3.9   Diffpack Implementation and Notation

In Diffpack, the generic base class FEM offers the standard finite element algorithms. Besides, the solution of systems of linear equations is administrated by the interface class LinEqAdmFE, which has access to different choices of solution methods, stopping criteria, preconditioners, etc. Therefore, a standard Diffpack simulator class is derived from FEM and contains an object of type

**Table 9.4.** An example demonstrating the effect of different stopping criteria on the number of iterations for test problem EB2 where $\epsilon = 10^{-8}$. The reference solution $\mathbf{x}$ needed in criterion $Sc0$ is obtained by applying the $FPCG$ method with the $Sc1(10^{-16})$ stopping criterion.

| Method | CG + fast Fourier trans. prec. | | | | CG + RILU ($\omega = 0.9$) prec. | | | |
|---|---|---|---|---|---|---|---|---|
| Criterion | $Sc0(\epsilon)$ | $Sc1(\epsilon)$ | $Sc2(\epsilon)$ | $Sc3(\epsilon)$ | $Sc0(\epsilon)$ | $Sc1(\epsilon)$ | $Sc2(\epsilon)$ | $Sc3(\epsilon)$ |
| $17 \times 17$ | 12 | 14 | 11 | 13 | 10 | 12 | 9 | 11 |
| $33 \times 33$ | 12 | 15 | 11 | 13 | 14 | 16 | 11 | 14 |
| $65 \times 65$ | 12 | 16 | 10 | 13 | 23 | 27 | 18 | 23 |
| $129 \times 129$ | 12 | 16 | 9 | 13 | 42 | 49 | 29 | 42 |
| $257 \times 257$ | 12 | 16 | 8 | 12 | 81 | 95 | 52 | 79 |
| $513 \times 513$ | 12 | 16 | 7 | 12 | 159 | 188 | 97 | 146 |

**Table 9.5.** Identifiers for the solution methods and their Diffpack names used in connection with `MenuSystem`.

| Identifier | Diffpack names | | Comments |
|---|---|---|---|
| | basic method | matrix type | |
| $BG$ | GaussElim | MatBand | |
| $J$ | Jacobi | MatSparse | |
| $GS$ | SOR | MatSparse | relaxation parameter=1 |
| $CG$ | ConjGrad | MatSparse | |
| $IPCG$ | ConjGrad | MatSparse | RILU relaxation parameter=0 |
| $MPCG$ | ConjGrad | MatSparse | RILU relaxation parameter=1 |
| $RPCG$ | ConjGrad | MatSparse | RILU relaxation parameter=.9 |
| $FPCG$ | ConjGrad | MatSparse | |
| $NMG$ | MLIter | MatSparse | multilevel method=NestedMultigrid |

`LinEqAdmFE`. The user then redefines the inherited member functions from `FEM`, such as `integrands` and perhaps also `makeSystem`, to implement the concrete finite element discretization. The class `MenuSystem` is used to enable the choice of different solution methods and problem dependent parameters. Table 9.5 gives a list of important identifiers associated with the use of `MenuSystem`.

## 9.4    Total CPU Time Consumption

The total consumption of CPU time by any numerical simulation is a machine-dependent amount, which also depends on many other parameters. The most important parameters are the size of the discretized problem, and the choice of the numerical method for solving the encountered linear system(s). Through analyzing measurements of the total CPU time consumption from a series of simulations, we intend to establish *simplified* performance models associated with different numerical methods. These simplified performance models will

**Table 9.6.** Identifiers of different stopping criteria and their Diffpack names. Note that for using Jacobi and Gauss-Seidel iterations in Diffpack, we must use `CMRelTrueResidual` instead of `CMRelResidual`, or `CMAbsTrueResidual` instead of `CMAbsResidual`.

| Identifier | Diffpack name | Comments |
|---|---|---|
| $Sc0(\epsilon)$ | CMRelRefSolution | |
| $Sc1(\epsilon)$ | CMRelResidual | |
| $Sc2(\epsilon)$ | CMAbsResidual | norm type=L2 |
| $Sc3(\epsilon)$ | CMAbsResidual | norm type=l2 |

be of the form:

$$t = C\,n^\alpha, \qquad (9.29)$$

where $t$ denotes the CPU consumption, and $n$ indicates the size of a discretized PDE. The values of the constants $C$ and $\alpha$ depend on a specific test problem and the choice of a particular hardware platform. Hopefully, these performance models will also provide a rough prediction of real CPU time consumption by more complicated PDE simulators.

### 9.4.1   Establishing Performance Models

As stated above, we want to establish simplified performance models (9.29) for a number of standard numerical methods. Each model will be based on a series of measurements of the total CPU time consumption. For simplicity, we normally only consider simple spatial domains of a rectangular or boxed shape, and use an equal number of grid points, denoted by $n$, in all the spatial directions. In addition, we use $N$ to denote the total number of grid points, which is typically $N = n^2$ for 2D test problems, and $N = n^3$ for 3D problems. If using $n$ is insufficient for representing the size of a discretized problem, such as in HT2 and HT3 due to unstructured grids, we express the performance models as a function of $N$ directly. In some test problems that involve a time-dependent PDE, such as HC2, WA2, and WA3, we also enforce $n$ as the number of time levels. We remark that the resulted relation between $\Delta t$ and $\Delta x$ may not be ideal according to theory, e.g., for HC2.

Using the above set-up, given a particular numerical method of interest, we typically run simulations for a specific test problem associated with a series of $n$ values. For each pair of a particular numerical method and a particular $n$, we execute the test case several times and report the best CPU measurement. By this strategy, we hope to eliminate the abnormal measurements that are due to the competition with other processes on a particular platform. Assuming that a simplified performance model takes the form of (9.29), we can estimate the values of the exponent $\alpha$ and the constant $C$ by using the method of least squares. More precisely, let us assume that we

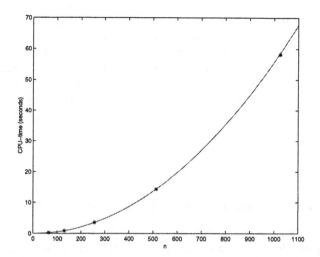

**Fig. 9.4.** An example of establishing the performance model; The *NMG* method is applied to the EB2 problem, $n$ has discrete values 65, 129, 257, 513, and 1025. The CPU measurements are marked by '*'. The curve represents the estimated performance model: $4.53 \times 10^{-5} \cdot n^{2.03}$, arising from a least-squares fit of the measurements.

have obtained a series of measurements $(n_i, t_i)_{i=1}^m$. To avoid the non-linearity of (9.29), we apply the logarithmic operator on both sides of (9.29). Consequently, we have the following equation:

$$\log(t) = \log(C) + \alpha \log(n). \tag{9.30}$$

The values of $\alpha$ and $C$ in (9.30) should be determined such that

$$\sum_{i=1}^{m} [\log(t_i) - \log(C) - \alpha \log(n_i)]^2 \tag{9.31}$$

is minimized. For example, the following Matlab program can be used to estimate $\alpha$ and $C$ for the measurements associated with the *NMG* method for the EB2 problem (see Table 9.10):

```
n=[65 129 257 513 1025];
t=[0.21 0.91 3.55 14.32 58.08];
[p,s]=polyfit(log(n),log(t),1); % linear model
alpha=p(1);
C=exp(p(2));
```

The resulting model is plotted together with the measurements in Figure 9.4.

### 9.4.2   The Software and Hardware Specification

The PDE simulators for all the model problems described in Section 9.2 are developed in the framework of Diffpack, version 3.5. Double precision is used in all the simulations. The timing is done using the standard UNIX system function `times`, and we use the sum of system time (`tms_stime`) and user time (`tms_utime`) as the reported CPU time. If not otherwise stated, the reported simulations are run on a PC that has an AMD Athlon 1GHz processor with 1.5 GB memory. The processor has a cache size of 256 KB. The C++ compiler used is `g++` of version 2.95.3 and the Linux operating system is of version 2.4.0. The g++ optimization options that have been used during compilation are as follows:

```
-Dgpp_Cplusplus -rdynamic -Wall -O3
```

### 9.4.3    The Best Performance Model for Each Test Problem

Before listing for every test problem the total CPU time consumption by different simulations, we first present in Table 9.7 the best performance model for each of the test problems: PE2, PE3, EB2, EB3, DEB2, EL2, HC2, WA2, WA3, and TF2. The reader should note that the performance results are subject to changes since Diffpack undergoes continuous improvement. The performance results obtained on another processor model, other than 1GHz Athlon, will of course be different. It should also be noted that the efficient *FPCG* method only applies to rectangular uniform grids. For more general domains and grids, methods such as *NMG* perform most efficiently.

Table 9.7. Summary of the best performance model for the test problems, based on measurements of the total CPU time consumption on an AMD Athlon 1GHz processor.

| Problem | CPU model | Method | Source |
|---------|-----------|--------|--------|
| PE2 | $1.70 \times 10^{-5} \cdot n^{2.03}$ | *FPCG* | Table 9.8 |
| PE3 | $3.54 \times 10^{-5} \cdot n^{3.09}$ | *FPCG* | Table 9.9 |
| EB2 | $4.53 \times 10^{-5} \cdot n^{2.03}$ | *NMG* | Table 9.10 |
| EB3 | $7.64 \times 10^{-5} \cdot n^{3.09}$ | *NMG* | Table 9.11 |
| DEB2 | $2.67 \times 10^{-5} \cdot n^{2.16}$ | *FPCG* | Table 9.12 |
| EL2 | $6.04 \times 10^{-6} \cdot n^{2.95}$ | *IPCG* | Table 9.13 |
| HC2 | $7.81 \times 10^{-6} \cdot n^{3.15}$ | *MPCG* | Table 9.14 |
| WA2 | $8.58 \times 10^{-6} \cdot n^{3.12}$ | *FPCG* | Table 9.15 |
| WA3 | $3.18 \times 10^{-5} \cdot n^{4.05}$ | *FPCG* | Table 9.16 |
| TF2 | $2.64 \times 10^{-5} \cdot n^{3.11}$ | *FPCG* | Table 9.17 |

### 9.4.4    The CPU Measurements and Performance Models

In Tables 9.8-9.21, we list the total CPU time consumption by different simulations for all the test problems. For test problems PE2, PE3, EB2, EB3,

DEB2, EL2, HC2, WA2, WA3, and TF2, we also present the estimated performance models as a function of $n$, which denotes the number of unknowns in each spatial direction. Note also that for test problems HC2, WA2, WA3, and TF2, we have deliberately chosen the number of time steps to be roughly equal to $n$. If the average CPU consumption per time step is of particular interest for these test problems, we can use

$$t_{\text{per time step}} = C\,n^{\alpha-1},$$

after the model of total CPU consumption (9.29) is established.

**Table 9.8.** Total CPU time consumption (in seconds) of the PE2 simulations. The stopping criterion $Scl(10^{-8})$ is used by the iterative methods.

| $n$ | 33 | 65 | 129 | 257 | 513 | 1025 | CPU model |
|---|---|---|---|---|---|---|---|
| $N$ | $33^2$ | $65^2$ | $129^2$ | $257^2$ | $513^2$ | $1025^2$ | |
| BG | 0.04 | 0.38 | 5.18 | 217.07 | | | $1.43 \times 10^{-8} \cdot n^{4.15}$ |
| J | | 1.35 | 45.61 | 876.19 | | | $1.78 \times 10^{-7} \cdot n^{4.60}$ |
| GS | | 0.67 | 22.88 | 438.04 | | | $1.09 \times 10^{-7} \cdot n^{4.55}$ |
| CG | 0.04 | 0.47 | 4.60 | 40.21 | 323.35 | | $9.25 \times 10^{-7} \cdot n^{3.16}$ |
| MPCG | 0.03 | 0.21 | 1.42 | 8.47 | 49.17 | 280.89 | $4.33 \times 10^{-6} \cdot n^{2.60}$ |
| FPCG | 0.02 | 0.08 | 0.35 | 1.38 | 5.59 | 22.31 | $1.70 \times 10^{-5} \cdot n^{2.03}$ |

**Table 9.9.** Total CPU time consumption (in seconds) of the PE3 simulations. The stopping criterion $Scl(10^{-8})$ is used by the iterative methods.

| $n$ | 9 | 17 | 33 | 65 | 129 | CPU model |
|---|---|---|---|---|---|---|
| $N$ | $9^3$ | $17^3$ | $33^3$ | $65^3$ | $129^3$ | |
| BG | 0.11 | 21.85 | 2169.32 | | | $7.07 \times 10^{-9} \cdot n^{7.61}$ |
| J | 0.07 | 3.71 | 136.64 | | | $2.10 \times 10^{-7} \cdot n^{5.83}$ |
| GS | 0.05 | 1.91 | 69.06 | | | $2.54 \times 10^{-7} \cdot n^{5.56}$ |
| CG | 0.03 | 0.35 | 5.10 | 70.36 | | $4.86 \times 10^{-6} \cdot n^{3.95}$ |
| MPCG | 0.03 | 0.31 | 3.30 | 34.93 | | $1.45 \times 10^{-5} \cdot n^{3.52}$ |
| FPCG | 0.03 | 0.22 | 1.80 | 14.37 | 116.27 | $3.54 \times 10^{-5} \cdot n^{3.09}$ |

**Table 9.10.** Total CPU time consumption (in seconds) of the EB2 simulations. The stopping criterion $Sc1(10^{-8})$ is used by the iterative methods.

| $n$ | 33 | 65 | 129 | 257 | 513 | 1025 | CPU |
|---|---|---|---|---|---|---|---|
| $N$ | $33^2$ | $65^2$ | $129^2$ | $257^2$ | $513^2$ | $1025^2$ | model |
| BG | 0.04 | 0.39 | 5.18 | 216.98 | | | $1.47 \times 10^{-8} \cdot n^{4.15}$ |
| J | | 1.33 | 44.78 | 862.85 | | | $1.44 \times 10^{-7} \cdot n^{4.64}$ |
| GS | | 0.67 | 22.08 | 432.63 | | | $1.12 \times 10^{-7} \cdot n^{4.54}$ |
| CG | 0.04 | 0.59 | 6.14 | 58.02 | 470.37 | | $8.55 \times 10^{-7} \cdot n^{3.23}$ |
| MPCG | 0.03 | 0.22 | 1.40 | 8.45 | 47.96 | 277.60 | $4.79 \times 10^{-6} \cdot n^{2.58}$ |
| FPCG | 0.03 | 0.16 | 0.73 | 3.19 | 12.78 | 51.77 | $2.74 \times 10^{-5} \cdot n^{2.09}$ |
| NMG | 0.05 | 0.21 | 0.91 | 3.55 | 14.32 | 58.08 | $4.53 \times 10^{-5} \cdot n^{2.03}$ |

**Table 9.11.** Total CPU time consumption (in seconds) of the EB3 simulations. The stopping criterion $Sc1(10^{-8})$ is used by the iterative methods.

| $n$ | 9 | 17 | 33 | 65 | 129 | CPU |
|---|---|---|---|---|---|---|
| $N$ | $9^3$ | $17^3$ | $33^3$ | $65^3$ | $129^3$ | model |
| BG | 0.21 | 21.85 | 2169.40 | | | $3.57 \times 10^{-8} \cdot n^{7.11}$ |
| J | 0.07 | 3.76 | 135.29 | | | $2.15 \times 10^{-7} \cdot n^{5.82}$ |
| GS | 0.06 | 1.93 | 70.09 | | | $3.92 \times 10^{-7} \cdot n^{5.44}$ |
| CG | 0.04 | 0.40 | 6.41 | 95.97 | | $3.85 \times 10^{-6} \cdot n^{4.08}$ |
| MPCG | 0.05 | 0.34 | 3.61 | 36.84 | | $1.74 \times 10^{-5} \cdot n^{3.49}$ |
| FPCG | 0.16 | 0.34 | 3.06 | 25.57 | 213.23 | $4.42 \times 10^{-5} \cdot n^{3.17}$ |
| NMG | 0.09 | 0.49 | 3.85 | 30.99 | 260.47 | $7.64 \times 10^{-5} \cdot n^{3.09}$ |

**Table 9.12.** Total CPU time consumption (in seconds) of the DEB2 simulations. The stopping criterion $Sc1(10^{-6})$ is used by all the methods.

| $n$ | 33 | 65 | 129 | 257 | 513 | CPU |
|---|---|---|---|---|---|---|
| $N$ | $33^2$ | $65^2$ | $129^2$ | $257^2$ | $513^2$ | model |
| J | | 2.36 | 81.94 | 1392.06 | | $2.10 \times 10^{-7} \cdot n^{4.68}$ |
| GS | | 1.19 | 41.02 | 685.46 | | $1.13 \times 10^{-7} \cdot n^{4.66}$ |
| CG | 0.19 | 3.08 | 29.66 | 527.62 | | $3.29 \times 10^{-7} \cdot n^{3.81}$ |
| RPCG | 0.02 | 0.23 | 1.76 | 13.34 | 105.27 | $9.78 \times 10^{-7} \cdot n^{2.96}$ |
| FPCG | 0.07 | 0.21 | 1.00 | 4.37 | 18.23 | $2.67 \times 10^{-5} \cdot n^{2.16}$ |

**Table 9.13.** Total CPU time consumption (in seconds) of the EL2 simulations. The stopping criterion $Sc1(10^{-4})$ is used by the iterative methods.

| $n$ | 33 | 65 | 129 | 257 | 513 | CPU |
|---|---|---|---|---|---|---|
| $N$ | $2 \cdot 33^2$ | $2 \cdot 65^2$ | $2 \cdot 129^2$ | $2 \cdot 257^2$ | $2 \cdot 513^2$ | model |
| BG | 0.22 | 2.76 | 109.85 | | | $2.18 \times 10^{-6} \cdot n^{4.56}$ |
| CG | 1.05 | 8.50 | 74.59 | 589.15 | | $2.13 \times 10^{-5} \cdot n^{3.09}$ |
| IPCG | 0.18 | 1.33 | 9.80 | 76.74 | 581.55 | $6.04 \times 10^{-6} \cdot n^{2.95}$ |

**Table 9.14.** Total CPU time consumption (in seconds) of the HC2 simulations. The stopping criterion $Sc2(10^{-10})$ is used by the iterative methods.

| $n$ | 17 | 33 | 65 | 129 | 257 | CPU model |
|---|---|---|---|---|---|---|
| $N$ | $17^2$ | $33^2$ | $65^2$ | $129^2$ | $257^2$ | |
| $\Delta t$ | $2^{-4}$ | $2^{-5}$ | $2^{-6}$ | $2^{-7}$ | $2^{-8}$ | |
| # steps | 16 | 32 | 64 | 128 | 256 | |
| BG | 0.04 | 0.42 | 4.56 | 57.47 | | $1.54 \times 10^{-6} \cdot n^{3.58}$ |
| J | 0.73 | 12.40 | 416.57 | 5817.21 | | $2.07 \times 10^{-6} \cdot n^{4.51}$ |
| GS | 0.37 | 6.29 | 210.35 | 2933.44 | | $1.06 \times 10^{-6} \cdot n^{4.50}$ |
| CG | 0.10 | 0.97 | 16.51 | 216.89 | | $1.76 \times 10^{-6} \cdot n^{3.83}$ |
| MPCG | 0.05 | 0.42 | 4.67 | 38.40 | 276.06 | $7.81 \times 10^{-6} \cdot n^{3.15}$ |

**Table 9.15.** Total CPU time consumption (in seconds) of the WA2 simulations. The stopping criterion $Sc2(10^{-8})$ is used by the iterative methods.

| $n$ | 17 | 33 | 65 | 129 | 257 | CPU model |
|---|---|---|---|---|---|---|
| $N$ | $17^2$ | $33^2$ | $65^2$ | $129^2$ | $257^2$ | |
| $\Delta t$ | $2^{-1}$ | $2^{-2}$ | $2^{-3}$ | $2^{-4}$ | $2^{-5}$ | |
| # steps | 16 | 32 | 64 | 128 | 256 | |
| BG | 0.05 | 1.02 | 22.61 | 646.07 | | $8.75 \times 10^{-8} \cdot n^{4.66}$ |
| CG | 0.19 | 2.42 | 69.33 | 1257.45 | | $6.40 \times 10^{-7} \cdot n^{4.40}$ |
| MPCG | 0.08 | 0.59 | 10.62 | 122.93 | 1284.16 | $1.53 \times 10^{-6} \cdot n^{3.73}$ |
| FPCG | 0.07 | 0.45 | 4.23 | 34.12 | 278.84 | $8.58 \times 10^{-6} \cdot n^{3.12}$ |

**Table 9.16.** Total CPU time consumption (in seconds) of the WA3 simulations. The stopping criterion $Sc2(10^{-8})$ is used by all the methods.

| $n$ | 9 | 17 | 33 | 65 | CPU model |
|---|---|---|---|---|---|
| $N$ | $9^3$ | $17^3$ | $33^3$ | $65^3$ | |
| $\Delta t$ | $2^{-1}$ | $2^{-2}$ | $2^{-3}$ | $2^{-4}$ | |
| # steps | 8 | 16 | 32 | 64 | |
| CG | 0.59 | 24.83 | 699.02 | | $4.14 \times 10^{-6} \cdot n^{5.44}$ |
| MPCG | 0.36 | 8.76 | 182.18 | 3669.09 | $2.58 \times 10^{-5} \cdot n^{4.50}$ |
| FPCG | 0.17 | 2.99 | 46.96 | 683.71 | $3.18 \times 10^{-5} \cdot n^{4.05}$ |

**Table 9.17.** Total CPU time consumption (in seconds) of the TF2 simulations. The stopping criterion $Sc2(10^{-10})$ is used by all the methods.

| $n$ | 17 | 33 | 65 | 129 | 257 | CPU model |
|---|---|---|---|---|---|---|
| $N$ | $17^2$ | $33^2$ | $65^2$ | $129^2$ | $257^2$ | |
| # steps | 36 | 70 | 142 | 289 | 588 | |
| RPCG | 0.17 | 1.33 | 21.64 | 309.13 | 3908.28 | $1.78 \times 10^{-6} \cdot n^{3.89}$ |
| FPCG | 0.19 | 1.35 | 11.82 | 99.36 | 800.04 | $2.64 \times 10^{-5} \cdot n^{3.11}$ |
| NMG | 0.34 | 1.95 | 20.64 | 173.45 | 1479.07 | $2.73 \times 10^{-5} \cdot n^{3.22}$ |

**Table 9.18.** CPU time consumption (in seconds) by 100 time steps of the HT2 simulator. In the table, $N_{\mathrm{para}}$ denotes the number of degrees of freedom in $H$ and $N_{\mathrm{elip}}$ denotes the number of degrees of freedom in $\Omega$. The performance model is $3.75 \times 10^{-2} \cdot N_{\mathrm{elip}}^{0.98}$.

| $N_{\mathrm{para}}$ | $N_{\mathrm{elip}}$ | total CPU |
|---:|---:|---:|
| 312 | 1192 | 44.89 |
| 1093 | 4661 | 159.33 |
| 4059 | 18433 | 575.11 |
| 15607 | 73313 | 2205.82 |
| 61167 | 292417 | 8632.96 |
| 242143 | 1168001 | 33566.70 |

**Table 9.19.** CPU time consumption (in seconds) by 100 time steps of the HT3 simulator. In the table, $N_{\mathrm{para}}$ denotes the number of degrees of freedom in $H$ and $N_{\mathrm{elip}}$ denotes the number of degrees of freedom in $\Omega$. The performance model is $7.05 \times 10^{-2} \cdot N_{\mathrm{elip}}^{0.93}$.

| $N_{\mathrm{para}}$ | $N_{\mathrm{elip}}$ | total CPU |
|---:|---:|---:|
| 27 | 53 | 4.26 |
| 125 | 321 | 16.26 |
| 729 | 2273 | 91.68 |
| 4913 | 17217 | 592.91 |
| 35937 | 134273 | 3782.52 |
| 274625 | 1061121 | 27614.80 |

**Table 9.20.** CPU time consumption (in seconds) by 100 time steps of the AD3 simulator. The stopping criterion $Scl(10^{-10})$ is used by the involved BiCGStab method.

| Grid ($N$) | $101 \times 21 \times 11$ | $201 \times 21 \times 11$ | $501 \times 11 \times 11$ | CPU model |
|---|---|---|---|---|
| CPU | 144.50 | 268.53 | 343.11 | $1.62 \times 10^{-2} \cdot N^{0.90}$ |

**Table 9.21.** CPU time consumption (in seconds) by 100 time steps of the Ns2 simulator. The stopping criterion $Scl(10^{-5})$ is used by the involved $CG$ method.

| Grid ($N$) | $51 \times 51$ | $101 \times 101$ | $201 \times 201$ | CPU model |
|---|---|---|---|---|
| CPU | 69.85 | 448.80 | 3175.04 | $1.22 \times 10^{-3} \cdot N^{1.39}$ |

### 9.4.5   Some Remarks About the Measurements

We notice from Tables 9.8-9.12 and 9.15-9.17 that FFT is a very efficient preconditioner for the CG method. However, this special preconditioner is only applicable to rectangular domains with special boundary conditions. Besides, the performance model of *FPCG* in Tables 9.10 and 9.11 may lead to an impression that the method is of order $\mathcal{O}(N)$. This impression is not correct because the total CPU times for those cases are dominated by the linear assembly process, whose CPU time consumption is of order $\mathcal{O}(N)$. Another wrong impression might be that *FPCG* is always more efficient than *NMG*, as is shown in Tables 9.10 and 9.11. We remark that this is due to the fact that the *NMG* method has to construct a hierarchy of linear systems on all grid levels, thus consuming more CPU time. As the tables in Section 9.5 will show, the *NMG* method has an inherently perfect order of complexity and is thus superior to the *FPCG* method when the problem size is large enough. Also, we can see from Tables 9.8-9.15 that banded Gaussian elimination suits quite well for small-sized 2D problems, whereas the Jacobi and Gauss-Seidel iterations are hardly efficient solution methods in any simulations.

### 9.4.6   Measurements Obtained on Another Platform

As a check of the CPU measurements that are obtained on the AMD Athlon 1GHz processor, we also run the simulations of test problems PE2 and PE3 on a HP SuperDome system, which has a ccNUMA[1] shared memory architecture with a total amount of memory of 88 GB. The processor in use is of type PA8600 with a clock frequency of 552 MHz, and the cache size is 1 MB. The operating system is HP-UX of version 11. The C++ compiler and the used compilation options are as follows (we used the 32-bit mode, although the system also supports a 64-bit mode):

```
aCC -DHP_ACC_Cplusplus +DA2.0N -O
```

The CPU measurements obtained on the HP system are listed in Tables 9.22 and 9.23. Compared with the CPU measurements in Tables 9.8 and 9.9, we notice that the performance associated with the two different types of CPUs agree quite well, with respect to the orders of the performance models. One exception is observed for the simulations using the banded Gaussian elimination. For such relatively memory intensive computations, the HP system is clearly superior to the PC due to better memory bandwidth and larger memory and cache sizes. For other computations, the Athlon processor easily outperforms the PA8600 processor.

---

[1] The term "ccNUMA" stands for cache coherent non-uniform memory access.

**Table 9.22.** Total CPU time consumption (in seconds) of the PE2 simulations. The CPU measurements are obtained on an HP system; see Section 9.4.6. The stopping criterion $Sc1(10^{-8})$ is used by the iterative methods.

| $n$ | 33 | 65 | 129 | 257 | 513 | 1025 | CPU |
|---|---|---|---|---|---|---|---|
| $N$ | $33^2$ | $65^2$ | $129^2$ | $257^2$ | $513^2$ | $1025^2$ | model |
| BG | 0.04 | 0.37 | 4.01 | 53.86 | | | $9.66 \times 10^{-8} \cdot n^{3.62}$ |
| J | 2.62 | 42.62 | 1301.55 | | | | $2.88 \times 10^{-7} \cdot n^{4.55}$ |
| GS | 1.43 | 22.93 | 674.84 | | | | $1.80 \times 10^{-7} \cdot n^{4.52}$ |
| CG | 0.06 | 0.35 | 4.59 | 47.30 | 411.50 | | $2.49 \times 10^{-7} \cdot n^{3.42}$ |
| MPCG | 0.05 | 0.24 | 2.03 | 12.38 | 74.90 | 418.71 | $7.63 \times 10^{-6} \cdot n^{2.57}$ |
| FPCG | 0.03 | 0.12 | 0.53 | 2.18 | 8.86 | 35.42 | $2.81 \times 10^{-5} \cdot n^{2.03}$ |

**Table 9.23.** Total CPU time consumption (in seconds) of the PE3 simulations. The CPU measurements are obtained on an HP system; see Section 9.4.6. The stopping criterion $Sc1(10^{-8})$ is used by the iterative methods.

| $n$ | 9 | 17 | 33 | 65 | CPU |
|---|---|---|---|---|---|
| $N$ | $9^3$ | $17^3$ | $33^3$ | $65^3$ | model |
| BG | 0.10 | 5.51 | 1922.02 | | $4.23 \times 10^{-9} \cdot n^{7.60}$ |
| J | 0.12 | 3.24 | 206.40 | | $3.57 \times 10^{-7} \cdot n^{5.74}$ |
| GS | 0.09 | 1.91 | 107.17 | | $4.88 \times 10^{-7} \cdot n^{5.46}$ |
| CG | 0.04 | 0.40 | 6.02 | 87.02 | $6.94 \times 10^{-6} \cdot n^{3.91}$ |
| MPCG | 0.05 | 0.40 | 4.82 | 52.11 | $1.97 \times 10^{-5} \cdot n^{3.54}$ |
| FPCG | 0.04 | 0.36 | 3.03 | 24.31 | $5.00 \times 10^{-5} \cdot n^{3.14}$ |

## 9.5   Solution of Linear Systems

The preceding section presents the information on total CPU times consumed by the different PDE simulators. We proceed our analysis further by examining how the CPU times have been spent. Roughly, the CPU time consumption of a PDE simulator may be divided into four parts:

1. Initialization – parameter input, grid generation etc;

2. Discretization – construction of the system of linear equations;

3. Solution of the system of linear equations;

4. Post-processing – analysis of computed results, visualization, etc.

For the simulators included in this report, Parts 1 and 4 normally take below 5% of the total CPU time consumption. Between the two remaining parts, Part 3 is usually the more CPU intensive one. This section thus examines the CPU time consumption of the solution of the linear systems, and performance models are established accordingly. Some issues on the efficiency of Part 2 will be addressed in the next section.

### 9.5.1    Summary

As discussed in Section 9.3.8, the choice of the solution method is important for the computational performance in each simulation. Table 9.24 summarizes the method that produces the best performance for each of the test problems: PE2, PE3, EB2, EB3, DEB2, EL2, HC2, WA2, WA3, and TF2.

**Table 9.24.** Summary of the best performance model for solving the linear systems in each test problem.

| Problem | CPU model | Method | Source |
|---------|-----------|--------|--------|
| PE2 | $1.91 \times 10^{-6} \cdot n^{2.10}$ | FPCG | Table 9.25 |
| PE3 | $2.28 \times 10^{-6} \cdot n^{3.17}$ | FPCG | Table 9.27 |
| EB2 | $2.22 \times 10^{-5} \cdot n^{2.02}$ | NMG | Table 9.26 |
| EB3 | $2.89 \times 10^{-5} \cdot n^{3.08}$ | NMG | Table 9.28 |
| DEB2 | $1.76 \times 10^{-5} \cdot n^{2.20}$ | FPCG | Table 9.29 |
| EL2 | $3.90 \times 10^{-6} \cdot n^{3.01}$ | IPCG | Table 9.30 |
| HC2 | $1.04 \times 10^{-5} \cdot n^{2.00}$ | MPCG | Table 9.31 |
| WA2 | $1.13 \times 10^{-5} \cdot n^{2.01}$ | FPCG | Table 9.32 |
| WA3 | $2.81 \times 10^{-5} \cdot n^{2.96}$ | FPCG | Table 9.33 |
| TF2 | $1.22 \times 10^{-5} \cdot n^{2.09}$ | FPCG | Table 9.34 |

### 9.5.2    Measurements

Tables 9.25-9.34 contain measurements of the CPU time consumption on solving the system of linear equations in every simulation. For iterative methods, we also list the number of iterations used. Theoretical estimates on the computational cost for different methods (Table 9.3) are applicable to test problems PE2, PE3, EB2, and EB3, which are derived from the elliptic model problem (9.1). We see that the performance models based on the actual measurements agree quite well with the theoretical estimates.

**Table 9.25.** Solution of the linear system in the PE2 simulations; CPU time (in seconds) and number of iterations. The stopping criterion $Scl(10^{-8})$ is used by the iterative methods. *) For the Poisson equation the fast Fourier transform preconditioner is an exact solver and consequently we get convergence in one iteration.

| $n$<br>$N$ | $33$<br>$33^2$ | | $65$<br>$65^2$ | | $129$<br>$129^2$ | | $257$<br>$257^2$ | | $513$<br>$513^2$ | | $1025$<br>$1025^2$ | | CPU<br>model |
|---|---|---|---|---|---|---|---|---|---|---|---|---|---|
| | CPU | #it. | CPU | #it. | CPU | #it. | CPU | # it. | CPU | #it. | CPU | #it. | |
| BG | 0.02 | ~ | 0.32 | ~ | 4.86 | ~ | 215.32 | ~ | | | | | $6.92 \times 10^{-10} \cdot n^{4.74}$ |
| J | 1.33 | 3794 | 45.54 | 15187 | 875.90 | 60751 | | | | | | | $8.72 \times 10^{-8} \cdot n^{4.76}$ |
| GS | 0.66 | 1898 | 22.81 | 7594 | 437.76 | 30376 | | | | | | | $4.26 \times 10^{-8} \cdot n^{4.76}$ |
| CG | 0.02 | 94 | 0.40 | 192 | 4.31 | 389 | 39.05 | 791 | 318.74 | 1601 | | | $6.03 \times 10^{-7} \cdot n^{3.23}$ |
| MPCG | 0.01 | 24 | 0.14 | 36 | 1.13 | 54 | 7.32 | 82 | 44.53 | 124 | 262.40 | 187 | $3.32 \times 10^{-6} \cdot n^{2.63}$ |
| FPCG* | 0.01 | 1 | 0.01 | 1 | 0.05 | 1 | 0.23 | 1 | 0.96 | 1 | 3.92 | 1 | $1.91 \times 10^{-6} \cdot n^{2.10}$ |

**Table 9.26.** Solution of the linear system in the EB2 simulations; CPU time (in seconds) and number of iterations. The stopping criterion $Scl(10^{-8})$ is used by the iterative methods.

| $n$<br>$N$ | $33$<br>$33^2$ | | $65$<br>$65^2$ | | $129$<br>$129^2$ | | $257$<br>$257^2$ | | $513$<br>$513^2$ | | $1025$<br>$1025^2$ | | CPU<br>model |
|---|---|---|---|---|---|---|---|---|---|---|---|---|---|
| | CPU | #it. | CPU | #it. | CPU | #it. | CPU | #it. | CPU | # it. | CPU | #it. | |
| BG | 0.02 | ~ | 0.32 | ~ | 4.85 | ~ | 215.61 | ~ | | | | | $6.93 \times 10^{-10} \cdot n^{4.74}$ |
| J | 1.31 | 3755 | 44.70 | 15030 | 862.55 | 60122 | | | | | | | $8.59 \times 10^{-8} \cdot n^{4.76}$ |
| GS | 0.65 | 1877 | 22.01 | 7512 | 432.33 | 30054 | | | | | | | $4.13 \times 10^{-8} \cdot n^{4.77}$ |
| CG | 0.03 | 116 | 0.52 | 254 | 5.84 | 541 | 56.82 | 1140 | 465.52 | 2348 | | | $6.10 \times 10^{-7} \cdot n^{3.29}$ |
| MPCG | 0.01 | 24 | 0.14 | 36 | 1.09 | 54 | 7.23 | 81 | 43.06 | 122 | 258.15 | 183 | $3.14 \times 10^{-6} \cdot n^{2.63}$ |
| FPCG | 0.01 | 15 | 0.09 | 16 | 0.43 | 16 | 1.97 | 16 | 7.93 | 16 | 32.33 | 16 | $1.84 \times 10^{-5} \cdot n^{2.08}$ |
| NMG | 0.02 | 7 | 0.09 | 7 | 0.42 | 7 | 1.65 | 6 | 6.78 | 5 | 27.73 | 6 | $2.22 \times 10^{-5} \cdot n^{2.02}$ |

**Table 9.27.** Solution of the linear system in the PE3 simulations; CPU time (in seconds) and number of iterations. The stopping criterion $Sc1(10^{-8})$ is used by the iterative methods.

| $n$<br>$N$ | 9<br>$9^3$ | | 17<br>$17^3$ | | 33<br>$33^3$ | | 65<br>$65^3$ | | 129<br>$129^3$ | | CPU<br>model |
|---|---|---|---|---|---|---|---|---|---|---|---|
| | CPU | # it. | CPU | # it. | CPU | # it. | CPU | # it. | CPU | # it. | |
| BG | 0.09 | ~ | 21.65 | ~ | 2166.13 | ~ | | | | | $4.27 \times 10^{-9} \cdot n^{7.76}$ |
| J | 0.04 | 231 | 3.51 | 940 | 134.98 | 3770 | | | | | $5.96 \times 10^{-7} \cdot n^{5.50}$ |
| GS | 0.03 | 117 | 1.71 | 472 | 67.40 | 1888 | | | | | $2.61 \times 10^{-7} \cdot n^{5.54}$ |
| CG | 0.00 | 27 | 0.14 | 57 | 3.45 | 118 | 57.19 | 241 | | | $4.63 \times 10^{-7} \cdot n^{4.48}$ |
| MPCG | 0.01 | 12 | 0.10 | 20 | 1.65 | 30 | 21.74 | 48 | | | $1.21 \times 10^{-6} \cdot n^{4.01}$ |
| FPCG* | 0.00 | 1 | 0.02 | 1 | 0.15 | 1 | 1.24 | 1 | 11.28 | 1 | $2.28 \times 10^{-6} \cdot n^{3.17}$ |

**Table 9.28.** Solution of the linear system in the EB3 simulations; CPU time (in seconds) and number of iterations. The stopping criterion $Sc1(10^{-8})$ is used by the iterative methods.

| $n$<br>$N$ | 9<br>$9^3$ | | 17<br>$17^3$ | | 33<br>$33^3$ | | 65<br>$65^3$ | | 129<br>$129^3$ | | CPU<br>model |
|---|---|---|---|---|---|---|---|---|---|---|---|
| | CPU | # it. | CPU | # it. | CPU | # it. | CPU | # it. | CPU | # it. | |
| BG | 0.10 | ~ | 21.61 | ~ | 2165.91 | ~ | | | | | $5.56 \times 10^{-9} \cdot n^{7.68}$ |
| J | 0.04 | 230 | 3.53 | 936 | 133.33 | 3753 | | | | | $6.47 \times 10^{-7} \cdot n^{5.47}$ |
| GS | 0.03 | 117 | 1.69 | 469 | 68.11 | 1878 | | | | | $2.35 \times 10^{-7} \cdot n^{5.57}$ |
| CG | 0.00 | 28 | 0.16 | 68 | 4.45 | 153 | 80.07 | 339 | | | $3.47 \times 10^{-7} \cdot n^{4.63}$ |
| MPCG | 0.01 | 12 | 0.10 | 19 | 1.65 | 30 | 21.14 | 47 | | | $1.29 \times 10^{-6} \cdot n^{4.00}$ |
| FPCG | 0.01 | 11 | 0.11 | 13 | 1.10 | 14 | 9.89 | 15 | 88.27 | 15 | $1.03 \times 10^{-5} \cdot n^{3.29}$ |
| NMG | 0.05 | 9 | 0.18 | 9 | 1.38 | 8 | 11.38 | 8 | 92.60 | 8 | $2.89 \times 10^{-5} \cdot n^{3.08}$ |

**Table 9.29.** Solution of the linear system in the DEB2 simulations; CPU time (in seconds) and number of iterations. The stopping criterion $Sc1(10^{-6})$ is used by the iterative methods.

| $n$ | 33 | | 65 | | 129 | | 257 | | 513 | | CPU |
|---|---|---|---|---|---|---|---|---|---|---|---|
| $N$ | $33^2$ | | $65^2$ | | $129^2$ | | $257^2$ | | $513^2$ | | model |
| | CPU | # it. | CPU | # it. | CPU | # it. | CPU | # it. | CPU | # it. | |
| J | 2.34 | 5019 | 81.89 | 19579 | 1391.86 | 77668 | | | | | $2.04 \times 10^{-7} \cdot n^{4.68}$ |
| GS | 1.18 | 2503 | 40.97 | 9797 | 685.26 | 38847 | | | | | $1.10 \times 10^{-7} \cdot n^{4.67}$ |
| CG | 0.17 | 500 | 3.03 | 1085 | 29.46 | 2245 | 526.82 | 9046 | | | $2.54 \times 10^{-7} \cdot n^{3.86}$ |
| RPCG | 0.01 | 20 | 0.19 | 33 | 1.56 | 62 | 12.54 | 120 | 102.08 | 242 | $5.88 \times 10^{-7} \cdot n^{3.04}$ |
| FPCG | 0.04 | 29 | 0.16 | 26 | 0.80 | 29 | 3.57 | 29 | 15.02 | 30 | $1.76 \times 10^{-5} \cdot n^{2.20}$ |

**Table 9.30.** Solution of the linear system in the EL2 simulations; CPU time (in seconds) and number of iterations. The stopping criterion $Sc1(10^{-4})$ is used by the iterative methods.

| $n$ | 33 | | 65 | | 129 | | 257 | | 513 | | CPU |
|---|---|---|---|---|---|---|---|---|---|---|---|
| $N$ | $2 \cdot 33^2$ | | $2 \cdot 65^2$ | | $2 \cdot 129^2$ | | $2 \cdot 257^2$ | | $2 \cdot 513^2$ | | model |
| | CPU | # it. | CPU | # it. | CPU | # it. | CPU | # it. | CPU | # it. | |
| BG | 0.17 | ~ | 2.52 | ~ | 108.58 | ~ | | | | | $9.09 \times 10^{-9} \cdot n^{4.74}$ |
| CG | 1.00 | 404 | 8.31 | 812 | 73.81 | 1627 | 586.04 | 3258 | | | $1.90 \times 10^{-5} \cdot n^{3.11}$ |
| IPCG | 0.13 | 28 | 1.13 | 55 | 9.02 | 108 | 73.64 | 218 | 568.83 | 434 | $3.90 \times 10^{-6} \cdot n^{3.01}$ |

**Table 9.31.** Solution of the linear system in the HC2 simulations; averaged CPU time (in seconds) and number of iterations at each time level. The stopping criterion $Sc2(10^{-10})$ is used by the iterative methods. The column "#it." lists the average number of needed iterations per time step.

| $n$ | 17 | | 33 | | 65 | | 129 | | 257 | | CPU |
|---|---|---|---|---|---|---|---|---|---|---|---|
| $N$ | $17^2$ | | $33^2$ | | $65^2$ | | $129^2$ | | $257^2$ | | model |
| | CPU | # it. | CPU | # it. | CPU | # it. | CPU | # it. | CPU | # it. | |
| BG | 0.00 | ~ | 0.01 | ~ | 0.03 | ~ | 0.22 | ~ | | | $1.61 \times 10^{-7} \cdot n^{2.91}$ |
| J | 0.04 | 358.94 | 0.38 | 805.81 | 6.48 | 1546.22 | 45.33 | 2531.72 | | | $2.10 \times 10^{-6} \cdot n^{3.51}$ |
| GS | 0.02 | 180.88 | 0.19 | 404.38 | 3.26 | 774.83 | 22.80 | 1267.51 | | | $1.04 \times 10^{-6} \cdot n^{3.51}$ |
| CG | 0.00 | 22.12 | 0.02 | 33.84 | 0.23 | 45.06 | 1.58 | 67.85 | | | $2.98 \times 10^{-7} \cdot n^{3.20}$ |
| MPCG | 0.00 | 7.25 | 0.01 | 8.09 | 0.04 | 7.88 | 0.19 | 6.57 | 0.62 | 4.89 | $1.04 \times 10^{-5} \cdot n^{2.00}$ |

**Table 9.32.** Solution of the linear system in the WA2 simulations; averaged CPU time (in seconds) and number of iterations at each time level. The stopping criterion $Sc2(10^{-8})$ is used by the iterative methods. The column "#it." lists the average number of needed iterations per time step.

| $n$ <br> $N$ | 17 <br> $17^2$ | | 33 <br> $33^2$ | | 65 <br> $65^2$ | | 129 <br> $129^2$ | | 257 <br> $257^2$ | | CPU <br> model |
|---|---|---|---|---|---|---|---|---|---|---|---|
| | CPU | # it. | CPU | # it. | CPU | # it. | CPU | # it. | CPU | # it. | |
| BG | 0.00 | ~ | 0.03 | ~ | 0.32 | ~ | 4.85 | ~ | | | $6.14 \times 10^{-8} \cdot n^{3.73}$ |
| CG | 0.01 | 109.81 | 0.07 | 203.41 | 1.06 | 384.25 | 9.75 | 729.20 | | | $2.43 \times 10^{-7} \cdot n^{3.62}$ |
| MPCG | 0.00 | 14.75 | 0.01 | 19.88 | 0.15 | 26.59 | 0.88 | 34.61 | 4.72 | 43.92 | $4.32 \times 10^{-6} \cdot n^{2.51}$ |
| FPCG | 0.00 | 7.69 | 0.01 | 6.97 | 0.05 | 6.69 | 0.19 | 6.07 | 0.79 | 5.87 | $1.13 \times 10^{-5} \cdot n^{2.01}$ |

**Table 9.33.** Solution of the linear system in the WA3 simulations; averaged CPU time consumption (in seconds) and number of iterations at each time level. The stopping criterion $Sc2(10^{-8})$ is used by the iterative methods. The column "#it." lists the average number of needed iterations per time step.

| $n$ <br> $N$ | 9 <br> $9^3$ | | 17 <br> $17^3$ | | 33 <br> $33^3$ | | 65 <br> $65^3$ | | CPU <br> model |
|---|---|---|---|---|---|---|---|---|---|
| | CPU | # it. | CPU | # it. | CPU | # it. | CPU | # it. | |
| CG | 0.06 | 95.38 | 1.48 | 158.12 | 21.28 | 269.50 | | | $3.30 \times 10^{-6} \cdot n^{4.52}$ |
| MPCG | 0.03 | 16.00 | 0.48 | 22.38 | 5.13 | 30.78 | 52.96 | 41.75 | $2.36 \times 10^{-5} \cdot n^{3.51}$ |
| FPCG | 0.01 | 6.38 | 0.12 | 6.50 | 0.91 | 6.25 | 6.34 | 5.39 | $2.81 \times 10^{-5} \cdot n^{2.96}$ |

**Table 9.34.** Solution of the linear system in the TF2 simulations; averaged CPU time (in seconds) and number of iterations at each time level. The stopping criterion $Sc2(10^{-10})$ is used by the iterative methods. The column "#it." lists the average number of needed iterations per time step.

| $n$ <br> $N$ | 17 <br> $17^2$ | | 33 <br> $33^2$ | | 65 <br> $65^2$ | | 129 <br> $129^2$ | | 257 <br> $257^2$ | | CPU <br> model |
|---|---|---|---|---|---|---|---|---|---|---|---|
| | CPU | # it. | CPU | # it. | CPU | # it. | CPU | # it. | CPU | # it. | |
| RPCG | 0.00 | 14.55 | 0.01 | 20.42 | 0.17 | 32.09 | 1.36 | 53.54 | 8.99 | 82.94 | $1.03 \times 10^{-6} \cdot n^{2.89}$ |
| FPCG | 0.00 | 8.93 | 0.01 | 10.73 | 0.07 | 11.09 | 0.32 | 10.70 | 1.28 | 9.91 | $2.67 \times 10^{-5} \cdot n^{2.10}$ |
| NMG | 0.06 | 7.92 | 0.10 | 7.52 | 0.28 | 6.75 | 1.21 | 5.89 | 4.78 | 5.26 | $1.22 \times 10^{-5} \cdot n^{2.09}$ |

### 9.5.3    Efficiency of the Linear Algebra Tools in Diffpack

The linear algebra tools in Diffpack have a carefully designed structure. By utilizing object-oriented programming techniques, Diffpack provides the users with a rich collection of solution methods, preconditioners, and stopping criteria. The users are able to make flexible combinations at run-time and the process of parameter fitting is greatly simplified. The generality is, of course, obtained at some cost of the computational efficiency. We will demonstrate through the following example that the linear algebra tools in Diffpack still maintain a relatively high performance level. We also mention that [1] contains several experiments concerning efficiency comparison between Diffpack and standard FORTRAN 77 routines.

As a competing code, we use a specially coded C program CGmilu for the solution of 3D elliptic equations on the unit cube, i.e., test problems PE3 and EB3. Here, CGmilu has hard-coded the 7-point-stencil of the finite difference discretization, meaning that the CPU time spent on constructing the linear system in CGmilu is negligible. We have thus found it fair to compare, in Table 9.35, the CPU time spent on the solution of the linear system by the Diffpack simulator with the total CPU time consumed by the CGmilu program. Assuming homogeneous Dirichlet boundary conditions on the entire boundary, CGmilu considers only inner nodal values as unknowns, so the resulting linear system has a reduced size compared with that in Diffpack. Moreover, CGmilu restricts to the (preconditioned) CG method where a specially coded matrix-vector product routine guarantees the extraordinary efficiency of the program. We also mention that the Diffpack simulator and CGmilu use the same stopping criterion $Scl(10^{-8})$, so both programs converge under the same number of CG iterations.

## 9.6    Construction of Linear Systems

In this section, we first analyze the process of constructing the linear system associated with the finite element discretization in Diffpack. Then, we point out some important implementation issues that can improve the performance.

### 9.6.1    The Process

In general, the CPU time consumed by the finite element discretization on constructing a linear system is proportional to the number of elements and thus proportional to the total number of unknowns. By examining the construction process, we can see that the CPU time is spent on the storage allocation of **A** and the following assembly process, i.e., loop through each element,

1. calculate the element matrix and vector,
2. enforce the essential boundary conditions, and
3. assemble the element contribution to the global linear system.

**Table 9.35.** Comparison between the Diffpack simulator and the specially coded C program CGmilu in PE3 and EB3 simulations. The measurements of the CPU time consumption (in seconds) are obtained on an AMD Athlon 1GHz processor.

| Solving the PE3 problem with $CG$ (no prec.) | | | |
|---|---|---|---|
| System size | $17 \times 17 \times 17$ | $33 \times 33 \times 33$ | $65 \times 65 \times 65$ | CPU model |
| Diffpack | 0.14 | 3.45 | 57.19 | $4.63\text{e-}7 \cdot n^{4.48}$ |
| System size | $15 \times 15 \times 15$ | $31 \times 31 \times 31$ | $63 \times 63 \times 63$ | CPU model |
| CGmilu | 0.05 | 2.16 | 35.54 | $2.39\text{e-}7 \cdot n^{4.58}$ |

| Solving the PE3 problem with $CG$+MILU prec. | | | |
|---|---|---|---|
| System size | $17 \times 17 \times 17$ | $33 \times 33 \times 33$ | $65 \times 65 \times 65$ | CPU model |
| Diffpack | 0.10 | 1.65 | 21.74 | $1.21\text{e-}6 \cdot n^{4.01}$ |
| System size | $15 \times 15 \times 15$ | $31 \times 31 \times 31$ | $63 \times 63 \times 63$ | CPU model |
| CGmilu | 0.04 | 0.88 | 11.78 | $9.38\text{e-}7 \cdot n^{3.96}$ |

| Solving the EB3 problem with $CG$ (no prec.) | | | |
|---|---|---|---|
| System size | $17 \times 17 \times 17$ | $33 \times 33 \times 33$ | $65 \times 65 \times 65$ | CPU model |
| Diffpack | 0.16 | 4.45 | 80.07 | $3.47\text{e-}7 \cdot n^{4.63}$ |
| System size | $15 \times 15 \times 15$ | $31 \times 31 \times 31$ | $63 \times 63 \times 63$ | CPU model |
| CGmilu | 0.06 | 2.76 | 49.72 | $2.14\text{e-}7 \cdot n^{4.68}$ |

| Solving the EB3 problem with $CG$+MILU prec. | | | |
|---|---|---|---|
| System size | $17 \times 17 \times 17$ | $33 \times 33 \times 33$ | $65 \times 65 \times 65$ | CPU model |
| Diffpack | 0.10 | 1.65 | 21.14 | $1.29\text{e-}6 \cdot n^{4.00}$ |
| System size | $15 \times 15 \times 15$ | $31 \times 31 \times 31$ | $63 \times 63 \times 63$ | CPU model |
| CGmilu | 0.03 | 0.83 | · 10.83 | $4.99\text{e-}7 \cdot n^{4.11}$ |

Many different matrix formats are supported by Diffpack. In particular, banded matrices and general sparse matrices are involved in the simulations of the current report. Typically, $\mathbf{A}$ has a banded structure, where the bandwidth is $\mathcal{O}(n)$ in 2D and $\mathcal{O}(n^2)$ in 3D, respectively. But for each row of $\mathbf{A}$ there are only a small number of non-zeros, independent of $n$. This means that a banded matrix format may waste too much storage on zero entries. The classical Gaussian elimination requires such a storage structure because those zero entries may later be filled with nonzero values during the LU-factorization. For the iterative methods, on the other hand, a compressed sparse row storage strategy, which only stores nonzero entries is preferable. This saves both storage and computational cost. Consequently, the cost of matrix-vector products becomes proportional to the number of rows of $\mathbf{A}$. However, the determination of the structure of a sparse matrix (locations of the non-zeros) requires some work, and a relatively complicated indexing procedure is needed to access individual entries of $\mathbf{A}$. As the CPU measurements indicate in Table 9.36, the assembly process runs more efficiently with the sparse matrix format. This is probably because the sparse format takes less storage space and has a better chance to be fitted into the cache.

**Table 9.36.** The CPU time (in seconds) spent on the `makeSystem` function in the PE2 simulations; Different matrix formats result in different amounts of CPU consumption.

| Grid | $33 \times 33$ | $65 \times 65$ | $129 \times 129$ | $257 \times 257$ |
|---|---|---|---|---|
| MatBand | 0.01 | 0.05 | 0.26 | 1.51 |
| MatSparse | 0.01 | 0.05 | 0.20 | 0.79 |

*Remark.* For dynamic problems such as Hc2 and Wa2, the CPU time spent on the storage allocation is only necessary at the first time level and should be avoided for later time levels.

### 9.6.2   Some Guidelines

The application of efficient solution methods raises stricter requirements on the efficiency of constructing the linear system. All the PDE simulators involved in this chapter have used the following guidelines for obtaining improved efficiency.

1. Reduction of arithmetic operations:
   - During the implementation of the virtual function `integrands`, identify constants independent of indices i and j, and pre-calculate them outside the loop.
   - Take advantage of the possible symmetry of the matrix **A**. For example, use the following procedure in `integrands`:

```
for (i=1; i<=nbf; i++) {
  for (j=1; j<=i; j++) {
    // ....
    elm_matvec.A(i,j) += a_ij;
  }
  // ....
  elm_matvec.b(i) += b_i;
}

for (i=1; i<nbf; i++)
  for (j=i+1; j<=nbf, j++)
    elm_matvec.A(i,j) = elm_matvec.A(j,i);
```

2. Reduction of (C++) overhead:
   - Use vector and matrix references instead of function calls, e.g., in `integrands`:

```
Vec(real)& N  = (Vec(real)&)fe.N();
Mat(real)& dN = (Mat(real)&)fe.dN();
```

   - If possible, use "inline" for frequently called functions, such as for the evaluation of a coefficient function.

3. Algorithmic improvement:

   - Use special (analytical) integration instead of standard numerical integration when possible. See Section 9.6.3 for a specific example.
   - For simulations of time-dependent problems, avoid stiffness matrix re-assembly when possible. See Section 9.6.4 for some examples.

We refer to [16, App. B.7] for more techniques for optimizing Diffpack codes.

### 9.6.3    The Mapped Laplace Equation in the WA3 Problem

We consider test problem WA3. Recall that the special transformation (9.7) maps the physical solution domain $\overline{\Omega}(t)$ (9.6) to a stationary computational domain $\Omega$, so we need to solve a new elliptic boundary-value problem which has a variable coefficient $K$ as given in (9.8). If we give a tripled index $(I, J, K)$ to each element matrix $\mathbf{A}^{I,J,K} = (A_{i,j}^{I,J,K})$, then we have

$$
\begin{aligned}
A_{i,j}^{I,J,K} = \int \int \int \frac{1}{H} \Bigg[ \frac{\partial N_i}{\partial x} & \left( (\eta + H)\frac{\partial N_j}{\partial x} - (z + H)\eta_x \frac{\partial N_j}{\partial z} \right) \\
+ \frac{\partial N_i}{\partial y} & \left( (\eta + H)\frac{\partial N_j}{\partial y} - (z + H)\eta_y \frac{\partial N_j}{\partial z} \right) \\
- (z + H)\eta_x \frac{\partial N_i}{\partial z} \frac{\partial N_j}{\partial x} & - (z + H)\eta_y \frac{\partial N_i}{\partial z} \frac{\partial N_j}{\partial y} \\
+ \frac{[H^2 + (z + H)^2(\eta_x^2 + \eta_y^2)]}{\eta + H} & \frac{\partial N_i}{\partial z} \frac{\partial N_j}{\partial z} \Bigg] \, dx\, dy\, dz \, .
\end{aligned}
$$

Suppose the element size in the $z$-direction is $\Delta z$. By examining the above formula for $A_{i,j}^{I,J,K}$, we see that the difference between $A_{i,j}^{I,J,K_1}$ and $A_{i,j}^{I,J,K_2}$ consists of two parts: one linearly varying part and one quadratically varying part, both depending on $(K_1 - K_2)\Delta z$. More precisely,

$$
A_{i,j}^{I,J,K} = A_{i,j}^{I,J,1} + (K - 1)D_{i,j}^{I,J} + (K - 1)^2 R_{i,j}^{I,J},
$$

where

$$
\begin{aligned}
D_{i,j}^{I,J} = -\Delta z \int \int \int \frac{1}{H} \Bigg[ \eta_x & \left( \frac{\partial N_i}{\partial x} \frac{\partial N_j}{\partial z} + \frac{\partial N_i}{\partial z} \frac{\partial N_j}{\partial x} \right) \\
+ \eta_y \left( \frac{\partial N_i}{\partial y} \frac{\partial N_j}{\partial z} + \frac{\partial N_i}{\partial z} \frac{\partial N_j}{\partial y} \right) & - 2\frac{(z + H)(\eta_x^2 + \eta_y^2)}{\eta + H} \frac{\partial N_i}{\partial z} \frac{\partial N_j}{\partial z} \Bigg] \, dx\, dy\, dz
\end{aligned}
$$

and

$$
R_{i,j}^{I,J} = \Delta z^2 \int \int \int \frac{1}{H} \frac{(\eta_x^2 + \eta_y^2)}{\eta + H} \frac{\partial N_i}{\partial z} \frac{\partial N_j}{\partial z} \, dx\, dy\, dz \, .
$$

This clearly simplifies the construction of the linear system since a 3D assembly process is almost replaced with a 2D assembly process where we only calculate $A_{i,j}^{I,J,1}$, $D_{i,j}^{I,J}$, and $R_{i,j}^{I,J}$. In this way, major efficiency improvement can be obtained (see Table 9.37). We mention that this efficient technique for constructing the linear system has already been used in the simulations for producing the measurements listed in Table 9.16.

**Table 9.37.** The performance difference between the standard FEM::makeSystem and a special makeSystem function for test problem WA3.

| makeSystem | $9 \times 9 \times 9$ | $17 \times 17 \times 17$ | $33 \times 33 \times 33$ | $65 \times 65 \times 65$ |
|---|---|---|---|---|
| standard | 0.03 | 0.23 | 1.88 | 15.09 |
| special | 0.01 | 0.06 | 0.47 | 3.72 |

### 9.6.4  Parabolic Problems

Let us consider a general parabolic problem of the form

$$\frac{\partial u}{\partial t} = \mathcal{L}(u) + f \quad \text{in } \Omega \times (0,T],$$
$$u = u_D \qquad \text{on } \partial\Omega \times (0,T],$$
$$u(\mathbf{x},0) = u^0(\mathbf{x}) \quad \text{in } \Omega,$$

where $\mathcal{L}$ is a linear elliptic operator.

Suppose we apply the finite element method for the spatial discretization and the $\theta$-rule for the temporal discretization. This results in a system of linear equations of the form (9.22) that needs to be solved at each time level. However, the time dependence of $\mathcal{L}$, $f$, and the essential boundary conditions $u_D$ will determine the computational cost of the assembly process. To clarify this, it is beneficial to rewrite $\mathbf{A}$ and $\mathbf{b}$ of (9.22) as:

$$\mathbf{A} = \mathbf{M} + \theta\Delta t\mathbf{K} + \mathbf{A}_{\text{mod}},$$
$$\mathbf{b} = \mathbf{A}_{\text{rhs}}\mathbf{u}^{n-1} + \mathbf{c} + \mathbf{b}_{\text{mod}},$$
$$\mathbf{A}_{\text{rhs}} = \mathbf{M} + (\theta - 1)\Delta t\mathbf{K},$$

where $\mathbf{M}$ is the consistent mass matrix, $\mathbf{K}$ is the stiffness matrix relating to $\mathcal{L}$, and $\mathbf{c}$ contains contributions from the $f$ term. Here, $\mathbf{M}$, $\mathbf{K}$, and $\mathbf{c}$ are constructed by the standard assembly process without regard to the essential boundary conditions. We note that $\mathbf{M}$ and $\mathbf{A}_{\text{mod}}$ are time independent and we have the relation

$$\mathbf{b}_{\text{mod}} = \mathbf{A}_{\text{mod}}\mathbf{u}_D,$$

where $\mathbf{u}_D$ is a vector containing essential boundary conditions on the boundary nodes and zeros otherwise. The following different situations in time-dependence of $\mathcal{L}$, $f$, and $u_D$ should be treated accordingly.

- For the simplest situation where $\mathcal{L}$, $f$, and $u_D$ are all independent of time, it is obvious that **A** remains the same for every time level. Moreover, only a matrix-vector product is needed to generate the right-hand side vector **b**.

- For the situation where only $f$ is time dependent, $\mathbf{b}_{\mathrm{mod}}$ is the same for all time levels. We note that **A** is also time independent. The only needed assembly is for generating **c**.

- For the special case where both $\mathcal{L}$ and $f$ are time independent whereas $u_D$ is time dependent, only two matrix-vector products are needed to construct the linear system.

- When $\mathcal{L}$ is time dependent and $f$ is not time dependent, it is possible to achieve a slightly better performance by skipping the recalculation of **c**.

For a Diffpack simulator treating such parabolic problems, it is therefore important to introduce three flags indicating the time-dependence of the linear operator $\mathcal{L}$, the right-hand side $f$, and the essential boundary condition $u_D$, respectively. The user can redefine the standard FEM::makeSystem function to obtain improved efficiency. Some results are shown in Table 9.38. We note that all the HC2 simulations have used this specially implemented makeSystem function.

**Table 9.38.** Analysis of the specially implemented makeSystem function in a heat conduction simulator class; The CPU times (in seconds) are associated with simulations on a uniform $257 \times 257$ grid.

| time-dependence | | | CPU on makeSystem | |
|---|---|---|---|---|
| $\mathcal{L}$ | $f$ | $u_D$ | first call | later calls |
| NO | NO | NO | 0.89 | 0.05 |
| NO | NO | YES | 1.01 | 0.09 |
| NO | YES | NO | 0.89 | 0.45 |
| NO | YES | YES | 1.03 | 0.47 |
| YES | NO | Y/N | 0.89 | 0.60 |
| YES | YES | Y/N | 0.89 | 0.91 |

## 9.7  Concluding Remarks

In this chapter, we have collected actual CPU time measurements for a number of prototypical PDE simulators. Detailed analyses of the measurements have been carried out and we have established some performance models, which agree quite well with theory. First of all, the actual measurements, which are obtained on a common hardware platform, can serve as a reference for assisting a Diffpack programmer to locate coding inefficiency or errors during code development. Secondly, we have demonstrated how to establish simplified performance models of different numerical methods. Thirdly, the established models may help a Diffpack user to choose an appropriate numerical method for solving a particular system of linear equations. Finally, it is hoped that these performance models can provide a rough prediction of CPU time consumption also for more complicated simulations of real-life problems.

# References

1. E. Arge, A. M. Bruaset, P. B. Calvin, J. F. Kanney, H. P. Langtangen, and C. T. Miller. On the numerical efficiency of C++ in scientific computing. In M. Dæhlen and A. Tveito, editors, *Numerical Methods and Software Tools in Industrial Mathematics*, pages 91–118. Birkhäuser, Boston, 1997.
2. G. B. Astrakhantsev. Methods of fictitious domains for a second-order elliptic equation with natural conditions. *U.S.S.R. Comput. Math. and Math. Phys.*, 18:114–121, 1978.
3. O. Axelsson and G. Lindskog. On the eigenvalue distribution of a class of preconditioning methods. *Numer. Math.*, 48:479–498, 1986.
4. J. B. Bell and G. R. Shubin. Higher-order Godunov methods for reducing numerical dispersion in reservoir simulation. In *SPE Reservoir Simulation Symposium*, pages 179–186, Dallas, Taxes, 1985.
5. A. M. Bruaset. *A Survey of Preconditioned Iterative Methods*. Pitman Research Notes In Mathematics Series 328. Longman Scientific & Technical, 1995.
6. X. Cai, H. P. Langtangen, B. F. Nielsen, and A. Tveito. A finite element method for fully nonlinear water waves. *J. Comput. Phys.*, 143:544–568, 1998.
7. P. Colella. Multidimensional upwind methods for hyperbolic conservation laws. *J. Comput. Phys.*, 87:171–200, 1990.
8. Diffpack Home Page. http://www.diffpack.com.
9. V. Faber and T. Manteuffel. Necessary and sufficient conditions for the existence of a conjugate gradient method. *SIAM J. Numer. Anal.*, 21:352–362, 1984.
10. D. Geselowitz and W. T. Miller. A bidomain model for anisotropic cardiac muscle. *Ann. Biomed. Eng.*, 11:191–206, 1983.
11. G. H. Golub and C. F. van Loan. *Matrix Computations*. Johns Hopkins University Press, 1989.
12. I. Gustafsson. A class of first order factorization methods. *BIT*, 12:142–156, 1978.
13. W. Hackbusch. *Multi-Grid Methods and Applications*. Springer-Verlag, Heidelberg, Berlin, 1985.
14. C. S. Henriquez, A. L. Muzikant, and K. Smoak. Fiber curvature and bath loading effects on activation in thin and thick cardiac tissue preparations. *J. Cardiovascular Electrophysiology*, 7:424–444, 1996.
15. M. R. Hestenes and E. Stiefel. Method of conjugate gradients for solving linear systems. *J. Res. Nat. Bur. Stand.*, 49:409–436, 1952.
16. H. P. Langtangen. *Computational Partial Differential Equations - Numerical Methods and Diffpack Programming*. Textbooks in Computational Science and Engineering. Springer, 2nd edition, 2003.

17. O. Munthe and H. P. Langtangen. Finite elements and object-oriented implementation techniques in computational fluid dynamics. *Computer Methods in Applied Mechanics and Engineering*, 190:865–888, 2000.

18. B. F. Nielsen and A. Tveito. On the approximation of the solution of the pressure equation by changing the domain. *SIAM J. Appl. Math.*, 57:15–33, 1997.

19. J. Saltzman. An unsplit 3d upwind method for hyperbolic conservation laws. *J. Comp. Phys.*, 115:153–168, 1994.

20. B. Smith, P. E. Bjørstad, and W. D. Gropp. *Domain Decomposition, Parallel Multilevel Methods for Elliptic Partial Differential Equations*. Cambridge University Press, 1996.

21. J. Sundnes, G. T. Lines, P. Grøttum, and A. Tveito. Electrical activity in the human heart. In H. P. Langtangen and A. Tveito, editors, *Advanced Topics in Computational Partial Differential Equations – Numerical Methods and Diffpack Programming*. Springer, 2003.

22. P. N. Swarztrauber. The methods of cyclic reduction, Fourier analysis and the FACR algorithm for the discrete solution of Poisson's equation on a rectangle. *SIAM Review*, 19:490–501, 1977.

23. R. S. Varga. *Matrix Iterative Analysis*. Prentice Hall, 1962.

24. G. B. Whitham. *Linear and Nonlinear Waves*. John Wiley & Sons, Inc., New York, 1974.

25. D. M. Young. *Iterative Solution of Large Linear Systems*. Academic Press, 1971.

# Chapter 10

# Electrical Activity in the Human Heart

J. Sundnes[1,2], G.T. Lines[1,2], P. Grøttum[2], and A. Tveito[1,2]

[1] Simula Research Laboratory
[2] Department of Informatics, University of Oslo

**Abstract.** The contraction of the heart is caused by a preceding cellular electro-chemical reaction. This reaction causes an electrical field to be created in the heart and the body. The measurement of this field on the body surface is called the electrocardiogram (ECG). The ECG is an important tool for the clinician in that it changes characteristically in a number of pathological conditions. A motivation for simulating the electrical activity in the heart is to gain a better understanding of the relationship between the ECG signal and the condition of the heart.

## 10.1 The Basic Physiology

The function of the heart muscle is to pump blood to the lungs for oxygenation and then back to the body. The blood enters the heart via the atria and is pumped out by the contraction of the ventricles. The right ventricle supplies blood to the lungs while the more powerful left ventricle pumps blood into the rest of the body. The circuit is illustrated in Figure 10.1.

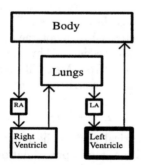

**Fig. 10.1.** The flow of blood through the body. The thick line around the left ventricle indicates the larger amount of muscle mass compared to the right ventricle. RA and LA represents the right and left atrium, respectively.

The tissue of the ventricles and atria are composed mainly of muscle fibers. These fibers are oblong-like cylinders with branches that connect to neighboring fibers. The contraction of the heart muscle is initiated by an electrical signal starting in the sinoatrial node, see e.g. Keener and Sneyd [14] for an explanation. The electrical signal then travels along a special type of cells known as Purkinje fibers. The primary function of these cells is to provide fast conduction of electrical pulses, and at their endings the surrounding muscle tissue is stimulated. A time delay caused by cells in the atrioventricular node (see e.g. [14]) ensures that the atria are stimulated before the ventricles. When the muscle cells are stimulated electrically they rapidly *depolarize*, i.e., the electrical potential inside the cell is changed. The depolarization causes the contraction of the cells, and the electrical signal is also passed on to neighboring cells. Thus the stimulus applied at the end points of the Purkinje fibers cause a wave of depolarization throughout the heart muscle. The direction of the depolarization wave is determined by the orientation of the muscle fibers, because the conduction of the signal is fastest in the direction of the fibers. This makes the heart tissue *anisotropic*, i.e. the electrical conductivity is not the same in all directions.

## 10.2   Outline of a Mathematical Model

The ECG is a measurement of the electrical potential on the surface of the body. In order to achieve realistic simulations of these measurements, we need to simulate how the electrical signal is created in the heart and how it is conducted through the heart tissue and the surrounding torso. The conduction of the electrical signal through the torso can be described with a simple mathematical model. The body may be regarded as a passive conductive medium, with a given conductance. Biological tissue is made up of cells, and the conductivity properties are thus likely to vary over a very short length scale, e.g. from the interior of a cell to the surrounding plasma. Problems with representing these short-scale variations will be avoided by treating the tissue as a continuum characterized by a volume averaged conductance.

The electrical field in the body is denoted by $E$. If we assume that this field is not changing in time, we have from Maxwell's equations that $\nabla \times E = 0$ and the vector field $E$ can thus be written in terms of a (scalar) potential function $u_T$,

$$E = -\nabla u_T. \tag{10.1}$$

The fluctuations of the electrical field in the human torso are so slow that the above relationship is also a good approximation in the non-static case. This situation is termed the *quasi-static condition*, see [22].

Assuming that Ohm's law applies to the tissues in the torso, the current $J_T$ may be expressed as

$$J_T = -M_T E = -M_T \nabla u_T, \tag{10.2}$$

where $M_T$ is the conductivity of the tissue. Because many types of biological tissue have anisotropic conductance properties, $M_T$ is not a scalar quantity.

There will also be discontinuous variations in $M_T$ corresponding to the different conductive properties of the tissues in the torso. There are no sources of electrical current in the tissue, hence conservation of current yields

$$\nabla \cdot J_T = 0. \tag{10.3}$$

Inserting (10.2) into (10.3) yields the equation for the electrical potential in the torso

$$\nabla \cdot (M_T \nabla u_T) = 0, \quad x \in T, \tag{10.4}$$

where the domain $T$ is the torso without the heart. Solving this equation to find the body surface potential from a known heart surface potential is called the forward problem.

The simple model used for the tissue outside the heart is not suitable for describing the heart tissue. Because the heart muscle cells belong to the class of so-called excitable cells (see e.g. [14]), the electrical properties of the heart muscle are more complex, and it cannot be modeled as a passive conductive medium. The electrical activity of heart tissue is described by the bidomain model, introduced by Geselowitz[7] in the late 70s. Because variations in the potential difference across the cell membrane are essential to the function of excitable tissue, the model treats the tissue as two separate continuous domains; the intracellular and the extracellular domain. The electrical potential in the two domains are denoted by $u_i$ and $u_e$, respectively. The total current in the two domains must be conserved, and current passing from one domain to the other must cross the cell membrane. If we define the positive direction for the membrane current to be out of the cell, we get the following equations for the potentials.

$$\nabla \cdot (M_i \nabla u_i) = \chi I_m, \tag{10.5}$$
$$\nabla \cdot (M_e \nabla u_e) = -\nabla \cdot (M_i \nabla u_i). \tag{10.6}$$

A detailed derivation of this model will be given below. The tensors $M_i$ and $M_e$ are the conductivities in the intra- and extracellular domain, respectively. The right hand side term $\chi I_m$ in (10.5) is the electrical current across the cell membrane. The membrane current $I_m$ is most easily expressed in terms of current per cell membrane area, while the other quantities of the equations are measured per volume. Hence it is necessary to introduce the scaling factor $\chi$, which is the ratio of membrane surface area to tissue volume. The membrane current term $I_m$ is a non-linear function of the form

$$I_m = I_m(v, \dot{v}, s),$$

where $v = u_i - u_e$ is the transmembrane potential, i.e. the potential difference across the cell membrane. The time derivative of $v$ is denoted by $\dot{v}$, and $s$ is a vector characterizing the state of the cell membrane. Models for this term will be discussed in detail below.

On realistic geometries it is not possible to solve the Eqs. (10.5)-(10.6) analytically, and even numerically the model is hard to solve. The purpose of this chapter is to describe the implementation of a simulator that simultaneously solves the equations (10.4), (10.5) and (10.6), to determine the electrical potentials throughout the heart muscle and the surrounding torso.

The rest of the chapter will be organized as follows. Section 10.3 presents the bidomain model in more detail. We make basic assumptions about the properties of the tissue and derive the equations from these assumptions, similar to what was done above for the surrounding torso. In Section 10.4 we complete the mathematical model by introducing suitable boundary conditions and formulating the complete problem in a compact form. Section 10.5 gives an introduction to the mathematical models for heart cells and excitable cells in general, and may be skipped by readers primarily interested in computational issues. In Section 10.6 we return to the bidomain equations and the forward problem to discuss how these equations may be discretized and solved numerically. Section 10.7 outlines one possible strategy for implementing the simulator, while sections 10.8 and 10.9 are concerned with simulation results and optimization of the numerical methods. Finally, we summarize our results in Section 10.10.

## 10.3   The Bidomain Model

Above we used Ohm's law and conservation of current to derive an equation for the electrical potential in the passive tissue surrounding the heart. We also introduced the more complex bidomain model describing the electrical activity in the heart muscle. In this section we will present a detailed derivation of the bidomain equations, based on conservation of current and basic assumptions regarding the behavior of the cell membrane.

### 10.3.1   A Continuous Model for the Heart Tissue

Modeling individual cells becomes impossible when simulating the activity of the entire heart, because the number of cells is simply too large ($\sim 10^{10}$). Instead we may follow the approach introduced above for the torso tissue, and study volume averaged quantities. If the quantity under study is, e.g., $u$ then the volume averaged quantity $\bar{u}$ at a point $P = (p_1, p_2, p_3)$ is given as

$$\bar{u} = \frac{1}{8r^3} \int_{-r}^{r} \int_{-r}^{r} \int_{-r}^{r} u(p_1 + x, p_2 + y, p_3 + z) \; dx \; dy \; dz$$

for some small $r$. This is a standard technique in continuum mechanics. By applying this technique to the heart muscle, one obtains a continuous approximation to the discrete reality, avoiding the complexity of having to study the cellular structure.

## 10.3.2    Derivation of the Bidomain Equations

As was briefly described in Section 10.2, the basis for the bidomain model is the division of the heart tissue into two domains; the extracellular domain and the intracellular domain. Each domain is continuous and encompasses the whole heart muscle, i.e., the domains have identical geometries and every point of the heart muscle is found in both domains. Inside each domain the current flow is purely resistive, i.e., it may be described by Ohm's law. This is a reasonable assumption in the extracellular domain where the current can flow in the continuous space between the cells. In the intracellular domain the flow model can be justified by the fact that the cells are connected and that current can pass from cell to cell via so-called gap junctions, see [14]. The magnitude of the resistance in both domains is characterized by conductivity parameters.

The quasi-static condition introduced above is assumed to be valid also for the electrical fields in the heart tissue. The electrical fields in the intra- and extracellular domains may hence be written in terms of the scalar potentials $u_i$ and $u_e$, respectively. Employing Ohm's law gives

$$J_i = -M_i \nabla u_i, \tag{10.7}$$
$$J_e = -M_e \nabla u_e, \tag{10.8}$$

with $J_i$ and $J_e$ denoting the intra- and extracellular current, respectively. The structure of the conductivity tensors $M_i$ and $M_e$ will be described in Section 10.5.10. If we assume no build-up of charge in any point, we have from conservation of charge that the outflow of charge from one domain is equal to the inflow of charge for the other domain. Mathematically we have

$$\int_{\partial D} J_i \cdot n \, dS = -\int_{\partial D} J_e \cdot n \, dS, \tag{10.9}$$

where $D$ is an arbitrary subdomain within the heart. The surface of $D$ is denoted by $\partial D$, while $n$ denotes the outward unit normal of $\partial D$. Using the Gauss theorem and the fact that (10.9) must hold over every volume we get

$$\nabla \cdot J_i = -\nabla \cdot J_e. \tag{10.10}$$

The divergence currents in (10.10) will pass between the domains. They are thus crossing the membrane and must be equal to the transmembrane current per unit volume. The transmembrane current $I_m$ is most easily expressed in terms of current per unit area of membrane surface. As described in Section 10.2, the transmembrane current per unit volume is then obtained by multiplying $I_m$ with a scaling factor $\chi$, which is the membrane surface area per unit volume of tissue. If we think of the heart cells as cylinders, $\chi$ expresses the surface area of all the cylinders – measured in $cm^2$ – that are contained within a $1 \, cm^3$ volume of tissue, see Figure 10.2.

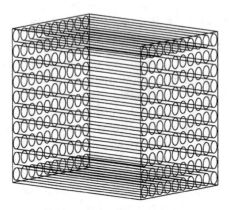

**Fig. 10.2.** The parameter $\chi$ expresses the amount of membrane area inside a unit volume cube.

The current leaving the extracellular domain is equal to $\chi I_m$ and opposite to the current leaving the intracellular domain, i.e., we have

$$\nabla \cdot J_i = -\chi I_m \quad \text{and} \quad \nabla \cdot J_e = \chi I_m . \tag{10.11}$$

We see that the bidomain equations (10.5)-(10.6), introduced in Section 10.2 are easily derived from (10.10) and (10.11). However, in the following it will be convenient to work with the transmembrane potential $v = u_i - u_e$, and to reformulate the bidomain equations by eliminating $u_i$ and use $v$ instead. After some simple manipulations using $u_i = u_e + v$, we get

$$\nabla \cdot (M_i \nabla v) + \nabla \cdot (M_i \nabla u_e) = \chi I_m, \tag{10.12}$$
$$\nabla \cdot ((M_i + M_e) \nabla u_e) = -\nabla \cdot (M_i \nabla v) . \tag{10.13}$$

To be able to solve (10.12) and (10.13), we need to express the membrane current $I_m$ in terms of the primary unknowns $v$ and $u_e$. We thus need a model for the electrical properties of the cell membrane. For now, we will simply state that the cell membrane may be modeled as a resistor and a capacitor in parallel. The transmembrane current $I_m$ is a sum of a capacitive current and an ionic current. The ionic current will depend on the state of the cell membrane, characterized by a vector $s$. Mathematically, we have the membrane current given by

$$I_m = C_m \frac{\partial v}{\partial t} + I_{\text{ion}}(v, s), \tag{10.14}$$

where $C_m$ is the capacitance of the membrane. A more thorough explanation of this cell membrane model will be given in Section 10.5. A large number of different models exist for the ionic current term $I_{\text{ion}}$, with huge differences in complexity and physiological correctness. Common to most of the models is that the state vector $s$ is determined from a system of ordinary differential equations (ODEs).

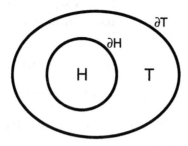

**Fig. 10.3.** Schematic view of the two domains and their boundaries.

## 10.4  A Complete Mathematical Model

To complete the mathematical model introduced in the preceding sections, we need a set of boundary conditions. We denote the heart domain by $H$ and the surrounding torso by $T$. In Figure 10.3 we see that the boundary of $T$ is made up of the surface of the heart and the body surface, while the boundary of $H$ is the surface of the heart. Following the notation from the figure, we need conditions for $v$ and $u_e$ on $\partial H$ and for $u_T$ on $\partial H$ and $\partial T$.

### 10.4.1  Boundary Conditions on the Heart-Torso Interface

For the surface of the heart there exist several choices of boundary conditions, see e.g. [11], but we shall here follow Colli Franzone et al [6]. On $\partial H$ the extracellular domain of the heart connects to the tissue surrounding the heart, and thus $u_e$ and $u_T$ will refer to the same quantity,

$$u_e(x) = u_T(x), \quad x \in \partial H. \tag{10.15}$$

The relationship between the current densities $J_T$, $J_e$ and $J_i$ on the boundary is not as easy to motivate. Following [6] we set

$$n \cdot J_T = n \cdot (J_e + J_i). \tag{10.16}$$

This states that both intra- and extracellular current can flow directly into $T$, implying that the intracellular space is in direct contact with $T$. Equation (10.16) may be rewritten as

$$n \cdot M_T \nabla u_T = n \cdot (M_i + M_e) \nabla u_e + n \cdot (M_i \nabla v). \tag{10.17}$$

A third boundary condition is needed to specify $v$. A convenient choice is

$$n \cdot (M_i \nabla v) = 0. \tag{10.18}$$

The physiological justification for this condition is not straightforward. One way to interpret (10.18) is to look at the right hand side of (10.13). This source term is only defined in $H$ and one may assume that the corresponding current only flows inside $H$. This requirement is fulfilled if (10.18) holds. Using (10.18) we can simplify (10.17) to

$$n \cdot M_T \nabla u_T = n \cdot (M_i + M_e) \nabla u_e . \tag{10.19}$$

There exist alternative sets of boundary conditions for the surface of the heart. An advantage of the set presented here is that it allows later simplifications which ease the implementation of the simulator.

### 10.4.2    Boundary Conditions on the Surface of the Body

We assume that no electrical current leaves the torso. Since the current is given by Ohm's law, the condition for the potential on the surface becomes

$$n \cdot (M_T \nabla u_T) = 0, \quad x \in \partial T . \tag{10.20}$$

This is a boundary condition of the Neumann type, hence the solution $u_T$ is only determined up to a constant. Because the value of the extracellular potential on $\partial H$ is only specified in terms of $u_T$, this applies to $u_e$ as well.

### 10.4.3    Summary of the Mathematical Problem

The equations and boundary conditions introduced in the previous sections may be summarized in a compact formulation of the mathematical problem.

$$\frac{\partial s}{\partial t} = F(t, s, v; x) \qquad x \in H, \tag{10.21}$$

$$\nabla \cdot (M_i \nabla v) + \nabla \cdot (M_i \nabla u_e) =$$
$$C_m \chi \frac{\partial v}{\partial t} + \chi I_{\text{ion}}(v, s) \quad x \in H, \tag{10.22}$$

$$\nabla \cdot (M_i \nabla v) + \nabla \cdot ((M_i + M_e) \nabla u_e) = 0 \qquad x \in H, \tag{10.23}$$

$$\nabla \cdot (M_T \nabla u_T) = 0 \qquad x \in T, \tag{10.24}$$

$$n \cdot (M_i \nabla v) = 0 \qquad x \in \partial H, \tag{10.25}$$

$$u_e = u_T \qquad x \in \partial H, \tag{10.26}$$

$$n \cdot ((M_i + M_e) \nabla u_e) = n \cdot (M_T \nabla u_T) \qquad x \in \partial H, \tag{10.27}$$

$$n \cdot (M_T \nabla u_T) = 0 \qquad x \in \partial T . \tag{10.28}$$

Eq. (10.21) determines the state of the cell membrane, which is needed to compute the ionic current $I_{\text{ion}}$. Models for this term will be discussed in detail in Section 10.5. In our continuous approximation to the discrete tissue, the state vector $s$ will in fact be a continuous field with variations in space and time. This is the reason for the inclusion of the position $x$ in the right hand side function $F(t, s, v; x)$. A numerical method for solving (10.21)-(10.28) will be presented in Section 10.6.

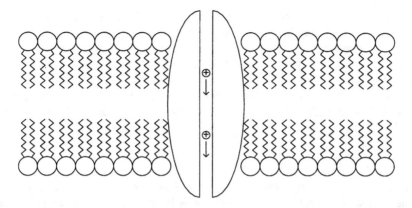

**Fig. 10.4.** A cross sectional view of the cell membrane with the bilipid layer and a gating protein. The protein shields the hydrophobic part of the phospholipids letting polar molecules and ions pass the membrane.

## 10.5   Physiology of the Heart Muscle Tissue

In the preceding section we formulated a complete model for the electrical potentials in the heart and the torso. In the derivation of the bidomain equations we simply stated that a suitable model for the membrane current is the sum of a capacitive current and an ionic current term, as described by (10.14). The purpose of this section is to give a short introduction to cell physiology, to justify this membrane model. We will also introduce and explain some well known models for the transmembrane ionic current. Finally we present how the cells are interconnected to form the muscle tissue, and how the structure of the tissue affects the conductivity properties.

### 10.5.1   Physiology of the Cell Membrane

A cell is delimited by the plasma membrane, which controls the flow of substances into and out of the cell. The membrane is made up of a double layer of phospholipids. These lipids have an ambivalent relationship to water. They have a head which contains polar molecular bonds and is therefore attracted to water. On the other hand, the tail is non-polar, thus being repelled by water. In the cell membrane the hydrophilic head of the lipid faces the interior and the exterior of the cell, with the hydrophobic tails forming the central part of the membrane, see Figure 10.4.

The hydrophobic core of the bilipid layer impedes the transport of polar molecules and ions, but gating proteins embedded in the membrane form channels in which these substances may pass. Figure 10.4 shows a schematic view of how a gating protein forms a channel, through which ions may pass the cell membrane. The channels are specialized and only a narrow group of molecules or even just a single substance may pass through a particular channel. The channels may be either passive, which means that their compound pass the membrane at a rate determined by the concentration gradient, or they may be active, using energy to pump molecules or ions against their concentration gradients.

The composition of the intra- and extracellular fluid differs, both chemically and electrically. There is thus a potential difference across the cell membrane which will be denoted as the transmembrane potential.

### 10.5.2   The Nernst Potential

The flow of ions through the cell membrane channels will depend upon the transmembrane potential. For a purely resistive current a good model would be

$$J = \sigma E. \tag{10.29}$$

This is again Ohm's law where $J$ is the current, $E$ is the electrical field and $\sigma$ some constant conductance. However, ion channels are far more complicated than an ordinary resistor. The flow depends on the ionic concentration in addition to the electrical potential. We stated above that the electrical field may be written as the gradient of a scalar potential, $E = \nabla \phi$. If we first assume that we have no concentration gradients, the flux of ions due to an electrical field is given by Planck's equation:

$$J_P = -m\frac{z}{|z|}c\nabla\phi, \tag{10.30}$$

where $m$ is the mobility of the ion in the liquid (see e.g. [14]). The charge of the ion is denoted by $z$, and $c$ is the concentration of the ion. The flux will also depend upon the concentration gradients. In an electrically neutral field the flux will not be zero since ions will tend towards areas with lower concentration. This is modeled by Fick's law:

$$J_F = -D\nabla c, \tag{10.31}$$

where $D$ is the diffusion coefficient. In the general case, we have both concentration gradients and a non-zero electrical field. The total current is then the sum of a diffusive and an electrically driven flow,

$$J = J_P + J_F.$$

The relation between the mobility $m$ and the diffusion constant $D$ is given by

$$m = D\frac{|z|F}{RT},$$

where $F$ is the Faraday constant, $R$ is the ideal gas constant, and $T$ is the absolute temperature. The total current may be written as

$$J = -D(\nabla c + \frac{zF}{RT}c\nabla\phi). \tag{10.32}$$

A natural question to ask is how the electrical field $\nabla\phi$ is generated. In a lab it can be applied externally, while in a living cell it is generated by the *specificity* of the ion channels. Specificity means that a channel (mainly) lets only one type of ions pass. The following example will illustrate how specificity can generate an electrical field.

We assume that two chambers filled with an electrically neutral liquid are separated by a membrane, which only allows $Na^+$-ions to pass. NaCl (ordinary table salt) is added in different concentrations to each chamber, say 100mM to chamber A and 10mM to chamber B. Both solutions will initially remain neutral since NaCl in itself is neutral. As the salt dissolves into $Na^+$ and $Cl^-$ ions, the concentration difference will cause a net flux of ions from A to B, but only for $Na^+$-ions since $Cl^-$ cannot cross the membrane. As the positive ions enter the B chamber, the solution in B will no longer be neutral. Likewise, the solution in chamber A will get a surplus of negative ions. We thus have a potential difference across the membrane.

Note that this would not have happened if also $Cl^-$ ions could cross the membrane. In that case ions of both types would flow into the B chamber at equal rate until the concentrations were the same on both sides. No potential difference would occur since the solutions would be neutral at all times.

The force of the electrical field generated by the potential difference will drive the ions in the opposite direction of the movement due to diffusion. An equilibrium will be reached where the flow of ions due to the concentration difference $(J_F)$ will be equal in size, but opposite in direction to the flow due to the potential difference $(J_P)$. In other words we will have $J_F = -J_P$.

For flow through membrane channels it is reasonable to only consider variation along the length of the channel. We arrange the coordinate system so that the $x$-axis is along the channel, and such that $x = 0$ is the interior boundary of the channel and $x = L$ is the outer boundary of the channel. In 1D with $J = 0$ (10.32) becomes

$$\frac{dc}{dx} + \frac{zF}{RT}c\frac{d\phi}{dx} = 0.$$

Dividing by $c$ and integrating from 0 to $L$ yields

$$\int_0^L \frac{1}{c}\frac{dc}{dx}dx + \int_0^L \frac{zF}{RT}\frac{d\phi}{dx}dx = 0$$

and finally

$$\ln(c)|_{c(0)}^{c(L)} = -\frac{zF}{RT}(\phi(L) - \phi(0)) = \frac{zF}{RT}v,$$

since the transmembrane potential is $v = \phi_i - \phi_e$. The value of the transmembrane potential at zero flux is then

$$v_{eq} = \frac{RT}{zF}\ln\left(\frac{c_e}{c_i}\right), \tag{10.33}$$

where $c_e$ and $c_i$ are the concentrations outside and inside the cell membrane, respectively. The potential $v_{eq}$, for which the net flux is zero, is referred to as the Nernst equilibrium potential.

### 10.5.3    Models for the Ionic Current

When the transmembrane potential differs from the Nernst equilibrium potential, we know that there will be a net flow of ions through the channel. To express the details of how the current relates to the transmembrane potential, we need to derive additional models. The simplest expression for the ionic current that satisfies the Nernst principle is a linear model, giving the current as

$$I = G(v - v_{eq}),$$

where $G$ is the constant conductivity. Another model for ionic current is derived from the constant field assumption. It states that the strength of the electrical field is independent of the ion flux and does not vary in space. Following the notation introduced above we can thus write $\nabla \phi = v/L$. As above we consider the 1D case and insert this into (10.32). We get

$$\frac{dc}{dx} - \frac{zFv}{RTL}c + \frac{J}{D} = 0. \tag{10.34}$$

This is an ODE in $c$ with known values at both endpoints, and $J$ as an additional unknown to be determined. Solving the equation gives the following expression for the flux,

$$J = \frac{D}{L}\frac{zFv}{RT}\frac{c_i - c_e \exp\left(\frac{-zFv}{RT}\right)}{1 - \exp\left(\frac{-zFv}{RT}\right)}.$$

Although this expression is far more complex than the simple linear expression above, it is easy to verify that the flux is zero when $v = v_{eq}$.

### 10.5.4    Electric Circuit Model for the Membrane

If we for a moment exclude the existence of the channels in the membrane, the most characteristic property of the membrane is that it separates charge. We may thus think of the membrane as a capacitor. A capacitor consists of two conducting plates separated by an insulating material. The intra- and extracellular fluid are in our case the conductive plates and the membrane is the insulating material. The potential $v$ over a capacitor is proportional to the amount of separated charge $q$. We have

$$v = q/C, \tag{10.35}$$

where $C$ is called the capacitance of the capacitor.

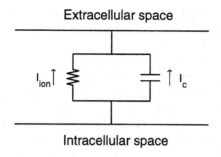

**Fig. 10.5.** The membrane is modeled as a resistor and a capacitor coupled in parallel.

Since the membrane does allow the ions to cross, it can not be viewed as a simple capacitor. It is more correct to expand the model with a resistor coupled in parallel with the capacitor. This circuit is shown in Figure 10.5.

The flow of ions will change the amount of charge separated by the membrane, and thus also the transmembrane potential according to (10.35). Consider the change over a time interval $\Delta t$. From (10.35) it follows that $\frac{\Delta v}{\Delta t} = \frac{1}{C}\frac{\Delta q}{\Delta t}$ and in the limit we get

$$\frac{dv}{dt} = \frac{1}{C}\frac{dq}{dt}. \tag{10.36}$$

The term $\frac{dq}{dt}$ is called the capacitive current and is denoted by $I_C$. The total transmembrane current $I_m$ is the sum of the ionic and capacitive current,

$$I_m = I_{\text{ion}} + I_C. \tag{10.37}$$

Combining this with (10.36) gives

$$I_m = I_{\text{ion}} + C\frac{dv}{dt}, \tag{10.38}$$

which is exactly the expression which is introduced in (10.14) and used in the derivation of the bidomain equations.

If the loop in Figure 10.5 is closed, i.e., there is no net transport of ions to or from the area, we have by conservation of current that

$$I_{\text{ion}} + C\frac{dv}{dt} = 0. \tag{10.39}$$

This means that all the ions passing the membrane accumulate and change the membrane potential accordingly.

## 10.5.5   Channel Gating

In Section 10.5.1 we introduced membrane channels in the form of transport proteins embedded in the membrane. For channels with constant conductance, the flow of ions is described by the expressions derived in Section

10.5.3. However, the conductance of ionic channels may change over time in response to changes in the transmembrane potential. The reason is that charged parts of the channel proteins cause the shape of the molecule to change when the voltage difference across the membrane changes. To illustrate how the behavior of these channels may be modeled mathematically, we consider a simple type of channel which may be either open or closed. We denote by $[O]_0$ the total concentration of such channels embedded in the cell membrane. The concentrations of channels in the open state and closed state are denoted by $[O]$ and $[C]$, respectively. We have $[O] + [C] = [O]_0$, and we assume that $[O]_0$ is constant. The change between the open and closed state is expressed by

$$C \underset{\beta}{\overset{\alpha}{\rightleftharpoons}} O, \qquad (10.40)$$

where $\alpha$ is the rate of channel opening and $\beta$ is the rate of channel closing. Applying the law of mass action (see e.g. [14]) to (10.40) we get

$$\frac{d[O]}{dt} = \alpha(v)[C] - \beta(v)[O] = \alpha(v)([O]_0 - [O]) - \beta(v)[O].$$

Dividing by the total concentration of channels yields

$$\frac{dg}{dt} = \alpha(v)(1 - g) - \beta(v)g, \qquad (10.41)$$

where $g = [O]/([C] + [O])$ is the portion of open channels.

To compute the current through a given channel type, we first compute the maximum current, i.e., the current driven by concentration differences and the transmembrane potential. The actual flow is then computed by multiplying this maximum flow with $g$, the portion of channels which are open.

In general $\alpha$ and $\beta$ depend on $v$, but the structure of the functions $\alpha(v)$ and $\beta(v)$ are not motivated by physiological considerations. Instead, their parameters are found by fitting them to empirically obtained data. To give a better understanding of the physical interpretation of $\alpha_g(v)$ and $\beta_g(v)$, (10.41) can be rewritten as

$$\frac{dg}{dt} = (g_\infty - g)/g_\tau, \qquad (10.42)$$

with

$$g_\infty = \alpha/(\alpha + \beta) \text{ and } g_\tau = 1/(\alpha + \beta).$$

If we for a moment assume that $g_\infty$ and $g_\tau$ are independent of $v$, the solution of (10.42) is

$$g(t) = g_\infty + (g_0 - g_\infty)e^{-t/g_\tau}, \qquad (10.43)$$

where $g_0$ is the initial value. We see that $g(t) \to g_\infty$ as $t \to \infty$, thus $g_\infty$ is the steady-state value of $g(t)$. The rate at which $g(t)$ approaches $g_\infty$ is controlled by the size of $g_\tau$. Since $g_\infty$ and $g_\tau$ are not constants the solution

(10.43) is not correct. However, if the time constant is small, so that the gating variable reaches its steady-state value almost momentarily, $g_\infty$ might be used as an approximation to $g(t)$. This equilibrium approach is used for important currents in some of the cell models which are presented below.

## 10.5.6   The Hodgkin-Huxley Model

The framework for channel gating presented this far was originally developed by Hodgkin and Huxley [12] in their work on nerve cells. For completion we state their model here. We will later describe more relevant models but they can all be viewed as modifications and refinements of the Hodgkin-Huxley model.

Three currents are assumed, a sodium current $I_{Na}$, a potassium current $I_K$, and a non-specified leakage current $I_L$. All the currents are linear in $v$. Denoting the capacity of the membrane by $C_m$, we have from (10.39)

$$-C_m \frac{dv}{dt} = I_{ion} = I_{Na} + I_K + I_L$$
$$= g_{Na}(v - v_{Na}) + g_K(v - v_K) + g_L(v - v_L). \quad (10.44)$$

The leakage current $I_L$ is not gated, i.e., $g_L$ is constant. The potassium current $I_K$ is gated by four identical subunits, each governed by an equation like (10.41). If we denote the probability that one of these gates is open by $n$, the probability that all four are open is given by $n^4$. The conductance for this channel is then $g_K = \bar{g}_K n^4$, where $\bar{g}_K$ is the maximal conductance of the channel. Hodgkin and Huxley found that the sodium current $I_{Na}$ shows a more complex behavior. It is first activated by the increased action potential, and then inactivated before the cell returns to its resting state. Modeling this behavior requires two different gate variables, $m$ and $h$, one for activating and one for inactivating the channel. Reproducing the fast activation of the channel was found to require three instances of the sub-unit $m$. Accordingly, $g_{Na} = \bar{g}_{Na} m^3 h$, where the dynamics of $m$ and $h$ are governed by equations like (10.41).

## 10.5.7   The Beeler-Reuter Model

The Hodgkin-Huxley model presented above gives an accurate description of the electrical behavior of a specific type of nerve cells. It is, however, not a suitable model for heart cells. One of the first models for this type of cells was the Beeler-Reuter model [1], published in 1976. The state vector of this model consists of eight variables:

$$s = (v, c, x, m, h, j, d, f).$$

Here $c$ is the intracellular calcium concentration, which is scaled for notational convenience: $c = 10^7 [Ca^{2+}]$. The latter six variables are gating variables,

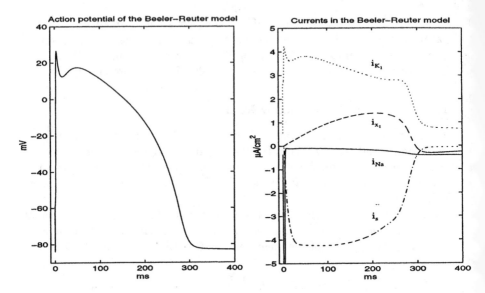

**Fig. 10.6.** Left: The action potentials of the Beeler-Reuter model. Right: The four currents of the model. The sodium current, $i_{Na}$, has been clipped, it peaks at $-129\mu A/cm^2$.

controlled by equations like (10.41). The gating variables $m, h$ and $j$ are related to the fast inward sodium current $I_{Na}$, while $d$ and $f$ are related to a slow inward current $I_s$, mainly consisting of $Ca^{2+}$ ions. The final gate variable $x$ is related to an outward current $I_x$. The total ionic current is given as a sum of these three currents plus an outward potassium current $I_K$ and an externally applied current, $I_{app}$. The equation for the transmembrane potential is then

$$\frac{dv}{dt} = -\frac{1}{C_m}(I_K + I_x + I_{Na} + I_{Ca} + I_s - I_{app}).$$

The update rule for the calcium concentration is given by

$$\frac{dc}{dt} = 0.07(1 - c) - I_s.$$

The equations for the gating variables have the same structure as (10.41), while the definitions of the currents and their gating coefficients ($\alpha$ and $\beta$) are listed in [1]. Figure 10.6 shows the action potential and the currents of the model in a simulation with standard parameters.

### 10.5.8   The Luo-Rudy Model

The Luo-Rudy model [18] is a more detailed description of the heart cell compared to the Beeler-Reuter model. It is based on the results from a number of studies [12,19,1,4]. An important forerunner for the work was the model of DiFrancesco and Noble [2].

The state vector consists of the transmembrane potential, five ionic concentrations and six gating variables,

$$s = (v, [Na^+]_i, [K^+]_i, [Ca^{2+}]_i, [Ca^{2+}]_{JSR}, [Ca^{2+}]_{NSR}, m, h, j, x, d, f).$$

The gating variables correspond closely to those of the Beeler-Reuter model, i.e., they are related to the same currents and have the structure as in (10.41), although the coefficients differ. In addition to the internal calcium concentration, the model also includes the internal sodium and potassium concentrations and the calcium concentrations in the sarcoplasmic reticulum (SR). The SR is a network inside the cell which is of importance to the dynamics of calcium. It is divided into two compartments, the junctional sarcoplasmic reticulum (JSR) and the network sarcoplasmic reticulum (NSR).

The dynamic change of the concentrations is determined by a number of ionic currents. The sign of the current is such that an outward current is positive. For example the rate of change for sodium is

$$\frac{d[Na^+]_i}{dt} = -K \cdot I_{Na,tot},$$

where $K$ is a constant that depends upon the geometry of the cell, as explained in [18]. Since a positive current reduces the internal concentration, $d[Na]_i/dt$ and $I_{Na,tot}$ have opposite signs.

The total transmembrane current is the sum of the three main ionic currents,

$$I_{ion} = I_{Na,tot} + I_{K,tot} + I_{Ca,tot}, \tag{10.45}$$

each representing the total current of one type of ions. These currents are again sums of different types of currents.

The sodium current $I_{Na,tot}$ is given by

$$I_{Na,tot} = I_{Na} + 3I_{NaCa} + 3I_{NaK} + I_{ns,Na} + I_{Nab} + I_{CaNa}.$$

Here $I_{Na}$ is the inward sodium current which is responsible for the fast depolarization. $I_{NaCa}$ is the current due to the sodium-calcium exchanger, which may be positive or negative. A positive $I_{NaCa}$ means that there is a sodium outflux, and the stoichiometry of the exchanger is 3:1 ($Na^+ : Ca^{2+}$). The sodium-potassium pump is modeled with the current $I_{NaK}$. When positive, which it is under all normal conditions, it extrudes three sodium ions in exchange for two potassium ions. A special non-specific channel activated by calcium is modeled by $I_{ns,Na}$. These channels are also permeable to potassium, but practically impregnable to $Ca^{2+}$ ions. A linear leakage current is also included, $I_{Nab}$. Finally, $I_{CaNa}$ is the sodium current passing through the so called L-type channels. These are primarily permeable to calcium but $Na^+$ and $K^+$ ions may also pass.

The potassium current is given by the following expression:

$$I_{K,tot} = I_K + I_{K1} + I_{Kp} - 2I_{NaK} + I_{ns,K} + I_{CaK} .$$

Two main components are the time dependent and time independent current, denoted by $I_K$ and $I_{K1}$ respectively. The so-called plateau current $I_{Kp}$ is similar to $I_{K1}$, but with a different voltage dependence for the permeability. As above, $I_{NaK}$ is the sodium-potassium pump, the negative sign is there since a positive current means an increase in the intracellular potassium. The current $I_{ns,K}$ passes through the non-specific channels activated by calcium. Finally, $I_{CaK}$ is the small potassium current passing through the L-type channels.

The third transmembrane ionic current, the calcium current, is given as

$$I_{Ca,tot} = I_{Ca} - 2I_{NaCa} + I_{p(Ca)} + I_{Ca,b} .$$

A major component of this current is $I_{Ca}$ which denotes the flow of ions through the L-type channels. The sarcolemma pump $I_{p(Ca)}$ extrudes calcium ions, like the Na-Ca exchanger $I_{NaCa}$. The final current $I_{Ca,b}$ is a linear leakage current.

The expressions for most of the currents are of the form presented in Section 10.5.3. For a complete definition we refer to the original publication of the model in [18]. Many of the expressions are directly taken from earlier models, but with parameters adjusted to fit into the new context (more currents and new data). An example of this is the fast inward sodium current $I_{Na}$ taken from the model of Beeler and Reuter. Other expressions are based on previous formulations but are structurally modified or expanded, e.g. the current due to the sodium-calcium exchanger $I_{NaCa}$. Finally, some of the expressions are original for this model, an example of this is the model of the current through the sodium-potassium pump.

Figure 10.7 shows the action potential for a simulation with standard parameters along with the most important currents. The time independent current $I_V$ is the sum of all currents that are not controlled by time dependent gates, except for the time independent Na-Ca exchanger which has been plotted separately. The current $I_V$ is then defined as

$$I_V = I_{K1} + I_{Kp} + I_{NaK} + I_{pCa} + I_{Nab} + I_{Cab} + I_{ns,Na} + I_{ns,K} .$$

Other currents in addition to those described here have been observed, but are not included in the model. They are primarily important under pathological conditions and may have to be included for such studies. These include a chloride current, and calcium-, sodium- and ATP-activated potassium currents.

*Currents Related to the Sarcoplasmic Reticulum* As mentioned above, the Luo-Rudy model includes the sarcoplasmic reticulum. In the Luo-Rudy model

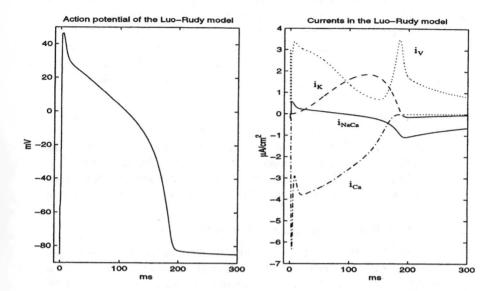

**Fig. 10.7.** Left: The action potentials of the Luo-Rudy model. Right: Four of the most important currents of the model. The time independent current, $I_V$, is the sum of all the time independent currents as described in the text. The sodium current, $I_{Na}$, has been left out, it peaks at $-367\mu A/cm^2$ after 1.5 ms and returns equally fast back to zero.

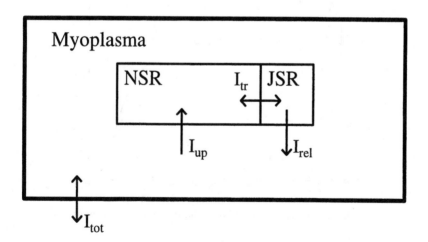

**Fig. 10.8.** The compartments of the Luo-Rudy model.

the SR is divided into two compartments, see Figure 10.8. The NSR represents the longitudinal tubules of the SR which takes up calcium from the myoplasm. This uptake current is termed $I_{up}$. The JSR is the cisternae which releases calcium back to the myoplasm, and this current is denoted by $I_{rel}$. The movement of calcium between these compartments is called the translocation current $I_{tr}$. For a definition of these currents, we again refer to [18].

*Buffering Mechanisms* A buffer is a substance that absorbs a compound (in our case ions) when the concentration of that compound is high, and returns the compound to the solution if the concentration decreases. In this way the buffer prevents large concentration variation of that compound. The buffers included in the Luo-Rudy model are calmodulin and troponin in the myoplasm and calsequestrin in the JSR, which all act as buffers for calcium.

### 10.5.9    A Different Model for Calcium Dynamics

Although the Luo-Rudy model is a highly complex and accurate description of the behavior of heart cells, many details of the model are still subject to debate. This has led to the development of even more advanced heart cell models, introducing a completely new model for the behavior of the L-type calcium channels and the release of calcium from the SR. The model to be used in our studies was published in 1999 by Winslow et al [26], and the main parts of the model are similar to a model presented by Jafri et al [13].

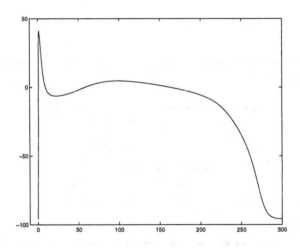

**Fig. 10.9.** The action potential of the model by Winslow et al.

A plot of the action potential is given in Figure 10.9. The state vector of the model consists of the transmembrane potential, six different ionic concentrations and 26 gate variables. The large number of gate variables compared to the Luo-Rudy model is mostly due to the new treatment of calcium currents. As in the Luo-Rudy model the transmembrane potential is affected by membrane currents of sodium, potassium, and calcium,

$$\frac{dV}{dt} = -I_{Ion} = -(I_{Na,tot} + I_{Ca,tot} + I_{K,tot}).$$

The sodium current consists of the following components

$$I_{Na,tot} = I_{Na} + 3I_{NaCa} + 3I_{NaK} + I_{Nab}.$$

These sodium currents are almost the same as in the Luo-Rudy model, except that the present model does not include sodium current through non-specific channels or through the L-type $Ca^{2+}$ channels. The mathematical expressions for the currents are similar to those of the Luo-Rudy model, with slightly adjusted coefficients.

The flow of potassium ions is given as

$$I_{K,tot} = I_{Kr} + I_{Ks} + I_{to1} + I_{K1} + I_{Kp} + I_{CaK} - 2I_{NaK}.$$

Important currents are $I_{Kr}$ and $I_{Ks}$, denoting respectively the rapid- and slow-activating component of the time dependent potassium current, which is denoted by $I_K$ in the Luo-Rudy model. Both currents are activated by the increased transmembrane potential, and they are not inactivated until the cell returns to its resting state. Their dynamics may thus be modeled with only one gate variable each. The transient outward current $I_{to1}$ contributes to the characteristic early drop in the transmembrane potential, seen in Figure 10.9. It is rapidly activated by the depolarization of the cell, and then inactivated after a relatively short time period. Like the channels for the sodium current $I_{Na}$, modeling this behavior requires two separate gating variables with different time constants. The currents $I_{K1}$, $I_{Kp}$ and $I_{NaK}$ are time independent currents similar to those present in the Luo-Rudy model, while the potassium current through the L-type channels, $I_{CaK}$, is quite different from earlier models. It is described further in Section 10.5.9 below.

The membrane current of calcium is given by

$$I_{Ca,tot} = I_{Ca} - 2I_{NaCa} + I_{p(Ca)} + I_{Ca,b}.$$

The most important component of this membrane current is $I_{Ca}$, the calcium flow through the L-type channels. The characteristics of this current are described in detail below. The other components of the calcium current are similar to those of the Luo-Rudy model.

*The L-type $Ca^{2+}$ Channel* The main difference between the present model and the earlier models like the Luo-Rudy model is the treatment of the complex calcium control mechanisms. The calcium currents are extremely important for the function of muscle cells, since it is the increased intracellular calcium concentration that initiates the contraction of the cell. During the action potential, the inward current of calcium ions through the L-type channels opposes the effect of the outward potassium currents and prolongs the plateau phase. Experiments have shown that the control mechanisms for the transmembrane flow of calcium are fairly complicated. As the transmembrane potential increases, the channels will first open but later inactivate, similar to the $I_{\mathrm{Na}}$ and $I_{\mathrm{to1}}$ currents described above. The added complexity comes from the fact that the L-type channels are also inactivated by an increase in the intracellular calcium concentration.

The Luo-Rudy model describes this calcium-induced inactivation as an instantaneous process, effectively closing all the channels at a given level of $\mathrm{Ca}^{2+}$ accumulation. However, according to [13], experimental evidence suggests that the inactivation is subject to a time delay. In the present model, the inactivation of L-type channels is modeled by so-called mode-switching. The channel is assumed to exist in two modes: the "normal" mode and the "Ca" mode. Independent of these modes, the channel consists of four separate sub-units, each capable of closing the channel. A schematic view of the states of the channel is given in Figure 10.10[1], where the states $C_0, \ldots, C_4$ and $C_{\mathrm{Ca0}}, \ldots, C_{\mathrm{Ca4}}$ represent the number of open sub-units in the normal and Ca mode respectively.

$$
\begin{array}{ccccccccccc}
C_0 & \underset{\beta}{\overset{4\alpha}{\rightleftarrows}} & C_1 & \underset{2\beta}{\overset{3\alpha}{\rightleftarrows}} & C_2 & \underset{3\beta}{\overset{2\alpha}{\rightleftarrows}} & C_3 & \underset{4\beta}{\overset{\alpha}{\rightleftarrows}} & C_4 & \underset{g}{\overset{f}{\rightleftarrows}} & O \\
\gamma \downarrow\uparrow \omega & & \gamma a \downarrow\uparrow \frac{\omega}{b} & & \gamma a^2 \downarrow\uparrow \frac{\omega}{b^2} & & \gamma a^3 \downarrow\uparrow \frac{\omega}{b^3} & & \gamma a^4 \downarrow\uparrow \frac{\omega}{b^4} & & \\
C_{\mathrm{Ca0}} & \underset{\beta'}{\overset{4\alpha'}{\rightleftarrows}} & C_{\mathrm{Ca1}} & \underset{2\beta'}{\overset{3\alpha'}{\rightleftarrows}} & C_{\mathrm{Ca2}} & \underset{3\beta'}{\overset{2\alpha'}{\rightleftarrows}} & C_{\mathrm{Ca3}} & \underset{4\beta'}{\overset{\alpha'}{\rightleftarrows}} & C_{\mathrm{Ca4}} & \underset{g'}{\overset{f'}{\rightleftarrows}} & O_{\mathrm{Ca}}
\end{array}
$$

**Fig. 10.10.** Schematic view of the states of the mode-switching L-type channel.

Opening and closing of each sub-unit is voltage dependent, both in normal and Ca mode, while the rate functions for the mode switching depends on the intracellular calcium concentration. In addition, the switch to the Ca mode occurs more easily at higher levels of the transmembrane potential. This effect is modeled by altering the rate functions for mode-switching as more sub-units open. When all four sub-units are open, a voltage independent transition to the open state occurs. Although the channel may open in both modes, the rate of opening in the Ca mode is very small compared to the normal mode, and the channel is thus effectively inactivated in the Ca mode. The rate of the mode switching to the Ca mode is substantially increased by the fact that the L-type channels are assumed to empty in a restricted subspace of

[1] This figure is reproduced in LaTeX from Figure 2 in [13].

the myoplasm, where the calcium concentration reaches values several orders of magnitude higher than that in the main parts of the myoplasm.

As mentioned above, the L-type channels are inactivated both by increased calcium concentration and by a delayed response to the increased transmembrane potential. The voltage dependent inactivation is not a part of the mode switching channel described above, and to have a full description of the L-type channels we need to include a separate inactivation variable. The behavior of this variable is similar to the voltage dependent gate variables for the time dependent sodium and potassium currents. For a complete description of the L-type channels, including the differential equations describing the transition between the states in Figure 10.10, we refer to Jafri et al [13]. The transmembrane flow of calcium through the L-type channels is given by

$$I_{Ca} = \bar{I}_{Ca} y \{O + O_{Ca}\},$$

where $y$ is the gate variable for the voltage dependent inactivation and $\bar{I}_{Ca}$ is a maximum current obtained by constant field theory, as described in Section 10.5.3.

As described above for the Luo-Rudy model, the L-type channels are not completely selective. Although the main current through the channels is of calcium ions, other ions are allowed to pass as well. The present model only considers flux of potassium ions, although it is known that small amounts of sodium ions may also be conducted. Measurements of the potassium current show that it is affected by the amount of calcium ions passing through the channel. When the channel is conducting a significant $Ca^{2+}$ current, it is in some sense occupied and the flow of potassium ions becomes very small. A mathematical model of this effect is achieved by letting the potassium conductivity be a decreasing function of the calcium current. The conductivity is computed as

$$P'_K = \frac{\bar{P}_K}{1 + \dfrac{I_{Ca}}{I_{Ca\text{-half}}}},$$

where $\bar{P}_K$ is the channel's conductivity to potassium in the absence of the calcium current, and $I_{Ca\text{-half}}$ is the level of $I_{Ca}$ that reduces the potassium current by 50 % . The complete expressions for both types of ionic currents passing through the L-type channel are given in [26].

*The Intracellular $Ca^{2+}$ Sub-system*  The complex L-type channel discussed above is not the only factor affecting the intracellular calcium concentration. It is also controlled by a complex intracellular sub-system of buffers, channels and sub-spaces. We have already mentioned that the L-type channels empty in a restricted sub-space of the myoplasm. In the description of the Luo-Rudy model we also introduced the SR, a network inside each cell holding large amounts of calcium ions in so-called cisternae. As described above, it is divided into two separate parts. The network sarcoplasmic reticulum (NSR) takes up $Ca^{2+}$ ions from the myoplasm, while the junctional sarcoplasmic reticulum (JSR) is the cisternae which releases $Ca^{2+}$ when stimulated by increased myoplasm calcium concentration. The release of calcium ions to the myoplasm is called the calcium induced calcium release (CICR).

In total, the model describes the change in four internal calcium concentrations. The calcium concentration in the myoplasm is denoted by $[Ca^{2+}]_i$, in the subspace at the mouth of the L-type channels it is denoted by $[Ca^{2+}]_{ss}$, and the calcium concentrations in the two parts of the sarcoplasmic reticulum are denoted by $[Ca^{2+}]_{NSR}$ and $[Ca^{2+}]_{JSR}$. In addition to these $Ca^{2+}$ concentrations, the model describes the dynamics of troponin buffers. These are considered slow buffers, meaning that they respond slowly to a change in the ionic concentration. The uptake and release of ions can hence not be viewed as an instantaneous process, but needs to be modeled with differential equations, see e.g. [26]. The concentration of calcium bound to low-affinity troponin binding sites is denoted by [LTRPNCa], and the concentration of calcium bound to high-affinity troponin binding sites is denoted by [HTRP-NCa]. Other buffers present in the subsystem are calsequestrin in the JSR and calmodulin in the myoplasm and subspace. Also included in the Luo-Rudy model, these are fast buffers and the reactions are thus assumed to always be in equilibrium. Their effects are calculated by a rapid buffering approach presented in [25]. The $Ca^{2+}$ sub-system includes six intracellular fluxes: CICR flux via the RyR channels $J_{rel}$, uptake of $Ca^{2+}$ to the NSR $J_{up}$, leak flux from the NSR to the myoplasm $J_{leak}$, transfer of $Ca^{2+}$ from NSR to JSR $J_{tr}$, transfer flux from the sub-space to the myoplasm $J_{xfer}$, and flux of $Ca^{2+}$ binding to troponin $J_{trpn}$ . In combination with the membrane currents $I_{Ca}$, $I_{p(Ca)}$, and $I_{NaCa}$ these currents control the $Ca^{2+}$ concentration in the various parts of the intracellular space.

Most of the intracellular calcium currents may be regarded as fairly simple diffusion processes, driven by concentration differences between the internal regions. This is, however, not the case for the release of calcium ions from the NSR, $J_{rel}$. This current responds to an increased myoplasm $Ca^{2+}$ concentration by releasing even more calcium into the myoplasm. To achieve this effect, the current needs to be controlled by some sort of channel gating dynamics. The channels controlling the $J_{rel}$ current, called Ryanodine Receptors or RyR channels, may exist in four modes, out of which two are open and two are closed. The channels empty in the same sub-space as the L-type channels, and respond to an increase in $[Ca^{2+}]_{ss}$ by opening and releasing $Ca^{2+}$ ions from the JSR into the sub-space. A detailed description of the transitions between the four states is given in [13]. The differential equations specifying the dynamics, as well as specifications of all intracellular fluxes, are given in [26].

The conservation equations for the intracellular calcium concentrations are

$$\frac{d[\text{Ca}^{2+}]_i}{dt} = \beta_i \left\{ J_{\text{xfer}} - J_{\text{up}} - J_{\text{trpn}} - (I_{\text{Ca,b}} - 2I_{\text{NaCa}} + I_{\text{p(Ca)}}) \frac{A_{\text{cap}} C_{\text{sc}}}{V_{\text{myo}} F} \right\},$$

$$\frac{d[\text{Ca}^{2+}]_{\text{ss}}}{dt} = \beta_{\text{ss}} \left\{ J_{\text{rel}} \frac{V_{\text{JSR}}}{V_{ss}} - J_{\text{xfer}} \frac{V_{\text{myo}}}{V_{ss}} - I_{\text{Ca}} \frac{A_{\text{cap}} C_{\text{sc}}}{2V_{ss} F} \right\},$$

$$\frac{d[\text{Ca}^{2+}]_{\text{JSR}}}{dt} = \beta_{\text{JSR}} \left\{ J_{\text{tr}} - J_{\text{rel}} \right\},$$

$$\frac{d[\text{Ca}^{2+}]_{\text{NSR}}}{dt} = J_{\text{up}} \frac{V_{\text{myo}}}{V_{\text{NSR}}} - J_{\text{tr}} \frac{V_{\text{JSR}}}{V_{\text{NSR}}}.$$

The parameters $\beta_i$, $\beta_{\text{ss}}$, and $\beta_{\text{JSR}}$ are functions of the respective calcium concentrations. They act as scale factors to calculate the amount of free calcium ions in the presence of the fast buffers calsequestrin and calmodulin. For a specification of these functions we again refer to the complete publication of the model in [26].

## 10.5.10  Structure of the Heart Tissue

The electrical behavior of each cell, described in the previous sections, is extremely important when solving the bidomain equations . However, the propagation of the electrical signal is also strongly affected by how the cells are connected to form the muscle tissue. The muscle cells are oblong-like cylinders which connect to neighboring cells through special pores called gap junctions. Electrical current is lead from cell to cell through these connections. The junctions are mostly located at the ends of the cylinders and more sparsely along the length, and the electrical conductance is therefore higher in the fiber axial direction than across the fiber axis. The web of muscle fibers is organized into sheets, and the conductivity across sheets is lower than the conductivity normal to the fiber direction in the sheet plane. Hence the heart tissue is anisotropic, with different conductivities in the three principal directions.

To simulate the propagation pattern of the wave front accurately, it is necessary to include the orientation of the muscle fibers in the model. Let

$$(a_l(x), a_t(x), a_n(x))$$

be a perpendicular set of vectors of unit length, with $a_l(x)$ lying along the fiber direction, $a_t(x)$ orthogonal to $a_l(x)$ in the sheet plane and $a_n(x)$ normal to the sheet plane, see Figure 10.11. Furthermore, let

$$M^* = \begin{bmatrix} \sigma_l & 0 & 0 \\ 0 & \sigma_t & 0 \\ 0 & 0 & \sigma_n \end{bmatrix} \tag{10.46}$$

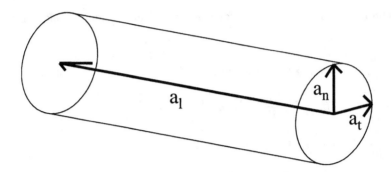

**Fig. 10.11.** The orientation of the local coordinate system of a single strand of fiber. The length of the arrows indicate the variation in conductivity in the different directions.

be the local conductivity tensor, i.e., expressed with respect to the basis formed by $(a_l, a_t, a_n)$. The conductivity along the fiber direction is $\sigma_l$, the conductivity across the fiber direction is $\sigma_t$, and $\sigma_n$ is the conductivity normal to the sheet plane.

For an electrical field $E^* = (e_1, e_2, e_3)^T$ expressed in this local coordinate system the corresponding current vector will, according to Ohm's law (10.29), be described by

$$J^* = M^* E^* = (\sigma_l e_1, \sigma_t e_2, \sigma_t e_3)^T = (j_1, j_2, j_3)^T .  \qquad (10.47)$$

This vector is mapped to global coordinates by

$$J = j_1 a_l + j_2 a_t + j_3 a_n = A J^* ,  \qquad (10.48)$$

where $A$ is the matrix with the vectors $a_l(x), a_t(x)$ and $a_n(x)$ as columns. An electrical field $E$ expressed in global coordinates is mapped to local coordinates with the inverse mapping, that is $E^* = A^{-1} E$. Since $A$ has perpendicular and normalized column vectors we have that $A^{-1} = A^T$, so the mapping is simply

$$E^* = A^T E .  \qquad (10.49)$$

Combining (10.47),(10.48) and (10.49) we get the relationship between an electrical field and current density, both expressed in global coordinates:

$$J = A M^* A^T E .  \qquad (10.50)$$

We thus define the global conductivity tensors by

$$M_i = A M_i^* A^T \quad \text{and} \quad M_e = A M_e^* A^T ,  \qquad (10.51)$$

where $M_i^*$ and $M_e^*$ are given by inserting the appropriate values in (10.46). The expressions in (10.51) may be simplified. A general entry in $M$ is

$$\begin{aligned} M_{ij} &= \sigma_l a_l^i a_l^j + \sigma_t a_t^i a_t^j + \sigma_t a_n^i a_n^j \\ &= (\sigma_l - \sigma_n) a_l^i a_l^j + (\sigma_t - \sigma_n)(a_t^i a_t^j) + \sigma_n (a_l^i a_l^j + a_t^i a_t^j + a_n^i a_n^j), \end{aligned}  \qquad (10.52)$$

so

$$M = (\sigma_l - \sigma_n)a_l a_l^T + (\sigma_t - \sigma_n)a_t a_t^T + \sigma_n I. \tag{10.53}$$

## 10.6   The Numerical Method

In Section 10.4 we presented a complete mathematical model for the electrical potential in the heart and the torso, with further details described in Section 10.5. Because of the complexity of the model it is impossible to obtain analytical solutions of the equations. In this section we will present a method for discretizing the mathematical model in space and time, to obtain a numerical solution of the problem.

### 10.6.1   Simplifications due to Boundary Conditions on $\partial H$

Before attempting to discretize the mathematical problem given by (10.21)-(10.28) in Section 10.4, we will use the boundary conditions on the surface of the heart to simplify the mathematical problem.

We assume now that $v$ is a known function. In general this is of course not true, since $v$ is one of the primary unknowns of the equation system. However, the assumption may be satisfied if the equations are solved by some sequential operator splitting technique, as will be demonstrated in Section 10.6.2 below. If $v$ is known it is convenient to rewrite (10.23) as

$$\nabla \cdot ((M_i + M_e)\nabla u_e) = -\nabla \cdot (M_i \nabla v). \tag{10.54}$$

We see that equations (10.54) and (10.24) share the same structure, with the difference being that the right hand side of (10.54) is non-zero. Due to the continuity of the extracellular and torso potentials on the heart surface, expressed by the conditions (10.26) and (10.27), it is possible to combine (10.54) and (10.24) into a single equation. This will simplify the implementation of the model considerably. We define $\Omega = T \cup H$ and a potential field $u(x)$ which coincides with the potentials on each of the subdomains;

$$u(x) = \begin{cases} u_T(x) & \text{if } x \in T, \\ u_e(x) & \text{if } x \in H. \end{cases}$$

The equation for $u$ becomes

$$\begin{aligned} \nabla \cdot (M\nabla u) &= f \ x \in \Omega, \\ n \cdot (M\nabla u) &= 0 \ \ x \in \partial\Omega (= \partial T), \end{aligned} \tag{10.55}$$

where

$$f(x) = \begin{cases} -\nabla \cdot (M_i \nabla v) & \text{if } x \in H, \\ 0 & \text{otherwise}, \end{cases} \tag{10.56}$$

and

$$M(x) = \begin{cases} M_T(x) & \text{if } x \in T, \\ M_i(x) + M_e(x) & \text{if } x \in H. \end{cases} \tag{10.57}$$

The simplification may be justified by considering the weak formulation of the combined equation (10.55), which may be written as

$$\int_H \nabla\phi((M_i + M_e)\nabla u)dx + \int_T \nabla\phi(M_T\nabla u)dx = -\int_\Omega \phi f dx. \tag{10.58}$$

From the Lax-Milgram representation theorem we know that (10.58) has a unique solution. Let us now regard the restriction of $u$ on $T$; $u|_T$. This function will be a classical (not just weak) solution of (10.23), see [8]. The same argument is valid for $u|_H$. Thus $u$ will solve (10.54) in $H$ and (10.24) in $T$, and have the correct behavior over the boundary $\partial H$.

## 10.6.2   Time Discretization

The mathematical model (10.21)-(10.28) applies for all moments in the time interval of the simulation $[0, T]$. However, solving this problem numerically restricts us to calculating the solution only at discrete time levels. If we use $N + 1$ equally spaced points in time denoted by $(t_0, \ldots, t_N)$ then $t_i = i\Delta t$ where $\Delta t = T/N$. The following notation will be used to denote the variables at each time step: $u^i = u(t_i, x), v^i = v(t_i, x)$ and $s^i = s(t_i, x)$.

Eqs. (10.21)-(10.28) are a coupled system and should ideally be solved simultaneously. However, due to the complexity of the system we have chosen to apply a sequential operator splitting technique. We assume that for sufficiently small time steps, the error introduced by this approach will be small. The integration of one time step is performed in the following order:

1. Assume that $u^l, v^l$ and $s^l$ are known at time $t_l$

2. Calculate $s^{l+1}$ by solving

$$\frac{\partial s}{\partial t} = F(t, s, v; x), \quad t_l < t \le t_{l+1} \text{ with } s(t_l, x) = s^l \text{ and } v(t_l, x) = v^l.$$

3. Calculate $v^{l+1}$ in $H$ by solving

$$C_m \chi \frac{v^{l+1} - v^l}{\Delta t} + \chi I_{\text{ion}}(v^l, s^{l+1}) = \nabla \cdot (M_i \nabla v^{l+1}) + \nabla \cdot (M_i \nabla u^l), \quad x \in H \tag{10.59}$$

4. Calculate $u^{l+1}$ in $\Omega$ by solving

$$\nabla \cdot (M \nabla u^{l+1}) = f^{l+1}, \quad x \in \Omega, \tag{10.60}$$

where $f^{l+1}$ is obtained by inserting $v^{l+1}$ in (10.56).

For the integration of the ODEs in Step 2 we use smaller time steps than for the discretization of the PDEs, because the time scale of the cellular reactions described by the ODEs is very short. Throughout the remainder of the chapter we will refer to these ODE steps as inner steps. There is one ODE problem for each point in space so in this spatially continuous description we have an infinite number of ODE systems. When $\Omega$ is discretized there will be one ODE problem for each point in the grid. If the size of the inner ODE time steps is chosen based on some adaptivity scheme, the work load of each ODE problem will depend upon its location. Large time steps can be used in quiescent areas, while small time steps are necessary in areas where the solution varies rapidly (i.e. close to the wavefront).

In Step 3 we have used a semi-implicit approach to discretize (10.22). The time derivative of $v$ is approximated by

$$\frac{\partial v}{\partial t} \approx \frac{v^{l+1} - v^l}{\Delta t},$$

and the diffusive term $\nabla \cdot (M_i \nabla v)$ is evaluated at time $t_{l+1}$, requiring the solution of a linear equation system for each time step. The non-linear ionic current term $I_{\text{ion}}(v, s)$ is computed using the previous update of the transmembrane potential. In this way the equation becomes linear with respect to $v$ and is thus much easier to solve. The error caused by the semi-implicit simplification will depend upon the step size $\Delta t$. Solving with a fully implicit scheme would have permitted larger time steps to be taken but each step would be very costly. Collecting terms of $v^{l+1}$ and $v^l$ in (10.59) yields

$$v^{l+1} - \frac{\Delta t}{C_m \chi} \nabla \cdot (M_i \nabla v^{l+1}) = v^l + g^l, \tag{10.61}$$

where

$$g^l = \frac{\Delta t}{C_m \chi} (\nabla \cdot (M_i \nabla u^l) - \chi I_{\text{ion}}(v^l, s^{l+1})). \tag{10.62}$$

## 10.6.3 Discretization in Space

To solve equations (10.21)-(10.28) numerically we also need to discretize the equations in the spatial domain. We have chosen to use the finite element method (FEM) for this purpose. The first step in the FEM is to obtain weak formulations of the partial differential equations. Consider first the combined equation (10.60). By multiplying with a test function $\phi \in H^1(\Omega)$ and integrating over $\Omega$, this equation may be rewritten as

$$\int_\Omega \phi \nabla \cdot (M \nabla u^{l+1}) dx = \int_\Omega \phi f^{l+1} dx. \tag{10.63}$$

The left hand side of this equation may be reformulated by applying Green's theorem. We get

$$\int_\Omega \nabla \phi \cdot M \nabla u^{l+1} dx = -\int_\Omega \phi f^{l+1} dx, \tag{10.64}$$

where the boundary integral in Green's theorem vanishes because of the boundary condition (10.28) on $\partial H$. For notational purposes it will be convenient to define the two inner products

$$(\phi, \psi)_\Omega \equiv \int_\Omega \phi\psi dx,$$

and

$$a_\Omega(\phi, \psi) \equiv \int_\Omega \nabla\phi \cdot (M\nabla\psi)dx.$$

Notice that $a_\Omega(\phi, \psi) = a_\Omega(\psi, \phi)$ since $M$ is symmetric. Equation (10.64) can now be reformulated as

$$a_\Omega(\phi, u^{l+1}) = -(\phi, f^{l+1})_\Omega. \tag{10.65}$$

The problem equivalent to the original PDE is to find $u^{l+1} \in H^1(\Omega)$ which satisfies (10.65) for all $\phi \in H^1(\Omega)$.

To discretize the weakly formulated equation, we introduce a grid on the domain $\Omega$, with corresponding basis functions $\varphi_i, i = 1, \dots, n$. We approximate the field $u^{l+1}$ by

$$u^{l+1} \approx \sum_{j=1} u_j\varphi_j,$$

where $u_j$ are real numbers, and insert this approximation into the weak formulation above. By a standard Galerkin procedure, i.e. using the basis functions $\varphi_i$ as test functions, we get a discrete approximation of the weakly formulated equation, given by

$$a_\Omega(\varphi_i, \varphi_j)u_j = -(f, \phi_i)_\Omega, \quad i, j = 1, \dots n. \tag{10.66}$$

We see that this is a system of $n$ linear equations in $n$ unknowns.

If we introduce the inner products

$$(\phi, v)_H \equiv \int_H \phi v dx$$

and

$$a_H(\phi, v) \equiv \int_H \phi v + \frac{\Delta t}{C_m\chi}\nabla\phi \cdot M_i\nabla v dx,$$

and follow the same procedure as above, a weak formulation of (10.61) may be written as

$$a_H(\phi, v^{l+1}) = (\phi, v^l + g^l)_H. \tag{10.67}$$

The term $g^l$, as defined in (10.62), includes double derivatives of $u$. To avoid this we invoke Green's theorem again in evaluating $(\phi, g^l)_H$. We get

$$\begin{aligned}
(\phi, g^l)_H &= \frac{\Delta t}{C_m\chi}(\phi, \nabla \cdot (M_i\nabla u^l) - \chi I_{ion}(v^l, s^{l+1}))_H \\
&= \frac{\Delta t}{C_m\chi}(\phi, \nabla \cdot (M_i\nabla u^l))_H - \frac{\Delta t}{C_m}(\phi, I_{ion}(v^l, s^{l+1}))_H \\
&= \frac{\Delta t}{C_m\chi}[\int_{\partial H} \phi(M_i\nabla u^l) \cdot n^T ds - \int_H \nabla\phi \cdot M_i\nabla u^l] \\
&\quad - \frac{\Delta t}{C_m}(\phi, I_{ion}(v^l, s^{l+1}))_H,
\end{aligned}$$

and we see that the final expression involves only first order derivatives.

By introducing a grid over $H$ with basis functions $\eta_i, i = 1, \ldots, m$, we may approximate the field $v^{l+1}$ by

$$v^{l+1} \approx \sum_{j=1}^{m} v_j \eta_j \, .$$

As described for the elliptic equation above, a discrete approximation of the weak PDE may now be written as

$$a_H(\eta_i, \eta_j) v_j = (\eta_i, v^l + g^l)_H, \quad i, j = 1, \ldots, m, \qquad (10.68)$$

where we use the expansion described above to avoid the second derivatives on the right hand side.

We see that using the FEM for spatial discretization requires the assembly and solution of two linear systems. In Step 3 we solve the $m \times m$ system (10.68), while the larger $n \times n$ system (10.66) is assembled and solved in Step 4.

## 10.6.4    Calculation of the Conductivity Tensors

The discretization procedure presented in Section 10.6.3 requires the computation of the conductivity tensors $M_i$, $M_e$ and $M_T$ at a given point in the grid. As shown in (10.53), the conductivity tensors are defined in terms of the direction of the muscle fibers of the heart. To define the tensor in a given point one must know the axis direction in that point. In practice, the fiber direction is only known at some discrete locations. We denote the direction at $n$ such locations by vectors $a_1, a_2 \ldots a_n$. The direction at an arbitrary point $x$ can be computed by a weighted average of these known directions. We have

$$a^i(x) = \frac{\sum_{j=1}^{n} a_j^i e^{-\lambda d(x)_j^2}}{\sum_{j=1}^{n} e^{-\lambda d(x)_j^2}},$$

where the superscript $i$ denotes the $i$'th component of the vector. The function $d_j(x)$ is the distance from the point $x$ to location number $j$, and $\lambda$ is a parameter controlling the relative weight of locations with different distances. The computed vector is normalized.

## 10.6.5    Solution of the ODEs

The discretization of the bidomain equations presented above requires the state vector $s$, which describes the state of the heart cells, to be updated for each time step. In the discrete problem, we have one such state vector for each node in the grid. For every time step, each state vector is updated by solving an initial value problem of the form

$$\frac{ds}{dt} = f(t, s), \quad T_l < t \leq T_{l+1}, \qquad (10.69)$$

with $s(t_l) = s^l$ and the time interval $T_{l+1} - T_l$ is equal to $\Delta t$, the time step of the PDE solver. For solving these initial value problems we have chosen the well-known Runge-Kutta methods. The basic idea of these methods is to divide each time step into a number of inner steps, referred to as the "stages" of the method. We then compute approximations to the derivatives at these stages, and the equations are stepped forward using a weighted average of these stage derivatives. A thorough description of Runge-Kutta methods and other ODE solvers is given in [9]. For ease of implementation, we have chosen to focus on explicit Runge-Kutta methods, where all stage derivatives are computed from explicit formulas involving only previously computed values. For a method with $p$ stages, we have

$$k^i = f(t + c_i \Delta t, s^l + \Delta t \sum_{j=1}^{i-1} a_{ij} k^j), \quad i = 1, 2, \ldots, p,$$

and the update of the solution is written as

$$s^{l+1} = s^l + \Delta t \sum_{i=1}^{p} b_i k^i.$$

The factors $c_i$, $b_i$, and $a_{ij}$, for $i = 1, \ldots, p$ and $j < i$, are given parameters characterizing the various Runge-Kutta methods.

The length of the time interval in which we are solving the ODEs, $T_{l+1} - T_l$, is equal to the time step of the PDE solvers and hence very small. However, because of the rapid variations in some of the state variables, it will not be possible to integrate this time interval with just one step of the ODE solver. Each "PDE step" will require a few inner steps of the ODE solver, and we want to minimize the number of these inner steps by choosing the step size adaptively. We want to choose the largest possible time step that satisfies a prescribed error tolerance, and we thus need an error estimate for the method. This is fairly easy to obtain with Runge-Kutta methods. The idea is to construct a Runge-Kutta method which obtains not only a numerical approximation $y_{l+1}$, but also an approximation $\hat{y}_{l+1}$ of higher order. The difference $\hat{y}_{l+1} - y_{l+1}$ may then be used as an error estimate to control the time step. By choosing the method coefficients carefully, it is possible to obtain this error estimate from only a single additional evaluation of the right hand side function $f(\cdot, \cdot)$. Out of the large number of such error estimating methods available, we chose to use a classic method of Fehlberg [5], which is of order four and uses a fifth order method for error estimation. The method is in six stages, where the sixth stage is only used for the fifth order method needed for the error estimation. Further details of the adaptive time stepping procedure are described in [9].

## 10.7     Implementation

The discrete formulation of the problem (10.21)-(10.28) has some distinct features that must be reflected in the design of the code:

1. Each time step consists of three main tasks, as described in Section 10.6.2.
2. The two coupled PDEs are defined over two different grids, $H$ and $\Omega$, where $H \subset \Omega$.

The complexity of the problem under study calls for a modular design. The problem can naturally be divided into separate pieces as described below. These pieces will have limited communication between them, typically only involving the exchange of solutions. This makes it feasible to let the implementation of the problem reflect the independence of the subproblems. Dividing the code into parts corresponding to these subproblems has several advantages. It makes the program easier to implement, maintain and extend. The latter point is important if new and improved models or numerical schemes are to be incorporated into the program. Given the rapid development of the description of the heart cell, extensibility is an obvious design goal.

### 10.7.1     Design

Looking at the time stepping algorithm in Section 10.6.2, it is evident that there are three major parts of the work per time step: The ODEs describing the dynamics of the concentrations and gates at the cellular level, the parabolic equation for the transmembrane potential, and finally the elliptic equation for the extracellular and torso potential. The coupling between the different parts is only required once per time step, and will only involve the exchange of the solutions. Thus separating the implementations of the subproblems will not imply any loss of efficiency. It is therefore natural to define three separate classes corresponding to these three subproblems. They will be denoted `Cells`, `Parabolic` and `Elliptic`. Information about the grid, fiber directions, and other resources that are shared between all the classes is contained in a separate class called `CommonRel`. This class also contains the methods for creating and modifying the grid and for computing the conductivity tensors.

The class `Heart` ties objects of the above four classes together. It contains one instance of each of the four classes and it initially sets up the necessary pointers between the objects. Their relationship is illustrated in Figure 10.12. In addition to pointers to `CommonRel`, the classes `Parabolic` and `Elliptic` also need pointers to each other in order to exchange solutions. The `Cells` class has a pointer to a `CellModel` which is a base class for general ODE description of a single cell.

The central routine in the `Heart` class is `timeLoop`. It calls the solving algorithms of the different classes consecutively and repeats this until the end

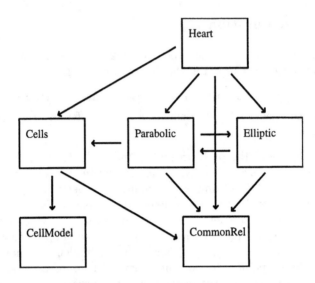

**Fig. 10.12.** The relationship between the classes. An arrow from class A to B indicates that an object of class A contains a pointer to an object of type B.

of the simulation time is reached. In pseudo-code the routine `timeLoop` looks as follows.

```
while(not finished)
{
  // increase time;
  cells->forward();
  parabolic->solveProblem();
  elliptic->solveProblem();
}
```

Each line in the loop corresponds to a step in the solution algorithm presented in Section 10.6.2.

### 10.7.2   The CommonRel Class

The `CommonRel` class contains data and functions that are shared between the three main classes `cells`, `parabolic`, and `Elliptic`. This includes grid information, fiber directions, and plotting routines. Since the multigrid method is used to solve the discretized equations, a series of grids is needed for each of the two PDEs. The fields for the conductivity tensors are also defined over these two series of grids:

torsogrid  is of type `GridFEAdT` and represents the grid hierarchy on the whole torso $\Omega$. It contains pointers to a series of grids, each corresponding to different levels of refinement. The `torsogrid` class is used in `Elliptic`.

musclegrid has the same type of structure as above, but for the heart domain
$H$. The relationship between torsogrid and musclegrid will be discussed
below. The musclegrid is used in the Parabolic class.

Mif,Mf are hierarchies of vector fields representing the entries conductivity
tensors. Because the conductivity tensors are symmetric we do not have
to store all the entries. In 2D there are three unique entries to be stored
while in 3D there are six. Mif is defined over the musclegrid and represents
the intracellular conductivity $M_i$. The vector field Mf is defined over the
torsogrid and represents the conductivity tensor $M$ in (10.57).

As can be seen from (10.56) the conductivity tensor $M_i$, which is only defined
in $H$, is also needed in the computation of the elliptic equation (10.55) which
is defined throughout the entire domain $\Omega$. Therefore a third field is defined,
Mi2f, which is defined over torsogrid and corresponds to Mif in $H$ and is zero
outside $H$.

*Managing the Two Grid Hierarchies.* Initially a grid for the complete domain
$\Omega$ is read from file. The subset of the grid representing the heart tissue
is assumed to have been tagged. The function makeInternalGrid takes the
initial grid and extracts the tagged elements to construct a grid for $H$. It
also returns a grid for $\Omega$ with an element numbering corresponding to that
of $H$, i.e. an element in the $\Omega$-grid that is inside $H$ has the same element
number as the corresponding element in the $H$-grid. This is accomplished by
the createPartOfGrid function of GridFEAdT.

The two grids are then refined uniformly a number of times as specified
by the input. When GridFEAdT refines a grid, the element numbering in the
new grid starts with the new elements in the first element of the coarse grid,
continues with the second element etc. In this way we know that the element
numbering in the two grid hierarchies coincide not only on the coarsest level
but also on all subsequent levels.

The numbering of the nodes in the refined grids follows a different scheme.
The new nodes in the refinements are given numbers following the numbers
in the coarse grid, and the effect of this is that the two grid hierarchies
will not have the same node numbering. Because the solution of the elliptic
problem appears on the right hand side of the parabolic equation, and vice
versa, we must be able to map solutions between the two grids. One could
use the predefined function FieldFE::interpolate but this is slow and should
be avoided if possible. Instead we construct two map vectors m2t and t2m,
mapping the node numbers of the two grids. By utilizing the close relationship
between the grids it is possible to construct this mapping efficiently (i.e. linear
work load with respect to the number of nodes). The Parabolic class has
a field extra defined over the heart grid which represents the extracellular
potential. Prior to assembling the linear system, this field is updated from
the solution of the elliptic problem, computed by the Elliptic object named
diffusion:

```
for (int n = 1; n <= userdef->getMuscleGrid().getNoNodes(); n++)
    extra(no_of_grids)().valueNode(n)=diffusion->valueNode(m2t(n));
```

A similar piece of code is found in Elliptic to map the transmembrane solution to a field defined over the torso grid.

### 10.7.3  The Cells Class

The class Cells implements the administration required for solving the ODEs, where each ODE problem is associated with a grid node. The following are the key variables of the class:

state This is an object of type FieldsFE, which represents a vector field. It is defined over the heart grid and it holds the state vector for the ODE system in each node.

aux Also of type FieldsFE, this vector field holds auxiliary variables which are necessary in some model formulations, e.g., in the Luo-Rudy cell model.

current This FieldFE object holds the values of the ionic currents. It is used for evaluating the ionic current term in the integrands function of Parabolic.

cellmodel is of type CellModel which is a generic class for ODE systems describing the heart cell. Derived classes hold the definition of the actual systems, i.e., the different ionic currents, update rules for the ionic concentrations, gating variables etc.

Note that CellModel is the class where the ODE system is specified, according to the choice of cell model, while Cells is used to administer a collection of ODE systems.

The data structure is updated by the following functions:

forward Loops over the nodes in the grid and for each node extracts the state vector and the auxiliary variables from state and aux, and possibly alters some parameters of the cell model. The state vector is then advanced one time step ahead by calling the forward function of CellModel. Adaptive time stepping is used in the numerical integration of the system. Finally state and aux are updated.

updateCurrent Loops over the nodes and updates current, based on the state, aux and the cellmodel. This function is called after forward.

### 10.7.4  The CellModel Class

The CellModel class is a base class for ODE systems related to models of the action potential. The class is derived from class ODE, which is a base class for ODE systems in general. The two major parts of the ODE class are;

**problem** This object is an instance of the class `ExplicitODEeqUDC` and holds the definition of the equations of the ODE system, i.e it can compute the right hand side of $\dot{s} = F(t, s)$ for a given state vector. It also has an object `state` which holds the current state of the system. We have let `CellModel` also be derived from `ExplicitODEeqUDC`. In this way we have the possibility to program the solver into the problem class, rather than having it as a separate class. This can be an advantage if we want to tailor the solution algorithm to a specific problem.

**solver** The object is an instance of type `ODESolver` which is a base class for a generic solver. A large number of solvers are implemented. The actual solver can be chosen at run-time.

Functions specific to `CellModel` include `ionicCurrent`, which returns the transmembrane current for a given state. The `forward` function takes the initial value and time interval $[T_l, T_{l+1}]$ as input, and returns the computed state at time $T_{l+1}$. A solver function in the `ODESolver` class is called for this task. In some subclasses of `CellModel` the function `forward` is redefined to suit the particular problem.

`CellModel` in itself does not contain any differential equations, but serves as a common interface to various models implemented as subclasses. Currently three subclasses have been implemented. `FitzHugh` implements the cell model of FitzHugh and Nagumo, which consists of two state variables. The `LuoRudy` class implements the Luo-Rudy cell model with 12 variable states. This model also needs a set of auxiliary variables to keep track of entities such as time of activation and the accumulation of internal calcium over a certain period. A modified solver is used for this problem. The `Winslow` class implements the model due to Winslow et al. The solver in this case is either from the standard library or, in later versions of the simulator, a specially optimized implicit Runge-Kutta method which will be described below.

### 10.7.5    The Parabolic Class

This class is designed to solve the parabolic PDE (10.22), i.e. to assemble and solve the linear system (10.68). Since we use the finite element method, the class is derived from the Diffpack base class `FEM`. The extensions of `FEM` in `Parabolic` is mainly related to the communication with the other classes in the program. Class `Parabolic` has pointers to `CommonRel`, `Elliptic` and `Cells` in order to get access to the geometric information, the solution of the elliptic problem, and the ionic current, respectively. The `GridFEAdT` handle in the class is set to point to the `musclegrid` hierarchy of `CommonRel`.

### 10.7.6    The Elliptic Class

This class implements the solver for the elliptic PDE (10.55), by assembling and solving the linear system (10.66). Like `Parabolic`, this class is derived from FEM and contains a pointer to CommonRel. It also contains a pointer to Parabolic in order to have access to the transmembrane potential, occurring in the right hand side function $f$ of (10.55). This class also contains a GridFEAdT handle, pointing to the torsogrid hierarchy of CommonRel.

## 10.8    Optimization of the Simulator

There are three main computational tasks in solving the present system of differential equations: Integrating the ODEs, assembling the linear systems, and solving the linear systems. To identify the most CPU-intensive parts, we performed a profiling of the simulator. The CPU consumption of each task was measured for a test case where the heart- and torso domains are nested cubes. The five main tasks are ranked as follows.

1. Solving ODE systems
2. Assembly of the linear parabolic system
3. Assembly of the linear elliptic system
4. Solving the linear elliptic system
5. Solving the linear parabolic system

The solution of the ODEs is by far the most demanding task, requiring close to 90% of the total CPU time of the program. It is hence the most obvious candidate for optimization. In Section 10.8.1 we present a new ODE solver which is more efficient than the explicit Runge-Kutta method used so far. In Section 10.8.2 the general-purpose assembly routines are replaced by local versions to cut overhead. The solver routine we use for the linear systems, the conjugate gradient method with multigrid as preconditioner, is already very efficient. To reduce the workload of the actual solving, we suggest using an adaptive grid. This is described in Section 10.8.3.

### 10.8.1    An Implicit ODE Solver

As stated above, the most time consuming part of the simulation is the solution of the ODE systems needed to determine the ionic current $I_{\mathrm{Ion}}(v, s)$. The system is stiff and thus the explicit solvers used initially suffer from severe stability problems, which means that the time steps used have to be very small. These two observations clearly indicate that a good strategy for optimizing the simulator is to implement a more stable ODE solver.

Analysis presented by Hairer and Wanner [10] suggests that a suitable class of solvers for stiff problems is implicit Runge-Kutta (IRK) methods. These methods offer greatly enhanced stability properties over explicit methods, but each time step is also considerably more costly because we need to solve a large system of non-linear algebraic equations. For our application we are only partly able to utilize the good stability properties, because the maximum step size of the ODE solver is limited by the PDE step size. Hence it is important to minimize the work per time step, and we chose to focus on a group of methods known as semi-explicit or diagonally implicit Runge-Kutta (DIRK) methods. These are methods which significantly reduce the work per time step compared to fully implicit methods, but the stability is still sufficient for our purpose.

Based on initial single-cell experiments, we chose to implement a DIRK method named ESDIRK23A, developed by Kværnø [15]. This is a four stage method of order three, with an embedded second order method for error estimation. Although the order of the method is lower than the explicit method used initially, the accuracy is sufficient for our application. The method has previously been successfully implemented in the ODE library Godess [21], but due to the special requirements of the coupling to the PDE solvers, it was difficult to use Godess directly.

To integrate one step, from $t_l$ to $t_{l+1}$, we first compute the four stage derivatives $k^1, \ldots, k^4$ by

$$
\begin{aligned}
k^1 &= f(t_l, s^l), \\
k^2 &= f(t_l + \Delta t c_2, s^l + \Delta t(a_{21} k^1 + \gamma k^2)) \\
k^3 &= f(t_l + \Delta t c_3, s^l + \Delta t(\hat{b}_1 k^1 + \hat{b}_2 k^2 + \gamma k^3)) \\
k^4 &= f(t_l + \Delta t c_4, s^l + \Delta t(b_1 k^1 + b_2 k^2 + b_3 k^3 + \gamma k^4))
\end{aligned}
$$

where $c_2, \ldots, c_4$, $\hat{b}_1, \hat{b}_2$, $b_1, \ldots, b_3$ and $\gamma$ are coefficients characterizing the method. We see that the first stage is explicit, while the rest of the stages require solving systems of non-linear equations to determine the stage derivatives. Finally, the solution and the error estimate at $t = t_{l+1}$ are computed from

$$
s^{l+1} = s^l + \sum_{i=1}^{4} b_i k^i,
$$

$$
e^{l+1} = \sum_{i=1}^{4} (b_i - \hat{b}_i) k^i,
$$

where we have introduced the coefficients $\hat{b}_3 = b_4 = \gamma$ and $\hat{b}_4 = 0$ for notational purposes. See e.g. [10] for a discussion of Runge-Kutta methods with embedded error estimates.

Initial tests of the new ODE solver indicated substantial efficiency improvements. For single-cell simulations the required number of time steps was reduced by almost 95%, even though a strict restriction on the time step was used, setting the maximum time step equal to the time step of the PDE solver. The CPU time for a single-cell simulation was reduced by more than 80%. We expect the efficiency improvements to drop slightly when the ODE solver is coupled to the full PDE system, because of potential problems related to reuse of the factorized Jacobian matrix (see [23] for details). Still, the results obtained for single-cell experiments are promising. The effect of the new ODE solver on the overall efficiency of the simulator is described below.

### 10.8.2    Local makeSystem Routines

Except for the solution of the ODE systems, the largest computational tasks of the problem are assembling the linear systems. In an effort to reduce the workload of this part of the simulator, the makeSystem routine in the FEM class has been redefined in the subclasses Parabolic and Elliptic.

The original version of makeSystem uses an object of the class ElmMatVec which contains the element matrix and vector for a single element. For each element in the grid the data structure of the ElmMatVec object must be initialized, and then passed on to numItgOverElm. This function loops over all the numerical integration points in the element, and calls the integrands for each point.

In the redefined version of makeSystem the ElmMatVec class is not used. Instead, the entire assembly is performed inside the body of the makeSystem procedure, without calls to numItgOverElm and integrands. In this way overhead connected to function calls and data initialization is reduced. The fact that our problem does not contain any essential boundary conditions also allows us to make simplifications compared to the original code of makeSystem. Furthermore, the code is optimized for one specific type of elements (tetrahedra), to exploit any simplifications which may be possible in this case.

The gain in efficiency comes at the price of a less elegant implementation. The high level of abstraction in the original formulation makes a clear distinction between the integrands function and the assembly of the linear system. In the new formulation the two are entwined and the resulting code is harder to maintain.

The original assembly code will reside in the base class FEM, with only the integrands functions specified in the Parabolic and Elliptic classes. The optimized makeSystem routines are implemented in subclasses of Parabolic and Elliptic. In this way the user can easily swap between the optimized and the flexible version. If the equations are to be changed this will first be done in the integrands function of either Parabolic or Elliptic before the optimized subclasses are updated. In this way it will be possible to verify that the optimizations have been performed correctly by comparing the results. This is crucial since the optimized functions are highly complex.

### 10.8.3    Adaptivity

An accurate representation of the sharp wave front of the depolarization requires very small elements. The distribution of the transmembrane potential away from the wave front displays comparatively little variation, hence the requirement on the element size is less strict. These characteristics of the solution may be exploited by using an adaptive grid.

Because we use the multigrid method to solve the equations, we already have a hierarchy of successively finer grids. An adaptive scheme may hence be implemented by adjusting the way the finer grids of this hierarchy are constructed. Instead of refining the grid uniformly to produce the next grid level, we want to restrict the refinements to the areas where it is most needed. In our context this is the area close to the wave front. One possible strategy is to go through the elements of the coarsest grid in the multigrid hierarchy, mark those elements which contain the wave front, and then subsequently refine all marked elements. This gives rise to a new grid on which the process can be repeated. The scheme is illustrated by an example in Figure 10.13. Note that in the first grid none of the two elements inside the square $[0, 2]^2$ are tagged since they do not contain the wave front, however they are still refined. The refinement of these elements is necessary to obtain a legal grid, i.e., a grid where no node lies on the edge of another element. Similar effects are also seen in the subsequent grids. In the finer grids the refinements are located in a narrow strip around the wave front. To determine if a given element contains the wave front, it is sufficient to compute the differences in transmembrane potential between all pairs of nodes in the element. If the wave front passes through an edge, the difference over the pair of nodes defining that edge will be large. The element is marked if the difference is above a given threshold. By choosing a threshold close to the difference between the polarized tissue ($-85$mV) and the depolarized tissue ($40$mV), a narrow strip of elements around the wave front will be refined. If the wave front needs to be represented with several layers of elements a smaller threshold can be chosen. In our simulations we have chosen a small threshold ($5$mV) to ensure that the wave front is always maximally refined.

We use the solution from the previous time step, $v^{l-1}$, to compute the change in the solution over the elements in the coarser grids, as shown in Figure 10.14. When the elements containing the wave front have been tagged, a refined grid is computed. The solution $v^{l-1}$ is then projected down to this refined grid, and the process is repeated until the finest grid level has been reached.

The use of the previous solution $v^{l-1}$ to estimate the location of the wave front is justified by the small time step size. With a typical propagation speed of 1mm/ms the wave front will move 0.125mm during a time step of 0.125ms. We will find in the next section that this is a sufficiently small time step and that the distance between the nodes of an element is normally larger than 0.125mm, thus the movement during a time step will not be larger than a single element. In addition to this, the low threshold value used and the requirements for a legal grid result in a high node density area several elements thick around the wave front. The wave front of the solution computed on the new grid will therefore also lie inside this refined area.

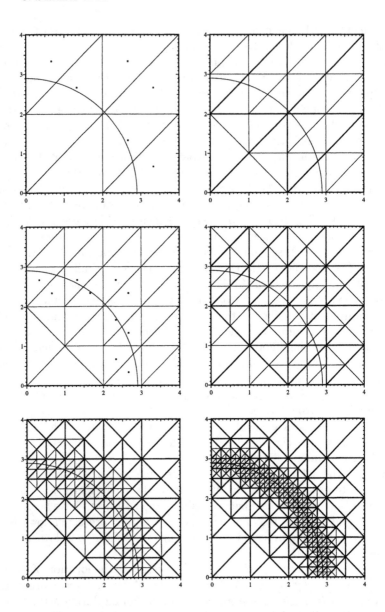

**Fig. 10.13.** Upper left: The coarsest grid with elements containing the wave front (curved line) are marked with dots. Right: The grid after refinement, original grid shown with thick lines. Middle: The new grid with tagged elements, and the resulting grid after another refinement. Bottom: Two more refinements.

## 10.9  Simulation Results

To validate the implementation of the model we would ideally like to compare the computed solution to an analytical solution and study the difference. Due

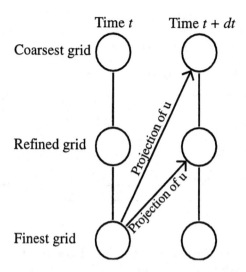

**Fig. 10.14.** The solution from the previous time step is projected onto the coarsest grid where the elements containing the wave front are refined. The process is repeated successively on the next grid, until a sufficient refinement is obtained.

to the complexity of the problem under study, it is not feasible to compute an analytical solution. Using more indirect techniques we have previously[16] found that a time step of 0.125ms and a spatial resolution of 0.2mm is necessary to obtain reliable results.

### 10.9.1    Test Results on a Simple 3D Geometry

As described above, simulations were performed on a simple three-dimensional geometry to identify the most CPU-intensive parts of the simulator. The geometry consists of two nested cubes. The smaller cube has sides of 1 cm and represents the heart tissue, while the larger cube has 2 cm long sides and represents in addition a surrounding bath. The coarsest grid for the heart tissue consists of one internal node and 26 boundary nodes ($3^3 - 1$). The larger grid, representing the heart tissue and the surrounding bath, contains these 27 nodes and in addition 26 torso boundary nodes. There are 48 tetrahedral elements inside the heart domain $H$ and 144 elements in $T$.

In the heart domain the edges of the elements are between 1/2 cm and $\sqrt{2}/2 \approx 0.71$cm. For each uniform refinement this length is halved. With five refinements the edge length is $(0.50, 0.71) \cdot 2^{-5} = (0.016, 0.022)$cm which is what we have found to be an acceptable resolution. On this fine grid there are 274,625 nodes in $H$ and 1,061,121 nodes in $\Omega$. The number of elements are 1,441,792 in $H$ and 6,291,456 in $\Omega$.

Stimulation is applied in the center of the cube, within a radius of 0.1cm. The duration of the simulation is 30ms or 240 time steps.

**Table 10.1.** The CPU consumption in seconds of the different parts of the problem. The second column is for the original implementation. The third shows the results when using a better ODE solver, and also the special purpose assembly routines. The rightmost column contains the results when using an adaptive grid. The optimizations from column three are also used in this last case.

| Task | Original | Optimized | Adaptivity |
|---|---|---|---|
| Solve ODEs | 546 971 | 69 324 | 32 569 |
| Assembly of linear parabolic system | 25 184 | 2 560 | 946 |
| Solving linear parabolic system | 2 823 | | 307 |
| Assembly of linear elliptic system | 14 899 | 1 080 | 987 |
| Solving linear elliptic system | 13 264 | | 646 |
| Total | 603 145 | 88 885 | 37 371 |

The simulation was carried out on a Linux PC with a Pentium III 900MHz processor and 1.5Gb of RAM. About 1.1Gb of memory was needed to run the simulation.

The CPU times displayed in Table 10.1 show the performance of the simulator before and after the optimizations described in Section 10.8.

The efficiency gain of the optimization is demonstrated by the numbers listed. The specialized SDIRK ODE solver uses only 12% of the CPU time used by the original explicit Runge-Kutta method, while the CPU time reduction for the optimization of the assembly routines is more than 90%.

Using adaptivity, the number of nodes varied greatly during the computation, as one would expect. It peaked at 122734 nodes, which is about 45% of the nodes used if we use uniform refinement throughout the heart domain. At the end of the simulation only nodes in the coarsest grids were used. The CPU consumption of this simulation is shown under the heading Adaptivity in Table 10.1.

On average 11% of the nodes in $H$ were used. This is reflected in the CPU consumption of the parabolic solve routine, which is reduced by almost 90%. Solving the elliptic problem is also reduced to a fraction corresponding to the reduction of nodes in $\Omega$. The assembly routines do not show the same reduction. The reason is that the stiffness matrices have to be recomputed for each time step because of the changes in the grid. Still there is speed gain, although only slightly in the elliptic case.

The time to solve the ODEs is reduced by a surprisingly small amount, about 47%. The reason is that the remaining nodes are clustered around the wave front, i.e. where the upstrokes take place and the inner time step of the ODE solver has to be very small. Discarding nodes in quiet areas gives little improvement since the ODEs in these areas are solved quickly anyway.

Comparing the solutions, we have found that the solution computed on the adaptive grid is very similar to the one obtained using uniform refinements. In particular, the depolarization wave uses the same amount of time to pass

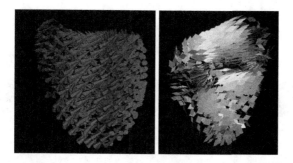

**Fig. 10.15.** The direction of the muscle fibers and the orientation of the sheet layers in the heart.

over the domain in the two cases. The propagation speed is therefore not altered as a consequence of using adaptivity.

### 10.9.2   Physiologically Correct Geometries

In order to perform simulations on more physiologically correct geometries, the geometry of the heart was taken from the data obtained by Nielsen et al [20]. These data only include the ventricles, but this is sufficient for our purposes because the electrical field propagating the body is dominated by the activity of the ventricles. The dataset is unique in that, in addition to the description of the internal and external boundaries of the heart, it also contains the orientation of the muscle fibers throughout the heart muscle. A visualization of these data is shown in Figure 10.15. Contours from the dataset were used to construct a finite element grid. The torso grid was obtained by tracing contours of photographs from the Visible Human data[24]. The surface of the resulting grid is shown in Figure 10.17.

Activation was initiated at several sites on the inner surface of the ventricles. These sites are taken from the activation maps published by Dürrer[3].

The potential has been recorded at several sites on the body surface. Figure 10.17 shows the location of these virtual electrodes, which correspond to V1-V6 in the 12 lead ECG.

A set of simulations for a time period of 300 ms has been performed. This includes the depolarization and the subsequent repolarization of the heart. Figure 10.16 shows a snapshot at three different time instances. At 30 ms the depolarization wave has propagated from the activation points out to the outer surface of the ventricles. At 70 ms the heart is nearly fully depolarized. At 200ms the heart is partially repolarized, 100ms later the whole heart is fully repolarized (not shown).

The corresponding ECG signals are shown in Figure 10.17. The signal can be divided into three phases. In the first phase the signal is strong and oscillating. This corresponds to the depolarization of the ventricles and it is called

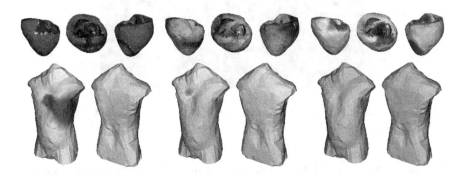

**Fig. 10.16.** Solution at 30ms, 70ms and 200ms. Top row shows the heart viewed from three different angles. Red indicates negative tissue in the resting state, blue indicates positive, activated, depolarized tissue. The torso is shown from two angels, red is negative potential, blue is positive.

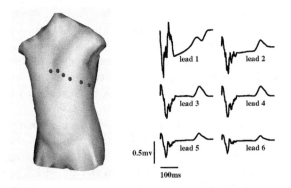

**Fig. 10.17.** The torso with six ECG leads and their corresponding readings.

the QRS complex. It is followed by a phase with an almost flat ECG reading, which is termed the ST-segment and corresponds to the plateau phase of the action potential. Finally we see the T-wave which corresponds to the repolarization of the heart. The simulation not only reproduces these distinct phases but does so with a realistic amplitude and duration. The biggest flaw of the simulated ECGs is that the detailed shape of the QRS-complex is not realistic. To achieve this, the propagation of the wave front must be more accurate. The current simulation has a very crude representation of the activation sites and the computational grid is also coarser than what is needed to reproduce a realistic propagation. A completely realistic result can therefore not be expected. This issue is further discussed in [17].

## 10.10    Concluding Remarks

We have seen that the electrical activity of the human heart may be modeled by a set of PDEs coupled with a complex system of ODEs. By coupling this system to an equation describing the surrounding torso, it is possible to simulate ECG signals, which are measurements of the electrical potential on the body surface. The model describing the heart may be adjusted to simulate pathological conditions such as infarction and arrhythmia, to obtain a better understanding of how such conditions may affect the ECG readings.

By a basic operator splitting technique in combination with the finite element method, we are able to discretize the complex system. To solve the resulting linear system we have constructed an optimal procedure, i.e. the work load of solving the system is proportional to the number of unknowns, by using the conjugate gradient method with a multigrid preconditioner.

The simulator was tested for a simple 3D geometry, and a profiling of the program was performed to identify the most CPU-intensive parts. Based on these results the simulator has been optimized to improve the performance, and simulations have been run on more physiologically correct geometries. We have found that the simulator produces results similar to real ECG-recordings, but the accuracy is not completely satisfactory. Experiments have been done to estimate the spatial and temporal resolution needed to produce sufficiently accurate results. We have found that for 3D simulations on realistic geometries, the required number of grid nodes is too large for the computation to be performed on a single processor. To achieve realistic results it will be necessary to utilize more powerful, parallel hardware.

The operator splitting technique used to discretize the problem results in a number of smaller sub-problems to be solved for each time step. In the implementation design, this modularity is utilized by implementing the solution of each sub-problem as a class. The main idea of the modular design is to increase the adaptability and extensibility of the simulator. The concept has already been utilized for the ODE-part of the problem, where new models and new solvers have been included without touching other parts of the code. The modular design combined with use of flexible Diffpack tools in the programming will also ease the porting of the code to parallel hardware, to enable accurate full-scale simulations.

# References

1. G.W. Beeler and H. Reuter. Reconstruction of the action potential of ventricular myocardial fibres. *J Physiol*, 268:177–210, 1977.

2. D. DiFrancesco and D. Noble. A model of the cardiac electrical activity incoportation ionic pumps and concentration changes. *Philos Trans R Soc Lond Biol*, 307:353–398, 1985.

3. D. Dürrer, R.T. van Dam, G. E. Freud, M. J. Janse, F.L. Meijler, and R.C. Arzbaecher. Total excitation of the isolated human heart. *Circulation*, 41:899–912, 1970.

4. L. Ebihara and E. A. Johnson. Fast sodium current in cardiac muscle: A quantitative descripton. *Biophys J*, 32:779–790, 1980.

5. E. Fehlberg. Klassische runge-kutta formeln fünfter und siebenter ordnung mit scrittweitenkontrolle. *Computing*, 4:93–106, 1969.

6. P. Colli Franzone, L. Guerri, and S. Rovida. Wavefront propagation in an activation model of the anisotropic cardiac tissue: asymptotic analysis and numerical simulations. *Journal of Mathematical Biology*, 28:121–176, 1990.

7. D. B. Geselowitz and W. T. Miller. A bidomain model for anisotropic cardiac muscle. *Annals of Biomedical Engineering*, 11:191–206, 1983.

8. W. Hackbusch. *Elliptic Differential Equations*. Springer-Verlag, 1992.

9. E. Hairer, S. P. Nørsett, and G. Wanner. *Solving Ordinary Differential Equations I, Nonstiff Problems*. Springer-Verlag, 1991.

10. E. Hairer and G. Wanner. *Solving Ordinary Differential Equations II, Stiff and Differential Algebraic Problems*. Springer-Verlag, 1991.

11. C. S. Henriquez. Simulating the electrical behavior of cardiac tissue using the bidomain model. *Critical Reviews in Biomedical Engineering*, 21:1–77, 1993.

12. A.L. Hodgkin and A. F. Huxley. A quantitative description of of membrane current and its aplication to conduction and excitation in nerve. *J Physiol*, 117:500–544, 1952.

13. M. S. Jafri, J. J. Rice, and R. L. Winslow. Cardiac $Ca^{2+}$ dynamics: The roles of ryanodine receptor adaption and sarcoplasmic reticulum load. *Biophysical Journal*, 74:1149–1168, 1998.

14. J. Keener and J. Sneyd. *Mathematical Physiology*. Springer-Verlag, 1998.

15. A. Kværnø. More, and to be hoped, better DIRK methods for the solution of stiff ODEs. Mathematical Sciences Div., Norwegian Institute of Technology, Trondheim, 1992.

16. G. T. Lines. *Simulating the electrical activity of the heart: a bidomain model of the ventricles embedded in a torso*. PhD thesis, Department of Informatics, Faculty of Mathematics and Natural Sciences, University of Oslo, 1999.

17. G. T. Lines, P. Grøttum, and A. Tveito. Modeling the electrical activity of the heart, a bidomain model of the ventricles embedded in a torso. Preprint 2000 4, Department of Informatics, University of Oslo, 2000.

18. C. H. Luo and Y. Rudy. A dynamic model of the cardiac ventricular action potenial. *Circulation Research*, 74:1071–1096, 1994.

19. R. E. McAllister, D. Noble, and R. W. Tsien. Reconstruction of the electrical activity of cardiac purkinje fibres. *J Physiol*, 251:1–59, 1975.

20. P. M. F. Nielsen, I. J. Le Grice, B. H. Smail, and P. J. Hunter. Mathematical model of geometry and fibrous structure of the heart. *American Journal of Physiology*, 260:1365–1378, 1991.

21. H. Olsson. *Runge-Kutta Solution of Initial Value Problems*. PhD thesis, Department of Computer Science, Lund Institute of Technology, Lund University, 1998.

22. R. Plonsey and D. B. Heppner. Considerations of quasi-stationarity in electro-physological systems. *Bulletin of Mathematical Biophysics*, 29:657–64, 1967.

23. J. Sundnes, G. T. Lines, and A. Tveito. Efficient solution of ordinary differential equations modeling electrical activity in cardiac cells. *Mathematical Biosciences*, 2001.

24. The visible human project. See URL *http://www.nlm.nih.gov/research/visible/visible_human.html* .

25. J. Wagner and J. Keizer. Effects of rapid buffers on $Ca^{2+}$ diffusion and $Ca^{2+}$ oscillations. *Biophysical Journal*, 67:447–456, 1994.

26. R. L. Winslow, J. Rice, S. Jafri, E. Marban, and B. O'Rourke. Mechanisms of altered excitation-contraction coupling in canine tachycardia-induced heart failure, II, model studies. *Circulation Research*, 84:571–586, 1999.

# Chapter 11

# Mathematical Models of Financial Derivatives

O. Skavhaug[1,2], B. F. Nielsen[1,2], and A. Tveito[1,2]

[1] Simula Research Laboratory
[2] Department of Informatics, University of Oslo

**Abstract.** In this chapter, we derive several mathematical models of financial derivatives, such as futures and options. The methodology used is commonly known as risk–neutral pricing, and was first presented by Merton, Black and Scholes in the 1970s. We start by presenting the basics of the Black–Scholes analysis, which leads to the Black–Scholes equation. Several option contracts such as plain European and American option contracts are derived. We also give an overview of some exotic option contracts. At last, we present mathematical models of the so–called Greeks, i.e., the partial derivatives of the value of the option contracts with respect to important model parameters.

## 11.1 Introduction

Financial derivatives play an important role in todays financial markets. Derivatives are financial instruments who's value depend on the value of some underlying, or primary, financial quantities such as stocks, interest rates, currencies, and commodities[1]. It is important to determine the fair price of such derivatives, and there exists a rich mathematical literature addressing this problem. It turns out that in many cases, the fair value of derivatives is governed by deterministic partial differential equations.

The primary purposes of financial derivatives, e.g., options and forwards, are speculating and hedging. Speculating is to expose a collection of financial instruments, called a portfolio, to a large amount of risk, and earn money if the market moves in a certain direction. Hedging is the opposite of speculating; to remove risk from a portfolio by taking opposite positions in the market, and ensure that these positions cancel each other out. There are always two parties involved when creating a financial derivative; a writer and a holder of the contract. These two parties have to agree on the value of the derivative. In this chapter, we derive models for such values, based on the principle of risk–neutral pricing.

In this chapter we will derive a number of mathematical models for financial derivatives. In order to solve most of these problems, numerical methods

---

[1] Also, the underlying quantity can be another derivative product, such as a second option contract.

must be utilized. In [14], various numerical solution strategies will be presented.

In 1973, Merton [11] and Black and Scholes [2] contributed significantly to the field of mathematical finance. They derived a model for how the price of an asset evolves in time. An asset may be a stock, a commodity, or a currency, to name a few. Based on this theory, they derived the Black-Scholes equation. This equation governs the fair price of many financial derivatives, see [2] and [11]. Their work gave rise to a new field of research that addresses the task of determining the fair price of financial derivatives, such as, for example, options.

This chapter is outlined as follows. In order to construct mathematical models we need to make certain basic assumptions for the markets. These assumptions are presented in Section 11.2.

In Section 11.3 we consider some simple derivatives; futures and forwards. These are so-called linear instruments, and we will see that their values are easily determined.

From a simple model of how the price of an asset evolves, we present the basis of the Black-Scholes analysis in Section 11.4. This approach leads to a deterministic partial differential equation governing the fair price of many derivatives.

Section 11.5 contains various models for different types of options. The fair price of the most common of these contracts, called plain vanilla options, can be determined by solving the Black-Scholes equation subject to simple final and boundary conditions. For European vanilla options, explicit analytical formulas are available.

We will also present models for exotic options in Section 11.5. The value of such options depend on complex changes in the value of the underlying financial instrument. In some cases, this behavior can be formalized in mathematical terms such that it is possible to determine the fair price of these exotic products by solving a partial differential equation, similar to the Black-Scholes equation, subject to contract-specific boundary and final conditions.

In Section 11.8, we derive methods for measuring the sensitivity of the fair price of derivatives with respect to changes in the model parameters. The value of an option depends on several parameters like the volatility of the underlying asset, the risk-free interest rate, etc. The partial derivatives of an option price with respect to these parameters defines the so-called Greeks. From a risk-management point of view, it is important to understand how changes in these parameters affect the value of options and other derivatives. We will see that several of the Greeks satisfy the Black-Scholes equation or similar PDEs.

Our presentation of the Black-Scholes models is based on the excellent book " Option Prices, Mathematical models and computation" [10], by Wilmott, Dewynne, and Howison.

## 11.2    Basic Assumptions

In order to derive mathematical models for the behavior of stock prices and financial derivatives, we must make some assumptions for the markets under consideration. We also need some technical assumptions that allow us to pose mathematical problems based on the behavior of the financial markets.

### 11.2.1    Arbitrage–Free Markets

The principle of an arbitrage–free market is important for understanding and deriving most of the models presented in this chapter.

First of all, we assume that all risk–free investments yield the same return, and that a risk–free investment[2] can be made. If it is possible to construct a portfolio, that is, a collection of different financial derivatives, with zero value today that pays out a *guaranteed* positive return in the future, this is called an *arbitrage opportunity* (see e.g. [12] for a thorough treatment of arbitrage).

It might seem strange that it is possible to construct a portfolio with zero value, but if *short selling* is allowed, it is quite easy to comprehend. Short selling is to sell a borrowed security. At a later time, the security must be bought back to close out the transaction[3].

Subject to some time restrictions short–selling is allowed in many markets today. Consider a portfolio consisting of one asset and short the same asset. A way of viewing this portfolio is that we have borrowed money from an asset, and used the money to purchase the asset for the same price. The portfolio will have the value $S + (-S) = 0$ at all times, where $S$ denotes the value of the asset. When deriving models for the fair value of financial derivatives, we construct portfolios similar to the one above.

There is another way of viewing arbitrage. Financial derivatives can be used to construct another derivative product synthetically. For example, a portfolio consisting of assets and forward contracts, see Section 11.3, can be made risk–free. Hence this portfolio is a synthetically constructed bank-deposit. Because they are basically the same derivative product, they must have the same value. If it is possible to construct a synthetic derivative from others to a cheaper price, this is an arbitrage opportunity.

Throughout this chapter we assume that the markets are arbitrage–free. This assumption can be made, because arbitrage opportunities are profitable, and thus quickly eliminated by market movements as soon as they occur.

The presentation given in this chapter is strongly influenced by [9] and [10], as they provide a intuitive approach to the foundation of mathematical finance.

---

[2] Usually offered by a bank system.

[3] Short selling can be used by investors that believes that the price of an asset will drop. The profit is the difference between the asset price at the time of the short sale and when the asset is bought back. If the asset increases its value, the investor will lose money on this investment.

## 11.2.2   The Efficient Market Hypothesis

The second assumption is the *efficient market hypothesis*. It states that at any given time, the asset price fully reflects all information available about the asset. Also, it states that the markets respond immediately to new information concerning the asset. Based on this hypothesis, the change of the asset price can be modeled as a *Markov process*, see [1] and [9].

A consequence of the efficient market hypothesis[4] is that technical analysis, i.e. to try to predict future asset prices based on historical information about the asset price, is questionable, see [12, 483].

## 11.2.3   Other Assumptions

To derive mathematical models of financial derivatives, we make the following technical and practical assumptions regarding the financial markets:

— To justify the derivation of continuous models for the price of an asset, and thus the price of derivatives based on these assets, we assume that *assets are divisible*. That is, fractions of financial instruments can be traded.

— Reading financial newspapers, we discover that the price for buying an asset is higher than the price for selling the same asset. This difference is called the bid–ask spread, and reflects that the market makers take some profit from each trade. We will assume that *the bid–ask spread is zero* and that *there are no transaction costs*.

— We assume that as long as the market is open, *transactions can take place continuously*.

— At certain times, assets may pay out dividends. Money is paid to the shareholders, and the value of the asset decreases correspondingly. If not stated otherwise, we will assume that *assets do not pay dividends* during the life-span of the derivative contracts.

## 11.3   Forwards and Futures

Different financial derivatives have different purposes for the traders in financial markets. Forwards and futures are simple financial derivatives that serve two main purposes. The most important is hedging, that is, reducing the risk a portfolio is exposed to. As mentioned earlier, a portfolio consisting of assets and futures or forwards can be made risk–free, i.e. the portfolio has a fixed payoff at some future time. The other use of futures and forwards is speculating. If we have a view on how the market will change, e.g., that the value of a given asset will rise, we can take a position in the market that will

---

[4] Note that this is a *hypothesis*, cf. [10].

pay out a positive profit if your view is correct. The downside is that if you are wrong, you may loose all your money, or worse[5].

We first consider forward contracts, which are so-called over-the-counter (OTC) derivatives. The name indicates that the contracts are direct agreements between two parties, e.g., a bank and a company. The reason is the high initial risk involved when buying and selling forward contracts.

### 11.3.1   The Forward Contract

We define a forward contract as follows.

**Definition 1.** A forward contract is a right and an obligation to buy (or sell) an underlying asset at a prescribed forward price at a certain time in the future.

When the contract is written, the forward price and the corresponding date are settled. Note that no money is interchanged prior to expiry, except from an initial margin used to reduce the credit risk. The point of interest is how to determine the risk-neutral forward price. Below, we will see that this price can be determined by an arbitrage argument.

### 11.3.2   The Forward Price

Let $F$ represent the forward price, $S = S(t)$ the price of the underlying asset, $T$ the expiry date of the contract, and $t_0$ some time prior to expiry. Furthermore, assume that the forward contract allows (and obligates) us to purchase the underlying asset for the forward price. At expiry, our profit will be the difference $S(T) - F$, which can either be positive, zero, or negative depending on the value of $S(T)$, cf. Figure 11.1.

At some time prior to expiry, the value of $S(T)$ is not known. Still, the value of $F$ can be determined by the following argument. Assume that we buy a forward contract at some initial time $t_0$. This costs nothing, because money is paid at expiry. However, we have taken a position in the market. This position is "canceled" by short selling the asset, i.e. we sell the asset not yet purchased, and put the money $S(t_0)$ in a bank. The total value of this portfolio is zero, $-S(t_0) + S(t_0) = 0$. At expiry, the bank deposit has increased to

$$S(t_0)e^{r(T-t_0)},$$

where $r$ is the risk–free interest rate, which is assumed to be constant during the life-time of the contract[6]. At expiry, we purchase the underlying asset for

---

[5] Consider an investment strategy that involves the obligation to buy certain assets at the market value some time in the future. If the market value of these assets increase too much, you may not be able to fulfill your commitments.

[6] Continuous interest rate is modeled by the exponential function.

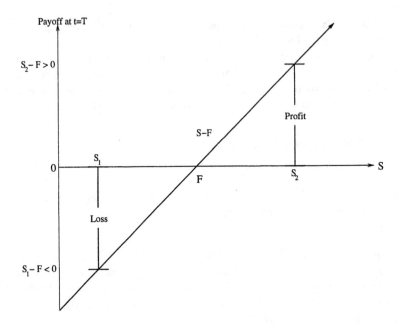

**Fig. 11.1.** The payoff at time $t = T$ for the buyer of the forward contract. The forward price is $F$. If the asset price has decreased to $S_1$, the buyer must pay $F - S_1$ to the writer (the seller of the contract), and loses money. On the other hand, if the value of the asset has increased to $S_2$, the buyer makes the profit $S_2 - F$.

the price $S(T)$ to cancel the short position, and the forward contract pays out $S(T) - F$. The net value of our portfolio is therefore

$$S(t_0)e^{r(T-t_0)} - S(T) + (S(T) - F) = S(t_0)e^{r(T-t_0)} - F.$$

Note that this value is independent of the value of the underlying asset at time $T$. The principle of arbitrage–free markets states that, because we started with no cash, we end up with no risk–free profit. We therefore obtain the following expression for the forward price,

$$F = S(t_0)e^{r(T-t_0)}. \tag{11.1}$$

Assume that this relation is violated, and that the value of the forward contract is below the price given by (11.1). Then the portfolio discussed above, which did not cost any money, would pay out $S(t_0)e^{r(T-t_0)} - F > 0$ at expiry, which violates the no-arbitrage assumption. On the other hand, if the value is higher than expressed in (11.1), the opposite position should be taken. That is, to short the forward and lend cash in the bank to purchase the underlying. This approach would give a risk-less profit of $F - S(t_0)e^{r(T-t_0)}$. Thus, because we assume that the market is arbitrage–free, we conclude that the forward price must be determined by (11.1).

The forward price $F$ given in equation (11.1) is a linear function of the price $S$ of the underlying asset. Derivatives that fulfill such a relationship are referred to as linear instruments.

### 11.3.3  The Future Contract

A future contract is similar to a forward agreement in that it is a settlement to trade an asset at a future time for a given price. However, future contracts are formalized and traded on exchanges. Thus, the buyer and seller do not need to know each other. The considerable default risk associated with forward contracts is reduced by introducing daily settlement agreements known as *marking to market*: The value of the future contract is evaluated several times during the life–time of the contract, and money is paid to the party that is in the profitable position, see [1] and [6]. It can be shown that if the interest rate $r$ is constant, the future price equals the forward price, see [10].

## 11.4  The Black-Scholes Analysis

In the celebrated work "The pricing of options and corporate liabilities" [2], Fischer Black and Myron Scholes derived a deterministic partial differential equation governing the fair price of an option. Robert Merton derived the same equation using another method the same year, and also generalized it in various ways. In 1997, Scholes and Merton were rewarded with the Nobel price for economics for this work[7].

The analysis that was carried out is surprisingly straight forward, and is the topic of this section. We start by defining European options, and distinguish between call and put options.

**Definition 2.** A European call option is a contract among two parties, with the buyer of the contract having the right, but not the obligation, to buy the underlying asset for a given strike price, $E$, at some time $T$ in the future, called the expiry date.

Similarly, we define a European put option.

**Definition 3.** A European put option is a contract among two parties, with the buyer of the contract having the right, but not the obligation, to sell the underlying asset for a given strike price, $E$, at some time $T$ in the future, called the expiry date.

Let $S = S(t)$ be the price of the underlying asset at time $t$, and consider the change $dS$ over a small period of time $dt$. We assume that the following stochastic differential equation (SDE) governs the change of the asset price

$$dS = \mu S dt + \sigma S dX, \tag{11.2}$$

---

[7] Fisher Black would undoubtedly have received this price together with Merton and Scholes if he had still been alive.

where $\mu$ is the drift of the asset, $\sigma$ is a constant referred to as the volatility of the asset, and $dX$ is a sample from a normal distribution with zero mean (i.e. a Wiener process). The drift is a measure of how the average value of the asset $S$ evolves in time, and we assume that it is constant.

Discretizing (11.2) using a Forward Euler method, we get a discrete random walk,

$$\Delta S^{k+1} = \mu S^k \Delta t + \sigma S^k \sqrt{\Delta t}\phi^k, \tag{11.3}$$

where $\phi^k$ is a sample from a standardized normal distribution. Starting with an initial state $S^0$, we can simulate asset prices $S^0, \ldots, S^M$ using (11.3) simply by updating $S^{k+1} = S^k + \Delta S^{k+1}$, $k = 1, \ldots, M - 1$. A simulation of a discrete random walk is shown in Figure 11.2.

A Random Walk

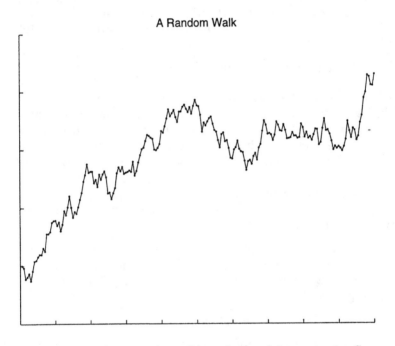

**Fig. 11.2.** A possible evolution of the asset price $S$.

The expectation of $dS$ is given by.

$$E(dS) = E(\mu S dt + \sigma dX S) = \mu S dt,$$

because the expectation of $dX$ is zero. That is, the expected change of the asset value, $E$, equals the drift of the asset and the current value.

### 11.4.1   Ito's Lemma

Derivatives are functions of stochastic variables (in our case functions of a variable on the form (11.2)). Ito's lemma expresses the change of a function $f(S)$ by changes of the variable $S$. Differentiation and integration of Wiener processes differ from classical calculus, see [16, 29–30]. Going in depth into this field is beyond the scope of this chapter, and the interested reader may see [6, 109–133] for details. It turns out that if we use the following rule–of–thumb:

$$dX^2 \to dt \quad \text{as } dt \to 0, \tag{11.4}$$

we can apply the rules of classical calculus when integrating Wiener processes, and we use this rule throughout this chapter.

Assume that $f = f(S)$ is a smooth function of $S$. Then a Taylor-series expansion of $f$ reads

$$df = f(S + dS) - f(S) \approx f(S) + f'(S)dS + \frac{1}{2}f''(S)(dS)^2 - f(S)$$

$$= f'(S)dS + \frac{1}{2}f''(S)(dS)^2.$$

By using (11.2), we get

$$df = f'(S)(\mu S dt + \sigma S dX) + \frac{1}{2}f''(S)(\mu S dt + \sigma S dX)^2$$

$$= f'(S)(\mu S dt + \sigma S dX) + \frac{1}{2}f''(S)(\mu^2 S^2 dt^2 + 2\sigma\mu S^2 dX dt + \sigma^2 S^2 dX^2).$$

Now, we use (11.4) and then ignore higher order terms in $dt$. We then get

$$df \approx f'(S)(\mu S dt + \sigma S dX) + \frac{1}{2}f''(S)\sigma^2 S^2 dt.$$

This expression follows from the fact that to leading order, $dX^2 = dt$ dominates the terms $dt^2$ and $dX dt$ for small values of $dt$. A re-arrangement of the above expression yields

$$df = \sigma S f'(S)dX + \left(\mu S f'(S) + \frac{1}{2}\sigma^2 S^2 f''(S)\right)dt, \tag{11.5}$$

which is referred to as Ito's lemma. If $f$ also depends on the time $t$, then the more general result,

$$df = \sigma S \frac{\partial f}{\partial S}dX + \left(\mu S \frac{\partial f}{\partial S} + \frac{1}{2}\sigma^2 S^2 \frac{\partial^2 f}{\partial S^2} + \frac{\partial f}{\partial t}\right)dt, \tag{11.6}$$

holds.

### 11.4.2   Elimination of Randomness

Assume that the two random walks (11.2) and (11.6) depend on the same Brownian motion $dX$. Then it turns out that we can utilize this fact to construct a new variable $g$ whose variation $dg$ is deterministic during the time period $dt$. To this end, define

$$g = f - \Delta S,$$

where $\Delta$ is constant during the time step $dt$. Now the change $dg$ is given by

$$
\begin{aligned}
dg &= df - \Delta dS \\
&= \sigma S \frac{\partial f}{\partial S} dX + (\mu S \frac{\partial f}{\partial S} + \frac{1}{2}\sigma^2 S^2 \frac{\partial^2 f}{\partial S^2} + \frac{\partial f}{\partial t}) dt - \Delta(\mu S dt + \sigma S dX) \\
&= \sigma S(\frac{\partial f}{\partial S} - \Delta) dX + \left(\mu S(\frac{\partial f}{\partial S} - \Delta) + \frac{1}{2}\sigma^2 S^2 \frac{\partial^2 f}{\partial S^2} + \frac{\partial f}{\partial t}\right) dt .
\end{aligned}
$$

By choosing

$$\Delta = \frac{\partial f}{\partial S}, \tag{11.7}$$

the above expression is reduced to

$$dg = (\frac{1}{2}\sigma^2 S^2 \frac{\partial^2 f}{\partial S^2} + \frac{\partial f}{\partial t}) dt,$$

which is completely deterministic, because the expression does not involve the stochastic term $dX$.

### 11.4.3   The Black-Scholes Equation

We are now in a position to derive the Black-Scholes equation. In addition to the assumptions made in Section 11.2, we assume that the volatility $\sigma$ of the asset and the risk–free interest rate $r$ are given constants. Let $V = V(S, t)$ be the value of an option as a function of the underlying asset $S$ and the time $t$. We construct a portfolio $\Pi$ as follows

$$\Pi = V - \Delta S. \tag{11.8}$$

This is a portfolio consisting of an option and short $\Delta$ shares of the corresponding underlying asset.

We assume that the asset price $S$ fulfills the SDE (11.2), and thus Ito's lemma (11.6) can be applied to the change of the value of the option $dV$ during the time step $dt$. The change of the value of this portfolio during this time step is then given by

$$
\begin{aligned}
d\Pi &= dV - \Delta dS \\
&= \sigma S(\frac{\partial V}{\partial S} - \Delta) dX + \left(\mu S(\frac{\partial V}{\partial S} - \Delta) + \frac{1}{2}\sigma^2 S^2 \frac{\partial^2 V}{\partial S^2} + \frac{\partial V}{\partial t}\right) dt .
\end{aligned}
$$

By choosing $\Delta = \frac{\partial V}{\partial S}$ we can eliminate the randomness and obtain a deterministic expression for the change of the value of the portfolio,

$$d\Pi = (\frac{1}{2}\sigma^2 S^2 \frac{\partial^2 V}{\partial S^2} + \frac{\partial V}{\partial t})dt. \tag{11.9}$$

Note that in the expression above, the drift parameter $\mu$ of the underlying asset has vanished.

Assume that instead of investing the amount $\Pi$ in the portfolio (11.8), we put the money in the bank, ensuring a risk-free profit of $r\Pi dt$ during the time step. By the principle of no arbitrage, this strategy is expected to pay out the same amount of money as the change in the value of the portfolio. This is known as risk-neutral pricing. Therefore,

$$d\Pi = r\Pi dt,$$

and inserting (11.8) and (11.9) in the above expression, we get

$$d\Pi = (\frac{1}{2}\sigma^2 S^2 \frac{\partial^2 V}{\partial S^2} + \frac{\partial V}{\partial t})dt = r\Pi dt$$
$$= r(V - \Delta S)dt = r(V - S\frac{\partial V}{\partial S})dt,$$

so

$$(\frac{1}{2}\sigma^2 S^2 \frac{\partial^2 V}{\partial S^2} + \frac{\partial V}{\partial t})dt = r(V - S\frac{\partial V}{\partial S})dt.$$

Finally, dividing by $dt$ gives the Black-Scholes equation

$$\frac{\partial V}{\partial t} + \frac{1}{2}\sigma^2 S^2 \frac{\partial^2 V}{\partial S^2} + rS\frac{\partial V}{\partial S} - rV = 0. \tag{11.10}$$

The above equation, subject to appropriate boundary and final conditions, fully specifies the fair price of various options, provided that the assumptions stated above hold. The equation (11.10) is backward parabolic, i.e. we solve it backwards in time[8]. Instead of an initial condition, we need a final condition at some future time, i.e., at the expiry date of the contract. Different option contracts lead to different boundary and final conditions. In the next chapter we discuss how these conditions can be determined.

Because an option contract is a right to buy or sell the underlying asset, but not an obligation to do so, the solution of (11.10) is more complicated than in the case of forward and future contracts. Especially, the relation between the value of the option and the price of the underlying asset is non-linear.

---

[8] A change of the $t$ variable, $\tau = -t$, changes this equation to forward parabolic in $\tau$.

## 11.5    European Call and Put Options

The list of available option contracts is long and in some cases rather obscure. Options can be constructed in various ways, ensuring a rise of the value of the option for almost any change of the underlying. Options can be divided into different classes, based on the basic structures of the different contracts.

Our first distinction is between vanilla and exotic options. Vanilla options are option contracts that do not include any unusual features. In our setting, this means contracts that fulfill Definition 2 or Definition 3 above. Some exotic options will be discussed in Section 11.7 below. Another way of categorizing options is to distinguish between European and American types of contracts. European options can only be exercised at a prescribed expiry date, whereas American options can be exercised at any time prior to this date. We start by considering European option contracts.

For simplicity, we will use the term European option for European vanilla put or call options. European options can only be exercised at the expiry date of the contract. This is the simplest type of options and the fair price of such a contract is easily determined by the Black-Scholes equation.

### 11.5.1    European Call Options

Recall Definition 2 of a call option. Because the European version of this contract may only be exercised at the time $T$ of expiry, the only decision the holder can make is whether or not to exercise it at this time. If the asset price is above the strike price, then the holder can make a profit from the contract. Exercising the option, the underlying asset is bought for the strike price. The market is willing to pay a higher price, and the profit is $S(T) - E$, where $E$ is the strike price and $S(T)$ the value of the asset at expiry. If the market price of the underlying is below the strike price stated in the option contract, the cheapest way to purchase the asset would be to buy it for the market price. In this case the option is worthless and thus the holder chooses not to exercise it.

To summarize, at expiry the value $C$ of the call option is given by,

$$C(S,T) = \max(S(T) - E, 0). \tag{11.11}$$

This payoff curve is the final condition needed in order to specify a mathematical model for the value of a European call option, and it is plotted in Figure 11.3.

The stochastic differential equation (11.2) has a fix point solution at $S = 0$, that is,

$$dS = \mu 0 dt + \sigma 0 dX = 0.$$

Therefore, if an asset ever becomes worthless, it will stay so forever. Consequently, the final condition (11.11) implies that $C(0, T) = 0$. Thus, at all times prior to expiry,

$$C(0, t) = 0. \tag{11.12}$$

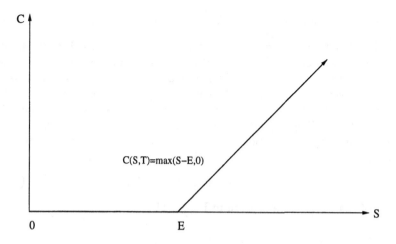

**Fig. 11.3.** The payoff at expiry for a European call option.

There is no theoretical upper limit for the value of the underlying asset. Therefore the contract is valid for all non-negative values of $S$. As $S$ increases, it becomes likely that the option will be exercised, and in the limit $S \to \infty$, the exercise price of the contract becomes less important compared to the value of the underlying. Thus,

$$C(S,t) \sim S(t) \quad \text{as } S \to \infty. \tag{11.13}$$

In order to obtain an exact expression for the value of the option as $S$ grows large, we must discount the exercise price stated in the contract. We then get,

$$C(S,t) = S(t) - Ee^{-r(T-t)} \quad \text{as } S \to \infty. \tag{11.14}$$

The fully specified model governing the value of a European call option is given below.

### The European Call Option

$$\frac{\partial C}{\partial t} + \frac{1}{2}\sigma^2 S^2 \frac{\partial^2 C}{\partial S^2} + rS\frac{\partial C}{\partial S} - rC = 0,$$
$$C(S,T) = \max(S - E, 0),$$
$$C(0,t) = 0,$$
$$C(S,t) = S - Ee^{-r(T-t)} \quad \text{as } S \to \infty.$$

Here, $r$ represents the interest rate, $E$ the exercise price, $T$ the expiry date, and $\sigma$ the volatility of the underlying asset.

## 11.5.2   European Put Options

A European put option is a right, but not the obligation, to sell the underlying at expiry for a given strike price $E$. If the market value of the asset is below the strike price at expiry, the holder of a European put option can buy the asset at market price $S$, and use the contract to sell it for $E$, giving the holder a profit of $E - S$. If the market value of the asset is above the strike price, the market is willing to pay a higher price for the asset than the contract specifies. In this case the put option is worthless. This gives the following final condition for a European put option,

$$P(S,T) = \max(E - S, 0).\qquad(11.15)$$

A plot of the payoff cure is given in Figure 11.4.

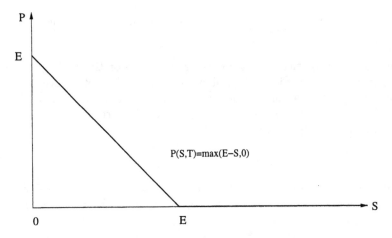

**Fig. 11.4.** The payoff curve at expiry for a European put option.

If the value of the underlying is zero, i.e. $S = 0$, we have a guaranteed payoff at expiry $P(0,T) = E$, and thus the boundary condition at $S = 0$ is the discounted value of the payoff, i.e.,

$$P(0,t) = Ee^{-r(T-t)}.$$

Because the put option is a right to sell an asset for a given price, the buyer of the contract will make a profit if the price of the underlying decreases sufficiently. On the other hand, if the underlying increases its value, the call option will become less worth, and in the limit vanish,

$$P(S,t) = 0 \quad \text{as } S \to \infty.$$

We have the following mathematical model for the value of European put options.

### The European Put Option

$$\frac{\partial P}{\partial t} + \frac{1}{2}\sigma^2 S^2 \frac{\partial^2 P}{\partial S^2} + rS\frac{\partial P}{\partial S} - rP = 0,$$
$$P(S,T) = \max(E - S, 0),$$
$$P(0,t) = Ee^{-r(T-t)},$$
$$P(S,t) = 0 \quad \text{as } S \to \infty.$$

Here, $r$ represents the interest rate, $E$ the exercise price, $T$ the expiry date, and $\sigma$ the volatility of the underlying asset.

### 11.5.3   Put–Call Parity

There is a simple relationship between the value of a European put and a European call option. To see this, consider a portfolio consisting of one asset $S$, one European put option $P$ written on the underlying asset $S$ (with the strike price $E$ and expiry date $T$), and short one European call option $C$ with the same exercise price, expiry date, and underlying. Then the payoff at expiry of this portfolio is,

$$S + P - C = S + \max(0, E - S) - \max(0, S - E) = E.$$

Having a fixed payoff, i.e. no risk, this portfolio must pay out the same as a bank deposit. Thus, at time $t$ prior to expiry, the value of the portfolio is the discounted value of $E$, i.e.

$$S + P - C = Ee^{-r(T-r)}. \tag{11.16}$$

This relation is called the put-call parity, and implies that it is only necessary to compute the value of either the put or the call option. The other value is then determined from (11.16).

### 11.5.4   Analytical Solutions

Explicit formulas for the fair price of European vanilla options are available in the literature, cf. [10]. Let

$$N(d) = \frac{1}{\sqrt{2\pi}} \int_{-\infty}^{d} e^{-\frac{1}{2}s^2}\, ds,$$

be the cumulative distribution function for the normal distribution. Then the solution of the European call problem is given by

$$C(S,t) = SN(d_1) - Ee^{r(T-t)}N(d_2),$$

where

$$d_1 = \frac{\log\left(\frac{S}{E}\right) + (r + \frac{1}{2}\sigma^2)(T-t)}{\sigma\sqrt{T-t}},$$

$$d_2 = \frac{\log\left(\frac{S}{E}\right) + (r - \frac{1}{2}\sigma^2)(T-t)}{\sigma\sqrt{T-t}}.$$

Having established analytical formulas for the value of European call options, the corresponding put option values are given by the put-call parity (11.16).

For most other option contracts, we do not have similar analytical expressions as above, and numerical techniques must be applied in order to determine the fair price $P$ or $C$ of the contract.

## 11.6   American Options

As mentioned above, American options differ from European agreements in that they can be exercised at any time up to and including the expiry date of the contract. Most exchange-traded options are American, and they potentially have a higher value than European due to the extra early exercise feature. An arbitrage argument determines this additional value. Suppose that the value of an American put option is below the payoff curve at expiry, $P(S,t) < \max(E - S, 0)$ at some time $t$ prior to expiry. If we buy the option for $P$, and then exercise it, selling the asset for $E$ while buying the asset for $S$ in the market, we earn $E - S - P > 0$. Thus a instantaneous profit is made. In the absence of such arbitrage possibilities, we conclude that the value of American put options must satisfy

$$P(S,t) \geq \max(E - S, 0) \quad \text{for all } t \leq T.$$

A similar argument holds for American call options,

$$C(S,t) \geq \max(S - E, 0) \quad \text{for all } t \leq T.$$

Note that European call options also fulfill this inequality, thus the value of American and European call options are identical. Stated in other words, it is never optimal to exercise the American call option prior to expiry[9]. To see this we present an arbitrage argument given in [13]. Given an American call option and the price $S$ of the underlying asset at some time $t$ before expiry $T$. Early exercise will pay $S(t) - E$, where $E$ is the exercise price of the underlying asset stated in the contract. However, if we instead sell the asset short and purchase it back at expiry for either the market price or by exercising the option, whichever is cheapest, the profit will be $S(t) - \min(E, S(T))$, which is clearly preferable.

---

[9] This statement only applies to assets that do not pay dividends.

### 11.6.1 American Put Options

An American put option is the right to sell an underlying asset for the exercise price within the expiry date. Thus, by buying an American put option, it is possible to have a guaranteed minimum price of the underlying asset at all times prior to the expiry date. Below, we will derive a mathematical model for the fair price of such contracts. As we will see, the model turns out to be more complicated than the European put option model.

An important point of interest for a holder of an American put contract, is whether or not to exercise the option. It turns out that if the underlying asset passes a given value, called the *critical stock price*, the option should be exercised. On the other hand, if this barrier is not crossed, the value of the option must satisfy the Black-Scholes equation. This critical stock price is not known a priori and moves during the life time of the option contract. In fact, the American put option problem is a free and moving boundary value problem, where the value of the critical stock price must be computed as part of the solution process. In order to fully specify the model for the American put problem, we must pose two conditions on the free boundary. A formal derivation of these conditions can be found in [6], and we will only present a motivation for these conditions below.

The first condition regards the continuity of the option price. If the option price is discontinuous for more than an infinitesimal period of time, a collection of options would make a guaranteed risk-less profit as soon as the value of the underlying asset passes a certain value cf [10]. Thus, at the critical stock price, the value of the option must equal the value of the payoff curve at expiry,

$$P(\bar{S}, t) = E - \bar{S},$$

where $\bar{S}$ denotes the critical stock price. A more subtle argument determines the second condition, that the Delta, $\Delta = \frac{\partial P}{\partial S}$, must be continuous for all values of $S$. Thus, this must also be valid at the unknown boundary, $\bar{S}$. Therefore,

$$\frac{\partial P(\bar{S}, t)}{\partial S} = -1, \tag{11.17}$$

where we have used that the slope of the payoff curve at expiry is $-1$. The interested reader should see [4] for a thorough treatment of the boundary conditions for American options.

It is common to pose the American put option problem for all values of $S > 0$. When the value of the underlying asset is less than $\bar{S}$, the value of the option contract is given by

$$P(S, t) = E - S, \quad 0 \le S < \bar{S}.$$

At expiry, the value of the American put option equals the value of the corresponding European contract, i.e.,

$$P(S, T) = \max(E - S, 0).$$

We need a condition for the value of the critical stock price $\bar{S}(T)$ at expiry. The value $\bar{S}(t)$ is the point that determines whether or not the option should be exercised. At expiry, this condition is clearly,

$$\bar{S}(T) = E,$$

because values of the underlying asset greater than $E$ at this time renders the put option contract worthless. On the other hand, if $S(T) < E$, the profit is $E - S(T)$.

Having established conditions on the boundary as well as final conditions, we have the following fully specified model of the value of an American put option.

**The American Put Option**

$$\frac{\partial P}{\partial t} + \frac{1}{2}\sigma^2 S^2 \frac{\partial^2 P}{\partial S^2} + rS\frac{\partial P}{\partial S} - rP = 0, \quad \bar{S} < S,$$
$$P(S,t) = E - S, \quad 0 < S < \bar{S},$$
$$P(S,T) = \max(E - S, 0),$$
$$\bar{S}(T) = E,$$
$$P(\bar{S}, t) = E - \bar{S},$$
$$\frac{\partial P(\bar{S})}{\partial S} = -1,$$
$$P(S,t) = 0 \quad \text{as } S \to \infty.$$

Here, $r$ represents the interest rate, $E$ the exercise price, $T$ the expiry date, $\bar{S}$ the unknown critical stock price, and $\sigma$ the volatility of the underlying asset.

Note that we have to compute $\bar{S}$ as part of solving this problem.

## 11.7    Exotic Options

Options that are more complex than the put and call options discussed above are called exotic options. We will briefly discuss two types of exotic options, namely correlation–dependent options and path–dependent options. As the name indicates, correlation–dependent options involve several underlying assets. The payoff curve at expiry for a path-dependent option depends on the history of the asset price during the life time of the option contract.

### 11.7.1    Correlation–Dependent Options

The simplest correlation dependent options are basket or multi-asset options [3] and [9]. The payoff is simply constructed as a combination of two or more underlying assets. This combination is called the basket. Basket options often involves currencies, where two or more currencies can be exchanged against a base currency or at a fixed exchange rate stated in the contract. Basket options are generally cheaper than the sum of the individual assets involved, due to the fact that the volatility of the basket is usually less than the individual volatilities.

Consider a European put basket consisting of $n$ underlying assets. Furthermore, let the payoff function at expiry have the form,

$$P(S_1, \ldots, S_n, T) = \max(E - \sum_{i=1}^{n} \alpha_i S_i, 0),$$

where $S_i$ denotes the value of the $i$th underlying asset and $\alpha_i > 0$ is a given weight. The governing equation for the fair value of this put option is the multi–dimensional Black-Scholes equation,

$$\frac{\partial P}{\partial t} + \frac{1}{2} \sum_{i=1}^{n} \sum_{j=1}^{n} \rho_{i,j} \sigma_i \sigma_j S_i S_j \frac{\partial^2 P}{\partial S_i \partial S_j} + r \sum_{i=1}^{n} S_i \frac{\partial P}{\partial S_i} - rP = 0, \qquad (11.18)$$

cf [5], [4] and [9]. Above, $\rho_{i,j} \in [0, 1]$ represents the correlation between the $i$th and $j$th underlying asset, which we assume is constant during the contract period. Special care has to be taken at the boundaries. As each individual asset $S_i$ increases towards infinity, the value of the basket asset must approach zero by the same reason as in the case of a single underlying asset. Thus,

$$\lim_{S_i \to \infty} P(S_1, \ldots, S_n, t) = 0 \quad i = 1, \ldots, n.$$

In order to determine the boundary conditions at $S_i = 0$, $i = 1, \ldots, n$, we first consider the model in case of two underlying assets, i.e., $n = 2$. Then, for $S_1 = 0$, the Black-Scholes equation (11.18) reduces to

$$\frac{\partial P}{\partial t} + \frac{1}{2} \sigma^2 S_2^2 \frac{\partial^2 P}{\partial S_2^2} + r S_2 \frac{\partial P}{\partial S_2} - rP = 0,$$

which is identical to (11.10) with $V = P$ and $S = S_2$. The only difference from the European put option model is that the payoff at expiry in this case is given by,

$$P(0, S_2, t) = P(S_2, t) = \max(E - \alpha_2 S_2, 0).$$

We have the same boundary condition as in the one–dimensional European put option problem. Thus, in order to determine the boundary condition at $S_1 = 0$, the problem above must be solved. The same argument holds for $S_2 = 0$, where the final condition is given by

$$P(S_1, 0, t) = P(S_1, t) = \max(E - \alpha_1 S_1, 0).$$

When these two problems at the boundary are solved, the solution of the two–dimensional problem can be computed.

We now turn to the case of three underlying assets. Then, at each $S_i = 0$, the problem degenerates to three two–dimensional problems by the same reasons as above. In order to find the solution of each two–dimensional problem, two one–dimensional problems must be computed.

In general, in order to solve an option problem with $n$ underlying assets, we must solve $n$ problems in $n - 1$ dimensions. This procedure is repeated until the case $n = 1$, where the problem can be solved with known boundary conditions. Thus, for a $n$–dimensional problem, the total number $p$ of problems that must be solved is

$$p(n) = \sum_{i=2}^{n} \prod_{j=i}^{n} j + 1.$$

For example, to find the option price for a contract consisting of five underlying asset, a total of 206 problems must be solved[10].

A natural way of implementing the procedure above is by recursion. See [14] for how this can be done. The fully specified model for European put basket options is given below.

### The European Put Basket Option

$$\frac{\partial P}{\partial t} + \frac{1}{2} \sum_{i=1}^{n} \sum_{j=1}^{n} \rho_{i,j} \sigma_i \sigma_j S_i S_j \frac{\partial^2 P}{\partial S_i \partial S_j} + r \sum_{i=1}^{n} S_i \frac{\partial P}{\partial S_i} - rP = 0,$$

$$P(S_1, \ldots, S_n, T) = \max(E - \sum_{i=1}^{n} \alpha_i S_i, 0),$$

$$P(S_1, \ldots, S_n, t) = g_i(S_1, \ldots, S_{i-1}, S_{i+1}, \ldots, S_n, t) \quad \text{at } S_i = 0,$$
$$P(S_1, \ldots, S_n, t) = 0 \quad \text{as } S_i \to \infty.$$

Here, $r$ represents the interest rate, $E$ the exercise price, $T$ the expiry date, $\rho_{i,j}$ the correlation between asset $S_i$ and $S_j$, $\alpha_i$ a given weight such that $\sum_{i=1}^{n} \alpha_i = 1$ and $\sigma_i$ the volatility of the underlying asset $S_i$.

Another type of correlation–dependent option is the spread option, where the value of the option is related to the difference in price between two underlying assets. For example, a spread call option pays out if the difference between two stocks is greater than the strike price stated in the contract. The pay off curve then reads,

$$C(S_1, S_2, T) = \max(0, (S_1 - S_2) - E),$$

---

[10] This is actually an overestimate, because the problems to some extent overlap each other.

where $E$ is the strike price. Spread options are written on individual stocks, baskets of stocks, or indices among others. Spread options can be constructed by taking a long and a short position in two similar vanilla option contracts, see [8].

## 11.7.2    Path–Dependent Options

Barrier options are contracts that depend on whether or not the underlying asset reaches a certain barrier, see [10, 164–169] and [4, 246–267]. The two main types of such options are knock-in and knock-out options. These contracts are similar to ordinary vanilla options, but includes an additional feature. Knock-out options become worthless if the underlying passes a barrier stated in the contract. Knock-in options come into existence if the underlying reaches the barrier. There are four different classes of barriers options; down-and-in, down-and-out, up-and-in, and up-and-out, and these types of barrier options can be either put or call option contracts. We discuss the down-and-out call barrier option, which means that the option becomes worthless if the underlying drops below the barrier. As long as the underlying asset stays above the barrier, the Black-Scholes equation determines the value of the option contract. At the barrier $B$, the option becomes worthless. This enters the model as a new boundary condition,

$$V(B,t) = 0. \tag{11.19}$$

There is a correlation between the value of a down-and-out and a down-and-in call option. Having two similar contracts, one down-and-out and one down-and-in, the barrier knocks either the first out or the latter in, and thus the sum of their values must equal the value of a European call option,

$$C_o + C_i = C,$$

where $C_o$ represents the down-and-out and $C_i$ the down-and-in barrier call option. Similar relationships hold for other barrier options, such as up-and-out and up-and-in call options, see [8].

When barrier options knock out or fail to knock in, it is quite common that the contracts specify a rebate that is paid to the holder of the barrier option. In case of down-and-out call barrier options, this can be modeled by changing (11.19) to,

$$V(B,t) = R,$$

where $R$ is the rebate paid if the option knocks out.

The barrier can be a constant, a continuous function of time, or only exist for some period of time within the life span of the option contract. Also, multiple barriers are possible.

Barrier options are not heavily dependent on the path of the underlying, and are sometimes referred to as weakly path dependent. Other option types are so–called strong path dependent. We will discuss two types; Asian options and look–back options.

The payoff for Asian options depends on some sort of averaging of the underlying asset price during the life span of the contract. Most Asian options are European style, i.e., they can only be exercised at the expiry date. The two main classes of Asian options are average value options and average strike options. The payoff for an average value call option is given by the function $\max(\hat{A} - E, 0)$, where $\hat{A}$ represents the average of the asset price during the contract period and $E$ as usual denotes the exercise price. In case of average strike options, the payoff curve at expiry is given by $\max(S(T) - \hat{A}, 0)$, where $S(T)$ is the price of the underlying at the expiry date.

There are several ways to specify the averages above. One possibility is to use an arithmetic average, that is

$$\hat{A} = \frac{1}{n} \sum_{i=1}^{n} S(t_i),$$

in the discrete case and

$$\hat{A} = \frac{1}{T} \int_{0}^{T} S(t)dt,$$

in the continuous case. Another possibility is to use geometric averaging. On discrete form this is given by

$$\hat{A} = \left[ \prod_{i=1}^{n} S(t_i) \right]^{\frac{1}{n}}.$$

In the continuous case, this average takes the following form,

$$\hat{A} = \exp\left( \frac{1}{T} \int_{0}^{T} \ln(S(t))dt \right).$$

In the expressions above, $S(t_i)$ is the asset price at the discrete time $t_i$.

In order to derive a mathematical model for the fair price of an Asian option, an analysis similar to the Black-Scholes analysis in Section 11.4 must be carried out. The reader should consult [4] for this derivation. The resulting governing PDE for the Asian option $V(S, A, t)$ turns out to be

$$\frac{\partial V}{\partial t} + \frac{1}{2}\sigma^2 S^2 \frac{\partial^2 V}{\partial S^2} + rS\frac{\partial V}{\partial S} + f(S,t)\frac{\partial V}{\partial A} - rV = 0. \tag{11.20}$$

Here $f(S,t)$ is related to the type of averaging stated in the option contract. More specifically, we have

$$A(t) = \int_{0}^{t} f(S, \tau)d\tau,$$

such that $A(T) = T\hat{A}$. Following [4], consider an arithmetic average strike European style call option $C(S, A, t)$. For this contract, the strike price $E$ is given by the average

$$E = \frac{1}{T} \int_0^T S(\tau)d\tau = \frac{A(T)}{T},$$

where $A(t) = \int_0^t S(\tau)d\tau$. The governing equation for the fair price of this contract is given by (11.20) with $f(S, t) = S$.

The fully specified model for this Asian option contract is given below.

**The Asian Arithmetic Average Strike European Style Call Option**

$$\frac{\partial C}{\partial t} + \frac{1}{2}\sigma^2 S^2 \frac{\partial^2 C}{\partial S^2} + rS\frac{\partial C}{\partial S} + S\frac{\partial C}{\partial A} - rC = 0, \qquad (11.21)$$

$$C(0, A, t) = 0,$$

$$\lim_{S\to\infty} \frac{\partial C}{\partial S}(S, A, t) = 1,$$

$$\lim_{A\to\infty} C(S, A, t) = 0,$$

$$C(S, A, T) = \max\left(0, S - \frac{A(T)}{T}\right).$$

Here, $r$ represents the interest rate, $\sigma$ the volatility of the underlying asset, $T$ the expiry date, and $A(T)/T$ the arithmetic average with $A(t) = \int_0^t S(\tau)d\tau$.

Note that one boundary condition on $A$ is needed because the first–order partial derivative with respect to $A$ appears in (11.21).

Look-back options [4] may also be modeled within the Black-Scholes framework. The payoff for look-back options depends on one of the extreme values of the underlying asset during the life time of the contract, or a period within this time called the look-back period. We shall consider a European style floating strike look-back put option $P(S, M, t)$. The payoff curve at expiry is given by

$$P(S, M, T) = \max(M - S, 0),$$

where $M$ is the maximum of the asset price over the look-back period. An important property of this option is that it is always in the money, i.e., it will always be exercised at the expiry date,

$$P(S, M, T) = \max(M - S, 0) = M - S,$$

because $S(T) \leq M$ by the construction of $M$. Due to this feature, this contract will pay out high profit if the underlying asset first increases its value during the look-back period, giving a high $M$, and then falls at the end of the contract period resulting in a relatively low $S(T)$.

It turns out that the standard Black-Scholes equation (11.10) models the option price, subject to the boundary conditions given below.

**The European Style Floating Strike Look-back Put Option**

$$\frac{\partial P}{\partial t} + \frac{1}{2}\sigma^2 S^2 \frac{\partial^2 P}{\partial S^2} + rS\frac{\partial P}{\partial S} - rP = 0, \quad 0 < S < M, \quad t > T_0,$$

$$P(0, M, t) = e^{-r(T-t)}M,$$

$$\frac{\partial P}{\partial S} = 0, \quad \text{at } S = M,$$

$$p(S, M, T) = \max(M - S, 0) = M - S.$$

Here, $T_0$ is the beginning of the look-back period, $r$ the interest rate, $\sigma$ the volatility of $S$, $M$ the maximum of the asset during the look-back period, and $T$ the expiry date of the contract.

## 11.8   Hedging

Hedging is a strategy for reducing investment risk by using options, short selling, futures, and other financial derivatives. Hedging can also be used to lock in profits. A hedged portfolio is constructed in order to reduce the overall volatility by reducing the risk of loss. When hedging a portfolio, we are interested in how the total value depends on variation of parameters such as the interest rate, the volatility of the different assets involved in the portfolio, the underlying asset prices, and time. These variations are known as the Greeks, and are given by

$$\text{Delta}, \quad \Delta = \frac{\partial V}{\partial S}, \tag{11.22}$$

$$\text{Gamma}, \quad \Gamma = \frac{\partial^2 V}{\partial S^2}, \tag{11.23}$$

$$\text{Theta}, \quad \Theta = -\frac{\partial V}{\partial t}, \tag{11.24}$$

$$\text{Vega}, \quad \wedge = \frac{\partial V}{\partial \sigma}, \tag{11.25}$$

$$\text{Rho}, \quad \rho = \frac{\partial V}{\partial r}. \tag{11.26}$$

### 11.8.1   The Delta Greek

We first consider the Delta Greek for European options. This Delta is the same as the $\Delta$ in (11.7) used to derive the deterministic Black-Scholes equation (11.10) earlier. Starting with put options, for a contract "in the money", i.e., a contract that probably will be exercised, the Delta Greek will move towards $-1$, because the slope of the payoff curve at expiry in this case is $-1$. For call options, the slope of this curve is 1, and thus $\Delta$ tends to 1 when call options are in the money. There are generally two ways of determining the Delta of an option contract. One way is to first compute the value $V(S, t)$ of the specific option contract, and then differentiate this value with respect to the underlying asset price: $\frac{\partial V}{\partial S}$. Another strategy is to differentiate the Black-Scholes equation to obtain a final value problem for the Delta Greek. In order to do so, it is preferable to rewrite the Black-Scholes equation on conservative form. Observing that

$$S^2 \frac{\partial^2 V}{\partial S^2} = \frac{\partial}{\partial S}\left(S^2 \frac{\partial V}{\partial S}\right) - 2S \frac{\partial V}{\partial S},$$

and that

$$S \frac{\partial V}{\partial S} = \frac{\partial}{\partial S}(SV) - V,$$

the Black-Scholes equation can be rewritten,

$$\frac{\partial V}{\partial t} + \frac{1}{2}\sigma^2 \frac{\partial}{\partial S}\left(S^2 \frac{\partial V}{\partial S}\right) + (r - \sigma^2)\frac{\partial}{\partial S}(SV) + (\sigma^2 - 2r)V = 0. \quad (11.27)$$

Differentiating the above equation with respect to $S$, and substituting $\frac{\partial V}{\partial S} = \Delta$, we obtain the following governing equation for $\Delta$,

$$\frac{\partial \Delta}{\partial t} + \frac{1}{2}\sigma^2 \frac{\partial}{\partial S}\left(S^2 \frac{\partial \Delta}{\partial S}\right) + r\frac{\partial}{\partial S}(S\Delta) - r\Delta = 0. \quad (11.28)$$

We consider the Delta for European put options, here denoted as $\Delta_P$. The boundary condition at $S = 0$ can be found by letting $S = 0$ in (11.28). Then

$$\frac{\partial \Delta_P}{\partial t} + r\Delta_P - r\Delta_P = 0 \quad \Rightarrow \quad \frac{\partial \Delta_P}{\partial t} = 0,$$

so the boundary value is constant in time. Hence, the boundary value at $S = 0$ equals the value of the Delta when the contract expires. At the expiry date $T$ we have,

$$\Delta_P(S, T) = \frac{\partial P}{\partial S}(S, T) = \frac{\partial}{\partial S}\big(\max(E - S, 0)\big) = \begin{cases} -1, & S \leq E, \\ 0, & S > E. \end{cases}$$

Thus, at $S = 0$, we have

$$\Delta_P(0, t) = \Delta_P(0, T) = -1.$$

The other boundary condition follows from the put-call parity (11.16). We define the Delta Greek for the European call option,

$$\Delta_C = \frac{\partial C}{\partial S}.$$

Then differentiating the put-call parity with respect to $S$ gives

$$\frac{\partial}{\partial S}(S + P - C) = 1 + \Delta_P - \Delta_C = 0. \tag{11.29}$$

Now, because the boundary condition for the call option as $S$ tends towards infinity is given by

$$\lim_{S \to \infty} C(S, t) = S - Re^{-r(T-t)},$$

the call Delta is clearly

$$\lim_{S \to \infty} \Delta_C = 1.$$

Inserting the above expression in (11.29) and taking the limit, we obtain

$$\lim_{S \to \infty} \Delta_P(S, t) = \lim_{S \to \infty} \Delta_C - 1 = 0.$$

The model for the European put Delta is summarized below.

### The European Put Delta

$$\frac{\partial \Delta_P}{\partial t} + \frac{1}{2}\sigma^2 \frac{\partial}{\partial S}\left(S^2 \frac{\partial \Delta_P}{\partial S}\right) + r\frac{\partial}{\partial S}(S\Delta_P) - r\Delta_P = 0,$$

$$\Delta_P(S, T) = \begin{cases} -1, & S \le E, \\ 0, & S > E, \end{cases}$$

$$\Delta_P(0, t) = -1,$$

$$\lim_{S \to \infty} \Delta_P(S, t) = 0.$$

Here, $\Delta_P$ is the European put Delta, $r$ is the interest rate, $E$ is the exercise price, $T$ is the expiry date, and $\sigma$ is the volatility of the underlying asset $S$.

Having established a model for the European put Delta, the corresponding call Delta is given by (11.29), i.e.

$$\Delta_C = 1 + \Delta_P. \tag{11.30}$$

### 11.8.2   The Gamma Greek

The next Greek of interest is the Gamma Greek (11.23), which is defined as the second–order partial derivative of the option with respect to the underlying asset:

$$\Gamma = \frac{\partial^2 V}{\partial S^2} = \frac{\partial \Delta}{\partial S}.$$

Differentiating (11.28) with respect to $S$, and substituting for Gamma gives,

$$\frac{\partial \Gamma}{\partial t} + \frac{1}{2}\sigma^2 \frac{\partial}{\partial S}\left(S^2 \frac{\partial \Gamma}{\partial S}\right) + (r + \sigma^2)\frac{\partial}{\partial S}(S\Gamma) = 0. \tag{11.31}$$

At the boundary $S = 0$, the above equation reduces to,

$$\frac{\partial \Gamma}{\partial t} + (r + \sigma^2)\Gamma = 0. \tag{11.32}$$

The solution of this equation is $\Gamma(t) = \Gamma_T e^{-(r+\sigma^2)(T-t)}$, where $\Gamma_T$ is independent of time and thus, $\Gamma_T = \frac{\partial}{\partial S}\Delta(S = 0, T)$. In case of the Gamma for a European put option we get $\Gamma_{P,T} = \frac{\partial}{\partial S}(-1) = 0$. Thus, at $S = 0$, we obtain the following boundary condition,

$$\Gamma_P(0, t) = 0.$$

In order to derive the boundary condition as $S \to \infty$, we apply a change of variables $x = 1/S$. Then,

$$\frac{\partial \Gamma}{\partial S} = \frac{\partial \Gamma}{\partial x}\frac{\partial x}{\partial S} = -\frac{1}{S^2}\frac{\partial \Gamma}{\partial x} = -x^2 \frac{\partial \Gamma}{\partial x},$$

and

$$\frac{\partial^2 \Gamma}{\partial S^2} = 2x^3 \frac{\partial \Gamma}{\partial x} + x^4 \frac{\partial^2 \Gamma}{\partial x^2}.$$

The equation for $\Gamma$ is now,

$$\frac{\partial \Gamma}{\partial t} + \frac{1}{2}\sigma^2(2x\frac{\partial \Gamma}{\partial x} + x^2\frac{\partial^2 \Gamma}{\partial x^2}) - (r + 2\sigma^2)x\frac{\partial \Gamma}{\partial x} + (r + \sigma^2)\Gamma = 0.$$

When $S \to \infty$, $x \to 0$, and the above equation reduces to

$$\frac{\partial \Gamma}{\partial t} + (r + \sigma^2)\Gamma = 0.$$

This equation is identical to (11.32) and by the same argument as above we conclude that

$$\Gamma_P(S, t) = 0 \quad \text{as } S \to \infty.$$

At expiry, the final condition for the European put Gamma $\Gamma_P$ is given by

$$\Gamma_P(S, T) = \frac{\partial}{\partial S}\begin{cases} -1, & S \le E, \\ 0, & S > E, \end{cases}$$

which is the Dirac delta function. More precisely, $\Gamma_P(S, T)$ is the distributional (or weak) derivative of a Heaviside function [7]. This is clearly a problem when solving this problem numerically, because we then have to solve a PDE with a Dirac delta function as final condition. Thus, in order to solve (11.31) numerically, we must apply a suitable discretization of $\Gamma_P(S, T)$ as a final condition for the associated discrete problem.

Alternatively, $\Gamma_P$ can be computed by applying a numerical differentiation procedure to the numerical solution of (11.28).

### 11.8.3   The Theta Greek

The Theta Greek is the negative time variation of the option value,

$$\Theta = -\frac{\partial V}{\partial t}.$$

Differentiating (11.27) with respect to $t$ and substituting for Theta yields,

$$\frac{\partial \Theta}{\partial t} + \frac{1}{2}\sigma^2 \frac{\partial}{\partial S}\left(S^2 \frac{\partial \Theta}{\partial S}\right) + (r - \sigma^2)\frac{\partial}{\partial S}(S\Theta) + (\sigma^2 - 2r)\Theta = 0.$$

Recall the put-call parity for European vanilla options,

$$S + P - C = Ee^{-r(T-t)}.$$

Differentiating this expression with respect to $t$ gives the put-call parity for Theta,

$$-\Theta_P + \Theta_C = rEe^{-r(T-t)}, \tag{11.33}$$

thus we only need an expression for either of the put or call Theta and use (11.33) to find the other. Consider the put Theta. At expiry, the final condition becomes

$$\Theta_P(S,T) = -\frac{\partial}{\partial t}\max(E - S, 0) = 0.$$

At $S = 0$, the put boundary condition is $P(0,t) = Ee^{-r(T-t)}$. This gives the following boundary condition for the put Theta,

$$\Theta_P(0,t) = -rEe^{-r(T-t)}.$$

The boundary condition as $S$ increases towards infinity is obtained by combining (11.33) with the boundary condition at infinity for the call Theta,

$$\lim_{S\to\infty} \Theta_C(S,t) = rEe^{-r(T-t)}.$$

Thus,

$$\lim_{S\to\infty} \Theta_P(S,t) = \lim_{S\to\infty}\left(\Theta_C(S,t) - rEe^{-r(T-t)}\right) = 0.$$

The fully specified model for the put Theta is given below.

### The European Put Theta

$$\frac{\partial \Theta_P}{\partial t} + \frac{1}{2}\sigma^2 \frac{\partial}{\partial S}\left(S^2 \frac{\partial \Theta_P}{\partial S}\right) + (r - \sigma^2)\frac{\partial}{\partial S}(S\Theta_P) + (\sigma^2 - 2r)\Theta_P = 0.$$

$$\Theta_P(S,T) = 0,$$

$$\Theta_P(0,t) = -rEe^{-r(T-t)},$$

$$\lim_{S\to\infty} \Theta_P(S,t) = 0.$$

Here, $\Theta_P$ denotes the European put Theta, $r$ the interest rate, $E$ the exercise price, $T$ the expiry date, and $\sigma$ the volatility of the underlying asset.

### 11.8.4    The Vega Greek

An important Greek is the Vega. It measures the variation of the option value in terms of changes of the volatility $\sigma$,

$$\Lambda = \frac{\partial V}{\partial \sigma} \, .$$

As before, we differentiate the conservative formulation of the Black-Scholes equation with respect to the desired parameter. We then obtain,

$$\frac{\partial \Lambda}{\partial t} + \frac{1}{2}\sigma^2 \frac{\partial}{\partial S}\left(S^2 \frac{\partial \Lambda}{\partial S}\right) + (r - \sigma^2)\frac{\partial}{\partial S}(S\Lambda) + (\sigma^2 - 2r)\Lambda = -\sigma S^2 \Gamma. \quad (11.34)$$

Note that we have a source term given by the Gamma Greek on the right hand side of the governing equation. The put-call parity for Vega is simple,

$$\Lambda_P = \Lambda_C, \quad (11.35)$$

hence we only need an expression for e.g. the Vega put Greek. At termination, $t = T$, the Vega is

$$\Lambda_P(S,T) = \frac{\partial}{\partial \sigma}(\max(E - S, 0)) = 0 \, .$$

Furthermore, at $S = 0$ the boundary condition reads,

$$\Lambda_P(0,t) = \frac{\partial}{\partial \sigma}\left(Ee^{-r(T-t)}\right) = 0 \, .$$

As $S$ grows large, the boundary condition for the call Vega Greek is

$$\lim_{S \to \infty} \Lambda_C(S,t) = \lim_{S \to \infty} \frac{\partial}{\partial \sigma}\left(S - Ee^{-r(T-t)}\right) = 0 \, .$$

Using the identity (11.35) then yields,

$$\lim_{S \to \infty} \Lambda_P(S,t) = \lim_{S \to \infty} \Lambda_C(S,t) = 0 \, .$$

The model is summarized below.

### The European Put Vega

$$\frac{\partial \Lambda_P}{\partial t} + \frac{1}{2}\sigma^2 \frac{\partial}{\partial S}\left(S^2 \frac{\partial \Lambda_P}{\partial S}\right) + (r - \sigma^2)\frac{\partial}{\partial S}(S\Lambda_P) + (\sigma^2 - 2r)\Lambda_P = -\sigma S^2 \Gamma_P \, .$$

$$\Lambda_P(S,T) = 0,$$
$$\Lambda_P(0,t) = 0,$$
$$\lim_{S \to \infty} \Lambda_P(S,t) = 0 \, .$$

Here, $\Lambda_P$ denotes the European put Vega, $\Gamma_P$ the European put Gamma, $r$ the interest rate, $E$ the exercise price, $T$ the expiry date, and $\sigma$ the volatility of the underlying asset.

### 11.8.5  The Rho Greek

The final Greek of interest is the variation with respect to the risk–free interest rate Rho,

$$\rho = \frac{\partial V}{\partial r}.$$

As before, the Black-Scholes equation is differentiated, and we substitute for Rho. This gives a slightly modified version of the Black-Scholes equation,

$$\frac{\partial \rho}{\partial t} + \frac{1}{2}\sigma^2 \frac{\partial}{\partial S}\left(S^2 \frac{\partial}{\partial S}\rho\right) + (r - \sigma^2)\frac{\partial}{\partial S}(S\rho) + (\sigma^2 - 2r)\rho = V - S\Delta. \quad (11.36)$$

Note that the left–hand side equals the conservative formulation of the Black-Scholes equation, and that the only difference is the additional source terms on the right–hand side. We have the put-call parity relation

$$\rho_P - \rho_C = -(T - t)Ee^{-r(T-t)}, \quad (11.37)$$

and it is thus sufficient to compute Rho for the European put options.

We now consider the Rho $\rho_P$, of European put options $P$. Because

$$P(S,T) = \max(E - S, 0),$$

the final condition for Rho is

$$\rho_P(S,T) = \frac{\partial}{\partial r}(P(S,T)) = 0.$$

Differentiating the left boundary condition with respect to $r$ for the European put option, we obtain the corresponding boundary condition for Rho,

$$\rho_P(0,t) = -(T - t)Ee^{-r(T-t)}.$$

Similarly, we have

$$\lim_{S\to\infty} \rho_P(S,t) = 0,$$

which gives the other boundary condition.

### The European Put Rho

$$\frac{\partial \rho_P}{\partial t} + \frac{1}{2}\sigma^2 \frac{\partial}{\partial S}\left(S^2 \frac{\partial}{\partial S}\rho_P\right) + (r - \sigma^2)\frac{\partial}{\partial S}(S\rho_P) + (\sigma^2 - 2r)\rho_P = P - S\Delta_P,$$

$$\rho_P(S,T) = 0,$$

$$\rho_P(0,t) = -(T - t)Ee^{-r(T-t)},$$

$$\lim_{S\to\infty} \rho_P(S,t) = 0.$$

Here, $\rho_P$ denotes the European put Rho, $P$ the value of the corresponding European put option, $\Delta_P$ the European put Delta, $r$ the interest rate, $E$ the exercise price, $T$ the expiry date, and $\sigma$ the volatility of the underlying asset.

## 11.9    Remarks

In this chapter, we have presented a few mathematical models of options that are commonly traded in financial markets throughout the world. The models are based on a simple model of how the price of an asset changes, and some basic assumptions regarding the financial markets. One should keep in mind that the models derived in this chapter are only as reasonable as the assumptions that are made. Much effort is put into refining these assumptions, and especially into refining the model (11.2) of how the underlying asset behaves. However, the framework presented here seems to be the most widely accepted.

There are several types of option contracts that are not treated in this chapter. Many excellent books treat such exotic options, and [3], [4], [9] and [10] are good starting points for further reading on this topic. Also, a rich literature about mathematical finance in general is available, see e.g. [1], [6] and [13].

In [14] we present various numerical techniques that can be applied when solving the models presented in this chapter. There are basically three different approaches that are commonly used. We will introduce these numerical methods, and hopefully provide some insight into how the models above can be solved efficiently and accurately numerically. When solving option problems utilizing finite element methods, we will describe how this can be done in an object-oriented fashion using Diffpack, see [3].

## References

1. N. H. Bingham and R. Kiesel. *Risk-Neutral Valuation, Pricing and Hedging of Financial Derivatives*. Springer-Verlag, 1998.
2. F. Black and M. Scholes. The Pricing of Options and Corporate Liabilities. *J. Political Economics*, 81:637–659, 1973.
3. L. Clewlow and C. Strickland. *Exotic Options, The State of the Art*. International Thomson Business Press, 1997.
4. Diffpack. See http://www.nobjects.no/ for further information.
5. D. Duffie. *Dynamic asset pricing theory*. Princeton University Press, 1996.
6. R. J. Elliot and P. E. Kopp. *Mathematics of Financial Markets*. Springer-Verlag, 1999.
7. D. H. Griffel. *Applied functional analysis*. Ellis Horwood, 1981.
8. J. C. Hull. *Options, futures and other derivatives*. Prentice Hall, 1997.
9. Y. K. Kwok. *Mathematical models of financial derivatives*. Springer-Verlag, 1998.
10. A. W. Lo and A. C. Mackinlay. *A Non-Random Walk Down Wall Street*. Princeton University Press, 1999.
11. Robert C. Merton. Theory of Rational Option Theory. *Bell Journal of Economics and Management Science*, 4, 1973.
12. S. N. Neftci. *An Introduction to the Mathematics of Financial Derivatives*. Academic Press, 2000.

13. S. H. Ross. *An introduction to mathematical finance.* Cambridge University Press, 1999.
14. O. Skavhaug, B. F. Nielsen, and A. Tveito. Numerical Methods for Financial Derivatives. In H. P. Langtangen and A. Tveito, editors, *Advanced Topics in Computational Partial Differential Equations – Numerical Methods and Diffpack Programming.* Springer, 2003.
15. P. Wilmott. *Derivatives, The theory and practice of financial engineering.* John Wiley & Sons, 1998.
16. P. Wilmott, J. Dewynne, and S. Howison. *Option Pricing, Mathematical models and computation.* Oxford Financial Press, 1993.

# Chapter 12

# Numerical Methods for Financial Derivatives

O. Skavhaug[1,2], B. F. Nielsen[1,2], and A. Tveito[1,2]

[1] Simula Research Laboratory
[2] Department of Informatics, University of Oslo

## 12.1 Introduction

Analytical solutions of the mathematical equations modeling the behavior of financial derivatives, like the price of option contracts, are seldom available. Only in the simplest cases, e.g., vanilla European put and call options, do analytical solutions exist. For most other option models, numerical techniques must be applied to compute solutions of the mathematical models. For exotic option contracts, computing the option prices numerically may be the only pricing mechanism available.

Various classes of numerical methods have been developed, and in this chapter we outline the main concepts of the three main frameworks. Thus, we do not set out to give a full overview of state–of–the–art numerical methods for solving problems arising in computational finance. The methods described here will hopefully serve as starting points for further investigations of the various methods that have been developed successfully in this field. The option models treated here have been presented in [8], and the reader should be familiar with these models prior to reading this chapter in detail.

This chapter is outlined as follows. In Section 12.2 we summarize the different models derived in [8]. These models can be solved by the numerical methods presented in the subsequent sections.

In Section 12.3 we present the basics of the Monte–Carlo simulation method. In this method, several simulations of the price of the underlying asset, during the duration of the contract, is carried out. On the basis of these simulations, the value of the option contract can be estimated. Some exotic option models cannot be written in terms of partial differential equations (PDEs). Hence, standard numerical techniques for solving (PDEs) cannot be applied. Still Monte–Carlo simulations can be carried out in order to obtain an estimate of the risk–neutral option price.

The basic setup for so–called lattice methods is presented in Section 12.4. The simplest lattice method, called the binomial method, is described. This method is based on a change in the model of how the price of an asset evolves in time. During the solution process, a tree of future asset values is generated, and later traversed in order to estimate the option price. Severe restrictions on the time step size may apply to binomial methods. Therefore, the trees generated generally tend to be memory consuming (at least for models involving several underlying assets). However, lattice algorithms have the same advantage as Monte–Carlo simulation methods; they do not need models posed as a PDEs. Therefore, lattice methods can be used to price exotic option contracts.

In Section 12.5 we turn our attention to finite difference methods (FDM) for PDEs. When the model of the value of an option can be expressed in terms of partial differential equations, FDMs can be used to compute the risk–neutral price of the contract. First the mathematical model is discretized by approximating the partial derivatives that occur in the Black–Scholes equation by finite differences. This leads to a system of algebraic equations on the form $Ax = b$ at every time step. These systems must be solved by suitable numerical methods. Both European style and American style options, see [8], are treated in this section.

The last numerical solution method for financial derivatives discussed in this chapter is the finite element method (FEM), presented in Section 12.6. In this method an approximate solution of the model is sought in a finite dimensional subspace of the function space used in the variational formulation of the problem. By rewriting the Black–Scholes equation on a suitable form, we can easily handle option contracts based on several underlying assets. Furthermore, related problems, that are governed by partial differential equations similar to the Black–Scholes equation, can easily be solved. The main focus of this section is on multi–asset option problems, with an arbitrary number of underlying assets.

In the case of European multi–asset option problems, we have to solve several sub–problems of the model in order to determine the necessary boundary conditions. When working within the Diffpack framework [3], we exploit one of the main strengths of this software library; the number of space dimensions of a problem can be set at run–time.

## 12.2    Model Summary

In this chapter we focus on how to solve the Black–Scholes equation,

$$\frac{\partial V}{\partial t} + \frac{1}{2}\sigma^2 S^2 \frac{\partial^2 V}{\partial S^2} + rS\frac{\partial V}{\partial S} - rV = 0, \tag{12.1}$$

by numerical methods. Here, $V$ denotes the value of the derivative product, $S$ is the price of the underlying asset, $r$ is the risk–free interest rate, and $\sigma$ is the volatility of $S$. In addition, we discuss how to numerically solve the partial differential equations governing the Greeks and some exotic option models that differ from the ordinary Black–Scholes model. As mentioned above, a derivation of these models can be found in [8]. We assume that the mathematical models described below have well behaved, unique solutions.

### The European Call Option

$$\frac{\partial C}{\partial t} + \frac{1}{2}\sigma^2 S^2 \frac{\partial^2 C}{\partial S^2} + rS\frac{\partial C}{\partial S} - rC = 0, \quad \text{in } \mathbb{R}^+ \times [0, T)$$
$$C(S, T) = \max(S - E, 0), \tag{12.2}$$
$$C(0, t) = 0,$$
$$C(S, t) = S - Ee^{-r(T-t)} \quad \text{as } S \to \infty.$$

Here, $r$ represents the interest rate, $E$ the exercise price, $T$ the expiry date of the contract, and $\sigma$ the volatility of the underlying asset price $S$.

### The European Put Option

$$\frac{\partial P}{\partial t} + \frac{1}{2}\sigma^2 S^2 \frac{\partial^2 P}{\partial S^2} + rS\frac{\partial P}{\partial S} - rP = 0, \quad \text{in } \mathbb{R}^+ \times [0, T)$$
$$P(S, T) = \max(E - S, 0), \tag{12.3}$$
$$P(0, t) = Ee^{-r(T-t)},$$
$$P(S, t) = 0 \quad \text{as } S \to \infty.$$

Here, $r$ represents the interest rate, $E$ the exercise price, $T$ the expiry date of the contract, and $\sigma$ the volatility of the underlying asset price $S$.

### The American Put Option

$$\frac{\partial P}{\partial t} + \frac{1}{2}\sigma^2 S^2 \frac{\partial^2 P}{\partial S^2} + rS\frac{\partial P}{\partial S} - rP = 0, \quad 0 < \bar{S}(t) < S, \quad t \in [0, T),$$
$$P(S, T) = \max(E - S, 0), \tag{12.4}$$
$$P(\bar{S}(t), t) = E - \bar{S}(t),$$
$$\frac{\partial P(\bar{S}(t))}{\partial S} = -1,$$
$$P(S, t) = 0 \quad \text{as } S \to \infty$$
$$\bar{S}(T) = E,$$
$$P(S, t) = E - S, \quad \text{for } 0 \leq S < \bar{S}(t).$$

Here, $r$ represents the interest rate, $E$ the exercise price, $T$ the expiry date of the contract, $\bar{S}$ the critical stock price (the unknown moving boundary), and $\sigma$ the volatility of the underlying asset price $S$.

### The European–style floating strike look–back Put Option

$$\frac{\partial P}{\partial t} + \frac{1}{2}\sigma^2 S^2 \frac{\partial^2 P}{\partial S^2} + rS\frac{\partial P}{\partial S} - rP = 0, \quad 0 < S < M, \quad T_0 < t < T,$$

$$P(S, M, T) = M - S, \tag{12.5}$$

$$P(0, M, t) = Me^{-r(T-t)},$$

$$\frac{\partial P}{\partial S} = 0 \quad \text{at } S = M.$$

Here, $r$ represents the interest rate, $E$ the exercise price, $T$ the expiry date of the contract, $M$ the maximum of the asset price over the look–back period, and $\sigma$ the volatility of the underlying asset price $S$.

### The European Put Basket Option

$$\frac{\partial P}{\partial t} + \frac{1}{2}\sum_{i=1}^{n}\sum_{j=1}^{n}\rho_{i,j}\sigma_i\sigma_j S_i S_j \frac{\partial^2 P}{\partial S_i \partial S_j} + r\sum_{i=1}^{n}S_i\frac{\partial P}{\partial S_i} - rP = 0,$$

$$P(S_1, \ldots, S_n, T) = \max(E - \sum_{i=1}^{n}\alpha_i S_i, 0), \tag{12.6}$$

$$P(S_1, \ldots, S_n, t) = g_i(S_1, \ldots, S_{i-1}, S_{i+1}, \ldots, S_n, t) \quad \text{at } S_i = 0,$$

$$P(S_1, \ldots, S_n, t) = 0 \quad \text{as } S_i \to \infty.$$

Here, $r$ represents the interest rate, $E$ the exercise price, $T$ the expiry date of the contract, $\rho_{i,j}$ the correlation between assets $S_i$ and $S_j$, $\{\alpha_i\}$ is a given set of weights such that $\sum_{i=1}^{n}\alpha_i = 1$, and $\sigma_i$ is the volatility of the underlying asset $S_i$ (measuring the standard deviation of the returns of the asset).

The boundary conditions $\{g_i\}$ are the solutions of the sub–problems posed in $n-1$ asset dimensions, corresponding to $S_i = 0$, $i = 1, \ldots, n$. Note that the correlation matrix $\Sigma$, given by $\Sigma_{i,j} = \rho_{i,j}$ $i, j = 1, \ldots, n$, is symmetric and positive definite with $\rho_{i,i} = 1$.

### The European Put Delta

$$\frac{\partial \Delta_P}{\partial t} + \frac{1}{2}\sigma^2 \frac{\partial}{\partial S}\left(S^2 \frac{\partial \Delta_P}{\partial S}\right) + r\frac{\partial}{\partial S}(S\Delta_P) - r\Delta_P = 0,$$

$$\Delta_P(S,T) = \begin{cases} -1, & S \leq E, \\ 0, & S > E, \end{cases} \tag{12.7}$$

$$\Delta_P(0,t) = -1,$$

$$\lim_{S\to\infty} \Delta_P(S,t) = 0.$$

Here, $\Delta_P$ is the European put Delta, $r$ is the interest rate, $E$ is the exercise price, $T$ is the expiry date of the contract, and $\sigma$ is the volatility of the underlying asset price $S$.

### The European Put Theta

$$\frac{\partial \Theta}{\partial t} + \frac{1}{2}\sigma^2 \frac{\partial}{\partial S}\left(S^2 \frac{\partial \Theta}{\partial S}\right) + (r - \sigma^2)\frac{\partial}{\partial S}(S\Theta) + (\sigma^2 - 2r)\Theta = 0.$$

$$\Theta_P(S,T) = 0, \tag{12.8}$$

$$\Theta_P(0,t) = -rEe^{-r(T-t)},$$

$$\lim_{S\to\infty} \Theta_P(S,t) = 0.$$

Here, $\Theta_P$ denotes the European put Theta, $r$ the interest rate, $E$ the exercise price, $T$ the expiry date of the contract, and $\sigma$ the volatility of the underlying asset price $S$.

### The European Put Vega

$$\frac{\partial \wedge_P}{\partial t} + \frac{1}{2}\sigma^2 \frac{\partial}{\partial S}\left(S^2 \frac{\partial \wedge_P}{\partial S}\right) + (r - \sigma^2)\frac{\partial}{\partial S}(S\wedge_P) + (\sigma^2 - 2r)\wedge_P = -\sigma S^2 \Gamma_P.$$

$$\wedge_P(S,T) = 0, \tag{12.9}$$

$$\wedge_P(0,t) = 0,$$

$$\lim_{S\to\infty} \wedge_P(S,t) = 0.$$

Here, $\wedge_P$ denotes the European put Vega, $\Gamma_P$ the European put Gamma, $r$ the interest rate, $E$ the exercise price, $T$ the expiry date of the contract, and $\sigma$ the volatility of the underlying asset price $S$.

## The European Put Rho

$$\frac{\partial \rho_P}{\partial t} + \frac{1}{2}\sigma^2 \frac{\partial}{\partial S}\left(S^2 \frac{\partial}{\partial S}\rho_P\right) + (r - \sigma^2)\frac{\partial}{\partial S}(S\rho_P) + (\sigma^2 - 2r)\rho_P = P - S\Delta_P,$$

$$\rho_P(S,T) = 0, \tag{12.10}$$

$$\rho_P(0,t) = -(T - t)Ee^{-r(T-t)},$$

$$\lim_{S \to \infty} \rho_P(S,t) = 0.$$

Here, $\rho_P$ denotes the European put Rho, $P$ the value of the European put option, $\Delta_P$ the European put Delta, $r$ the interest rate, $E$ the exercise price, $T$ the expiry date, and $\sigma$ the volatility of the underlying asset price $S$.

## 12.3   Monte–Carlo Methods

Let $f(S)$ be the payoff curve at the expiry date[1] for an option contract. Because the Black–Scholes equation is based on a risk–neutral price valuation, the value $V$ of an option at some time $t$ prior to the expiry date $T$ can be stated as the risk–neutral expectation of the discounted payoff at the expiry date $f(S)$, see [2]. When the interest rate $r$ is assumed to be constant, we get a simple expression for this risk–neutral option price,

$$V_{S,t} = e^{-r(T-t)}E[f(S)]. \tag{12.11}$$

The expected payoff at expiry is denoted $E[f(S)]$. If we are able to compute an approximation of this expectation, we can estimate the risk–neutral price of the option contract given the current value $S(t)$ of the underlying asset. This is the main idea behind the Monte–Carlo method for option models.

Recall from [8] the assumption that the evolution of $S$ is governed by the stochastic differential equation,

$$dS = rSdt + \sigma SdX, \tag{12.12}$$

i.e., a geometric Brownian motion, where $dX$ is a sample from a normal distribution with zero mean and variance $\sqrt{dt}$. It is possible to use (12.12) to simulate the future stock price $S(T)$, and thus valuate the payoff function $f(S)$. On discrete form, (12.12) can be written,

$$\Delta S^{k+1} = rS^k \Delta t + \sigma S^k \sqrt{\Delta t}\phi. \tag{12.13}$$

Here, $\phi$ is a sample from a standardized normal distribution. Note that, if $S^k$ is given, we can easily determine the stock price at time step $t_{k+1}$, because

---

[1] For the sake of convenience, we often use the term *expiry* instead of *the expiry date*.

$S^{k+1} = S^k + \Delta S^{k+1}$. The generated values $S(t_k)$, $k = 1, \ldots, K$, are called a realization of the discrete random walk, and represent a possible path of future asset prices up to the expiry date of the option contract. At expiry, the value of the contract is a known function of the underlying asset, and we compute the possible future option price by evaluating the payoff function for the generated asset value $S^K$ at time $t_K = T$.

Running several realizations of the discrete random walk, denoting the corresponding generated asset values at expiry by $S_l^K$, $l = 1, 2, \ldots, M$, the mean value of the payoff at expiry is given by,

$$\tilde{f}(S) = \frac{1}{M} \sum_{l=1}^{M} f(S_l^K), \tag{12.14}$$

where $M$ is the total number of realizations of the discrete random walk. This mean value is used as an approximation of the expectation of the payoff at expiry, and the discounted value is given by (12.11). By the law of large numbers[2], the approximation obtained by this method converges with the rate $\mathcal{O}(1/\sqrt{M})$ towards the risk–neutral option value as $M$ increases. This approach is known as the Monte–Carlo simulation (MCS) method.

One of the advantages of MCS methods is the simplicity of the algorithm. This makes implementation of this method straight forward. The main component of a computer simulator is the generator of the discrete random walks, i.e., a random number generator. The rest of the algorithm consists of computing the mean value of the payoff function and applying formula (12.11).

Another strength of MCS methods is that the often complicated conditions in exotic contracts can be evaluated in the simulation process of the random walk. This is even the case when it is hard to write these conditions as explicit functions, e.g., for look–back and barrier options. For these option contracts, the terminal payoff functions depend on the history of the asset price, which is available during the simulation of the random walk. At each time step the conditions stated in the contract are checked. If an event that affects the value of the option has taken place, the option contract is updated correspondingly, and the time stepping continues. In the case of knock out barrier options, we check, at each time step, whether or not the value of the underlying asset has passed the point that knocks the option out.

Consider options on baskets, i.e., option contracts that depend on several underlying assets, see (12.6), and assume for simplicity that the options in the basket are non–correlated. This model is multi–dimensional, where the number of dimensions $n$ equals the number of underlying assets. Using Monte–Carlo methods, this is handled by running $n$ discrete random walks at each step in the main Monte–Carlo iteration. Thus, we can simulate the movements of all the involved assets, and are able to evaluate the payoff function at expiry. In order to obtain a good estimate of the option price, the

---

[2] The Central Limit Theorem, see [6].

above iteration is carried out several times, and the mean value of the payoff function is used as the expectation in (12.11). The numerical approximation of the option value is in this case given by,

$$V_{S_1,\ldots,S_n,t} = e^{-r(T-t)}\hat{E}[f(S_1,\ldots,S_n)],$$

where $\hat{E}$ denotes the computed average of the payoff function,

$$\hat{E}[f(S_1,\ldots,S_n)] = \frac{1}{M}\sum_{l=1}^{M} f(S_{1,l}^K,\ldots,S_{n,l}^K).$$

We have assumed that the random walks in the MCS algorithm do not depend on each other, i.e., that the assets are not correlated. Thus, if we have a parallel computer, we can compute the random walks concurrently. The nearly optimal speed–up by such an approach may compensate for the relatively slow convergence rate of the method.

Note that a Monte–Carlo simulation only computes the price of the option for the current value $S$ of the underlying asset, i.e., a point value. Thus, in order to compute the Greeks, i.e., the partial derivatives of the solution with respect to the different parameters in the model, we first compute several point values of the option price, corresponding to different parameter values. Thereafter, we can differentiate the option price numerically, by applying finite difference approximations, using these point values.

One of the major drawbacks of Monte–Carlo simulations in option pricing is that a large number of random walk simulations must be computed in order to obtain an accurate approximation. The reason for this is that the convergence rate is "only" $\mathcal{O}(1/\sqrt{M})$. There exist several refinements of MCS methods that address this problem, e.g., quasi Monte–Carlo methods are known to converge faster than the original method. The main motivation behind quasi MCS algorithms is that when random numbers are generated, the values cluster. Thus, methods for drawing quasi–random numbers that do not cluster have been developed. See, e.g., [2, pp. 129–133] and [9, pp. 688–692] for further information on quasi–random numbers.

## 12.4    Lattice Methods

The simplest versions of lattice methods are called binomial methods. The basic assumption in these algorithms is that the value of the underlying asset follows a binomial process. This means that at a given time step $t_i$, we assume that the value of the underlying asset $S(t_i)$ either goes up by a factor $u$ or down by a factor $v$. Thus, at time step $t_{i+1}$, the new possible values of $S$ become

$$S(t_{i+1}) = uS(t_i) \quad \text{or} \quad S(t_{i+1}) = vS(t_i).$$

Here $u$ and $v$ incorporate the average behavior and volatility of the asset. Note that $u$ and $v$ should be chosen in accordance with the assumption that the underlying asset follows a geometric Brownian motion (12.12),

$$dS = rSdt + \sigma S\sqrt{\Delta t}\phi.$$

Following [2] and references therein, this can be achieved by choosing,

$$u = \exp\left((r - \frac{1}{2}\sigma^2)\Delta t + \sigma\sqrt{\Delta t}\right), \qquad (12.15)$$

$$v = \exp\left((r - \frac{1}{2}\sigma^2)\Delta t - \sigma\sqrt{\Delta t}\right). \qquad (12.16)$$

Consider a European call option $C$, where the underlying asset does not pay out dividends during the life–time of the option contract. We construct a portfolio with value $\pi$, consisting of a short position of one call option and a long position of $\Delta$ shares[3] of the underlying asset, i.e., $\pi = -C + \Delta S$. We want this portfolio to be risk–free during the time step $\Delta t$. I.e., we seek the number $\Delta$ such that the value of this portfolio is unchanged by movements of the underlying asset during the time period $\Delta t$. Thus, we require that

$$-C_u + \Delta uS = -C_v + \Delta vS,$$

where $C_u$ and $C_v$ denotes the values of the option after an upwards and downwards movement of the underlying asset, respectively. Hence, $\Delta$ is given by

$$\Delta = \frac{C_u - C_v}{(u - v)S}. \qquad (12.17)$$

During this time step the portfolio is risk–free, and the principle of no arbitrage possibilities assures that the return of the portfolio must equal that of a risk–free bank deposit, i.e.,

$$(-C_u + \Delta uS) = e^{r\Delta t}(-C + \Delta S). \qquad (12.18)$$

Combining (12.17) and (12.18), we get

$$C = e^{-r\Delta t}\left(pC_u + (1 - p)C_v\right), \qquad (12.19)$$

where

$$p = \frac{e^{r\Delta t} - v}{u - v},$$

is interpreted as the probability of a rise in the value $S$ of the underlying asset. For the choice of $u$ and $v$ in (12.15) and (12.16), $p = \frac{1}{2}$, see [2, 10–52]

---

[3] This notation may be a bit confusing, using $\Delta$ both as a number and for the time step size $\Delta t$. The meaning should, however, be clear from the context.

and [4, 187–202]. Thus, the probability of a rise in the asset price equals the probability of a fall[4].

Assume that the time to expiry of a call option is $\Delta t$, and that we wish to find the risk–neutral price of the call option. We then construct a binomial tree consisting of three nodes. One node for the current value of the under-lying asset, and two nodes for the two possible values of the underlying at expiry. At this time, the corresponding values of the option are stated in the contract:

$$C_u = \max(0, E - uS),$$
$$C_v = \max(0, E - vS).$$

Equation (12.19) can then be used to compute the numerical solution of the value of the call option at the time $\Delta t$ prior to expiry.

More generally, consider a contract for a call option at time $t$ with expiry date $T$, such that $T - t = k\Delta t$, where $k > 1$ is a positive integer. In case of $k = 4$, the binomial tree schematically takes the form given in Figure 12.1.

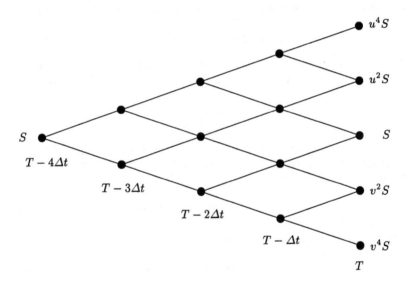

**Fig. 12.1.** The binomial tree with four time steps to expiry.

The nodes in this tree define the possible paths of the values that the underlying asset may take from the initial time $t$ until the expiry date $T$. Starting at the rightmost position in the tree, we easily compute the value of

---

[4] We may choose other formulas for $u$ and $v$ in (12.15) and (12.16), in which case the probabilities of a rise and a fall may change, see, e.g., [4].

the option contract at the nodes at expiry, using the given payoff function. We then compute the values of the option contract at the nodes at time $T - \Delta t$ using formula (12.19), because all the values on the right hand side of the equation, that is at time $T$, are known. This procedure is repeated at each time step in the tree structure above (backwards in time), until the value of the option contract has been computed at all nodal points. When the algorithm finishes, the approximate price of the option contract at time $t$, $C(S(t))$ has been computed.

Note that, as MCS methods, binomial methods compute single point values of the option price, i.e., the value of the option contract for the current value of the underlying asset $S(t)$.

Recall that the basic assumption for this method is that the asset price $S(t_i)$ can change to one of two different values after a time step $\Delta t$. That is $S(t_{i+1}) = uS(t_i)$, corresponding to a rise in the value, or $S(t_{i+1}) = vS(t_i)$ corresponding to a fall. In order to justify this assumption, the time step size has to be small. Thus, if there is a long time to expiry, the binomial tree generally grows large, demanding much computer memory.

Extensions of the binomial method lead to lattice methods. Here, the assumption regarding the possible changes in the value of the asset during a time step $\Delta t$ is extended from two to several steps. Instead of the two values $u$ and $v$ in the binomial method, we get

$$S(t_{i+1}) = c_j S(t_i),$$

where $\{c_j\}$, $j = 1, \ldots, J$ are the allowed changes. The trees that are generated in these generalized methods, called lattice trees, have more branches to the right hand side of each node than in the tree associated with the binomial algorithm, and more computational effort is needed in order to obtain the approximate solution of the option price. However, due to less strict assumptions regarding the movement of the underlying asset, generally larger time steps can be chosen and still obtain an accurate solution.

American option problems can also be incorporated within the framework of binomial and lattice methods. For further information about binomial and lattice methods, see [2], [4], and [9].

## 12.5    Finite Difference Methods

For a number of option contracts, we can discretize the mathematical models, posed as PDE models, using finite difference methods, obtaining numerical solutions of the option prices. Using Diffpack's FDM module, we can construct a framework where by changing final and boundary conditions, different types of options are handled.

The Black–Scholes PDE (12.1) is backwards parabolic in time. Thus, in order to solve this equation numerically, a final condition is needed instead of the usual initial condition. We then step backwards in time to compute solutions at earlier time steps.

The value $S$ of the underlying asset can be any number in $\mathbb{R}^+$. However, in order to apply a finite difference method, the solution domain must be of finite size. Thus, we truncate the original domain. The problem is re–posed on a fixed domain, $(0, S_\infty)$, where $S_\infty$ is a sufficiently large value of the underlying asset. At $S_\infty$, we apply the boundary condition at infinity in the original problem.

The magnitude of $S_\infty$ must be chosen with care. If chosen too small, it will affect the numerical solution, because the boundary condition forces the price of the option contract to a fixed value at the boundary. On the other hand, $S_\infty$ should not be chosen too large. If we fix the number of nodes in the asset direction, the size of the discretization parameter $\Delta S$ grows linearly as a function of the size of the domain, $S_\infty$. Hence, the error of the numerical method will increase as the size of the solution domain grows.

For a given value of the underlying asset, it is unlikely that the asset price will rise above a certain level before expiry. Thus, we assume that when the value of the underlying asset has grown to a given factor of the exercise price, $S_\infty = kE$, we can apply the boundary conditions at infinity of the original problem. Numerical experiments should be used to determine $k$ for a given problem.

### 12.5.1  European Options

The simple mathematical models of European put and call option contracts, so–called vanilla option contracts, $V$, can be solved numerically by applying standard finite difference approximations to the terms in the Black–Scholes equation (12.1).

A finite difference solution is given at certain discrete points in time and space. We use the term "space" when referring to the asset dimension. Let

$$V_j^n \approx V(j\Delta S, n\Delta t), \quad j = 0, \ldots, M+1, \quad n = 0, \ldots, N+1, \qquad (12.20)$$

be a finite difference approximation of the solution $V$ of a vanilla option contract, i.e., a European put or call option, at the mesh points. Here

$$\Delta S = \frac{S_\infty}{M+1},$$

is the spatial step size and

$$\Delta t = \frac{T}{N+1},$$

is the temporal step size. An explicit finite difference scheme for the Black–Scholes equation can be written,

$$\frac{V_j^n - V_j^{n-1}}{\Delta t} + \frac{1}{2}\sigma^2 S_j^2 \frac{V_{j+1}^n - 2V_j^n + V_{j-1}^n}{(\Delta S)^2} + rS_j \frac{V_{j+1}^n - V_j^n}{\Delta S} - rV_j^n = 0,$$

$$(12.21)$$

where an upwind finite difference approximation is used for the transport term.

Recall put–call parity from [8], that relates the solutions of a European put and call option to each other,

$$P = C - S + Ee^{-r(T-t)}. \tag{12.22}$$

Hence, it suffices to compute a numerical solution of the European call option problem, and use (12.22) to compute the value of the corresponding put option.

Consider a call option, and let $C_j^n = V_j^n$ in (12.21). The final condition of this option is given by,

$$C_j^{N+1} = \max(S_j - E, 0), \quad j = 0, \dots, M+1. \tag{12.23}$$

The discrete versions of the boundary conditions are incorporated in the scheme by requiring,

$$C_0^n = 0,$$
$$C_{M+1}^n = Ee^{-r(T-n\Delta t)}.$$

Note that the explicit scheme poses stability restrictions on the time step size. For this problem, the requirement on $\Delta t$ turns out to be,

$$\Delta t \le \frac{(\Delta S)^2}{\sigma^2 S_\infty^2 + rS_\infty(\Delta S) + r(\Delta S)^2}.$$

To avoid this restriction on the size of the time steps, an implicit finite difference scheme can be used. In this case, the finite difference scheme of the vanilla option price $V$ is given by,

$$\frac{V_j^n - V_j^{n-1}}{\Delta t} + \frac{1}{2}\sigma^2 S_j^2 \frac{V_{j+1}^{n-1} - 2V_j^{n-1} + V_{j-1}^{n-1}}{(\Delta S)^2} + rS_j\frac{V_{j+1}^{n-1} - V_j^{n-1}}{\Delta S} - rV_j^{n-1} = 0, \tag{12.24}$$

which results in a linear system of algebraic equations at each time step. Because the spatial problem is one–dimensional, the resulting linear system is tri–diagonal. Thus, tri–diagonal Gaussian elimination is an efficient linear system solver for this problem.

A more general solution strategy is to apply the $\theta$-rule. Using the same notation as above, this method can be written,

$$\frac{V_j^{n+1} - V_j^n}{\Delta t} + (1 - \theta)\mathcal{L}(V_j^{n+1}) + \theta\mathcal{L}(V_j^n) = 0,$$

where

$$\mathcal{L}(V_j^n) = \frac{1}{2}\sigma^2 S_j^2 \frac{V_{j+1}^n - 2V_j^n + V_{j-1}^n}{(\Delta S)^2} + rS_j\frac{V_{j+1}^n - V_j^n}{\Delta S} - rV_j^n.$$

Choosing $\theta = 0$ gives an explicit scheme (forward Euler), $\theta = 0.5$ gives a Crank–Nicholson scheme and $\theta = 1$ gives an implicit scheme (backward Euler).

## 12.5.2   American Options

We consider the American put option contract only, because the American call price equals the European call price when the underlying asset does not pay dividends during the life–time of the option contract, see [8]. Recall that the American put is regarded as a free and moving boundary value problem. There are several strategies for solving such problems.

The most widely used approach is the Successive Over-relaxation Method (SOR) method, described in, e.g., [10]. At a given time step, we first solve the scheme for the European put option contract, and place the result in a temporary solution $Q_j^n$. The constraint given by the early exercise feature of the contract, that $P_j^n \geq \max(0, E - j\Delta S)$ is enforced by applying this constraint to the solution $Q_j^n$,

$$P_j^n = \max \left( \max(0, E - j\Delta S), Q_j^n \right).$$

This completes the computation of the American put option at a given time step, and we proceed with the time–loop. When computing $Q_j^n$, explicit, implicit, or $\theta$-rule methods can be used, like in the case of European options.

Another way of solving the American put problem is by penalty methods, see [7] and [11]. The main idea here is to add a term to either the numerical scheme or to the Black–Scholes equation before discretizing. This penalty term forces the solution to stay above the pay–off curve at expiry for all time.

Following [7], a penalty formulation of the American put option problem reads,

$$\frac{\partial V_\epsilon}{\partial t} + \frac{1}{2}\sigma^2 S^2 \frac{\partial^2 V_\epsilon}{\partial S^2} + rS\frac{\partial V_\epsilon}{\partial S} - rV_\epsilon$$
$$+\frac{\epsilon C}{V_\epsilon + \epsilon - q(S)} = 0, \quad S \geq 0, \ t \in [0, T),$$
$$V_\epsilon(S, T) = \max(E - S, 0),$$
$$V_\epsilon(0, t) = E,$$
$$V_\epsilon(S, t) = 0 \quad \text{as } S \to \infty,$$

where $\epsilon > 0$ is a small regularization parameter.

An important point is that the penalty term renders the implicit and $\theta$-rule schemes non–linear. Thus, the corresponding algebraic equations must be solved by using a non–linear solution method like e.g., Newton's method.

An alternative to this approach is to use an explicit finite difference approximation of the mathematical model. In this case, the stability requirements for the time step size slows the solution algorithm down. A semi–implicit scheme, placing the non–linear term on the time step already computed, can be used in order to obtain an efficient numerical algorithm. In this case, weaker stability conditions are required for the time step size.

## 12.6    Finite Element Methods

In this section, we discuss how to solve option problems using finite element methods. After a reformulation of the Black–Scholes equation on conservative form, it is quite easy to write a simulator for multi–asset, or basket, option models using Diffpack. We concentrate on European basket options in this section.

In order to apply FEM to financial derivatives, we re–write the multi–asset version of the Black–Scholes equation on the following, conservative form,

$$\frac{\partial P}{\partial t} + \nabla \cdot (K\nabla P) + \nabla \cdot [(r\mathbf{S} - \mathbf{v})P] + (\nabla \cdot \mathbf{v} - r(n+1))P = 0, \quad (12.25)$$

where

$$K_{ij} = \frac{1}{2}\rho_{i,j}\sigma_i\sigma_j S_i S_j,$$

and

$$\mathbf{v}_i = S_i\Big(\sigma_i^2 + \sum_{j\neq i}^{n} \frac{1}{2}\rho_{i,j}\sigma_i\sigma_j\Big).$$

Above, $n$ is the number of underlying assets. The solution domain for this problem is $\Omega = \mathbb{R}^{n,+}$. For the sake of convenience, we transform the time variable $t$ into time to expiry $\tau$, which is simply given by $\tau = T - t$. This results in a forward parabolic PDE,

$$\frac{\partial P}{\partial \tau} - \nabla \cdot (K\nabla P) - \nabla \cdot [(r\mathbf{S} - \mathbf{v})P] - (\nabla \cdot \mathbf{v} - r(n+1))P = 0. \quad (12.26)$$

We discretize in time by a standard finite difference $\theta$-rule, dividing the computations into $T_K$ spatial weak problems. Then, at each time step $\tau_k > 0$, we obtain the following weak formulation:

Find $P^k \in H^1(\Omega)$, $k = 1, \ldots, T_K$, such that

$$a(P^k, v) = f_{k-1}(v) \quad \forall v \in H^1(\Omega),$$

where

$$a(P^k, v) = \int_\Omega P^k v + \theta\Delta\tau\Big((K\nabla P^k)\cdot\nabla v -$$
$$\nabla \cdot [(r\mathbf{S} - \mathbf{v})P^k]v - (\nabla\cdot\mathbf{v} - r(n+1))P^k v\Big),$$

and

$$f_k(v) = \int_\Omega P^k v + (1-\theta)\Delta\tau\Big((K\nabla P^k)\cdot\nabla v -$$
$$\nabla \cdot [(r\mathbf{S} - \mathbf{v})P^k]v - (\nabla\cdot\mathbf{v} - r(n+1))P^k v\Big).$$

At the initial time step, $P^0(\mathbf{S}) = \phi(\mathbf{S})$, denotes the payoff function at expiry. The interval of $\theta$ is $[0, 1]$, and $\theta = 0$ gives the forward Euler scheme, $\theta = \frac{1}{2}$ gives the Crank–Nicholson scheme, and $\theta = 1$ is the backward Euler scheme, see [5].

As with finite difference methods, the solution domain must be truncated. Thus, we solve the option problem on $\Omega^h = [0, L_1] \times \ldots \times [0, L_n]$, where $n$ is the number of underlying assets stated in the contract and $L_i$ is the length of the solution domain in the direction of the $i$'th asset. Typically, this length depends on the exercise price $E$ and the $\alpha_i$'s given in (12.6).

The finite element method is given by finding a solution to the above weak formulation in a finite dimensional subspace spanned by the element basis functions, $H^{1,h}(\Omega^h) = \mathrm{span}(N_1, \ldots, N_m) \subset H^1(\Omega)$. Here, the basis functions $N_i$ are piecewise, continuous polynomials over the triangulation of the solution domain $\Omega^h$. See [1] for further reading about finite elements and triangulation. The finite element formulation is now given by:

Find $P^k \in H^{1,h}(\Omega)$, $k = 1, \ldots, T_k$, such that,

$$a(P^k, v) = f_{k-1}(v), \quad \forall v \in H^{1,h}(\Omega).$$

The standard Galerkin method is given by expanding $P^k$ in terms of the basis functions, $P^k = \sum_{j=1}^{n} \alpha_j N_j$, and using the same basis functions as test functions, $v = N_i$, $i = 1, \ldots, n$. The discrete problem now reads:

Let $P^k = \sum_{i=1}^{m} \alpha_i^k N_i$. For $k = 1, \ldots, T_k$, find $\alpha^k$ that satisfies $A\alpha^k = f_{k-1}$, where $A_{i,j}$ is given by

$$A_{i,j} = \int_{\Omega} N_i N_j + \theta \Delta\tau \Big( (K\nabla N_i) \cdot \nabla N_j - \nabla \cdot [(r\mathbf{S} - \mathbf{v})N_i] N_j$$
$$- (\nabla \cdot \mathbf{v} - r(n+1)) N_i N_j \Big), \tag{12.27}$$

and $f_{k-1,j} = f_{k-1}(N_j)$.

### 12.6.1   Implementing a FEM Solver

In this section we describe how to implement a multigrid FEM solver for European put basket options, using Diffpack.

We start by determining the the size of the solution domain. For simplicity, we assume that this domain is a square hyper–cube in $n$ dimensions, where $n$ is the number of assets in the basket. The important parameters that determines the size of the solution domain are the exercise price $E$ and $\alpha_i$, $i = 1, \ldots, n$. A suitable formula is $L = jE\max(\alpha_i)$ for some suitable integer value $j$.

The multi–dimensional Black–Scholes equation is straight forward to implement on the interior of the solution domain. The tricky parts are the boundary conditions. Recall that the Black–Scholes equation degenerates to a problem of one lesser dimension when $S_i = 0$, $i = 1, \ldots, n$. The reason for this is that $S_i = 0$ is a fix point solution of the stochastic differential equation that governs the $i$'th asset,

$$dS_i|_{S_i=0} = rS_i dt + \sigma_i S_i dX|_{S_i=0} = 0, . \tag{12.28}$$

Hence, at this boundary we have to solve the corresponding degenerated problem.

In order to implement such a method in practice, the simulator class needs pointers to the sub–problems at the boundaries. We want the simulator to handle arbitrary many underlying assets, and at the class level, we do not wish to distinguish between the general problem and the sub–problems.

We write a class `BlackScholesMG` which is a subclass of `FEM`, and also declare the class `VecSimplest(BlackScholesMG)`. We add a pointer to this vector in the definition of the `BlackScholesMG` class,

```
VecSimplest(BlackScholesMG)* boProb;
```

When the simulator is initialized, we initialize the sub–problems also. This is done in the `adm` function,

```
void BlackScholesMG::adm(MenuSystem& menu) {
  SimCase::attach(menu); // Enables later access to menu
  define(menu);          // Define/build the menu
  menu.prompt();         // Prompt user, read menu answers into memory
  scan();                // Read menu answers and init
  makeBoProbl(menu);     // Create the lower dimensional problems
}
```

The `makeBoProbl` function extracts the parameters relevant to the different sub–problems. After creating the `VecSimplest(BlackScholesMG)` and re–dimensioning it to the correct number of sub–problems, `nsd`, we modify the `gridfile` string to generate grids for the sub–problems. We use box elements with one degree of freedom per node.

```
void BlackScholesMG::makeBoProbl(MenuSystem& menu) {
    if (nsd > 1 ) { // Create sub-problems
        boProb = new VecSimplest(BlackScholesMG)();
        VecSimplest(BlackScholesMG)& bp = *boProb;
        bp.redim(nsd);

        // Construct the gridfile-string
        Ptv(real) max, min;
        grid().getMinMaxCoord(min, max);
        String igf = menu.get("gridfile");
        int n = mgtools->getGrid(1).getLattice().getDivisions(1);
        String gf;
        gf = oform("P=PreproBox | d=%d [0,%.1f]", nsd-1, max(1));
        for (int i=2;i<nsd;i++) gf += oform("x[0,%.1f]", max(1));

        int elmn = pow(2, (nsd-1));
        gf += oform(" | d=%d e=ElmB%dn%dD [%d", nsd-1, elmn, nsd-1, n);
```

```
for (int i=2;i<nsd;i++) gf += oform(",%d", n);
gf += oform("] [1");
for (int i=2;i<nsd;i++) gf += oform(",1");
gf += oform("]");
menu.set("gridfile", gf);
```

The MenuSystem in Diffpack is used to distribute the new gridfile string to
the sub–problems.

The parameters for the sub–problems are extracted and passed them to
the simulators using the MenuSystem as well. The correlations are stored in
a $n \times n$ Mat(real) named rho. For the $i$'th subproblem, we must eliminate
the $i$'th row and column from this matrix. The volatilities $\sigma_j$ and weights $\alpha_j$
from the payoff function at expiry ,$j = 1, \ldots, n$, are stored in two Ptv(real)
called sigma and alpha, and we must eliminate the $i$'th entry in these vectors.
We build strings for the MenuSystem during this procedure.

```
// Store initial values. Reset at end of function
String isigma = menu.get("volatility");
String ialpha = menu.get("alpha");
String irho   = menu.get("rho");

String sval; // Volatilites of boundary problem
String aval; // Alphas in boundary problem
String rval; // Correlations in boundary problem

for (int i=1;i<=nsd; i++) { // Derive parameters for sub-problems
    sval=""; aval=""; rval="";
    for (int j=1;j<i;j++)    {
        sval += oform("%f ", sigma(j));
        aval += oform("%f ", alpha(j));
        for (int k=1;k<i;k++)     rval += oform("%f ",rho(j,k));
        for (int k=i;k<nsd;k++) rval += oform("%f ",rho(j,k+1));
    }
    for (int j=i;j<nsd;j++) {
        sval += oform("%f ", sigma(j+1));
        aval += oform("%f ", alpha(j+1));
        for (int k=1;k<i;k++)     rval += oform("%f ",rho(j+1,k));
        for (int k=i;k<nsd;k++) rval += oform("%f ",rho(j+1,k+1));
    }
    menu.set("volatility", sval);
    menu.set("alpha", aval);
    menu.set("rho", rval);
    bp(i).initBP(menu); // read params and init sub-problems
}

// Reset values
menu.set("gridfile", igf);
menu.set("volatility", isigma);
menu.set("alpha", ialpha);
menu.set("rho", irho);
}
}
```

The call bp(i).initBP(menu), initializes the boundary problems with the new parameter values. At the end of this method, we test if the dimension of the problem is higher that one. If so, we call the makeBoProbl function again.

```
void BlackScholesMG::initBP(MenuSystem& menu) {
   lineq.rebind(new LinEqAdmFE()); lineq->scan(menu);

   mgtools.rebind(new MGtools(*this));
   mgtools->attach(*lineq); mgtools->scan(menu);

   if (lineq->getSolver().description().contains("multilevel")) {
      MLIter& ml = CAST_REF(lineq->getSolver(), MLIter);
      ml.attach(mgtools->getMLSolver());
   }
   if (lineq->getPrec().description().contains("multilevel")) {
      PrecML& ml = CAST_REF(lineq->getPrec(), PrecML);
      ml.attach(mgtools->getMLSolver());
   }

   mgtools->initStructure();
   no_of_grids = mgtools->getNoOfSpaces();
   grid.rebind(mgtools->getGrid(no_of_grids));
   dof.rebind(mgtools->getDof(no_of_grids));

   nsd   = grid->getNoSpaceDim();
   theta = menu.get("theta").getReal();
   r     = menu.get("interest rate").getReal();
   sigma.redim(nsd); sigma.scan(menu.get("volatility"));
   alpha.redim(nsd); alpha.scan(menu.get("alpha"));
   rho.redim(nsd, nsd); rho.scan(menu.get("rho"));

   tip.rebind(new TimePrm()); tip().scan(menu.get("time params"));
   FEM::scan(menu); PG.scan(menu);

   // Create finite element fields
   u.rebind(new FieldFE(*grid, "u"));
   u_prev.rebind(new FieldFE(*grid, "u_prev"));
   linsol.redim(dof->getTotalNoDof());
   lineq->attach(linsol);

   tip->initTimeLoop(); setIC();
   if (nsd>1) makeBoProbl(menu); // More subproblems to be created
}
```

This is a recursive procedure that generates the necessary boundary problems needed in order to solve the full problem in $n$ dimensions.

In the solveProblem function, we initialize the time loop and insert the initial condition[5] in the finite element fields u_prev and u. At a given time step, we must first compute the solution of the boundary problems before we can compute the numerical solution of the original problem. This is done by calling the step function for each of the sub–problems.

---

[5] When we made a change of variables from the time $t$ to time to expiry $\tau$, the original final condition for $t$ became a initial condition for $\tau$.

```
void BlackScholesMG::solveProblem() {
   tip->initTimeLoop();
   setIC();
   database->dump(*u, tip.getPtr());
   while (!tip->finished()) {
      tip->increaseTime();
      cout << "t=" << tip->time()<<endl;
      if (nsd > 1) { // solve sub-problems first
         for (int i=1;i<=nsd;i++) {
            (*boProb)(i).step(); // Compute sub-problems
         }
      }
      solveAtThisTimeStep();     // Compute the solution
      *u_prev = *u;
   }
}
```

Before we can solve a given sub–problem, the sub–problems of that sub–problem must be computed. This is not necessary when the sub–problem is one–dimensional.

```
void BlackScholesMG::step() {
   if (!tip->finished()) {
      tip->increaseTime();
      if (nsd > 1 ) { // sub-problems exist, must handle these first
         for (int i = 1;i<=nsd;i++) {
            (*boProb)(i).step(); // compute boundary at this time step
         }
      }
      solveAtThisTimeStep();
      *u_prev = *u;
   }
}
```

In order to assemble the linear system of equations, we must incorporate the boundary conditions, i.e., we must obtain the value of the solution of the sub–problems at given points. When a node i is a boundary node at $S_k = 0$, we must incorporate this essential boundary condition in the linear system of equations. First, we extract the coordinate x of the node. This coordinate and the boundary indicator b are passed to the function g,

```
real BlackScholesMG::g(const Ptv(real)& x, int b) {
   if (nsd==1)  // Base case, return known BC
      return exp(-r*tip->time());
   else { // nsd > 1, get values from boundary solutions
      int k=b-nsd; // probl. number (DP convension)
      Ptv(real) xr; xr.redim(nsd-1); // reduced vector dim
      for (int i=1; i<k  ; i++) xr(i) = x(i);
      for (int i=k; i<nsd ; i++) xr(i) = x(i+1);
      return  (*boProb)(k).u().valuePt(xr); // Solution of sub-probl
   }
}
```

We remove[6] the entry k from x and place the result in xr. At this boundary, $S_k = 0$, and we pass this coordinate to the sub–problem k, which returns the boundary value needed.

As usual, the system of linear equations is assembled in the the the integrands function. In case of basket options written on arbitrary many underlying assets, this function is quite complicated, e.g., we must assemble all the second order cross derivatives.

```
void BlackScholesMG::integrands
    (ElmMatVec& elmat, const FiniteElement& fe) {
    int i, j;
    const int nbf = fe.getNoBasisFunc(); // Num. of basis funcs
    const real detJxW = fe.detJxW();     // det J times num.itg.-weight
    const int nsd = fe.getNoSpaceDim();

    const Ptv(real) x = fe.getGlobalEvalPt();

    // Petrov-Galerkin
    Ptv(real) vel = velo(x); Mat(real) k = K(x);
    Ptv(real) tens; Ptv(real) pg; pg.redim(nsd);
    u_prev->derivativeFEM(pg, fe);
    tens.redim(nsd); tens.fill(1.);
    PG.calcWeightingFunction(fe, vel, tens, tip->Delta(), false);

    real dt = tip->Delta(); real up_pt = u_prev->valueFEM(fe);

    // Help variables for the loop below
    real nablaq, nabla;
    Ptv(real) prod; prod.redim(nsd);
    for (i = 1; i <= nbf; i++) {
        for (j = 1; j <= nbf; j++) {
            nabla=nablaq=.0; prod.fill(.0);
            for (int p = 1; p <= nsd; p++) {
                nabla +=(r*x(p)-vel(p))*PG.dW(i, p)*fe.N(j);
                for (int q=1;q<=nsd;q++) // Dot products (loop over nsd)
                    prod(p) += k(p, q)*fe.dN(j, q); // Inc. cross derivs
                nablaq += prod(p)*PG.dW(i, p);
            }
            // theta from time discr. rule (theta=0.5 is C-N)
            elmat.A(i, j) += PG.W(i)*fe.N(j)*detJxW;
            elmat.A(i, j) += theta*dt*nablaq*detJxW;
            elmat.A(i, j) += theta*dt*nabla*detJxW;
            elmat.A(i, j) += theta*dt*(r*(nsd+1)-dv())
                            *PG.W(i)*fe.N(j)*detJxW;
        }
        // Right hand side below this point
        nabla=nablaq=.0; prod.fill(.0);
        for (int p = 1; p <= nsd; p++) {
            nabla +=(r*x(p)-vel(p))*PG.dW(i, p)*up_pt;
            for (int q=1;q<=nsd;q++) prod(p) += k(p, q)*pg(q);
            nablaq += prod(p)*PG.dW(i, p);
        }
        elmat.b(i) += PG.W(i)*up_pt*detJxW;
```

---

[6] Due to the numbering convention of the boundaries in Diffpack, we subtract nsd from b in order to find the dimension k corresponding to $S_k = 0$.

```
        elmat.b(i) += (1-theta)*dt*nablaq*detJxW;
        elmat.b(i) += (1-theta)*dt*nabla*detJxW;
        elmat.b(i) += (1-theta)*dt*(r*(nsd+1)-dv())*PG.W(i)*up_pt*detJxW;
    }
}
```

Note that we allow for Petrov–Galerkin discretization, due to the small diffusion term compared to convection as the asset prices decrease towards zero. The functions `velo`, `K` and `dv` returns the coefficients given in (12.27) above, i.e., $v$, $K$, and $\nabla \cdot v$, respectively.

### 12.6.2    Extensions

Having outlined how to implement a FEM solver for European basket options, we address how to extend this framework in order to handle American options and the Greeks, see (12.4) and (12.7) – (12.10) for model descriptions. For simplicity, we assume that we have only one underlying asset.

In order to handle American style contracts, we can use the same penalty approach as described in Section 12.5.2. In the general case, this results in a non–linear model that has to be solved using a non–linear solver such as Newton's method. This approach to solving American style option contracts has been mathematically proved to generate a numerical solution that fulfills the early exercise feature in case of FDM only. However, this method should also be useful in a finite element setting.

Consider the Greeks, which are the partial derivatives of the option price with respect to the parameters in the model. The Delta Greek $\Delta$ can be computed in two different ways. Either by numerical differentiating or by solving the model given in (12.7). The first approach is easy to handle using Diffpack functionality. Assume that we have computed the FEM solution and, for simplicity, that we have a one dimensional problem,

$$V^k(S) = \sum_{j=1}^{n} \alpha_j^k N_j,$$

where $V^k(S)$ denoted the FEM solution of the option contract at time step $k$. The Delta Greek can then be computed by calling `FEM::makeFlux` with $V^k(S)$ as input. We then seek a solution, $\Delta^k(S) = \sum_{j=1}^{n} \beta_j^k N_j$, such that

$$\int_\Omega \Delta^k(S) N_i d\Omega = \int_\Omega \frac{\partial}{\partial S} V^k(S) N_i d\Omega = \int_\Omega \sum_{j=1}^{n} \alpha_j^k \frac{\partial N_j}{\partial S} N_i,$$

for $i = 1, \ldots, n$. In case of linear basis functions, $\frac{\partial N}{\partial S}$ is piecewise constant, yet $\Delta^k(S)$ is a linear approximation. Having computed this $\Delta^k(S)$, the Gamma Greek can be computed by calling `makeFlux` once more, using $\Delta^k(S)$ as input.

The other Greeks must be solved by other means. Consider the Rho Greek, i.e., the partial derivative of the risk–neutral price of an option with respect to the risk–free interest rate,

$$\rho = \frac{\partial V}{\partial r}.$$

We approximate this partial derivative with a finite difference scheme. Assume that we have computed two solutions of the option problem using two different values for $r$, say $V_{r_1}^k(S)$ and $V_{r_2}^k(S)$, and define $\Delta r = r_2 - r_1$. Then a first order finite difference approximation is given by,

$$\rho^k(S) = \frac{V_{r_2}(S) - V_{r_1}(S)}{\Delta r}.$$

The scheme must be valuated point–wise, at all nodal points in the finite element mesh. This approach can be applied to the remaining Greeks as well, substituting the interest rate with the parameters corresponding to the definition of the individual Greeks.

The other way of computing the Greeks, is to solve the mathematical models listed in Section 12.2. We first consider Delta (12.7) and Theta (12.8) Greeks. These models are similar to the Black–Scholes model, and thus the same solution method can be used in order to solve these models numerically. The only changes that have to be employed are the coefficients in the PDE and some adjustments of the boundary and final conditions.

The last two Greeks listed in Section 12.2 are the Vega and Rho Greeks. In addition to the adjustments that have to be made for the two previous Greeks, these models depend on other financial derivatives. We must therefore solve weakly coupled systems of PDEs. In case of the Vega Greek, we can first compute the Gamma Greek, and then use this solution when discretizing and solving the Vega Greek problem. Thus, such a simulator for the Vega Greek must be equipped with a pointer to a simulator that solves the Gamma Greek model.

Similarly, when writing a simulator for the Rho Greek, we need two pointers to simulators that compute the risk–neutral price of the option contract and the corresponding Delta Greek. We then discretize the Rho Greek model, using the numerical solution of the option price and the Delta Greek as known finite element fields.

# References

1. D. Braess. *Finite elements.* Cambridge University Press, 1997.
2. L. Clewlow and C. Strickland. *Implementing derivative models.* John Wiley & Sons, 1998.
3. Diffpack. See http://www.nobjects.no/ for further information.
4. Y. K. Kwok. *Mathematical models of financial derivatives.* Springer-Verlag, 1998.
5. H. P. Langtangen. *Computational Partial Differential Equations. Numerical Methods and Diffpack Programming.* Springer-Verlag, 1999.

6. R. J. Larsen and M. L. Marx. *Mathematical Statistics and Its Applications.* Prentice-Hall, 1986.

7. B. F. Nielsen, O. Skavhaug, and A. Tveito. Penalty and Front-Fixing Methods for the Numerical Solution of American Option Problems. *The Journal of Computational Finance,* 5(4):69–97, 2002.

8. O. Skavhaug, B. F. Nielsen, and A. Tveito. Mathematical Models of Financial Derivatives. In H. P. Langtangen and A. Tveito, editors, *Advanced Topics in Computational Partial Differential Equations – Numerical Methods and Diffpack Programming.* Springer, 2003.

9. P. Wilmott. *Derivatives, The theory and practice of financial engineering.* John Wiley & Sons, 1998.

10. P. Wilmott, J. Dewynne, and S. Howison. *Option Pricing, Mathematical models and computation.* Oxford Financial Press, 1993.

11. R. Zvan, P. A. Forsyth, and K. R. Vetzal. Penalty Methods for American Options With Stochastic Volatility. *Journal of Computational and Applied Mathematics,* 91(2):199–218, 1998.

# Chapter 13

# Finite Element Modeling of Elastic Structures

T. Thorvaldsen[1,2], H. P. Langtangen[1,2], and H. Osnes[3]

[1] Simula Research Laboratory
[2] Dept. of Informatics, University of Oslo
[3] Dept. of Mathematics, University of Oslo

**Abstract.** This chapter describes the structural version of the finite element method, commonly used in analysis of structures like buildings, ships, aircrafts, etc. The structural version of the finite element method is based on energy minimization and appear in the literature to differ from the finite element method based on the Galerkin formulation of partial differential equations. We show that these two versions of the finite element method are mathematically equivalent for problems in structural analysis. Then we describe the structural elements available in Diffpack: the 2D/3D bar elements, the 2D beam element, the 2D frame element, and the 2D triangular Kirchhoff plate element. Programming of structural elements in Diffpack is outlined, with emphasis on adding new structural elements. Finally, we describe some Diffpack simulators for computing the deformation and stress state in structures built of bar, beam, frame, and plate elements.

## 13.1 Introduction

Modern engineering concerned with building elastic structures, such as ships, buildings, cars, and aircrafts, utilizes computer simulations to a large extent in the design phase of the structures. The simulations are typically used to predict the deformation and internal stress state in the structures, resulting from some prescribed environmental loads. The purpose is to ensure that the structure can sustain the given loads with acceptable upper values of stresses and deformations. The activity of analyzing structures by computer simulations, based on mathematical models, is often called *structural analysis*. We shall in the present chapter give an introduction to this field and its tools.

The mathematical models used in structural analysis are based on a continuum description of the constructions. For simplicity, we limit the attention to elastic structures, i.e., structures that recover their shapes after the loads are removed. This property is relevant for constructions whose deformations are small, and this is very often the case (e.g., buildings, ships, cars, and aircrafts).

### 13.1.1 Two Versions of the Finite Element Method

*Dimension Reduction in Elasticity Models.* Chapter 5 in Langtangen [6] gives an introduction to the mathematical model for elasticity and finite element discretization methods for two- and three-dimensional problems. The elasticity model is expressed as a vector partial differential equation (PDE), and can describe elastic structures in general. However, many structures have special geometric shapes that can be taken advantage of to derive simpler mathematical models. The most prominent example is the *beam*; a beam has significantly larger extent in one direction compared to the other two, a property that makes it possible to formulate an approximative *one-dimensional* PDE describing stresses and deformations in a beam undergoing bending. That is, a one-dimensional mathematical model describes a three-dimensional physical problem. Similarly, a thin *plate* has much larger extent in two of the space directions than in the third. This geometric feature can be utilized to formulate a *two-dimensional* PDE for three-dimensional bending of thin plates. However, the classical PDE for elasticity must address beams and plates as three-dimensional domains. Thus, the specialized PDEs for beams and plates are much more computationally efficient because the number of space dimensions in the mathematical model is reduced. Besides beams and plates, there are two-dimensional models for shells (aircrafts and ships are typically modeled using shell elements) and one-dimensional models for *bars* (beams without bending resistance, also called trusses).

*Finite Elements as Physical Parts of a Structure.* Structural analysis is concerned with looking at a structure as an assembly of beams, plates, shells, and three-dimensional elastic components. The behavior of each of these components (i.e., structural elements) can be described by some PDE model, and each PDE can be discretized by the Galerkin finite element method. However, practitioners in structural analysis seldom work in terms of PDEs and Galerkin formulations. Instead they view the structure as a gathering of discrete structural elements, and describe the deformation and stress state of each element by a typical discrete finite element expansion. Next they apply an energy principle to derive a linear system, whose solution provides the unknown parameters in the finite element expansion (typically nodal displacements and rotations). This way of thinking is quite dissimilar from a PDE-based Galerkin finite element method, and there are hence two different "schools" of the finite element method: the one in structural analysis, and the one based on PDE models. Some would alternatively speak about the engineering versus the mathematical finite element method.

*The Mathematical Equivalence.* Looking at the literature, it may be difficult to see the links between these two approaches. The formulations are mathematically equivalent, but different notation, terminology, physical principles, and applications obscure the equivalence. The first part of this chapter tries to describe this equivalence, with a two-fold purpose: (i) understanding the similarity improves the understanding of the finite element method in general, and (ii) people with a background in one of the approaches may quickly learn the other procedure from an understanding of the equivalence. The latter point is important if you come from structural analysis, and need to do fluid dynamics applications, or vice versa. It should be mentioned that the two versions of the finite element method are not equally *convenient* for solving a given problem; the structural analysis version is clearly the most suitable tool for analyzing structures, whereas the PDE version of the finite element method is probably preferable for most other applications.

## 13.1.2   Two Element Concepts

*Different Meaning of the Term "Element".* The term "element" has a slightly different meaning in structural analysis compared to the PDE-based finite element method. An element in Diffpack is a geometric object; it has a geometric shape, a mapping from a reference element to the shape in the physical domain, a set of basis functions, and a set of degrees of freedom (usually nodal unknowns). An important feature in Diffpack is that the element is decoupled from the PDE. As an example, the 2D biquadratic element can be applied in a range of different physical contexts, including fluid flow, heat transfer, and elasticity. In structural analysis, the element contains information about the mathematical model and its physical parameters (material properties and loads) *in addition to* the geometric information regarding shape, basis functions, and degrees of freedom. An element in structural analysis is therefore strongly tied to the mathematical model. Moreover, all integrals are (normally) integrated analytically, so one has mathematical formulas for the element matrix and vector (i.e., there is no use of `integrands` functions, in Diffpack terminology). To distinguish the two concepts of an element, we shall refer to *geometric elements* (such as the standard Diffpack elements [6]) and *structural elements*. The latter type is used in the analysis of structures and constitute the subject of this chapter. We present in particular the various structural elements implemented in Diffpack at the time of this writing.

*Historical Background.* The modern finite element method was invented by engineers in the 1950s. They introduced the concept of a structural element – as a simple discrete description of a bar, beam, plate, or shell component. The total structure was viewed as a gathering of such elements. The principle of minimum potential energy was used to determine the (discrete) free displacement parameters in the construction. The free displacement parameters could be the displacement and rotation of the two ends of a beam element, or the displacement and rotation of the corners of a plate element. Assembling the structural elements in a correct energy minimizing way, implies a linear system, coupling all free displacement parameters in the structure. This became the finite element method for structural analysis.

Later, in the 1960s, mathematicians understood that the approximations used in structural analysis could also be applied to PDEs in general. They realized that this approach was the same as the classical Galerkin method

for PDEs (invented early in the 20th century) – just with some more computationally feasible basis functions. The main difference between the classical
Galerkin method and the computational method in structural analysis, is that
the Galerkin method employ global basis functions, whereas one in structural
analysis work with basis functions of a local nature. As mentioned in Chapter 2 in [6], basis functions of local nature have two convenient properties:
they are almost orthogonal (avoiding ill-conditioned matrices, and making
the matrices sparse) and they can be adapted to complicated geometries.

As already mentioned, the finite element method based on a PDE model
for elastic structural components, is equivalent to the finite element method
based on the principle of minimum potential energy. This equivalence is exemplified for two different element types in the next two sections.

## 13.2  An Introductory Example; Bar Elements

The purpose of this and the next section is to explain the basic concepts in
the structural analysis version of the finite element method, and show how
this version relates to the finite element method for PDEs. We assume that
the reader is familiar with the latter approach, e.g., from Chapter 2 in [6]. It
will also be advantageous for the reader to have studied Chapters 5.1.1–5.1.3
in [6].

*How to Solve Problems in Structural Mechanics.*  As previously said, analysis
of structures typically follow two alternative paths. The first path consists
of modeling the complete structure as a 3D elastic body and solving the associated 3D elasticity PDE. In the second path one views the structure as
an assembly of beams, plates, shells, and other structural components. Each
component has its own simplified mathematical model. These models can be
expressed either as lower-dimensional PDEs (e.g., a one-dimensional beam
equation) or relations between a few force and displacement degrees of freedom (e.g., a beam finite element). The most general and accurate approach
is to solve the 3D elasticity equations throughout the whole structure, but
this is often very much more computer resource consuming than utilizing an
assembly of simplified, lower-dimensional (beam, plate, shell, etc.) models.
With today's computer resources, the latter approach is frequently the only
feasible one in engineering contexts.

For the examples in this section, and the next, we use a one-dimensional
partial differential equation for showing the equivalence between the two finite
element formulations. In real cases doing structural analysis of a construction,
a three-dimensional PDE is employed.

*Sample Problem: Deformation of a Bar.*  Let us now demonstrate the two
versions of the finite element method in the simplest possible example, namely
a bar. A bar structure corresponds to a two-point boundary value problem

in PDE terminology. We consider a long and slender bar, or rod, subject to a body force $F$ along the axis of the bar (gravity, for instance), and an end force $P$, as depicted in Figure 13.1. Our aim is to compute the deformation

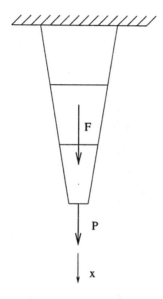

**Fig. 13.1.** Sketch of a bar, with varying cross section, subject to a body force $F$ and an end force $P$.

and stress state in the bar. We shall do this using a one-dimensional mathematical model for this three-dimensional physical problem (note that we have deformations in all three space directions; the cross section gets contracted as we elongate the bar).

If the cross section, the material properties, or body forces are not constant along the bar, the one-dimensional models we develop below are not a mathematically correct simplification of the classical three-dimensional elasticity PDE. The one-dimensional model in this case is, nevertheless, a reasonable approximation if the geometry, material properties, and loads are slowly varying along the bar.

### 13.2.1   Differential Equation Formulation

*Derivation of the Differential Equation.*   Let us derive a differential equation for the bar deformation by looking at a small section (of length $\Delta x$) of the bar, as depicted in Figure 13.2. The small section is in equilibrium, which implies that the net sum of forces vanishes. There are two types of forces acting: normal stress $\sigma$ in the cross section, and a body force $F$. The sum $S$

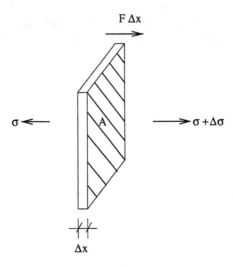

**Fig. 13.2.** Equilibrium of a small length $\Delta x$ of a bar element.

of all forces equals

$$S = F\Delta x + A(\sigma + \Delta\sigma) - A\sigma.$$

The body force $F$ is defined as force per unit length, so we need to multiply by $\Delta x$ to get the total body force of the small section of consideration. The force in the cross section is obtained as the product of the stress $\sigma$ and area $A$, where $\sigma$ is assumed constant over $A$. Equilibrium of the small section now ensures $S = 0$. Thus, the relation above may be expressed as

$$A\frac{\Delta\sigma}{\Delta x} + F = 0.$$

In the limit $\Delta x \to 0$ we obtain the equation[1]

$$A\frac{d\sigma}{dx} + F = 0. \tag{13.1}$$

We can now integrate to find $\sigma$ along the bar ($F$ is considered known). However, boundary conditions are needed, as always, when working with differential equations. The stress $\sigma$ is uniquely determined by (13.1) if the boundary condition specifies $\sigma$ in one of the end points of the bar, but sometimes the boundary condition involves the displacement $u$ in the $x$ direction. Therefore, in general cases additional equations are needed to close the model.

In elasticity problems, the differential equation models are based on force equilibrium, the stress-strain relation, and the strain-displacement relation.

---

[1] This is the counterpart to $\sigma_{rs,s} + \varrho b_r = 0$ in three-dimensional elasticity, see Chapter 5.1.1 in [6].

The former is in this case expressed in (13.1), whereas the latter two read

$$\sigma = E\varepsilon, \quad \varepsilon = \frac{du}{dx},$$

where $E$ is Young's modulus, $\varepsilon$ is the normal strain in the $x$ direction, and $u(x)$ is the displacement in the $x$ direction. These formulas are valid for a uniform straight bar, but we apply them as a reasonable approximation for non-uniform bars if the geometry, material properties, and loads are slowly varying. Eliminating $\sigma$ and $\varepsilon$ gives the governing differential equation[2]

$$\frac{d}{dx}\left( EA\frac{du}{dx}\right) + F = 0, \quad a < x < b. \tag{13.2}$$

This equation is to be solved on an interval $[a, b]$ with boundary conditions of the type $u(a)$ or $u(b)$ known, or $\sigma(a)$ or $\sigma(b)$ known (one condition at each boundary point, $x = a, b$).

*The Finite Element Formulation.*  The finite element solution to (13.2) consists in approximating the primary unknown $u$ by $\hat{u} = \sum_{j=1}^{n} u_j N_j(x)$, where $u_j$ are nodal values and $N_j(x)$ are prescribed basis functions. Inserting $\hat{u}$ in (13.2) gives a residual $R$. In the Galerkin method, $R$ is multiplied by $n$ weighting functions $N_i(x)$ and integrated over the domain. Second-order derivatives are integrated by parts to relax the continuity requirements of the $N_i$ functions. The result is a linear system for $u_1, \ldots, u_n$. The linear system can be viewed as a sum of contributions from each element, and these element contributions are known as the element matrix and element vector. The element matrix and element vector are the two problem-dependent quantities that must be supplied to a finite element program; the rest of the computations are independent of the physical application. Normally, element matrices and vectors are computed in a reference element on $[-1, 1]$, utilizing numerical evaluation of integrals.

We shall go through these details using a notation that is compatible with the one used in the structural analysis version of the finite element method. The derivation of expressions for the element matrix and element vector will be based on working with a reference element exclusively. This reference element has now length $L$. It would be mathematically more pleasing to introduce (say) $\xi$ as coordinate in the reference element, but it is common practice to use $x$ both for the physical system and for the $[0, L]$ reference element. On the reference element we have two nodes; node 1 at $x = 0$ and node 2 at $x = L$. We introduce $v_{xi}$ as the unknown $x$-displacement of $u$ at node $i$. The displacement function $u(x)$ is expanded, on this reference element, in the usual finite element sense, as

$$u \approx \hat{u} = N_1(x)v_{x1} + N_2(x)v_{x2},$$

---

[2] This is the counterpart to $((\lambda + \mu)u_{s,s})_{,r} + (\mu u_{r,s})_{,s} + \varrho b_r = 0$ in three-dimensional elasticity, see Chapter 5.1.1 in [6].

where $N_j(x)$ are the linear basis functions on the element:

$$N_1(x) = \frac{L-x}{L}, \quad N_2(x) = \frac{x}{L}. \tag{13.3}$$

The basis functions are collected in a row vector

$$\boldsymbol{N} = \begin{bmatrix} N_1, N_2 \end{bmatrix},$$

and the nodal displacements are collected in a column vector

$$\boldsymbol{v} = \begin{bmatrix} v_{x1} \\ v_{x2} \end{bmatrix}.$$

The approximation $\hat{u}$ can now be written as an inner product $\hat{u} = \boldsymbol{N}\boldsymbol{v}$. Higher-order elements can easily be constructed by introducing more nodes and associated basis functions. In general, we may have displacement degrees of freedom $v_{x1}, v_{x2}, \ldots, v_{xq}$ collected in a column vector $\boldsymbol{v}$, and basis functions $N_1, N_2, \ldots, N_q$ collected in a row vector $\boldsymbol{N}$, leading to the formula $\hat{u} = \boldsymbol{N}\boldsymbol{v}$ also in the more generic case.

Inserting $\hat{u} = \boldsymbol{N}\boldsymbol{v}$ in the differential equation (13.2), now viewed as a differential equation governing the physics on the reference element, results in an element residual, which we multiply by $\boldsymbol{N}^T$, and integrate from 0 to $L$:

$$\int_0^L \boldsymbol{N}^T \frac{d}{dx}\left( EA\frac{d}{dx}\boldsymbol{N}\boldsymbol{v} \right) dx + \int_0^L \boldsymbol{N}^T F dx = 0.$$

It is common to introduce a special symbol $\boldsymbol{B}$ for the derivatives of $\boldsymbol{N}$, here $\boldsymbol{B} = d\boldsymbol{N}/dx$. Applying integration by parts and omitting boundary terms (assuming $u$ is prescribed at both ends), we get

$$\left( \int_0^L EA\boldsymbol{B}^T\boldsymbol{B}dx \right) \boldsymbol{v} = \int_0^L \boldsymbol{N}^T F dx. \tag{13.4}$$

This is the finite element equation for the reference element. The element matrix is given as

$$\boldsymbol{A}^{(e)} = \int_0^L EA\boldsymbol{B}^T\boldsymbol{B}dx,$$

and the formula for the element vector reads

$$\boldsymbol{b}^{(e)} = \int_0^L \boldsymbol{N}^T F dx.$$

For linear elements and constant $EA$ we get

$$\boldsymbol{A}^{(e)} = \frac{EA}{L}\begin{bmatrix} 1 & -1 \\ -1 & 1 \end{bmatrix}, \quad \boldsymbol{b}^{(e)} = F\frac{L}{2}\begin{bmatrix} 1 \\ 1 \end{bmatrix}.$$

The element matrix and the element vector are both, of course, identical to the ones obtained when working with a reference element $[-1, 1]$, except for the factor $L$.

*The Element Assembly Process.* The element matrices and vectors for each element can be assembled, yielding a linear system, which can be solved for the nodal displacements. We assume that the reader is familiar with the assembly procedure such that we can solely focus on deriving expressions for element matrices and vectors [6].

*Variational Problem.* It is well known that certain differential equation problems can be equivalently expressed as minimization problems. (See Chapter 2.10 in [6], or a book on variational calculus for details of the mathematical background.) Our model equation (13.2) can be recast as a minimization problem, where the functional to be minimized reads

$$J(v) = \int_a^b \frac{1}{2} EA \left( \frac{\partial v}{\partial x} \right)^2 dx - \int_a^b Fv dx . \tag{13.5}$$

(No boundary conditions on $\sigma = Eu'$ is assumed, otherwise an additional boundary term enters (13.5).) The minimization can be expressed as

$$\min_v J(v) = J(u) \quad \text{or} \quad J(u) \le J(v),$$

where $v$ is an arbitrary function in a suitable space of functions (in this case $H^1([a,b])$). The solution of (13.2) is the value $u$ of $v$ that minimizes $J(v)$.

Readers who are familiar with finite element theory (e.g., from Chapter 2.10 in [6]) will know that there is a minimization problem $J(u) \le J(v)$ for $J(v) = \frac{1}{2} a(v, v) - F(v)$ when the weak formulation of the underlying differential equation problem is on the form $a(u, v) = F(v)$ (for all $v$ in a suitable function space). In the present case[3],

$$a(u, v) = \int_0^L EA \left( \frac{\partial u}{\partial x} \right) \left( \frac{\partial v}{\partial x} \right) dx, \quad F(v) = \int_0^L Fv dx .$$

Approximating $\hat{u} \approx \sum_{j=1}^n u_j N_j$ and inserting $\hat{u}$ in the functional to be minimized gives

$$J(\hat{u}) = \int_a^b \frac{1}{2} EA \left( \frac{\partial \hat{u}}{\partial x} \right)^2 - \int_a^b F \hat{u} dx .$$

Now $J(\hat{u})$ is a function of $n$ variables to be minimized. The finite element equations result from requiring $\partial J / \partial u_i = 0$, $i = 1, \ldots, n$. This approach is mathematically equivalent with the discretization based on the Galerkin method. However, there are many PDE problems that can not be formulated as a minimization problem, so the Galerkin approach is more general. From the global finite element equations $\partial J / \partial u_i = 0$ we can easily derive

---

[3] Note that we here use the same symbol $F$ for two different things: the linear functional $F(v)$ and the body force $F$, but we are not going to work further with the $F(v)$ symbol, so we can stick to the notation from Chapter 2.10 in [6].

element matrices and vectors. Alternatively, we can perform the minimization at the element level, as shown in the next paragraph, using the notation and reasoning common in the structural analysis version of the finite element method.

Starting with the minimization of the functional (13.5), we can arrive at the element matrix and vector by inserting a function $v$, expanded as a finite element function on an element in (13.5), and performing the minimization at the element level. That is, we want to minimize

$$J^{(e)}(v) = \int_0^L \frac{1}{2} EA \left( \frac{d}{dx} Nv \right)^2 dx - \int_0^L N^T F v^T dx,$$

or written with the $B$ matrix as usual:

$$J^{(e)}(v) = \int_0^L \frac{1}{2} EA(Bv)^2 dx - \int_0^L N^T F v^T dx.$$

This $J^{(e)}(v)$ is now a function of the scalar parameters $v$. The minimization of $J^{(e)}(v)$ leads to the equations $\partial J^{(e)}(v)/\partial v = 0$ at the element level. In case of linear elements we have two such parameters, and minimizing $J^{(e)}(v_{x1}, v_{x2})$ implies

$$\frac{\partial J^{(e)}}{\partial v_{x1}} = 0, \quad \frac{\partial J^{(e)}}{\partial v_{x2}} = 0.$$

These two equations give the same result as in (13.4).

### 13.2.2   Energy Formulation

*The Potential Energy.*   In the structural analysis version of the finite element method the derivation of the element matrix and vector is often based on an energy formulation. This energy formulation corresponds to minimizing the potential energy in the structure. The potential energy is, not surprisingly, identical to the $J^{(e)}(v)$ expressions in the previous subsection, and the mathematical details of the minimization procedure are thus the same. However, it is usual in structural analysis to formulate the functional to be minimized directly from the physics of the problem, i.e., no differential equation and associated Galerkin method appear in the derivation of the element matrix and vector. This different way of reasoning, combined with other sets of symbols, tends to expose the two versions of the finite element method more unlike than what they really are. The equation of equilibrium is equivalent to minimizing

$$\Pi = U - W.$$

The quantity $\Pi$ is called *potential energy*, and the two terms $U$ and $W$ can be interpreted as work done by internal stresses and work done by body forces, respectively. It is common to refer to $U$ as the *strain energy* of the structure.

*The Bar Element Case.* We shall exemplify the energy formulation for the bar element. For a small elongation $\Delta u$ of the bar element, corresponding to a normal strain $\Delta \varepsilon$ in the $x$ direction, the work done on an infinitesimal element $dV$ is $\sigma \Delta \varepsilon dV = \sigma \Delta \varepsilon A dx$, where $\sigma$ is the normal stress and $A$ the cross section area. The work of $F$ (where $F$ is defined as force per unit length) for a cross section is $F dx \Delta u$. For the whole bar the work expressions require integration along the bar. Hence, the contributions to $U$ and $W$ become

$$\Delta U = \int_0^L A\sigma \Delta \varepsilon \, dx, \quad \Delta W = \int_0^L F \Delta u \, dx \, .$$

Now, $\sigma$ depends on $\varepsilon$ because $\sigma = E\varepsilon$ (according to Hooke's law in one dimension). In the infinitesimal limit we get

$$dU = \int_0^L EA\varepsilon d\varepsilon \, dx, \quad dW = \int_0^L F du \, dx \, .$$

Integrating with respect to the deformation leads to the total potential energy

$$\Pi = U - W = \int_0^L \left[ \frac{1}{2} EA\varepsilon^2 - Fu \right] dx \, .$$

The idea is then to insert $u \approx \hat{u} = \boldsymbol{N}\boldsymbol{v}$ in $\Pi$, and minimize $\Pi(\boldsymbol{v})$ with respect to the parameters $\boldsymbol{v}$. We have

$$\varepsilon = \frac{du}{dx} \approx \frac{d\hat{u}}{dx} = \frac{d}{dx}\boldsymbol{N}\boldsymbol{v} = \boldsymbol{B}\boldsymbol{v},$$

resulting in

$$U = \int_0^L \frac{1}{2} EA\boldsymbol{v}^T \boldsymbol{B}^T \boldsymbol{B}\boldsymbol{v} \, dx = \boldsymbol{v}^T \frac{1}{2} EA \int_0^L \boldsymbol{B}^T \boldsymbol{B} \, dx\boldsymbol{v}$$

and

$$W = \int_0^L \boldsymbol{v}^T \boldsymbol{N}^T F dx = \boldsymbol{v}^T F \int_0^L \boldsymbol{N}^T dx \, .$$

In the above expressions we have assumed that $EA$ and $F$ are independent of $x$.

*Minimizing the Potential Energy.* Minimization of

$$\Pi = U - W = \boldsymbol{v}^T \frac{1}{2} EA \int_0^L \boldsymbol{B}^T \boldsymbol{B} \, dx\boldsymbol{v} - \boldsymbol{v}^T F \int_0^L \boldsymbol{N}^T dx$$

is known as *the principle of minimum potential energy*, leading to

$$\frac{\partial \Pi}{\partial \boldsymbol{v}^T} = EA \int_0^L \boldsymbol{B}^T \boldsymbol{B} \, dx\boldsymbol{v} - F \int_0^L \boldsymbol{N}^T dx = 0$$

or

$$EA \int_0^L B^T B \, dx v = F \int_0^L N^T dx \,. \tag{13.6}$$

Equation (13.6) is the so-called stiffness relation between the *element stiffness matrix*

$$A^{(e)} = EA \int_0^L B^T B \, dx$$

and the *element load vector*

$$b^{(e)} = F \int_0^L N^T dx \,.$$

*Different Finite Element Formulations.* We have now seen how these formulas can be developed along different lines:

 - the Galerkin method applied to a differential equation,
 - minimization of a functional associated with the differential equation,
 - the principle of minimum potential energy.

A fourth method is the principle of virtual work, which is also a widely used approach in structural analysis. This principle can be viewed as a physical counterpart to Galerkin's method. All these methods can be applied globally to establish the complete system of algebraic equations in a finite element model. Alternatively, we can employ the methods locally on an element to form the algebraic equations at the element level.

## 13.3   Another Example; Beam Elements

Bar elements correspond to a second-order differential equation in space. The parameters in $v$ are then values of the primary unknown (displacement) at the nodes, and the basis functions $N$ are piecewise continuous with jumps in the derivatives. The expressions for the element matrix and vectors typically contain first derivatives of $N$.

In structural analysis, the elements also model beams and plates. The underlying differential equations are of forth order, which leads to second-order derivatives of $N$ in the expressions for the element matrix and vector. This means that $N$ must have continuous first-order derivatives, and furthermore, a demand for more degrees of freedom in an element. Typically, one has both the displacement and its derivative(s) (rotations) as unknowns at the nodes. To illustrate the implication of such an extension of the degrees of freedom, it may be helpful to repeat the bar example in the case of a beam element.

### 13.3.1   Differential Equation Formulation

*The Beam Element.*   A beam is a long and slender elastic structural component, like the bar, but with more general deformation patterns and loads; we can have moments and/or shear forces at both ends, as well as distributed and concentrated loads (between the end points) in the transverse direction. Moments and shear forces give rise to *bending*, which is the characteristic deformation of a beam. For simplicity in the mathematical model, tension and body forces are neglected now. The beam element is thus solely subject to bending[4]. Furthermore, the beam element covers the interval $[0, L]$, in a local coordinate system, and has one node at each end point of the interval. For simplicity we only allow bending in one plane, i.e., we have a two-dimensional deformation problem.

*Degrees of Freedom and Basis Functions.*   The natural degrees of freedom for bending are the vertical displacement at the nodes. However, the governing differential equation for a beam is of forth order. Thus, there are four constants, arising from integration, that need to be fixed. In a finite element formulation, this requires (at least) third-order polynomials for $N$ and correspondingly more degrees of freedom than in the bar element case. The additional degrees of freedom for a beam element are rotations at the nodes. For small displacements, which is a fundamental assumption in the beam models we consider herein, rotation equals the first derivative of the vertical displacement.

*The Differential Equation.*   The prevailing differential equation for a beam reads

$$\frac{d^2}{dx^2}\left(EI\frac{d^2w}{dx^2}\right) = q, \tag{13.7}$$

where $w(x)$ is the deflection (vertical displacement) of the beam, $q$ is a vertical load along the beam, $E$ is Young's modulus, and $I$ is the second moment of inertia of the cross section (i.e., a geometric feature of the cross section). The product $EI$ is termed flexural rigidity, and may vary slowly along the beam. Because the spatial operator in the equation is of fourth order, two boundary conditions are needed at each end point. Four types of conditions are possible:

- prescribed vertical displacement $w$,
- prescribed rotation $w'$,
- prescribed moment $M = -EIw''$,
- prescribed shear force $Q = EIw'''$.

---

[4] Combining bending and elongation, i.e., shear forces, tension forces, moments, and body forces result in a superposition of the beam and bar element, often called the *frame* element, see Section 13.5.3.

*The Finite Element Formulation.* To derive expressions for the element matrix and vector, we view (13.7) as the governing equation for the deflection of a beam element, and set $EI$ and $q$ constant. The primary unknown $w$ is expanded, as usual, in basis functions and the associated degrees of freedom at the element level:

$$w \approx \hat{w} = N_1 w_1 + N_2 \theta_1 + N_3 w_2 + N_4 \theta_2, \qquad (13.8)$$

where $w_i$ are nodal vertical displacements and $\theta_i$ are nodal rotations (derivatives of $w$). The basis functions are cubic Hermite polynomials with continuous derivatives up to first order:

$$N_1 = 1 - \frac{3x^2}{L^2} + \frac{2x^3}{L^3}, \qquad (13.9)$$

$$N_2 = -x + \frac{2x^2}{L} - \frac{x^3}{L^2}, \qquad (13.10)$$

$$N_3 = \frac{3x^2}{L^2} - \frac{2x^3}{L^3}, \qquad (13.11)$$

$$N_4 = \frac{x^2}{L} - \frac{x^3}{L^2}. \qquad (13.12)$$

As usual, $N_i$ equals 1 for degree of freedom no. $i$ and zero for all other degrees of freedom. More precisely, $N_1$ equals 1 at node 1, its derivative is zero at node 1, and its value and derivative are zero at node 2; $N_2$ equals 0 at node 1, its derivative is 1, and its value and derivative are zero at node 2; $N_3$ equals 1 at node 2, its derivative is zero at node 2, and its value and derivative are zero at node 1; $N_4$ equals 0 at node 2, its derivative is 1 at node 2, and its value and derivative are zero at node 1.

We insert (13.8) into (13.7), multiply by $N_i$, $i = 1, 2, 3, 4$, and integrate by parts *twice*. The result becomes (omitting boundary terms, for simplicity)

$$EI \int_0^L \frac{d^2 N_i}{dx^2} \frac{d^2 N_j}{dx^2} dx \begin{bmatrix} w_1 \\ \theta_1 \\ w_2 \\ \theta_2 \end{bmatrix} = q \int_0^L \begin{bmatrix} N_1 \\ N_2 \\ N_3 \\ N_4 \end{bmatrix} dx \, .$$

Using the ordinary notation in structural finite element analysis,

$$\boldsymbol{N} = [\, N_1, N_2, N_3, N_4 \,],$$

$$\boldsymbol{v} = [\, w_1, \theta_1, w_2, \theta_2 \,]^T,$$

and[5]

$$\boldsymbol{B} = \frac{d^2}{dx^2} \boldsymbol{N},$$

---

[5] Note that the $\boldsymbol{B}$ matrix now contains the second derivative of the basis functions (and not, as for the bar elements, the first derivative).

we end up with expressions on the same form as for the bar element:

$$EI \int_0^L \boldsymbol{B}^T \boldsymbol{B} \, dx \, \boldsymbol{v} = q \int_0^L \boldsymbol{N}^T dx \, .$$

By evaluating the integrals, we get

$$\boldsymbol{A}^{(e)} \boldsymbol{v} = \boldsymbol{b}^{(e)},$$

with the element stiffness matrix $\boldsymbol{A}^{(e)}$ and element load vector $\boldsymbol{b}^{(e)}$ as

$$\boldsymbol{A}^{(e)} = \frac{EI}{L^3} \begin{bmatrix} 12 & 6L & -12 & 6L \\ 6L & 4L^2 & -6L & 2L^2 \\ -12 & -6L & 12 & -6L \\ 6L & 2L^2 & -6L & 4L^2 \end{bmatrix}, \quad \boldsymbol{b}^{(e)} = \frac{qL}{2} \begin{bmatrix} 1 \\ \frac{L}{6} \\ 1 \\ \frac{L}{6} \end{bmatrix} \, .$$

## 13.3.2 Energy Formulation

*The Potential Energy.* Now moving to the finite element formulation employed in structural analysis, it can be shown that for the beam element carrying a load $q$ per unit length, the strain energy stored due to bending of the beam is

$$U = \int_0^L \frac{1}{2} \frac{M^2}{EI} dx,$$

where $M$, being a function of $x$, is the bending moment along the beam. For the work done by the external load $q$ it can be shown that

$$W = \int_0^L qw \, dx \, .$$

Now, the bending moment $M$ is related to $w$ through

$$M = -EI \frac{d^2 w}{dx^2},$$

or $M = -EI\boldsymbol{B}\boldsymbol{v}$, if we switch to discrete quantities and the corresponding vector notation. The stored potential energy in the beam is then

$$\Pi = U - W = \frac{1}{2} \boldsymbol{v}^T \int_0^L \boldsymbol{B}^T EI \boldsymbol{B} \, dx \boldsymbol{v} - \boldsymbol{v}^T \int_0^L q \boldsymbol{N}^T dx \, .$$

*Minimizing the Potential Energy.* The principle of minimum potential energy states that stable equilibrium is achieved when $\Pi$ is a minimum with respect to all parameters in $\boldsymbol{v}$:

$$\frac{\partial \Pi}{\partial \boldsymbol{v}^T} = \int_0^L \boldsymbol{B}^T EI \boldsymbol{B} \, dx \boldsymbol{v} - \int_0^L q \boldsymbol{N}^T dx = 0,$$

which alternatively can be written as

$$EI \int_0^L B^T B \, dx v = q \int_0^L N^T dx,$$

assuming that $EI$ and $q$ are constants. The element stiffness matrix is

$$A^{(e)} = EI \int_0^L B^T B \, dx$$

and the element load vector reads

$$b^{(e)} = q \int_0^L N^T dx \, .$$

These expressions are the same as we derived using the differential equation and the Galerkin method.

## 13.4     General Three-Dimensional Elasticity

Bar, beam, and plate models are simplified versions of the general, linear, three-dimensional model for deformation of elastic bodies. Here we illustrate how the full three-dimensional elasticity model fits into the framework from the previous sections.

### 13.4.1     Differential Equation Formulation

*General Three-Dimensional Elasticity.*   General three-dimensional elasticity is governed by Navier's equation

$$\nabla \left[ (\lambda + \mu) \nabla \cdot u \right] + \nabla \cdot [\mu \nabla u] + F = 0 \qquad (13.13)$$

inside the elastic body $\Omega$. Thermal deformations are neglected in this model (see Chapter 5.1.1 in [6] for a derivation of (13.13), and how to incorporate thermal deformations). The parameters in (13.13) have the following meaning: $u$ is the displacement field, $\lambda$ and $\mu$ are Lamé's elasticity parameters, and $F$ denotes body forces[6]. The stress tensor $\sigma$ is related to the displacement field through

$$\sigma = \lambda \nabla \cdot u I + \mu (\nabla u + (\nabla u)^T) \, .$$

Three scalar boundary conditions must be prescribed at each boundary point when solving (13.13). These conditions can be displacement components or components of the stress vector $\sigma \cdot n$, $n$ being an outward unit normal vector to the boundary $\partial \Omega$.

---

[6] We remark that the body force term in this chapter differs from what is used in [6] (where $\varrho b$ is used). This is to avoid conflicts with the element load vector $b^{(e)}$ and the right hand side $b$ of the global linear system.

*The Engineering Finite Element Notation.*  The details of applying a Galerkin finite element method to (13.13) are explained in detail in Chapter 5.1.2 in [6]. Of particular relevance to structural analysis, as discussed in the present chapter, is the "engineering finite element notation" in Chapter 5.1.3 in [6]. For convenient reference, we quote and explain the most important formulas from [6] (but omit thermal deformation).

The stress tensor is not written as a tensor, but as a vector

$$\sigma = [\sigma_{xx}, \sigma_{yy}, \sigma_{zz}, \sigma_{xy}, \sigma_{yz}, \sigma_{zx}]^T.$$

It is common in this notation to use $\sigma_{xy}$ rather than $\sigma_{12}$. The strain tensor is written as a vector as well,

$$\varepsilon = [\varepsilon_{xx}, \varepsilon_{yy}, \varepsilon_{zz}, \gamma_{xy}, \gamma_{yz}, \gamma_{zx}]^T,$$

where one applies the "engineering shear strain" $\gamma_{xy} = 2\varepsilon_{xy}$. Hooke's law can now be expressed as

$$\sigma = D\varepsilon, \tag{13.14}$$

with

$$D = \frac{E(1-\nu)}{(1+\nu)(1-2\nu)} \begin{bmatrix} 1 & \frac{\nu}{1-\nu} & \frac{\nu}{1-\nu} & 0 & 0 & 0 \\ & 1 & \frac{\nu}{1-\nu} & 0 & 0 & 0 \\ & & 1 & 0 & 0 & 0 \\ & & & \frac{1-2\nu}{2(1-\nu)} & 0 & 0 \\ & \text{symmetric} & & & \frac{1-2\nu}{2(1-\nu)} & 0 \\ & & & & & \frac{1-2\nu}{2(1-\nu)} \end{bmatrix} \tag{13.15}$$

for three-dimensional elasticity. The parameters $E$ and $\nu$ are known as Young's modulus and Poisson's ratio, respectively.

The common cases of plane strain

$$\varepsilon_{xz} = \varepsilon_{yz} = \varepsilon_{zz} = 0, \quad \sigma_{zz} = \nu(\sigma_{xx} + \sigma_{yy})$$

and plane stress

$$\sigma_{xz} = \sigma_{yz} = \sigma_{zz} = 0, \quad \varepsilon_{zz} = -\frac{\nu}{E}(\sigma_{zz} + \sigma_{yy})$$

can be handled by a constitutive law of the form (13.14), but now with

$$\sigma = [\sigma_{xx}, \sigma_{yy}, \sigma_{xy}]^T, \quad \varepsilon = [\varepsilon_{xx}, \varepsilon_{yy}, \gamma_{xy}]^T,$$

and

$$D = \frac{E}{1-\nu^2} \begin{bmatrix} 1 & \nu & 0 \\ \nu & 1 & 0 \\ 0 & 0 & \frac{1-\nu}{2} \end{bmatrix} \tag{13.16}$$

for plane stress, and

$$D = \frac{E(1-\nu)}{(1+\nu)(1-2\nu)} \begin{bmatrix} 1 & \frac{\nu}{1-\nu} & 0 \\ \frac{\nu}{1-\nu} & 1 & 0 \\ 0 & 0 & \frac{1-2\nu}{2(1-\nu)} \end{bmatrix}$$

for plane strain.

*The Finite Element Formulation.* Let us approximate the displacement field, i.e., the primary unknown in the elasticity problem, with

$$\hat{u} = \sum_{j=1}^{n} u_j N_j(x),$$

where $u_j$ is the displacement (and/or rotation) vector at node no. $j$.

The discrete strain vector can now be expressed by

$$\varepsilon = \sum_{j=1}^{n} B_j u_j, \tag{13.17}$$

with

$$B_i = \begin{bmatrix} N_{i,x} & 0 & 0 \\ 0 & N_{i,y} & 0 \\ 0 & 0 & N_{i,z} \\ N_{i,y} & N_{i,x} & 0 \\ 0 & N_{i,z} & N_{i,y} \\ N_{i,z} & 0 & N_{i,x} \end{bmatrix}.$$

In plane stress and strain, where only the $x$ and $y$ components of $u_i$ enter the equations, the $B_i$ matrix takes the form

$$B_i = \begin{bmatrix} N_{i,x} & 0 \\ 0 & N_{i,y} \\ N_{i,y} & N_{i,x} \end{bmatrix}.$$

Axisymmetric elasticity problems in $(z, r)$ coordinates fits into the framework above, using the following definitions of $\varepsilon$, $\sigma$, $B_i$, and $D$.

$$\varepsilon = [\varepsilon_{zz}, \varepsilon_{rr}, \varepsilon_{\theta\theta}, \gamma_{rz}]^T,$$
$$\sigma = [\sigma_{zz}, \sigma_{rr}, \sigma_{\theta\theta}, \sigma_{rz}]^T,$$
$$B_i = \begin{bmatrix} 0 & N_{i,z} \\ N_{i,r} & 0 \\ \frac{1}{r} N_i & 0 \\ N_{i,z} & N_{i,r} \end{bmatrix},$$

$$D = \frac{E(1-\nu)}{(1+\nu)(1-2\nu)} \begin{bmatrix} 1 & \frac{\nu}{1-\nu} & \frac{\nu}{1-\nu} & 0 \\ \frac{\nu}{1-\nu} & 1 & \frac{\nu}{1-\nu} & 0 \\ \frac{\nu}{1-\nu} & \frac{\nu}{1-\nu} & 1 & 0 \\ 0 & 0 & 0 & \frac{1-2\nu}{2(1-\nu)} \end{bmatrix}.$$

All integrals $\int_\Omega () \, d\Omega$ are in the axisymmetric case transformed to $2\pi \int () r \, dr \, dz$.

The Galerkin formulation of (13.13) can in the engineering notation be expressed as

$$\int_\Omega B_i^T \sigma \, d\Omega = \int_{\partial\Omega} N_i t \, d\Gamma. \tag{13.18}$$

The symbol $t$ represents the traction $\sigma \cdot n$.

Inserting (13.17) in (13.14) and then the resulting (13.14) in (13.18) yields

$$\sum_{j=1}^{n} \int_{\Omega} B_i^T D B_j \, d\Omega \, u_j = \int_{\partial\Omega} N_i t \, d\Gamma, \qquad (13.19)$$

for $i = 1, \ldots, n$. For each $i$ in (13.19) we have a vector equation with three components (or two in plane stress/strain or axisymmetry). The element matrix can be viewed as a block matrix with $n_e \times n_e$ blocks $\int_{\Omega_e} B_i^T D B_j \, d\Omega$, $i, j = 1, \ldots, n_e$, where $n_e$ is the number of nodes in the element. The element vector can also be written on block form, with block no. $i$ taking the form of the right-hand side in (13.19). A nice feature of this formulation is that plane stress, plane strain, axisymmetric, and general three-dimensional problems can straightforwardly be treated in a unified notation and implementation.

At the element level it is also common to collect all the element degrees of freedom in a vector

$$v = [u_1 \; u_2 \; \cdots \; u_{n_e}]^T,$$

where $u_i$ is the displacements and/or rotations in node $i$.

Furthermore, one collects the $B_i$ matrices in a matrix $B$ such that

$$\varepsilon = Bv.$$

The element stiffness matrix becomes, with this latter notation,

$$A^{(e)} = \int_{\Omega_e} B_i^T D B_j \, d\Omega,$$

whereas the element load vector can be written

$$b^{(e)} = \int_{\partial\Omega_e} N_i t \, d\Gamma.$$

### 13.4.2   Energy Formulation

*The Potential Energy.* The strain energy due to a small deformation $\Delta u$ with corresponding strains $\Delta\varepsilon$ is

$$\Delta U = \int_{\Omega} \sigma^T \Delta\varepsilon \, d\Omega.$$

Inserting Hooke's law $\sigma = D\varepsilon$, looking at an infinitesimal deformation $du$ (with $d\varepsilon$), and integrating with respect to $\varepsilon$ gives

$$U = \int_{\Omega} \frac{1}{2} \varepsilon^T D\varepsilon \, d\Omega.$$

The work done by the body forces $F$ and surface tractions $t$ is

$$W = \int_{\Omega} u^T F \, d\Omega + \int_{\partial\Omega} u^T t \, d\Gamma \, .$$

The principle of minimum potential energy implies minimizing the functional $\Pi = U - W$ with respect to the structure's degrees of freedom. Here we shall present a more general form of the potential energy, where we also allow for initial stress and strain, as well as point loads at the nodes:

$$\Pi = \int_{\Omega} (\frac{1}{2}\varepsilon^T D\varepsilon - \varepsilon^T D\varepsilon_0 + \varepsilon^T \sigma_0) \, d\Omega$$

$$- \int_{\Omega} u^T F \, d\Omega - \int_{\partial\Omega} u^T t \, d\Gamma - x^T P \, . \qquad (13.20)$$

In this expression $u$ is the displacement field, $D$ is the material property matrix, $\varepsilon$ are strains, $\varepsilon_0$ are initial strains, $\sigma_0$ are initial stresses, $F$ are body forces, $t$ are surface tractions, $x$ nodal degrees of freedom for the global system, and $P$ are loads applied to the degrees of freedom by external agencies. The symbols $\partial\Omega$ and $\Omega$ denote the surface and the volume of the structure, respectively. Usually initial stresses are neglected, and the primary application of initial strains is the modeling of thermal strains, i.e., isotropic expansion due to heating. We will come back to this later.

*The Finite Element Formulation.* We *assume* that the displacement field $u$ may be expressed by basis functions $N$ and local element degrees of freedom $v$:

$$u = Nv \, . \qquad (13.21)$$

This kind of assumption is the same as in many other formulations of the finite element method, and is the salient point in the method. We move from the *continuous* displacement field to a *discrete* representation (without changing notation in any way). The approximation is expressed by degrees of freedom $v$ and basis functions $N$, and thus the choice of basis functions is essential for how good the discrete estimated solution is.

Further, strains are given as derivatives of the displacements. Using the assumption over, we get

$$\varepsilon = \partial u = Bv, \qquad (13.22)$$

where

$$B = \partial N \, . \qquad (13.23)$$

The symbol $\partial$ denotes an operator, which for the general 3D case reads

$$\partial = \begin{bmatrix} \frac{\partial}{\partial x} & 0 & 0 \\ 0 & \frac{\partial}{\partial y} & 0 \\ 0 & 0 & \frac{\partial}{\partial z} \\ \frac{\partial}{\partial y} & \frac{\partial}{\partial x} & 0 \\ 0 & \frac{\partial}{\partial z} & \frac{\partial}{\partial y} \\ \frac{\partial}{\partial z} & 0 & \frac{\partial}{\partial x} \end{bmatrix}.$$

Bar, beam, and plate elements will have other expressions for $\partial$; the purpose here is to introduce a unified notation.

Inserting (13.21)–(13.23) into (13.20), and using the principle of minimum potential energy, we end up with a linear system. Applying the principle to a single element gives the element matrix

$$A^{(e)} = \int_{\Omega_e} B^T D B \, d\Omega.$$

The element load vector yields

$$b^{(e)} = \int_{\Omega_e} B^T D \varepsilon_0 d\Omega - \int_{\Omega_e} B^T \sigma_0 d\Omega + \int_{\Omega_e} N^T F d\Omega + \int_{\partial\Omega_e} N^T t d\Gamma. \quad (13.24)$$

We remark that contributions to the element load vector are only from loads acting *on* the element, that is, *between* the element nodes.

For the rest of this chapter we are going to focus on the structural element types implemented in Diffpack, that is, bar elements, beam elements, frame elements, and triangular thin plate elements.

## 13.5   Degrees of Freedom and Basis Functions

*Basis Function Requirements.* In the previous section we assumed that the continuous displacement field ($u$) in an element could be expressed by basis functions and local element degrees of freedom. For the structural elements implemented in Diffpack the basis functions are polynomials that interpolate displacements and rotations between the element nodal points. The polynomials depend on the element type, i.e., the number of nodes, and number and type of degrees of freedom.

For elements having displacement degrees of freedom only, the basis functions must fulfill

- $N_i = 1$ at node $i$,
- $N_i = 0$ at all other nodal points.

For elements having both displacement and rotational degrees of freedom, we have the following requirements:

- Displacement degrees of freedom:
  - $N_i = 1$ at node $i$,
  - $N_i = 0$ at all other nodal points,
  - $\frac{\partial N_i}{\partial x_j} = 0$ at all nodal points; $j = 1, 2, 3$.
- Rotational degrees of freedom:
  - $\frac{\partial N_i}{\partial x_j} = 1$ at node $i$; $j = 1, 2, 3$,
  - $\frac{\partial N_i}{\partial x_j} = 0$ at all other nodal points; $j = 1, 2, 3$,
  - $N_i = 0$ at all nodal points.

### 13.5.1   Bar Elements

*Bar Element in 1D.* For one-dimensional problems the local $(x)$ and the global $(\bar{x})$ axis point in the same direction. The local degrees of freedom are $v = \begin{bmatrix} v_{x1}, v_{x2} \end{bmatrix}^T$, and the global degrees of freedom are $\bar{v} = \begin{bmatrix} \bar{v}_{x1}, \bar{v}_{x2} \end{bmatrix}^T$. The nodes are located at the end points. For this element, with length $L$, the basis functions are given in (13.3).

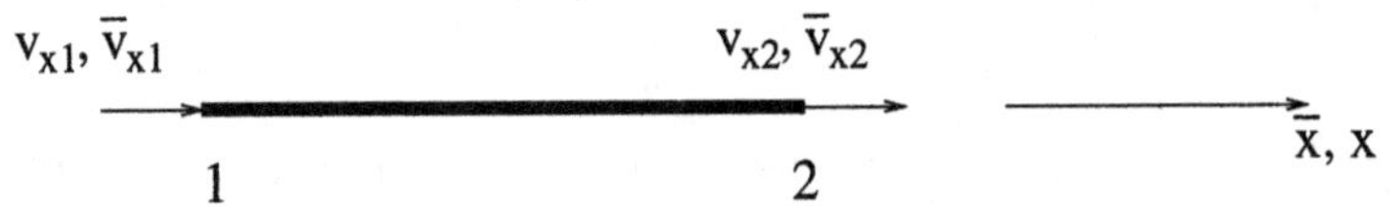

**Fig. 13.3.** 1D bar element; node numbering and degrees of freedom.

*Bar Element in 2D.* In two dimensions the bar element is described by a total of four translational degrees of freedom, referred to the global system; two at each of the two nodes, see Figure 13.4. The global degrees of freedom

$$\bar{v} = \begin{bmatrix} \bar{v}_{x1}, \bar{v}_{y1}, \bar{v}_{x2}, \bar{v}_{y2} \end{bmatrix}^T$$

can be expressed by a transformation matrix and the local degrees of freedom:

$$v = T\bar{v}. \tag{13.25}$$

The suitable transformation matrix is in this case

$$T = \begin{bmatrix} \cos\theta & \sin\theta & 0 & 0 \\ 0 & 0 & \cos\theta & \sin\theta \end{bmatrix},$$

where $\theta$ is the angle between positive global $\bar{x}$ and positive local longitudinal axis. In this way, we map the 1D bar element from one to two space dimensions. The mapping consists of a pure rotation plus a translation. For computational purposes, only the rotations matter.

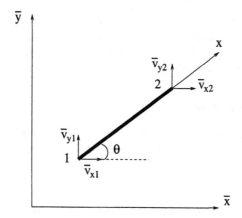

**Fig. 13.4.** Bar element in 2D; node numbering and global degrees of freedom.

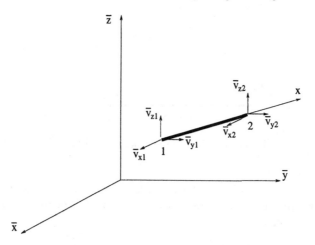

**Fig. 13.5.** Bar element in 3D; node numbering and global degrees of freedom.

*Bar Element in 3D.* In three dimensions the local element $x$ axis generally forms different angles with all three global axes, see Figure 13.5. This element is described by a total of six translational degrees of freedom; three at each of the two nodes:

$$\bar{v} = \left[ \bar{v}_{x1}, \bar{v}_{y1}, \bar{v}_{z1}, \bar{v}_{x2}, \bar{v}_{y2}, \bar{v}_{z2} \right]^{T} .$$

As for the 2D case, we need a transformation matrix for mapping the local element to the global coordinate system:

$$T = \begin{bmatrix} l_1 & m_1 & n_1 & 0 & 0 & 0 \\ 0 & 0 & 0 & l_1 & m_1 & n_1 \end{bmatrix},$$

where $l_1 = \cos(\bar{x}, x)$, $m_1 = \cos(\bar{y}, x)$, and $n_1 = \cos(\bar{z}, x)$, that is, the angles between positive global axes and the positive local longitudinal axis.

All computations are in the local coordinate system, and we only need the basis functions expressed in local coordinates.

### 13.5.2  Beam Elements

As shown in Figure 13.6, the two-dimensional beam element is expressed by displacement (translational) and rotational degrees of freedom:

$$v = \left[\, v_{z1},\, v_{\theta 1},\, v_{z2},\, v_{\theta 2} \,\right]^T .$$

For this element type, no change in element length is permitted, i.e., axial deformations are neglected. The beam element has two nodes; one at each

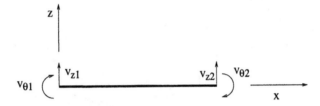

**Fig. 13.6.** Beam element in 2D; node numbering and local degrees of freedom.

end of the element. Node 1 is localized at $x = 0$ and node 2 at $x = L$ (as for bar elements). In global coordinates the number of degrees of freedom for the element is six,

$$\bar{v} = \left[\, \bar{v}_{x1},\, \bar{v}_{z1},\, \bar{v}_{\theta 1},\, \bar{v}_{x2},\, \bar{v}_{z2},\, \bar{v}_{\theta 2} \,\right]^T$$

resulting in a $4 \times 6$ transformation matrix for the correspondence between local and global degrees of freedom:

$$T = \begin{bmatrix} -\sin\theta & \cos\theta & 0 & 0 & 0 & 0 \\ 0 & 0 & 1 & 0 & 0 & 0 \\ 0 & 0 & 0 & -\sin\theta & \cos\theta & 0 \\ 0 & 0 & 0 & 0 & 0 & 1 \end{bmatrix},$$

where $\theta$ (as for the 2D bar element) is the angle between the positive global $\bar{x}$ axis and the positive local longitudinal axis.

The appropriate basis functions for this element (having four degrees of freedom in local coordinates) are given in (13.9)–(13.12).

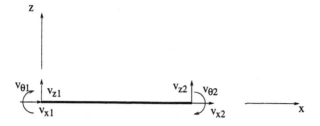

**Fig. 13.7.** Frame element in 2D; node numbering and local degrees of freedom.

### 13.5.3    Frame Elements

The frame element extends the beam element with elongational degrees of freedom, i.e., the element allows bending *and* changes in the element length [7]. We can view the 2D frame element as a superposition of the 2D bar and the 2D beam element. The displacement field for the frame element can be expressed as the sum of the basis functions and degrees of freedom for the beam element and the basis functions and degrees of freedom for the bar element:

$$u \approx \hat{u} = N_1^{beam} v_{z1} + N_2^{beam} v_{\theta 1} + N_3^{beam} v_{z2} + N_4^{beam} v_{\theta 2}$$
$$+ N_1^{bar} v_{x1} + N_2^{bar} v_{x2},$$

where the basis functions for the beam element are given in (13.9)–(13.12) and the basis functions for the bar element are found in (13.3). By ordering the degrees of freedom in a suitable way we get the frame element type described by a total of six basis functions and six degrees of freedom, referred to local coordinates; two translational and one rotational degree of freedom at each of the two nodes, see Figure 13.7. We form

$$N = \left[ N_1^{bar}, N_1^{beam}, N_2^{beam}, N_2^{bar}, N_3^{beam}, N_4^{beam} \right]^T$$

and

$$v = \left[ v_{x1}, v_{z1}, v_{\theta 1}, v_{x2}, v_{z2}, v_{\theta 2} \right]^T .$$

As for the beam element, the frame element is expressed by (a total of) six degrees of freedom in global coordinates, as depicted in Figure 13.8. By enlarging the transformation matrices for the bar element $(2 \times 2 \rightarrow 6 \times 6)$ and the beam element $(4 \times 6 \rightarrow 6 \times 6)$, we get the transformation matrix for the relation between the local and global degrees of freedom:

$$T = \begin{bmatrix} \cos\theta & \sin\theta & 0 & 0 & 0 & 0 \\ -\sin\theta & \cos\theta & 0 & 0 & 0 & 0 \\ 0 & 0 & 1 & 0 & 0 & 0 \\ 0 & 0 & 0 & \cos\theta & \sin\theta & 0 \\ 0 & 0 & 0 & -\sin\theta & \cos\theta & 0 \\ 0 & 0 & 0 & 0 & 0 & 1 \end{bmatrix} .$$

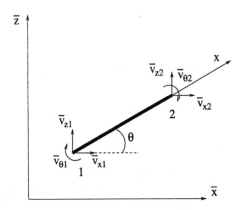

**Fig. 13.8.** Beam and/or frame element in 2D; node numbering and global degrees of freedom.

*Three-dimensional Element Types.* For the completeness of the presentation of (bar and) beam elements, it should also be mentioned that there is a three-dimensional beam element type. We do not go into details about this element type, but refer to [2].

### 13.5.4   DKT Plate Elements

Finally, we will take a look at the degrees of freedom and basis functions for the triangular plate bending element. The thickness of the plate is assumed to be small compared to the other element dimensions, i.e., the plate is thin. Kirchhoff has established a theory for thin plates, such that the three-dimensional stress and deformation state can be described by a two-dimensional mathematical model. For a thorough description of thin plates and the assumptions made, consult [7]. It is convenient to work with the element directly in *global* coordinates.

*Element with Twelve Degrees of Freedom.* We start by assuming that the displacements can be expressed by

$$u = z\beta_x(x,y), \qquad v = z\beta_y(x,y) \qquad \text{and} \qquad w = w(x,y),$$

where $\beta_x$ and $\beta_y$ are the rotations of the plate in the $xz$- and $yz$-planes, respectively, and $z$ is the coordinate value in the transverse direction. Furthermore, we assume that the rotations can be described by basis functions and rotational degrees of freedom at each node. The element has corner nodes and also nodes on each of the sides, that is, a total of six:

$$\beta_x = \sum_{i=1}^{6} N_i \beta_{x_i} \qquad \text{and} \qquad \beta_y = \sum_{i=1}^{6} N_i \beta_{y_i}.$$

The basis functions may be written as

$$N_1 = 2(1 - \xi - \eta)(\tfrac{1}{2} - \xi - \eta),$$
$$N_2 = \xi(2\xi - 1),$$
$$N_3 = \eta(2\eta - 1),$$
$$N_4 = 4\xi\eta,$$
$$N_5 = 4\eta(1 - \xi - \eta),$$
$$N_6 = 4\xi(1 - \xi - \eta),$$

where $\xi$ and $\eta$ are area coordinates. This gives us an element with a total of twelve rotational degrees of freedom; $\beta_{x_i}$ and $\beta_{y_i}$, $i = 1, 2, 3, 4, 5, 6$, see Figure 13.9a [4,7]. These are the primary unknowns for the element.

*Element with Nine Degrees of Freedom.* A computationally more efficient element, having $w$ and its two derivatives as primary unknowns at the corners of the triangular element, see Figure 13.9b, can be derived along the lines of Batoz, Bathe, and Ho [4]. A new notation is introduced:

$$\beta_x = \mathbf{H}_x^T(\xi, \eta)\bar{v}, \qquad (13.26)$$
$$\beta_y = \mathbf{H}_y^T(\xi, \eta)\bar{v}. \qquad (13.27)$$

The new basis functions are given by

$$H_{x1} = \frac{3}{2}((x_{31}/l_{31}^2)N_5 - (x_{12}/l_{12}^2)N_6),$$

$$H_{x2} = \frac{3}{4}((x_{31}y_{31}/l_{31}^2)N_5 + (x_{12}y_{12}/l_{12}^2)N_6),$$

$$H_{x3} = N_1 - ((\tfrac{1}{4}x_{31}^2 - \tfrac{1}{2}y_{31}^2)/l_{31}^2)N_5 - ((\tfrac{1}{4}x_{12}^2 - \tfrac{1}{2}y_{12}^2)/l_{12}^2)N_6,$$

$$H_{x4} = \frac{3}{2}((x_{12}/l_{12}^2)N_6 - (x_{23}/l_{23}^2)N_4),$$

$$H_{x5} = \frac{3}{4}((x_{12}y_{12}/l_{12}^2)N_6 + (x_{23}y_{23}/l_{23}^2)N_4),$$

$$H_{x6} = N_2 - ((\tfrac{1}{4}x_{12}^2 - \tfrac{1}{2}y_{12}^2)/l_{12}^2)N_6 - ((\tfrac{1}{4}x_{23}^2 - \tfrac{1}{2}y_{23}^2)/l_{23}^2)N_4,$$

$$H_{x7} = \frac{3}{2}((x_{23}/l_{23}^2)N_4 - (x_{31}/l_{31}^2)N_5),$$

$$H_{x8} = \frac{3}{4}((x_{23}y_{23}/l_{23}^2)N_4 + (x_{31}y_{31}/l_{31}^2)N_5),$$

$$H_{x9} = N_3 - ((\tfrac{1}{4}x_{23}^2 - \tfrac{1}{2}y_{23}^2)/l_{23}^2)N_4 - ((\tfrac{1}{4}x_{31}^2 - \tfrac{1}{2}y_{31}^2)/l_{31}^2)N_5,$$

and

$$H_{y1} = \frac{3}{2}((y_{31}/l_{31}^2)N_5 - (y_{12}/l_{12}^2)N_6),$$

$$H_{y2} = -N_1 + ((\frac{1}{4}y_{31}^2 - \frac{1}{2}x_{31}^2)/l_{31}^2)N_5 + ((\frac{1}{4}y_{12}^2 - \frac{1}{2}x_{12}^2)/l_{12}^2)N_6,$$

$$H_{y3} = -\frac{3}{4}((x_{31}y_{31}/l_{31}^2)N_5 + (x_{12}y_{12}/l_{12}^2)N_6),$$

$$H_{y4} = \frac{3}{2}((x_{12}/l_{12}^2)N_6 - (x_{23}/l_{23}^2)N_4),$$

$$H_{y5} = -N_2 + ((\frac{1}{4}y_{12}^2 - \frac{1}{2}x_{12}^2)/l_{12}^2)N_6 + ((\frac{1}{4}y_{23}^2 - \frac{1}{2}x_{23}^2)/l_{23}^2)N_4,$$

$$H_{y6} = -\frac{3}{4}((x_{12}y_{12}/l_{12}^2)N_6 + (x_{23}y_{23}/l_{23}^2)N_4),$$

$$H_{y7} = \frac{3}{2}((y_{23}/l_{23}^2)N_4 - (y_{31}/l_{31}^2)N_5),$$

$$H_{y8} = -N_3 + ((\frac{1}{4}y_{23}^2 - \frac{1}{2}x_{23}^2)/l_{23}^2)N_4 + ((\frac{1}{4}y_{31}^2 - \frac{1}{2}x_{31}^2)/l_{31}^2)N_5,$$

$$H_{y9} = -\frac{3}{4}((x_{23}y_{23}/l_{23}^2)N_4 + (x_{31}y_{31}/l_{31}^2)N_5).$$

The basis functions $N_i$ ($i = 1, 2, 3, 4, 5, 6$) are the same as given previously, and $x_{ij} = x_i - x_j$, $y_{ij} = y_i - y_j$, and $l_{ij}^2 = (x_{ij}^2 + y_{ij}^2)$, where $(x_i, y_i)$ are the coordinates of the element node in the global system. Moreover $\bar{v}$ is the vector of degrees of freedom:

$$\bar{v} = \left[ w_1, w_{1,y}, -w_{1,x}, w_2, w_{2,y}, -w_{2,x}, w_3, w_{3,y}, -w_{3,x} \right]^T,$$

(where $w_{i,x} = \frac{\partial w_i}{\partial x}$, and $w_{i,y} = \frac{\partial w_i}{\partial y}$; $i = 1, 2, 3$). Different material types will be discussed in the next section. However, it should be mentioned here that the basis functions for the plate element are the same for both isotropic and orthotropic materials.

For the DKT plate element the stiffness matrix and the load vector are given directly in global coordinates. We therefore do not need any transformation matrix in this case.

## 13.6   Material Types and Elasticity Matrices

As we saw in Section 13.4, the expressions for the stiffness matrices and load vectors contain the matrix of material properties $D$. This matrix depends on the problem of consideration, i.e., the element type and the material type.

Now we are going to take a closer look at both isotropic and orthotropic materials employed for the structural elements implemented in Diffpack. In this context, orthotropic materials refer to fiber composites with unidirectional, continuous fibers. We restrict ourselves to linear elastic materials.

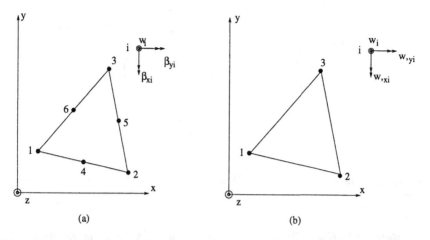

**Fig. 13.9.** Kirchhoff triangular discrete plate bending element (DKT); node numbering and global degrees of freedom. (a) Element with twelve rotational degrees of freedom as primary unknowns; (b) element with $w$ and its derivatives, a total of nine degrees of freedom, as primary unknowns.

### 13.6.1    Linear Elastic, Isotropic Material

The $D$ matrix in general, isotropic, linear elastic 3D problems is given in (13.15). It should be mentioned that this matrix contains two independent material coefficients; $\nu$ and $E$.

For thin plates it is assumed that the theory of Kirchhoff is valid, which means that 2D *plane stress* conditions are presumed [7]. Thus, the material stiffness matrix is given in (13.16). Still there are two independent material parameters.

Finally, for the bar, beam, and frame elements we have a length dimension, but no transverse dimension. Therefore, we do not need to take $\nu$ into account. The elastic properties are thus expressed by $E$ only.

### 13.6.2    Orthotropic Material; Fiber Composite

Orthotropic materials are made of two or more distinct materials put together. For orthotropic materials the parameters are direction dependent. Fiber composites are built of several layers, where each layer is composed of fiber reinforcements embedded in a continuous phase termed the matrix [1]. In general, all layers can be made of different materials. Moreover, the direction of the fibers in each layer, according to common axes, influences on the properties of the whole composite. This makes the expressions for the effective material properties for the fiber composite more complicated, and more than two material coefficients are required.

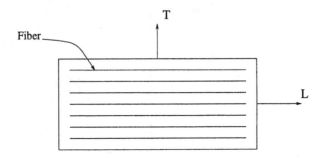

**Fig. 13.10.** The fiber direction in a composite build of unidirectional, continuous fibers.

*The Material Stiffness Matrix for a Single Layer.* In the same way as for the isotropic thin plate, we assume plane stress conditions to prevail. Thus, for a single layer of unidirectional, continuous fibers, referred to local axes for the layer, it can be shown that the material stiffness matrix may be expresses as

$$Q_k = \begin{bmatrix} Q_{11} & Q_{12} & 0 \\ Q_{12} & Q_{22} & 0 \\ 0 & 0 & Q_{66} \end{bmatrix},$$

where $k$ is the layer number, and

$$Q_{11} = \frac{E_L}{1 - \nu_{LT}\nu_{TL}},$$

$$Q_{22} = \frac{E_T}{1 - \nu_{LT}\nu_{TL}},$$

$$Q_{12} = \frac{\nu_{LT}E_T}{1 - \nu_{LT}\nu_{TL}} = \frac{\nu_{TL}E_L}{1 - \nu_{LT}\nu_{TL}},$$

$$Q_{66} = G_{LT}.$$

The fiber direction is called the longitudinal direction ($L$), and the direction normal to the fibers is the transverse direction ($T$), see Figure 13.10. In the expressions above, $E_L$ and $E_T$ are the elastic moduli in the longitudinal and the transverse directions, respectively. The quantities $\nu_{LT}$ and $\nu_{TL}$ are Poisson's ratio numbers; $\nu_{LT}$ (*major Poisson ratio*), giving transverse strain caused by longitudinal stress, and $\nu_{TL}$ (*minor Poisson ratio*), giving the longitudinal strain resulting from a transverse stress. The four material constants are related through

$$\frac{\nu_{LT}}{E_L} = \frac{\nu_{TL}}{E_T}.$$

Finally, $G_{LT}$ is the shear modulus. Therefore, four independent material parameters (for each layer) are required for such orthotropic problems in 2D.

Using transformation matrices [1], it can be shown that the matrix for a single layer, referred to global $(\bar{x}, \bar{y})$ axes, is

$$\bar{Q}_k = \begin{bmatrix} \bar{Q}_{11} & \bar{Q}_{12} & \bar{Q}_{16} \\ \bar{Q}_{12} & \bar{Q}_{22} & \bar{Q}_{26} \\ \bar{Q}_{16} & \bar{Q}_{26} & \bar{Q}_{66} \end{bmatrix}, \tag{13.28}$$

where

$$\bar{Q}_{11} = Q_{11} \cos^4 \theta + Q_{22} \sin^4 \theta + 2(Q_{12} + 2Q_{66}) \sin^2 \theta \cos^2 \theta,$$
$$\bar{Q}_{22} = Q_{11} \sin^4 \theta + Q_{22} \cos^4 \theta + 2(Q_{12} + 2Q_{66}) \sin^2 \theta \cos^2 \theta,$$
$$\bar{Q}_{66} = (Q_{11} + Q_{22} - 2Q_{12} - 2Q_{66}) \sin^2 \theta \cos^2 \theta + Q_{66}(\cos^4 \theta + \sin^4 \theta),$$
$$\bar{Q}_{12} = (Q_{11} + Q_{22} - 4Q_{66}) \sin^2 \theta \cos^2 \theta + Q_{12}(\cos^4 \theta + \sin^4 \theta),$$
$$\bar{Q}_{16} = (Q_{11} - Q_{12} - 2Q_{66}) \cos^3 \theta \sin \theta - (Q_{22} - Q_{12} - 2Q_{66}) \cos \theta \sin^3 \theta,$$
$$\bar{Q}_{26} = (Q_{11} - Q_{12} - 2Q_{66}) \cos \theta \sin^3 \theta - (Q_{22} - Q_{12} - 2Q_{66}) \cos^3 \theta \sin \theta.$$

In these expressions $\theta$ is the angle between the positive global $\bar{x}$ axis and positive local $x$ axis for the layer. Again, $k$ is the layer number. The stress-strain relation for one single layer of the composite can then be expressed as [1, page 190]

$$\sigma_k = \bar{Q}_k \varepsilon.$$

To get the material properties for the whole laminate, we sum up for all the layers. (Actually we integrate over the plate thickness.) We get back to this in Section 13.9.2.

## 13.7    Element Matrices in Local Coordinates

In the context of the finite element method in structural mechanics it is common practice to start by establishing stiffness matrices for the elements with respect to local element coordinates. These matrices depend on the element and material types. After having defined the stiffness matrices in the local system, it is time to transform the contributions into values related to the global structural system. In the present section we will derive expressions for various local element stiffness matrices.

### 13.7.1    Elements with Linear, Isotropic Material Properties

*Bar Element.* As we have seen in Section 13.2 the element stiffness matrices typically contain contributions from the material properties, derivatives of the basis functions (expressed by a $B$ matrix), and an integral over the geometry of the element. For the bar element the $B$ matrix may be expressed by

$$B = \frac{d}{dx} N = \begin{bmatrix} -\frac{1}{L} & \frac{1}{L} \end{bmatrix},$$

where $N$ is the vector containing the basis functions, which are defined in (13.3). Thus, in agreement with the derivations in Section 13.2, the element stiffness matrix is written as

$$A^{(e)} = \int_0^L B^T E A B \, dx = \frac{EA}{L} \begin{bmatrix} 1 & -1 \\ -1 & 1 \end{bmatrix},$$

where $A$ is the element cross section area (assumed constant), $E$ is Young's modulus of elasticity, and $L$ is the element length.

*Beam Element.* In a similar way as for the bar element, we stated in Section 13.3 that the $B$ matrix for the beam element may be expressed as

$$B = \frac{d^2}{dx^2} N = \left[ \left( -\frac{6}{L^2} + \frac{12x}{L^3} \right), \left( \frac{4}{L} - \frac{6x}{L^2} \right), \left( \frac{6}{L^2} - \frac{12x}{L^3} \right), \left( \frac{2}{L} - \frac{6x}{L^2} \right) \right]. \quad (13.29)$$

The components in vector $N$ giving the basis functions, are found in (13.9)–(13.12). In this case the element stiffness matrix yields

$$A^{(e)} = \int_0^L B^T E I B \, dx = \frac{EI}{L^3} \begin{bmatrix} 12 & -6L & -12 & -6L \\ -6L & 4L^2 & 6L & 2L^2 \\ -12 & 6L & 12 & 6L \\ -6L & 2L^2 & 6L & 4L^2 \end{bmatrix}.$$

The bending stiffness $EI$ is assumed to be constant for the element. Moreover, $I = \int_{\partial\Omega} z^2 d\Gamma$ is the second moment of area, and $L$ is the element length.

*Plane Frame Element.* Finally, we have the plane frame element, which is a combination of the bar and beam elements. The element stiffness matrix for this kind of element may be obtained by a superposition of the corresponding matrices for the bar and beam elements. However, the latter matrices must be modified due to the extended number of degrees of freedom for the frame element (a total of six degrees of freedom). The bar element matrix with size $2 \times 2$ has to be enlarged to $6 \times 6$. Furthermore, the beam element matrix with size $4 \times 4$ must be enlarged to $6 \times 6$. Thus, the matrix reads

$$A^{(e)} = \frac{AE}{L} \begin{bmatrix} 1 & 0 & 0 & -1 & 0 & 0 \\ 0 & 0 & 0 & 0 & 0 & 0 \\ 0 & 0 & 0 & 0 & 0 & 0 \\ -1 & 0 & 0 & 1 & 0 & 0 \\ 0 & 0 & 0 & 0 & 0 & 0 \\ 0 & 0 & 0 & 0 & 0 & 0 \end{bmatrix}$$

$$+ \frac{EI}{L^3} \begin{bmatrix} 0 & 0 & 0 & 0 & 0 & 0 \\ 0 & 12 & -6L & 0 & -12 & -6L \\ 0 & -6L & 4L^2 & 0 & 6L & 2L^2 \\ 0 & 0 & 0 & 0 & 0 & 0 \\ 0 & -12 & 6L & 0 & 12 & 6L \\ 0 & -6L & 2L^2 & 0 & 6L & 4L^2 \end{bmatrix}.$$

### 13.7.2   Elements with Orthotropic Material Properties

A bar element, which is mathematically one-dimensional (in local element coordinates), is only capable of carrying tension and compression loads in the longitudinal direction. Two-dimensional beam and frame elements have no transverse dimension, and may only bend in one plane. For bar, beam, and frame elements the fibers must be directed in the same direction, that is, along the element (longitudinal) axis. This indicates that we do not need separate orthotropic bar, beam, or frame elements. Instead we can use the isotropic element types, with $E = E_L$. (We then have to use the same type of fibers for the whole element.)

### 13.7.3   DKT Plate Elements

For the DKT plate elements it is more convenient to express the stiffness matrix in global coordinates directly. The element stiffness matrices for the DKT plate elements are described in Section 13.9.2.

## 13.8   Element Load Vectors in Local Coordinates

The general expression for the element load vector is given in (13.24). It is seen that it contains contributions from initial stresses and strains, body forces, and surface tractions. Besides body forces and surface tractions we only take temperature changes into account. Temperature effects can be expressed either as initial stresses or initial strains [7]. We choose to express these loads by initial strains, and put $\sigma_0$ equal to zero. To indicate that the initial strains ($\varepsilon_0$) are caused by temperature changes, the vector is named $\tau$ [6].

The load vector depends on the physical problem under consideration, and different load cases (naturally) give rise to dissimilar contributions. Now we are going to take a closer look at the load vector for the different structural elements implemented. The load contributions that are considered and implemented, are limited to:

- change in element length, influenced by temperature changes,
- linearly distributed loads,
- concentrated loads (also bending moments), and,
- gravity forces.

The different load types are shown in Figures 13.11-13.16. External point loads (forces and moments) acting at nodal points, are not included in the local element load vectors. Instead, these effects are directly implemented into the global system load vector.

### 13.8.1  Bar Elements

*Temperature Loads.*  Bar elements are only capable of carrying loads acting in the element longitudinal direction. Let us first take a look at the case where the element is exposed to a temperature change $\Delta T = T - T_0$, $T_0$ being the reference temperature, see Figure 13.11. In practice, initial strains (like temperature) are often constant in the longitudinal direction and varies in the transverse direction. Variations in the transverse direction result in bending, which is not allowed for this element type. Moreover, the element is mathematically one-dimensional. Thus, in this case we set the temperature change $\Delta T$ constant in all directions. The associated thermal normal strain is $\alpha \Delta T$, i.e., $\tau = [\alpha \Delta T]$, where $\alpha$ is the thermal expansion coefficient. For the bar element it can be shown that this contribution ends up with [7]

$$b_\tau^{(e)} = \int_0^L \frac{1}{L} \begin{bmatrix} -1 \\ 1 \end{bmatrix} E\alpha \Delta T A\, dx = EA\alpha \Delta T \begin{bmatrix} -1 \\ 1 \end{bmatrix},$$

where $A$ is the area of the cross section.

The temperature of the bar element is taken to be constant. However, we may allow for different temperature in different elements. Such a piecewise constant temperature field seems unphysical. The natural finite element representation of the temperature field is a thermal counterpart to the bar element, i.e., a linearly varying scalar field along the element, corresponding to linear elements for a two-point boundary value problem (the temperature and the longitudinal displacement fulfill the same differential equation along the element). A constant temperature in the bar element is therefore a pure numerical approximation. A scalar element temperature value $\Delta T$ is conveniently treated as another physical parameter for the element (along with $L$, $E$, $A$, $\alpha$). Alternatively, one could add two nodal temperature values to the element's parameter set, to describe a linearly varying temperature field, but this must be done consistently such that two elements sharing a node have the same temperature value at this node.

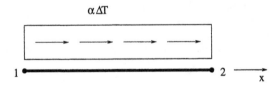

**Fig. 13.11.** Temperature load on a bar.

*Linearly Varying Tangential Loads.*  Bar elements can also carry surface tractions. First, we consider a linearly varying tangential load:

$$q_x = \frac{L - x}{L} q_{x1} + \frac{x}{L} q_{x2}. \tag{13.30}$$

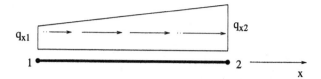

**Fig. 13.12.** Linearly distributed axial load.

Here, $q_x$ is the traction load per unit length in the axial direction, and $q_{x1}$ and $q_{x2}$ are the values at the nodes 1 and 2, respectively, which is illustrated in Figure 13.12. In this case we get the expression

$$b_{q_x}^{(e)} = \int_0^L N^T q_x \, dx = \frac{L}{6} \begin{bmatrix} 2q_{x1} + q_{x2} \\ q_{x1} + 2q_{x2} \end{bmatrix}.$$

*Concentrated Tangential Loads.* In addition to linearly distributed surface tractions, the element can be exposed to concentrated tangential loads. We examine a load $F_x$, acting at $x = \tilde{x}$ ($0 < \tilde{x} < L$), see Figure 13.13. In this case the total force is concentrated at a single point. Therefore, the traction force is

$$t = F_x \delta(x - \tilde{x}) e_x,$$

where $\delta(x - \tilde{x})$ is the standard delta function centered at $x = \tilde{x}$, and $e_x$ denotes the unit vector in the axial direction. Thus, the load contribution in the axial direction is

$$b_{F_x}^{(e)} = \int_0^L F_x \delta(x - \tilde{x}) N^T \, dx = N^T|_{x=\tilde{x}} F_x = F_x \begin{bmatrix} \frac{L-\tilde{x}}{L} \\ \frac{\tilde{x}}{L} \end{bmatrix}.$$

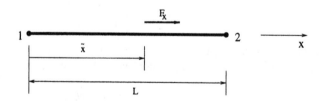

**Fig. 13.13.** Concentrated axial load, $0 < \tilde{x} < L$.

*The Gravity Force.* The gravity force, which is the only body force that is taken into consideration in this chapter, is expressed as the product of the total mass of the bar, $m$, and the acceleration due to the gravity $g$, $G = mg$. Remark that, for bar elements, we only get contribution from the part acting

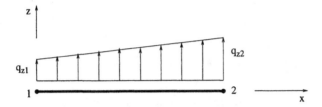

**Fig. 13.14.** Linearly distributed transverse load.

in the longitudinal direction of the element, i.e., $G_x = mg\cos\phi$, where the angle $\phi$ depends on the element orientation compared to the direction of the gravitational force. We also remark that the angle $\phi$ is different from the angle indicating the orientation of the element in global axes, see Section 13.5. We assume that the gravity force can be treated as a concentrated load acting in the mass center of the bar element, $x = \tilde{x}$. Again, by using the delta function, we get:

$$b_{G_x}^{(e)} = N^T|_{x=\tilde{x}}AG_x = AG_x \begin{bmatrix} \frac{L-\tilde{x}}{L} \\ \frac{\tilde{x}}{L} \end{bmatrix}.$$

### 13.8.2    Beam Elements

For beam elements axial deformations are neglected. Thus, loads giving change in element length is not possible. Beam elements can, however, carry concentrated moment loads, as well as concentrated and distributed transverse loads, and body forces.

*Linearly Distributed Transverse Loads.* First we consider a linearly distributed transverse load. The load is directed in the $z$ direction, as illustrated in Figure 13.14:

$$q_z = \frac{L-x}{L}q_{z1} + \frac{x}{L}q_{z2},$$

where $q_z$ is load per unit length in the transverse direction, and $q_{z1}$ and $q_{z2}$ are the values at the nodes 1 and 2, respectively. The load vector becomes

$$b_{q_z}^{(e)} = \int_0^L N^T q_z dx = \frac{L}{60} \begin{bmatrix} 21q_{z1} + 9q_{z2} \\ -3Lq_{z1} - 2Lq_{z2} \\ 9q_{z1} + 21q_{z2} \\ 2Lq_{z1} + 3Lq_{z2} \end{bmatrix}.$$

*Concentrated Transverse Loads.* The beam element can also be subject to concentrated transverse loads. We examine a concentrated force $P_z$ (in the $z$

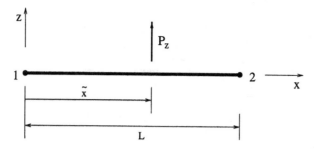

**Fig. 13.15.** Concentrated transverse force, $0 < \tilde{x} < L$.

direction) acting at $x = \tilde{x}$, see Figure 13.15. By employing the delta function for evaluating the integral (see the previous subsection), we end up with

$$
\boldsymbol{b}_{P_z}^{(e)} = \boldsymbol{N}|_{x=\tilde{x}}P_z = P_z
\begin{bmatrix}
1 - \frac{3\tilde{x}^2}{L^2} + \frac{2\tilde{x}^3}{L^3} \\
-\tilde{x} + \frac{2\tilde{x}^2}{L} - \frac{\tilde{x}^3}{L^2} \\
\frac{3\tilde{x}^2}{L^2} - \frac{2\tilde{x}^3}{L^3} \\
\frac{\tilde{x}^2}{L} - \frac{\tilde{x}^3}{L^2}
\end{bmatrix}.
$$

*Concentrated Moment Loads.* The beam element can also be loaded with concentrated bending moments, see Figure 13.16. A concentrated moment $M$, at $x = \tilde{x}$ $(0 < \tilde{x} < L)$, may be modeled as two point loads working in opposite directions at a distance $h$. Employing the delta function for each of the point loads and letting $h \to 0$, we end up with an expression for the derivative of the basis functions multiplied by the concentrated moment $M$:

$$
\boldsymbol{b}_{M}^{(e)} = \left[\frac{d\boldsymbol{N}}{dx}\right]_{x=\tilde{x}}^{T} M = M
\begin{bmatrix}
6(-\frac{\tilde{x}}{L^2} + \frac{\tilde{x}^2}{L^3}) \\
-1 + \frac{4\tilde{x}}{L} - \frac{3\tilde{x}^2}{L^2} \\
6(\frac{\tilde{x}}{L^2} - \frac{\tilde{x}^2}{L^3}) \\
\frac{2\tilde{x}}{L} - \frac{3\tilde{x}^2}{L^2}
\end{bmatrix}.
$$

*The Gravity Force.* The gravity force may also contribute to the element load vector. However, when dealing with beam elements, only the transverse component, $G_z = mg\sin\phi$, contributes. We assume, as for the bar element, that the gravity force may be implemented as a concentrated force acting at the element mass center $x = \tilde{x}$. We end up with

$$
\boldsymbol{b}_{G_z}^{(e)} = \boldsymbol{N}^T|_{x=\tilde{x}}AG_z = AG_z
\begin{bmatrix}
1 - \frac{3\tilde{x}^2}{L^2} + \frac{2\tilde{x}^3}{L^3} \\
-\tilde{x} + \frac{2\tilde{x}^2}{L} - \frac{\tilde{x}^3}{L^2} \\
\frac{3\tilde{x}^2}{L^2} - \frac{2\tilde{x}^3}{L^3} \\
\frac{\tilde{x}^2}{L} - \frac{\tilde{x}^3}{L^2}
\end{bmatrix}.
$$

Here $G_z$ acts in the positive local $z$ direction (same as $P_z$).

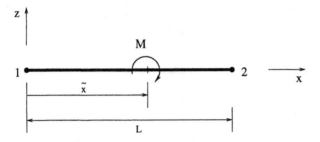

**Fig. 13.16.** Concentrated bending moment, $0 < \tilde{x} < L$.

### 13.8.3   Frame Elements

The frame element can be loaded with all load types described in the previous subsections for the bar and beam elements. The expressions are similar, so they are just listed in the following [7].

The contribution to the element load vector from temperature changes is

$$
\boldsymbol{b}_T^{(e)} = EA\alpha\Delta T
\begin{bmatrix}
-1 \\
0 \\
0 \\
1 \\
0 \\
0
\end{bmatrix}.
$$

Furthermore, the load vector due to a linearly distributed axial load $q_x$ is expressed as

$$
\boldsymbol{b}_{q_x}^{(e)} = \frac{L}{6}
\begin{bmatrix}
2q_{x1} + q_{x2} \\
0 \\
0 \\
q_{x1} + 2q_{x2} \\
0 \\
0
\end{bmatrix}.
$$

A concentrated axial force $F_x$, at $x = \tilde{x}$, yields

$$
\boldsymbol{b}_{F_x}^{(e)} = F_x
\begin{bmatrix}
\frac{L-\tilde{x}}{L} \\
0 \\
0 \\
\frac{\tilde{x}}{L} \\
0 \\
0
\end{bmatrix},
$$

while linearly distributed transverse loads gives

$$
b_{q_z}^{(e)} = \frac{L}{60}
\begin{bmatrix}
0 \\
21q_{z1} + 9q_{z2} \\
-3Lq_{z1} - 2Lq_{z2} \\
0 \\
9q_{z1} + 21q_{z2} \\
2Lq_{z1} + 3Lq_{z2}
\end{bmatrix}.
$$

The element load vector obtained by a concentrated transverse load $P_z$ takes the form

$$
b_{P_z}^{(e)} = P_z
\begin{bmatrix}
0 \\
1 - \frac{3\tilde{x}^2}{L^2} + \frac{2\tilde{x}^3}{L^3} \\
-\tilde{x} + \frac{2\tilde{x}^2}{L} - \frac{\tilde{x}^3}{L^2} \\
0 \\
\frac{3\tilde{x}^2}{L^2} - \frac{2\tilde{x}^3}{L^3} \\
\frac{\tilde{x}^2}{L} - \frac{\tilde{x}^3}{L^2}
\end{bmatrix},
$$

while a concentrated moment $M$ yields

$$
b_M^{(e)} = M
\begin{bmatrix}
0 \\
6(-\frac{\tilde{x}}{L^2} + \frac{\tilde{x}^2}{L^3}) \\
-1 + \frac{4\tilde{x}}{L} - \frac{3\tilde{x}^2}{L^2} \\
0 \\
6(\frac{\tilde{x}}{L^2} - \frac{\tilde{x}^2}{L^3}) \\
\frac{2\tilde{x}}{L} - \frac{3\tilde{x}^2}{L^2}
\end{bmatrix}.
$$

In the context of the frame element, both axial and transverse components of the gravity force contribute to the local load vector. Assuming again that the force due to gravity is applied at the mass center $x = \tilde{x}$, and that its components in the longitudinal and transverse directions are denoted $G_x$ and $G_z$, respectively, the load vector may be expressed by

$$
b_G^{(e)} = A
\begin{bmatrix}
G_x(\frac{L-\tilde{x}}{L}) \\
G_z(1 - \frac{3\tilde{x}^2}{L^2} + \frac{2\tilde{x}^3}{L^3}) \\
G_z(-\tilde{x} + \frac{2\tilde{x}^2}{L} - \frac{\tilde{x}^3}{L^2}) \\
G_x(\frac{\tilde{x}}{L}) \\
G_z(\frac{3\tilde{x}^2}{L^2} - \frac{2\tilde{x}^3}{L^3}) \\
G_z(\frac{\tilde{x}^2}{L} - \frac{\tilde{x}^3}{L^2})
\end{bmatrix}.
$$

### 13.8.4   DKT Plate Elements

As mentioned in Section 13.7.3, for the element stiffness matrix, we give the element contributions to the global linear system in global coordinates directly. The element load vectors for the DKT plate elements are given in Section 13.9.2.

## 13.9    Element Matrices and Vectors in Global Coordinates

### 13.9.1    Bar, Beam, and Frame Elements

Until now, the element stiffness matrix and the element load vector have been expressed in local coordinates. The stiffness relation for the element referred to the *local* system is

$$S = A^{(e)}v + b^{(e)},$$

where $S$ is a vector containing the element forces (i.e., the nodal forces), $A^{(e)}v$ are the elastic forces, and $b^{(e)}$ is the load vector.

The expressions need to be transformed to global axes before being put into the system stiffness matrix and system load vector. This includes enlarging the element stiffness matrix and the element load vector to the right size, and allowing for rotating the element according to the global axes. In Section 13.5 we introduced a transformation matrix $T$ (in (13.25)), which relates the local ($v$) and global ($\bar{v}$) degrees of freedom. (We remark that $T$ depends on the element type and dimension.) This transformation is a general rule that can be employed for the element matrix and the element load vector as well. For the element stiffness relation referred to the global system we have

$$\bar{S} = T^T S = T^T (A^{(e)}v + b^{(e)}) = T^T A^{(e)} T\bar{v} + T^T b^{(e)} = A^{\overline{(e)}}\bar{v} + b^{\overline{(e)}}.$$

The element stiffness matrix and the element load vector in global coordinates are hence expressed as

$$A^{\overline{(e)}} = T^T A^{(e)} T$$

and

$$b^{\overline{(e)}} = T^T b^{(e)},$$

respectively. The element contributions, referred to common global axes, are put into the system stiffness matrix and the system load vector.

### 13.9.2    DKT Plate Elements

*The Stiffness Matrix.* In contrast to bar, beam, and frame elements, it is more convenient to derive the stiffness matrix for plate elements directly in global coordinates. In this way the computational efficiency is increased, because we do not have to transform from local to global coordinates, and vice versa. Furthermore, for plate elements we are not primarily interested in forces and displacements referred to local coordinates.

The basis functions in (13.26) and (13.27) may be written as

$$\{H_x\} = [1, \quad \xi, \quad \eta, \quad \xi^2, \quad \xi\eta, \quad \eta^2]G_x,$$
$$\{H_y\} = [1, \quad \xi, \quad \eta, \quad \xi^2, \quad \xi\eta, \quad \eta^2]G_y,$$

where the matrices $G_x$ and $G_y$, both with size $6 \times 9$, are independent of the area coordinates[7].

The strains are given by

$$\varepsilon = z B \bar{v},$$

where $z$ is the coordinate normal to the (plate) plane, and $\bar{v}$ are the element degrees of freedom referred to the global system. Employing the expressions for $G_x$ and $G_y$ it can be shown that the $B$ matrix may be expressed as [4,5]

$$B = \begin{bmatrix} B_1 \\ B_2 \\ B_3 \end{bmatrix},$$

where

$$B_1 = \frac{1}{2A} [1 \quad \xi \quad \eta] \begin{bmatrix} y_{31}\{G_x\}_2 + y_{12}\{G_x\}_3 \\ 2y_{31}\{G_x\}_4 + y_{12}\{G_x\}_5 \\ y_{31}\{G_x\}_5 + 2y_{12}\{G_x\}_6 \end{bmatrix},$$

$$B_2 = \frac{1}{2A} [1 \quad \xi \quad \eta] \begin{bmatrix} x_{31}\{G_y\}_2 + x_{12}\{G_y\}_3 \\ 2x_{31}\{G_y\}_4 + x_{12}\{G_y\}_5 \\ x_{31}\{G_y\}_5 + 2x_{12}\{G_y\}_6 \end{bmatrix},$$

$$B_3 = \frac{1}{2A} [1 \quad \xi \quad \eta]$$

$$\times \begin{bmatrix} x_{31}\{G_x\}_2 + x_{12}\{G_x\}_3 + y_{31}\{G_y\}_2 + y_{12}\{G_y\}_3 \\ 2x_{31}\{G_x\}_4 + x_{12}\{G_x\}_5 + 2y_{31}\{G_y\}_4 + y_{12}\{G_y\}_5 \\ x_{31}\{G_x\}_5 + 2x_{12}\{G_x\}_6 + y_{31}\{G_y\}_5 + 2y_{12}\{G_y\}_6 \end{bmatrix}.$$

Here $\{G_x\}_i$ and $\{G_y\}_i$ indicate row number $i$ in $G_x$ and $G_y$, respectively, and $A$ is the area of the triangle,

$$2A = x_{21}y_{31} - x_{31}y_{21}.$$

Furthermore, $x_{ij} = x_i - x_j$ and $y_{ij} = y_i - y_j$, where $(x_i, y_i)$ are the coordinates of the element node in the global system.

*Isotropic Material.* The material property matrix is found in (13.16). The element stiffness matrix can be calculated as

$$A^{(e)} = \int_{\Omega_e} B^T D B z^2 d\Omega = \int_{\partial\Omega_e} B^T D_{iso} B d\Gamma, \qquad (13.31)$$

where $\Omega_e$ is the element with physical thickness (three-dimensional volume), and $\partial\Omega_e$ is the plate surface area (the two-dimensional plate finite element).

---

[7] Note that there is no connection between $G_x$ and $G_y$ and the gravity vector (from Section 13.8).

The matrix $\boldsymbol{D}_{iso}$ results from integrating over the thickness $t$ of the plate element:

$$\boldsymbol{D}_{iso} = \int_{-t/2}^{t/2} \boldsymbol{D}z^2 dz = \frac{Et^3}{12(1-\nu^2)} \begin{bmatrix} 1 & \nu & 0 \\ \nu & 1 & 0 \\ 0 & 0 & \frac{1-\nu}{2} \end{bmatrix}.$$

The element stiffness matrix in (13.31) has size $9 \times 9$.

*Orthotropic Material; Fiber Composite.* In the orthotropic case, the only difference in the expression for the stiffness matrix is another material property matrix. For fiber composites with unidirectional, continuous fibers, the material properties for a single layer are found in (13.28). The stiffness matrix for the whole composite (made of $n$ layers) now yields:

$$A^{(e)} = \int_{\partial \Omega_e} \boldsymbol{B}^T \boldsymbol{D}_{orto} \boldsymbol{B} d\Gamma,$$

where

$$\boldsymbol{D}_{orto} = \sum_{k=1}^{n} \left( \begin{bmatrix} \bar{Q}_{11} & \bar{Q}_{12} & \bar{Q}_{16} \\ \bar{Q}_{12} & \bar{Q}_{22} & \bar{Q}_{26} \\ \bar{Q}_{16} & \bar{Q}_{26} & \bar{Q}_{66} \end{bmatrix}_k \int_{h_{k-1}}^{h_k} z^2 dz \right)$$

$$= \frac{1}{3} \sum_{k=1}^{n} (\bar{Q}_{ij})_k (h_k^3 - h_{k-1}^3),$$

and $(h_k - h_{k-1})$ is the thickness of layer number $k$ [1]. The element stiffness matrix has the same size as in the isotropic case.

*The Element Load Vector.* The plate elements can allow nodal loads in addition to a distributed transverse load acting on the entire element. We have two types of nodal loads; loads in the vertical direction, and moment loads $M_x$ and $M_y$, as shown in Figure 13.17 for element node 3. For the distributed transverse load, we employ a concentrated load vector. The total load acting on an element, with load per unit square $q$ and area $A$, is $Aq$. We assume that the major part of the load acts in the vertical direction. One third of the total load is thus distributed to each node. Doing this is an approximation, which is called *lumping* [6]. The load contribution is shown i Figure 13.18, and yields:

$$b_q^{(e)} = \frac{Aq}{3} \{ 1, 0, 0, 1, 0, 0, 1, 0, 0 \}^T.$$

In a computer simulator, the load contributions are put directly into the system load vector.

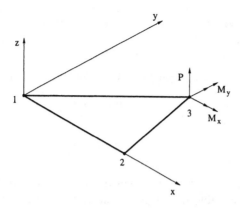

**Fig. 13.17.** Nodal loads for plate elements.

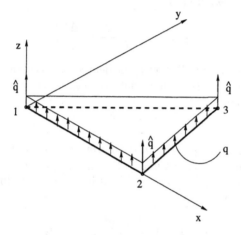

**Fig. 13.18.** Uniformly distributed normal load for plate elements, with $\hat{q} = \frac{Aq}{3}$.

## 13.10  Element Forces, Stresses, and Strains

The finite element formulation leads to a linear equation system for the global set of degrees of freedom. Having solved this system, we can calculate forces, stresses, and strains for each element.

### 13.10.1  Bar, Beam, and Frame Elements

*Element Forces.* The element forces, referred to global axes, may now be determined from the relation

$$\mathbf{S}^{(\overline{e})i} = \boldsymbol{A}^{(\overline{e})i}\bar{\boldsymbol{v}}^i + \boldsymbol{b}^{(\overline{e})i}.$$

For getting the element's nodal displacements (and/or rotations) in local coordinates, we use

$$v^i = T^i \bar{v}^i .$$

The element forces in local coordinates are calculated from

$$\mathbf{S}^{(e)i} = \mathbf{A}^{(e)i} v^i + \mathbf{b}^{(e)i},$$

where

$$\mathbf{b}^{(e)i} = (T^T)^{-1} \mathbf{b}^{(\bar{e})i} .$$

*Stresses and Strains for the Bar Element (Local Coordinates).* The bar elements are only capable of carrying axial loads. Therefore, we only have stresses and strains in the longitudinal direction for this element type. The axial strain may be written as

$$\varepsilon_{xx} = \frac{v_{x2} - v_{x1}}{L} = \frac{\Delta L}{L},$$

where $v_{x1}$ and $v_{x2}$ are the displacement degrees of freedom at nodes 1 and 2, respectively. For linear elastic materials we get the stress (in the longitudinal direction) $\sigma_{xx}$ from

$$\sigma_{xx} = E\varepsilon_{xx} . \tag{13.32}$$

*Stresses and Strains for the Beam and Frame Elements (Local Coordinates).* For beam elements the strains due to bending may be determined using the basis functions (actually the second derivative of the basis functions) and the element degrees of freedom:

$$\varepsilon = \{\varepsilon_{xx}\} = -z\frac{d^2}{dx^2}u = -z\mathbf{B}v,$$

where $z$ is the coordinate value in the transverse direction and $\mathbf{B}$ is given in (13.29). The stresses are found from (13.32).

As said in a previous section, the frame element is a superposition of the bar and beam elements. The tension/contraction and bending states are in this case decoupled, and stresses and strains may be found from the expressions for the bar and beam elements.

## 13.10.2   DKT Plate Elements

*Stresses and Strains.* For plate elements the strains due to bending read

$$\varepsilon = z\kappa,$$

where

$$\kappa = \mathbf{B}\bar{v} .$$

For isotropic materials the stresses are calculated employing

$$\sigma = D\varepsilon = zD\kappa,$$

where $D$ is given in (13.16). Further, for orthotropic materials the material properties varies through the thickness of the plate. Therefore, the stresses for a given layer $(k)$ are expressed by

$$\sigma_k = \bar{Q}_k\varepsilon = z\bar{Q}_k\kappa,$$

where $\bar{Q}_k$ is found in (13.28). In this latter case stresses and strains refer to *global* $(\bar{x}, \bar{y})$ axes. We are also able to determine the stresses and strains for each layer referred to *local* axes [1].

*Bending Moments.* The bending moments at each node are found by integrating the stresses over the thickness, that is,

$$M = \begin{bmatrix} M_x \\ M_y \\ M_{xy} \end{bmatrix} = \int_{-t/2}^{t/2} \sigma z dz = \hat{D}\kappa,$$

where $\hat{D}$ depends on the material properties, $D_{iso}$ and $D_{orto}$ for the isotropic and orthotropic cases, respectively, and $t$ is the plate thickness.

## 13.11   Implementation of Structural Elements

In this section we shall take a closer look at the different structural element types implemented in Diffpack. The focus is on describing the hierarchy of the element types. At the end of this section we give a recipe on how to implement new structural element types in Diffpack. A more thorough description of member functions and functionality is presented in the header file for each element class.

*The Meaning of "Material".* In an element formulation based on the structural elements, the elements may have different elastic properties, second moment of area, cross section area, thickness, length, loadings, etc. For the structural elements in Diffpack we demand that *all properties are identical* for any two elements to be of the same *material.* For example, if two beam elements are indistinguishable, except for a transverse distributed load, the two elements are treated as two different materials. Dissimilar element types are always defined as different materials. In an analysis model we therefore often end up having one material for each element.

This convention, for defining a material, may seem very rigorous and strict; for the geometric element types in Diffpack we do not define a material this way. But, for the structural elements, it is central to gather all information about each element in one place, and to make sure that this information is easy to get. We therefore utilize this convention in this chapter and in the implementation.

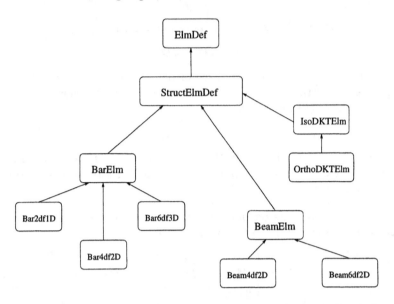

**Fig. 13.19.** Class hierarchy of structural element types. Arrows indicate class inheritance.

### 13.11.1    Class StructElmDef

ElmDef is the common abstract base class for all element type classes in Diffpack, including the structural elements. This class should be familiar to readers having implemented geometric Diffpack element types.

The structural elements are implemented as a new group of elements in the element class hierarchy, all being subclasses of class ElmDef, as depicted in Figure 13.19. The structural elements are implemented as a new "branch" owing to the fact that these elements differ a bit compared to all the other types.

The common base class for the structural elements is called StructElmDef. StructElmDef is an abstract base class, defining a common interface to the structural element classes. The base class has the following subclasses:

- BarElm: abstract base class for bar elements in 1D, 2D, and 3D,
- BeamElm: abstract base class for beam and frame elements,
- IsoDKTElm: triangular Kirchhoff (DKT) plate element class for isotropic materials.

Initially we declare a handle to class StructElmDef for each material. The proper subclass element type is initialized and bound to this handle at runtime.

## 13.11.2    Class BarElm

Class `BarElm` works as an abstract base class for bar elements in 1D, 2D, and 3D, and has three subclasses:

- `Bar2df1D`: one-dimensional bar element with two nodes, and a total of two degrees of freedom in global coordinates,
- `Bar4df2D`: two-dimensional bar element with two nodes, and a total of four degrees of freedom in global coordinates,
- `Bar6df3D`: three-dimensional bar element with two nodes, and a total of six degrees of freedom in global coordinates.

In local coordinates all the bar element types have two nodes and a total of two degrees of freedom. As explained in a previous section, we map the bar elements into higher dimensions employing transformation matrices.

Bar elements can only be extended or compressed in the element longitudinal direction. These element types are thus only capable of carrying axial loads (and not normal or moment loads). For treating two bar elements as one material, the elastic modulus $E$, the area of the cross section $A$, the length $L$, and all loads must be identical. As explained in Section 13.7.2, bar elements are used for isotropic materials only.

## 13.11.3    Class BeamElm

Class `BeamElm` is an abstract base class for beam and frame elements, collecting the common data and functions for these element types. Only 2D element types are implemented as subclasses of `BeamElm` until now, but a 3D element type may easily be implemented in a similar way. The class names reflect the number of degrees of freedom in *local* coordinates:

- `Beam4df2D`: two-dimensional beam element with two nodes, a total of four degrees of freedom in local coordinates, and a total of six degrees of freedom in global coordinates,
- `Beam6df2D`: two-dimensional frame element with two nodes, and a total of six degrees of freedom both in local and global coordinates.

`Beam4df2D` has no degrees of freedom in the (beam's) axial direction (see Section 13.5.2), and is thus unable to carry axial and temperature loads changing the element length. The `Beam4df2D` element is not very well suited for numerical computations in a general-purpose structural analysis program, because it cannot be elongated. (No such element is implemented in, e.g., ANSYS [2].) `Beam6df2D`, on the other hand, is much more suited for computer-based calculations.

To treat two beam or frame elements as the same material, the elastic modulus $E$, the area of the cross section $A$, the length $L$, the second moment of area $I$, the beam height $h$, and all loads acting have to be identical. As for bar elements, beam and frame elements are used for isotropic materials only.

### 13.11.4   Class IsoDKTElm

Class IsoDKTElm is used for modeling thin plates, divided into triangular elements, having isotropic material properties. The behavior of the elements is based on Kirchhoff's thin plate theory [7]; the thickness has to be small compared to the in-plane dimensions. The element has a total of nine degrees of freedom, see Figure 13.9b. Recall from Section 13.9.2 that the element stiffness matrix is given directly in global coordinates. Furthermore, the loads are put directly into the system load vector, i.e., the right-hand side of the linear system.

Plate elements are used for structures that are exposed to transverse loads, both uniformly distributed loads and point loads. In addition, the element can bear moment loads. The element is not capable of carrying loads acting *in* the plate plane[8]. Possible load contributions are shown in Figures 13.17 and 13.18.

To treat two isotropic plate elements as the same material, the elastic modulus $E$, the Poisson's ratio $\nu$, the thickness $t$, and all loads acting have to be identical.

### 13.11.5   Class OrthoDKTElm

The OrthoDKTElm element is also used for modeling thin plates, and is very similar to IsoDKTElm. The only difference is the material properties. That is, the material is a fiber composite with unidirectional, continuous fibers. Because the only distinction is the material properties, OrthoDKTElm is implemented as a subclass of IsoDKTElm. For the orthotropic plate element we have more material constants; the properties of the material are direction dependent, and the element is built up of several layers. The calculation of the matrices defining the material characteristics are thus more comprehensive than for isotropic problems.

To treat two orthotropic elements as the same material, the order, the orientation, *and* the thickness of each layer have to be identical. Furthermore, the elastic moduli $E_L$ and $E_T$, the Poisson's ratios $\nu_{LT}$ and $\nu_{TL}$, and the shear modulus $G_{LT}$ – *layer by layer* – have to be identical. All loads also have to be equal. OrthoDKTElm can bear the same loads as described for IsoDKTElm.

### 13.11.6   Class StructElms

Class StructElms keeps track of all the structural elements created in a simulator at run time. We emphasize that this class is not a part of the structural element class hierarchy, but an independent class used by the simulators. All

---

[8] If we have loads acting in the plane, membrane elements are used. For cases where we have both transverse loads and loads in the plane, we rather use shell elements. Membrane and shell elements are not discussed in this chapter. See [2] for more information.

correspondence between the simulator and the element objects goes through `StructElms`.

As an example of the correspondence, we assume that the simulator's `StructElms` handle and all the element objects have been created, and that we now want to set the elasticity modulus for material m. This is done with the following code line:

```
// have already created Handle(StructElms) structelms
structelms->getStructElmDefRef(m).setE(E);
```

In the same way the rest of the information about the various materials (or element objects) are set, and later, extracted from `StructElms`.

For more details about the use and functionality of this class we refer to the header file for the class, and to the simulator programs described in the next section.

### 13.11.7   How to Implement New Structural Elements

The structural element types that are implemented are not capable of solving all types of problems in structural mechanics, so implementation of new structural elements may be required in a specific problem. In this subsection we outline how to implement new structural element types, i.e., how to extend the element class hierarchy.

For the structural elements already implemented, we have analytical expressions for the element stiffness matrix and load vector. Thus, we assume that such expressions are available for a potential new element type. If not, we are not able to use the same design as for the already implemented structural elements. Structural element types are given names according to the *type* (beam, plate, shell, etc.), the *number of degrees of freedom*, and *dimension*.

To implement a new structural element type, the following tasks should be addressed:

- Study the already existing element hierarchy to decide where to put the new element type. All structural element types have to be subclasses of `StructElmDef`. If we want to implement, e.g., a three-dimensional beam element, it is probably advantageous to let it become a subclass of class `BeamElm`. In this way we are (most likely) able to make use of already existing functionality. Moreover, it makes the code structure more perspicuous.

- Create a new header file (`.h` file) and a `.cpp` file. These files should contain the Diffpack code needed for the new element.

  Remember to include the `StructElmDef.h` or `StructElmDefs.h` header file, depending on where in the structural element class hierarchy you put your new element.

- Make sure that all virtual functions are defined. We do not have to re-define all member functions in the new subclass if the functions already defined in the base class are proper for our use.

- All new member functions, not (already) declared in the implemented classes, have to be declared in class StructElmDef.

- Extend the element type list.

  All the structural element types in Diffpack are listed in class ElmDefStruct_prm[9]. New elements must also be listed here. See the Diff-pack FAQ entry on parameter classes [3], and also ElmDefStruct_prm.h (and if possible ElmDefStruct_prm.cpp) for further details.

The reader is encouraged to learn about the structural elements already implemented before making new ones.

## 13.12   Some Example Programs

In this section we take a look at some simulators made for solving structural analysis problems, employing the Diffpack structural elements. All simulator classes are subclasses of class FEM, as depicted in Figure 13.20. Although the structural element types differ with respect to material and geometrical quantities, as well as sizes of matrices and vectors used in the element formulation, most of the functionality in the simulator classes is similar. By making a hierarchy, we are able to reuse code in several classes.

The new simulator classes are given names in accordance with the structural element type being used in the modeling. BarSim1 is a simulator program for bar elements. Further, BeamSim1 and BeamSim2 are simulators for frame elements and beam elements, respectively. The BarBeamSim1 simulator program is used for problems where we employ both bar elements and frame elements (in two dimensions) in the modeling. BarSimAnsys and BarBeamSimAnsys are simulators for bar and frame elements using an ANSYS input file, and are thus called ANSYS simulators in this chapter[10]. Finally, DKTSim1 is used when modeling with isotropic plate bending elements, and DKTSim2 when employing orthotropic plates (i.e., fiber composites). All the simulator programs are found in subdirectories of

    src/structure/

We are not going to describe the simulator programs in detail in this chapter. Our main purpose is just to show how to build a simulator for structural problems. More advanced simulators may be needed for more complicated

---

[9] All the other element types are defined in class ElmDef_prm [3,6].

[10] Actually we only use the ANSYS input file for generating the Diffpack grid file; see Section 13.13.7 for more details.

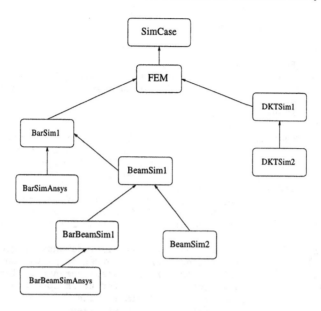

**Fig. 13.20.** Class hierarchy of simulators for structural analysis; arrows indicating class inheritance.

problems. However, having the basic simulators at hand, new ones should be fairly easy to implement.

As just said, the simulator programs are, among themselves, rather similar. Therefore, we only go through the basic steps in a solution process for one of the simulator programs. Our pedagogical choice is the bar element simulator. Readers familiar with writing and running applications in Diffpack will also realize that we follow the same programming style as in [6]; most of the functionality and member function names are adopted. For these reasons the practitioner should quite easily grasp how to solve structural problems with Diffpack.

At the end of this section, we give some comments about the indicators in ANSYS simulators, which differ a bit from the boundary indicators in Diffpack input files.

### 13.12.1    The Bar Element Simulator

Class **BarSim1** is a simulator made for computing displacements, nodal forces, stresses, and strains in structures modeled by bar elements. This simulator can be used both in 1D, 2D, and 3D cases. We will now point out the major steps in the solution process.

*The Header File.* First we list the essential parts of the header file:

```
class BarSim1 : public FEM
{
public:
  Handle(GridFE)        grid;
  Handle(DegFreeFE)     dof; // Transmits the field values of u to a
                             // vector of unknowns in the linear system
  Handle(FieldsFE)      u;
  Handle(SaveSimRes)    database;
  Vec(real)             solution;
  Handle(LinEqAdmFE)    lineq;    // Contains the linear system
                                  // for the problem under consideration
                                  // and different solvers for
                                  // linear systems

  Handle(StructElms)    structelms;
  int                   no_of_materials; // number of materials
  int                   ndofno;          // number of dof pr node
  Vec(real)             nodalloads;      // vector for nodal loads

  Mat(real)             v_glob_coor; // elm displacements in glob.coor.
  Mat(real)             s_glob_coor; // elm forces in glob. coor.
  Mat(real)             v_loc_coor;  // elm displacements in loc.coor.
  Mat(real)             s_loc_coor;  // elm forces in loc. coor.
  Mat(real)             strains;     // elm strains
  Mat(real)             stresses;    // elm stresses

  BarSim1();
  ~BarSim1() {}

  virtual void adm(MenuSystem& menu);
  virtual void define(MenuSystem& menu, int level = MAIN);
  virtual void scan ();
  virtual void solveProblem();

protected:
  virtual void fillEssBC();
  void saveResults();

  // creating the system
  virtual void makeSystem(DegFreeFE& dof, LinEqAdmFE& lineq,
                          bool compute_A = true,
                          bool compute_RHS = true);
  virtual void calcElmMatVecAna(int e, ElmMatVec& elm_matvec);
```

The major difference compared to other simulators is the data member for organizing the element objects, Handle(StructElms) structelms. The handle is initialized in the scan function, writing

```
    structelms.rebind(new StructElms());
    structelms->init(*grid, no_of_materials);
```

We refer to Section 13.11.6 for an example of using structelms for setting and getting values from the element objects.

*The Menu System and Reading From File.* Reading values from an input file employing the menu system in Diffpack [6]. There is no preprocessor class available in this case. Information about elements, nodes, material properties, and boundary conditions thus has to be written manually into the grid file. (See Section 13.13.2 or [6] for examples.)

*Building Up and Solving the Linear Equation System.* Structural elements make use of analytical expressions for the element stiffness matrix and the element load vector. Thus, we need to reimplement the `makeSystem` member function in this case, when building up the global linear system $Kx = b$. Further, instead of using the `calcElmMatVec` member function, we employ the "special" member function `calcElmMatVecAna`, where "Ana" indicates use of analytic expressions. The code for the two member functions is given below:

```
void BarSim1::makeSystem(DegFreeFE& dof, LinEqAdmFE& lineq,
                         bool compute_A, bool compute_RHS)
{
  lineq.initAssemble(dof,compute_A,compute_RHS);

  elm_matvec.attach(dof);

  // General algorithm.
  // Run through all the elements and add the element's
  // contribution to the linear system
  const int nel = dof.grid().getNoElms();
  for (int e = 1; e <= nel; e++)
    {
      // allocate space in elm_matvec for this element, set A=b=0
      elm_matvec.refill(e);
      calcElmMatVecAna(e, elm_matvec);
      elm_matvec.enforceEssBC();
      lineq.assemble(elm_matvec);
    }

  if(have_nodal_loads){
      Vector(real)& rhs = lineq.b();
      rhs.add(rhs,nodalloads);
      }
}

void BarSim1::calcElmMatVecAna(int elm_no, ElmMatVec& elm_matvec)
{
  int m;
  Mat(real) elmcoor;

  m = grid->getMaterialType(elm_no);
  grid->getElmCoor(elmcoor,elm_no);

  // Element stiffness matrix
  structelms->getStructElmRef(m).
              elmStiffnessMat(elm_matvec.A,elmcoor);

  // Contribution to the load vector except for nodal loads
  structelms->getStructElmRef(m).elmLoadVec(elm_matvec.b,elmcoor);
}
```

Remark that we do not use the `integrands` function [6]. The information needed is stored in the element objects.

Before solving the linear system, the global degrees of freedom are marked with boundary indicators [6]. In this case the boundary indicators are given by (general 3D case):

- Indicator 1: $r_x = 0$
- Indicator 2: $r_y = 0$
- Indicator 3: $r_z = 0$

The indicators denote restrictions on displacement in the $x$, $y$, and $z$ directions. The quantity $r = [r_x, r_y, r_z]^T$ is the global (displacement) degrees of freedom for each node found in the global solution vector $x^{11}$.

*Calculating Results.* The simulator calculates the displacements and nodal forces for each element, both with respect to global and local coordinates. Furthermore, the axial stresses and strains, for each element, are computed.

*Deformed Configuration.* We are also able to make a sketch of the deformed structure (function plotDeformedGrid). The figure may later be visualized using, for example, Plotmtv. (See Section 13.13.2 for an example.)

*Automatic Report Generation and Writing Results.* The simulator is also equipped with the opportunity for automatic report generation in LaTeX, HTML, and ASCII format. All results calculated by the simulator program are (automatically) put into tables [6].

### 13.12.2   Indicators in the ANSYS Simulators

Diffpack is equipped with a filter, called ansys2dp_2, which makes it possible to generate a general Diffpack grid file [6] from an ANSYS [2] input file. When using ANSYS input files for generating the grid file, we have to adjust the Diffpack functions reading parameter values, to fit the ANSYS programming style. This is because indicators are used to a larger extent in ANSYS.

For ANSYS simulators, loads are also pointed out using indicators. We employ the following indicator names in the ANSYS simulators:

**UX,UY,UZ** : Mark nodes with zero displacement in the $x$, $y$, and $z$ directions, respectively.

**ROTZ** : Marks nodes where the nodal rotation is zero.

**FX,FY,FZ** : Mark nodal forces.

**MZ** : Marks nodal moment.

---

[11] The other simulators use other indicators. For example, in the beam element simulators, indicator 3 denotes restriction on rotation instead; see the header files for the indicators in each case.

**TEMP** : Marks temperature loads. Notice that the ANSYS function BFE generally takes four different temperature value arguments, depending on element type (see ANSYS element documentation in each case [2]). In Diffpack we need the thermal expansion coefficient $\alpha$ and the temperature change $\Delta T$, i.e., other parameters than given in the ANSYS function. Anyway, no temperature load values are transferred when generating the Diffpack grid file using the ansys2dp_2 filter; indicators only mark nodes with temperature loads. The parameter values $\Delta T$ and $\alpha$ are instead given in the Diffpack input file.

**PRESX** : Marks element pressure load in the $x$ direction (i.e., distributed axial load).

**PRESY** : Marks element pressure load in the $y$ (or $z$) direction (i.e., distributed transverse load).

No other indicator names are supported by the Diffpack simulator programs. We assume that the reader is familiar with ANSYS and recognize the quantities employed in ANSYS.

## 13.13   Test Problems

We have made a range of test problems. The examples are relative simple, but complicated enough to verify the simulators and the structural element library. The reader may use these cases as starting points for solving more comprehensive structural problems. For all the test cases, there are ready-made input files (including grid files) in the Verify directory. Notice that not all test problems in this directory are described in detail in this chapter.

All the simulators use the same executable app file. For creating the proper simulator object (at run time), each simulator is given an unambiguous name. The simulator name has to be provided as command-line input when running the application[12]. The test problem summary table referred to in Section 13.7 lists the correct simulator name in each case.

We start with a one-dimensional bar element problem, and list the commands for running the simulator. In Section 13.13.2 we also explain the contents of the input files. With the information given in the two first simulator examples, it should be quite easy to understand and work through the rest of the test problems.

### 13.13.1   Bar Elements in 1D

Figure 13.21 shows a structure modeled by two Bar2df1D elements. Element 1 has elastic modulus $E_1 = 1000000$ N/m$^2$, cross section $A_1 = 0.1$ $m^2$, and

---

[12] Take a look at file main.cpp to see the details of creating the right simulator object.

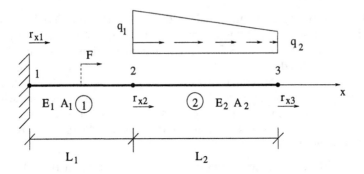

**Fig. 13.21.** Bar elements in 1D; the element numbers in the circles, and the node numbers given at each node.

length $L_1 = 5$ $m$. For element 2 we have $E_2 = 2000000$ N/m$^2$, $A_2 = 0.1$ $m^2$, and $L_2 = 7$ $m$. Element 1 has a concentrated axial force, $F = 1000$ N, acting at the element mid-point, i.e., at $x = L_1/2$. Element 2 has a linearly distributed axial load, with $q_1 = 100$ N/m and $q_2 = 50$ N/m. We can also have temperature loads and volume loads for this element type, but no such loading is applied. Because the construction is built in at node 1, we mark the global degree of freedom at this node ($r_{x1}$) with boundary indicator 1 (see Section 13.12.1). The components of the system solution vector is (in this case) denoted $r_{xi}$, $i =$ node number. That leaves us with only two unknown degrees of freedom.

To run the application we write:

```
app --casename Bar1D < Verify/bar1.i -sim bar
```

Table 13.1 gives displacements and cross section axial forces in each bar. Positive displacement value indicates elongation of the bar elements.

**Table 13.1.** Bar elements in 1D, Bar2df1D. Displacements (in $m$) at nodes ($r_{xi}$, $i = 2, 3$), and axial forces (in N) in elements ($S_{xe}$, $e = 1, 2$).

| $r_{x2}$ | $r_{x3}$ | $S_{x1}$ | $S_{x2}$ |
|---|---|---|---|
| 0.05125 | 0.05942 | 1025 | 233.33 |

## 13.13.2    Bar Elements in 2D

A 2D frame structure modeled by bar elements, i.e., Bar4df2D, is depicted in Figure 13.22. All elements have the same elastic modulus and cross section;

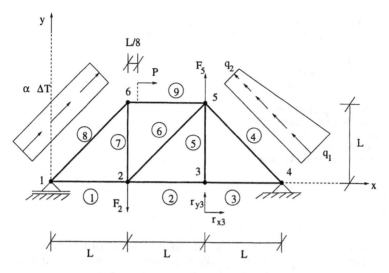

**Fig. 13.22.** Bar elements in 2D.

$E = 10000 \text{ N/m}^2$ and $A = 0.1 \ m^2$. All horizontal and vertical elements have length $L = 2 \ m$. Element 4 is loaded with a linearly distributed axial load, with $q_1 = 100$ N/m and $q_2 = 50$ N/m. Element 8 has temperature loads acting, with thermal expansion coefficient $\alpha = 0.001$ /°C and temperature change $\Delta T = 50$ °C. Element 9 carries a concentrated axial load, $P = 100$ N, acting in the positive $x$ direction, at a distance $\hat{x} = L/8$ from node 6 (see the figure). Further, nodes 2 and 5 are loaded with pure nodal loads; $F_2 = 50$ N in the negative $y$ direction, and $F_5 = 50$ N in the positive $y$ direction. Every node has two degrees of freedom as shown for node 3. Node 1 is marked with boundary indicator 2 ($r_{y1} = 0$), and node 4 with boundary indicators 1 ($r_{x4} = 0$) and 2 ($r_{y4} = 0$).

The simulator is run as explained for the 1D bar case. (We actually only change the case name and the name of the input file.):

```
app --casename Bar2D < Verify/bar2.i -sim bar
```

*The Input and Grid Files.* In this case we give the input file explicitly, and go through it in detail, line by line. The file **Verify/bar2.i** reads:

```
set gridfile = Verify/bar2.grid
set number of materials = 5
set E modulus = 1=[10000] 2=[10000] 3=[10000] 4=[10000] 5=[10000]
set cross section area = 1=[0.1] 2=[0.1] 3=[0.1] 4=[0.1] 5=[0.1]
set temperature loads = 1 4=[0.001 50]
set distributed axial loads = 1 2=[100 50]
set concentrated axial loads = 1 5=[100 0.25]
set nodal loads = 2 2=[0 -50]  5=[0 50]
set print results ('YES' or 'NO') = YES
```

```
set filename for print = Axi3.res
set filename for grid plot = gridplot.gb

sub LinEqAdmFE
 sub Matrix_prm
  set matrix type = MatSparse
 ok
 sub LinEqSolver_prm
  set basic method = ConjGrad
 ok
 sub Precond_prm
  set preconditioning type = PrecNone
 ok
ok
h
ok
```

There is no preprocessor available. Therefore, we need to write all information about geometry, boundary indicators, and elements explicitly in a file ourselves. Let us take a closer look at file `Verify/bar2.grid` before moving on with the rest of the input file:

```
Finite element mesh (GridFE):

  Number of space dim. =  2
  Number of elements   =  9
  Number of nodes      =  6

  All elements are of the same type : true
  Max number of nodes in an element: 2
  Only one material                 : false
  Lattice data                      ? 0


  2 boundary indicator(s):
    r_x=0 r_y=0


  Nodal coordinates and nodal boundary indicators,
  the columns contain:
   - node number
   - coordinates
   - no of boundary indicators that are set (ON)
   - the boundary indicators that are set (ON) if any.
#
   1 ( 0.00000e+00, 0.00000e+00)  [1] 2
   2 ( 2.00000e+00, 0.00000e+00)  [0]
   3 ( 4.0000e+00,  0.00000e+00)  [0]
   4 ( 6.0000e+00,  0.00000e+00)  [2] 1 2
   5 ( 4.0000e+00,  2.00000e+00)  [0]
   6 ( 2.0000e+00,  2.00000e+00)  [0]


  Element types and connectivity
  the columns contain:
   - element number
```

```
        - element type
        - material number
        - the global node numbers of the nodes in the element.

#
   1  Bar4df2D  1         1   2
   2  Bar4df2D  1         2   3
   3  Bar4df2D  1         3   4
   4  Bar4df2D  2         4   5
   5  Bar4df2D  1         3   5
   6  Bar4df2D  3         2   5
   7  Bar4df2D  1         2   6
   8  Bar4df2D  4         1   6
   9  Bar4df2D  5         6   5
```

Grids with 2D bar elements must be specified with number of space dimensions as 2. We have nine elements and six nodes, as depicted in Figure 13.22 (element numbers in circles and node number at each node). All elements are of the same type. (Otherwise we would have had to write `false`.) We use different materials (see Section 13.11). The lattice data is set to `false`, because an underlying finite difference grid is not relevant here (as well as in all other structural analysis problems). Further, we have two different boundary indicators with names 'r_x=0' and 'r_y=0'; by giving the boundary indicators logical names, it is easier to understand their meaning. The first indicator is number 1, and the other indicator number 2. Next we give the coordinates and boundary indicators for each node. For example, we place node 4 at point $(6, 0)$, and mark it with indicators 1 and 2. At the end, we point out the element type and connectivity, as also explained in the file. Here we indicate what elements having what material properties, i.e., the material number.

Having written one grid file, we are able to reuse its structure in other cases. Later we will also see how to give all the information about elements, node coordinates and boundary indicators to the simulator program using a preprocessor [6]. The preprocessor then generates the grid file automatically.

We now return to the input file (`bar2.i`). There are five different materials in this model. For each material we have to set the elastic modulus $E$ and the area of the cross section $A$. Next, we indicate the number of materials carrying a given load type, and denote what materials bear each particular load. There is one material having temperature loads, and that is material number 4. The thermal expansion coefficient is $\alpha = 0.001$ and the temperature change $\Delta T = 50$. One material, material 2, has a distributed axial load, with $q_1 = 100$ and $q_2 = 50$. Next, material 5 is the only material which has a concentrated load $P = 100$, located at $\hat{x} = 0.25$. Here $\hat{x}$ is a local element coordinate for the element corresponding to material 5, i.e., element 9. Finally, nodes 2 and 5 carry pure nodal loads in the negative and positive $y$ direction, respectively. Then we indicate if the results calculated in the simulator should be written to file or terminal screen. If we want to put the results on a file, we have to provide the file name. If no filename is given (i.e., writing `NONE`, which also is default) all results are automatically printed to the screen. For later making

**Table 13.2.** Cross section axial forces in N, `Bar4df2D`.

| Element | Axial force |
|---------|-------------|
| 1 | -16.67 |
| 2 | -83.33 |
| 3 | -83.33 |
| 4 | 70.71 |
| 5 | 0.00 |
| 6 | 94.28 |
| 7 | -16.67 |
| 8 | 73.57 |
| 9 | -58.33 |

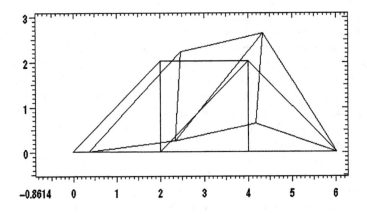

**Fig. 13.23.** Bar elements in 2D; undeformed and deformed structure.

a plot of the deformed structure, we have to set a file name for plotting the deformed grid. The rest of the input file gives the menu level structure for the solving procedure; for more details about this part we refer to Chapter 3 in [6].

*Results and Plotting.* Table 13.2 shows cross section axial forces for each element. Positive value indicates elongation, and negative value compression. Furthermore, Figure 13.23 shows the deformed and the undeformed structure. To generate this plot using Plotmtv, we write:

```
gb2mtv gridplot.gb > gridplot.mtv
```

and then

```
plotmtv gridplot.mtv
```

The file name `gridplot.gb` is set in the input file.

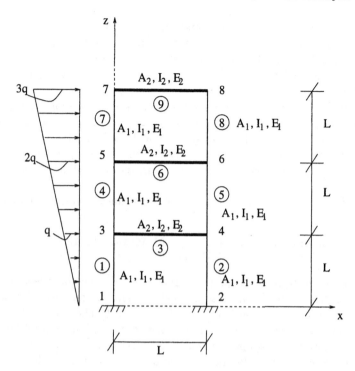

**Fig. 13.24.** Three-storey framework.

## 13.13.3   Beam Elements in 2D; Three-Storey Framework

Next we study a framework structure, as shown in Figure 13.24. The storey framework is loaded with a linearly distributed load directed in the horizontal ($x$) direction. We model the "walls" with one beam element per "floor". The linear load then increases by $q = 20$ N/m per element.

The vertical "wall" elements have cross section $A_1 = 0.1 \ m^2$, second moment of area $I_1 = 0.5 \ m^4$, and elastic modulus $E_1 = 200000$ N/m$^2$. The horizontal "floor" elements have cross section $A_2 = 0.5 \ m^2$, second moment of area $I_2 = 0.6 \ m^4$, and elastic modulus $E_2 \gg E_1$. We therefore assume that the "floor" elements can be regarded as perfectly rigid, i.e., $E_2 I_2 \to \infty$, which means that all nodal rotations are zero. This approximation reduces the number of degrees of freedom, and makes it possible to use the beam element (Beam4d2D) in the modeling. All elements have length $L = 5 \ m$.

*Special Treatment of the Degrees of Freedom.*   As noted in a previous section, special care has to be taken for the system degrees of freedom when using this element type. For the beam element, indicators 1 ($r_{xi}$) and 2 ($r_{zi}$) denote restrictions on the displacement in the $x$and $z$directions, respectively, while 3 ($r_{\theta i} = 0$) indicates zero rotation ($i$ = node number). We go through this in detail for the current case: Nodes 1 and 2 can not be translated or rotated, and are both marked with boundary indicators 1, 2, and 3. Because we neglect axial deformations in this case, the "floor" elements are translated in the $x$ direction. Furthermore, the "wall" elements are neither compressed nor elongated in the $z$ direction. Thus, all vertical degrees of freedom must be marked with indicator 2, and we have to take special care of the dependent degrees of freedom; $r_{x3} = r_{x4}$, $r_{x5} = r_{x6}$, and $r_{x7} = r_{x8}$. The nodal rotations connected to the "floor" elements are also neglected, and (therefore) marked with indicator 3. That leaves only three unknown, independent degrees of freedom, i.e., the horizontal displacement degrees of freedom at nodes 3, 5, and 7.

This is how we define dependencies between the degrees of freedom in the input file:

```
set dependent dofs = 3 [4 1|3 1|1.0] [6 1|5 1|1.0]
                       [8 1|7 1|1.0]
```

In this case we have three pairs of dependent degrees of freedom (dofs); one pair inside each of the three brackets ([ ]). The number of dependent degrees of freedom is given before the first bracket. Then, for each pair, we start by giving the node and the nodal degree of freedom that is dependent on another node (values appearing after the first bracket and before the first |). Next, we give the node and the nodal degree of freedom for this other node (between the first and second |). At the end we provide the value of the proportionality factor (between the second | and the ] bracket). This hopefully becomes apparent with an example. Let us explain the first pair in the line above: The horizontal degrees of freedom at nodes 3 and 4 are dependent. We decide to keep $r_{x3}$ when solving the system of equations, and let $r_{x4}$ be the dependent degree of freedom. (There is no problem in choosing it the other way around.) Thus, the degree of freedom in direction 1 ($x$) at node 4, is dependent of the degree of freedom in direction 1 ($x$) at node 3. The degrees of freedom are equal, which gives the proportionality value 1.0. The input file giving this and the rest of the commands is named beam1.i.

The results from the simulation are shown in Table 13.3. Figure 13.25

**Table 13.3.** Three-storey framework. All numbers given in meters ($m$).

| Horizontal displacement at node | | |
|---|---|---|
| 3 and 4 | 5 and 6 | 7 and 8 |
| 0.0453125 | 0.0802083 | 0.0942708 |

shows a sketch of the deformed and undeformed structure. In this figure the elastic modulus is reduced, giving larger displacements and rotations, thus making it easier to observe the deformations.

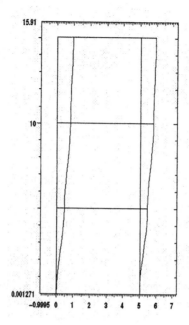

**Fig. 13.25.** Three-storey framework; undeformed and deformed structure.

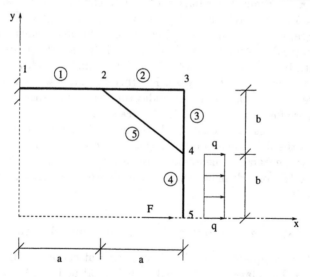

**Fig. 13.26.** Construction model with a tension rod.

### 13.13.4   Bar and Frame Elements in 2D

A structure model where we use both bar and frame elements, that is, both
`Bar4df2D` and `Beam6df2D`, is depicted in Figure 13.26. The structure is modeled

**Table 13.4.** Nodal displacements (in $m$) and rotations (in radians), construction model with a tension rod.

| Degree of freedom | Displacement/rotation |
|:---:|:---:|
| $r_{x2}$ | 0.000375 |
| $r_{y2}$ | 0.125000 |
| $r_{\theta 2}$ | -0.050000 |
| $r_{x3}$ | -0.000081 |
| $r_{y3}$ | 0.444574 |
| $r_{\theta 3}$ | -0.066745 |
| $r_{x4}$ | 0.259950 |
| $r_{y4}$ | 0.445110 |
| $r_{\theta 4}$ | -0.068140 |
| $r_{x5}$ | 0.551180 |
| $r_{y5}$ | 0.445110 |
| $r_{\theta 5}$ | -0.074807 |

by five elements, where elements 1–4 are frame elements, and element 5 is a bar element. All elements have the same elastic modulus $E = 2000000 \text{ N/m}^2$. The bar element's cross section is half the area of the cross section of the frame elements; $A_{\text{beam}} = 2A_{\text{bar}} = 0.2 \ m^2$. The second moment of area is the same for all frame elements, $I = 0.01 \ m^4$. The horizontal elements have length $a = 5 \ m$, and the vertical elements length $b = 4 \ m$. Node 5 carries a nodal force, $F = 10$ N, in the $x$ direction. In addition, element 4 is loaded with a uniform distributed normal load $q = 5$ N/m, see the figure. Node 1 is marked with boundary indicators 1, 2, and 3. Boundary indicator 3 denotes rotational restraint in this 2D case. For this simulator we thus employ the same convention for the boundary indicators as in BeamSim1 and BeamSim2. All details are found in the input file barframe1.i.

Table 13.4 gives the nodal displacement and rotational values. Figure 13.27 shows the deformed and undeformed structure. We are not able to illustrate how the frame elements bend. Nevertheless, the sketch gives an indication on nodal point displacements. In this figure we use $\hat{E} = 200000 \text{ N/m}^2$, i.e. $E/\hat{E} = 10$.

### 13.13.5    Twisting of a Square Plate; Isotropic Material

We now study a square plate simply supported at corners A, B, and D, see Figure 13.28. The plate carries a load, $F = 5$ lb, at point C. We want to find the nodal moment values, and also the vertical deflection at the center and at C.

Figures 13.28a–13.28d show four different meshes. In all cases the plate is 8" × 8". Further, the elastic modulus is $E = 10000$ psi, the thickness $t = 1$ in, and Poisson's ratio $\nu = 0.3$. Files plate1a–plate1d are used as input. The results from the simulations are given in Table 13.5.

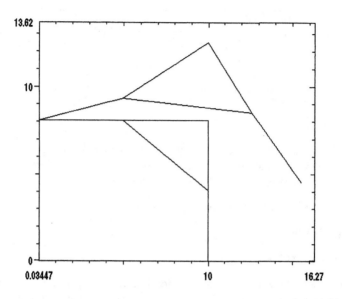

**Fig. 13.27.** Construction model with a tension rod; undeformed and deformed structure.

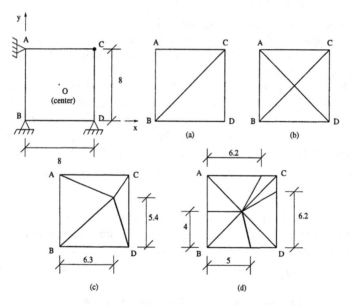

**Fig. 13.28.** Twisting of a square plate.

**Table 13.5.** Twisting of a square isotropic plate, different meshes.

| Mesh | Deflection at | | Nodal moments | |
|------|------|------|------|------|
| | O | C | $M_x$ and $M_y$ | $M_{xy}$(lb − in/in) |
| (a) | - | 0.24960 | 0.0 | 2.5 |
| (b) | 0.06240 | 0.24960 | 0.0 | 2.5 |
| (c) | - | 0.24960 | 0.0 | 2.5 |
| (d) | 0.06240 | 0.24960 | 0.0 | 2.5 |

*Contour Plot..* In addition to the plot showing the deformed grid, we are, for the plate elements, also able to make contour plots, i.e., indicate where the plate has the same value of deflection. We write:

```
simres2mtv_struct -f plate1 -s -n w -a
```

Then launch Plotmtv:

```
plotmtv plate1.w.mtv
```

The filter `simres2mtv_struct` is a modification of the `simres2mtv` filter [6, page 230]. This new filter, for the plate bending elements, works similarly.

### 13.13.6   Simply Supported Fiber Composite Plate

We shall next look at a simply supported rectangular plate with orthotropic material properties. In this case we have the opportunity to compare the Diffpack results with an analytical solution indicated by Timoshenko and Woinowsky-Krieger [8].

We have a composite material with four layers, where the layers are oriented [0/90/90/0]; for the two mid layers $\theta = 90$ and for the two outer layers $\theta = 0$, see Section 13.6.2[13]. The thickness of each layer is $t_{\text{layer}} = 0.025\,m$, and the material parameters are the same for each layer: $E_L = 19755000\,\text{N/m}^2$, $E_T = 197550\,\text{N/m}^2$, $\nu_{LT} = 0.35$, and $G_{LT} = 3352000\,\text{N/m}^2$. Furthermore, the plate carries a uniform load $q = 1\,\text{N/m}^2$. Moreover, the plate size is $a \times b$. In all cases we use 800 elements.

*Simplifying Command Input.* If all layers have the same properties, we can avoid a lot of typing in the input file. Using the elastic modulus in the longitudinal direction $(E_L)$ as an example, we write

```
set E_L = 1=S[25000000]
```

By putting a capital 's' in from of the bracket ([), all layers are given the same value. If this is not the case, we have to give the values for each layer separately.

---

[13] The notation is explained in detail in [1].

*The DKT Plate Element Preprocessor.* In this case we use the preprocessor `PreproBoxDKT` when generating the grid and elements, and denoting the boundary indicators:

```
set gridfile = P=PreproBoxDKT | d=2 [0,10]x[0,6.25] |
               d=2 e=OrthoDKTElm div=[20,20] g=[1,1]
set redefine boundary indicators = n=4 names= w=0 w_x=0 w_y=0 free
                                    1=(1 2 3 4) 2=() 3=() 4=()
```

The input file is called `plate5.i`. Table 13.6 gives the maximum deflection, located at the center of the plate.

**Table 13.6.** Deflection at the center of a simply supported orthotropic plate subjected to a uniform load.

| a | b | Deflection |
|---|---|---|
| 10.00 | 6.25 | 0.0281 |
| 10.00 | 5.68 | 0.0229 |
| 10.00 | 5.21 | 0.0188 |
| 10.00 | 4.80 | 0.0154 |
| 10.00 | 4.46 | 0.0127 |
| 10.00 | 4.17 | 0.0106 |
| 10.00 | 3.90 | 0.0087 |
| 10.00 | 3.68 | 0.0073 |
| 10.00 | 3.47 | 0.0061 |
| 10.00 | 3.29 | 0.0052 |
| 10.00 | 3.13 | 0.0044 |
| 10.00 | 2.50 | 0.0020 |
| 10.00 | 2.08 | 0.0010 |
| 10.00 | 1.56 | 0.0003 |
| 10.00 | 1.25 | 0.0001 |

### 13.13.7   Test Problems Using ANSYS Input Files

A regular ANSYS input file (giving nodes, elements, material types, loads, etc) can be converted to a Diffpack grid file. To do this, we use the ansys2dp_2 filter. Having an ANSYS input file at hand, the Diffpack grid file is created by running:

```
ansys2dp_2 -f ansysfile.ANS -g gridfile
```

where `ansysfile.ANS` is the name of the ANSYS input file, and `gridfile` is the Diffpack grid file name. The grid file is automaticly given the extension `.grid`, resulting in `gridfile.grid` as the real file name. More information about the use of the ansys2dp_2 filter may be found by typing

```
ansys2dp_2
```

Having generated the Diffpack grid file, we can run our application in the same way as before.

Using the bar element test case given in `ansysbar1.i` as example, we first generate the Diffpack grid file by writing

```
ansys2dp_2 -f ansysbar1.ANS -g ansysbar1
```

Next, for running the simulation, we type[14]:

```
app --casename AnsysCase1 < Verify/ansysbar1.i -sim ansys1
```

As said in Section 13.12.2, only nodal coordinates and boundary indicators, are transferred from ANSYS to Diffpack. Therefore, all material properties, geometric values, and load values have to be set in the Diffpack input file.

### 13.13.8   Test Problems Summary

Table 7 is a summary table containing information about all the currently available test problems made.

## 13.14   Summary

In this chapter we have presented the basic features of the finite element method commonly employed in structural engineering. This formulation is preferable when working with structural analysis of ships, cars, aircrafts, etc. We have shown that the energy formulation is equivalent with the variational formulation and the Galerkin method based on partial differential equations for some simple examples. Even though there exists two "schools", and the various formulations at first glance seem quite different in the literature, we end up with the same algebraic expressions for the element matrix and vector.

Detailed descriptions of the element matrix and load vector for the structural element types implemented in Diffpack are given. For the structural elements we obtain analytical expressions for the element matrix and vector. Moreover, the structural elements differ from the geometric element types in that the structural element classes store information about the mathematical model and its physical parameters, in addition to the geometric information regarding shape, basis functions, and degrees of freedom. Further, structural analysis is concerned with looking at a structure as an assembly of beams, plates, shells, and so on; the mathematical model can thus be viewed as a gathering of "physical" parts of the structure. It would be possible to model each part by PDEs, but structural elements are more convenient. From a

---

[14] The ANSYS simulators are given simulator names `ansys1` and `ansys2`, see the table referred to in Section 13.7.

**Table 13.7.** Test problem summary.

| Test problem | Element type(s) | Input file(s) | Geometry file(s) | Simulator name |
|---|---|---|---|---|
| Bars in 1D | Bar2df1D | bar1.i | bar1.grid | bar |
| Bars in 2D | Bar4df2D | bar2.i | bar2.grid | bar |
| Bars in 3D | Bar6df3D | bar3.i | bar3.grid | bar |
| Three-storey framework | Beam4df2D | beam1.i | beam1.grid | beam |
| Point loaded framework | Beam4df2D | beam2.i | beam2.grid | beam |
| Frame elements | Beam6df2D | beam3.i | beam3.grid | frame |
| Bar and frame elements | Bar4df2D & Beam6df2D | barbeam1.i | barbeam1.grid | barframe |
| Twisting of a isotropic plate | IsoDKTElm | plate1a.i – plate1d.i | plate1a.grid – plate1d.grid | isoplate |
| Rhombic cantilever | IsoDKTElm | plate2.i | plate2.grid | isoplate |
| Simply supported isotropic plate | IsoDKTElm | plate3.i | use a preprocessor | isoplate |
| Twisting of a orthotropic plate | OrthoDKTElm | plate4a.i – plate4d.i | plate4a.grid – plate4d.grid | orthoplate |
| Simply supported orthotropic plate | OrthoDKTElm | plate5.i | use a preprocessor | orthoplate |
| Bars in 2D | Bar4df2D | ansysbar1.i | ansysbar1.grid | ansys1 |
| Bars in 3D | Bar6df3D | ansysbar2.i | ansysbar2.grid | ansys1 |
| Beam frame | Beam6df2D | ansysbeam1.i | ansysbeam1.grid | ansys2 |
| Bar and beam frame | Bar4df2D & Beam6df2D | ansysbarbeam1.i | ansysbarbeam1.grid | ansys2 |

mathematical point of view, using structural elements is equivalent to discretizing local PDEs by the Galerkin finite element method.

Some application programs are described. The main purpose is to show how to build simulators in Diffpack employing the structural elements. The simulator programs follow the same programming style as other simulators in Diffpack, and have adopted most of the functionality and member function names. Thus, experienced Diffpack users should be able to run simulations for analyzing construction models quite easily.

A couple of simple, yet illustrating, test problems are shown. By going through these problems, the practitioner should be able to approach more complicated cases. The summary table referred to in Section 13.7 lists the currently available test problems.

The structural element types implemented in Diffpack at this time of writing, are not capable of solving all problems in structural mechanics. There is a need for more sophisticated element types, e.g., shell elements and membrane elements [2]. Moreover, in recent years composite materials are used in a larger and larger scale in bikes, cars, boats, etc. For solving such problems, more advanced element types, allowing orthotropic materials, are required.

# References

1. B. D. Agarwal and L. J. Broutman. *Analysis and performance of fiber composites.* John Wiley & Sons, Inc., 2nd edition, 1990.
2. ANSYS software package. Manuals, version 5.5, ANSYS Inc.
3. Diffpack software package. *http://www.diffpack.com.*
4. K. J. Bathe J. L. Batoz and L. W. Ho. A study of three-node triangular plate bending elements. *Int. J. Num. Meth. Engng.*, 15(12):1771–1812, 1980.
5. C. Jeyachandrabose and J. Kirkhope. An alternative formulation for the dkt plate bending element,. *Int. J. Num. Meth. Engng.*, 21(7):1289–1293, 1985.
6. H. P. Langtangen. *Computational Partial Differential Equations – Numerical Methods and Diffpack Programming.* Textbook in Computational Science and Engineering. Springer, 2nd edition, 2003.
7. D. S. Malkus R. D. Cook and M. E. Plesha. *Concepts and applications of finite element analysis.* John Wiley & Sons, Inc., 1989.
8. S. Timoshenko and S. Woinowsky-Krieger. *Theory of plates and shells.* McGraw-Hill Book Company, Inc., 1959.

# Chapter 14

# Simulation of Aluminum Extrusion

K. M. Okstad[1] and T. Kvamsdal[2]

[1] SINTEF Applied Mathematics, N–7465 Trondheim, Norway. Currently at
FEDEM Technology AS, Vestre Rosten 78, N–7075 Tiller, Norway
[2] SINTEF Applied Mathematics, N–7465 Trondheim, Norway

**Abstract.** In this chapter we investigate the simulation of aluminum extrusion processes by means of a new heat and fluid flow solver. The extrusion problem is here formulated as two sub-problems, separately solved via a staggered solution strategy, 1) a non-Newtonian Navier-Stokes fluid flow problem where the viscosity depends on both the effective strain rate and the temperature, and 2) a linear heat conduction problem. The latter problem is further divided into the sub-problems of the different material regions that make up the extrusion package. The constitutive properties of the aluminum are expressed by the Zener-Hollomon material model. A mixed interpolation of the unknown variables is employed in the finite element discretization, using second order polynomials for the velocity and temperature fields and linear polynomials for the pressure. The simulator is implemented in C++, using the Diffpack library. The results of the numerical simulations are compared with physical experiments conducted at the extrusion laboratory at SINTEF in Trondheim, Norway.

## 14.1 Introduction

Aluminum alloys are steadily gaining ground in new application areas, partly due to the benefits these metals show during the user phase of a finished product (recycling, etc.). Building industries and automotive industries are possibly the most important areas where aluminum products are used, and examples of such products are depicted in Figure 14.1.

Extrusion is an important process in the manufacture of these products. This involves pressing a cylindrical piece of massive aluminum (called the Billet) through a Die with a prescribed velocity, such that a profile with the desired cross section is output, see Figure 14.2. The aluminum billet is preheated up to 400-500°C such that it is more easily deformed. Nevertheless, a cross section reduction ratio of order 20-100 for the billet is common, so it is clear that huge forces are involved (the ram force is typically in the range 10-65 MN).

Quantitative knowledge about the maximum ram force, flow pattern, and temperature distribution in the aluminum and extrusion tool is important for choosing an optimal procedure for extrusion of aluminum. Physical experiments have been the main approach in learning such extrusion processes.

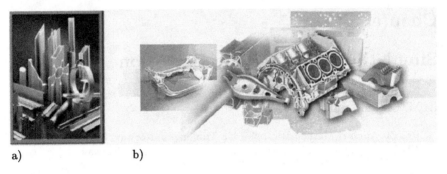

a)                          b)

**Fig. 14.1.** Some extruded aluminum products: a) For the building industry. d) For the automotive industry.

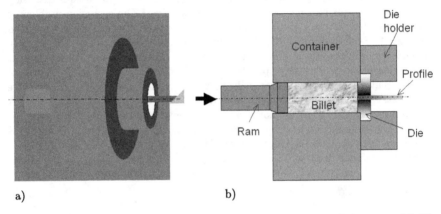

a)                                    b)

**Fig. 14.2.** The aluminum extrusion press: a) 3D illustration of the press. b) 2D section through the press showing the aluminum billet and profile.

However, this methodology is often costly and cumbersome to perform, and a large amount of experiments are usually required before an improvement in the design or process is reached. A tool for performing the experiments numerically will yield significant savings for the industry as the number of such numerical experiments then will be limited only by the available computer resources, and the performance of the numerical solver. Some of the physical quantities are also much easier to calculate from a numerical simulation than to measure from a physical experiment. A reliable numerical simulator may enable the analyst to learn more about the process than ever would be possible through physical experiments. Numerical simulation of aluminum extrusion in order to quantify the various mechanical properties is therefore of great interest to the industry.

The extrusion process is quite complicated, and various simplifications in the numerical models are introduced in order to keep the computing time to a reasonable level. Our aim herein is to investigate the accuracy of a

newly developed thermo-mechanical simulation tool called *Extrud*, through
a comparison of the ram force and the temperature distribution (in certain
critical points) with physical experiments.

In the present version of *Extrud*, the extrusion problem is formulated as
two sub-problems which are solved separately in a staggered solution strategy:
1) A non-Newtonian Navier-Stokes flow problem where the viscosity depends
both on the effective strain rate and the temperature. 2) A linear heat conduction problem including both the aluminum domain and the extrusion tool
package. The constitutional relationship is described by the simple Oldroyd-
Bingham and Zener-Hollomon material models. However, other models may
be implemented easily when desired. A mixed interpolation of the unknown
variables is employed, using second order polynomials for the velocity- and
temperature fields, and linear polynomials for the pressure.

Among the most important simplifications in the present implementation
are: 1) No elasto-plastic effects are taken into account. 2) The contact between
the aluminum and the die is either full stick or full slip, i.e., no friction laws are
implemented. 3) The extrusion die is assumed to be fixed, i.e., no deformation
of the bearing channel is taken into account.

In what follows, the mathematical formulation of the extrusion problem is
described in detail, and then the numerical finite element model is presented.
Finally, we presents some numerical results of the temperature distribution
(in a critical point) compared with physical experiments.

## 14.2     Mathematical Formulation

### 14.2.1     Basic Definitions

The system in study is a fluid body with mass density $\rho$ occupying a de-
formable domain $\Omega = \Omega(t) \subset \mathbb{R}^{n_d}$ in its present configuration. We assume
that $\Omega$ is connected and that its boundary is smooth, i.e., $\partial\Omega$ is Lipschitzian,
see [1]. The corresponding unit outer boundary normal, $n \in \mathbb{N}$, will then
exist "almost everywhere" (a.e.) on $\partial\Omega$.

In this study, it is assumed that the time evolution of the fluid domain is
known. We also assume that the fluid considered (hot aluminum) is incompressible such that $\rho$ is a known constant. The problem is then limited to
finding the fluid velocity $u$, the pressure $p$ and the temperature $T$ within the
deforming fluid domain.

The deformation of the fluid body is assumed to be large. Therefore,
the change of the domain $\Omega$ must be taken into account when evaluating
the integrals of the conservation laws governing the problem. These laws
are posed on the current deformed configuration and then transported to a
fixed reference configuration $\Omega^0$. For this purpose, we introduce a continuous
mapping as in [2],

$$x \; : \; \Omega^0 \times \mathbb{R} \longrightarrow \Omega(t), \tag{14.1}$$

which maps any point $X \in \Omega^0$ of the reference configuration to its image $x(X,t) \in \Omega(t)$. This may be viewed as an Eulerian description with a moving coordinate system.

Since the configuration movement is different from the movement of the fluid particles themselves, we also need to distinguish between the fluid particle velocity, $u$, and the *configuration* or *grid* velocity $u^g$. The latter is given by

$$u^g(X,t) := \frac{\partial}{\partial t} x(X,t).$$ 
(14.2)

### 14.2.2    Governing Equations

The governing equations of the extrusion problem may be derived from the conservation laws of fluid mechanics and the Reynolds transport theorem (see e.g. [3]), stating that *the rate of change of the integral of a single-valued point function over a material volume is equal to the time rate of change of this integral over the corresponding control volume, plus the flux of this point function across the surface of the control volume*. The material volume follows the fluid particles, whereas the control volume normally is fixed in time and coincides with the material volume at a given time instant. However, herein the control volume is moving independently of the fluid particles and the *relative* velocity of the corresponding material volume is $u^{rel} := u - u^g$.

Through this we end up with the following three partial differential equations for the fluid domain:

$$\text{Conservation of mass:} \qquad \frac{\partial J}{\partial t} = -J\,\nabla \cdot u^{rel}. \qquad (14.3)$$

$$\text{Conservation of linear momentum:} \qquad \rho \frac{Du}{Dt} = \nabla \cdot \sigma + f. \qquad (14.4)$$

$$\text{Conservation of energy:} \qquad \rho \frac{De}{Dt} = \sigma : \varepsilon - \nabla \cdot q. \qquad (14.5)$$

Here, $J$ is the determinant of the deformation gradient of the grid, i.e., $J = \det[\frac{\partial x^i}{\partial X^j}]$ where $x^i$ and $X^j$ denote the Cartesian components of $x$ and $X$, respectively; $\sigma$ the symmetric Cauchy stress tensor, $\varepsilon$ the symmetric strain rate tensor, $f$ the applied body force density (herein assumed to be zero), $e$ the internal energy density and $q$ the heat flux through a given cross section. The *material derivative* operator $\frac{D}{Dt}$ is defined through

$$\frac{D\bullet}{Dt} := \begin{cases} \dfrac{\partial \bullet}{\partial t} + (\nabla \otimes \bullet)^T \cdot u^{rel} & \text{(for vector fields),} \\[2mm] \dfrac{\partial \bullet}{\partial t} + \nabla \bullet \cdot u^{rel} & \text{(for scalar fields).} \end{cases} \qquad (14.6)$$

The strain rate and stress tensors are assumed as, respectively,

$$\varepsilon(\boldsymbol{u}) = \frac{1}{2}\left(\boldsymbol{\nabla}\otimes\boldsymbol{u} + (\boldsymbol{\nabla}\otimes\boldsymbol{u})^{\mathrm{T}}\right), \tag{14.7}$$

$$\boldsymbol{\sigma} = -p\boldsymbol{I} + 2\mu(\bar{\varepsilon},T)\varepsilon, \tag{14.8}$$

where $\boldsymbol{I}$ denotes the $n_{\mathrm{d}} \times n_{\mathrm{d}}$ identity tensor and $\mu$ is the dynamic viscosity, which is a function of the effective strain rate $\bar{\varepsilon} := \sqrt{\frac{2}{3}\varepsilon : \varepsilon}$, and the temperature $T$. The colon operator (:) denotes the tensor inner product: $\varepsilon : \varepsilon \equiv \varepsilon_{ij}\varepsilon_{ij}$. In this work, we use the Zener-Hollomon material model [4], where

$$\mu(\bar{\varepsilon},T) := \frac{1}{3\bar{\varepsilon}\,\alpha_{\mathrm{z}}} \sinh^{-1}\left(\left(\frac{Z}{A}\right)^{\frac{1}{m}}\right) \quad \text{with} \ \ Z = \bar{\varepsilon}\exp\left(\frac{Q}{RT}\right), \tag{14.9}$$

with $\alpha_{\mathrm{z}}$, $A$, and $m$ being material parameters, $Q$ the activation energy, and $R$ the universal gas constant. Note that the temperature $T$ in (14.9) is given in Kelvin [K] temperature scale. To avoid numerical difficulties in areas where $\bar{\varepsilon} \to 0$, expression (14.9) is replaced by a quadratic approximation for small values of $\bar{\varepsilon}$, in the same way as in the *ALMA* code [5].

We assume that the internal energy density $e$ is a function of the temperature only, and that Fourier's law of heat conduction applies, i.e.,

$$\frac{De}{Dt} = c_{\mathrm{v}}\frac{DT}{Dt}, \tag{14.10}$$

$$\boldsymbol{q} = -\boldsymbol{k}\cdot\boldsymbol{\nabla}T, \tag{14.11}$$

where $c_{\mathrm{v}}$ is the specific heat capacity and $\boldsymbol{k}$ is the second-order thermal conductivity tensor. Herein, we assume that both $c_{\mathrm{v}}$ and $\boldsymbol{k}$ are constant.

### 14.2.3  Boundary Conditions

The problem domain $\Omega$ is now assumed to consist of four non-overlapping sub-domains $\Omega_A$, $\Omega_B$, $\Omega_C$, and $\Omega_D$ such that $\overline{\Omega} = \overline{\Omega}_A \cup \overline{\Omega}_B \cup \overline{\Omega}_C \cup \overline{\Omega}_D$, see Figure 14.3. We assume that all four sub-domains have Lipschitzian boundaries. The interfaces between the sub-domains are $\Gamma_{AB} := \partial\Omega_A \cap \partial\Omega_B$, $\Gamma_{AC} := \partial\Omega_A \cap \partial\Omega_C$, $\Gamma_{AD} := \partial\Omega_A \cap \partial\Omega_D$, $\Gamma_{BC} := \partial\Omega_B \cap \partial\Omega_C$, and $\Gamma_{CD} := \partial\Omega_C \cap \partial\Omega_D$. We also define the external boundary of each sub-domain, respectively, $\partial\Omega_{A_{\mathrm{ext}}} := \overline{\partial\Omega_A \setminus (\Gamma_{AB} \cup \Gamma_{AC} \cup \Gamma_{AD})}$, $\partial\Omega_{B_{\mathrm{ext}}} := \overline{\partial\Omega_B \setminus (\Gamma_{AB} \cup \Gamma_{BC})}$, $\partial\Omega_{C_{\mathrm{ext}}} := \overline{\partial\Omega_C \setminus (\Gamma_{AC} \cup \Gamma_{BC} \cup \Gamma_{CD})}$, and $\partial\Omega_{D_{\mathrm{ext}}} := \overline{\partial\Omega_D \setminus (\Gamma_{AD} \cup \Gamma_{CD})}$, such that $\partial\Omega = \partial\Omega_{A_{\mathrm{ext}}} \cup \partial\Omega_{B_{\mathrm{ext}}} \cup \partial\Omega_{C_{\mathrm{ext}}} \cup \partial\Omega_{D_{\mathrm{ext}}}$. We assume that all these boundary- and interface-surfaces are of nonzero Lebegues measure.

The four sub-domains are assumed to have the following characteristics:

$\Omega_A$ : Deforming fluid domain (aluminum billet and profile).
- Known field variables: $\rho$, $c_{\mathrm{v}}$, $\boldsymbol{u}^{\mathrm{g}}$, $\boldsymbol{f}$, $\boldsymbol{k}$
  and the function $\mu(\bar{\varepsilon},T)$.

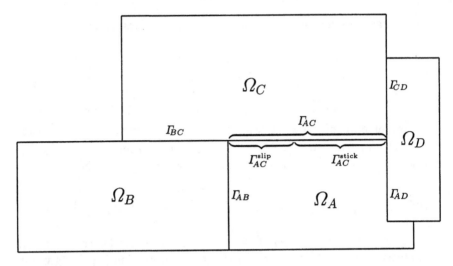

**Fig. 14.3.** Schematic partitioning of the whole problem domain into four subdomains.

- Primary unknowns: $u$, $p$, and $T$

$\Omega_B$ : Rigid but moving solid domain (ram).
  - Known field variables: $\rho$, $c_v$, $u = u^g = u_B$ and $k$
  - Primary unknown: $T$

$\Omega_C$ : Fixed solid domain (container).
  - Known field variables: $\rho$, $c_v$, $u = u^g = 0$ and $k$
  - Primary unknown: $T$

$\Omega_D$ : Fixed solid domain (die).
  - Known field variables: $\rho$, $c_v$, $u = u^g = 0$ and $k$
  - Primary unknown: $T$

Thus, the only difference between the rigid domains $\Omega_B$, $\Omega_C$, and $\Omega_D$ is that $\Omega_B$ is moving, whereas $\Omega_C$ and $\Omega_D$ are fixed.

In the present work we assume either full stick or full slip for the fluid velocity along the interfaces $\Gamma_{AB}$, $\Gamma_{AC}$, and $\Gamma_{AD}$. The interface $\Gamma_{AC}$ is thus subdivided into $\Gamma_{AC}^{slip}$ and $\Gamma_{AC}^{stick}$ (see Figure 14.3), such that $\Gamma_{AC} = \Gamma_{AC}^{slip} \cup \Gamma_{AC}^{stick}$ and $\mathring{\Gamma}_{AC}^{slip} \cap \mathring{\Gamma}_{AC}^{stick} = \emptyset$. Similarly, we divide interface $\Gamma_{AD}$ into $\Gamma_{AD}^{slip}$ and $\Gamma_{AD}^{stick}$ such that $\Gamma_{AD} = \Gamma_{AD}^{slip} \cup \Gamma_{AD}^{stick}$ and $\mathring{\Gamma}_{AD}^{slip} \cap \mathring{\Gamma}_{AD}^{stick} = \emptyset$. We then assume full stick for the fluid velocity at the interfaces $\Gamma_{AB}$, $\Gamma_{AC}^{stick}$, and $\Gamma_{AD}^{stick}$, whereas along $\mathring{\Gamma}_{AC}^{slip}$ and $\mathring{\Gamma}_{AD}^{slip}$ only the normal component of the velocity is zero. The Dirichlet boundary conditions of the fluid flow problem is therefore

$$u = \bar{u} \quad \text{on} \quad \partial\Omega_u, \tag{14.12}$$

where $\partial\Omega_u := \Gamma_{AB} \cup \Gamma_{AC} \cup \Gamma_{AD}$ and $\bar{u}$ is a function defined through

$$\bar{u} : \partial\Omega_u \to \mathbb{R}^{n_d}, \quad \begin{cases} \bar{u}(x) = u_B, & \forall x \in \Gamma_{AB}, \\ \bar{u}(x) = 0, & \forall x \in \Gamma_{AC}^{stick} \cup \Gamma_{AD}^{stick}, \\ n(x) \cdot \bar{u}(x) = 0, & \forall x \in \dot{\Gamma}_{AC}^{slip} \cup \dot{\Gamma}_{AD}^{slip}. \end{cases} \quad (14.13)$$

We also assume that we can have some non-homogeneous Neumann boundary conditions

$$n \cdot \sigma = \bar{t} \quad \text{on } \partial\Omega_t, \quad (14.14)$$

where the prescribed traction $\bar{t}$ could be the result of a puller force on the extruded profile, and $\partial\Omega_t := \overline{\partial\Omega_A \setminus \partial\Omega_u} = \partial\Omega_{A_{ext}}$.

The heat flux through a domain boundary is assumed to be governed by

$$n \cdot q = \alpha(T_a - T_a^{amb}) \quad \text{on } \partial\Omega_a, \, a = A \ldots D, \quad (14.15)$$

where $\alpha$ is the heat transfer coefficient and $T_a$ denotes the temperature field in sub-domain $\Omega_a$. The field $T_a^{amb}$ represents the temperature "outside" the domain in question, thus

$$T_a^{amb} : \partial\Omega_a \to \mathbb{R}, \quad \begin{cases} T_a^{amb}(x) = T_b(x), & \forall x \in \Gamma_{ab}, \, b = A \ldots D \neq a, \\ T_a^{amb}(x) = T_{air}, & \forall x \in \partial\Omega_{a_{ext}}, \, a = A \ldots D, \end{cases} \quad (14.16)$$

where $T_{air}$ is the ambient air temperature. Heat exchange is assumed to take place at all external boundaries and interfaces, but the heat transfer coefficient $\alpha$ may be set equal to zero for adiabatic boundaries, if any.

## 14.2.4  Variational Formulation

To establish the variational formulation, we introduce the following spaces:

$$\mathcal{U}(\Omega) = \{ v : \Omega \to \mathbb{R}^{n_d}, \, v \in \boldsymbol{H}^1(\Omega) : v = \bar{u} \text{ a.e. on } \partial\Omega_u \}, \quad (14.17)$$
$$\mathcal{V}(\Omega) = \{ v : \Omega \to \mathbb{R}^{n_d}, \, v \in \boldsymbol{H}^1(\Omega) : v = 0 \text{ a.e. on } \partial\Omega_u \}, \quad (14.18)$$
$$\mathcal{Q}(\Omega) = \{ q : \Omega \to \mathbb{R}, \quad q \in L_2(\Omega) \}, \quad (14.19)$$
$$\mathcal{T}(\Omega) = \{ T : \Omega \to \mathbb{R}, \quad T \in H^1(\Omega) \}, \quad (14.20)$$

where $\boldsymbol{H}^1(\Omega)$ and $H^1(\Omega)$ denote the usual Sobolev spaces of vector- and scalar-valued functions on $\Omega$, respectively, and $L_2(\Omega)$ is the space of all scalar-valued functions on $\Omega$ for which the $L_2$ norm is bounded, see [1] for details. Since we assume no Dirichlet boundary conditions for the heat conduction problems, the test and trial spaces for the temperature field are equal, as is the case also for the pressure field.

The governing equations given in Sections 14.2.2 and 14.2.3 can now be used to formulate a set of boundary value problems for each sub-domain, as shown by Okstad, Kvamsdal, and Abtahi in [6]. The associated variational formulation is then obtained by multiplying these partial differential equations by test functions $\delta u \in \mathcal{V}(\Omega)$, $\delta q \in \mathcal{Q}(\Omega)$, and $\delta T \in \mathcal{T}(\Omega)$, respectively, and integrating over the appropriate domains. Through integration by parts and the use of the divergence theorem, we then end up with the following set of variational problems:

1. Fluid flow problem in sub-domain $\Omega_A$.

   Given $\rho \in \mathbb{R}$, $p_0 \in \mathcal{Q}(\Omega_A)$, $u^g \in H^1(\Omega_A)$, $T \in \mathcal{T}(\Omega_A)$, $\bar{u} : \partial\Omega_u \to \mathbb{R}^{nd}$, $\bar{t} : \Omega_t \to \mathbb{R}^{nd}$, and $\mu : \mathbb{R}^{nd} \times \mathbb{R} \to \mathbb{R}$, find the fluid velocity $u \in \mathcal{U}(\Omega_A)$ and the fluid pressure $p \in \mathcal{Q}(\Omega_A)$, such that for any $\{\delta u, \delta p\} \in \mathcal{V}(\Omega_A) \times \mathcal{Q}(\Omega_A)$, the following is satisfied:

$$\int_{\Omega_A} \left( \frac{1}{J}\frac{\partial J}{\partial t} + \nabla \cdot u^{\text{rel}} \right) \delta q \, dV = 0. \tag{14.21}$$

$$\rho \int_{\Omega_A} \frac{\partial u}{\partial t} \cdot \delta u \, dV + \rho \int_{\Omega_A} \left[ (\nabla \otimes u)^{\text{T}} \cdot u^{\text{rel}} \right] \cdot \delta u \, dV$$

$$- \int_{\Omega_A} p \nabla \cdot \delta u \, dV + \int_{\Omega_A} 2\mu(u, T)\, \varepsilon(u) : \varepsilon(\delta u) \, dV \tag{14.22}$$

$$= \int_{\partial\Omega_t} \bar{t} \cdot \delta u \, dA.$$

2. Heat conduction problem in sub-domain $\Omega_A$.

   Given $\rho \in \mathbb{R}$, $c_v \in \mathbb{R}$, $\alpha \in \mathbb{R}$, $k \in \mathbb{R}^{nd} \times \mathbb{R}^{nd}$, $T_0 \in \mathcal{T}(\Omega_A)$, $u^{\text{rel}} \in H^1(\Omega_A)$, $T_A^{\text{amb}} : \partial\Omega_A \to \mathbb{R}$, $\sigma : \Omega_A \to \mathbb{R}^{nd} \times \mathbb{R}^{nd}$, and $\varepsilon : \Omega_A \to \mathbb{R}^{nd} \times \mathbb{R}^{nd}$, find the temperature $T \in \mathcal{T}(\Omega_A)$, such that for any $\delta T \in \mathcal{T}(\Omega_A)$, the following is satisfied:

$$\rho\, c_v \int_{\Omega_A} \frac{\partial T}{\partial t} \delta T \, dV + \rho\, c_v \int_{\Omega_A} \nabla T \cdot u^{\text{rel}} \delta T \, dV$$

$$+ \int_{\Omega_A} k \cdot \nabla T \cdot \nabla \delta T \, dV + \alpha \int_{\partial\Omega_A} T \delta T \, dA \tag{14.23}$$

$$= \int_{\Omega_A} (\sigma : \varepsilon)\, \delta T \, dV + \alpha \int_{\partial\Omega_A} T_A^{\text{amb}} \delta T \, dA.$$

3. Heat conduction problems in sub-domains $\Omega_B$, $\Omega_C$, and $\Omega_D$.

   Given $\rho \in \mathbb{R}$, $c_v \in \mathbb{R}$, $\alpha \in \mathbb{R}$, $k \in \mathbb{R}^{nd} \times \mathbb{R}^{nd}$, $T_0 \in \mathcal{T}(\Omega_B)$, and $T_B^{\text{amb}} : \partial\Omega_B \to \mathbb{R}$, find the temperature $T \in \mathcal{T}(\Omega_B)$, such that for any

$\delta T \in \mathcal{T}(\Omega_B)$, the following is satisfied:

$$
\rho\, c_v \int_{\Omega_B} \frac{\partial T}{\partial t}\, \delta T\, \mathrm{d}V + \int_{\Omega_B} k \cdot \boldsymbol{\nabla} T \cdot \boldsymbol{\nabla} \delta T\, \mathrm{d}V + \alpha \int_{\partial\Omega_B} T \delta T\, \mathrm{d}A
$$
$$
= \alpha \int_{\partial\Omega_B} T_B^{\mathrm{amb}} \delta T\, \mathrm{d}A. \tag{14.24}
$$

and similarly for the sub-domains $\Omega_C$ and $\Omega_D$ with index $B$ replaced by $C$ and $D$, respectively.

## 14.3 Finite Element Implementation

### 14.3.1 Time Discretization

The first issue we need to address in a numerical implementation of a solver for the mathematical problem described in Section 14.2, is to construct a time discretization scheme for the evolution of the solution in time. The solution is computed at a sequence of discrete times $t_n, n = 1, \ldots$, using some finite difference approximations of the various time derivatives involved. In the following we let the subscript $n$ indicate quantities evaluated at time step $n$. The question now is how to approximate the time derivatives appearing in the variational problems (14.21)–(14.24).

For the mass conservation equation (14.21), a finite difference approximation of the time derivative of $J$ is implicitly defined as

$$
\frac{1}{J_n} \left( \frac{\partial J}{\partial t} \right)_n := \boldsymbol{\nabla} \cdot \boldsymbol{u}_n^{\mathrm{g}}. \tag{14.25}
$$

This can be viewed as a discrete geometric conservation law for the mass conservation equation [2], and implies that the time-discrete mass conservation equation for a finite element approximation of an incompressible flow reduces to

$$
\int_{\Omega(t_n)} \boldsymbol{\nabla} \cdot \boldsymbol{u}_n\, \delta q\, \mathrm{d}V = 0, \quad \forall\, \delta q \in \mathcal{Q}(\Omega). \tag{14.26}
$$

For the fluid- and heat-flow problems considered in this work, the time derivatives are so small that a simple backward Euler scheme is assumed to be a sufficient approximation. Thus, at time step $n$ we let

$$
\left( \frac{\partial \boldsymbol{u}}{\partial t} \right)_n = \frac{\boldsymbol{u}_n - \boldsymbol{u}_{n-1}}{\Delta t_n}, \tag{14.27}
$$
$$
\left( \frac{\partial T}{\partial t} \right)_n = \frac{T_n - T_{n-1}}{\Delta t_n}, \tag{14.28}
$$

where $\Delta t_n := t_n - t_{n-1}$ is the time step size. A trapezoidal scheme has also been implemented, but the simulations carried out so far showed only negligible differences in the numerical results for the two schemes.

With the above choice, the time discrete sub-problems may be obtained, see [6] for the details.

### 14.3.2    Spatial Discretization

The linearized version of the variational sub-problems may be solved at a given time $t_n$ by introducing a finite element discretization over each sub-domain and replacing the test- and trial spaces (14.17)–(14.20) with finite dimensional equivalents. The subsequent treatment then follows standard finite element procedures in order to obtain the solution at a given time.

In the present work we use a mixed interpolation of the fluid flow variables, with quadratic $C^0$-continuous test and trial functions for the velocities and linear $C^0$-continuous functions for the pressure (Taylor-Hood elements). The temperature field is also interpolated with quadratic functions in the sub-domain $\Omega_A$, whereas in the other sub-domains both quadratic and linear interpolations are possible.

The nonlinear algebraic equations of the fluid problem are solved using a Newton-Raphson procedure with fixed temperature variables. Although it is possible also to update the temperatures within the nonlinear iteration loop, a fully coupled nonlinear solution is not computed since the temperature dependent term of the linearized viscous stress tensor is neglected.

### 14.3.3    Global Solution Procedure and Mesh Movement

The global solution procedure employed in the *Extrud* solver is summarized in Figure 14.4. The overall solution at a given time step is obtained by iterating between the various linear or linearized sub-problems in a prescribed manner. Herein, we use a staggering scheme of Gauss-Seidel type, meaning that when we solve a given sub-problem, we use the solution at the *present* time step in the neighboring sub-domains that already have been solved at this time step, whereas we use the solution at the *previous* step in the sub-domains that have not yet been solved.

The mesh movement step, i.e., the computation of updated grid-position and -velocity, is the first thing that is performed within a certain time step. Since the location and velocity of the moving boundary $\Gamma_{AB}$ is known at each time step, we only need to compute the location and velocity of the interior nodes of the fluid domain $\Omega_A$ in a consistent manner such that the mesh quality (element shapes) is maintained as well as possible.

Herein, we use two different methods to accomplish this. The first and simplest method is just to let the grid velocity field and the incremental displacement field of the grid be defined by a linearly decreasing function from the interface $\Gamma_{AB}$ over a user-specified distance from the initial ram position.

---

Set initial conditions $\Rightarrow u_0, p_0, T_0$.

FOR EACH time step $n = 1, \ldots$ DO

FOR EACH stagger iteration DO
1. <u>Mesh movement</u>. Update the nodal coordinates of the fluid and ram meshes ($\Omega_A$ and $\Omega_B$) and compute mesh velocity $\Rightarrow x_n, u_n^g$.
2. <u>Fluid flow</u>.

   WHILE fluid problem not converged DO
   2.1 Solve the linearized fluid flow problem in $\Omega_A \Rightarrow \Delta u_n, \Delta p_n$
   2.1 Update the nonlinear flow solution $\Rightarrow u_n, p_n$
      IF update of temperature in $\Omega_A$ is desired THEN
   2.2   Solve the linear heat conduction problem in $\Omega_A$
      based on the latest fluid flow solution $\Rightarrow (T_A)_n$
   END IF
   END DO
3. <u>Heat conduction</u>.
   3.1 Solve the linear heat conduction problem in $\Omega_A \Rightarrow (T_A)_n$
   3.2 Solve the linear heat conduction problem in $\Omega_B \Rightarrow (T_B)_n$
   3.3 Solve the linear heat conduction problem in $\Omega_D \Rightarrow (T_D)_n$
   3.4 Solve the linear heat conduction problem in $\Omega_C \Rightarrow (T_C)_n$
END DO

END DO

---

**Fig. 14.4.** Global solution procedure in *Extrud*.

A more sophisticated approach is to consider the aluminum domain as a linear elastic continuum. The incremental nodal displacements, $\Delta d$, of the grid are then solved from a finite element discretization of the linear elasticity equations over the current grid configuration, in which the prescribed incremental displacement of the boundary $\Gamma_{AB}$ is enforced as Dirichlet conditions. The grid velocity and updated grid position are then taken as

$$u_n^g = \frac{\Delta d_n}{\Delta t_n}, \tag{14.29}$$

$$x_n = x_{n-1} + \Delta d_n. \tag{14.30}$$

We refer to [7], Section 5.1 for details on the elasticity equations and the associated finite element implementation.

## 14.4 Object-Oriented Implementation

### 14.4.1 Introduction

When implementing a numerical procedure for solving some mathematical problem, one should aim to make the implementation as generic as possible such that it can be reused for different problems without changing the core of the program. An important aspect in this regard is to separate the problem-dependent data and methods from those of the common numerical solution procedure. Maintaining such a separation may be difficult if relying on traditional procedure-oriented programming languages, such as Fortran-77, C, etc. A program written in these languages may run very well as long as it is used on problems for which it was originally designed. But as soon as one tries to use it on a somewhat different kind of problem, even if the implemented numerical procedure is applicable to this problem too, one often realizes that this cannot be done without significantly changing some data structures in the program which can affect all relevant parts of the program. In contrast, using object-oriented techniques, such data structures can be encapsulated in problem-dependent objects on which the numerical solver operates through a well-defined interface. Solving a new problem is then not more difficult than creating a new sub-class of the problem definition class with the additional data and/or functions.

In the present implementation of *Extrud*, this philosophy has been pursued from the beginning. All problem-specific data are stored in a class hierarchy, with the most general problem type as a base class and then adding more data and functions down the hierarchy as the problems get more specialized. Similarly, a class hierarchy of numerical solvers has also been created with the common features collected in base classes, and then adding more specialized features as we go down the hierarchy. In this way we have a computational tool that not only can solve the extrusion problem, albeit the initial goal of this study, but also other kinds of fluid flow problems or pure heat-conduction problems.

To illustrate this way of programming, we will now discuss in detail some major classes of our implementation, which is based upon the Diffpack FEM C++ library [7,8]. We assume in the following that the reader is familiar with basic use of this library and refer to the excellent book [7] for details.

### 14.4.2   Class Hierarchy for the Problem-Dependent Data

Figure 14.5 shows an overview of the classes containing problem-dependent data, and methods and how they are correlated. Since the problem we want to solve is a coupled fluid and heat flow problem, we have two base classes for representing each of these branches, respectively. These two classes do not contain much data themselves, but serves merely as interfaces to the lower-level problem-classes through the virtual member functions. Let us see the definition of the CFDProblem class:

```
class CFDProblem
{
protected:
   real      rho, mu; // Density and dynamic viscosity
   Handle(Fields) u;  // Analytical velocity field, if any
```

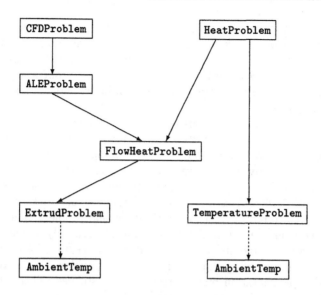

**Fig. 14.5.** Class hierarchy for the problem dependent data and methods. The solid arrows indicate class inheritance ("is a"-relationship), whereas the dashed arrows indicate reference to internal data objects ("has a"-relationship).

```
  Handle(Field)  p; // Analytical pressure field, if any
public:
              CFDProblem () { rho = mu = 1.0; }
    virtual ~CFDProblem () {}

    virtual void define (MenuSystem& menu, int level) = 0;
    virtual void scan (MenuSystem& menu, Handle(GridFE)& grid) = 0;

    virtual void initialVelocity (FieldsFE& u0);
    virtual void initialPressure (FieldFE& p0);

    virtual bool movingMesh () const { return false; }

    virtual real inletVelocity (BasisFuncGrid& g, int node, int dof,
                                real t);
    virtual void traction (Ptv(real)& trac, FiniteElement& fe, real t);
    virtual void bodyforce (Ptv(real)& force, FiniteElement& fe, real t);

    real getDensity   () const { return rho; }
    real getViscosity () const { return mu; }

    Fields* Uex () { return u.getPtr(); }
    Field*  Pex () { return p.getPtr(); }
};
```

The base class for CFD-problems contains the density and dynamic viscosity parameters as protected data members and associated non-virtual member functions for retrieving the values. Thus, this class is restricted to problems where these parameters are constant in space and time (incompressible Newtonian flow). However, herein we use it as a base also for non-Newtonian problems. The viscosity parameter is then provided from a separate class outside the problem class hierarchy, as shown later, and the getViscosity functions is not used.

The virtual member functions of this class define the interface through which the problem properties are obtained within the numerical solvers. This includes definition of initial conditions, obtaining the current value of the inflow velocity at a given node, evaluating the surface traction on the outflow boundary if prescribed, and evaluating the body force at an interior point. Sub-classes of CFDProblem may now be derived where these functions are re-implemented to reflect the characteristics of that problem. Some problems may have very simple implementations, which return only a constant value, whereas others may require more sophisticated procedures for obtaining the values. However, by hiding all these details in the problem-classes, the numerical solver implementation may be retained undisturbed. This is the major advantage of objected-oriented implementation.

The CFDProblem base class also contains pointers to fields representing the analytical solution. This opens for implementation of functions for evaluating error norms of the computed solution, and is useful for evaluating the accuracy of the numerical procedures on problems, where the solution is known. In an earlier investigation (see [9,10]) we have shown how such error norm evaluation and estimation may be implemented in an object-oriented setting based on Diffpack.

The other base class for problem data, HeatProblem, is defined as follows:

```
class HeatProblem : public virtual HandleId
{
protected:
  real        c; // Heat capacity
  Handle(Field) q; // Analytical temperature field, if any

public:
            HeatProblem () { c = 1.0; }
  virtual ~HeatProblem () {}

  virtual void define (MenuSystem& menu, int level) = 0;
  virtual void scan (MenuSystem& menu, Handle(GridFE)& grid) = 0;

  virtual void initialTemperature (FieldFE& q0);
  virtual void initialTemperatureTimeDerivative (FieldFE& dq0);

  virtual real dirichlet (GridFE& g, int node, int boInd, real t);
  virtual real neumann (FiniteElement& fe,
                        int side, int boInd, real t);
  virtual real robin (real& alpha, FiniteElement& fe,
                      int side, int boInd, real t);
  virtual real diffusionCoeff (FiniteElement& fe, real t);
  virtual real heatSource (FiniteElement& fe, real t);
```

```
real getHeatCapacity () const { return c; }

Field* Qex () { return q.getPtr(); }
};
```

As shown above, the overall structure of this class is somewhat similar to that of `CFDProblem`, with virtual member functions for retrieving problem-specific properties on the boundaries and inside the domain. The `dirichlet` and `neumann` functions are the equivalent to `inletVelocity` and `traction` functions of `CFDProblem`, respectively. In addition, we have the `robin` function returning both the heat flux and the heat transfer coefficient across the boundary. This function is needed in our application since the problem we want to solve consists of multiple temperature domains.

The `HeatProblem` base class is restricted to problems with constant heat capacity and thus this parameter is stored as a single variable in the class. We also have the option of deriving sub-classes with an analytical solution field for accuracy assessment, as in the `CFDProblem` class.

Now, moving down in the class hierarchy depicted in Figure 14.5, we have the sub-class `ALEProblem` of `CFDProblem`. This may be regarded as a base class for CFD-problems for which the grid is deforming during the simulation and this necessitates an ALE-formulation. This class only adds a few extra virtual member-functions for describing how the mesh-domain should deform and contains no data itself.

The next level consists of the `FlowHeatProblem` base class which is defined through multiple inheritance from `CFDProblem` and `HeatProblem`. Thus, this is our base class for coupled fluid and heat flow problems.

Finally, we arrive at our application-specific problem class at the lowest level, which we derive as a sub-class of `FlowHeatProblem` as follows:

```
class ExtrudProblem : public FlowHeatProblem
{
  VecSimple(real) Uram; // ram velocity at distinct time instants
  VecSimple(real) tram; // time instants at which the ram vel. changes
  int             dram; // direction of ram movement
  real            tbar; // outflow traction value
  real      k, T0, Tair; // diffusion coeff., initial- and air-temp.
  bool           withALE; // true, if a moving mesh should be used

public:
  AmbientTemp ambient; // ambient temp. fields and transfer coeff.

  ExtrudProblem ();
  ~ExtrudProblem () {}

  virtual void define (MenuSystem& menu, int level);
  virtual void scan (MenuSystem& menu, Handle(GridFE)& grid);

  virtual void initialTemperature (FieldFE& q0) { q0.fill(T0); }

  virtual bool movingMesh () const { return withALE; }
```

```
    virtual real ALEdis (int dof, real t, real dt);

    virtual real inletVelocity (BasisFuncGrid& g, int n, int dof,
                                real t);
    virtual void traction (Ptv(real)& trac, FiniteElement& fe, real t);
    virtual real robin (real& alpha, FiniteElement& fe,
                        int side, int boInd, real t);
    virtual real diffusionCoeff (FiniteElement& fe, real t) { return k; }
    real airTemperature () const { return Tair; }
};
```

This class re-implements some of the member functions of CFDProblem and
HeatProblem (but not all of them) and contains also more data for describing
the problem we want to solve. More specifically, it stores data for description
of the ram movement as a function of time. The present implementation thus
allows the ram to move with piece-wise constant velocity, and the ALEdis
member function (inherited from the ALEProblem class) uses this data to
update the position of the moving boundary that represents the interface
between the ram and the aluminum domains.

The ExtrudProblem class contains all the necessary data and functions
for describing the aluminum domain in our coupled fluid and heat flow simulator. However, the extrusion problem also consists of other temperature-only
domains (the extrusion tool), for which we use the following problem class
derived directly from HeatProblem:

```
class TemperatureProblem : public HeatProblem
{
    real k, T0, Tair; // diffusion coeff., initial- and air-temp.
public:
    AmbientTemp ambient; // ambient temp. fields and transfer coeff.

    TemperatureProblem ();
   ~TemperatureProblem () {}

    virtual void define (MenuSystem& menu, int level);
    virtual void scan (MenuSystem& menu, Handle(GridFE)& grid);

    virtual void initialTemperature (FieldFE& q0);
    virtual real diffusionCoeff (FiniteElement& fe, real t);
    virtual real robin (real& alpha, FiniteElement& fe,
                        int side, int dof, real t);
    real airTemperature () const { return Tair; }
};
```

As shown above, all data and functions of this class are also present in
ExtrudProblem, so one could argue that it would be better to derive the
latter class from TemperatureProblem and thereby avoiding defining these
functions twice. However, that is not possible since both FlowHeatProblem
and TemperatureProblem are derived from the base class HeatProblem.

Both problem classes described above contain a data member of the class
AmbientTemp. This class serves as a collection of all the temperature fields of

the various domains surrounding the problem domain of interest. Its definition is as follows:

```
class AmbientTemp
{
  VecSimplest(Handle(Field)) T;      // ambient temperature fields
  VecSimplest(real)          alpha;  // heat transfer coefficients
  VecSimplest(int)           bind;   // boundary ind. for each field
public:
  AmbientTemp ();
  ~AmbientTemp () {}

  void redim (const int nDomain);
  void attach (Field* f, real coeff, int bInd, int index);
  real valuePt (real& coeff, int bInd, Ptv(real)& x, real t);
};
```

In addition to a vector of temperature fields, it contains vectors of associated heat transfer coefficients and boundary indicators defining the extent of the boundary through which the heat transfer shall take place for each temperature field. The `valuePt` member function is then used within the `robin` function of the owner class when evaluating the boundary conditions.

### 14.4.3   Class Hierarchy for the Numerical Solvers

Having defined the class structure for the problem data, we now proceed with the numerical solvers themselves. They are organized in class hierarchies based on the same object-oriented philosophy as for the problem classes; collecting the most general features in 'general' base classes and deriving more special purpose solvers as sub-classes. In addition, we need some administrative classes for the coupling of the flow-, temperature-, and mesh-movement solvers.

The overall class hierarchy of this coupled system is shown in Figure 13.4. This chart might be viewed as four distinct parts; the left-most part is concerned with the fluid problem, the right-most part with the heat flow problems, the upper part with the mesh-movement problem, and finally the center part with the coupling and overall solution procedure. We will now look into these parts in more detail starting with the center part.

The overall solution process of the extrusion simulation is handled through the class `ExtrudSolver` which is defined as follows:

```
typedef VecSimplest(Handle(HeatProblem)) HeatProblems;
typedef VecSimplest(Handle(FieldSolver)) HeatSolvers;

class ExtrudSolver : public SimCase
{
protected:
  FlowHeatProblem&   alum;   // aluminum domain (fluid and heat)
  HeatProblems&      tool;   // extrusion tool domains (heat only)
  MatSimple(real)    alpha;  // heat transfer coeff. betw. domains
```

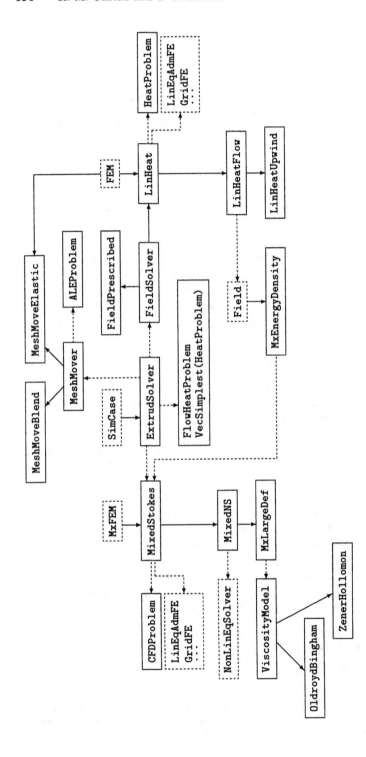

**Fig. 14.6.** Class hierarchy for the coupled fluid and heat flow solvers with moving mesh. The solid arrows indicate class inheritance ("is a"-relationship), whereas the dashed arrows indicate references to internal data objects ("has a"-relationship). Classes with dotted line boxes are part of the Diffpack library. Here we only show a few of the most important Diffpack classes. The three dots '...' in two of the dotted boxes represent other Diffpack-classes which are used here.

```
Handle(TimePrm)       tip;     // time integration parameters

Handle(MeshMover)     meshSolver;
Handle(MixedStokes)   flowSolver;
HeatSolvers           heatSolver;
public:
  ExtrudSolver (FlowHeatProblem& alu_, HeatProblems& tool_);
 ~ExtrudSolver () {}

  void define (MenuSystem& menu, int level = MAIN);
  void scan (MenuSystem& menu);

  virtual void adm (MenuSystem& menu);
  virtual void solveProblem ();
};
```

The class contains pointers (or references) to each of the sub-problem domains which are objects of the classes defined in Section 14.4.2, and pointers to associated sub-domain solvers which are defined below. In addition, there is a separate sub-solver for the mesh-movement problem. It should be noted that the temperature domain- and solver-objects are kept in vectors of pointers (or Handle's) to these objects. This makes it very simple to create a solver having any number of temperature domains, i.e., one only needs to set the dimension of these vectors accordingly.

The ExtrudSolver class is derived from the Diffpack class SimCase and re-implements its adm and solveProblem member functions. These two functions administer the input phase and the solution phase of the simulation, respectively. The constructor of ExtrudSolver takes the sub-problem domains as arguments. Thus, they are to be created outside the scope of this class, which consequently does not need to know about what sub-class of FlowHeatProblem and HeatProblem those objects are.

Next, let us look at the fluid solver branch. The bottom level class here is MixedStokes and its definition is as follows:

```
class MixedStokes : public MxFEM
{
protected:
  CFDProblem&         problem;    // problem-dependent data and functions
  Handle(GridFE)      mesh;       // geometry discretization
  Handle(DegFreeFE)   dof;        // nodal field values <--> global dofs
  Handle(TimePrm)     tip;        // time stepping parameters
  Handle(FieldsFE)    u_h;        // finite element velocity field
  Handle(FieldFE)     p_h;        // finite element pressure field
  Handle(FieldsFE)    unknowns;   // contains both u_h and p_h
  LinEqAdmFE          lineq;      // interface to linear solvers and data
  Vec(real)           solvec;     // linear solution vector
  ...

  virtual real fillEssBC (real t);
  virtual void calcElmMatVecMx (int elm, ElmMatVec& elmat,
                                MxFiniteElement& mfe);
  virtual void integrandsMx (ElmMatVec& elmat,
```

```
                          const MxFiniteElement& mfe);
        virtual void integrands4sideMx (int side, int boInd, ElmMatVec& em,
                             const MxFiniteElement& mfe);

public:
  MixedStokes (CFDProblem& p);
  ~MixedStokes ();

  void solutionNorm (real& U, real& P, Norm_type type = L2);

  virtual void define (MenuSystem& menu, int level = MAIN);
  virtual void scan (MenuSystem& menu);
  virtual void adm (MenuSystem& menu);

  virtual void initSolver ();
  virtual void setIC ();
  virtual void updateDataStructures () {}
  virtual bool solveAtThisTimeLevel ();
  virtual void solveProblem ();

  virtual void attachGridVel (Fields* u) {}
  virtual void attachTemp (Field* T) {}
  virtual Field* energyRateDensity () { return NULL; }

  virtual real energyFEM (FiniteElement& feU, FiniteElement& feP);
  virtual real strainFEM (Ptm(real)& eps, FiniteElement& feU);
  virtual void stressFEM (Ptm(real)& sig, FiniteElement& feU,
                             FiniteElement& feP);

  void tractionFEM (Ptv(real)& trac, FiniteElement& feU,
                             FiniteElement& feP);
  bool sectionalForceFEM (Ptv(real)& SF, const int dir = 1);
  bool sectionalForceVW  (Ptv(real)& SF, const int dir = 1);
  ...
};
```

Most of the members of this class are what we normally find in a Diffpack simulator class and need no further explanation. However, since we use the fluid solver as one part of a coupled-problem solver, we have some additional member functions to facilitate this coupling, e.g., attachGridVel and attachTemp, which are used to attach the grid velocity field and the temperature field to the fluid solver, such that those fields might be evaluated as if they belong to this solver class directly. But when used in a coupled solver, they merely belong to and are updated by other sub-solvers. Finally, the class also contains some member functions for computation of derived solution quantities, such as stresses and strain as well as sectional forces.

We derive our fluid solvers from the Diffpack class MxFEM since we rely on a mixed interpolation of the fluid velocity and the pressure variables. The core of the solver is hidden within the member functions integrandsMx and integrands4sideMx, which calculate the entries of the finite element matrix equations. In the above class, those are resulting from the Stokes flow equations which are the simplest and crudest approximation of a fluid flow problem. But it is a purely linear system such that the solution may be obtained by solving one linear system of equations per time step.

Although the MixedStokes class might be used as a solver itself, its main purpose is to serve as a base class for more specific fluid solvers based on the mixed finite element methods. On the next level, we create the sub-class MixedNS which implements a Navier–Stokes solver for Newtonian fluids. The main ingredients of its definition is as follows:

```
class MixedNS : public MixedStokes, public NonLinEqSolverUDC
{
protected:
  Handle(FieldsFE)        u_t;     // velocity field time derivative
  Handle(FieldsFE)        u_p;     // velocity field at prev. time step
  Handle(NonLinEqSolver)  nlsolver; // interface to nonlinear solvers

  virtual void makeAndSolveLinearSystem ();
  virtual void integrandsMx (ElmMatVec& elmat,
                             const MxFiniteElement& mfe);
  virtual real normOfResidual (Norm_type type = L2);
public:
  MixedNS (CFDProblem& p);
 ~MixedNS ();
  virtual void define (MenuSystem& menu, int level = MAIN);
  virtual void scan (MenuSystem& menu);
  virtual void updateDataStructures ();
  virtual bool solveAtThisTimeLevel ();
};
```

Here we make use of two existing Diffpack classes, NonLinEqSolver and NonLinEqSolverUDC, for handling the non-linear solution procedure which now is needed at each time step. Accordingly, we must also re-implement some of the member functions.

From the Navier–Stokes solver class, we now derive another sub-class for handling non-Newtonian fluids, i.e., the viscosity is no longer invariant but depends on the current fluid velocity and optionally the temperature:

```
class MxLargeDef : public MixedNS
{
protected:
  Handle(Fields)     v_g, v_p; // current and previous grid velocity
  Handle(Field)         T_h; // temperature field
  Handle(Field)      energyd; // strain energy density
  Handle(ViscosityModel) mat; // the material law

  virtual void integrandsMx (ElmMatVec& elmat,
                             const MxFiniteElement& mfe);

public:
  MxLargeDef (CFDProblem& p);
 ~MxLargeDef ();

  virtual void attachGridVel (Fields* u) { v_g.rebind (u); }
  virtual void attachTemp (Field* T) { T_h.rebind (T); }
  virtual Field* energyRateDensity () { return energyd.getPtr(); }
```

```
virtual void define (MenuSystem& menu, int level = MAIN);
virtual void scan (MenuSystem& menu);
virtual void updateDataStructures ();

virtual real energyFEM (FiniteElement& feU, FiniteElement& feP);
virtual real strainFEM (Ptm(real)& eps, FiniteElement& feU);
virtual void stressFEM (Ptm(real)& sig, FiniteElement& feU,
                        FiniteElement& feP);
};
```

As mentioned earlier in Section 14.4.2, the CFDProblem class does not handle non-linear viscosity. Instead we introduce a separate class, ViscosityModel, to represent the viscosity of the fluid problem. It has the following definition:

```
class ViscosityModel : public HandleId
{
protected:
  real a, b, eps_g; // Data for parabola
  bool temperature; // True for temperature dependent models

  void initParabola (real eps, real T);

public:
          ViscosityModel () { a=b=eps_g=0; temperature=false; }
  virtual ~ViscosityModel () {}

  virtual real value      (real eps, real T) const = 0;
  virtual real derivative (real eps, real T) const = 0;

  virtual void scan (const String&) {}

  bool isTemperatureDependent () const { return temperature; }
};
```

This class contains two purely virtual member functions for returning the current value and derivative of the effective stress as a function of the effective strain rate and temperature, i.e.,

$$\text{value} = \bar{\tau}(\bar{\varepsilon}, T), \tag{14.31}$$

$$\text{derivative} = \frac{\partial}{\partial \bar{\varepsilon}} \bar{\tau}(\bar{\varepsilon}, T), \tag{14.32}$$

as these two values are needed in the linearized expressions for the constitutive terms. So far we have implemented two different viscosity models as separate sub-classes of ViscosityModel, namely, the Oldroyd–Bingham and the Zener–Hollomon models, see Section 14.2.2.

Recall the definition of class MxLargeDef, we see that this class also contains some additional fields representing the grid velocity, temperature, and strain energy density. The fields are used to facilitate the coupling of the fluid solver to the mesh-movement and temperature solvers. Again, we note that these fields are declared with the Field base class (and not FieldFE) such that they can be rebinded to fields of any type. Thus, it does not make any difference to our fluid solver if the temperature field is just a constant field with some prescribed value, or a finite element field that is updated outside the fluid solver - the fluid solver implementation is completely unaffected by that. The strain energy density field energyd is rebinded to an object of class MxEnergyDensity, which is just a Field-encapsulation of the energyFEM function of the fluid solver class. This field is referred to, by the temperature solver on the aluminum domain, as the strain energy density, appearing as a 'heat source' term.

Now, having finished the fluid solvers, we can set up a similar hierarchy for the temperature solvers (see the right-hand side of Figure 13.4). However, since the main methodology is basically the same, we will omit the details. The temperature solver classes are derived from the Diffpack class FEM which contains the necessary administrative tools for assembling the finite element equations. The class LinHeat is a standard solver for linear heat conduction problems based on the finite element method. From that class, we derive the sub-class LinHeatFlow, which also have the energy density field from the fluid solver and the mesh velocity fields as members. Thus, this class is used for the temperature problem in the aluminum domain, whereas the LinHeat class is used for all the other domains. We have also derived a third solver LinHeatUpwind which only re-implements the integrands function to account for an upwind formulation, for testing purposes.

For the validation of the present implementation of the extrusion solver, there was a desire to run only the non-Newtonian fluid solver separately where the temperature field was known *a priori*. To facilitate such executions, we let the fluid solver be derived from a general FieldSolver class, from which we also derive the class FieldPrescribed. The latter class just read, the temperature field from a file at each time step instead of solving a finite element problem. So, if we want to let the temperature field be a known quantity, we just have to rebind the FieldSolver handle of the ExtrudSolver object to a FieldPrescribed object instead of one of the temperature solvers.

Finally, the third and last branch of the extrusion simulation system is the mesh-movement 'solver', represented by the upper part of Figure 13.4. It consists of the base class MeshMover and the sub-classes MeshMoveBlend and MeshMoveElastic. In the first sub-class, the mesh displacement in the direction of the ram-movement is prescribed to decrease linearly from the ram-aluminum boundary to zero over a specified distance, whereas in the other two directions it is zero. This is normally sufficient for extrusion-type problems. However, for more complex geometries, it might be necessary to allow a more free mesh deformation. For that purpose, we use the MeshMoveElastic class, which solves for the updated nodal positions and velocity from a finite element elasticity problem as described in Section 14.3.3. Thus, this class is also derived from the FEM class of Diffpack (see also Section 5.1.4 in [7] for a similar implementation).

## 14.5    Numerical Experiments

### 14.5.1    The Jeffery–Hamel Flow Problem

When developing software for numerical simulation of physical problems, it is very useful to validate the implementation on simple benchmark problems where the solution is known. For this purpose we use the Jeffery–Hamel flow problem, which is governed by the incompressible Navier–Stokes equations on a wedge-shaped domain with a source or sink at the origin (see Figure 14.7a). The analytical solution to this problem has been derived in [11] (pp. 184–189) and gives the radial velocity $u_r$ (the angular velocity $u_\theta$ being zero) and the pressure $p$ as, respectively,

$$u_r = \frac{U_0}{r} f(\eta), \quad \eta = \frac{\theta}{\alpha}, \tag{14.33}$$

$$p = p_0 - \frac{1}{2} \rho \left( \frac{U_0}{r/r_0} \right)^2 \left[ f^2 \pm \frac{1}{Re\,\alpha} f'' \right], \tag{14.34}$$

where $U_0$ is the radial velocity at a given point $r = r_0$ and $\theta = 0$, $\rho$ the mass density of the fluid, and $p_0$ an arbitrary constant, here selected such that $p = 0$ for $r = r_0$ and $\theta = 0$. The Reynolds number $Re$ is given by

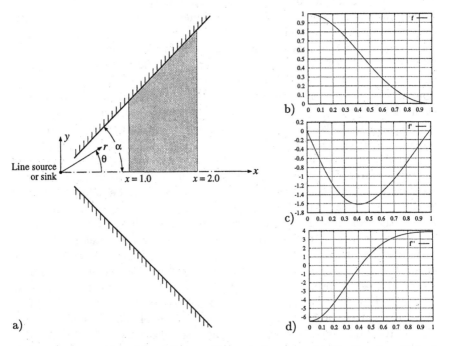

**Fig. 14.7.** The Jeffery–Hamel flow problem: a) Geometry. b) The function $f(\eta)$. c) First derivative $f'(\eta)$. d) Second derivative $f''(\eta)$.

$Re = U_0 r_0 / \alpha$. The $+$ sign in front of the last term in (14.34) is for flow in the positive $x$-direction and the $-$ sign for flow in the negative $x$-direction.

The function $f(\eta)$ is determined through the differential equation

$$f''' + 2 Re \alpha f f' + 4 \alpha^2 f' = 0, \qquad (14.35)$$

with boundary conditions $f(0) = 1$, $f(1) = 0$ and $f'(0) = 0$. This third order, non-linear equation has been solved numerically using MATLAB and the solution for our choice of parameters, $Re = 10$ and $\alpha = \pi/4$, is plotted in Figure 14.7b. The associated first and second derivatives are plotted in Figures 14.7c and 14.7d, respectively.

The Cartesian components of the velocity field are $u_x = u_r \cos \theta$ and $u_y = u_r \sin \theta$, respectively. With $x = r \cos \theta$ and $y = r \sin \theta$, we may write the velocity field in terms of Cartesian coordinates as

$$\mathbf{u}(x,y) = \begin{Bmatrix} u_x \\ u_y \end{Bmatrix} = \frac{U_0}{r^2} f(\eta) \begin{Bmatrix} x \\ y \end{Bmatrix}, \qquad (14.36)$$

The distribution of these components as well as the pressure over the shaded quadrilateral section shown in Figure 14.7a is plotted in Figure 14.8.

This portion is now analyzed with the flow in the positive $x$-direction. The exact velocity (14.36) is imposed as Dirichlet conditions along both the inflow and outflow boundaries. The pressure is then determined up to a constant and here it is prescribed equal to zero at the point $\{x,y\} = \{1.0, 0.0\}$. This problem is simulated using the Navier–Stokes solver MixedNS in which the viscosity is constant. Since this is a stationary problem, we can omit the

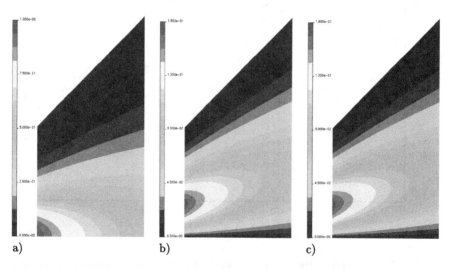

a)                    b)                    c)

**Fig. 14.8.** Analytical solution of the Jeffery–Hamel flow problem. a) Horizontal velocity component. b) Vertical velocity component. c) Pressure.

time-derivatives in (14.21)–(14.22) and solve the non-linear equations only once. We have done this on a series of successively refined meshes to study the convergence of the FE solution toward the exact solution shown in Figure 14.8. The obtained convergence rates are presented in Figure 14.9.

Here we have plotted the error in the computed solution in terms of the energy norm, the $L_2$-velocity norm, and the $L_2$-pressure norm, respectively. These are defined trough, respectively,

$$\eta_{\mathrm{E}} := 100\% \times \frac{\sqrt{\|\nabla \otimes e_u\|_{L_2}^2 + \|\nabla \cdot \hat{u}\|_{L_2}^2}}{\|\nabla \otimes u\|_{L_2}}, \tag{14.37}$$

$$\eta_{\mathrm{u}} := 100\% \times \frac{\|e_u\|_{L_2}}{\|u\|_{L_2}}, \tag{14.38}$$

$$\eta_{\mathrm{p}} := 100\% \times \frac{\|e_p\|_{L_2}}{\|p\|_{L_2}}, \tag{14.39}$$

where $e_u := u - \hat{u}$ is the point-wise error in the FE velocity field $\hat{u}$, and similarly, $e_p := p - \hat{p}$ is the point-wise error in the computed pressure $\hat{p}$. It is seen that we obtain the expected rate of convergences in these physical quantities according to the order of interpolation.

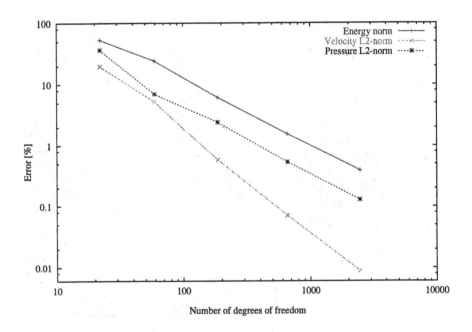

**Fig. 14.9.** The Jeffery–Hamel flow problem: Convergence of the error in energy norm, and $L_2$-norms of velocity and pressure, respectively.

## 14.5.2    The Extrusion Problem

Being satisfied with the results of the Navier–Stokes verification, we now proceed with a real extrusion problem, where we only can compare numerical results against experimental data. The problem we want to simulate is displayed in Figure 14.10. The diameter of the container is 100 mm and the extruded profile is rectangular with the width and thickness 78.5 mm and 1.7 mm, respectively. The ratio between the cross sectional area of the container and the profile (approximately 59:1) equals that between the container diameter and the profile thickness. Hence, the reduction ratio for a 2D longitudinal section located at the center of the container and normal to the rectangular profile is equal to the reduction ratio for the 3D case.

The numerical simulations are performed with material properties corresponding to aluminum alloy AA6060.35[1], ram speed of 10 mm/s, initial bolt temperature 486° C, initial container temperature 432° C and initial ram temperature 30° C. This corresponds to the physical experiment number 6 reported by Lefstad and Flatval [12].

The bearing channel is 5 mm long and has a 3° release from the inlet. With such a short bearing channel, the effect of the friction in the channel is assumed to be negligible.

---

[1] Hydro Aluminum internal designation.

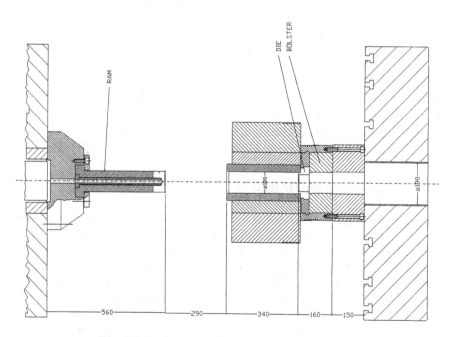

**Fig. 14.10.** Drawing of the extrusion equipment.

### 14.5.3   Simulations of the Temperature

The location of the points where the temperature is measured is shown in
Figure 14.11. Here we report results obtained at point 4 in the die such that
we may investigate *Extrud*'s capability to model heat conduction including
heat transfer between domains of different mechanical character. The finite
element model used in the 2D simulations with *Extrud* is shown in Figure 14.12.

In order to investigate the sensitivity of the obtained numerical results in
terms of the choice of global solution strategy, we perform the following three
simulations with *Extrud*.

-  One stagger iteration only: $(N_{\text{stag}} = 1, T_{\text{upd}} = 0)$.

-  Four stagger iterations: $(N_{\text{stag}} = 4, T_{\text{upd}} = 0)$.

-  Four stagger iterations and update of the temperature between each
   Newton-Raphson iterations in the fluid sub-problem solver: $(N_{\text{stag}} = 4,$
   $T_{\text{upd}} = 1)$.

Here, $N_{\text{stag}}$ denotes the number of stagger iterations and $T_{\text{upd}}$ is understood
as a logical variable indicating whether the temperature should be updated
for each Newton–Raphson iteration or not, see Figure 14.4. The results are
presented in Figure 14.13a, and we see no significant differences occur. Thus,
in the comparisons below we report the results obtained with $N_{\text{stag}} = 4$ and
$T_{\text{upd}} = 0$.

In Figure 14.13b we compare the results obtained, respectively, with *Extrud* and the *ALMA* code [5] with the physical experiments. The temperatures
obtained by *Extrud* and *ALMA* at point 4 (die) compares both well with the
temperatures measured in the laboratory experiments. The differences between results of *Extrud* and *ALMA* are assumed to be due to the different
finite element schemes used in the two codes.

In Figure 14.14 the temperature distribution computed by *Extrud* is displayed. Notice the sharp temperature gradients in the bearing channel.

## 14.6   Concluding Remarks

The mathematical background and numerical implementation of an object-
oriented solver for the simulation of aluminum extrusion has been presented.
The solver, *Extrud*, deals with the extrusion problem as a non-Newtonian
Navier–Stokes fluid problem, coupled with linear heat conduction in multiple temperature domains. The numerical solution is based on a mixed FE
method for the fluid flow and a standard isoparametric FE method for the
heat conduction.

The implementation is based on existing classes in Diffpack for assembling
and solving finite element equations, and therefore, the amount of new code

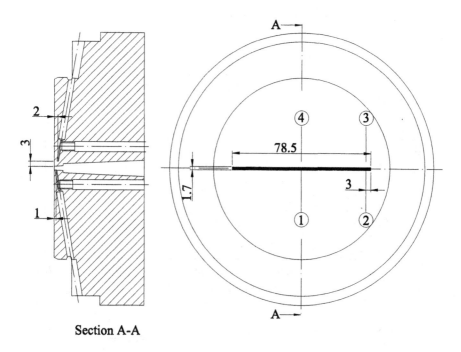

Section A-A

**Fig. 14.11.** The locations of the points where the temperature is measured.

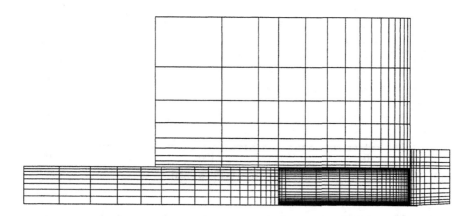

**Fig. 14.12.** The finite element model used in the *Extrud* 2D simulations.

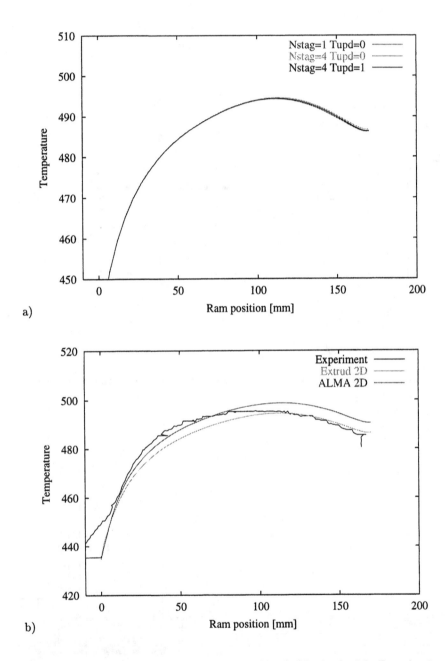

**Fig. 14.13.** Temperatures at point 4 in the die. a) Obtained with *Extrud* using three different solution strategies. b) Obtained with *Extrud* and *ALMA* compared with physical experiments.

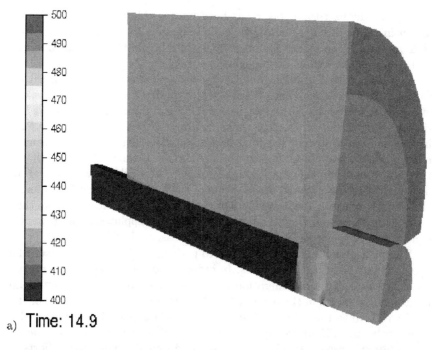

a) Time: 14.9

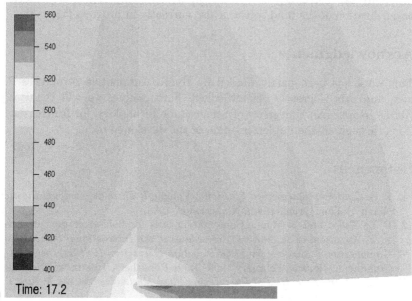

b) Time: 17.2

**Fig. 14.14.** Temperatures computed by *Extrud.* a) Overall view of the 3D temperature distribution in the extrusion tool. b) Temperature distribution in the vicinity of the bearing channel at the final time step.

needed for the extrusion solver has been limited. Moreover, the implementation in terms of class hierarchies is done in such a way that much of the code may easily be reused for other types of fluid problems. The code may be used on 2D as well as 3D models.

It is the authors' opinion that an object-oriented approach has great advantages over traditional procedural-oriented methods in developing this kind of numerical software. Although there might be a substantial beginners barrier on using object-oriented methods, and a need of a comprehensive FEM library such as Diffpack, the final implementation can end up being much more modular, readable, compact and extendible. It should be noted though that an optimal implementation depends on a thorough knowledge and understanding of the different components of the Diffpack library. Therefore, a challenge is lying on the developers of these kind of libraries, to make them well-documented and flexible to use, such that the application programmers can exploit their potential fully with only a reasonable effort.

The present solver has been shown to perform well on a simple extrusion problem with idealized conditions. In order to do a more accurate analysis, one has to take into consideration friction between the die and aluminum, elasto-plastic effects in the extruded profile, etc. Further development on *Extrud* will involve these issues. Moreover, it is clear that simulations with a sufficiently fine FE discretization for these kinds of problems are computationally expensive, especially over the fluid domain. Therefore, efforts on parallelization of the fluid solver is also currently in progress [13].

## Acknowledgments

This work has been partly funded by Hydro Automotive Structures Raufoss, and this is greatly acknowledged. Furthermore, we will thank Kjell Holthe, Norwegian University of Science and Technology, for fruitful discussions throughout the implementation of the developed code.

## References

1. P. G. Ciarlet. *Mathematical Elasticity, Volume I: Three-dimensional Elasticity.* North–Holland, Amsterdam, Netherlands, 1988.
2. P. Le Tallec and S. Mani. Conservation laws for fluid-structure interaction. In T. Kvamsdal et al., editors, *Proceedings of the International Symposium on Computational Methods for Fluid-Structure Interaction (FSI'99)*, pages 61–78, Trondheim, Norway, February 1999. Tapir Publishers, Trondheim, Norway.
3. R. M. Olson. *Essentials of Engineering Fluid Mechanics.* Harper & Row Publishers, New York, 1980.
4. C. Zener and J. H. Hollomon. Effects of Strain Rates upon the Plastic Flow of Steel. *Journal of Applied Physics*, 15:22, 1944.
5. K. Holthe. Numerical simulation of aluminium extrusion. Volume 1: ALMA, Theory Manual. SINTEF Report STF71 F91028, SINTEF Structural Engineering, Trondheim, Norway, April 1991.

6. K. M. Okstad, T. Kvamsdal, and S. Abtahi. A Coupled Heat and Flow Solver for Extrusion Simulation. In *Proceedings of the 3rd ECCOMAS Conference on Numerical Methods in Engineering*, volume on CD-ROM, Barcelona, Spain, September 2000.
7. H. P. Langtangen. *Computational Partial Differential Equations - Numerical Methods and Diffpack Programming*. Textbooks in Computational Science and Engineering. Springer, 2nd edition, 2003.
8. The Diffpack world-wide-web Home Page. http://www.diffpack.com.
9. K. M. Okstad and T. Kvamsdal. Object-Oriented Programming in Field Recovery and Error Estimation. *Engineering with Computers*, 15:90–104, 1999.
10. K. M. Okstad and T. Kvamsdal. Object-Oriented Field Recovery and Error Estimation in Finite Element Methods. In A. M. Bruaset, H. P. Langtangen, and E. Quak, editors, *Advances in Software Tools for Scientific Computing*, pages 283–317. Springer–Verlag, Berlin Heidelberg, 1999.
11. F. M. White. *Viscous Fluid Flow*. McGraw–Hill, 1974.
12. M. Lefstad and R. Flatval. Ekstrudering av skinne og U-profil i laboratoriepresse ved sintef som underlag for simuleringsprogrammet Ekstrud3D. Technical Report STF24 F99602, SINTEF Materials Technology, Trondheim, Norway, 1999.
13. T. Kvamsdal, R. Holdahl, and K. M. Okstad. Parallel Incompressible non-Newtonian Flow Code for Aluminium Extrusion. In *Proc. of abstracts 6th U.S. National Congress on Computational Mechanics*, Dearborn, Michigan, August 2001.

# Chapter 15

# Simulation of Sedimentary Basins

A. Kjeldstad[1,2], H. P. Langtangen[3,4], J. Skogseid[5], and K. Bjørlykke[2]

[1] Norske Shell
[2] Department of Geology, University of Oslo
[3] Simula Research Laboratory
[4] Department of Informatics, University of Oslo
[5] Norsk Hydro

**Abstract.** In this chapter we focus on mathematical and numerical models of coupled partial differential equations governing geological processes. We consider a model governing some of the fundamental space-time geological processes in sedimentary basins. The mathematical model couples fluid and heat transfer in time in a deforming porous medium. A finite element discretization technique with a fully implicit approach represent the most robust and reliable solution. The basin model is used in a case study simulating the cooling of a high temperature (1100 degrees C) horizontal magmatic intrusion (sill), including the aspects of maturity of hydrocarbons. The model is able to handle sharp temperature gradients and give a quantitative description of the conductive and advective temperature transfer, which causes changes in the pore pressure, fluid circulation, as well as effective and thermal stresses, in the proximity of the sill.

## 15.1 Introduction

Geological processes are often highly complex, first of all as result of the heterogenous assembly and geometries of rocks and sediments. Modeling such processes by systems of partial differential equations (PDEs) can in many cases prove to be a fruitful approach to increase the geological understanding. The challenge is to derive a model which mirrors the major geological processes yet with a comprehensible complexity regarding software development and simulation time. This requires skills in both geology and numerical modeling.

The evolution of sedimentary basins are of interest in geoscience in general since the basins hold large amounts of sediments deposited over millions of years, thus conserving the sediments as a history book of the past geological events. Sedimentary basins are often formed during episodes of extension, when the earth's crust and mantle are stretched and thinned, causing upper crustal brittle faulting and subsidence [14]. Major extension episodes are often associated with magmatic activity [7,18], and sometimes result in continental separation. During extension magmatic bodies with high temperature (1100 degrees C) are likely to penetrate the sediments, causing changes in the heat transfer and fluid flow, as well as the stress field.

Basin subsidence and sediment accumulation is further associated with loading of the substratum, causing a compressive stress and deformations. Vertical downward deformations of sediments in a basin, where there is a volume and porosity reduction, is often referred to as compaction or consolidation. These processes can, however, be encountered by excess pore pressures if the pore fluids cannot escape due to low permeable seals e.g. shales. The excess pressure can cause the pore fluids to move towards low pressured areas, provided that a permeable path way exists. All these processes of heat transfer, fluid flow, and sediment deformation are usually coupled and play a primary role in the evolution of the basin.

The effect and importance of each individual process are often under debate in the geological community. Complexity is increased when the processes are non-linear and coupled, and hence processes become increasingly difficult to understand, without the use of quantitative models such as PDEs. Solving such quantitative models can lead to great challenges both in the numerical solution and the verification process.

The numerical challenges in basin modeling lie in convergence and stability of the PDE solvers, how the PDEs are coupled numerically, and how the initial conditions are handled. The coupled PDEs can be solved sequentially or in a fully implicit manner. The sequential approach is often preferred since it may save simulation time because the linear systems are smaller than in the fully implicit approach. However, if the PDEs are strongly coupled, a fully implicit solution is often much more robust and stable than a sequential approach [4,12].

The verification process of the numerical model is another challenge, as very few analytical solutions are available, especially for coupled processes. Physical experiments and field observations are the most useful tools for verification in such cases. Since the verification of the code is usually the most time consuming part of the model development, it is essential that the model can be expanded without altering previously tested code. To this end, we have applied a software development technique based on object-oriented programming as outlined in [11, ch. 3.5].

Simulation of the evolution of sedimentary basins is a matter of simulating fluid flow, deformations, tectonic stress fields, and heat transfer in porous geological media over very long time periods. There are a number of proposed mathematical models governing deformations, fluid flow and heat transfer in porous media [8,12]. The models are derived from continuum mechanics as well as geotechnical and geological studies. The proposed models couples multi-phase fluid flow and heat transfer in non-deforming porous medium, or single-phase flow in deforming porous medium. All these processes are normally coupled and should not be solved independently [12]. We will here propose a model that couples both heat and fluid flow in a linearly elastic deforming porous medium. The reason to use an elastic model, despite our knowledge that deformation in sediments is extremely complicated and to a large degree permanent (i.e. plastic), is mainly that elastic models are computationally efficient, and it appears that the predicted deformations are sufficiently accurate for many applications in basin modeling.

Some work has been published on heat transfer and fluid flow around cooling magmatic vertical bodies such as batholites and dikes in basins, see i.e. [16]. However, little attention has been given to horizontal magmatic bodies, such as sills, emplaced in a layered sedimentary succession in a rift basin associated with volcanic activity. On volcanic margins such sills are particular frequent and may occur in large sill complexes at different depths [18,1]. Such complexes will alter the heat transfer gradient significantly in parts of the basin. Because heat transfer is an important parameter in the maturation of organic rich shales, sills can play a major role in generation of and in the primary migration of hydrocarbons out of the source rock. Here, we use the cooling of a sill intrusion as a case study for our modeling application. We aim to demonstrate some of the capabilities of the basin model and to quantitatively address these processes going on around the sill.

To summarize, our aim is to propose a basin model, which couples the stress field, fluid flow, and heat transfer in a deforming porous medium over time. The model should handle different complex initial conditions and basin scale grids with a large number of nodes. Furthermore, it must handle both high rates of loading, i.e., high sedimentation rates, and rapid changes in heat transfer from magmatic and volcanic activities, as frequently encountered in sedimentary basins.

We first present the mathematical model and the numerical discretization techniques. Then, in a case study, we model the thermal and effective stress field, fluid flow and heat transfer, and the oil generation maturity index [19] around a cooling 23 m thick sill.

## 15.2    The Geomechanical and Mathematical Problem

Our mathematical model consists basically of three partial differential equations governing stress-induced deformations, single-phase fluid flow, and heat transfer. These equations are based on the continuum hypothesis, the assumption of mass conservation, Newton's 2nd law of motion for the fluid flow, and the first law of thermodynamics. Below we describe the fundamental equations and their associated geological processes.

### 15.2.1    Elastic Deformations

Deformations or consolidation of sediments couples fluid flow and the stress in the porous sediments. In this respect the concept of effective stress is important both in basin modeling and geotechnical engineering. The coupling of the elastic porous matrix, the effective stress, and the pore pressure is developed on the lines of Biot's theory [2]. When tension is positive, the effective stress is expressed as

$$\sigma'_{rs} = \sigma_{rs} + P\delta_{rs}. \tag{15.1}$$

Here, $\sigma_{rs}$ is the total stress tensor, $\sigma'_{rs}$ is the effective stress tensor, $P$ is the pore pressure, and $\delta_{rs}$ is the identity tensor (Kronecher delta). The effective stress plays a fundamental role in any fluid-saturated deforming porous medium. If any load is put on top a of porous medium, parts of the resulting internal stress is immediately taken up by the pore pressure. This slows down the deformation or compaction process. The rate of vertical deformations can increase if the pore pressure is allowed to drain out of the porous medium. An increase or decrease in the pore pressure inside the porous medium leads to isotropic expansion or contraction. The effective stress indicates the average contact stress over a given area in the continuum model of the geological medium. However, actual contact stress between the individual grains can be much higher, since the grain contact area can be as low as one to five percent of a given geometrical area inside the sediment.

In basin modeling we often use the term excess pressure. The excess pressure, $P_e$, is referred to as the pressure above the hydrostatic pressure:

$$P_e = P - \varrho_f g z. \tag{15.2}$$

Here, $\varrho_f$ is the fluid density, $z$ is the depth and $g$ is the acceleration of gravity. Thus, $P$ can also be modeled as the excess pressure when gravity is neglected. During a constant rate of deposition of sediments in a basin, the mechanical deformations stop if the permeability is low and the pore water cannot escape. This results in an increase of excess pressure leading to a reduction in effective stress relative to a hydrostatic condition, where the pore water is allowed to escape. In basin modeling, the excess pressure can often be of more interest, since excess pressure affects fluid flow, bore-hole stability during drilling, and hydrocarbon production from reservoirs.

The mineral framework expands as the temperature rises in a particular area. Heating or cooling the porous medium leads to isotropic expansion or contraction. The strains associated with this deformation are referred to as *thermal strains* and are denoted by $\varepsilon_{rs}^{(T)}$. They cause an associated stress and deformation field, and this might be a significant phenomenon in basin modeling [14]. Thus, both the pore pressure and temperature affect the stress and deformation of the sediments. An empirical model for $\varepsilon_{rs}^{(T)}$ is

$$\varepsilon_{rs}^{(T)} = \beta_s(T - T_0)\delta_{rs}, \tag{15.3}$$

where $\beta_s$ is a thermal expansion coefficient, $T$ is the temperature, and $T_0$ is a reference temperature where thermal strains vanish.

The momentum equation for the solid deformation is based on Newton's 2nd law. In a geological time scale it is relevant to ignore the acceleration. The resulting equilibrium equation can then be expressed as

$$\sigma_{rs,s} + \varrho b_r = 0 \,. \tag{15.4}$$

Here, $\sigma_{rs}$ is the total stress tensor for the solid part of the sediment, $\varrho$ is the average total density of the sediment, and $b_r = -g\delta_{r3}$ represents acceleration due to gravity. A comma notation is used to indicate partial differentiation, whereas a double index signals the use of Einstein's summation convention [13]. For example, $\sigma_{rs,s}$ is the divergence of the stress tensor $(\nabla \cdot \boldsymbol{\sigma})$ and $\varrho_{f,t} \equiv \partial \varrho_f / \partial t$.

Assuming isotropic elastic properties, the relation between the effective stress and strain is given by Hooke's law:

$$\sigma'_{rs} = \lambda \varepsilon^{(E)}_{q,q} \delta_{rs} + 2\mu \varepsilon^{(E)}_{rs} \,. \tag{15.5}$$

Here, $\varepsilon^{(E)}_{rs}$ is the part of the total strain $\varepsilon_{rs}$ that is caused by the stresses (hence, $\varepsilon^{(E)}_{rs}$ is commonly called elastic strain). Provided that the medium can be modeled as isotropic, the material law (15.5) involves only two parameters, the Lame's constant $\lambda$ and the shear modulus $\mu$. These parameters are space dependent in heterogeneous sediments, but the common case is to view them as piecewise constant, i.e., constant in each sedimentary layer. Isotropic medium is usually not the case. Nevertheless, elastic isotropic conditions are widely used as an approximation for mechanical deformation of sediments [10], and we adopt the assumption here for simplicity and lack of anisotropic material values.

It is common to divide the total strain $\varepsilon_{rs}$ into a thermal component $\varepsilon^{(T)}_{rs}$ and an elastic component $\varepsilon^{(E)}_{rs}$:

$$\varepsilon_{rs} = \varepsilon^{(T)}_{rs} + \varepsilon^{(E)}_{rs} \,. \tag{15.6}$$

The total strains $\varepsilon_{rs}$ are related to the displacement field through

$$\varepsilon_{rs} = \frac{1}{2}(u_{r,s} + u_{s,r}) \,. \tag{15.7}$$

The $u_r$ quantity is the deformation vector for the solid part and $u_{r,s}$ are the deformation gradients [11]. Combining equations (15.3), (15.5), and (15.6) yields Hooke's generalized thermo-elastic law, which is inserted into (15.1). This yields the relation between the effective stresses, temperature, and pore pressure. We can now summarize the constitutive law for the sediments:

$$\sigma_{rs} = \lambda u_{q,q}\delta_{rs} + \mu(u_{r,s} + u_{s,r}) - P\delta_{rs} - \beta_s(3\lambda + 2\mu)(T - T_0)\delta_{rs} \,. \tag{15.8}$$

In basin modeling the temperature scale is in Celsius and a reference temperature is, e.g., the seafloor temperature of $T_0 \approx 0°C$.

The momentum equation for the solid is obtained by inserting (15.8) into (15.4). However, loading by sedimentary lobes and fans into a basin in time is complicated as their geometry and position changes. At each time level an incremental load must be put on top of the basin to model the sedimentation. To avoid complicated boundary conditions for the stress vector $\sigma_{rs}n_s$, where $n_s$ is the unit outward surface normal vector, we differentiate (15.8) with respect to time. This gives the dynamic momentum equation

$$((\lambda + \mu)\dot{u}_{q,q}),_r + (\mu\dot{u}_{r,q}),_q = \dot{P},_r + \beta_s(3\lambda + 2\mu)\dot{T},_r. \tag{15.9}$$

Here, the gravitational force vanishes as it does not change with time. The dot notation, like in $\dot{P} = \partial P/\partial t$, signals partial differentiation with respect to time.

The time derivative is discretized by a standard, first-order, *backward Euler* scheme. Looking at (15.9) we realize that $\Delta t$ will appear in all terms such that the time step parameter vanishes from the discrete equations. We are then left with the increments $\Delta u_r = u_r^{\ell+1} - u_r^\ell$. The superscript $\ell$ represents the time level. Since (15.9) will be coupled to other time dependent PDEs, which do not have increments as an unknown, we prefer to work with variables as $u_r^{\ell+1}$, $P^{\ell+1}$, and $T^{\ell+1}$ as unknowns. The results from the previous time step, $u_r^\ell$, $P^\ell$, and $T^\ell$ are then treated as known coefficients in the equation system. Furthermore, the total displacement field, $u_r^\ell$, from the previous incremental loads is then kept in (15.9). We then only need a simple boundary condition when applying the incremental loads at the present time level, and solve for the new displacement field $u_r^{\ell+1}$. Thus, we do not need a load function for the whole sedimentary fan or lobe complex. We simply apply the incremental load at the specific areas of interest at the current time level. We also find (15.9) easy to couple numerically with other time-dependent PDEs.

The use of linear elastic theory as material model for sedimentary deformations is a highly simplified model, since their deformation behavior is non-linear. However, the non-linear deformations can be linearized by taking, e.g., the secant stress-strain modulus from laboratory experiments for the sediment at the appropriate stress interval [10]. Different sedimentary layers are then assigned the appropriate secant modulus, and the non-linear deformations are approximated. Our model can only be used to study the deposition of one sedimentary lobe or wedge with its stresses and deformations fields. It cannot model the loading or infill of the whole basin with sediments, since only the load from the layer is put on top of the domain and not the physical layer itself. However, the model is useful to study tectonic stress fields and specific depositional episodes over short geological time periods. For studies of infill history of the whole basin and detailed porosity reduction studies over tens or hundreds of million of years we refer to [21].

### 15.2.2    Fluid Flow

When a basin is filled with sediments over time, the sedimentary succession gets compacted and deformed. The porosity is reduced and the sand or clay particles are compressed in an increasingly tighter configuration, causing volumetric decrease of the sediments and an increase in the effective stress. However, the total volume of sediments cannot decrease without the escape of pore water, found between the grains. For a low permeable medium excess pore pressure may develop because the pore water cannot escape. There is a strong physical coupling between effective stress and pore pressure, which again controls fluid flow. The volumetric deformations of the sediments is a dominant source for generating excess pore pressure in sedimentary basins.

Other factors controlling the pore pressure are the compressibility of the pore water and the mineral skeleton in the sediments, expressed by the bulk modulus for the pore water and for the solid matrix, denoted by $K_f$ and $K_s$, respectively. Increasing heat transfer can also result in an excess pore pressure, as the pore water and mineral framework in sediments thermally expand. This is modeled by thermal expansion coefficients of the pore water and solid phase, denoted by $\beta_f$ and $\beta_s$, respectively. The expansion of the pore water as a result of increased heat transfer causes changes in the density of the pore water, thus making hot pore water lighter than the adjacent cooler pore water in the gravity field. We assume that the density of the solid and fluid phases in the sediments is a function of both temperature and pore pressure, written as $\varrho_f(P,T)$ and $\varrho_s(P,T)$ for the fluid and solid phases, respectively.

The above factors can be incorporated in the pressure equation governing the fluid flow. The starting point is the mass conservation for the different phases in the sediment:

$$\frac{\partial \phi_i \varrho_i}{\partial t} - \nabla \cdot \phi_i \varrho_i \mathbf{V}^{(i)} = 0. \tag{15.10}$$

Here, the index $i$ is the phase component in the porous medium, e.g., fluid or quartz mineral, each having its own parameters. The symbols $\phi_i$, $\varrho_i$, and $\mathbf{V}^{(i)}$ are the volume fraction, the density and the velocity for each component, respectively.

Assuming only a two-component porous medium, the two mass conservation equations are written as (15.10). Thus, the solid phase mass conservation equation is

$$\frac{\partial (\varrho_s(1-\phi))}{\partial t} + \nabla \cdot (1-\phi)\varrho_s \mathbf{V}^{(s)} = 0, \tag{15.11}$$

and for the fluid phase we have

$$\frac{\partial \phi \varrho_f}{\partial t} + \nabla \cdot \varrho_f \phi \mathbf{V}^{(f)} = 0. \tag{15.12}$$

Here, $\varrho_f$ and $\varrho_s$ are the densities for the fluid and solid phases, respectively. The volume fraction for the fluid phase is $\phi_f = \phi$ and $\phi_s = 1 - \phi$ for the solid, $\phi$ being the porosity in the sediment. Combining (15.11) and (15.12), one can derive the equation of continuity for the fluid phase:

$$(\phi - 1)\varrho_{s,t} + \phi\varrho_{f,t} + (\varrho_f V_r^{(f)})_{,r} + (u_{k,k})_{,t} = 0. \tag{15.13}$$

(Here we have used that $V_i^{(s)} \approx u_{i,t}$ which leads to $V_{i,i}^{(s)} = u_{i,it}$, where $u_{i,it}$ is the volumetric change and $u_i$ is the deformation field for the solid part.) The velocity vector $V_r^{(f)}$ is related to the pore pressure through Darcy's law (this is the counterpart to Newton's 2nd law for single-phase flow in a porous medium):

$$\frac{\mu_f}{k} V_r^{(f)} + P_{,r} + \varrho_f g_r = 0.\tag{15.14}$$

Here, $V_r^{(f)}$ represents an average velocity over a large number of pores, $\mu_f$ is the fluid viscosity, $k$ is the permeability (may be tensor if $1/k$ is interpreted as the inverse matrix $k^{-1}$), $P$ is the pore pressure, and $g_r = -g\delta_{r3}$ is the acceleration due to gravity. The time derivative of the fluid density can be replaced by time derivatives of the pressure and temperature using the equation of state $\varrho_f = \varrho_f(P,T)$ of the fluid. We have

$$\phi \frac{1}{\varrho_f} \frac{\partial \varrho_f}{\partial t} = \phi \frac{1}{\varrho_f} \frac{\partial \varrho_f}{\partial P}\bigg|_T \frac{\partial P}{\partial t} + \phi \frac{1}{\varrho_f} \frac{\partial \varrho_f}{\partial T}\bigg|_P \frac{\partial T}{\partial t}.\tag{15.15}$$

The compressibility $K_f$ and the heat expansion $\beta_f$ for the fluid phase are defined through

$$\frac{1}{\varrho_f} \frac{\partial \varrho_f}{\partial P}\bigg|_T = \frac{1}{K_f},$$

$$\frac{1}{\varrho_f} \frac{\partial \varrho_f}{\partial T}\bigg|_P = -\beta_f.$$

Similarly, the time derivatives for the solid density can be replaced by the time derivatives for the pore pressure and temperature using the equation of state for the solid $\varrho_s = \varrho_s(P,T)$. We then have

$$(1-\phi) \frac{1}{\varrho_s} \frac{\partial \varrho_s}{\partial t} = (1-\phi) \frac{1}{\varrho_s} \frac{\partial \varrho_s}{\partial P}\bigg|_T \frac{\partial p}{\partial t} + (1-\phi) \frac{1}{\varrho_s} \frac{\partial \varrho_s}{\partial T}\bigg|_P \frac{\partial T}{\partial t},\tag{15.16}$$

and the compressibility $K_s$ and the heat expansion $\beta_s$ for the solid phase are defined through

$$\frac{1}{\varrho_s} \frac{\partial \varrho_s}{\partial P}\bigg|_T = \frac{1}{K_s},$$

$$\frac{1}{\varrho_s} \frac{\partial \varrho_s}{\partial T}\bigg|_P = -\beta_s.$$

Thus, the final pressure equation can be expressed as follows:

$$\left[(1-\phi)K_s^{-1} + \phi K_f^{-1}\right]\dot{P} - \left[(1-\phi)\beta_s - \phi\beta_f\right]\dot{T} + \dot{u}_{q,q}$$

$$= \frac{k_{rs}}{\mu_v}(P_{,j} + \varrho_f(1 - \beta_f T)g_i)_{,j}.\tag{15.17}$$

The term $-\beta_f T g_i$ in (15.17) takes into account the density contrasts in the gravity field caused by the heat transfer. All the parameters used in (15.17) and their units are given in Table 15.1.

### 15.2.3   Heat Transfer

Energy transport in a porous medium is based on the assumption of energy balance in the solid and fluid phases. In the following, all the parameters used are defined in the text or in Table 15.1. The first law of thermodynamics states that the rate of change of the total energy of a material volume is equal to the work done on the volume by external forces plus the heat input to the volume. In our model we assume the work done by external forces and the kinetic energy due to the velocities are negligible compared to the heat input to the volume (because we assume small velocities and a significant heat input). Hence, the first law of thermodynamics can be expressed mathematically as:

$$\int_\Omega \sum_i \frac{\partial \phi_i \varrho_i e_i}{\partial t} d\Omega \;=\;$$
$$-\int_{\partial\Omega} \sum_i \left( \phi_i \kappa_i \nabla T + \phi_i \varrho_i \mathbf{V}^{(i)} e_i \right) d\Gamma + \int_\Omega \sum_i \phi_i \varrho_i h_i d\Omega . \tag{15.18}$$

The rate of change of energy is integrated over a volume $\Omega$, and under isotropic conditions it must be equal to the amount of energy transported through the surface $\partial\Omega$ for each time unit plus the energy generated inside $\Omega$. In (15.18) the $e_i$ is the specific internal energy for each component. The quantity $h$ is an internal source of energy, e.g., radioactive decay in the sediments per unit mass. The integral statement (15.18) can be transformed into a volume integral that covers all terms. This integral will be zero, and since the domain of integration $\Omega$ is arbitrary, the integrand must vanish. This leads to the partial differential equation

$$\sum_i e_i \left( \frac{\partial \phi_i \varrho_i}{\partial t} - \nabla \cdot \phi_i \varrho_i \mathbf{V}^{(i)} \right) + \sum_i \phi_i \varrho_i \frac{\partial e_i}{\partial t} = - \sum_i \left( \nabla \cdot \phi_i \kappa_i \nabla T \right)$$
$$+ \sum_i \phi_i \varrho_i \mathbf{V}^{(i)} \cdot \nabla e_i + \sum_i \phi_i \varrho_i h_i . \tag{15.19}$$

The first term in (15.19) is simply the mass conservation equation (15.10), in which each term in the summation equals zero. The energy equation can then be simplified to

$$\sum_i \phi_i \varrho_i \frac{\partial e_i}{\partial t} = - \sum_i \left( \nabla \cdot \phi_i \kappa_i \nabla T \right) + \sum_i \phi_i \varrho_i \mathbf{V}^{(i)} \cdot \nabla e_i + \sum_i \phi_i \varrho_i h_i . \tag{15.20}$$

To close the model, the internal energy $e_i$ must be related to the temperature $T$. The heat absorbed will in general depend upon how it was transferred;

with constant volume or with constant pressure. The energy is converted to temperature over a constant volume $V$ by

$$\frac{\partial e_i}{\partial t} = \frac{\partial e_i}{\partial T}\bigg|_V \frac{\partial T}{\partial t} = C_i \frac{\partial T}{\partial t}, \quad C_i \equiv \frac{\partial e_i}{\partial T}\bigg|_V. \qquad (15.21)$$

Here, $C_i$ has the unit Joule/Kg$^{\circ}$C which is the specific heat for each component in the sediment.

The energy equation for a two component porous medium with a fluid and a solid phase is then written as:

$$\left((1-\phi)\varrho_s C_s + \phi\varrho_f C_f\right)\dot{T} + \varrho_f C_f T_{,j}\phi V_j^{(f)}$$
$$= ([(1-\phi)\kappa_s + \phi\kappa_f]T_{,j})_{,j} + (1-\phi)\varrho_s h_s. \qquad (15.22)$$

The equations (15.9), (15.17), and (15.22) constitute a fully coupled non-linear system in time and space, where the parameters used in our simulations are given in Table 15.1. The term $\phi V_j^{(f)}$ in (15.22) is the Darcy flow, defined in (15.14), which depends on the pore pressure. Thus, the term $C_f T_{,j}\phi V_j^{(f)}$ couples the temperature and the pore pressure and makes the energy equation nonlinear.

### 15.2.4   Initial Conditions

The geological history, geometry, and many lithologies in a sedimentary basin can result in a complex initial condition. To compute suitable initial conditions for the temperature, pore pressure, and stress field we use a stationary version of the basic equations (15.9), (15.17), and (15.22), i.e., the time derivatives are assumed to be negligible. The initial conditions are then governed by

$$-\varrho_f C_f T_{,j}\phi V_j^{(f)} + ([(1-\phi)\kappa_s + \phi\kappa_f]T_{,j})_{,j} = -(1-\phi)\varrho_s h_s \qquad (15.23)$$

$$(\frac{k_{rs}}{\mu_f}(P_{,s} + \varrho_f g_s))_{,s} = 0, \qquad (15.24)$$

$$((\lambda + \mu)u_{q,q})_{,r} + (\mu u_{r,q})_{,q} - P_{,r} - \beta_s(3\lambda + 2\mu)T_{,r} = -\varrho g_r. \qquad (15.25)$$

Thus, the initial pore pressure (15.24) and temperature (15.23) are simply governed by the Laplace and Poisson equations, respectively. The solutions of these equations are then directly inserted into the static equilibrium equation (15.25), and the initial deformation can be solved. Any hydrostatic pressure and gravitational force can also be initialized in (15.24) and (15.25), respectively, since they are independent of time. This approach removes the usual need for analytical expression for the initial condition; a numerically computed initial condition offers greater flexibility.

**Table 15.1.** Parameters for the governing partial differential equation system (15.9), (15.17), and (15.22).

| Parameter | Unity | Symbol |
|---|---|---|
| Lame constant | Pa | $\lambda$ |
| Shear modulus | Pa | $\mu$ |
| Bulk modulus for the sediment matrix | Pa | $K_s$ |
| Bulk modulus for the water | Pa | $K_f$ |
| Porosity in the sediment | | $\phi$ |
| Density for the sediment solid | Kg/m$^3$ | $\varrho_s$ |
| Density for the pore water | Kg/m$^3$ | $\varrho_f$ |
| Density for the sediment | Kg/m$^3$ | $\varrho$ |
| Heat expansion for the sediment matrix | 1/°C | $\beta_s$ |
| Heat expansion for the pore water | 1/°C | $\beta_f$ |
| Permeability matrix for the sediment | m$^2$ | $k_{rs}$ |
| Viscosity of the pore water | Kg/ms | $\mu_v$ |
| Specific heat for the pore water | Joule/Kg°C | $C_f$ |
| Specific heat for the sediment | Joule/Kg°C | $C_s$ |
| Heat conductivity for the sediment solid | f/m°C | $\kappa_s$ |
| Heat conductivity for the pore water | W/m°C | $\kappa_f$ |
| Heat conductivity for sediment | W/m°C | $\kappa$ |
| Internal heat source | W/kg | $h$ |
| Gravity acceleration | m/s$^2$ | $g_r$ |

### 15.2.5  Boundary Conditions

We here present some of the basic types of boundary conditions that are relevant for our mathematical model. The geometry of the basin is for simplicity assumed to be a two-dimensional box-shaped domain $\Omega$. Restricting to a plane strain assumption means that $u_3 = 0$ and that there is no variation in the third direction of any quantity ($\partial/\partial x_3 = 0$). We let the sea floor $\partial\Omega_1$ be the top of the domain. Furthermore, the sides are named $\partial\Omega_2$ and $\partial\Omega_4$, respectively, whereas the base is the $\partial\Omega_3$ boundary. In sedimentary basins the $\partial\Omega_3$ boundary is often referred to as the basement, i.e., the transition between sediments and crystalline rocks.

The boundary conditions are usually the same for the initial problems (15.23), (15.24), and (15.25) as for the time dependent coupled problem (15.9), (15.17), and (15.22). At the $\partial\Omega_3$ boundary, the essential displacement condition is $u_r = 0$ for the equations (15.9) and (15.25). This is a good assumption in locations where the basement consists of hard crystalline rock with a substantially higher stiffness than the sediments.

Symmetry of a scalar field at a boundary implies that the normal derivative vanishes. This type of boundary condition is important since it reduces the grid size and simulation time. The symmetry boundary condition of a displacement field, is $\mathbf{u} \cdot \mathbf{n} = 0$, where $\mathbf{n}$ is the unit outward normal vector to the surface $\partial\Omega$. Vanishing tangential stress implies $\boldsymbol{\sigma} \cdot \mathbf{n} - (\mathbf{n} \cdot \boldsymbol{\sigma} \cdot \mathbf{n})\mathbf{n} = \mathbf{0}$. Symmetric boundary conditions for the pore pressure and temperature can be written as $\nabla P \cdot \mathbf{n} = 0$ and $\nabla T \cdot \mathbf{n} = 0$, respectively.

At the top of the domain, $\partial\Omega_1$, we use a natural boundary condition with a traction force $\boldsymbol{\tau} = \boldsymbol{\sigma} \cdot \mathbf{n}$. This traction force could be related to deposition and loading of sediments.

In hydrostatic conditions, where $\varrho_f g_i \neq 0$ in (15.24) and (15.17), the pore pressure boundary conditions are set at the seafloor $\partial\Omega_1$, with the appropriate pressure depending on the water depth. However, if only the excess pressure is modeled we have $\varrho_f g_i = 0$ in (15.24), (15.17), and $P = 0$ at the seafloor. Also, a geothermal gradient is imposed by setting a fixed expected temperature $T$ at the seafloor $\partial\Omega_1$ and the bottom boundary $\partial\Omega_3$. Symmetric boundary conditions are preferred at the sides $\partial\Omega_2$ and $\partial\Omega_4$ for the pore pressure and the temperature, since only parts of a sedimentary basin is contained within $\Omega$. Finally, a heat or fluid flux can also be applied at the boundaries, by setting $\nabla P \cdot \mathbf{n} \neq 0$ and $\nabla T \cdot \mathbf{n} \neq 0$, respectively.

## 15.3  Numerical Methods

We use the weighted residual method to discretize the spatial part of (15.9), (15.17), and (15.22)–(15.25). We refer to [11] for a detailed description of the numerical methods used to solve the spatial problem (15.23), (15.24), and (15.25). The focus here is on the numerical methods used to solve the spatial and time-dependent coupled problem (15.9), (15.17), and (15.22).

### 15.3.1  Discretization Technique

In time we use a *backward Euler* discretization scheme. This results in a fully implicit method. At each time step a spatial problem must be solved, and to this end we apply the finite element method.

A spatial unknown $u$ evaluated at time level $\ell$ is denoted by $u^\ell$. This $u^\ell$ is approximated by $\hat{u}^\ell = \sum_j u_j^\ell N_j$. Here, $N_j$ denotes the prescribed basis functions and $u_j^\ell$ represents the unknown coefficients. We let $W_i$ denote a weighting function, with $W_i = N_i$ corresponding to the Galerkin method. The rapid changes in the Darcy flux in the convective term in (15.22) make a strong demand on the discretization methods. The local mesh Peclet number $Pe_\Delta$ can indicate whether a Galerkin approach is appropriate or not, here $Pe_\Delta = (\phi V^{(f)})h/\lambda_c$, where $(\phi V^{(f)})$ and $h$ are the characteristic Darcy flux and elements length, respectively, and $\lambda_c$ is the thermal diffusivity. In linear convection-diffusion model problems, $Pe_\Delta > 2$ leads to non-physical oscillations for a Galerkin solution. Thus, for a large flux $(\phi V^{(f)})$ relative to $\lambda_c$ on a coarse grid one must generally apply Petrov Galerkin "upwind" weighting functions [5], where $W_i \neq N_i$. We refer to [11] for more details.

The unknowns to be solved for are the displacement vector $u_r$ and the scalars $P$ and $T$. We let $u_r$, $P$, and $T$ be approximated by

$$u_r^\ell \approx \hat{u}_r^\ell = \sum_{j=1}^{n} u_j^{r,\ell} N_j(x_1, \ldots, x_d), \tag{15.26}$$

$$P^\ell \approx \hat{P}^\ell = \sum_{j=1}^{n} P_j^\ell N_j(x_1, \ldots, x_d), \tag{15.27}$$

$$T^\ell \approx \hat{T}^\ell = \sum_{j=1}^{n} T_j^\ell N_j(x_1, \ldots, x_d), \tag{15.28}$$

where $u_j^{r,\ell}$, $P_j^\ell$, and $T_j^\ell$ are the unknown coefficients, $n$ is the number of nodes and the subscript $d$ is the number of spatial dimension. We insert the quantities $\hat{u}_r^\ell$, $\hat{P}^\ell$, and $\hat{T}^\ell$ into the governing equations, multiply by $W_i$, and integrate over the domain. Second-order derivatives are integrated by parts. We refer to [11, ch. 2.1] for more details on how to derive a discrete weak formulation for PDEs like (15.9), (15.17), and (15.22).

### 15.3.2  Nonlinear Solution Technique

Due to the complexity of the nonlinear coupled system of PDEs, a *Successive Substitution* technique, also called *Picard iteration*, is employed to solve the nonlinear algebraic equations in (15.22). Newton iterations generally give the best convergence performance if the initial guess is good, but Successive Substitutions is often more robust, i.e., less dependent on the initial guess. The discrete coupled system (15.9), (15.17), and (15.22) with the finite element method in space and a backward Euler scheme in time gives a non-linear algebraic system, which can be written on the form

$$\mathbf{K}(\mathbf{x}^{\ell+1})\mathbf{x}^{\ell+1} = \mathbf{b}(\mathbf{x}^{\ell}). \tag{15.29}$$

Applying the Successive Substitution method implies a recursive set of linear problems

$$\mathbf{K}(\mathbf{x}^{\ell+1,k})\mathbf{x}^{\ell+1,k+1} = \mathbf{b}(\mathbf{x}^{\ell,k}), \qquad k = 1, 2, 3, \cdots \tag{15.30}$$

where $\ell$ is the time level and $k$ is the iteration level. The vector

$$\mathbf{x}^{\ell} = (u_1^{1,\ell}, \ldots, u_1^{d,\ell}, u_2^{1,\ell}, \ldots, u_2^{d,\ell}, \ldots, u_n^{1,\ell}, \ldots, u_n^{d,\ell},$$
$$P_1^{\ell}, \ldots, P_n^{\ell}, T_1^{\ell}, \ldots, T_n^{\ell})^T \tag{15.31}$$

contains the unknown coefficients. The iterations continue until a sufficiently small convergence criteria is met: $\|\mathbf{x}^{k+1} - \mathbf{x}^k\| \leq \epsilon$. The Successive Substitution technique requires a start vector $\mathbf{x}^{\ell,0}$ at each time level. A preferred choice is the solution from the previous time step $\mathbf{x}^{\ell,0} = \mathbf{x}^{\ell-1}$.

### 15.3.3   The Linear System

The fully linearized system (15.30) can be partitioned into block matrices and block vectors, where each block is associated with a node or coupling of two nodes. The relation between one matrix block, the corresponding block of unknowns, and the corresponding right-hand side block can be written as

$$
\begin{bmatrix}
A_{ij}^{1,1} & \cdots & A_{ij}^{1,d} & P_{ij}^{1,d+1} & T_{ij}^{1,d+2} \\
\vdots & \ddots & \vdots & \vdots & \vdots \\
A_{ij}^{d,1} & \cdots & A_{ij}^{d,d} & P_{ij}^{d,d+1} & T_{ij}^{d,d+2} \\
C_{ij}^{d+1,1} & \cdots & C_{ij}^{d+1,d} & P_{ij}^{d+1,d+1} & T_{ij}^{d+1,d+2} \\
0 & \cdots & 0 & 0 & T_{ij}^{d+3,d+3}
\end{bmatrix}
\begin{bmatrix}
u_j^{1,\ell+1,k+1} \\
\vdots \\
u_j^{d,\ell+1,k+1} \\
P_j^{\ell+1,k+1} \\
T_j^{\ell+1,k+1}
\end{bmatrix}
$$
$$
=
\begin{bmatrix}
b_i^{1,\ell} \\
\vdots \\
b_i^{d,\ell} \\
b_i^{d+1,\ell} \\
b_i^{d+2,\ell}
\end{bmatrix}.
\tag{15.32}
$$

Here, the block matrix represents block $\mathbf{K}_{ij}$ in the coefficient matrix $\mathbf{K}$, reflecting the coupling of degrees of freedom in node $i$ and $j$. The two vectors in (15.32) are the block of unknowns $\mathbf{x}_j^{\ell+1,k}$ at node $j$ and the right-hand side block $\mathbf{b}_i$ of the algebraic at node $i$ (respectively). The $A_{ij}^{pq}$ and $C_{ij}^{pq}$ entries come from the terms governing deformation in the porous medium, while the $P_{ij}^{pq}$ and $T_{ij}^{pq}$ entries come from the pore pressure and the temperature terms in the governing equations.

For a fixed $i$ and $j$ (node numbers), $\mathbf{K}_{ij}$ is a $(d+2)\times(d+2)$ matrix. At a global level $\mathbf{K}$ in (15.29) can be viewed as an $n(d+2)\times n(d+2)$ matrix with $n\times n$ blocks of $\mathbf{K}_{ij}$.

The entries in the coefficient matrix $\mathbf{K}$ and $\mathbf{b}$ in (15.32) are derived from the weak discrete formulation of (15.9), (15.17), and (15.22) with the help of Green's lemma [11] to integrate second-order spatial derivatives by part. Below we will formulate each of the entries in (15.32) and indicate which part of the PDE they represent.

Let us start with (15.4). This equation is differentiated with respect to time since time-derivatives of displacements (and hence stresses) appear in other equations. Employing the Galerkin method on $\int_\Omega \dot\sigma_{rs,s} N_i d\Omega$ with integration by parts, where $\sigma_{rs}$ is defined in (15.8), results in the discrete weak formulation of a time-differentiated version of (15.4):

$$\int_\Omega \hat\sigma^{\ell+1}_{rs} N_{i,s} d\Omega = \int_\Omega \hat\sigma^\ell_{rs} N_{i,s} d\Omega + \int_{\partial\Omega} N_i\left(t^{\ell+1}_r - t^\ell_r\right)d\Gamma, \qquad (15.33)$$

where the $t^\ell_r = \sigma^\ell_{rs} n_s$ is the known stress vector on the surface $\partial\Omega$. Observe that $\Delta t$ cancels. The notation $\hat\sigma_{rs}$ implies that the approximations (15.26)-(15.28) are inserted into $\sigma_{rs}$.

The entries in $\mathbf{K}_{ij}$ from the part $\int_\Omega [\lambda \hat u^{\ell+1}_{q,q}\delta_{rs} + \mu(\hat u^{\ell+1}_{r,s} + \hat u^{\ell+1}_{s,r})]N_{i,s}d\Omega$ is expressed as [11]:

$$A^{rs}_{i,j} = \sum_{j=1}^n \sum_{s=1}^d \int_\Omega \left[\mu\left(\sum_k N_{i,k} N_{j,k}\right)\delta_{rs} + \mu N_{i,s} N_{j,r} + \lambda N_{i,r} N_{j,s}\right]d\Omega,$$

$$r = 1,\ldots,d, \quad i = 1,\ldots,n. \tag{15.34}$$

The pore pressure entries from the part $-\int_\Omega \hat P^{\ell+1} N_{i,s} d\Omega$ in (15.33) are simply

$$P^{s,d+1}_{ij} = -\int_\Omega (N_j N_{i,s}) d\Omega, \qquad s = 1, 2, \ldots, d. \tag{15.35}$$

Similarly, the temperature entries in $\mathbf{K}_{ij}$ from (15.33) are

$$T^{s,d+2}_{ij} = -\beta_s(3\lambda + 2\mu)\int_\Omega (N_j N_{i,s}) d\Omega, \qquad s = 1, 2, \ldots, d. \tag{15.36}$$

The right-hand side of (15.32) is the $b^{r,\ell}_i$ vector, and is expressed as

$$b^{r,\ell}_i = \int_\Omega \left[\lambda(\sum_{s=1}^d \hat u^\ell_{s,s})N_{i,r} + \mu\left((\sum_{s=1}^d \hat u^\ell_{r,s} N_{i,s}) + (\sum_{s=1}^d \hat u^\ell_{s,r} N_{i,s})\right)\right.$$
$$\left. -\hat T^\ell N_{i,r} - \hat P^\ell N_{i,r}\right]d\Omega + \int_{\partial\Omega_N} N_i(t^{\ell+1}_r - t^\ell_r)d\Gamma. \tag{15.37}$$

The discrete weak formulation of the pressure equation (15.17) can be expressed with the help of Green's lemma as

$$
\int_{\Omega} \left[ \left( \hat{u}_{q,q}^{\ell+1} + \frac{1-\phi}{K_s} \hat{P}^{\ell+1} + \frac{\phi}{K_f} \hat{P}^{\ell+1} - (1-\phi)\beta_s \hat{T}^{\ell+1} - \phi\beta_f \hat{T}^{\ell+1} \right) N_i \right.
$$
$$
\left. - \Delta t \frac{k}{\mu_v} (\nabla \hat{P}^{\ell+1} + \varrho_f(1-\beta_f \hat{T})g_k) \cdot \nabla N_i \right] d\Omega =
$$
$$
\int_{\Omega} \left[ \hat{u}_{q,q}^{\ell} + \frac{1-\phi}{K_s} \hat{P}^{\ell} + \frac{\phi}{K_f} \hat{P}^{\ell} - (1-\phi)\beta_s \hat{T}^{\ell} - \phi\beta_f \hat{T}^{\ell} \right] N_i d\Omega
$$
$$
+ \Delta t \int_{\partial\Omega} (Q_f) N_i d\Gamma.
$$
$$(15.38)$$

Here, $Q_f = k/\mu_f(P_{,r} n_r)$ is the fluid flux at the boundary, $k$ is the permeability matrix and $n_r$ is the surface normal vector.

The entries in $\mathbf{K}_{i,j}$ originating from the term $-\int_{\Omega} \hat{u}_{q,q}^{\ell+1} N_i d\Omega$ in (15.38) can be written as

$$
C_{ij}^{d+1,s} = - \int_{\Omega} (N_{j,s} N_i) d\Omega, \qquad s = 1, 2, \dots, d. \tag{15.39}
$$

In (15.38) the pressure terms in $\mathbf{K}_{i,j}$ can be expressed as

$$
P_{ij}^{d+1,d+1} = \int_{\Omega} \left[ \Delta t \frac{1}{\mu_v} k \left( \sum_{s=1}^{d} N_{j,s} N_{i,s} \right) \right.
$$
$$
\left. + \frac{\phi}{K_f} N_i N_j + \frac{1-\phi}{K_s} N_i N_j \right] d\Omega. \tag{15.40}
$$

Similarly, the temperature terms in (15.38) on the left-hand side are

$$
T_{ij}^{d+1,d+2} = \int_{\Omega} \left[ -(\phi-1)\beta_s N_i N_j - \phi\beta_s N_i N_j + \Delta t \frac{k}{\mu_v} \beta_s g \varrho_f N_j N_{i,d} \right] d\Omega. \tag{15.41}
$$

The right-hand side in (15.38) is the $b_i^{r+1,\ell}$ vector in (15.32), which reads

$$
b_i^{d+1,\ell} = -\Delta t \int_{\partial\Omega} (Q_f) N_i d\Gamma + \int_{\Omega} \left[ \left( \sum_{s=1}^{d} \hat{u}_{s,s}^{\ell} \right) N_i + \frac{\phi}{K_f} \hat{P}^{\ell} N_i \right.
$$
$$
\left. + \frac{1-\phi}{K_s} \hat{P}^{\ell} N_i - \left( \phi\beta_f + (1-\phi)\beta_s \right) \hat{T}^{\ell} N_i + \Delta t \frac{k}{\mu_v} \varrho_f g N_{i,d} \right] d\Omega. \tag{15.42}
$$

The energy equation (15.22) for a porous medium, is written in a discrete formulation as

$$\int_\Omega \left((1-\phi)\varrho_s C_s + \phi\varrho_f C_f\right)\hat{T}^{\ell+1}W_i + \Delta t\varrho_f C_f \nabla \hat{T}^{\ell+1}\phi\cdot V^k$$

$$-\Delta t((1-\phi)\kappa_s + \phi\kappa_f)\nabla\hat{T}^{\ell+1}\cdot\nabla W_i)d\Omega =$$

$$\int_\Omega \left[\left((1-\phi)\varrho_s C_s + \phi\varrho_f C_f\right)\hat{T}^\ell + \Delta t(1-\phi)\varrho_s h_s\right]W_i d\Omega \qquad (15.43)$$

$$+\Delta t \int_{\partial\Omega} (Q_T)W_i d\Gamma .$$

Note that $W_i \neq N_i$ when we use a Petrov-Galerkin upwind discretization technique. The heat flux at the boundary can be written as $Q_T = \kappa(T_{,r}n_r)$.

The entries for $T_{ij}^{d+2,d+2}$ from the left-hand side of (15.43) are:

$$T_{ij}^{d+2,d+2} = \int_\Omega \left[\left((1-\phi)C_s\varrho_s + \phi C_f\varrho_f\right)W_iN_j + \Delta t\kappa\left(\sum_{s=1}^{d} W_{i,s}N_{j,s}\right)\right.$$

$$\left. +\Delta t\varrho_f C_f\left(\sum_{s=1}^{d} N_{j,s}W_iV_s^{(f),k}\right)\right]d\Omega .$$

$$(15.44)$$

The subscript $k$ is the iteration level in the nonlinear transportation term. This is the only term making the matrix $\mathbf{K}_{ij}$ non-symmetric. The fluid velocity vector in (15.44) is written as:

$$V_s^{(f),k} = \frac{k_s}{\mu_v}(P_j^k N_{j,s} - \varrho_f g_s), \qquad s = 1,2,\ldots,d. \qquad (15.45)$$

Finally, the last entry in (15.32) is the $b_i^{r+2,\ell}$ vector, which is the right-hand in the expression (15.43),

$$b_i^{r+2,\ell} = -\Delta t \int_{\partial\Omega} (Q_T)W_i d\Gamma$$

$$+ \int_\Omega \left[\left((1-\phi)C_f\varrho_s + \phi C_f\varrho_s\right)\hat{T}^\ell W_i + \Delta t(1-\phi)\varrho_s hW_i\right]d\Omega .$$

$$(15.46)$$

In basin modeling, as well as in geotechnical engineering, the effective and thermal stresses are usually of more interest than the displacement field. Having computed the latter, the effective stress tensor (15.8) can be calculated at each time level. Since the stresses are derivatives of the displacement components, the computed stress tensor field will be discontinuous over the element boundary. For the purposes of plotting and analysis it is desirable to have a continuous stress field. Thus, components in the stress tensor are smoothed and plotted as a scalar continuous stress field, together with the computed pore pressure and temperature fields. Having computed the pore pressure, the fluid phase flow vectors can also be visualized.

When the calculated stress tensor is established, it is desirable that we understand the critical state in the sediment or simply the stress level. A simple model of the stress level is the widely used von Mises yield function, also called the equivalent stress $m$:

$$m = \sqrt{\frac{3}{2}\hat{\sigma}_{rs}\hat{\sigma}_{rs}}, \tag{15.47}$$

where $\hat{\sigma}_{rs}$ is the stress deviator, $\hat{\sigma}_{rs} = \sigma_{rs} - \frac{1}{3}\sigma_{kk}\delta_{rs}$.

The level of maximum shear stress and the probability of fracturing or failure of the sediment is of interest to both geomechanical engineers and geologists. Fracture and failure criteria frequently involve principal values (eigenvalues) of the effective stress tensor. The Mohr-Coulomb criterion is a widely used failure criterion:

$$(\sigma_1' - \sigma_3') = 2c\cos\phi - (\sigma_1' + \sigma_3')\sin\phi. \tag{15.48}$$

Here, $c$ is the cohesion and $\phi$ is the internal friction angle and the effective principal stresses are $\sigma_1' \geq \sigma_2' \geq \sigma_3'$. We compute the effective principal stresses in each element, and if (15.48) is approximately fulfilled, the sediments in the element are close to failure.

## 15.4    Implementing a Solver for a System of PDEs

We shall in this section describe a flexible software implementation that is capable of solving the numerical problem presented in the previous section. Emphasis is placed on producing a software which is easy to test and expand. When dealing with large equation systems it is of great importance that the development and testing of the software is undertaken in a piece-wise manner. This is cost effective, since verification takes the bulk of the time in this type of numerical software development.

The solution strategy will have two main objectives. First, an expansion of a solvers code should be possible without any modification to already verified and working components. Second, the lines of code can be minimized and the flexibility increased by keeping the common variable coefficients for the solvers in separate classes, which then can be accessed by all the solvers. A manager class is also needed to act as the solver class for the system, and to call the different solvers in a proper sequence.

The basin model consist of three stand-alone PDE solvers for (15.23), (15.24), and (15.25), as well as one compound solver for the system of PDEs (15.9), (15.17), and (15.22). All PDE solvers are organized as C++ classes in Diffpack as explained in [11]. The building blocks for the PDE solvers are found in the Diffpack library. Examples on such components are linear systems solvers, finite element grids, finite element assembly algorithms, arrays etc. at different abstraction levels [11].

The organization of the PDE classes making up the basin simulator is shown in Figure 15.1. In Diffpack, pointers are used in the manager class Administrator to gain access to the different member functions in the PDE solver classes. This makes it easy to add new PDE solvers to the model, as they are independent and only have pointers to the class Administrator. The class Administrator is responsible for creating the different solver objects and for holding data common for all the solvers, such as finite element grid objects and time stepping parameters.

Prior to solving the initial effective stresses in (15.25), the initial pore pressure and temperature fields governed by (15.24) and (15.23) must be found. The temperature governed by the Poisson equation (15.23) is solved in the class Heat. Class Administrator then calls class Pressure to solve the initial pore pressure problem governed by the Laplace equation (15.24). Afterwards the deformation and stress state is computed by class Elasticity. We refer to [11] for information on the basic building blocks for Laplace or Poisson equations. Furthermore, the class Elasticity is a simple expansion of the Elasticity1 solver in [11], where only the pore pressure is added as an extra term.

In geological modeling, the grid may have to be sectioned up in different lithological layers. These layers will have different values of the physical parameters given in Table 15.1. The finite element method is well suited for solving PDEs in realistic geometries with, e.g., variable thickness in the layers, anticlines, and rotated fault blocks, and Diffpack offers a flexible way of working with such layered media. The finite element grid can be divided into subdomains, and class GeoDataField reads the values for the variable coefficients for different subdomains and stores them as FieldFormat objects. Class FieldFormat [11] in Diffpack enables flexible handling of various formats of a variable coefficient in a PDE. A variable coefficient is represented as a scalar field, or more precisely, as a Diffpack Field class. Here, the parameters are assumed to be constant in the finite element grid subdomains. This is often a realistic or adequate geological assumption. For example, in the paralic environment (i.e., along the delta, coastal plain shore line, or shelf systems) there are often relative abrupt changes in lithology. Fluctuation in the relative sea level, over geological time, results in lithology changes, e.g., from shale to sand over short distances in the sedimentary sequences deposited. Therefore, piece-wise constant parameters in the finite element grid can represent a good geological approximation to the changing lithology. All the solver classes can independently access the value of variable coefficients in the class GeoDataFieldAtPt at a point in the finite element grid during the simulation. This means that there are no geological parameters in the solvers; the solvers access the `GeoDataField` object through pointers to class `Administrator`.

Class `DynamicEquilibrium` solves the coupled model (15.9), (15.17), and (15.22) in a fully implicit fashion. We can use the same boundary conditions in `DynamicEquilibrium` as in the solvers `Heat`, `Pressure`, and `Elasticity`. The initial conditions for the time-dependent system are obtained from the latter three solvers. Class `DynamicEquilibrium` gets the initial conditions and the boundary conditions through class `Administrator`.

The subclasses `Terzaghi1Dconsolidation` and `Cylinders` are expansions of their respective base classes containing analytical equations used to verify parts of a numerical solution (using the Diffpack set-up from [11, ch. 3.5]).

## 15.5    Verification

An important part of the development of any numerical model is the verification of the implementation. Numerical discretization errors are common for all the numerical models. The numerical errors are estimated by testing the numerical model against analytical solutions, whenever available. To the authors' knowledge, there are no analytical solution available for the coupled model (15.9), (15.17), and (15.22). However, one-dimensional analytical solutions are available for parts of the coupled model and for the spatial problem (15.23), (15.24), and (15.25). We will here focus on verifying the models against some of these analytical solutions. The correlation with field data is described in the section concerning specific case studies.

### 15.5.1    Cylinder with Concentric Circular Hole

The initial condition given by (15.23)-(15.25) is tested on a porous cylinder with a concentric circular hole. To simplify the problem we let $(1 - \phi)\varrho_s h = 0$ and we set thermal diffusivity to unity in (15.23). In the pressure equation (15.24) we let $k_{rs}/\mu_v = 1$. In both (15.24) and (15.25) the equations are simplified by $g_i = 0$. Since the geometry is axis-symmetric, an analytical solution can easily be established. Furthermore, the cylinder geometry implies that it is reasonable to introduce a plain strain assumption. As a result, the solution is dependent only on the radial coordinate. The analytical boundary value problem for the temperature and pore pressure are given by the Laplace equation in cylindrical coordinates:

$$\left\{ \begin{array}{l} \dfrac{1}{r}\dfrac{d}{dr}\left(r\dfrac{d(T(r), P(r))}{dr}\right) = 0 \\ P(a) = P_0, \quad P(b) = 0 \\ T(a) = T_0, \quad T(b) = 0 \end{array} \right\}. \tag{15.49}$$

Here, the $P_0$ and $T_0$ are the boundary values at the inner wall of the cylinder, $r = a$, for the pore pressure and temperature, respectively. At the outer

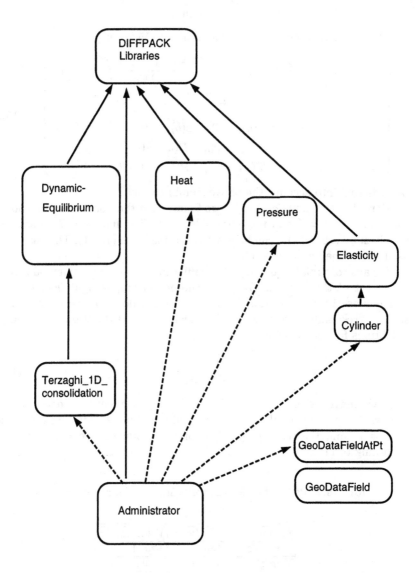

**Fig. 15.1.** The main class hierarchy constituting the basin simulator. Solid arrows illustrates inheritance ("is-a" relationship) of code and dashed arrows are pointers ("has-a" relationship).

wall, $r = b$, the temperature and pore pressure are set to zero. By simply integrating (15.49), these solutions are derived:

$$(T(r), P(r)) = \frac{(T_0, P_0)}{\ln(a/b)} \ln(\frac{b}{r}). \qquad (15.50)$$

The corresponding analytical solution for (15.25) is

$$
\left\{
\begin{aligned}
u(r) &= \frac{(1+\nu)\beta_s}{(1-\nu)r} \int_a^r T(r) r \, dr + \frac{(1+\nu)(1-2\nu)}{(1-\nu)Er} \int_a^r P(r) r \, dr \\
&\quad + c_1 r + \frac{c_2}{r} \\
&u(a)_r = 0, \quad u(b)_r = 0 \\
&P(a) = P_0, \quad P(a) = 0 \\
&T(a) = T_0, \quad P(b) = 0.
\end{aligned}
\right\} \qquad (15.51)
$$

when taking into account the above simplifications.

Here, the constants $c_1$ and $c_2$ are found from the boundary conditions at the inner wall $r = 1$ and the outer side $r = 2$. The functions $T(r)$ and $P(r)$ are given by (15.50) and can be directly inserted into (15.51). The parameters used are defined in Table 15.2.

Because of radial symmetry, it is sufficient to solve the numerical problem on a part of the cylinder. We have chosen the upper right quarter of the cylinder as the computational domain. In this domain there are two new sides, corresponding to symmetry lines, where we set the standard symmetry conditions for scalar fields,

$$
\begin{aligned}
\nabla T \cdot \mathbf{n} &= 0, \\
\nabla P \cdot \mathbf{n} &= 0.
\end{aligned}
\qquad (15.52)
$$

The symmetric boundary condition for the displacement is $\mathbf{u} \cdot \mathbf{n} = 0$ together with vanishing tangential stress. At the outer and inner wall the displacement is $\mathbf{u} = \mathbf{0}$, and the temperature and pressure boundary conditions are set according to (15.51).

**Table 15.2.** Parameters for the verification of the numerical solution.

| Parameter | unity | Value | Symbol |
|---|---|---|---|
| Poisson's constant | | 0.25 | $\nu$ |
| Young's module | Pa | 10000 | $E$ |
| Heat expansion | 1/°C | 0.005 | $\beta_s$ |

Class `Cylinder` holds additional information to compute the analytical result and to compare it with the numerical solution. A common model for the discretization error is $e(h) = Ch^r$, where $e$ is the error, $h$ is the grid size, and $C$ and $r$ are constants that can be estimated either from theoretical considerations or from numerical experiments. We have for two numerical experiments $s$ and $s+1$ the relations $e_s = Ch_s^r$ and $e_{s+1} = Ch_{s+1}^r$. Dividing these relations eliminates $C$, and by solving with respect to $r$ we get the estimate of the *convergence rate*

$$r = \frac{\ln\left(e(h_s)/e(h_{s+1})\right)}{\ln\left(h_s/h_{s+1}\right)}. \tag{15.53}$$

We have used four-node box-shaped elements with bi-linear basis functions. The error is expected to be proportional to $h^2$ in the $L^2$ norm [11]. Thus, the convergence rate should be close to 2. We refer to [11] for how to compute error estimates experimentally. In Table 15.3 the errors in the $L^2$ norm are given for different grid sizes. The results confirm the convergence of the numerical solution and the convergence rate $r \sim 2$.

**Table 15.3.** Numerical errors in the $L^2$-norm, for different element sizes $h$. Note that the reduction in the error as the grid size decreases is of order $h^2$, hence the convergence rate is $r \sim 2$.

|       | $h$ | $h/2$ | $h/4$ | $h/8$ | $h/16$ |
|-------|-----|-------|-------|-------|--------|
| $L^2$ | $1.082 \cdot 10^{-4}$ | $2.7579 \cdot 10^{-5}$ | $6.9150 \cdot 10^{-6}$ | $1.7155 \cdot 10^{-6}$ | $4.1769 \cdot 10^{-7}$ |
| $r$   |     | $1.972$ | $1.996$ | $2.001$ | $2.038$ |

## 15.5.2   One-Dimensional Consolidation

We have found no analytical solutions available for the coupled system (15.9), (15.17), and (15.22). Verification in this case is based on testing part of the system against analytical solutions and later the whole system with field data. For example, mechanical deformations is governed by (15.9) and (15.17), and in one-dimensional consolidation and deformation the pore pressure equation in time can be expressed as:

$$P(z,t) = \sum_{i=1}^{N} P_o \frac{2}{i\pi}(1 - \cos(i\pi))\sin(\frac{i\pi z}{2}) \exp\left(-\frac{kE(1-\nu)t}{\mu_f(1+\nu)(1-2\nu)}\right). \tag{15.54}$$

Here, the exponent is the consolidation constant $c_v$ for one-dimensional, confined consolidation, and $P_o$ is the uniform initial pore pressure in the layer. Instantly loading a sedimentary layer results in the pore pressure taking up all the applied load. Over time the pore pressure in the layer will decrease, as the pore water is drained out of the layer. In (15.54) the rate of pore pressure decay is dependent on the stiffness of the layer, controlled by the Young's modulus $E$ and Poisson relation $\nu$, as well as the permeability $k$ and the fluid viscosity $\mu_f$.

The boundary-value problem to be solved reads

$$
\left\{
\begin{array}{c}
0 = \nabla \cdot [\mu \nabla \dot{\mathbf{u}}] + \nabla[(\mu + \lambda)\nabla \cdot \dot{\mathbf{u}}] - \nabla \dot{P} \\[2mm]
\nabla \cdot \dot{\mathbf{u}} = \nabla \cdot (\dfrac{k}{\mu_f}\nabla(P)) \\[2mm]
P(0) = 0 \\[1mm]
P(2) = 0 \\[1mm]
P(z,0) = 1 \\[1mm]
u_2(z,0) = 0
\end{array}
\right\} .
\tag{15.55}
$$

The grid is a box with 9×9 four-node bi-linear elements on a uniform mesh. There is no temperature in this problem (in the simulator we just set $T = 0$). We have symmetric boundary conditions on the two sides to achieve a one-dimensional condition. In all the results presented in this section, only cross sections of the 2×2 box are plotted, from the coordinates (0.5,0) to (0.5,2). The solutions are found with Galerkin spatial discretization and finite difference backward Euler discretization in time.

**Table 15.4.** Physical parameters for the solution (15.54) of (15.55).

| Parameter | Unity | Value | Symbol |
|---|---|---|---|
| Poissons's constant | | 0.25 | $\nu$ |
| Young,s module | Pa | $4.0 \cdot 10^4$ | $E$ |
| Permeability | $m^2$ | $3.0 \cdot 10^{-10}$ | $k$ |
| Viscosity | Kg/m·s | 0.001 | $\mu_v$ |

The first test had a uniform initial pore pressure of $P_0 = 1$ in the layer. The time steps were $\Delta t = 1$. Some oscillations were obtained at the first time step, but at the second and the third time step the oscillations were gradually smoothed out (Figure 15.2a). The analytical solution (15.54) is visualized at the same time steps as a reference. Similar results were encountered in [4] and depicted in Figure 15.2b. In [4] they used a 1D consolidation program, TINY, which was tested in a similar way. In Figure 15.2b only half of the layer is used because of symmetry. Reference [4] also gave a relation on how far the drainage had reached into the sediment layer in time:

$$
l = \sqrt{12 c_v \Delta t}.
\tag{15.56}
$$

Here, $c_v$ is defined as the exponent in (15.54), and $l$ is the length the drainage has reached within $\Delta t$. The oscillations occur when $l \leq h$, where $h$ is the element length [4]. Thus, the stability criteria is $l > h$ to avoid these oscillations.

To compare the numerical error in (15.55) with the analytical solution (15.54), and to see if it converges as the time steps get smaller, three numerical experiments were performed. With a *backward Euler* scheme in time, the truncation error is of order $\Delta t$. Using bilinear elements in space, we expect the error to behave like $e(h, \Delta t) = C_1 h^2 + C_2 \Delta t$, where $h$ is the element size, and $C_1$ and $C_2$ are constants.

The initial uniform pore pressure was set to $P_o = 10$ in the layer. The tests were conducted with three different time steps $\Delta t = 5, 10, 20$. The plots were produced at times 20, 60, and 100. When plotting the solution curves from (15.55) together with the analytical solution (15.54) in Figures 15.3a, 15.3c, and 15.4a, we see that the pore pressure curves converges as the time step is reduced. The relative error is reduced from 1.47 at $\Delta t = 20$ to 0.31 at $\Delta t = 5$. It is important to note that when the time steps are reduced by 50 percent from $t = 20$ to $t = 10$, the relative errors in Figure 15.3b are reduced by 50 percent in Figure 15.3d. This has a direct relationship to the expected discretization error, where the error is expected to be proportional to $\Delta t$. However, this also shows that the spatial error $C_1 h^2$ is negligible compared to the temporal discretization error $C_2 \Delta t$.

The results from numerical verifications of (15.54) indicates stability limitations and numerical errors. The stability criterion in time is $l \leq h$ and this limits how small $\Delta t$ can be, provided that $h$ is kept constant. However, the relative error in Figures 15.3 and 15.4 is of order $\Delta t$ and shows that small time steps are preferred to minimize the relative error. One can, e.g., choose $\Delta t = h/\sqrt{12c_v}$.

An upwinding technique [11] is used to stabilize the temperature solution in (15.22). However, this upwind weighting, as in finite difference techniques, curbs numerical oscillation by introducing numerical diffusion, thus smearing the temperature front [8]. The numerical diffusion must not be too excessive, as this can lead to a lack of conservation of energy.

## 15.6    A Magmatic Sill Intrusion Case Study

During extension and subsequent continental breakup the lithospheric thermal gradient, and thus the heat transfer will be increased as a function of the amount of lithosphere thinning [14]. Magmatic intrusions will in addition transport heat into the crust and sedimentary basin, introducing local sharp rise in the heat transfer. During the opening of the NE Atlantic Ocean in Paleocene-early Eocene time (~55 million years ago), large volumes of igneous rocks were emplaced at crustal level. The thickest units are found at the continental-ocean transition where a 100% igneous crust of ~ 20-25 km thickness is suggested. Beneath the adjacent thinned continental crust 5-7 km

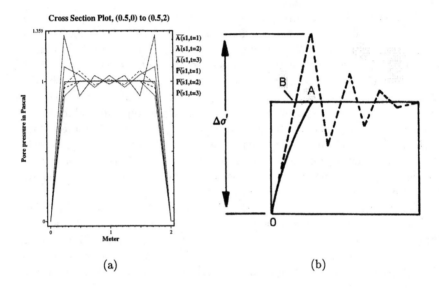

(a)                                              (b)

**Fig. 15.2.** Curve plots (a) of analytical solutions, $A(s1, t)$, from (15.54), and the numerical solution, $P(s1, t)$, from the initial boundary problem (15.55); $t$ is the time for the individual graphs and $s1$ is a local spatial coordinate along a line from $(0.5, 0)$ to $(0.5, 2)$. The numerically computed pore pressure in $P(s1, t)$ is oscillating, whereas the analytical solution has no oscillations. Plot (b) illustrates the same non-physical oscillations computed by the TINY program [4] (the dashed line is the numerical solution, and the solid line is the analytical one). In plot (b), $A$ is the distance to first interior node and $B$ is how far the pore pressure reduction has reached into the layer. Note the similar results from numerical solution, $P(s1, t)$, from (15.55) in plot (a) and results from the TINY program in plot (b).

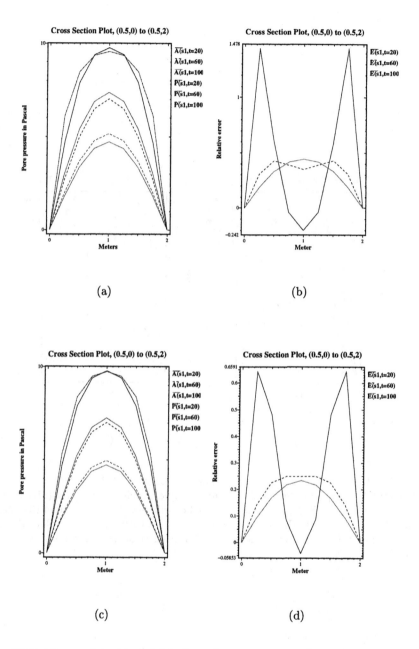

**Fig. 15.3.** Curve plots (a) and (c) show the pore pressure from the theoretical solution (15.54), $A(s1, t)$, and the initial boundary value problem (15.55), $P(s1, t)$, with $\Delta t = 20$ and $\Delta t = 10$ for times $t$, respectively, and $s1$ is a local spatial coordinate along a line from $(0.5, 0)$ to $(0.5, 2)$. Plots (b) and (d) show the relative error, $E(s1, t)$, at each time step in plot (a) and (c), respectively. The error computations indicate that the discretization error in time is proportional to $\Delta t$.

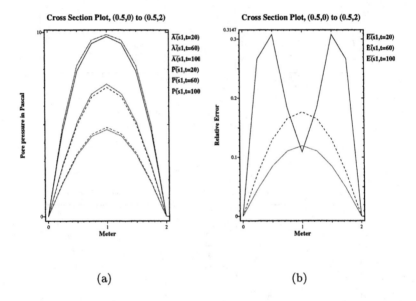

(a)                                      (b)

**Fig. 15.4.** Curve plot (a) shows the pore pressure from the analytical solution (15.54), $A(s1, t)$, and the initial boundary value problem (15.55), $P(s1, t)$, with $\Delta t = 5$ for times $t$. Here, $s1$ is a local spatial coordinate along a line from $(0.5, 0)$ to $(0.5, 2)$. Plot (b) shows the relative error, $E(s1, t)$, at each time step in plot (a). The error computations indicate that the discretization error in time is proportional to $\Delta t$.

thick bodies of magmatic underplating characterize a 50-70 km wide zone, whereas numerous dikes and sills are found as much as $\sim$ 150 km landward of the continental breakup axis. The magmatic episode was characterized by extensive volcanic activity (i.e., basaltic flows at the surface). With respect to heat transfer, however, 1-5 m thick lava flows at the surface cool rapidly, and does not contribute significantly to the thermal budget. Magmatic underplating and sill intrusions, on the other hand, are of major importance. The associated spike in heat transfer may also have had effects on the exploration potential for hydrocarbons in some basins. The sediment and the pore water will expand and change the effective stress in the basin, whereas thermal expansion also leads to density anomalies in the pore water, which may lead to advection. However, this requires that the permeability, the height of the permeable layer, and the temperature gradient are sufficiently large. Free convection is not considered likely to occur with normal thermal gradients [3].

The main objective of the present section is to demonstrate an application of significant geological relevance. As stated in the introduction, the model can be used to study fluid and heat transfer in the sediments. We have selected a case where a horizontal magmatic sill intrusions is causing an abrupt physical process, which sends a steep heat gradient into the sediments. This causes heat transport and changes in fluid flow pattern in the adjacent sediments, which causes the thermal and the effective stresses to change in the sill and the adjacent sediments. Thus, this case study requires all the couplings in (15.9), (15.17), and (15.22). Moreover, it can also give an important geological contribution in trying to understand the physical effects the sill intrusions have on adjacent sediments and maturity of kerogen (a topic covered later in the chapter).

The steep temperature gradients from a magmatic intrusion body may effect the maturity of kerogen, and thus the onset of primary migration. During primary migration the oil starts to leak out of the organic rich shales. Kerogen maturity is measured using vitrinite reflectance. Vitrinite is usually plant tissue of terrestrial origin. Its color changes from light brown to black with exposure to temperatures between 60 and 200°C. The scale of vitrinite reflectance ranges from 0-8, depending on how much of the organic matter (kerogen) is converted into oil and gas. If the value is 8 then there is almost only pure carbon left as a result of the high temperatures.

In a coastal marine setting the kerogen assemblage usually comprises of organic matter from terrestrial sources and marine phytoplankton. These kerogens comprise a spectrum of different chemical compounds with large organic molecules that are stable at low temperatures. As the temperatures increases the thermal vibration increases in the molecules. When the temperature or energy reaches a certain level, the vibrations causes the molecular bonds to 'crack'. This cracking process causes the molecules to separate into heavier carbon compounds and lighter carbon compounds. The lighter compounds (hydrocarbons), such as oil and wet gas, are produced once the vitrinite reflectance passes 0.5, while at values of 2 and above only dry gas is generated. The energy level that is required to cause cracking of certain hydrocarbon compounds is called activation energy. Light hydrocarbons (e.g. alkans and aromatics) with a high hydrogen/carbon (H/C) ratio are continuously released from the kerogen as the heat increases. This means that the H/C ratio in kerogen itself decreases, the vitrinite reflectance increases, the kerogen darkens, and at the very highest temperatures it ends up as pure carbon. We refer to [20] for further description of the maturation process.

To quantify the vitrinite reflectance we employ the $EASYR_o$ model [19]. This model is easy to use in conjunction with our coupled model. The chemical model of vitrinite maturation is described by the first order Arrhenius reactions,

$$\frac{dw_i}{dt} = -w_i A \exp\left(\frac{-E_i}{RT(t)}\right), \tag{15.57}$$

where $i$ is the reaction component number, $w_i$ is the weight of the unreacted component, $E_i$ is the activation energy, $A$ is a factor, $R$ is the universal gas constant, and $T$ is the temperature. Each component is an organic compound with different thermo-chemical properties.

The weight loss due to the cracking process is given by the rate of change of the components $w_i$ integrated over time:

$$w_i = w_{i0} - \int_0^t \frac{dw_i}{dt} dt. \tag{15.58}$$

Here, $w_{i0}$ is the initial amount of the total reactant of the individual compound. In the $EASYR_o$ model there are 20 reactive components.

The vitrinite reflectance value is then calculated from

$$EASYR_o = \exp\left(-1.6 + 3.7F\right), \tag{15.59}$$

where the function $F$ is given by the relation

$$F = 1 - \frac{w}{w_0}. \tag{15.60}$$

Here, $w$ is the total weight of unreacted components and the $w_0$ is the initial total weight of all components. The $EASYR_o$ quantity ranges between 0.2 at $F = 0$ and 4.7 at $F = 0.85$. It is important to note that the $EASYR_o$ model only takes into account 85% of the kerogen (i.e. $F_{max} = 0.85$). This means that the modeled $EASYR_o$ can be too small compared with observed values, for the very high temperatures.

The model used was tested against vitrinite reflectance values measured in the shale prone, low permeability Bottenheia Formation near a horizontal sill intrusion on Svalbard. The measured thickness of the sill was 23 m and samples of vitrinite were taken at 10, 15, 20, and 25 m vertical distance from the sill (pers. comm. Karlsen D. 2000)(Figure 15.5). In the modeling it is assumed that the heat transfer was dominantly conductive, and the initial temperature in the sill was 1100°C. The model output shown is made at a time step when the temperature in the sill had dropped to 200°C, and the heat aureole had reached its maximum extension. At each time step the rate of weight loss is calculated in (15.57) for each of the 20 components. The total sum of all the components weight loss to the present simulation time is then integrated and subtracted from the initial weight $w_{i0}$ of components in (15.58). The $EASYR_o$ quantity is calculated from the total weight $w$ and $w_0$ from all the 20 components. Close to the intrusion the model underestimate the vitrinite values because of the high temperature and 100 % of the kerogen has reacted and the $EASYR_o$ model only takes into account 85% of the kerogen. However, the 1D plot of the modeled maturity resembles the measurements well, so the model is useful in quantifying the maturity in the volumes of source rock around the sill. Furthermore, this correlation with field data can also be used in the calibration process of the basin model, since the observed data fits the simulation results quite well.

### 15.6.1    Case Definition

In our selected case we retain the 23 m thickness of the sill from Svalbard. The basin simulator is used to model the two-dimensional cooling of a sill intrusion and its influence on maturation of a possible source rock in its vicinity. The simulations involves the fluid/heat transfer, thermal and effective stresses. At each time step the $EASYR_o$ vitrinite reflectance is updated, giving the vitrinite reflectance in the host sediment.

The geological model has three sedimentary layers, where the sill is placed in the center (Figure 15.6). There are 1960 eight-node box shaped elements and a total of 6021 nodes in the finite element grid, and the model is 300 m wide and 400 m high. The depth to the base of the magmatic sill is 1500 m below the seafloor.

The overburden stress and the hydrostatic pressure at these depths are high relative to the thermal stress and pore pressures caused by the heat transfer. This makes it hard to visualize the changes in the pore pressure and stress as result of heat transfer from the sill alone. Thus, we conduct some of the simulations without any hydrostatic pressure or overburden. However, when the hydrostatic pressure and overburden are used in the simulations, the term $g_i$ is switched on in (15.24) and (15.25). The 1300 m overburden is simulated by a traction force on the top of the finite element grid. In our simulations internal heat sources (radioactive heat) are small which leaves $(1-\phi)h\varrho = 0$. At both sides symmetry boundary conditions are imposed both for the temperature and the pore pressure. At the bottom, the displacement is set to $u_2 = 0$, allowing motion only in the lateral directions, with vanishing tangential stress. The upper boundary of the domain has no displacement restrictions.

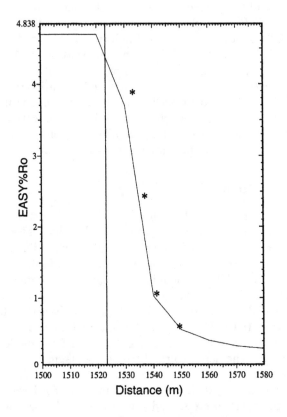

**Fig. 15.5.** Modeled vitrinite reflectance $EASYR_o$, solid curve, compared with the values, labeled with *, at various distances above the edge of a 23 m thick sill intrusion, vertical line, on Svalbard. The assumed total heat conduct for the sediment and pore water is $\kappa = 1W/m^2$, permeability is $k = 1.0\times10^{-15}m^2$| and porosity is $\phi = 30\%$.

We have conducted three types of simulations in time, where the initial temperature of the sill intrusion is set to 1100 °C in all of them. In the first two simulations we only model the excess pressures caused by the heat transfer from the sill. The first simulation assumes a medium with low permeability. In the second simulation the permeabilities are higher and fluid flow dominates the heat transfer (i.e. advection). Only the excess pressures are modeled in these simulations. In the third simulation the hydrostatic pressure due to the overburden is implemented, providing the thermal and effective stress situation around the sill in a presumed natural stress environment.

The parameters used in the simulations are taken from [15,17,12]. The parameters for the geological model are given in Table 15.5 and 15.6, whereas the presumed lithologies are shown in Figure 15.6. Pore pressures, temperatures, and stress fields that differ from the hydrostatic pressure, the normal geothermal gradient, and the overburden stress caused by the gravity field, respectively, are referred to as anomalies.

### 15.6.2    Results and Discussion

In the first simulation we plot the temperature field around the sill 12 years after the magma has intruded the sediments (Figure 15.7). The temperature in the core of the intrusion is 400°C and the heat aureole is symmetric around the sill. The sill is heating up sediments more than 50 m above and below the intrusion. The heat conducted outwards from the sill heats up both the sediment and the pore fluid causing both the mineral skeleton and the pore fluid to expand. Immediately after emplacement of the sill, the steep heat gradient causes heat expansion and density anomalies in the pore water. This leads to pore pressure increase as result of boiling and vapor phase transition provided that the hydrostatic pressure is below 30 MPa [6]. However, in this study we look at the long term effects of heat transfer and fluid flow. We also assume that the sills are at depths with higher hydrostatic pressures than 30 MPa, thus the boiling and vapor phase transition is not taken into account. The pore pressure field propagates outwards at a rate which mainly depends on the coefficient of consolidation for plain strain

$$c_v = \frac{kE}{2(1 - 2\nu)(1 + \nu)}. \tag{15.61}$$

The fluctuation of stiffness in the sediments is far less than the changes in permeability, which makes the permeability the main governing parameter, since the other parameters are close to constant. After the initial high pressure fluid explosion phase by vapor pressure the temperature decreases and the buoyancy driven forces gradually become dominant [6]. Provided that the permeability is sufficiently high, the density anomalies in the pore water can instigate buoyancy driven fluid flow.

In our simulation the permeability is too low to instigate any buoyancy driven fluid flow, (permeability $\leq 0.01$mD; milli Darcy). The low permeability coupled with the heat expansion of the pore water causes a circular zone

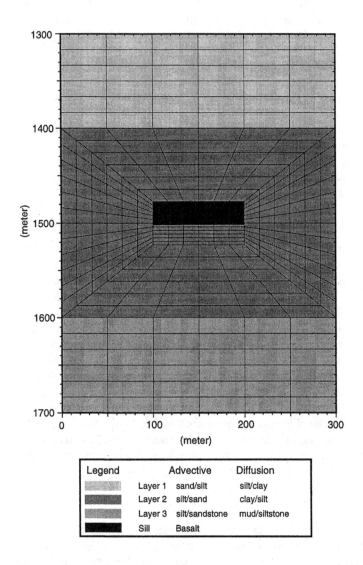

**Fig. 15.6.** Geological model used for the 23 m thick and 100 m wide sill, masked with a finite element grid. The model has three sedimentary layers. Different lithologies are related to advective and diffusion dominated regimes.

of increased pore pressure around the sill (Figure 15.8). The highest positive pressure anomalies are located close to the sill and in the modeling the maximum pore pressure anomaly is 0.7 MPa after 1.25 years. The increasing temperature causes volumetric expansion of the sediments, which contributes to a reduction in the pore pressure. A drop in the temperature has the opposite effect. However, the thermal expansion of the pore water is dominant and results in only excess pressure. Thus, the relationship between the thermal expansion $\beta_f$ of the pore water and the permeability of the sediments are the driving forces for the pore pressure field. At the start of the simulation the fluid flow and pore pressure are at their peak level (Figure 15.8). As the heat is conducted outwards and the temperatures decline, the pore pressure also decreases. The fluid flow velocity (8 cm/yr) is still slower than the rate of conduction. The increase in the pore pressure is, thus, dependent on the permeability of the sediments and the temperature. In low permeability sediments the pore pressure will be higher as the pore water is unable to escape fast enough to drain the pore pressure as the aureole expands outwards. In the very lowest permeability sequences it is expected that the pore pressure will become so high that it leads to fracturing of the sediment which will then increase the outwards flow rate and lead to a drainage of the excess pore pressures.

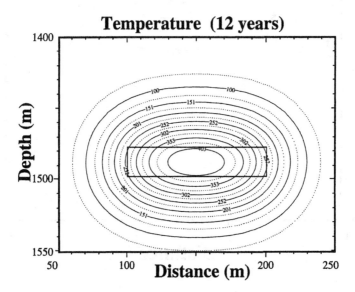

**Fig. 15.7.** Temperature in a conduction dominated regime at 12 years after emplacement. A box shows the geometry of the sill. Note the symmetric areol caused by the temperature field.

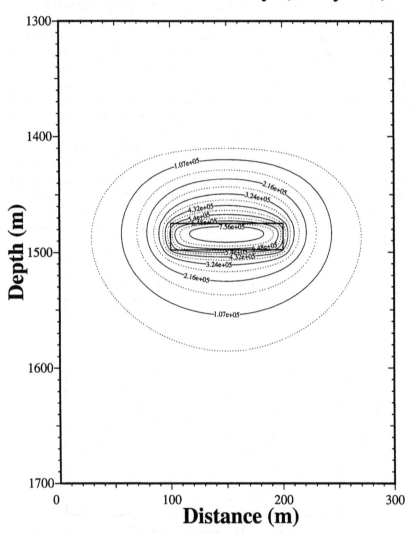

**Fig. 15.8.** Fluid pressure (Pa) anomaly (excess pressure) generated by the temperature field at 1.25 years after emplacement. Note that the pressure and temperature field are closely linked (Figure 15.7). The sill is located within the box.

**Table 15.5.** Parameters for the coupled model (15.9), (15.17), and (15.22) in a conductive dominated heat transport. Units are defined in Table 15.1.

| Parameter | Layer1 | Layer2 | Layer3 | Intrusion |
|---|---|---|---|---|
| $E$ | 1.0 GPa | 1.0 GPa | 1.6 GPa | 1.6 GPa |
| $\mu$ | 0.35 | 0.35 | 0.35 | 0.35 |
| $K_s$ | 30.0 GPa | 30.0 GPa | 30.0 GPa | 30.0 GPa |
| $K_f$ | 4.3 GPa | 4.3 GPa | 4.3 GPa | 4.3 GPa |
| $\phi$ | 35% | 35% | 25% | 10% |
| $\varrho_s$ | 2300 | 2300 | 2300 | 2300 |
| $\varrho_f$ | 1020 | 1020 | 1020 | 1020 |
| $\beta_s$ | $1.0 \times 10^{-5}$ | $1.0 \times 10^{-5}$ | $1.0 \times 10^{-5}$ | $1.0 \times 10^{-5}$ |
| $\beta_f$ | $1.0 \times 10^{-3}$ | $1.0 \times 10^{-3}$ | $1.0 \times 10^{-3}$ | $1.0 \times 10^{-3}$ |
| $k_x$ | $1.0 \times 10^{-13}$ | $1.0 \times 10^{-16}$ | $5.0 \times 10^{-17}$ | $1.0 \times 10^{-18}$ |
| $k_z$ | $1.0 \times 10^{-13}$ | $1.0 \times 10^{-16}$ | $5.0 \times 10^{-17}$ | $1.0 \times 10^{-18}$ |
| $\mu_f$ | $1.0 \times 10^{-3}$ | $1.0 \times 10^{-3}$ | $1.0 \times 10^{-3}$ | $1.0 \times 10^{-3}$ |
| $C_f$ | 4190 | 4190 | 4190 | 4190 |
| $C_s$ | 753 | 753 | 753 | 800 |
| $\kappa_s$ | 1.5 | 1.5 | 1.5 | 1.5 |
| $\kappa_f$ | 0.6 | 0.6 | 0.6 | 0.6 |
| $g_r$ | 9.81 | 9.81 | 9.81 | 9.81 |

The $EASYR_o$ vitrinite reflectance value is a direct consequence of maximum temperature and duration of heating. Hence, the vitrinite reflectance field shows the same geometry as the temperature field (Figure 15.9). The maximum extent of the vitrinite reflectance field is approximately 20-25 m both above and below the sill. Based on Figure 15.9 we estimate the matured part of the host sediment to be roughly twice that of the sill itself.

In the second simulation, the permeability was increased so that fluid flow rather than conduction is the dominant mechanism of heat transfer. After the initial emplacement the fluids are expelled outwards from the sill. Less than a year after emplacement the fluid flow is buoyancy driven. More fluids are transported upwards from the sill than downwards. This is due to the density contrasts caused by the heating, leading to density changes (expansion) of the fluid and associated buoyancy driven fluid flow upwards. Through time an asymmetric temperature field develops around the sill (Figure 15.10) and a steep pore pressure gradient builds up between the top and bottom of the sill (Figure 15.11).

Immediately after emplacement there is a high excess pressure center located in and around the sill. The fluids are expelled radially outwards from the sill. Provided that the permeability is sufficiently high, there is a gradual transition to buoyancy driven advection as the excess pore pressures propagate outwards. The total fluid transport upwards generates a fluid mass deficiency below the sill. The excess pressure center gradually moves upward

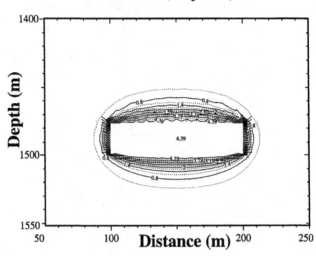

**Fig. 15.9.** *EASYR<sub>o</sub>* vitrinite reflectance 12 years after emplacement of the sill with conduction dominated heat transfer. Note that the area affected is comparable with the thickness of the intrusion itself. The sill is located within the box.

from the center of the sill to just above it, as fluids are expelled upwards. Simultaneously, the excess pore pressure gradually decreases and becomes negative below the sill as a result of the fluid mass deficiency (Figure 15.8). The negative pore pressure subsequently drives the fluid flow downwards, mainly following the surface of the sill body because the sill itself has lower permeability (Figure 15.12). A sharp pressure gradient develops within the sill (Figure 15.11). The transition from initial outwards expulsion to buoyancy driven advection of the pore fluid takes less than one year. The result is a plume shaped 'mushroom' areol, which is often seen when a hot fluid body rises in an advective system. The fluid flow velocity is locally at 17 m/yr, for a short period of time after emplacement. As the sill cools, the fluid circulation decreases further and the heat transfer becomes more dominated by conduction.

According to [6] a 1D case study shows, a short time after emplacement of a vertical magmatic dike intrusion, high excess pore pressure and fluid expelling outwards from the intrusions wall. Later, depending on the permeability, buoyancy driven fluid flow is instigated and the fluid starts to flow backward towards the negative excess pore pressured areas, close to the dike wall, to compensate for the fluid mass deficiency. Although our geometry, dimensions, and intrusion type differs from these conditions, we believe the same physical process takes place. The fluid flow as a result of the the buoyancy driven advection is mainly dependent on the temperature gradient in the areol, sediment permeability, and the thermal expansion coefficient of the pore water. In our conduction-dominated case the permeability is too low to instigate any buoyancy driven fluid flow. Thus, more than one year after the sill intrusion the pore fluids are still expelled slowly outwards (Figure 15.8).

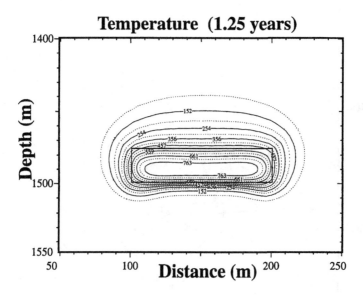

**Fig. 15.10.** Temperature field after 1.25 years in the advection dominated heat transfer simulation. Note the asymmetric temperature above and below the sill, and the plume shaped areol. The sill is located within the box.

The $EASYR_o$ vitrinite reflectance pattern 12 years after sill emplacement in a permeable setting has an asymmetric distribution above and below the sill (Figure 15.13), which can be directly related to the temperature distribution pattern (Figure 15.10). Above the sill the vitrinite anomaly extends as far as 30 m whereas below the sill the gradient is very steep and the anomaly extends only 5 m away from the sill. The total energy and heat output is not different from the conduction-dominated case, but the distribution is changed and hence, the vitrinite reflectance pattern. Higher permeability only changes the distribution of the heat, and thus the distribution of the vitrinite reflectance field. The total volume of source rock to reach maturity is expected to be the same as for the diffusion dominated case.

The third and final simulation uses both hydrostatic pressure and overburden as given by (15.24) and (15.25) (Table 15.6). To account for the overburden at 1300 m depth a traction force equivalent to 30 MPa is placed at the top boundary, and boundary condition for the pore pressure is set to 15MPa to account for the overlying water column. For simplicity we use the

# Fluid Pressure anomaly (1.25 years)

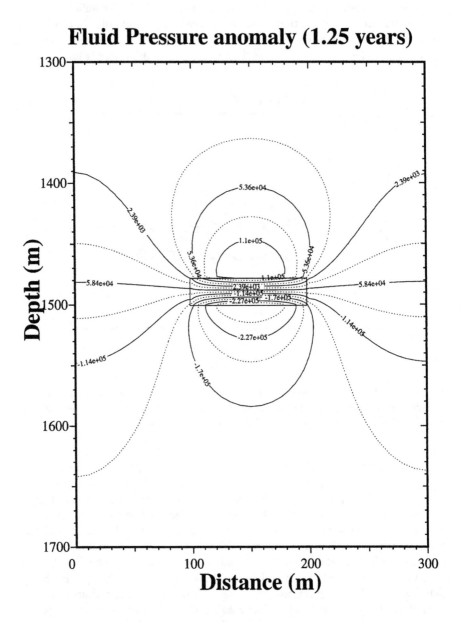

**Fig. 15.11.** Pore pressure (Pa) anomaly generated by the temperature field in Figure 15.10. Note the high excess pore pressure above the sill and the negative pore pressure below the sill. Also note the difference to conduction dominated pressure field (Figure 15.8).

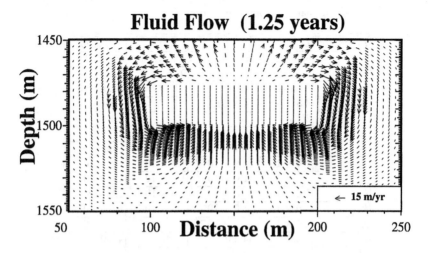

**Fig. 15.12.** Fluid flow due to the pore pressure anomaly (Figure 15.11) in a convectional flow pattern. Fluids are moving downward along the sides and under the sill.

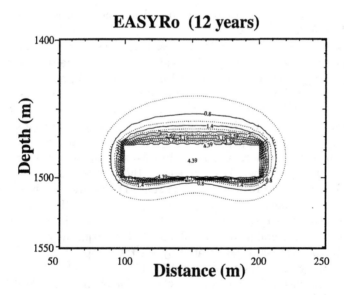

**Fig. 15.13.** *EASYR₀* vitrinite reflectance, after 12 years, in a convective environment. Note that the geometry follows that of the temperature field (Figure 15.10).

**Table 15.6.** Parameters for the coupled model (15.9), (15.17), and (15.22) in an advection dominated heat transport simulation.

| Parameter | Layer1 | Layer2 | Layer3 | Intrusion |
|---|---|---|---|---|
| $E$ | 600 MPa | 600 MPa | 600 MPa | 800 MPa |
| $\mu$ | 0.35 | 0.35 | 0.35 | 0.35 |
| $K_s$ | 30.0 GPa | 30.0 GPa | 30.0 GPa | 30.0 GPa |
| $K_f$ | 4.3 GPa | 4.3 GPa | 4.3 GPa | 4.3 GPa |
| $\phi$ | 25% | 25% | 25% | 0.1% |
| $\varrho_s$ | 2300 | 2300 | 2300 | 2300 |
| $\varrho_f$ | 1020 | 1020 | 1020 | 1020 |
| $\beta_s$ | $1.0 \times 10^{-5}$ | $1.0 \times 10^{-5}$ | $1.0 \times 10^{-5}$ | $1.0 \times 10^{-5}$ |
| $\beta_f$ | $1.0 \times 10^{-3}$ | $1.0 \times 10^{-3}$ | $1.0 \times 10^{-3}$ | $1.0 \times 10^{-3}$ |
| $k_x$ | $1.0 \times 10^{-13}$ | $1.0 \times 10^{-13}$ | $5.0 \times 10^{-13}$ | $1.0 \times 10^{-17}$ |
| $k_z$ | $1.0 \times 10^{-13}$ | $1.0 \times 10^{-13}$ | $5.0 \times 10^{-13}$ | $1.0 \times 10^{-17}$ |
| $\mu_f$ | $1.0 \times 10^{-3}$ | $1.0 \times 10^{-3}$ | $1.0 \times 10^{-3}$ | $1.0 \times 10^{-3}$ |
| $C_f$ | 4190 | 4190 | 4190 | 4190 |
| $C_s$ | 753 | 753 | 753 | 800 |
| $\kappa_s$ | 1.5 | 1.5 | 1.5 | 1.5 |
| $\kappa_f$ | 0.6 | 0.6 | 0.6 | 0.6 |
| $g_r$ | 9.81 | 9.81 | 9.81 | 9.81 |

equivalent effective stress (15.47), $m$, to indicate the stress level and here $\hat{\sigma}_{rs}$ is the effective stress deviator.

One year after emplacement the stress field close to the sill is dominated by the thermal stress (Figure 15.14). Heat is displaced radially outwards from the sill as a result of thermal stress within the sill itself, and the maximum principal stress follows the same direction [9]. This maximum stress is higher than the overburden stress levels, and is seen as an upward displacement over the sill such that the thermal stress elevates the surrounding sediments. Further away from the sill, approximately 30-40 m both above and below, the overburden stress becomes dominant. In this zone the displacement is more lateral and small, but there is an increase in the strains and, hence, the shear stress. We refer to this region as a *shock zone*. The position of the shock zone and its size will change with changes in the heat level in the sill.

The equivalent stress values are at their highest within and close to the sill, whereas, above and below the sill a zone of low equivalent stress is found (Figure 15.15). This is demonstrated by a stress drop from $3.19 \times 10^7$ Pa in the center of the sill to $1.86 \times 10^7$ Pa outside it (Figure 15.15). However, above the sill there is again an increase, to $2.39 \times 10^7$ Pa, towards the *shock zone*. At the sides of the sill there are a steady lateral decline in equivalent stresses.

To summarize our results we can state that low permeability (meaning less than $5.0 \times 10^{-14} \text{m}^2$, i.e., 0.5mD) favors conduction. Such low permeabilities are normally found in silt or slightly consolidated clays. At higher perme-

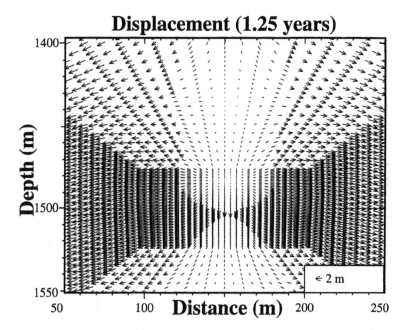

**Fig. 15.14.** Displacements caused by thermally and effective stress 1.25 years after emplacement of the sill. Note the thermal generated radial outwards displacements. The *shock zone* is seen 30-40 m above and below the sill were the displacements are close to horizontal.

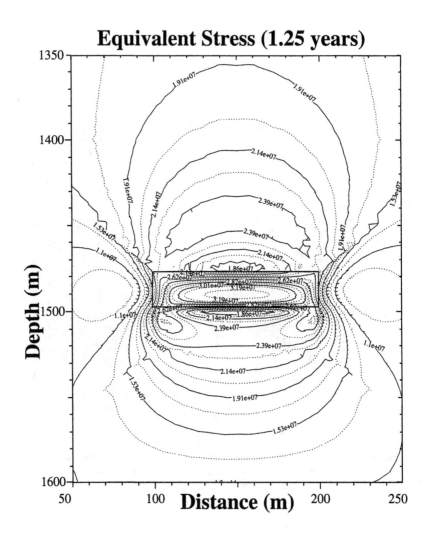

**Fig. 15.15.** Equivalent stress as a result of effective and thermal stresses with hydrostatic and gravitational forces. Note the magnitude of the equivalent stress inside the sill, the low values just above and below the sill, and the increase towards the *shock zone*.

abilities heat transport is dominated by advection. We emphasize, however, that advection transport of heat is only expected to occur around very hot intrusions. At normal geothermal gradients (20-40 °C/km) advective heat transport cannot occur [3].

Our main findings from this study are as follows:

- When heat transfer is dominantly controlled by conduction the heat aureole is symmetric and extends above and below the sill to a height and depth equivalent to the sill's vertical thickness.

- When heat transfer is advection dominated the aureole extends further above than below the sill. The height of the aureole above the sill exceeds the vertical thickness of the sill whereas the depth of the aureole below the sill, is less than the vertical thickness of the sill.

- Advection is dominant at permeabilities greater than 0.5mD in our test cases.

- A sill can mature $(0.5 < EASYR_o)$ roughly twice its own volume of host rock if the background maturity level is low $(0.3\ EASYR_o)$. If background maturity is high or the sill thickness is large, say 100 m, a greater volume of sediments can be heated to oil maturity levels or higher.

- The model gives a first order view of the thermal stress and effective stresses both within the cooling sill and in the surrounding sediments.

- Simulation results indicate that from a numerical standpoint the coupled model performs adequately, i.e., the method converges and produces stable solutions.

The coupled model provides a highly simplified description of the stress, fluid, and heat transfer around the sill. It does not consider such aspects as the dynamics of the emplacement process itself; the pressure exerted by the crystallization processes in the cooling magma or the diagenetic changes in the sediments around the sill. Temperature dependent processes, such as quartz cementation, which will be promoted by the increased temperatures in the host sediment, will reduce porosity and permeability in the affected zone. Changes in the permeability and fluid flow will also be associated with fracturing. Inclusion of all these processes is beyond the scope of the model, but their effects should be taken into consideration when interpreting the geological results of the simulations. In addition it should be noted that there is an almost infinite possible variation in both the lithology and mineralogy of the sediments surrounding a sill. This variation can be on a very fine scale (mm) and the layering used in the model is a simplified and idealized case. Our model does not include high pressurization effects such as steam generation from boiling, which would be the case at depths less than 2-3 km below sea level.

However, the model can give a first order estimate and be a useful tool in the study of the basic phenomena of fluid and heat transfer around an intrusion, especially at hydrostatic pressures above a value corresponding to 2 km. The fully coupled heat transfer and fluid flow, and the stress distribution associated with sill intrusions, are difficult or impossible to study in the field. The model is a tool, which can be used to visualize some of the basic physical and mechanical processes involved.

## 15.7   Concluding Remarks

We have investigated a fully implicit method for a finite element model that couples fluid and heat transfer through time in a deforming porous medium. Key points have been a safe and flexible implementation strategy in an (object-oriented programming) environment and adequate verification of the numerical scheme. The latter was achieved with the help of both analytical solutions and case studies. In the physical case study we chose to focus on a magmatic intrusion into horizontally layered sediments in a sedimentary basin. The goal was to present a basin model that is able to simulate the basic mechanical processes ongoing through time. The coupled model (15.9), (15.17), and (15.22) can be applied in a wide range of situations. In geotechnical engineering it can be used for subsidence stability analysis for buildings and foundations, and in sedimentary basins analysis it can be applied to study processes associated with sedimentary loading, such as fluid flow and effective stress. Another possible application is the simulation of a geothermal power plant, where (15.22) can be used to model the transport of injected water through the permeable crust at great depth and the export of the surplus heat to the power plant. The model will estimate how fast the wells and the reservoir will cool as the geothermal heat is depleted, which are important considerations in the evaluation of whether to build and how to run the power plant.

Our experience shows that the design of the computer code is essential in order to easily implement and verify the numerical model. The separation of the basin model into classes or building blocks, with the help of a class Administrator, allowed flexibility in the selection of the various initial conditions for the basin model at particular times. The fully implicit solutions of (15.9) and (15.17) show that the grid size $l$ must satisfy $l \leq c_v \Delta t$ to avoid non–physical oscillations. These limitations force us to select time steps such that $\Delta t < c_v/l$. With the aid of both field data and a case study, the basin model proved to be sufficiently stable and robust. The case study showed that the model could handle steep thermal gradients and rapid changes in the stress and pore pressure level, and we conclude that the implicit solution strategy with successive substitution is successful in handling a violent and rapidly changing environment within a sedimentary basin.

The case study also provided data on the mechanisms of heat transport and fluid flow/pore pressure distribution around a cooling, horizontal magmatic body. The maturity model $EASYR_o$, in combination with the modeled temperature field, is able to indicate the likely maturity level in the sediments around the sill. The amount of energy released into the host sediment is the same whether the heat transfer is conductive or advective. Two different cases showed that in places where permeability is below 0.5mD there is a symmetric distribution of heat and vitrinite reflectance and that these extend outwards in a zone equivalent to the sill thickness. Increasing the permeability results in an asymmetric distribution where the area above the sill is most affected. In this case the zone of increased reflectance above the sill is approximately 1.5 times the sill thickness but below the sill it is equivalent to only half the sill thickness. Here, the cooler fluids are transported up under the sill. We conclude that under the modeled conditions, a sill is able to increase the maturity level $EASYR_o$ to $> 0.5$ in a volume of rock equivalent to at least twice its own volume, thus producing hydrocarbons from the corresponding volume. The thermal stress level exceeds the overburden stress and thus creates an expanding stress field that deflects the gravity field in a shock zone above and below the sill, where shear stresses are prominent, and the possibility of fracturing is highest.

These results indicate that the basin model could be a useful tool in the risk assessment and analysis of petroleum prospects that are intruded by sill complexes or are dependent on sill influenced maturation of source rocks. In a sill complex the number of sills can be very large with thickness ranging from a few meters to several hundred meters [1]. The emplacement of sills into a sedimentary basin will directly impact local geothermal gradients and hence the timing of maturation, the intensity of maturation of potential source rocks, and the onset of primary migration.

*Acknowledgement.* This chapter is made possible with financial support from The Research Council of Norway (NFR) and Volcanic Basin Petroleum Research (VBPR). The authors wish to thank VBPR and Anders Malthe-Sørenssen for fruitful discussions on the interpretation of the numerical data from the heat transfer around the sill intrusives. We also thank Sylfest Glimsdal, Linda Ingebrigtsen, and Halvard Moe for their very useful feedback on the manuscript.

# References

1. C. Berndt. *Continental Breakup Volcanism on the Norwegian Margin.* PhD thesis, University of Oslo, 2000.
2. M. A. Biot. General theory of three dimensional consolidation. *J. Appl. Phys.*, 12(155-164), 1941.
3. K. Bjørlykke, A. Mo, and E. Palm. Modelling thermal convection in sedimentary basins and its relevance to diagenetic reactions. *Marine and Petroleum Geology*, 5, 1988.
4. A. M. Britt and M. J. Gunn. *Critical State Soil Mechanics Via Finite Element.* Ellis Horwood Limited, 1987.

5. A. N. Brooks and T. J. R. Hughes. A streamline-upwind/Petrov-Galerkin finite element formulation for advection domainated flows with particular empha sis on the incompressible Navier-Stokes equations. *Comp. Methods Appl. Mech. Engrg.*, pages 199–259, 1982.

6. P. T. Delaney. Rapid intrusion of magma into wet rock: Groundwater flow to pore pressure increases. *Journal of Geophysical Research*, pages 7739–7756, 1982.

7. O. Eldholm, J. Thiede, and E. Taylor. Evolution of the vøring vulcanic margin. In *Eldholm, O., Thiede, J., Taylor, E. et al., Proc. ODP, Sci. Results. College Station (Ocean Drilling Program)*, volume 104, pages 1033–1065, 1989.

8. P. S. Huyakorn and G. F. Pinder. *Computational methods in subsurface flow.* Academic Press, 1983.

9. R. B. Knapp and D. Norton. Preliminary numerical analysis of processes related to magma crystallization and stress evolution in cooling pluton environments. *American Journal of Science*, 1981.

10. T. W. Lambe and R. V. Whitman. *Soil Mechanics.* John Wiley and Sons, 1969.

11. H. P. Langtangen. *Computational Partial Differential Equations – Numerical Methods and Diffpack Programming.* Springer-Verlag, 1999.

12. R. W. Lewis and B. A. Schrefler. *The Finite Element Method in the Deformation and Consolidation of Porous Media.* John Wiley and Sons, 1987.

13. G. E. Mase. *Continuum Mechanics.* McGraw-Hill, 1970.

14. D. P. McKenzie. Some remarks on the development of sedimentary basins. *Earth Planet. Sci. Lett.*, 40, 1978.

15. K. Midttømme. *Thermal conductivity of sedimentary rocks - selected methological, mineralogical and textural studies.* PhD thesis, NTNU Norges teknisknaturvitenskaplig universitet, 1997.

16. Hanson B. R. The hydrodynamic of contact metamorphism. *GSA Bulletin*, 1995.

17. J. H. Schon. *Physical Properties of rocks: Fundamentals and Principles of Petrophysics*, volume 18. Pergamon, 1996.

18. J. Skogseid, T. Pedersen, O. Eldholm, and V. B. Larsen. Vøring Basin: subsidence and tectonic evolution. *Norwegian Petroleum Society (NPF) Special Publication.*, 1:55–82, 1992b.

19. J. J. Sweeney and A. K. Burnham. Evaluation of a simple model of vitrinite reflectance based on chemical kinetics. *Am. Ass. Petrol. Geol. Bull.*, 74:1559–1570, 1990.

20. B. P. Tissot and D. H. Welte. *Petroleum Formation and Occurrence.* Springer Verlag, 1984.

21. M. Wangen. A finite element formulation in lagrangian coordinates for heat and fluid flow compacting sedimentary basins. *Int.J. for Numerical and Analytical Metods for Geomechanics*, 17(6):401–432, 1993.

# Editorial Policy

§1. Volumes in the following three categories will be published in LNCSE:

i)   Research monographs
ii)  Lecture and seminar notes
iii) Conference proceedings

Those considering a book which might be suitable for the series are strongly advised to contact the publisher or the series editors at an early stage.

§2. Categories i) and ii). These categories will be emphasized by Lecture Notes in Computational Science and Engineering. **Submissions by interdisciplinary teams of authors are encouraged.** The goal is to report new developments – quickly, informally, and in a way that will make them accessible to non-specialists. In the evaluation of submissions timeliness of the work is an important criterion. Texts should be well-rounded, well-written and reasonably self-contained. In most cases the work will contain results of others as well as those of the author(s). In each case the author(s) should provide sufficient motivation, examples, and applications. In this respect, Ph.D. theses will usually be deemed unsuitable for the Lecture Notes series. Proposals for volumes in these categories should be submitted either to one of the series editors or to Springer-Verlag, Heidelberg, and will be refereed. A provisional judgment on the acceptability of a project can be based on partial information about the work: a detailed outline describing the contents of each chapter, the estimated length, a bibliography, and one or two sample chapters – or a first draft. A final decision whether to accept will rest on an evaluation of the completed work which should include

– at least 100 pages of text;
– a table of contents;
– an informative introduction perhaps with some historical remarks which should be accessible to readers unfamiliar with the topic treated;
– a subject index.

§3. Category iii). Conference proceedings will be considered for publication provided that they are both of exceptional interest and devoted to a single topic. One (or more) expert participants will act as the scientific editor(s) of the volume. They select the papers which are suitable for inclusion and have them individually refereed as for a journal. Papers not closely related to the central topic are to be excluded. Organizers should contact Lecture Notes in Computational Science and Engineering at the planning stage.

In exceptional cases some other multi-author-volumes may be considered in this category.

§4. Format. Only works in English are considered. They should be submitted in camera-ready form according to Springer-Verlag's specifications. Electronic material can be included if appropriate. Please contact the publisher. Technical instructions and/or TEX macros are available via http://www.springer.de/math/authors/help-momu.html. The macros can also be sent on request.

# General Remarks

Addresses:

Timothy J. Barth
NASA Ames Research Center
NAS Division
Moffett Field, CA 94035, USA
e-mail: barth@nas.nasa.gov

Michael Griebel
Institut für Angewandte Mathematik
der Universität Bonn
Wegelerstr. 6
53115 Bonn, Germany
e-mail: griebel@iam.uni-bonn.de

David E. Keyes
Department of Applied Physics
and Applied Mathematics
Columbia University
200 S. W. Mudd Building
500 W. 120th Street
New York, NY 10027, USA
e-mail: david.keyes@columbia.edu

Risto M. Nieminen
Laboratory of Physics
Helsinki University of Technology
02150 Espoo, Finland
e-mail: rni@fyslab.hut.fi

Dirk Roose
Department of Computer Science
Katholieke Universiteit Leuven
Celestijnenlaan 200A
3001 Leuven-Heverlee, Belgium
e-mail: dirk.roose@cs.kuleuven.ac.be

Tamar Schlick
Department of Chemistry
Courant Institute of Mathematical
Sciences
New York University
and Howard Hughes Medical Institute
251 Mercer Street
New York, NY 10012, USA
e-mail: schlick@nyu.edu

Springer-Verlag, Mathematics Editorial IV
Tiergartenstrasse 17
69121 Heidelberg, Germany
Tel.: *49 (6221) 487-8185
e-mail: peters@springer.de
http://www.springer.de/math/
peters.html

# Lecture Notes
in Computational Science
and Engineering

**Vol. 1**  D. Funaro, *Spectral Elements for Transport-Dominated Equations.* 1997. X, 211 pp. Softcover. ISBN 3-540-62649-2

**Vol. 2**  H. P. Langtangen, *Computational Partial Differential Equations.* Numerical Methods and Diffpack Programming. 1999. XXIII, 682 pp. Hardcover. ISBN 3-540-65274-4

**Vol. 3**  W. Hackbusch, G. Wittum (eds.), *Multigrid Methods V.* Proceedings of the Fifth European Multigrid Conference held in Stuttgart, Germany, October 1-4, 1996. 1998. VIII, 334 pp. Softcover. ISBN 3-540-63133-X

**Vol. 4**  P. Deuflhard, J. Hermans, B. Leimkuhler, A. E. Mark, S. Reich, R. D. Skeel (eds.), *Computational Molecular Dynamics: Challenges, Methods, Ideas.* Proceedings of the 2nd International Symposium on Algorithms for Macromolecular Modelling, Berlin, May 21-24, 1997. 1998. XI, 489 pp. Softcover. ISBN 3-540-63242-5

**Vol. 5**  D. Kröner, M. Ohlberger, C. Rohde (eds.), *An Introduction to Recent Developments in Theory and Numerics for Conservation Laws.* Proceedings of the International School on Theory and Numerics for Conservation Laws, Freiburg / Littenweiler, October 20-24, 1997. 1998. VII, 285 pp. Softcover. ISBN 3-540-65081-4

**Vol. 6**  S. Turek, *Efficient Solvers for Incompressible Flow Problems.* An Algorithmic and Computational Approach. 1999. XVII, 352 pp, with CD-ROM. Hardcover. ISBN 3-540-65433-X

**Vol. 7**  R. von Schwerin, *Multi Body System SIMulation.* Numerical Methods, Algorithms, and Software. 1999. XX, 338 pp. Softcover. ISBN 3-540-65662-6

**Vol. 8**  H.-J. Bungartz, F. Durst, C. Zenger (eds.), *High Performance Scientific and Engineering Computing.* Proceedings of the International FORTWIHR Conference on HPSEC, Munich, March 16-18, 1998. 1999. X, 471 pp. Softcover. 3-540-65730-4

**Vol. 9**  T. J. Barth, H. Deconinck (eds.), *High-Order Methods for Computational Physics.* 1999. VII, 582 pp. Hardcover. 3-540-65893-9

**Vol. 10**  H. P. Langtangen, A. M. Bruaset, E. Quak (eds.), *Advances in Software Tools for Scientific Computing.* 2000. X, 357 pp. Softcover. 3-540-66557-9

**Vol. 11**  B. Cockburn, G. E. Karniadakis, C.-W. Shu (eds.), *Discontinuous Galerkin Methods.* Theory, Computation and Applications. 2000. XI, 470 pp. Hardcover. 3-540-66787-3

**Vol. 12**  U. van Rienen, *Numerical Methods in Computational Electrodynamics.* Linear Systems in Practical Applications. 2000. XIII, 375 pp. Softcover. 3-540-67629-5

**Vol. 13**  B. Engquist, L. Johnsson, M. Hammill, F. Short (eds.), *Simulation and Visualization on the Grid.* Parallelldatorcentrum Seventh Annual Conference, Stockholm, December 1999, Proceedings. 2000. XIII, 301 pp. Softcover. 3-540-67264-8

**Vol. 14**  E. Dick, K. Riemslagh, J. Vierendeels (eds.), *Multigrid Methods VI.* Proceedings of the Sixth European Multigrid Conference Held in Gent, Belgium, September 27-30, 1999. 2000. IX, 293 pp. Softcover. 3-540-67157-9

**Vol. 15**  A. Frommer, T. Lippert, B. Medeke, K. Schilling (eds.), *Numerical Challenges in Lattice Quantum Chromodynamics.* Joint Interdisciplinary Workshop of John von Neumann Institute for Computing, Jülich and Institute of Applied Computer Science, Wuppertal University, August 1999. 2000. VIII, 184 pp. Softcover. 3-540-67732-1

**Vol. 16**  J. Lang, *Adaptive Multilevel Solution of Nonlinear Parabolic PDE Systems.* Theory, Algorithm, and Applications. 2001. XII, 157 pp. Softcover. 3-540-67900-6

**Vol. 17**  B. I. Wohlmuth, *Discretization Methods and Iterative Solvers Based on Domain Decomposition.* 2001. X, 197 pp. Softcover. 3-540-41083-X

**Vol. 18**  U. van Rienen, M. Günther, D. Hecht (eds.), *Scientific Computing in Electrical Engineering.* Proceedings of the 3rd International Workshop, August 20-23, 2000, Warnemünde, Germany. 2001. XII, 428 pp. Softcover. 3-540-42173-4

**Vol. 19**  I. Babuška, P. G. Ciarlet, T. Miyoshi (eds.), *Mathematical Modeling and Numerical Simulation in Continuum Mechanics.* Proceedings of the International Symposium on Mathematical Modeling and Numerical Simulation in Continuum Mechanics, September 29 - October 3, 2000, Yamaguchi, Japan. 2002. VIII, 301 pp. Softcover. 3-540-42399-0

**Vol. 20**  T. J. Barth, T. Chan, R. Haimes (eds.), *Multiscale and Multiresolution Methods.* Theory and Applications. 2002. X, 389 pp. Softcover. 3-540-42420-2

**Vol. 21**  M. Breuer, F. Durst, C. Zenger (eds.), *High Performance Scientific and Engineering Computing.* Proceedings of the 3rd International FORTWIHR Conference on HPSEC, Erlangen, March 12-14, 2001. 2002. XIII, 408 pp. Softcover. 3-540-42946-8

**Vol. 22**  K. Urban, *Wavelets in Numerical Simulation.* Problem Adapted Construction and Applications. 2002. XV, 181 pp. Softcover. 3-540-43055-5

**Vol. 23**  L. F. Pavarino, A. Toselli (eds.), *Recent Developments in Domain Decomposition Methods.* 2002. XII, 243 pp. Softcover. 3-540-43413-5

**Vol. 24**  T. Schlick, H. H. Gan (eds.), *Computational Methods for Macromolecules: Challenges and Applications.* Proceedings of the 3rd International Workshop on Algorithms for Macromolecular Modeling, New York, October 12-14, 2000. 2002. IX, 504 pp. Softcover. 3-540-43756-8

**Vol. 25**  T. J. Barth, H. Deconinck (eds.), *Error Estimation and Adaptive Discretization Methods in Computational Fluid Dynamics.* 2003. VII, 344 pp. Hardcover. 3-540-43758-4

**Vol. 26**  M. Griebel, M. A. Schweitzer (eds.), *Meshfree Methods for Partial Differential Equations*. 2003. IX, 466 pp. Softcover. 3-540-43891-2

**Vol. 27**  S. Müller, *Adaptive Multiscale Schemes for Conservation Laws*. 2003. XIV, 181 pp. Softcover. 3-540-44325-8

**Vol. 28**  C. Carstensen, S. Funken, W. Hackbusch, R. H. W. Hoppe, P. Monk (eds.), *Computational Electromagnetics*. Proceedings of the GAMM Workshop on "Computational Electromagnetics", Kiel, Germany, January 26-28, 2001. 2003. X, 209 pp. Softcover. 3-540-44392-4

**Vol. 29**  M. A. Schweitzer, *A Parallel Multilevel Partition of Unity Method for Elliptic Partial Differential Equations*. 2003. V, 194 pp. Softcover. 3-540-00351-7

**Vol. 30**  T. Biegler, O. Ghattas, M. Heinkenschloss, B. van Bloemen Waanders (eds.), *Large-Scale PDE-Constrained Optimization*. 2003. VI, 349 pp. Softcover. 3-540-05045-0

**Vol. 31**  M. Ainsworth, P. Davies, D. Duncan, P. Martin, B. Rynne (eds.) *Topics in Computational Wave Propagation*. Direct and Inverse Problems. 2003. VIII, 399 pp. Softcover. 3-540-00744-X

**Vol. 32**  H. Emmerich, B. Nestler, M. Schreckenberg (eds.) *Interface and Transport Dynamics*. Computational Modelling. 2003. XV, 432 pp. Hardcover. 3-540-40367-1

**Vol. 33**  H. P. Langtangen, A. Tveito (eds.) *Advanced Topics in Computational Partial Differential Equations*. Numerical Methods and Diffpack Programming. 2003. XIX, 658 pp. Softcover. 3-540-01438-1

**Vol. 34**  V. John, *Large Eddy Simulation of Turbulent Incompressible Flows*. Analytical and Numerical Results for a Class of LES Models. 2004. XII, 261 pp. Softcover. 3-540-40643-3

# Texts in Computational Science and Engineering

**Vol. 1**  H. P. Langtangen, *Computational Partial Differential Equations*. Numerical Methods and Diffpack Programming. 2nd Edition 2003. XXVI, 855 pp. Hardcover. ISBN 3-540-43416-X

**Vol. 2**  A. Quarteroni, F. Saleri, *Scientific Computing with MATLAB*. 2003. IX, 257 pp. Hardcover. ISBN 3-540-44363-0

*For further information on these books please have a look at our mathematics catalogue at the following URL:* `http://www.springer.de/math/index.html`

Printing and Binding: Strauss GmbH, Mörlenbach